Ionospheres
Physics, Plasma Physics, and Chemistry
Second Edition

This combination of text and reference book provides a comprehensive description of the physical, plasma, and chemical processes controlling the behavior of ionospheres, upper atmospheres and exospheres. It describes in detail the relevant processes, mechanisms, and transport equations that are required to solve fundamental research problems. Our current understanding of the structure, chemistry, dynamics, and energetics of the terrestrial ionosphere is summarized in two chapters, while that of other solar system bodies is outlined in a separate chapter. The final chapter of the book is devoted to relevant in-situ and remote measurement techniques.

This second edition incorporates the results, model developments, and interpretations from the last ten years, both in the text and through the addition of new figures and references. In particular, it includes new material on neutral atmospheres, on the terrestrial ionosphere at low, middle, and high latitudes, and on planetary atmospheres and ionospheres, where results from recent space missions have yielded a wealth of new data. Extensive appendices (including an additional four for this edition) provide information about physical constants, mathematical formulas, transport coefficients, and other important parameters needed for ionospheric calculations.

The book forms an extensive and lasting volume for researchers studying ionospheres, upper atmospheres, aeronomy, and plasma physics, and is an ideal textbook for graduate-level courses. Problem sets are provided, for which solutions are available to instructors online at www.cambridge.org/9780521877060.

Cambridge Atmospheric and Space Science Series
Editors: J. T. Houghton, M. J. Rycroft, and A. J. Dessler

This series of upper-level texts and research monographs covers the physics and chemistry of different regions of the Earth's atmosphere, from the troposphere and stratosphere, up through the ionosphere and magnetosphere, and out to the interplanetary medium.

ROBERT W. SCHUNK is Professor of Physics and the Director of the Center for Atmospheric and Space Sciences at Utah State University. He is also a co-founder and the President of Space Environment Corporation, a small high-tech company in Logan, Utah. He has over 35 years of experience in theory, numerical modeling, and data analysis in the general areas of plasma physics, fluid mechanics, kinetics, space physics, and planetary ionospheres and atmospheres. He has been a Principal Investigator on numerous NASA, NSF, Air Force, and Navy grants and has chaired many national committees, international organizations and review panels. Professor Schunk received the D. Wynne Thorne Research Award from USU in 1983, the Governor's Medal for Science & Technology from the State of Utah in 1988, gave the AGU Nicolet Lecture in 2002, is a Fellow of the AGU, and was inducted into the International Academy of Astronautics in 2006.

ANDREW F. NAGY has been on the faculty of the University of Michigan since 1963, serving as a professor of Space Science and Electrical Engineering, Associate Vice President for Research (1987–1990), and Director of the Space Physics Research Laboratory (1990–1992). He has over 40 years of experience in both theoretical and experimental studies of the upper atmospheres, ionospheres, and magnetospheres of the Earth and planets, and has been principal and co-investigator and interdisciplinary scientist on a variety of space missions. Professor Nagy has chaired or been a member of over 40 national and international committees and boards. He was President of the Space Physics and Aeronomy Section of the AGU. He is a Fellow of the AGU, a member of the International Academy of Astronautics, has given the AGU Nicolet Lecture (1998), and received the NASA Public Service Medal (1983) and the Attwood (1998) and the Distinguished Faculty Achievement Awards (2003) from the University of Michigan.

Cambridge Atmospheric and Space Science Series

EDITORS

John T. Houghton

Michael J. Rycroft

Alexander J. Dessler

Titles in print in this series

M. H. Rees
Physics and chemistry of the upper atmosphere

R. Daley
Atmosphere data analysis

J. K. Hargreaves
The solar–terrestrial environment

J. R. Garratt
The atmosphere boundary layer

S. Sazhin
Whistler-mode waves in a hot plasma

S. P. Gary
Theory of space plasma microinstabilities

I. N. James
Introduction to circulating atmospheres

T. I. Gombosi
Gaskinetic theory

M. Walt
Introduction to geomagnetically trapped radiation

B. A. Kagan
Ocean–atmosphere interaction and climate modelling

D. Hastings and H. Garrett
Spacecraft–environment interactions

J. C. King and J. Turner
Antarctic meteorology and climatology

T. E. Cravens
Physics of solar system plasmas

J. F. Lemaire and K. I. Gringauz
The Earth's plasmasphere

T. I. Gombosi
Physics of space environment

J. Green
Atmospheric dynamics

G. E. Thomas and K. Stamnes
Radiative transfer in the atmosphere and ocean

R. W. Schunk and A. F. Nagy
Ionospheres: Physics, plasma physics, and chemistry

I. G. Enting
Inverse problems in atmospheric constituent transport

R. D. Hunsucker and J. K. Hargreaves
The high-latitude ionosphere and its effects on radio propagation

M. C. Serreze and R. G. Barry
The Arctic climate system

N. Meyer-Vernet
Basics of the solar wind

V. Y. Trakhtengerts and M. J. Rycroft
Whistler and Alfvén mode cyclotron masers in space

R. W. Schunk and A. F. Nagy
Ionospheres: Physics, plasma physics, and chemistry, second edition

IONOSPHERES

Physics, Plasma Physics, and Chemistry

Second Edition

ROBERT W. SCHUNK

Utah State University

and

ANDREW F. NAGY

University of Michigan

CAMBRIDGE
UNIVERSITY PRESS

CAMBRIDGE
UNIVERSITY PRESS

University Printing House, Cambridge CB2 8BS, United Kingdom

One Liberty Plaza, 20th Floor, New York, NY 10006, USA

477 Williamstown Road, Port Melbourne, VIC 3207, Australia

314-321, 3rd Floor, Plot 3, Splendor Forum, Jasola District Centre, New Delhi - 110025, India

79 Anson Road, #06-04/06, Singapore 079906

Cambridge University Press is part of the University of Cambridge.

It furthers the University's mission by disseminating knowledge in the pursuit of education, learning and research at the highest international levels of excellence.

www.cambridge.org
Information on this title: www.cambridge.org/9781108462105

© R. W. Schunk and A. F. Nagy 2000, 2009

This publication is in copyright. Subject to statutory exception and to the provisions of relevant collective licensing agreements, no reproduction of any part may take place without the written permission of Cambridge University Press.

First published 2000
Second edition 2009
First paperback edition 2018

A catalogue record for this publication is available from the British Library

ISBN 978-0-521-87706-0 Hardback
ISBN 978-1-108-46210-5 Paperback

Additional resources for this publication at www.cambridge.org/9780521877060

Cambridge University Press has no responsibility for the persistence or accuracy of URLs for external or third-party internet websites referred to in this publication, and does not guarantee that any content on such websites is, or will remain, accurate or appropriate.

*To our parents for their past guidance, encouragement
and support, and to our children and AFN's wife
(Allison, Lisa, Michael, Robert, and Susan)
for their love and understanding.*

Contents

Chapter 1	**Introduction**	*page* 1
1.1	Background and purpose	1
1.2	History of ionospheric research	3
1.3	Specific references	8
1.4	General references	9

Chapter 2	**Space environment**	11
2.1	Sun	11
2.2	Interplanetary medium	17
2.3	Earth	22
2.4	Inner planets	31
2.5	Outer planets	37
2.6	Moons and comets	39
2.7	Plasma and neutral parameters	43
2.8	Specific references	46
2.9	General references	48

Chapter 3	**Transport equations**	50
3.1	Boltzmann equation	50
3.2	Moments of the distribution function	53
3.3	General transport equations	55
3.4	Maxwellian velocity distribution	58
3.5	Closing the system of transport equations	60
3.6	13-moment transport equations	62
3.7	Generalized transport systems	65
3.8	Kinetic, Monte Carlo, and particle-in-cell methods	66
3.9	Maxwell equations	67
3.10	Specific references	68
3.11	Problems	69

Chapter 4	**Collisions**	72
4.1	Simple collision parameters	73
4.2	Binary elastic collisions	74

vii

4.3	Collision cross sections	80
4.4	Transfer collision integrals	85
4.5	Maxwell molecule collisions	89
4.6	Collision terms for Maxwellian velocity distributions	92
4.7	Collision terms for 13-moment velocity distributions	98
4.8	Momentum transfer collision frequencies	102
4.9	Specific references	109
4.10	Problems	110

Chapter 5	**Simplified transport equations**	113
5.1	Basic transport properties	114
5.2	The five-moment approximation	119
5.3	Transport in a weakly ionized plasma	120
5.4	Transport in partially and fully ionized plasmas	125
5.5	Major ion diffusion	126
5.6	Polarization electrostatic field	128
5.7	Minor ion diffusion	130
5.8	Supersonic ion outflow	132
5.9	Time-dependent plasma expansion	135
5.10	Diffusion across **B**	137
5.11	Electrical conductivities	139
5.12	Electron stress and heat flow	143
5.13	Ion stress and heat flow	148
5.14	Higher-order diffusion processes	149
5.15	Summary of appropriate use of transport equations	153
5.16	Specific references	155
5.17	General references	156
5.18	Problems	156

Chapter 6	**Wave phenomena**	159
6.1	General wave properties	159
6.2	Plasma dynamics	164
6.3	Electron plasma waves	168
6.4	Ion-acoustic waves	170
6.5	Upper hybrid oscillations	172
6.6	Lower hybrid oscillations	174
6.7	Ion-cyclotron waves	175
6.8	Electromagnetic waves in a plasma	177
6.9	Ordinary and extraordinary waves	179
6.10	L and R waves	183
6.11	Alfvén and magnetosonic waves	185
6.12	Effect of collisions	186
6.13	Two-stream instability	188
6.14	Shock waves	191
6.15	Double layers	196
6.16	Summary of important formulas	201
6.17	Specific references	203
6.18	General references	204
6.19	Problems	204

Chapter 7	**Magnetohydrodynamic formulation**	206
7.1	General MHD equations	206
7.2	Generalized Ohm's law	211
7.3	Simplified MHD equations	213
7.4	Pressure balance	214
7.5	Magnetic diffusion	216
7.6	Spiral magnetic field	217
7.7	Double-adiabatic energy equations	219
7.8	Alfvén and magnetosonic waves	221
7.9	Shocks and discontinuities	225
7.10	Specific references	228
7.11	General references	229
7.12	Problems	229

Chapter 8	**Chemical processes**	231
8.1	Chemical kinetics	231
8.2	Reaction rates	236
8.3	Charge exchange processes	240
8.4	Recombination reactions	243
8.5	Negative ion chemistry	245
8.6	Excited state chemistry	246
8.7	Optical emissions; airglow and aurora	248
8.8	Specific references	250
8.9	General references	252
8.10	Problems	252

Chapter 9	**Ionization and energy exchange processes**	254
9.1	Absorption of solar radiation	254
9.2	Solar EUV intensities and absorption cross sections	258
9.3	Photoionization	260
9.4	Superthermal electron transport	264
9.5	Superthermal ion and neutral particle transport	270
9.6	Electron and ion heating rates	272
9.7	Electron and ion cooling rates	276
9.8	Specific references	284
9.9	General references	287
9.10	Problems	287

Chapter 10	**Neutral atmospheres**	289
10.1	Rotating atmospheres	290
10.2	Euler equations	291
10.3	Navier–Stokes equations	292
10.4	Atmospheric waves	294
10.5	Gravity waves	295
10.6	Tides	300
10.7	Density structure and controlling processes	304
10.8	Escape of terrestrial hydrogen	311
10.9	Energetics and thermal structure of the Earth's thermosphere	314

10.10	Exosphere	321
10.11	Hot atoms	325
10.12	Specific references	328
10.13	General references	331
10.14	Problems	332

Chapter 11	**The terrestrial ionosphere at middle and low latitudes**	335
11.1	Dipole magnetic field	337
11.2	Geomagnetic field	341
11.3	Geomagnetic variations	344
11.4	Ionospheric layers	346
11.5	Topside ionosphere and plasmasphere	356
11.6	Plasma thermal structure	360
11.7	Diurnal variation at mid-latitudes	365
11.8	Seasonal variation at mid-latitudes	367
11.9	Solar cycle variation at mid-latitudes	368
11.10	Plasma transport in a dipole magnetic field	369
11.11	Equatorial F region	371
11.12	Equatorial spread F and bubbles	373
11.13	Sporadic E and intermediate layers	379
11.14	F_3 layer and He^+ layer	381
11.15	Tides and gravity waves	381
11.16	Ionospheric storms	386
11.17	Specific references	391
11.18	General references	395
11.19	Problems	396

Chapter 12	**The terrestrial ionosphere at high latitudes**	398
12.1	Convection electric fields	399
12.2	Convection models	405
12.3	Effects of convection	410
12.4	Particle precipitation	419
12.5	Current systems	423
12.6	Large-scale ionospheric features	425
12.7	Propagating plasma patches	430
12.8	Boundary and auroral blobs	432
12.9	Sun-aligned arcs	434
12.10	Cusp neutral fountain	434
12.11	Neutral density structures	437
12.12	Neutral response to convection channels	438
12.13	Supersonic neutral winds	443
12.14	Geomagnetic storms	445
12.15	Substorms	448
12.16	Polar wind	450
12.17	Energetic ion outflow	465
12.18	Neutral polar wind	470
12.19	Specific references	472
12.20	General references	479
12.21	Problems	480

Chapter 13	**Planetary ionospheres**	482
13.1	Mercury	482
13.2	Venus	482
13.3	Mars	492
13.4	Jupiter	496
13.5	Saturn, Uranus, Neptune, and Pluto	498
13.6	Satellites and comets	502
13.7	Specific references	509
13.8	General references	514
13.9	Problems	515
Chapter 14	**Ionospheric measurement techniques**	517
14.1	Spacecraft potential	517
14.2	Langmuir probes	519
14.3	Retarding potential analyzers	522
14.4	Thermal ion mass spectrometers	525
14.5	Magnetometers	529
14.6	Radio reflection	532
14.7	Radio occultation	534
14.8	Incoherent (Thomson) radar backscatter	538
14.9	Specific references	544
14.10	General references	546
Appendix A	**Physical constants and conversions**	548
A.1	Physical constants	548
A.2	Conversions	548
Appendix B	**Vector relations and operators**	550
B.1	Vector relations	550
B.2	Vector operators	551
B.3	Specific references	553
Appendix C	**Integrals and transformations**	554
C.1	Integral relations	554
C.2	Important integrals	555
C.3	Integral transformations	556
Appendix D	**Functions and series expansions**	558
D.1	Important functions	558
D.2	Series expansions for small arguments	559
Appendix E	**Systems of units**	560
Appendix F	**Maxwell transfer equations**	562
Appendix G	**Collision models**	567
G.1	Boltzmann collision integral	567
G.2	Fokker–Planck collision term	571

G.3   Charge exchange collision integral   572
G.4   Krook collision models   572
G.5   Specific references   574

Appendix H   **Maxwell velocity distribution**   575

H.1   Specific reference   580

Appendix I   **Semilinear expressions for transport
                    coefficients**   581

I.1   Diffusion coefficients and thermal conductivities   581
I.2   Fully ionized plasma   582
I.3   Partially ionized plasma   583
I.4   Specific references   583

Appendix J   **Solar fluxes and relevant cross sections**   584

J.1   Specific references   593

Appendix K   **Atmospheric models**   594

K.1   Introduction   594
K.2   Specific references   599

Appendix L   **Scalars, vectors, dyadics, and tensors**   600

Appendix M   **Radio wave spectrum**   605

Appendix N   **Simple derivation of continuity equation**   606

Appendix O   **Numerical solution for F region ionization**   608

O.1   Specific reference   613

Appendix P   **Monte Carlo methods**   614

P.1   Specific references   617

             Index   618

Chapter 1

Introduction

1.1 Background and purpose

The ionosphere is considered to be that region of an atmosphere where significant numbers of free thermal (<1 eV) electrons and ions are present. All bodies in our solar system that have a surrounding neutral-gas envelope, due either to gravitational attraction (e.g., planets) or some other process such as sublimation (e.g., comets), have an ionosphere. Currently, ionospheres have been observed around all but two of the planets, some moons, and comets. The free electrons and ions are produced via ionization of the neutral particles both by extreme ultraviolet radiation from the Sun and by collisions with energetic particles that penetrate the atmosphere. Once formed, the charged particles are affected by a myriad of processes, including chemical reactions, diffusion, wave disturbances, plasma instabilities, and transport due to electric and magnetic fields. Hence, an understanding of ionospheric phenomena requires a knowledge of several disciplines, including plasma physics, chemical kinetics, atomic theory, and fluid mechanics. In this book, we have attempted to bridge the gaps among these disciplines and provide a comprehensive description of the physical and chemical processes that affect the behavior of ionospheres.

A brief history of ionospheric research is given later in this introductory chapter. An overview of the space environment, including the Sun, planets, moons, and comets, is presented in Chapter 2. This not only gives the reader a quick look at the overall picture, but also provides the motivation for the presentation of the material that follows. Next, in Chapter 3, the general transport equations for mass, momentum, and energy conservation are derived from first principles so that the reader can clearly see where these equations come from. This is followed by a derivation of the collision terms that appear in the transport equations, including those relevant to resonant charge exchange, nonresonant ion-neutral and electron-neutral interactions, and Coulomb collisions (Chapter 4). These general collision terms and transport equations are complicated and in many situations it is possible to use simpler sets of transport equations. Therefore, in Chapter 5, several simplified systems of transport equations are derived, including the Euler, Navier–Stokes, diffusion, and thermal conduction equations. This is followed by a discussion of the wave modes, plasma instabilities, and shocks that can occur in the ionospheres (Chapter 6). In Chapter 7, the magnetohydrodynamic (MHD) equations are derived and then used to describe MHD waves, shocks, and pressure balance.

In Chapter 8, chemical kinetics and a variety of reactions relevant to the ionospheres are discussed and presented, including those involving metastable species and negative ions. Optical emissions are also briefly discussed in this chapter. The relevant ionization and energy exchange processes are detailed in Chapter 9, including those pertaining to both photons and particles. The chapter concludes with a summary of the heating and cooling expressions that are needed for practical applications. Chapter 10 is devoted to a discussion of neutral atmospheres. The Euler and Navier–Stokes equations for neutral gases are presented at the beginning of the chapter, and this is followed by a discussion of atmospheric waves and tides. The rest of Chapter 10 deals with atmospheric structure, escape fluxes, the exosphere, and hot atoms. In Chapters 11 and 12, the general material given in the previous chapters is applied to elucidate the unique characteristics associated with the terrestrial ionosphere at low, middle, and high latitudes. Although much of this material is still of a fundamental nature, an overview of what has been accomplished to date is also provided. Chapter 13 summarizes what is currently known about all of the other ionospheres in the solar system. The most commonly used experimental techniques for measuring ionospheric densities, temperatures, and drifts are briefly described in Chapter 14. Finally, several Appendices are included that contain physical constants, mathematical formulas, some important derivations, and useful tables.

This book is the outgrowth of two decades of numerous joint research endeavors and publications by the authors. Some of the material was used in courses taught by the authors at Utah State University and at the University of Michigan. This book should be useful to graduate students, postdoctoral fellows, and established scientists who want to fill gaps in their knowledge. It also serves as a reference book for obtaining important equations and formulas. A subset of the material can be used for a graduate level course about the upper atmosphere and ionosphere, and plasma physics. At the University of Michigan a one-semester graduate course on the ionosphere and upper atmosphere has been based on Chapters 2, 3, parts of 5, 8, 9, 10, most of 11 and 12, and 13 and 14. At Utah State University, a one-semester course on plasma physics has been based on Chapters 3–7, and a course on aeronomy has been based on Chapters 2, 3, 5, 8–12. To facilitate the use of this book as a text, problems are provided at the end of most of the chapters.

Several people were helpful in the preparation of this book, and we wish to acknowledge them here. The help came in a variety of forms (e.g., providing some unpublished material, reading, or proofing part of the manuscript, etc.), and it certainly improved the book. AFN would especially like to thank (in alphabetical order)

J. R. Barker, T. E. Cravens, J. L. Fox, B. E. Gilchrist, T. I. Gombosi, J. W. Holt, A. J. Kliore, M. W. Liemohn, H. Rishbeth, and C. T. Russell. RWS would like to thank Melanie Oldroyd for typing a preliminary form of some of the chapters. We would both like to thank Shawna Johnson for drawing some of the figures, for digitizing figures, and for overseeing the production of the book. We would both also like to thank Elizabeth Wood for preparing the manuscript in LaTeX. Some of the material in the book comes from lecture notes collected over many years and thus may contain material without appropriate references to their sources, which we have forgotten. This is inadvertent and we apologize to such authors. Also, to keep the bibliographies from becoming unrealistically long, we limited our referencing to only those papers from which figures were taken, to either the latest or original reference for the material discussed, and to review papers. Hence, we omitted many deserving, appropriate, and relevant references. We hope that the readers and scientists working in the field will understand and appreciate our dilemma.

The units used in the book are a mixture of MKSA and Gaussian-cgs because of the corresponding usage by practitioners in the field. Most of the equations and formulas throughout the book are in MKSA units, and some tables and numbers are given in Gaussian-cgs units when this is the common practice. The conversion from one system to the other is briefly discussed in Appendix E.

1.2 History of ionospheric research

The earliest exposure of humankind to a phenomenon originating in the upper atmosphere is the visual aurora. The visual displays of colored light appear in the form of arcs, bands, patches, blankets, and rays, and often the features move rapidly across the night sky. It has been suggested that the earliest records of the aurora can be traced to the Stone Age.[1] References to the aurora appear in the Old Testament, in writings of Greek philosophers, including Aristotle's *Meteorologica*, and possibly in ancient Chinese works from before 2000 BC. In most of these early writings, the auroral displays were interpreted to be manifestations of God. The name *aurora borealis* (northern dawn) appears to have been coined by Galileo at some time prior to 1621.[1] The first recorded observation of the southern hemispheric aurora (*aurora australis*) was by Cook in 1773.

A serious scientific study of auroras began at about 1500 AD.[1] However, the early theories put forth by noted scientists were completely wrong. Edmund Halley, who predicted the reappearance of what is now known as Halley's comet, suggested that the auroras were "watery vapors, which are rarefied and sublimed by subterraneous fire, [and] might carry along with them sulphureous vapors sufficient to produce this luminous appearance in the atmosphere." In 1746, the Swiss mathematician Leonard Euler suggested that "the aurora was particles from the Earth's own atmosphere driven beyond its limits by the impulse of the sun's light and ascending to a height of several thousand miles. Near the poles, these particles would not be dispersed by the Earth's rotation."[2] Benjamin Franklin, who was a respected scientist in his time, thought that the aurora was related to atmospheric circulation patterns.[3] Basically, Franklin argued that the atmosphere in the polar regions must be heavier and lower than in the equatorial region because of the smaller centrifugal force, and therefore, the vacuum–atmosphere interface must be lower in the polar regions. He then further argued that the electricity brought into the polar region by clouds would not be able to penetrate the ice, and hence, would break through the low atmosphere and run along the vacuum toward the equator. The electricity would be most visible at high latitudes, where it is dense, and much less visible at lower latitudes, where it diverges. Franklin claimed such an effect would "give all the appearances of an Aurora Borealis."[1,3]

Numerous other theories of the aurora have been proposed over the last 150 years, including reflected sunlight from ice particles, reflected sunlight from clouds, sulfurous vapors, combustion of inflammable air, luminous magnetic particles, meteoric dust ignited by friction with the atmosphere, cosmic dust, currents generated by compressed cosmic ether, thunderstorms, electric discharges between the Earth's magnetic poles, and electric discharges between fine ice needles. A comprehensive and fascinating account of the aurora in science, history, and the arts is given in Reference 1, and additional theories are presented there.

Although early auroral theories did not fare very well, observations made during the latter half of the 1700s and throughout the 1800s elucidated many important auroral characteristics. In 1790, the English scientist Cavendish used triangulation and estimated the height of auroras at between 52 and 71 miles.[4] In 1852, the relationship among geomagnetic disturbances, auroral displays, and sunspots was clearly established; the frequency and amplitude of these features varied with the same 11-year periodicity.[5,6] In 1860, Elias Loomis drew the first diagram of the region where auroras are most frequently observed and noted that the narrow ring is not centered on the geographic pole, but that its oval form resembles lines of equal magnetic dip, thereby establishing the relationship between the aurora and the geomagnetic field. In 1867, the Swedish physicist Ångström made the first measurements of the auroral spectrum.[7] However, a significant breakthrough in auroral physics was not achieved until the end of the nineteenth century, when cathode rays were discovered and identified as electrons by the British physicist J. J. Thomson. Subsequently, the Norwegian physicist Kristian Birkeland proposed that the aurora was caused by a beam of electrons emitted by the Sun. Those electrons reaching the Earth would be affected by the Earth's magnetic field and guided to the high-latitude regions to create the aurora.

Until the discovery of sunspots by Galileo in 1610, the Sun was generally thought to be a quiet, featureless object. Galileo not only discovered the dark spots but also noted their westward movement, which was the first indication that the Sun rotates. In subsequent observations, it was quickly established that the number of sunspots varies with time. It was not until more than two centuries later, however, that an amateur astronomer in Germany, Heinrich Schwabe, noted an apparent 10-year periodicity in his 17 years of sunspot observations.[8] Shortly after Schwabe's discovery, professional astronomers set out to determine whether or not the cycle was real. The leader of this effort was Rudolf Wolf of the Zürich observatory. Wolf conducted an extensive search of past data and was able to establish that the number of sunspots varied with an 11-year cycle that had been present since at least 1700.[9] In 1890, Maunder called attention to the 70-year period from 1645 to 1715, when almost no sunspots were observed.[10] This period, which is known as the Maunder Minimum Period, raises the question whether the sunspot cycle is a universal feature or just a recent phenomenon.

As defined at the beginning of this chapter, the terrestrial ionosphere begins at an altitude of about 60 km and extends beyond 3000 km, with the peak electron concentration occurring at approximately 300 km. The first suggestion of the existence of what is now called the ionosphere can be traced to the 1800s. Carl Gauss and Balfour Stewart hypothesized the existence of electric currents in the atmosphere to explain the observed variations of the magnetic field at the surface of the Earth. Gauss argued:[11]

> It may indeed be doubted whether the seat of the proximate causes of the regular and irregular changes which are hourly taking place in this [terrestrial magnetic] force, may not be regarded as external in reference to the Earth ... But the atmosphere is no conductor of such [galvanic] currents, neither is vacant space. But our ignorance gives us no right absolutely to deny the possibility of such currents; we are forbidden to do so by the enigmatic phenomena of the Aurora Borealis, in which there is every appearance that electricity in motion performs a principal part.

It had been well established that there was a direct correlation between the solar cycle and magnetic disturbances on the Earth. To account for this strong correlation, Stewart speculated that electrical currents must flow in the Earth's upper atmosphere, and that the Sun's action is responsible for turning air into a conducting medium.[12] It was also concluded that the conductivity of the upper atmosphere is higher at sunspot maximum than at sunspot minimum. This view, however, was not widely accepted and strong counterarguments were presented in 1892 by Lord Kelvin.

The existence of the ionosphere was clearly established in 1901 when G. Marconi successfully transmitted radio signals across the Atlantic. This experiment indicated that radio waves were deflected around the Earth's surface to a much greater extent than could be attributed to diffraction. The following year, A. E. Kennelly and O. Heaviside suggested that free electrical charges in the upper atmosphere could reflect radio waves.[13] That same year, the first *physical* theory of the ionosphere was proposed.[14]

> The observed effect, which if confirmed is very interesting, seems to me to be due to the conductivity ... of air, under the influence of ultra-violet solar radiation. No doubt electrons must be given off from matter ... in the solar beams; and the presence of these will convert the atmosphere into a feeble conductor.

In 1903, J. E. Taylor independently suggested that solar ultraviolet radiation was the source of electrical charges, which implied solar control of radio propagation.[15] The first rough measurements of the height of the reflecting layer were made by Lee de Forest and L. F. Fuller at the Federal Telegraph Company in San Francisco from

1912 to 1914. The reflecting layer's height was deduced using a transmitter–receiver spacing of approximately 500 km, which was determined by the circuits of the Federal Telegraph Company.[16] However, the de Forest–Fuller results were not well known, and generally accepted measurements of the height of the reflecting layer were made in 1924 by Breit and Tuve[17] and by Appleton and Barnett.[18] The Breit–Tuve experiments involved a "pulse sounding" technique, which is still in use today, while Appleton and Barnett used "frequency change" experiments, which demonstrated the existence of downcoming waves by an interference technique. These experiments led to a considerable amount of theoretical work, and in 1926 the name "ionosphere" was proposed by R. A. Watson-Watt in a letter to the United Kingdom Radio Research Board, but it did not appear in the literature until three years later.[19] Radio soundings of the ionosphere initially seemed to indicate that the ionosphere consisted of distinct layers; we now know that this is generally not the case and we refer to different regions. These regions are called the D, E, and F regions. The names of these regions originated with Appleton, who stated that in his early work he wrote E for the reflected electric field from the first layer that he recognized. Later, when he recognized a second layer at higher altitudes, he wrote F for the reflected field. Subsequently, he conjectured that there may be another layer at lower altitudes so he decided to name the first two layers E and F and the possible lower one D, thus allowing the alphabetical designation of other undiscovered layers.[20]

The rocket technology available at the end of World War II was used by scientists to study the upper atmosphere and ionosphere, paving the way for space exploration via satellites. The first rocket-borne scientific payload, which carried instrumentation to make measurements directly in the upper atmosphere and ionosphere, was launched in 1946 on a *V-2* from White Sands, New Mexico. The University of Michigan payload consisted of a Langmuir probe and a thermionic pressure gage; although the *V-2* failed during this flight it marked the beginning of direct exploration of the ionosphere. The first book devoted to the ionosphere was published in 1952 by Rawer.[21]

The rocket technology, coupled with a major advance in ground-based instrumentation, led scientists to realize that a dramatic increase in our knowledge of the terrestrial environment was possible. To take advantage of these new capabilities, the International Geophysical Year (IGY), 1957–1958, was organized.[22,23] This cooperative effort was to begin with the next maximum of the solar cycle. As part of the IGY, scientists proposed to launch artificial satellites, and eventually *Sputnik 1* was launched on October 4, 1957.

Many consider the launch of *Sputnik 1* the beginning of the Space Age, but to some degree it started much earlier. Rockets have been with us ever since the ancient Chinese used them for fireworks. Later variations of "rockets" were used, basically for military purposes, to send payloads from one location to impact at another. Newton developed the scientific basis to describe how an object could be placed in orbit around the Earth, and visionaries like Jules Verne and H. G. Wells dreamt such thoughts.

The modern era of rocket propulsion began in Russia in the 1880s, where Konstantin Tsiolkovsky worked out the fundamental laws of rocket propulsion and published his work proving the feasibility of achieving orbital velocities by rockets at the turn of the century. He had earlier described the phenomenon of weightlessness in space, predicted Earth satellites, and suggested the use of liquid hydrogen and oxygen as propellants. Robert H. Goddard, a high school physics teacher in Massachusetts, was not aware of Tsiolkovsky's work and independently began studying rocket propulsion after World War I. On March 16, 1926, he launched the first liquid fuel rocket, which burned for only 2.5 seconds and landed a couple of hundred feet away from the "launch site." He continued to work, supported by the Guggenheim Foundation, in seclusion from the press, which ridiculed him. The third rocket pioneer was Hermann Oberth of Germany, also a school teacher. His work gained a great deal of attention and support and eventually led to the development of the *V-2* rocket (the first operational liquid fuel rocket).

After World War II, part of the German team responsible for the development of the *V-2*, including Wernher von Braun, came to the United States, while others went to the Soviet Union. Some of the captured *V-2* rockets that were brought to the United States were used to carry scientific payloads; these flights started in May 1946 from White Sands, New Mexico. A year later, the first Soviet *V-2* was launched from Kapustin-Yar. The limited supplies of *V-2*s and the estimated large expense of reproducing them led to the development of new sounding rockets for scientific research. The first one of these was the liquid fueled *Aerobee*; other rockets, many of them having a military heritage, followed later. The ascent of the Cold War spurred the development of Intercontinental Ballistic Missiles (ICBMs), but much of this effort was highly secret. Reading the history of repetitive studies, interservice jealousies, politicking, backbiting, and bickering provides a fascinating view of this secret world of the 1950s.[24,25]

The first US study regarding the feasibility of artificial satellites can be traced to 1945 when a Navy committee concluded that they were possible, but nothing developed at that time. After a failed Navy–Air Force collaborative effort, the Air Force conducted an independent study and concluded that the United States could launch a 500 pound satellite by 1951. Again, this suggestion was not pursued. However, with pressure from the American Rocket Society and the scientists involved in planning for the IGY, serious consideration was at last given to the launch of a small satellite for scientific purposes. Specifically, on July 29, 1955, a White House announcement indicated that the United States would launch "small unmanned Earth-circling satellites as part of the US participation in the IGY." Two days later, the Soviet Union announced it would also launch artificial satellites as part of the IGY in the late summer or early autumn of 1957. However, this announcement was basically ignored by the US press and public, possibly because it was believed that the Russians did not have the required technology. In the United States, all three military services proposed to launch the first satellite, which was to be placed in orbit during the 1957–8 time period. The Air Force proposed to use the *Atlas* ICBM, the Army proposed to use the *Jupiter C* Intermediate Range Ballistic Missile (IRBM), and the Navy proposed to develop a new rocket that did not have a military heritage (the *Vanguard*). The Vanguard Project was chosen primarily because it would not interfere with the existing military missile programs and because it seemed more appropriate to use a nonmilitary missile for a scientific mission. Despite not being selected, the Army's design of the *Jupiter C* IRBM contained a fourth stage, which appeared to have no specific military function. In September 1956, when the *Jupiter C* was ready to be launched, the Pentagon was so concerned that the Army might take "the glory" away from the Navy's Vanguard Project that von Braun was personally ordered to make sure that the fourth stage was not live. The launch was successful, and with a live fourth stage, the *Jupiter C* could have placed a satellite in orbit.

On October 4, 1957, the Soviet Union launched *Sputnik ("Traveling Companion") 1*, which was an 83 kg satellite. *Sputnik 2*, a 507 kg satellite followed on November 3. This created a tremendous public and political reaction in the United States.

Vanguard was still given a first chance, but the launch attempt on December 6, 1957, was a televised public failure (the second launch attempt on February 5, 1958, was also a failure). In the meantime the Army was given the green light to proceed with a *Jupiter C* launch, and an 8 kg satellite named *Explorer I* was successfully placed in orbit on January 31, 1958. *Explorer I* carried a small Geiger counter supplied by James Van Allen of the University of Iowa. The instrument was supposed to record the presence of cosmic rays, which are very fast particles from deep space; but surprisingly the instrument showed no response when the satellite was at high altitudes. There seemed to be no logical explanation, but a second instrument flown two months later confirmed the result. A graduate student (Carl McIlwain) working with Van Allen solved the problem. He suggested that the satellite encountered a region of very intense energetic particle fluxes, which saturated the Geiger tube and caused the counting circuits to read zero. Thus, the Van Allen radiation belts were discovered.[26]

The large international cooperative efforts, the vast amount of geophysical data collected, and the launch of artificial satellites, which began during the IGY, led to the birth of solar–terrestrial physics. The subsequent major infusion of money into this area by several countries led to a rapid advance in our knowledge of the Earth's environment. In the early phase of these explorations, every measurement yielded new and exciting results. A phase has now been reached where detailed measurements are available and theoretical models are generally able to explain and reproduce the observed large-scale features of the terrestrial ionosphere. This does not imply that a complete understanding has been achieved and there is nothing more to learn. On the contrary, the time has been reached when the problems that need further study can be clearly defined and then attacked in a systematic manner.

1.3   Specific references

1. Eather, R. H., *Majestic Lights*, American Geophysical Union, Washington, DC, 1980.
2. Euler, L., Recherches physiques sur la cause des queues comètes de la lumière boreale et de la lumière zodiacale, *Hist. Acad. Roy. Sci. Belles Lett. Berlin* **2**, 117, 1746.

3. Franklin, B., *Political, Miscellaneous and Philosophical Pieces*, ed. by B. Vaughan, London: J. Johnson, 504, 1779.

4. Cavendish, H., On the height of the luminous arch which was seen on February 23, 1784, *Phil. Trans. Roy. Soc.*, **80**, 101, 1790.

5. Sabine E., On periodical laws discoverable in the mean effects of the larger magnetic disturbances, *Phil. Trans. Roy. Soc.*, **142**, 103, 1852.

6. Wolf, R., *Acad. Sci.*, **35**, 364, 1852.

7. Angström, A., Spectrum des Nordlichts, *Ann. Phys.*, **137**, 161, 1869.

8. Schwabe, S. H., Die Sonne, *Astron. Nachr.*, **20**, 280, 1843.

9. Wolf, R., Neue Untersuchungen über die Periode der Sonnenflecken, *Astron. Mitt. Zürich*, No. 1, 8, 1856.

10. Maunder, E. W., The prolonged sunspot minimum, 1645–1715, *MNRAS*, **50**, 251, 1890.

11. Gauss, C. F., General theory of terrestrial magnetism, English translation in *Scientific Memoirs*, (ed. R. Taylor), vol. 2, 184, London, 1841.

12. Stewart, B., Aurora Borealis, in *Encyclopaedia Britannica*, 9th Edn., 36, 1882.

13. Ratcliffe, J. A., The ionosphere and the engineer, *Proc. Inst. Elec. Eng. (London)*, **114**, 1, 1967.

14. Lodge, O., Mr. Marconi's results in day and night wireless telegraphy, *Nature*, **66**, 222, 1902.

15. Taylor, J. E., Characteristics of electric earth-current disturbances, and their origin, *Proc. Phys. Soc. (London)*, **LXXI**, 225, 1903.

16. Villard, O. G., The ionospheric sounder and its place in the history of radio science, *Radio Sci.*, **11**, 847, 1976.

17. Breit, G. and M. A. Tuve, A radio method of estimating the height of the conducting layer, *Nature*, **116**, 357, 1925.

18. Appleton, E. V. and M. A. F. Barnett, Local reflection of wireless waves from the upper atmosphere, *Nature*, **115**, 333, 1925.

19. Watson-Watt, R. A., Weather and wireless, *Q. J. Roy. Meteorol. Soc.*, **55**, 273, 1929.

20. Silberstein, R., The origin of the current nomenclature for the ionospheric layers, *J. Atmos. Terr. Phys.*, **13**, 382, 1959.

21. Rawer, K., *Die Ionosphere*, Groningen: P. Noordhoff Ltd., 1952.

22. Van Allen, J. A., Genesis of the International Geophysical Year, in *History of Geophysics*, ed. by C. S. Gillmor, American Geophysical Union, **4**, 49, 1984.

23. Nicolet, M., Historical aspects of the IGY, in *History of Geophysics*, ed. by C. S. Gillmor, American Geophysical Union, **44**, 44, 1984.

24. Emme, E. M., *Aeronautics and Astronautics, 1915–1960*, Washington, DC: US Government Printing Office, 1961.

25. Thomas, S., *Men in Space*, vol. 2, 218, Philadelphia: Chilton Press, 1961.

26. Van Allen, J. A., Radiation belts around the earth, *Sci. Amer.*, **200**, 39, 1959.

1.4   General references

Akasofu, S.-I., *Exploring the Secrets of the Aurora*, Springer, Netherlands, 2007.

Eather, R. H., *Majestic Lights*, American Geophysical Union, Washington, DC, 1980.

Hess, W. N., *The Radiation Belt and Magnetosphere*, Blaisdell Publishing Co., Waltham, MA, 1968.

Newell, H. E., *Beyond the Atmosphere*, NASA SP-4211, Washington, DC, 1980.

Rishbeth, H., H. Kohl, and L. W. Barclay, A history of ionospheric research and radio communications, in *Modern Ionospheric Science*, European Geophysical Society, Katlenburg-Lindau, FRG, 1996.

Van Allen, J. A., *Origins of Magnetospheric Physics*, Smithsonian Institution Press, Washington, DC, 1983.

Chapter 2

Space environment

Before discussing the various ionospheres in detail, it is necessary to describe the physical characteristics of the bodies in the solar system that possess ionospheres as well as the plasma and electric–magnetic environments that surround the bodies because they determine the dynamical processes acting within and on the ionospheres. It is also useful to give a brief overview of the characteristics of the different ionospheres, including those associated with planets, moons, and comets. This not only allows the reader to see easily the diversity of ionospheric characteristics and features, but also provides motivation for the fundamental physics and chemistry covered in later chapters. In what follows, the sequence of the discussion is the Sun, the interplanetary medium, the Earth, the inner and outer planets, and then moons and comets.

2.1   Sun

The Sun is a star of average mass (1.99×10^{30} kg), radius (6.96×10^5 km), and luminosity (3.9×10^{26} watts) whose remarkable steady output of radiation over several billion years has allowed life to develop on Earth. The Sun is composed primarily of hydrogen and helium, with small amounts of argon, calcium, carbon, iron, magnesium, neon, nickel, nitrogen, oxygen, silicon, and sulfur. The solar energy is generated from the nuclear fusion of hydrogen into helium in a very hot central *core*, which is about 16 million kelvins. This energy is first transmitted through the *radiative zone* and then the *convective zone*, which is the outer 2.00×10^5 km of the Sun. The Sun's surface is irregular because of the strong convection in this outer zone, displaying both small-scale and large-scale convective cells or *granules*. The small-scale cells are about 1000 km in diameter, with individual cells lasting for approximately

10 minutes. On the large scale, there are networks of cells (*supergranules*) that have dimensions of about 30 000 km and can last as long as an Earth day.

The Sun's atmosphere, which extends out to beyond 10 solar radii, is composed of three regions, consisting of the photosphere, chromosphere, and corona. The *photosphere* is a very thin, cool layer from which the visible radiation is emitted. The temperature in this layer decreases with radial distance from about 6000 K at its sunward boundary to a minimum of about 4500 K near the photosphere–chromosphere boundary. The *chromosphere* is also a relatively thin layer (~4000 km) in which the temperature increases rapidly from the temperature minimum of 4500 K to about 25 000 K near the base of the outer atmosphere. This third region, or *corona*, contains a very tenuous, hot (~10^6 K), ionized plasma that typically extends several radii from the Sun.

Close to the Sun the solar magnetic field is basically dipolar, but there is an offset between the rotational and dipole axes (Figure 2.1). Hot plasma can be trapped on these closed field lines and its presence can be detected via the electromagnetic radiation that it emits. However, away from the Sun, the high coronal temperatures cause a continuous outflow of plasma from the corona, which is called the *solar wind*. As this hot plasma flows radially away from the Sun, it tends to drag the dipolar magnetic field lines with it into interplanetary space. At times, the solar wind can be very nonuniform because the magnetic field in the corona can be highly structured, as shown schematically in Figure 2.2. Hot coronal plasma can be trapped on strong magnetic field loops, and a very intense X-ray emission is associated with these *coronal loops*. Depending on the strength of the magnetic field, some hot plasma can slowly escape from these loops, forming *coronal streamers* that extend into space. These streamers are the source of the slow component of the solar wind.

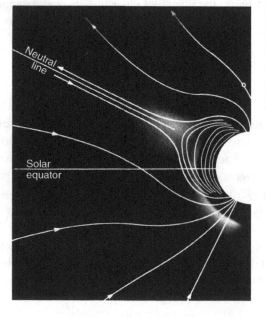

Figure 2.1 A photograph of the white-light corona above the east limb of the Sun on June 5, 1973. The solid lines correspond to a suggested magnetic field geometry that is consistent with the plasma distribution emitting the white light.[1]

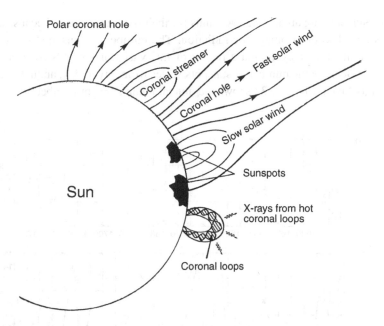

Figure 2.2 Schematic diagram of the magnetic field topology in the solar corona and the associated coronal features. The solid curves with arrows are the magnetic field lines.[2]

However, at other places in the corona, the Sun's magnetic field does not loop, but extends in the radial direction. In these regions, the hot plasma can easily escape from the corona, which leads to the high-speed component of the solar wind. As a result of this rapid escape, the plasma densities and associated electromagnetic radiation are low, and consequently, these regions have been named *coronal holes*. Typically, coronal holes are transient features that vary from day to day, but during quiet solar conditions, extensive coronal holes can exist at the Sun's polar regions. In the polar regions, the magnetic field lines extend into deep space because the solar magnetic field is basically dipolar, and hence, hot plasma can readily escape along these field lines.

The Sun rotates with a period of about 27 days, but because the Sun's surface is not solid there is a differential rotation between the equator (25 days) and the poles (31 days). This rotation and plasma convection act to produce intense electric currents and magnetic fields via a dynamo action. However, the magnetic fields that are generated display a distinct temporal variation. Specifically, there is an overall increase and decrease in magnetic activity that follows a 22-year cycle, which coincides with the change in polarity of the Sun's magnetic poles. One of the primary manifestations of solar magnetic activity is the appearance of *sunspots*, which are dark regions on an active Sun (Figure 2.2). Sunspots, which can last from several hours to several months, are located in the photosphere and are a result of stormy localized magnetic fields (several thousand gauss). The stormy magnetic fields choke the flow of energy from below, and consequently, sunspots are cooler than the surrounding area, which accounts for their dark appearance because cooler regions emit less electromagnetic radiation. The number of sunspots is known to vary with an 11-year cycle and a record of this variation extends for more than 300 years. Because the number of sunspots varies from day to day, annual averages are usually taken. Figure 2.3a shows the annual mean sunspot numbers from 1610

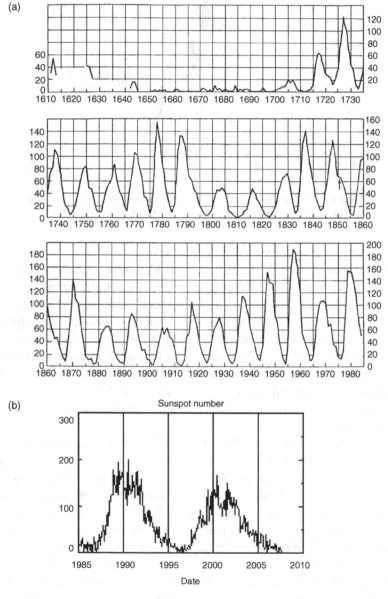

Figure 2.3 (a) Annual mean sunspot numbers from 1610 to 1985. The numbers before 1650 are not reliable.[3, 4] (b) Sunspot numbers from 1985 to 2008. (From NASA Marshall Space Flight Center.)

1632 UT   1724 UT   1809 UT

Figure 2.4 A rising prominence, as seen in a sequence of photographs taken on September 8, 1948.[5]

to 1985 and Figure 2.3b shows sunspot numbers from 1985 to 2008. Clearly evident in the figure is the 11-year sunspot cycle. However, during the 1600s there was very little solar activity and this period is known as the Maunder Minimum Period.

Sometimes there are powerful explosions in the atmosphere above sunspots, which are called *solar flares*. These bright flashes of light last only a few minutes to a few hours, but the explosions send bursts of energetic particles into space. Another kind of solar explosion stems from a *prominence* (Figure 2.4). The prominence extends far into the Sun's upper atmosphere and follows the loop of a closed magnetic flux tube, with the ends of the loop rooted in sunspots. The strong, curved magnetic field traps hot plasma, and because of intense heating, thermal conduction fronts can race through the loops, raising the temperature to 20–30 million kelvins. At times, one of the ends of the magnetic flux loop breaks free, sending streams of energetic plasma into space. Another form of mass release is called a *coronal mass ejection* (CME). Coronal mass ejections were once thought to be initiated by flares, but it is now known that most CMEs are not associated with flares. Coronal mass ejections expand as they move away from the Sun, at speeds as high as $1000 \ km \ s^{-1}$. Large CMEs contain as much as 10^{16} g of plasma. Figure 2.5 shows snapshots of a CME moving away from the Sun on October 24, 1989. In this figure a black disk apparently 1.6 times the diameter of the Sun blocks the bright sunlight so that the CME can be observed.

The loss of energy from the Sun is due to both electromagnetic radiation and particle outflow, with radiated energy being by far the dominant loss process. Table 2.1 shows the wavelength ranges for the different solar spectral regions. The radiated energy per second in all wavelengths is approximately constant and at the Earth it is 1370 watts m^{-2}, which is called the *solar constant*. The main energy contributions are from the infrared (52%), visible (41%), and ultraviolet (<7%) spectral regions, and the energy associated with these regions is steady. The radio and X-ray emissions display large fluctuations, but they are minor contributors to the total radiated energy. The energy loss due to particle outflow (solar wind and CMEs) is also very small, as shown in Table 2.2. However, as will be discussed later, the solar wind and coronal mass ejections have a dramatic effect on planetary ionospheres and atmospheres. Likewise, extreme ultraviolet (EUV) radiation, which amounts to only about 0.1% of the total radiated energy, is a critical source of plasma in planetary ionospheres.

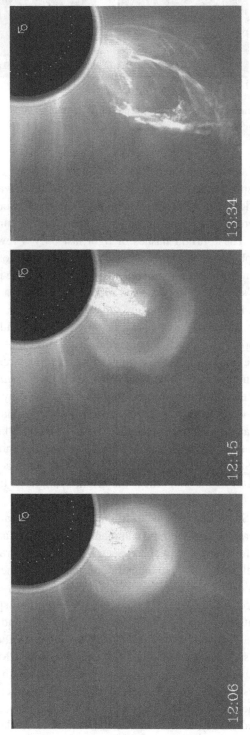

Figure 2.5 Snapshots of a coronal mass ejection that occurred on October 24, 1989. The images were obtained from the High Altitude Observatory in Boulder, Colorado. A black disk, apparently 1.6 times the Sun's diameter, blocks the bright sunlight so that the CME can be observed.[6]

Table 2.1 Solar spectral regions.

Radio	$\lambda > 1\,\text{mm}$
Far infrared	$10\,\mu\text{m} < \lambda < 1\,\text{mm}$
Infrared	$0.75\,\mu\text{m} < \lambda < 10\,\mu\text{m}$
Visible	$0.3\,\mu\text{m} < \lambda < 0.75\,\mu\text{m}$
Ultraviolet (UV)	$1200\,\text{Å} < \lambda < 3000\,\text{Å}$
Extreme ultraviolet (EUV)	$100\,\text{Å} < \lambda < 1200\,\text{Å}$
Soft X-rays	$1\,\text{Å} < \lambda < 100\,\text{Å}$
Hard X-rays	$\lambda < 1\,\text{Å}$

Note: $\text{Å} = 10^{-10}\,\text{m}$.

Table 2.2 Energy and mass loss from the Sun.[7]

Radiated power	3.8×10^{26} watts
Solar wind power	4.1×10^{20} watts
CME power	7.0×10^{18} watts
Mass loss (radiation)	4.2×10^9 kg s^{-1}
Mass loss (particles)	1.3×10^9 kg s^{-1}

2.2 Interplanetary medium

Prior to the 1950s it was generally believed that interplanetary space was a vacuum, except for the occasional bursts of energetic particles associated with solar flares. However, because of satellite measurements, it is now known that the solar wind is a continuous source of plasma for this region. The solar wind outflow starts in the lower corona and the velocity steadily increases as the plasma moves radially away from the Sun. At a distance of a few solar radii, the solar wind becomes *supersonic*, which means its outward bulk velocity becomes greater than the characteristic wave speeds in the medium. At about the same distance, the rarefied solar wind plasma becomes *collisionless*; that is, the collisional mean free path exceeds the characteristic scale length for density changes. In a collisionless plasma, electric currents flow with little resistance. As a consequence, the solar magnetic field, which resembles a dipole close to the Sun, gets "frozen" into the solar wind and is carried with it into space, becoming the *interplanetary magnetic field* (IMF).

As the magnetic field is drawn outward by the radial solar wind, the Sun's slow rotation (2.7×10^{-6} rad s^{-1}) acts to bend the field lines into spirals that extend deep into space (Figure 2.6). At the Earth's orbit, the spiral angle is approximately $43°$ with respect to a line that connects the Sun and Earth. In three dimensions, the spirals can be described by the *ballerina skirt* model.[8] The skirt represents a sheet of current that flows in an azimuthal direction around the Sun, but the skirt has a wavy structure that is similar to a ballerina's skirt (Figure 2.7). The magnetic fields on the opposite sides of this *heliospheric current sheet* have opposite polarity, and

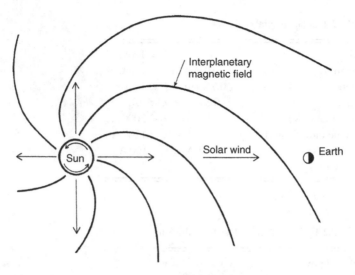

Figure 2.6 Schematic diagram of the Sun–Earth system in the Sun's ecliptic plane. The solar wind is in the radial direction away from the Sun and the magnetic field lines bend into spirals as the Sun slowly rotates.

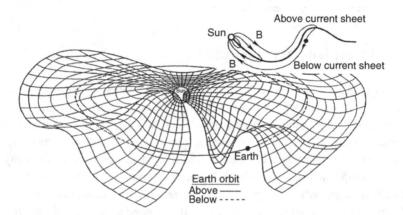

Figure 2.7 Schematic diagram of the three-dimensional structure of the current sheet that flows in an azimuthal direction around the Sun. The inset at the top of the figure shows the opposite polarities of the magnetic fields on the two sides of the current sheet.[9] (Courtesy of S.-I. Akasofu, Geophysical Institute, University of Alaska).

as the different folds of the skirt drape the various bodies in the solar system, they are exposed to different IMF polarities. The polarity of the whole system reverses at the beginning of each new 11-year cycle because of the reversal in polarity of the Sun's magnetic poles.

The formation of shocks in the interplanetary medium can have important consequences for the various ionospheres because of the strong impulsive force associated with them. Shocks can form when a fast solar wind stream overtakes a slower moving solar wind, as shown schematically in Figure 2.8. This figure shows the rotating

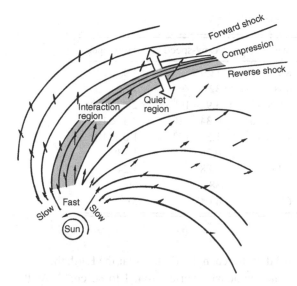

Figure 2.8 Schematic diagram showing the conditions leading to the formation of forward and reverse shocks in the solar wind.[9] Courtesy of D. S. Intriligator, Carmel Research Corporation.

Sun in the *ecliptic plane*, which is the plane containing the orbits of the planets, and the associated radial solar wind and spiral magnetic field lines. When the solar wind is slow, the spirals are tightly coiled. However, when a coronal hole rotates around, the high-speed stream associated with it also leads to spiral magnetic field lines, but they are not as tightly coiled because of the higher outward velocity. As the high-speed stream overtakes the slow solar wind that is ahead of it, there is a density compression at its leading edge and a rarefaction at its trailing edge. If the velocity difference between the high- and low-speed streams is greater than the local sound speed, *forward and reverse shocks* form. In a frame of reference that is fixed to the high-speed stream, the forward and reverse shocks are seen to propagate in opposite directions away from the compression zone. However, these shocks propagate in a plasma that is streaming away from the Sun, and hence, in an inertial reference frame, the forward shock appears to be moving away from the Sun at a faster speed than the reverse shock.

Shocks can also form in association with coronal mass ejections. Some CMEs can become magnetically isolated from the Sun and then they become *plasmoids* or *magnetic clouds*. As the plasmoid moves rapidly toward the Earth, a shock wave can be driven ahead of it in the ambient plasma. The ambient plasma is then deflected around the plasmoid in such a way that the IMF drapes around the plasmoid. When this happens the magnetic field lines in the plasmoid form closed loops and an isolated magnetic cloud results.

The solar wind can vary markedly on an hourly basis and is highly structured throughout the solar system because of time variations, shocks, CMEs, and flares. Despite this marked variation, it is useful to provide average values for the parameters describing the interplanetary medium near the Earth, for which there is a large body of measurements. The Earth's orbit is approximately 217 solar radii from the Sun, which is defined to be one *astronomical unit* (1 AU $\approx$ 150 million km). The solar

Table 2.3 Solar wind parameters near the Earth.[10]

Parameter	Average	Low-speed	High-speed
$n(\text{cm}^{-3})$	8.7	11.9	3.9
$u(\text{km s}^{-1})$	468	327	702
$nu(\text{cm}^{-2}\,\text{s}^{-1})$	3.8×10^8	3.9×10^8	2.7×10^8
$T_\text{p}(\text{K})$	1.2×10^5	0.34×10^5	2.3×10^5
$T_\text{e}(\text{K})$	1.4×10^5	1.3×10^5	1.0×10^5
$(1/2m_\text{p}u^2/nu)(\text{erg cm}^{-2}\,\text{s}^{-1})$	0.70	0.35	1.13
β	2.17	1.88	1.24
$V_\text{A}(\text{km s}^{-1})$	44	38	66
$V_\text{S}(\text{km s}^{-1})$	63	44	81

wind plasma generally takes 2 to 3 days to reach the Earth. Near the Earth the speed ranges from 200 to 900 km s^{-1} and the density varies from 1 to 80 cm^{-3}. As the plasma moves away from the Sun, it expands and cools, with the electron temperature decreasing from about one million kelvin in the corona to about 100 000 K near the Earth. The interplanetary magnetic field also decreases with distance from the Sun, from about 1 gauss at the Sun's surface to about 3×10^{-5} gauss near the Earth.

Table 2.3 compares the plasma characteristics near the Earth for low-speed, high-speed, and average solar wind conditions. Given in this table are the plasma density ($n = n_\text{e} = n_\text{p}$), the drift velocity ($u = u_\text{e} = u_\text{p}$), the number flux ($nu$), the proton temperature (T_p), the electron temperature (T_e), the energy flux ($0.5m_\text{p}u^2/nu$), the ratio of kinetic to magnetic pressure (β), the Alfvén wave speed (V_A), and the ion-acoustic (sound) speed (V_S), where subscripts e and p refer to electrons and protons, respectively. The parameters β, V_A, and V_S are defined as

$$\beta = n_\text{p}k(T_\text{e} + T_\text{p})/(B^2/2\mu_0) \tag{2.1}$$

$$V_\text{A} = B/(\mu_0 n_\text{p} m_\text{p})^{1/2} \tag{2.2}$$

$$V_\text{S} = [k(T_\text{e} + 3T_\text{p})/m_\text{p}]^{1/2} \tag{2.3}$$

where B is the magnetic field, k is Boltzmann's constant, m_p is the proton mass, and μ_0 is the permeability of free space. Note that these parameters will be rigorously derived in later chapters. The fact that the β of the plasma is greater than unity for all solar wind conditions means that the magnetic field is relatively weak and is carried along with the flow. In and close to the solar corona, however, the β of the plasma is much less than unity, which indicates that the magnetic field is strong and directs the flow. Another interesting result shown in Table 2.3 is that the solar wind velocity is much greater than both the Alfvén wave speed and the ion-acoustic speed, which means that the solar wind is supersonic at 1 AU.

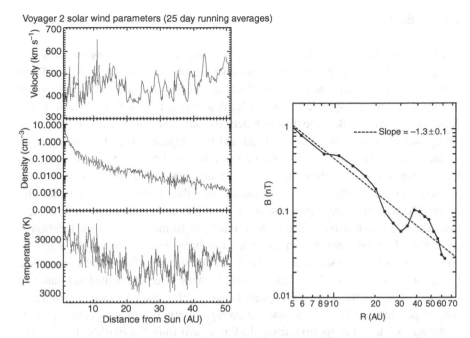

Figure 2.9 Variation of the solar wind velocity (top left panel), density (middle left panel), temperature (bottom left panel), and interplanetary magnetic field (right panel) with distance from the Sun. The left panels are from *Voyager 2* measurements (courtesy of J. D. Richardson) and the right panel is from *Voyager 1*.[11]

An indication of how the solar wind varies at distances beyond the Earth's orbit has been provided by the *Pioneer, Voyager, Galileo, Cassini,* and *Ulysses* spacecraft. Figure 2.9 shows the variation of the solar wind velocity, density, temperature, and IMF from 1 AU to past Pluto's orbit. The plasma parameters correspond to 25 day running averages and the magnetic field to yearly averages, as measured by instruments on the *Voyager 2* and *Voyager 1* spacecraft, respectively. Beyond the Earth's orbit, the solar wind speed is between 400 and 500 km s^{-1} regardless of the distance from the Sun. The solar wind density, on the other hand, displays a continuous decrease with distance. The density decrease is fairly rapid between the Earth (1 AU) and Saturn ($\sim$10 AU), with the density decreasing from about 10 to 0.08 cm^{-3} over this distance. Beyond 10 AU, the density decrease is not as rapid. The temperature displays a large variation at all locations even though the profile was constructed from 25 day averages. However, there is an overall temperature decrease, from about 50 000 K to 5000 K, between 1 and 20 AU (past Uranus's orbit), and then it is difficult to discern a clear trend. The variation of the observed magnitude of the magnetic field, B, beyond a few AU is basically consistent with the expected $1/r$ variation, where r is the radial distance from the Sun. The deviations from this simple behavior during the 18 years of the *Voyager 1* measurements are the result of solar cycle variations, changes in the solar wind velocity, and the increasing ecliptic latitude of the spacecraft location.

2.3   Earth

Comparisons of the physical characteristics of the planets and moons are shown in Tables 2.4 and 2.5, respectively. One of the important features to note is that the Earth possesses a strong intrinsic magnetic field. Because collisionless plasmas cannot readily flow across magnetic fields, the Earth's field acts as a hard obstacle to the solar wind, and the bulk of the flow is deflected around the Earth, leaving a magnetic cavity that is shaped like a comet head and tail (Figure 2.10). The head occurs on the sunward side of the Earth where the solar wind pressure acts to compress the geomagnetic field, while the solar wind flow past the Earth acts to produce an elongated tail on the side away from the Sun that extends well past the orbit of the Moon. When the supersonic solar wind hits the Earth's magnetic field, a free-standing shock wave, called a *bow shock*, is formed. The shock location is determined by a balance between the solar wind dynamic pressure and the magnetic pressure of the compressed geomagnetic field. The shock surface drapes around the Earth, and its shape and orientation vary with both the direction of the interplanetary magnetic field and the solar wind speed. However, the shock surface is symmetric with respect to the ecliptic plane, and the average location of the *nose* (closest point) of the shock surface is approximately 12 Earth radii from the Earth's surface. The bow shock is unusual in that it is a *collisionless shock*; the shock is a result of particle "collisions" with oscillating electric fields, in contrast to shocks around supersonic aircraft, which are caused by particle–particle collisions.

As the solar wind passes through the bow shock, it is decelerated, heated, and deflected around the Earth in a region called the *magnetosheath*. The magnetosheath thickness is approximately $3R_E$ (R_E denotes the Earth's radius) near the subsolar point, but it increases rapidly in the downstream direction. After being decelerated by the bow shock, the heated solar wind plasma is accelerated again from subsonic to supersonic flow as it moves past the Earth. The boundary layer that separates the magnetized solar wind plasma in the magnetosheath from that confined by the Earth's magnetic field is called the *magnetopause*. The magnetopause is generally very thin (~100 km), and its location is determined approximately by a balance between the dynamic pressure of the "shocked" solar wind and the magnetic pressure of the compressed geomagnetic field. Along the Earth–Sun line on the day side, the magnetopause radial position is approximately $9R_E$. An extensive current flows along the magnetopause, which acts to separate the solar wind's magnetic field from the geomagnetic field. On the front of the magnetopause, the current flow is primarily from dawn to dusk, but it acquires an increasing meridional (north–south) component as it flows around and past the Earth.

The domain where the Earth's magnetic field dominates is called the *magneto-sphere*. This large region, which encompasses the entire three-dimensional volume inside the magnetopause, is populated by thermal plasma and energetic charged particles of both solar wind and terrestrial origin. Although the bulk of the solar wind is deflected around the Earth in the magnetosheath, some of it can cross the magnetopause and enter the magnetosphere. Direct entry of solar wind plasma occurs

Table 2.4 Physical characteristics of the planets.

Planet	Mass (10^{23} kg)	Equatorial radius (10^3 m)	Equatorial surface gravitational acceleration[a] (m s^{-2})	Average distance from Sun (10^9 m)	Orbital eccentricity	Length of year (days or years)	Period of rotation (days)	Magnetic dipole moment ($T\text{m}^3$)
Mercury	3.30	2439.7	3.7	57.9	0.206	87.96 d	58.65	~ 3.5–$4.4(12)^b$
Venus	48.7	6052	8.87	108.2	0.007	224.7 d	-243	<4.3(11)
Earth	59.8	6378	9.77	149.6	0.017	365.3 d	1	8.06(15)
Mars	6.42	3397	3.693	228	0.093	686.98 d	1.026	<2(11)
Jupiter	18987	71492	20.87	778	0.048	11.86 y	0.41	1.6(20)
Saturn	5685	60268	7.207	1427	0.054	29.46 y	0.44	4.7(18)
Uranus	868	25559	8.43	2871	0.047	84.01 y	-0.718	3.8(17)
Neptune	1024	24764	10.71	4498	0.009	164.8 y	0.67	2.8(17)
Pluto	0.129	1151	0.4	5906	0.248	248.0 y	6.39	—

[a] Values are $GM/(R_{eq})^2$ and do not include rotational and wind effects.
[b] 4.4(12) $= 4.4 \times 10^{12}$, etc.

Table 2.5 Physical characteristics of selected satellites.

Satellite	Mean radius (km)	Mass (kg)	Mean position (km)	Mean position (planetary radii)	Orbital period (h)	Surface gravity (m s^{-2})
Callisto	2400	1.077 (23)^a	1.883 (6)	26.34 R_J	400.54	1.25
Ganymede	2631	1.48 (23)	1.070 (6)	14.97 R_J	171.72	1.43
Europa	1569	4.80 (22)	6.709 (5)	9.38 R_J	85.22	1.30
Io	1815	8.94 (22)	4.216 (5)	5.9 R_J	42.46	1.81
Titan	2575	1.346 (23)	1.222 (6)	20.28 R_S	383	1.35
Triton	1350	2.14 (22)	3.548 (5)	14.33 R_N	−141	0.78

^{a} 1.077(23) = 1.077 × 10^{23}

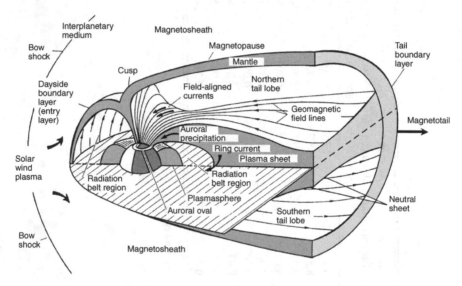

Figure 2.10 Schematic diagram of the Earth's bow shock and magnetosphere showing the various regions and boundaries.[9] (Courtesy of J. R. Roederer, Geophysical Institute, University of Alaska.)

on the day side in the vicinity of the *polar cusp* (or *cleft*). At low altitudes (∼300 km), the cusp occupies a narrow latitudinal band that is centered near noon, but is extended in longitude. Within this band, the solar wind particles can travel along geomagnetic field lines and deposit their energy in the upper atmosphere. Solar wind particles also get into the tail of the magnetosphere by mechanisms that have not yet been fully established. These solar wind particles, along with plasma that has escaped the Earth's upper atmosphere and has convected to the tail, populate a region known as the *plasma sheet*. However, the plasma sheet particles have an average energy 10 times larger than that found in the magnetosheath and a density that is lower by a factor of 10 to 100. The particles in the plasma sheet are not trapped, but have direct

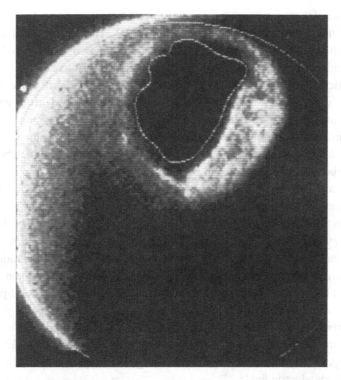

Figure 2.11 Earth's northern auroral oval as observed in the atomic oxygen emission at 130.4 nm with the *Dynamics Explorer 1* satellite at 1242 UT on November 11, 1981. The boundary of the polar cap is shown by the dotted curve.[12]

access to the Earth's upper atmosphere on the night side along specific magnetic field lines. At low altitudes, these field lines converge to a spatial region that is narrow in latitude, but longitudinally extended around the Earth, joining the day side cusp to form what is known as the *auroral oval*. Note that auroral ovals exist in both the northern and southern polar regions. As the plasma sheet particles stream toward the Earth along geomagnetic field lines, they get accelerated and then collide with the Earth's upper atmosphere, which acts to produce the auroral displays (Figure 2.11).

In addition to the plasma sheet flow toward the Earth that occurs on magnetic field lines that connect to the auroral ovals, there is a large-scale current flow across the plasma sheet from dawn to dusk, which is called the *neutral current sheet*. This dawn-to-dusk current acts to separate the two regions of oppositely directed magnetic fields in the magnetospheric tail; the magnetic field is toward the Earth above the neutral current sheet (northern hemisphere) and away from the Earth below the current sheet (southern hemisphere). Although these stretched magnetic field lines extend deep into the magnetospheric tail, near the magnetopause they get connected to the magnetic field embedded in the shocked solar wind. This magnetic connection acts to generate voltage drops across the magnetospheric tail larger than 100 000 volts, electric currents greater than 10^7 amps, and more than 10^{12} watts of power. The potential drop across the magnetospheric tail maps down to the *polar cap*, which is the region poleward of the auroral oval. The electric field that is generated points from dawn to dusk across the polar cap. As will be discussed later, this electric field has a major effect on the Earth's upper atmosphere.

The energetic particles near the center of the plasma sheet also drift closer to the Earth as a result of magnetospheric electric fields and then get trapped on closed geomagnetic field lines, thereby forming the *Van Allen radiation belts*. As these trapped high energy particles spiral along the closed geomagnetic field lines toward the Earth, they encounter an increasing magnetic field strength, are reflected, and then bounce back and forth between the northern and southern hemispheres. These trapped energetic electrons and protons (and at times a significant number of oxygen ions) also drift in an azimuthal direction around the Earth because of gradients in the geomagnetic field, with the electrons and protons drifting in opposite directions. The drift of the lower-energy (10–300 keV) particles results in a large-scale ring of current that encircles the Earth, which is called the *ring current*. A final aspect of the radiation belt and ring current that is important to note is that it prevents the dynamo-generated electric fields at high latitudes from penetrating to middle and low latitudes. Specifically, in response to penetrating high-latitude electric fields, the electrons and protons in the ring current polarize and set up an oppositely directed electric field that effectively cancels the penetrating high-latitude electric field. Hence, except for brief transient time periods, the mid- and low-latitude regions are generally not affected by magnetospheric electric fields.

Closer to the Earth is the *plasmasphere*, which is a torus-shaped volume that surrounds the Earth and contains a relatively cool ($\sim$5000 K), high-density ($\sim$$10^2$ cm^{-3}) plasma that has its origin in the Earth's ionosphere (Figure 2.12). The plasma in this region co-rotates with the Earth, but it can also flow along geomagnetic field lines from one hemisphere to the other. In the equatorial plane, the plasmasphere has a radial extent of about 4–8 R_E depending on magnetic activity, and its boundary, called the *plasmapause*, is typically marked by a large and sharp decrease in plasma density as one leaves the plasmasphere. The plasmapause is essentially the boundary between the plasma that co-rotates with the Earth and the plasma that is influenced by magnetospheric electric fields.

The Earth's atmosphere is the primary source of plasma close to the planet. It occupies a relatively thin, spherical envelope that extends from the Earth's surface to beyond 1000 km. Below about 90 km the atmosphere is mixed and the relative composition of the major constituents (N_2 and O_2) is essentially constant, although

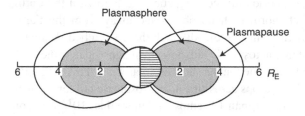

Figure 2.12 Schematic illustration of the plasmasphere and its bounding surface, which is called the plasmapause.[13]

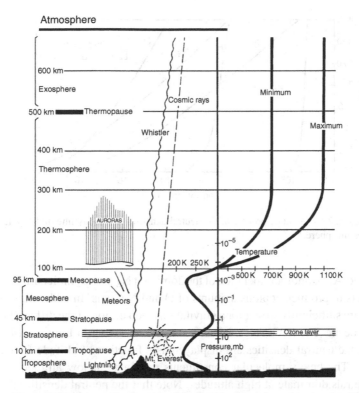

Figure 2.13 Schematic diagram of the Earth's atmosphere showing the different domains. The dark solid curves show atmospheric temperature profiles for solar maximum and minimum conditions.[9]

the atmospheric density decreases rapidly with altitude. However, the temperature in the lower atmosphere displays important variations with altitude that act to produce stratified layers, as shown in Figure 2.13. The layer closest to the Earth is the *troposphere*, which extends up to about 10 km and is the region normally associated with atmospheric weather. In this region the atmospheric temperature decreases with altitude up to a minimum value, which defines its upper boundary (the *tropopause*). Above this boundary is the *stratosphere*, which extends from about 10 to 45 km and is the region where the ozone layer exists. In the stratosphere the atmospheric temperature basically increases with altitude up to a local maximum, which defines its top boundary (the *stratopause*). In the next layer, which extends from about 45 to 95 km and is called the *mesosphere*, the atmospheric temperature decreases again to a local minimum at its upper boundary (the *mesopause*). The mesopause corresponds to the coldest region of the atmosphere, with the temperature getting as low as 180 K. Also, near and below the mesopause is the region where meteors can typically be seen streaking across the sky.

The *thermosphere* is the region of the Earth's upper atmosphere that extends from about 95 to 500 km. In this region, the atmospheric temperature first increases with altitude to an overall maximum value ($\sim$1000 K) and then becomes constant

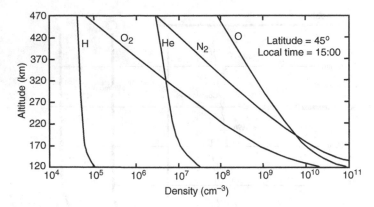

Figure 2.14 Altitude profiles of the neutral densities in the daytime mid-latitude thermosphere.[14]

with altitude. Also, photodissociation of the dominant N_2 and O_2 molecules is important and acts to produce copious amounts of O and N atoms. In addition, diffusion processes are sufficiently strong for a gravitational separation of the different neutral species to occur. The net effect of these processes is shown in Figure 2.14, where profiles of the neutral densities are displayed as a function of altitude for daytime conditions. The heavy molecular constituents dominate at low altitudes and the atomic neutrals dominate at high altitudes. Note that the neutral densities decrease exponentially with altitude at rates that are determined by the neutral masses. At about 500 km, the neutral densities become so low that collisions become unimportant and, hence, the upper atmosphere can no longer be characterized as a fluid. This transition altitude is called the *exobase*, and the region above it is called the *exosphere*, where the neutrals behave like individual ballistic particles.

The dynamics of the upper atmosphere during quiet geomagnetic activity are primarily controlled by solar heating on the day side. The thermospheric wind tends to blow horizontally from the subsolar heated region around the Earth to the coldest region on the night side. As the wind develops, Coriolis forces that are associated with the Earth's rotation act to deflect the flow. In addition, at high latitudes heating due to magnetospheric electric fields and particle precipitation acts either to retard or enhance the predominately anti-solar flow. The net effect of the magnetospheric processes is to decrease the anti-solar winds on the day side and increase them both in the polar caps and on the night side. Typically, the horizontal wind speeds in the upper thermosphere range from 100 to 300 m s^{-1} for quiet geomagnetic conditions, but they can approach 900 m s^{-1} over the polar caps during active magnetic conditions when the magnetospheric electric fields are large and the auroral precipitation is intense. Also, during active times, the upwelling associated with the magnetospheric heating processes can be sufficiently large to impede the gravitational separation of the neutral species. Such major changes in the thermospheric circulation and density structure have a significant effect on the charged particles embedded in the neutral gas.

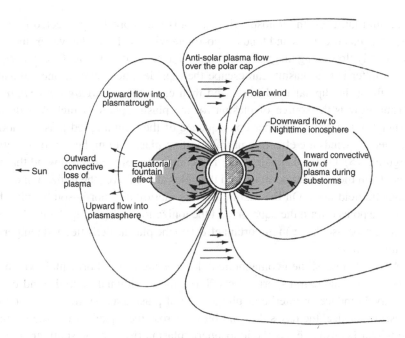

Figure 2.15 Schematic diagram showing the Earth's magnetic field and the plasma flow regimes in the ionosphere.[15]

The *ionosphere* is the ionized portion of the upper atmosphere. It extends from about 60 to beyond 1000 km and completely encircles the Earth. The main source of plasma for the ionosphere is photoionization of neutral molecules via solar EUV and soft X-ray radiation, although other production processes may dominate in certain regions. The ions produced then undergo chemical reactions with the neutrals, recombine with the electrons, diffuse to either higher or lower altitudes, or are transported via neutral wind effects. However, the diffusion and transport effects are strongly influenced by the Earth's intrinsic magnetic field, which is dipolar at ionospheric altitudes (Figure 2.15).

At high latitudes, the geomagnetic field lines extend deep into space in an anti-sunward direction. Along these so-called open field lines, ions and electrons are capable of escaping from the topside ionosphere in a process termed the *polar wind*. This loss of plasma can have an appreciable effect on the density and temperature structure. In addition, the dynamo electric field that is generated through the solar wind–magnetosphere interaction is mapped down to ionospheric altitudes and typically causes a two-cell plasma flow pattern, with antisunward flow over the polar cap and return flow equatorward of the auroral oval. This horizontal flow is continuous and its speed can be as high as 4 km s^{-1}. As a result, the high-latitude plasma is subjected to widely changing conditions as it drifts into different regions, including the sunlit hemisphere, the day side auroral oval, the polar cap, the nocturnal auroral oval, and the dark sub-auroral region. When it is in the auroral oval, the plasma is heated and ionization is produced due to precipitating energetic electrons.

At mid-latitudes, the ionospheric plasma is not appreciably affected by magnetospheric electric fields and tends to co-rotate with the Earth. However, the plasma can readily flow along magnetic field lines like beads on a string. One consequence of the latter is that plasma can escape the topside ionosphere in one hemisphere, flow along the dipolar field lines, and then enter the *conjugate ionosphere*. This plasma flow is the source of plasma for the plasmasphere, which was discussed earlier (Figure 2.12). Another consequence of the field-aligned plasma motion is that neutral winds are effective in transporting plasma to higher or lower altitudes (Figure 2.15). On the day side, there is a component of the neutral wind that blows away from the subsolar point toward the poles and it drives the ionization down the magnetic field lines. On the night side, this meridional (north–south) wind blows from the poles toward the equator, and the ionization is driven up the field lines. All of these processes have an important effect on the plasma densities and temperatures at mid-latitudes.

At low latitudes, the geomagnetic field lines are nearly horizontal, which introduces some unique transport effects. First, the meridional neutral wind can very effectively induce an interhemispheric flow of plasma along these horizontal field lines. At solstice, the day side wind blows across the equator from the summer to the winter hemisphere. As the ionospheric plasma rises on the summer side of the equator, it expands and cools, while on the winter side it is compressed and heated as it descends. Another interesting transport effect at low latitudes is the so-called *equatorial fountain*. In the daytime equatorial ionosphere, eastward electric fields associated with neutral wind-induced ionospheric currents drive a plasma motion that is upward. The plasma lifted in this way then diffuses down the magnetic field lines and away from the equator because of the action of gravity. The combination of electromagnetic drift and diffusion produces a fountainlike pattern of plasma motion, and this motion acts to produce plasma density enhancements on both sides of the magnetic equator, which are known as the *Appleton anomaly*.

Although different physical processes dominate in the different latitudinal domains, the electron density variation with altitude still displays the same basic structure at all latitudes. Specifically, the electron density profile exhibits a layered structure, with distinct D, E, F_1, and F_2 regions (Figure 2.16). In the D and E regions, chemical processes are the most important, molecular ions dominate, and N_2, O_2, and O are the most abundant neutral species. Additionally, in the D region (60–100 km), there are both positive and negative ions, water cluster ions, and three-body chemical reactions. The cluster ions dominate the D region at altitudes below about 85 km and their formation occurs via hydration starting from the primary ions NO^+ and O_2^+. In the E region (100–150 km), the basic chemical reactions are not as complicated, and the major ions are NO^+, O_2^+, and N_2^+. The total ion density is of the order of 10^5 cm^{-3}, while the neutral density is greater than 10^{11} cm^{-3}. Therefore, the E region plasma is *weakly ionized*, and collisions between charged particles are not important. In the F_1 region (150–250 km), ion–atom interchange and transport processes start to become important and in the F_2 region the ionization maximum occurs as a result of a balance between plasma transport and chemical

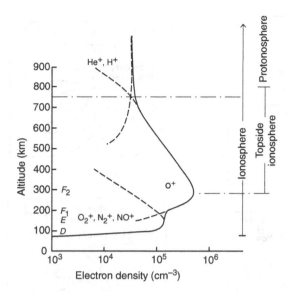

Figure 2.16 Representative ion density profiles for the daytime mid-latitude ionosphere showing the layered structure $(D, E, F_1, F_2$ layers).[16]

loss processes. In these regions, the atomic species (O^+ and O) dominate. The peak ion density in the F_2 region (10^6 cm^{-3}) is roughly a factor of 10 greater than that in the E region, while the neutral density (10^8 cm^{-3}) is still two orders of magnitude greater than the ion density. The plasma in this region is *partially ionized*, and collisions between the different charged particles and between the charged particles and neutrals must be taken into account. The *topside ionosphere* is generally defined to be the region above the F region peak, while the *protonosphere* is the region where the lighter atomic ions (H^+ and He^+) dominate. Although the neutrals still outnumber the ions in the protonosphere, the plasma is effectively *fully ionized* and only collisions between charged particles need to be considered. In both the topside ionosphere and protonosphere, plasma transport processes dominate.

2.4 Inner planets

2.4.1 Mercury

Figure 2.17 shows representative magnetospheres in the solar system. These sketches provide a rough idea of the scales and extent of these magnetospheres. The planet Mercury is unique among the inner planets in that it has a strong intrinsic magnetic field (Table 2.4). Given this strong magnetic field, a bow shock and a magnetosphere are formed around Mercury. The region of post shock, decelerated solar wind flow is called the magnetosheath, just as in the terrestrial case, and a long tail is also present. The planet is less than half the size of the Earth, so the different magnetospheric regions have appropriately scaled dimensions (e.g., the magnetopause stand-off distance is about 1460 km). Direct information about Mercury is extremely limited; until very recently all the available data were from three flybys of the planet by the

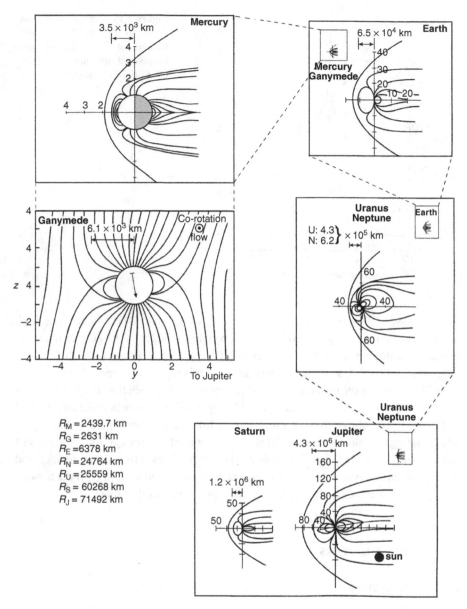

Figure 2.17 Schematic diagram showing the magnetospheres that are known to exist in the solar system. The axes are scaled relative to body radii. (Adapted from Reference 17.)

Mariner 10 spacecraft in 1974, but the *Messenger* spacecraft which is on its way to Mercury will provide a wealth of new information. It will fly by Mercury three times starting in 2008 and will go into orbit around the planet in 2011.

Mercury does not have a conventional, gravitationally bound atmosphere. *Mariner 10* optical observations indicated an upper limit on the day side surface density of about $1 \times 10^6 \, \text{cm}^{-3}$. Helium and atomic hydrogen were positively identified and atomic oxygen tentatively identified, with subsolar densities of about 4.5, 8 and $7 \times 10^3 \, \text{cm}^{-3}$, respectively.[18] In 1985, Earth-based optical observations established the presence of sodium and potassium; the sunlit column densities were estimated to be ~ 1–2×10^{11} and $\sim 1 \times 10^9$ atoms cm^{-2}, respectively.[19] Note, that these column densities are comparable to or less than the estimated sunlit helium column density of $\sim 3 \times 10^{11}$ atoms cm^{-2}. Given these very low neutral gas densities, Mercury does not have a conventional ionosphere; an ion exosphere is expected to be present.

2.4.2 Venus

As indicated in Table 2.4 Venus has no intrinsic magnetic field (of any significance), therefore its interaction with the solar wind is dissimilar to that of the Earth. The obstacle to the supersonic solar wind is Venus's ionosphere and atmosphere, and a well-established bow shock is present. Some details of the solar wind interaction and ionospheric processes and regions are sketched in Figure 2.18. Analogous to the magnetopause, a so-called *ionopause* is formed at Venus. This ionopause is a *tangential discontinuity*, (see Table 7.1) across which the total (kinetic, dynamic, and magnetic) pressure is constant and the normal components of the velocity and magnetic field are zero. At Venus, it is formed at a location where the kinetic pressure of the ionospheric plasma is approximately equal to the

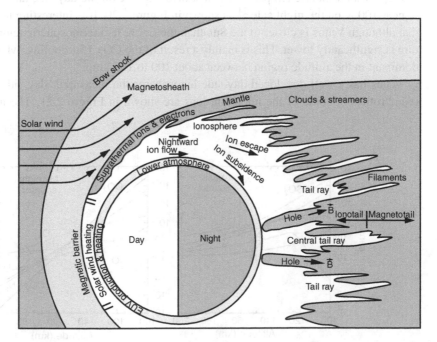

Figure 2.18 A schematic (not to scale) drawing of the plasma environment of Venus, showing some of the important regions and processes.[20]

dynamic pressure of the unperturbed solar wind. On the day side the solar wind dynamic pressure is transformed to magnetic pressure between the bow shock and the ionopause. This "piled" up magnetic field region, just outside the ionopause, is called the *magnetic barrier*. The shocked solar wind is deflected and flows around the ionopause; the region between the bow shock and the ionopause is called the *magnetosheath* or *ionosheath*. The interplanetary magnetic field (IMF) gets draped around the planet and a long tail extending to tens of Venus radii is created behind the planet.

The first indications and suggestions that the atmosphere of Venus is composed of CO_2 were based on ground-based observations made in the early 1930s.[21] These suggestions were confirmed by *in situ* measurements at Venus made from the *Venera 4* entry probe in 1967.[22] The upper atmosphere–ionosphere region of Venus is the most studied one of all the bodies in our solar system, except for the Earth, and now Titan. The surface pressure on Venus is about 100 times greater than that on the Earth and the surface temperature is about 750 K. CO_2 is by far the most abundant gas species near the surface. However, in the upper atmosphere above about 150 km, atomic oxygen becomes the dominant neutral. At even higher altitudes, helium, nonthermal atomic oxygen, and eventually atomic hydrogen become the main neutral species. Figure 2.19 shows representative upper atmospheric neutral density values from an empirical model, which is based on neutral mass spectrometer measurements.[23] At Venus the *exobase*, (see Section 10.10), the altitude above which collisions between the neutral atoms become negligible, is around 180 km. The upper atmospheric temperature is just below 300 K on the day side and drops to near 100 K on the night side,[24] as shown in Figure 2.20. It is interesting to note that although Venus is closer to the Sun than the Earth, its thermospheric temperature is significantly lower. This is mainly a result of the CO_2 15μ cooling, which is dominant in the altitude region between about 100 to 160 km.[25]

Venus has a well-developed day side ionosphere; the measured, day side altitude profiles of some of the important ions are shown in Figure 2.21. The major

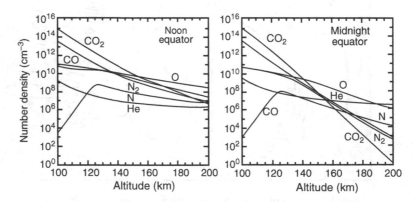

Figure 2.19 Representative neutral gas densities at Venus.[23]

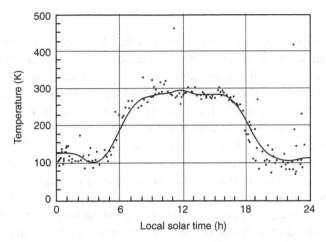

Figure 2.20 Measured kinetic temperatures of the upper atmosphere of Venus.[24]

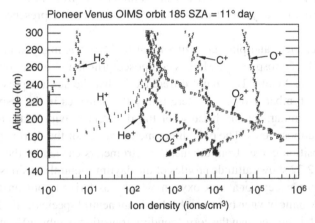

Figure 2.21 Measured ion densities in the Venus day side ionosphere.[26]

ion, near the peak altitude of about 140 km, is O_2^+, which was a surprise initially, because the major neutral species is CO_2 and there is essentially no O_2 in the upper atmosphere. However, it was soon realized that photochemical processes can easily explain the observed result, making Venus an excellent example of the importance of chemistry in controlling ionospheric behavior. A significant night side ionosphere was also observed at Venus.[27] This was also a surprise originally, because the night on Venus lasts about 58 Earth days. It was soon recognized that pressure gradients drive ionospheric plasma from the day side to the night side, helping to maintain a night side ionosphere. Low-energy electron impact ionization, somewhat similar to auroral precipitation, also contributes to the night side ionosphere. As indicated in Figure 2.18 the night side ionosphere is a very complex region with tail rays, filaments, streamers, and patches of plasma clouds.

2.4.3 Mars

Mars has no significant intrinsic magnetic field (Table 2.4), although some rem-
nant crustal magnetic anomalies of small spatial scales are present, mostly in the
southern hemisphere.[28] A well-defined bow shock has been observed around Mars,
and its magnetosheath has been extensively explored by the Phobos spacecraft[29]
and more recently, *Mars Express*.[30] The region of the piled up magnetic field outside the ionopause is possibly more complex at Mars than at Venus, although this
apparent difference may simply be the result of better and more data for Mars. This
region at Mars is commonly referred to as the *magnetic pile-up boundary* or *region*.
Only very limited and inconclusive information is currently available concerning
the ionopause location at Mars. The electron reflectometer, carried by the *Mars
Global Surveyor*, indicates a transition in the measured photoelectron fluxes; it
has been suggested that this change is related to the presence of an ionopause.[31]
If this interpretation is correct the day side ionopause is in the altitude region
between about 300 and 500 km. Radio occultation data from the *Mars Global Surveyor* and *Mars Express*, as well as the topside sounder data from *Mars Express*,
have not so far provided any direct definitive information on the presence of an
ionopause.

The upper atmosphere and plasma environment of Mars has many similarities
to that of Venus. The atmosphere of Mars is composed principally of carbon dioxide, as is the case for Venus. The major difference is that the surface pressure at
Mars is only about 6 mbar. However, interestingly the densities in the respective
thermospheres are similar, mainly because of the different gravity and temperatures in the lower atmospheres. The only direct measurement of the thermospheric
neutral gas composition comes from the mass spectrometers carried by the *Viking
Landers*.[32] Figure 2.22 shows altitude profiles of the daytime neutral densities based
on these observations, except for atomic oxygen, which was derived from ion density
measurements.[33] Atomic oxygen becomes the dominant neutral species at an altitude
near 200 km, which is higher than the corresponding transition height at Venus. The

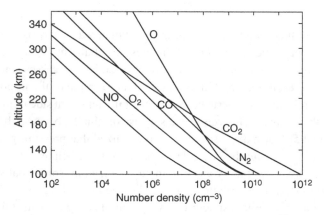

Figure 2.22 Representative neutral gas densities at Mars.[33]

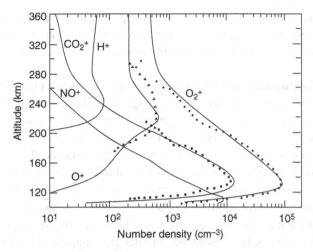

Figure 2.23 Measured and calculated ion densities for the day side ionosphere of Mars.[33]

exospheric neutral gas temperatures have been estimated to vary between about 175 and 300 K. These low temperatures are also caused by the CO_2 15μ cooling, but appear to vary with solar cycle more than the temperatures at Venus.[25] One of the only two directly measured ion density profiles for Mars is shown in Figure 2.23, and they are similar to the Venus profiles shown in Figure 2.21. A theoretical fit to the data is also shown in Figure 2.23. The UV spectrometer carried by the *Mars Express* spacecraft observed apparent auroral emissions, which when correlated with the nature of electron and ion fluxes indicated possible similarities with terrestrial aurorae.[34]

2.5 Outer planets

In any discussion of the outer planets we need to recognize the fact that the amount of information available is rather limited. All four of the giant planets (Jupiter, Saturn, Uranus, and Neptune) have strong intrinsic magnetic fields, but the field orientations with respect to the spin axes and the ecliptic plane varies. Jupiter and Saturn are the best explored of these giant planets, mainly because of the *Galileo* and *Cassini* missions. Both planets have strong intrinsic magnetic fields, which are relatively closely aligned (10° for Jupiter[35] and less than 1° for Saturn[36]) with their rotation axes. The interaction of these planets with the solar wind is to some degree similar to that of the Earth. The obstacle is the magnetic field, and a strong bow shock and magnetopause are formed. The region between the bow shock and the magnetopause, where the shocked solar wind moves around the obstacle or magnetopause, is called the magnetosheath, as in the terrestrial case. Some of the major differences between the terrestrial magnetosphere and the Jovian, as well as the Kronian one, are due to the relatively rapid rotation of these planets, resulting in high centrifugal forces, and the presence of numerous large moons within the magnetosphere, some of which are important sources of magnetospheric plasma.

All the other giant planets also have strong bow shocks and magnetospheres; only, as indicated earlier, the orientation of the magnetic field does result in some important differences. For example, the magnetic dipole axis at Uranus is tilted by ~58.6° relative to the rotation axis.[37] Its rotation axis lies essentially in the ecliptic plane and points roughly toward the Sun at the present time. This unusual combination of circumstances means that the actual dipole tilt with respect to the so-called *GSM coordinate system* (the x-axis in this system points from the planet to the Sun; z is positive to the north and is perpendicular to x and in the plane that contains x and the magnetic dipole axis) is similar to that of the Earth, so the resulting magnetosphere has an "Earth-type" bipolar geomagnetic tail. However, because of the 17.9 hour rotation period of the planet, the magnetosphere changes from a "closed" to an "open" configuration every 8.9 hours.

The giant planets do not have solid surfaces as do the inner planets. Altitude scales are generally referred to a reference pressure level, which is now generally accepted to be the 1 bar level. This pressure level corresponds to a radial distance of 71 492 km from the center of Jupiter at the equator. Note that these planets are oblate, given their rapid rotation rate (e.g., there is a nearly 10% difference between the polar and equatorial radius at Saturn). The atmospheres of these planets consist predominantly of molecular hydrogen and some lesser amounts of helium and atomic hydrogen. In the lower atmosphere CH_4 and other hydrocarbons are also present as minor constituents. The latest estimates of the thermospheric temperatures at Jupiter, Saturn, Uranus, and Neptune are about 900, 400, 800, and 750 K, respectively. However, these values are very uncertain. At this time the energy sources responsible for these relatively high temperatures have not been established; candidate sources include Joule heating, gravity wave dissipation, and precipitating particle energy deposition. The latest estimates of the densities and the neutral gas temperature at Jupiter are shown in Figure 2.24, as a representative example for the giant planets.

Radio occultation observations by the *Pioneer, Voyager, Galileo*, and *Cassini* spacecraft have established the presence of ionospheres at all the giant planets. All these occultation measurements, because of the nature of the encounter geometries, are from near the terminator. Figure 2.25 shows representative electron density observations from the *Galileo* spacecraft at Jupiter.

Pluto used to be considered a planet and its companion Charon as its satellite. However, the new classification places both of them in the category of "dwarf planets." Evaporation of surface frost caused by solar radiation and sputtering by energetic particles are the likely causes of the atmosphere believed to be currently surrounding Pluto. The information available on the nature of Pluto comes from a very limited set of remote sensing observations. The surface temperature is estimated to fall between 30 and 44 K, with a most probable value of 36 K. The temperature

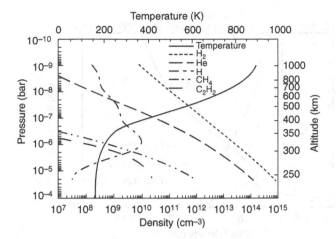

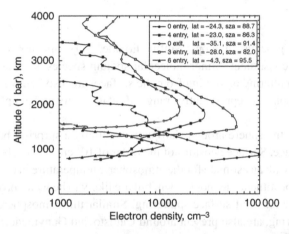

Figure 2.24 A model of Jupiter's atmosphere, showing neutral gas densities and temperatures. (Courtesy of Tariq Majeed.)

Figure 2.25 *Galileo* radio occultation measurements of ionospheric electron densities at Jupiter. (Courtesy of A. J. Kliore.)

in the upper atmosphere is believed to be nearly isothermal with a value around 100 K. The atmosphere is likely to consist mainly of N_2, CH_4, and CO, along with many other minor constituents.[38] An associated ionosphere with a peak density of less than 10^3 cm^{-3} is expected to be present. Pluto's low gravity implies that the atmosphere is only weakly bound and thus a significant neutral escape rate is likely to be present. The *New Horizon* spacecraft is on its way to Pluto and will fly by in July 2015.

2.6   Moons and comets

A number of the giant planets' moons are known to have atmospheres surrounding them. Io has been observed to have "volcanic" eruptions and thus it must have

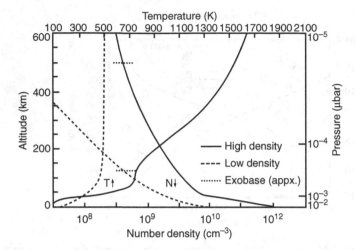

Figure 2.26 Model atmosphere values for Io; two sets of profiles for high and low total SO_2 densities.[39]

a highly time-variable gaseous envelope; sulfur dioxide, SO_2, appears to be its dominant atmospheric constituent. Minor molecular sodium species, such as Na_2S or Na_2O, released by sputtering or venting from the surface, are also believed to be present. Figure 2.26 displays a representative range of total density and temperature values.[39]

The presence of an atmosphere around Europa was initially a surprise, because of its frozen water surface. Surface densities of the order of 10^7 cm^{-3} have been indirectly deduced along with an estimate for the atmospheric temperature in the range of 350–600 K.[40] The constituents are not known, but are likely to be water products, such as O_2 and OH, the result of surface sputtering. Similar thin atmospheres, the result of surface sputtering, are also present around Callisto and Ganymede.[41]

From the *Cassini* mission we now know more about the atmosphere and ionosphere of Saturn's moon Titan than any other moon in the solar system. The surface pressure at Titan is about one and a half times that of the Earth (1.467 bar).[42] This dense atmosphere consists mostly of molecular nitrogen, N_2, and some lesser amount of methane, CH_4. There is also some molecular hydrogen, H_2, and a variety of hydrocarbons (e.g., C_2H_2, C_2H_4) as well as more complex organic molecules.[43, 44] The measured upper atmospheric N_2 and CH_4 densities are shown in Figure 2.27. The neutral temperature derived from these density profiles is around 150 K. These results also indicate that the homopause and exobase (see Sections 10.7 and 10.10) are at approximately 850 and 1430 km, respectively.

Cassini also found that Enceladus, a relatively small moon of Saturn, has a neutral gas plume emanating from near its southern polar cap, which is likely to be associated with observed surface cracks, referred to as "tiger paws." The main constituent was observed to be water and the densities reached around 10^6 cm^{-3},[45] as shown in Figure 2.28.

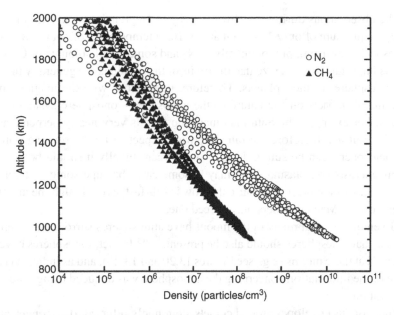

Figure 2.27 Titan density profiles measured by the ion–neutral mass spectrometer carried by the *Cassini* spacecraft.[43] (Courtesy of J. H. Waite.)

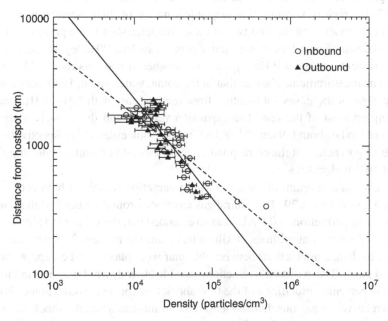

Figure 2.28 Water vapor density measured at Enceladus by the ion–neutral mass spectrometer[43] carried by the *Cassini* spacecraft. (Courtesy of J. H. Waite.)

The observations during the *Voyager* flyby of Neptune's moon Triton indicated a surface pressure of only 15–19 mbar, a surface temperature of about 38 K and an atmosphere consisting predominantly of N_2 and some small amounts of CH_4.[46]

It is important to recognize that the orbits of these moons are generally inside the magnetospheres of their planets. Therefore, interactions with the magnetospheres have major impacts on the nature of the atmosphere–ionosphere system of these moons. For example, the Saturnian magnetic field is very nearly perpendicular to Titan's orbit and therefore the ramside, with respect to the co-rotating Saturnian magnetosphere, can be sunlit, dark, or in between. Finally it should be noted that the magnetospheric plasma flows may in some cases be supersonic, but generally no bow shocks are present because the flow is sub-fast, even though the alfvenic or magnetosonic Mach numbers may exceed one.

Given the fact that many of the moons have atmospheres surrounding them, one expects that ionospheres should also be present.[47, 48] In fact, ionospheres have been observed at these moons (e.g., see Figures 13.20 and 13.26), and as indicated earlier, some of the information concerning the atmosphere was deduced using ionospheric observations.

The gaseous envelopes around comets, commonly referred to as *comas*, are different from conventional atmospheres in a number of important ways. The most important distinguishing characteristics of comas are (1) the lack of any significant gravitational force, (2) relatively fast radial outflow velocities ($\sim$1 km s^{-1}), and (3) the rapidly varying, time-dependent nature of their physical properties. A direct consequence of the first two of these characteristics is the presence of a very extended neutral envelope around active comets, such as P/Halley. Direct spacecraft measurements at comet P/Halley were made when it was less than 1 AU from the Sun. The measurements showed that in the coma, water vapor, H_2O, accounted for about 80% of the gases sublimating from the nucleus, with NH_3, CH_4, and CO_2 making up most of the rest. The expansion velocity and the mass loss rate were measured to be about 0.9 km s^{-1} and 6.9×10^{29} molecules s^{-1}, respectively.[49] It is interesting to note that the corresponding mass loss rate of comet Hale-Bopp[50] was about 10^{31} molecules s^{-1}.

A schematic diagram of the solar wind interaction region with an active comet is shown in Figure 2.29. The neutral gas envelope around comet P/Halley, when it was near its perihelion, $\sim$0.6 AU, was so extended that the solar wind already began to "see" the comet at millions of kilometers from the nucleus.[52] This occurred via charge exchange interactions between the solar wind plasma and escaping cometary neutral gases. These distant interactions acted to slow down the solar wind to about twice supersonic velocities and the bow shock that formed around comet P/Halley was a relatively weak one. Direct measurements indicated that the shock was located at a distance of about 1.15×10^6 km from the nucleus.[53] A tangential discontinuity or ionopause (sometimes called *contact surface* or *diamagnetic cavity boundary*) is formed near the nucleus of comets that have significant gas production. When comet

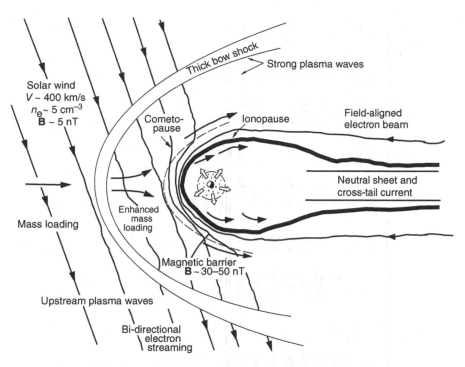

Figure 2.29 Schematic diagram showing the various regions associated with the interaction of the solar wind with a cometary atmosphere and ionosphere.[51]

P/Halley was near its perihelion, the distance of this tangential discontinuity from the nucleus was about 4700 km.[49] The magnetized, shocked solar wind flows around this contact discontinuity and never enters it, resulting in a diamagnetic cavity. The region between the bow shock and the tangential discontinuity is again called the magnetosheath, although some authors have introduced some new terminology that yields further subdivisions.

The *Rosetta* spacecraft was launched on March 2, 2004 and will arrive in the vicinity of comet 67P/Churyumov-Gerasimenko in May 2014. In November of that year an instrumented package will be deployed and land on the comet. The mother spacecraft will continue "orbiting" the comet as it approaches and then travels away from the Sun, until December 2015. This mission promises to provide a wealth of new information on periodic comets.

2.7   Plasma and neutral parameters

As is evident from the descriptions given in the previous subsections, the ionospheres found in our solar system display widely different characteristics. Table 2.6 provides

Table 2.6 Typical ionospheric and thermospheric parameters.

Body	Height (km)	Ion species	n_e (cm^{-3})	T_e (K)	T_i (K)	Neutral species	N_n (cm^{-3})	T_n (K)	λ_D (cm)	$N_{\lambda D}$	ω_{pe} (s^{-1})	ω_{ce} (s^{-1})	r_e (cm)	ω_{pi} (s^{-1})	ω_{ci} (s^{-1})	r_i (cm)
Venus	140	O_2^+	10^5	1000	500	O	1(10)	300	0.7	1.4(5)^a	1.8(7)	—	—	7.4(4)	—	—
Earth	300	O^+	10^5	2000	2000	O	1(8)	1000	1	4.2(5)	1.8(7)	4.7(6)	6.3	1.0(5)	1.6(2)	889
Mars	140	O_2^+	10^5	500	250	CO_2	1(10)	200	0.5	4.9(4)	1.8(7)	—	—	7.4(4)	—	—
Jupiter	~2000	$H^+(H_3^+?)^b$	10^4–10^5	2000^b	1000^b	H_2	1(6)	900	1	2.0(5)	1.3(7)	6.8(7)	0.36	3.0(5)	3.7(4)	11.6
Saturn	~2000	$H^+(H_3^+?)^b$	10^3–10^4	2000^b	1000^b	H_2	1(8)	400	3	6.2(5)	4.0(6)	3.3(6)	7.5	9.3(4)	1.8(3)	240
Uranus	~2000	$H^+(H_3^+?)^b$	10^3	2000^b	1000^b	H_2	1(10)	800	7	1.4(6)	1.8(6)	2.9(6)	8.6	4.2(4)	1.6(3)	274
Neptune	~2000	$H^+(H_3^+?)^b$	10^3	2000^b	1000^b	H_2	5(6)	750	7	1.4(6)	1.8(6)	1.7(6)	14.9	4.2(4)	9.3(2)	476

[a] $1.4(5) = 1.4 \times 10^5$

[b] No data available; estimate

a summary of representative plasma and neutral parameters that describe the primary ionization peak in the various ionospheres. The table includes the height of the peak, the dominant ion species, the electron density (n_e), the electron temperature (T_e), the ion temperature (T_i), the dominant neutral species, the neutral density (N_n), and the neutral temperature (T_n). All of these parameters are for typical daytime conditions.

Also included in Table 2.6 are some important plasma frequencies and scale lengths, which are given by

$$\lambda_D = \left(\frac{\varepsilon_0 k T_e}{n_e e^2} \right)^{1/2}, \tag{2.4}$$

$$N_{\lambda_D} = \frac{4\pi}{3} \lambda_D^3 n_e, \tag{2.5}$$

$$\omega_{p_\alpha} = \left(\frac{n_\alpha e^2}{\varepsilon_0 m_\alpha} \right)^{1/2}, \tag{2.6}$$

$$\omega_{c_\alpha} = \frac{eB}{m_\alpha}, \tag{2.7}$$

$$r_\alpha = \frac{(2k T_\alpha / m_\alpha)^{1/2}}{\omega_{c_\alpha}}. \tag{2.8}$$

where ε_0 is the permittivity of free space and α corresponds to either electrons or ions. The *Debye length* (λ_D) is the minimum distance over which a plasma can exhibit collective behavior. That is, for plasma phenomena that vary over scale lengths less than λ_D, the ions and electrons can be treated as individual particles. The degree to which collective behavior occurs is determined by the number of plasma particles in a *Debye sphere* (N_{λ_D}). When this number is much greater than unity, collective behavior dominates. The *gyroradius* (r_α) is the radius at which charged particles gyrate about magnetic field lines. For plasma scale lengths much less than the gyroradius, the charged particles behave as if they were not magnetized, i.e., they are not tied to magnetic field lines. The *cyclotron frequency* (ω_{c_α}) is the frequency at which charged particles gyrate about magnetic field lines. For plasma phenomena with frequencies much greater than ω_{c_α}, the gyrating motion is not important. The *plasma frequency* (ω_{p_α}) describes the ability of the charged particles to oscillate in response to time varying electric fields. If the frequency of the electric field is greater than the plasma frequency, the charged particles cannot keep up with the changing electric field.

These plasma frequencies and scale lengths will be rigorously defined in later chapters. Typical values of these parameters for the various ionospheres are given now because they will help explain why different mathematical approaches are used for different ionospheric phenomena.

2.8   Specific references

1. Hundhausen, A. J., An interplanetary view of coronal holes, in *Coronal Holes and High Speed Wind Streams*, ed. by J. B. Zirker, 225, Boulder, CO: Colorado Associated University Press, 1977.

2. *International Solar Terrestrial Physics Program*, Washington, DC: NASA Headquarters, 1985.

3. Waldmeier, M., *The Sunspot-Activity in the Years 1610–1960*, Zurich: Schulthess and Co., 1961.

4. Eddy, J. A., The Maunder Minimum, *Science*, **192**, 1189, 1976.

5. Brandt, J. C., *Introduction to the Solar Wind*, San Francisco, CA: Freeman and Company, 1970.

6. Charbonneau, P., and O. R. White, *The Sun: A Pictorial Introduction*, Boulder, CO: High Altitude Observatory, 1998.

7. Schwenn, R., Transport of energy and mass to the outer boundary of the Earth system, in *STEP Major Scientific Problems*, Urbana: University of Illinois, 13, 1988.

8. Alfvén, H., *Cosmic Plasma*, Netherlands: D. Reidel, 1981.

9. *Solar-Terrestrial Research for the 1980s*, Washington, DC: National Research Council, National Academy Press, 1981.

10. Feldman, W. C., J. R. Asbridge, S. J. Bane, and J. T. Gosling, Plasma and magnetic fields from the Sun, in *The Solar Output and Its Variation*, ed. by O. R. White, 351, Boulder, CO: Colorado Associated University Press, 1977.

11. Burlaga, L. F. *et al.*, Heliospheric magnetic field strength out to 66 AU: Voyager 1 1978–1996, *J. Geophys. Res.*, **103**, 23727, 1998.

12. Frank, L. A., Dynamics of the near-earth magnetotail-recent observations, in *Modeling Magnetospheric Plasma*, ed. by T. E. Moore and J. H. Waite, *Geophys. Monograph*, **44**, 261, 1988.

13. Chappell, C. R., Conference on Magnetospheric–Ionospheric Coupling, *Trans. A. G. U.*, **55**, 776, 1974.

14. Hedin, A. E., J. Salah, J. Evans, *et al.*, A global thermospheric model based on mass spectrometer and incoherent scatter data, MSIS, *J. Geophys. Res.*, **82**, 2139, 1977.

15. Burch, J. L., The magnetosphere, in *The Upper Atmosphere and Magnetosphere*, 42, National Research Council, Washington, DC: National Academy Press, 1977.

16. Banks, P. M., R. W. Schunk, and W. J. Raitt, The topside ionosphere: a region of dynamic transition, *Annl. Rev. Earth Planet. Sci.*, **4**, 381, 1976.

17. Williams, D. J., B. Mauk, and R. W. McEntire, Properties of Ganymede's magnetosphere as revealed by energetic particle observations, *J. Geophys. Res.*, **103**, 17523, 1998.

18. Broadfoot, A. L., D. E. Shemansky, and S. Kumar, Mariner 10: Mercury atmosphere, *Geophys. Res. Lett.*, **3**, 577, 1976.

19. Potter, A., and T. Morgan, Discovery of sodium in the atmosphere of Mercury, *Science*, **229**, 651, 1985.

20. Brace, L. H., and A. J. Kliore, The structure of the Venus ionosphere, *Space Sci. Rev.*, **55**, 81, 1991.

21. Adams, W. S., and T. Dunham, Absorption bands in the infra-red spectrum of Venus, *Publ. Astron. Soc. Pac.*, **44**, 243, 1932.

22. Vinogradov, A. P., Y. A. Surkov, and C. P. Florensky, The chemical composition of Venus atmosphere based on the data of the interplanetary station Venera 4, *J. Atmos. Sci.*, **25**, 535, 1968.

23. Hedin, A. E., H. Niemann, W. Kasprzak, and A. Sieff, Global empirical model of the Venus thermosphere, *J. Geophys. Res.*, **88**, 73, 1983.

24. Niemann, H. B., W. Kasprzak, A. Hedin, D. Hunten, and N. Spencer, Mass spectrometric measurements of the neutral gas composition of the thermosphere and exosphere of Venus, *J. Geophys. Res.*, **85**, 7817, 1980.

25. Fox, J. L., and S. W. Bougher, Structure, luminosity and dynamics of the Venus thermosphere, *Space Sci. Rev.*, **55**, 357, 1991.

26. Taylor, H. A., H. Brinton, S. Bauer, *et al.*, Global observations of the composition and dynamics of the ionosphere of Venus: implications for the solar wind interaction, *J. Geophys. Res.*, **85**, 7765, 1980.

27. Kliore, A. J., E. S. Levy, D. L. Cain, G. Fjeldbo, and S. I. Rasod, Atmosphere and ionosphere of Venus from the Mariner 5 S-band radio occultation measurements, *Science*, **158**, 1683, 1967.

28. Acuna, M. H., J. E. P. Connerney, P. Wasilewski, *et al.*, Magnetic field and plasma observations at Mars: initial results of the Mars Global Surveyor mission, *Science*, **279**, 1676, 1998.

29. Nagy, A. F., D. Winterhalter, K. Sauer, *et al.*, The plasma environment of Mars, *Space Sci. Rev.*, **111**, 33, 2004.

30. Dubinin, E., M. Fränz, J. Woch, *et al.*, Plasma morphology at Mars; ASPERA-3 observations, *Space Sci. Rev.*, **126**, 209, 2006.

31. Mitchell, D. L., R. Lin, C. Mazelle, *et al.*, Probing Mars' crustal magnetic field and ionosphere with the MGS electron reflectometer, *J. Geophys. Res.*, **106**, E10, 23, 419, 2001.

32. Nier, A. O., and M. B. McElroy, Composition and structure of Mars' upper atmosphere: results from the neutral mass spectrometers on Viking 1 and 2, *J. Geophys. Res.*, **82**, 4341, 1977.

33. Chen, R. H., T. E. Cravens, and A. F. Nagy, The Martian ionosphere in light of the Viking observations, *J. Geophys. Res.*, **83**, 3871, 1978.

34. Lundin, R., D. Winningham, S. Barabash, *et al.*, Plasma acceleration above Martian magnetic anomalies, *Science*, **311**, 980, 2006.

35. Acuna, M. H., K. W. Behannon, and J. E. P. Connerney, Jupiter's magnetic field and magnetosphere, *Physics of the Jovian Magnetosphere*, ed. by A. J. Dessler, 1, Cambridge University Press, 1983.

36. Dougherty, M. K., N. Achilleos, N. Andre, *et al.*, *Science*, **307**, 1266, 2005.

37. Connerney, J. E., M. H. Acuna, and N. F. Ness, The magnetic field of Uranus, *J. Geophys. Res.*, **92**, 15329, 1987.

38. Lara, L.M., W.-H. Ip, and R. Rodrigo, Photochemical models of Pluto's atmosphere, *Icarus*, **130**, 16, 1997.

39. Summers, M.E., and D.F. Strobel, Photochemistry and vertical transport in Io's atmosphere and ionosphere, *Icarus*, **120**, 290, 1996.

40. Kliore, A.J., P.P. Hinson, F.M. Flasar, A.F. Nagy, and T.E. Cravens, The ionosphere of Europa from Galileo radio occultations, *Science*, **277**, 355, 1997.

41. Kliore, A.J., A. Anabtawi, R.G. Herrera, *et al.*, The ionosphere of Callisto from Galileo radio occultation, observations, *J. Geophys. Res.*, **107**, 11.1407/2002JA009365, 2002.

42. Fulchignoni, M., F. Ferri, F. Angrilli, *et al.*, In situ measurements of the physical characteristics of Titan's environment, *Nature*, **437**, 785, doi:10.1038/nature04314, 2005.

43. Waite, J.H., H. Niemann, R.V. Yelle, *et al.*, Ion neutral mass spectrometer results from the first flyby of Titan, *Science*, **308**, 982, 2005.

44. Waite, J.H., D.T. Young, T.E. Cravens, *et al.*, The process of tholin formation in Titan's upper atmosphere, *Science*, **316**, 870, 2007.

45. Waite, J.H., M.R. Combi, W.-H. Ip, *et al.*, Cassini ion neutral mass spectrometer: Enceladus plume composition and structure, *Science*, **311**, 1419, 2006.

46. Strobel, D.F., M. Simmers, F. Herbert, and B. Sandel, The photochemistry of methane in the atmosphere of Triton, *Geophys. Res. Lett.*, **17**, 1729, 1990.

47. Kliore, A.J., A.F. Nagy, E.A. Marouf, *et al.*, First results from the Cassini radio occultations of the Titan ionosphere, *J. Geophys. Res.*, **113**, doi:10.1029/2007JA012965, 2008.

48. Wahlund, J.-E., R. Bostram, G. Gnstafson, *et al.*, Cassini measurements of cold plasma in the ionosphere of Titan, *Science*, **308**, 986, 2005.

49. Krankowsky, D., P. Lammerzahl, I. Herrwerth, *et al.*, In situ gas and ion measurements at comet Halley, *Nature*, **321**, 326, 1986.

50. Weaver, H.A., P.D. Feldman, M.F. A'Hearn, and C. Arpigny, The activity and size of the nucleus of comet Hale-Bopp (c/1995 01), *Science*, **275**, 1900, 1997.

51. Flammer, K.R., The global interaction of comets with the solar wind, in *Comets in the Post-Halley Era*, ed. by R.L. Newburn, M. Neugebauer, and J. Rahe, by Kluwer Academic Press, Dordrecht: 1191, 1991.

52. Somogyi, A.J., K.I. Gringauz, K. Szego, *et al.*, First observations of energetic particles near comet Halley, *Nature* **321**, 285, 1986.

53. Gringauz, K.I., T.I. Gombosi, A.P. Remizov, *et al.*, First in situ plasma and neutral gas measurements at comet Halley, *Nature* **321**, 282, 1986.

2.9   General references

Cravens, T.E., *Physics of Solar System Plasmas*, Cambridge, UK: Cambridge University Press, 1997.

Esposito, L.W., *et al.*, eds., Exploring Venus as a Terrestrial Planet, *Geophysical Monograph Series*, Washington, DC: American Geophysical Union, 2007.

Lemiare, J. F. and K. I. Gringauz, *The Earth's Plasmasphere*, Cambridge, UK: Cambridge University Press, 1998.

Mendillo, M., *et al.* eds., Atmospheres in the Solar System: Comparative Aeronomy, *Geophysical Monograph Series*, **130**, Washington, DC: American Geophysical Union, 2002.

Nagy, A. F., *et al.* Comparative Aeronomy, *Space Science Series of ISSI*, **29**, Springer, 2008.

Sharma, A. S., *et al.* eds., Disturbances in Geospace: The Storm-Substorm Relationship, *Geophysical Monograph Series*, Washington, DC: American Geophysical Union, 2004.

Chapter 3

Transport equations

A wide variety of plasma flows can be found in the various planetary ionospheres. For example, gentle near-equilibrium flows occur in the terrestrial ionosphere at mid-latitudes, while highly nonequilibrium flow conditions exist in the terrestrial polar wind and in the Venus ionosphere near the solar terminator. The highly nonequilibrium flows are generally characterized by large temperature differences between the interacting species, by flow speeds approaching and exceeding thermal speeds, and by flow conditions changing from collision-dominated to collisionless regimes. In an effort to model the various ionospheric flow conditions, several different mathematical approaches have been used, including collision-dominated and collisionless transport equations, kinetic and semikinetic models, and macroscopic particle-in-cell techniques. However, the transport equation approach has received the most attention, primarily because it can handle most of the flow conditions encountered in planetary ionospheres. Therefore, the main focus of this chapter is on transport theory, although other mathematical approaches are briefly discussed at the end of the chapter. Typically, numerous assumptions are made to simplify the transport equations before they are applied, and therefore, it is instructive to trace the derivation of the various sets of transport equations in order to establish their intrinsic strengths and limitations. Before diving into the rigorous derivation of the transport equations, it is useful to review the simple derivation of the continuity equation given in Appendix N.

3.1   Boltzmann equation

The Boltzmann equation is not only the starting point for the derivation of the different sets of transport equations but also forms the basis for the kinetic and semikinetic theories. With Boltzmann's approach, one is not interested in the motion of individual

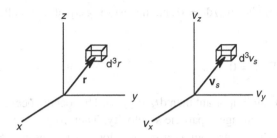

Figure 3.1 Volume element $\mathrm{d}^3 r$ about position vector $\mathbf{r}$ in configuration space (left) and volume element $\mathrm{d}^3 v_s$ about velocity $\mathbf{v}_s$ in velocity space (right). Note that each volume element $\mathrm{d}^3 r$ must contain a sufficient number of particles for a complete range of velocities.

particles in the gas, but instead with the distribution of particles. Accordingly, each species in the gas mixture is described by a separate velocity distribution function $f_s(\mathbf{r}, \mathbf{v}_s, t)$, where $\mathbf{r}, \mathbf{v}_s$ and t are *independent variables*. The distribution function corresponds to the number of particles of species s that, at time t, are located in a volume element $\mathrm{d}^3 r$ about $\mathbf{r}$ and simultaneously have velocities in a velocity-space volume element $\mathrm{d}^3 v_s$ about $\mathbf{v}_s$ (Figure 3.1). Alternatively, f_s can be viewed as a probability density in the $(\mathbf{r}, \mathbf{v}_s)$ phase space. The evolution of f_s is determined by the flow in phase space of particles under the influence of external forces and by the net effect of collisions. The rate of change of f_s due to an explicit time variation and a flow in phase space is given by

$$\frac{\mathrm{d}f_s}{\mathrm{d}t} = \lim_{\Delta t \to 0} \frac{f(\mathbf{r} + \Delta\mathbf{r}, \mathbf{v}_s + \Delta\mathbf{v}_s, t + \Delta t) - f_s(\mathbf{r}, \mathbf{v}_s, t)}{\Delta t}. \tag{3.1}$$

Since Δt is a small quantity, $f(\mathbf{r} + \Delta\mathbf{r}, \mathbf{v}_s + \Delta\mathbf{v}_s, t + \Delta t)$ can be expanded in a Taylor series:

$$\frac{\mathrm{d}f_s}{\mathrm{d}t} = \lim_{\Delta t \to 0} \frac{1}{\Delta t} \bigg[f(\mathbf{r}, \mathbf{v}_s, t) + \frac{\partial f_s}{\partial t} \Delta t + \Delta\mathbf{r} \cdot \nabla f_s$$

$$+ \Delta\mathbf{v}_s \cdot \nabla_v f_s + \cdots - f_s(\mathbf{r}, \mathbf{v}_s, t) \bigg], \tag{3.2}$$

where ∇ is the gradient operator in configuration space and ∇_v is a similar gradient operator in velocity space. Taking the limit of $\Delta t \to 0$ yields

$$\frac{\mathrm{d}f_s}{\mathrm{d}t} = \frac{\partial f_s}{\partial t} + \mathbf{v}_s \cdot \nabla f_s + \mathbf{a}_s \cdot \nabla_v f_s, \tag{3.3}$$

where all the higher-order terms in the Taylor series drop out as $\Delta t \to 0$ and

$$\frac{\Delta\mathbf{r}}{\Delta t} \to \frac{\mathrm{d}\mathbf{r}}{\mathrm{d}t} \to \mathbf{v}_s, \tag{3.4}$$

$$\frac{\Delta\mathbf{v}_s}{\Delta t} \to \frac{\mathrm{d}\mathbf{v}_s}{\mathrm{d}t} \to \mathbf{a}_s. \tag{3.5}$$

The vector $\mathbf{a}_s$ is the acceleration of the particles (force/mass).

If collisions are not important, then $df_s/dt = 0$ and the resulting equation is called the *Vlasov equation*

$$\frac{\partial f_s}{\partial t} + \mathbf{v}_s \cdot \nabla f_s + \mathbf{a}_s \cdot \nabla_v f_s = 0. \tag{3.6}$$

On the other hand, if collisions are important then $df_s/dt \neq 0$. This occurs because collisions act to instantaneously change a particle's velocity. Therefore, particles instantaneously appear in, and disappear from, regions of velocity space as a result of collisions, and hence, they correspond to production and loss terms for f_s. Letting $\delta f_s/\delta t$ represent the effect of collisions, the equation describing the evolution of f_s becomes

$$\frac{\partial f_s}{\partial t} + \mathbf{v}_s \cdot \nabla f_s + \mathbf{a}_s \cdot \nabla_v f_s = \frac{\delta f_s}{\delta t}, \tag{3.7}$$

which is known as the *Boltzmann equation*.

The main external forces acting on the charged particles in planetary ionospheres are the Lorentz and gravitational forces. With allowance for these forces, the acceleration becomes

$$\mathbf{a}_s = \mathbf{G} + \frac{e_s}{m_s}(\mathbf{E} + \mathbf{v}_s \times \mathbf{B}), \tag{3.8}$$

where $\mathbf{G}$ is the acceleration due to gravity, $\mathbf{E}$ is the electric field, $\mathbf{B}$ is the magnetic field, e_s is the species charge, and m_s is the species mass. On the other hand, gravitational, Coriolis, and centripetal forces can be important for planetary neutral atmospheres (Chapter 10).

For binary elastic collisions between particles, the appropriate collision operator is the *Boltzmann collision integral* (Appendix G)

$$\frac{\delta f_s}{\delta t} = \iint d^3 v_t \, d\Omega \, g_{st} \sigma_{st}(g_{st}, \theta)\left(f_s' f_t' - f_s f_t\right), \tag{3.9}$$

where
$d^3 v_t$ = velocity–space volume element for the target species t,
g_{st} = $|\mathbf{v}_s - \mathbf{v}_t|$ is the relative speed of the colliding particles s and t,
$d\Omega$ = element of solid angle in the colliding particles' center-of-mass reference frame,
θ = center-of-mass scattering angle,
$\sigma_{st}(g_{st}, \theta)$ = differential scattering cross section, defined as the number of molecules scattered per solid angle $d\Omega$, per unit time, divided by the incident intensity,
$f_s' f_t'$ = $f_s(\mathbf{r}, \mathbf{v}_s', t) f_t(\mathbf{r}, \mathbf{v}_t', t)$, where the primes indicate the distribution functions are evaluated with the particle velocities after the collision.

In Equation (3.9) the first term in the brackets corresponds to the particles scattered into a given region of velocity space (production term) and the second term corresponds to the particles scattered out of the same region of velocity space (loss term).

The Boltzmann collision integral can be applied to both self-collisions ($t = s$) and collisions between unlike particles. It can be applied to Coulomb collisions, to elastic ion–neutral collisions, and to collisions between different neutral species. In addition, it can be applied to a resonant charge exchange interaction between an ion and its parent neutral because the charge exchange process is pseudo-elastic. The net energy loss in the interaction is small.

3.2   Moments of the distribution function

In the ideal situation one would like to solve the Boltzmann equation for each of the species in the gas mixture and thereby obtain the individual velocity distribution functions, but this can only be done for relatively simple situations. As a consequence, one is generally restricted to obtaining information on a limited number of low-order velocity moments of the species distribution function. For example, since $f_s(\mathbf{r}, \mathbf{v}_s, t)$ represents the number of particles at time t that are located in a volume element $d^3 r$ about $\mathbf{r}$ and simultaneously have velocities in a volume element $d^3 v_s$ about $\mathbf{v}_s$, then an integration over all velocities yields the number of particles in the volume element $d^3 r$ at time t, which is the species *number density*, $n_s(\mathbf{r}, t)$

$$n_s(\mathbf{r}, t) = \int d^3 v_s f_s(\mathbf{r}, \mathbf{v}_s, t). \tag{3.10}$$

Likewise, the *average* or *drift velocity* of a species, $\mathbf{u}_s(\mathbf{r}, t)$, can be obtained by integrating the product $\mathbf{v}_s f_s(\mathbf{r}, \mathbf{v}_s, t)$ over all velocities and then dividing by the density

$$\mathbf{u}_s(\mathbf{r}, t) = \frac{\int d^3 v_s \mathbf{v}_s f_s(\mathbf{r}, \mathbf{v}_s, t)}{\int d^3 v_s f_s(\mathbf{r}, \mathbf{v}_s, t)}. \tag{3.11}$$

This process can be continued so that if $\xi_s(\mathbf{v}_s)$ is any function of velocity of the particles of type s, then the average value of $\xi_s(\mathbf{v}_s)$ at any position $\mathbf{r}$ and time t is given by

$$\langle \xi_s(\mathbf{v}_s) \rangle = \frac{1}{n_s} \int d^3 v_s f_s(\mathbf{r}, \mathbf{v}_s, t) \xi_s(\mathbf{v}_s). \tag{3.12}$$

The procedure of multiplying the species distribution function by powers or products of velocity and then integrating over all velocities is called taking *velocity moments*. However, the definition of all higher-order velocity moments is not unique. For example, the temperature is a measure of the spread about some average velocity, and this average velocity must be selected before the temperature can be defined.

Likewise, all of the higher-order velocity moments of f_s must be defined relative to an average velocity. In the early work of Chapman, Enskog, Burnett, and others,[1] the velocity moments of the distribution function were defined relative to the average velocity of the gas mixture

$$\mathbf{u} = \sum_s n_s m_s \mathbf{u}_s \Big/ \sum_s n_s m_s. \tag{3.13}$$

Such a definition is appropriate for highly collisional gases, where the individual species drift velocities and temperatures do not significantly differ from the average drift velocity and temperature of the gas mixture.

As an alternative to defining the transport properties with respect to the average gas velocity, Grad proposed that the transport properties of a given species be defined with respect to the average drift velocity of that species, $\mathbf{u}_s$.[2] This definition is more appropriate for planetary atmospheres and ionospheres, where large relative drifts between interacting species can occur. In terms of the species average drift velocity, the *random* or *thermal* velocity is defined as

$$\mathbf{c}_s = \mathbf{v}_s - \mathbf{u}_s. \tag{3.14}$$

At this point it is necessary to decide what velocity moments are needed beyond the first two moments n_s and $\mathbf{u}_s$. In general, this will depend on how far the flow is from equilibrium. For most applications, the following moments are sufficient:

Temperature:

$$\frac{3}{2} k T_s = \frac{1}{2} m_s \langle c_s^2 \rangle = \frac{m_s}{2 n_s} \int d^3 v_s f_s (\mathbf{v}_s - \mathbf{u}_s)^2, \tag{3.15}$$

Heat flow vector:

$$\mathbf{q}_s = \frac{1}{2} n_s m_s \langle c_s^2 \mathbf{c}_s \rangle = \frac{m_s}{2} \int d^3 v_s f_s (\mathbf{v}_s - \mathbf{u}_s)^2 (\mathbf{v}_s - \mathbf{u}_s), \tag{3.16}$$

Pressure tensor:

$$\mathbf{P}_s = n_s m_s \langle \mathbf{c}_s \mathbf{c}_s \rangle = m_s \int d^3 v_s f_s (\mathbf{v}_s - \mathbf{u}_s)(\mathbf{v}_s - \mathbf{u}_s), \tag{3.17}$$

Higher-order pressure tensor:

$$\mu_s = \frac{1}{2} n_s m_s \langle c_s^2 \mathbf{c}_s \mathbf{c}_s \rangle$$

$$= \frac{m_s}{2} \int d^3 v_s f_s (\mathbf{v}_s - \mathbf{u}_s)^2 (\mathbf{v}_s - \mathbf{u}_s)(\mathbf{v}_s - \mathbf{u}_s), \tag{3.18}$$

Heat flow tensor:

$$\mathbf{Q}_s = n_s m_s \langle \mathbf{c}_s \mathbf{c}_s \mathbf{c}_s \rangle = m_s \int d^3 v_s f_s (\mathbf{v}_s - \mathbf{u}_s)(\mathbf{v}_s - \mathbf{u}_s)(\mathbf{v}_s - \mathbf{u}_s), \tag{3.19}$$

where k is the Boltzmann constant. The pressure tensors $\mathbf{P}_s$ and $\boldsymbol{\mu}_s$ are second-order tensors, each with nine elements, and the heat flow tensor $\mathbf{Q}_s$ is a third-order tensor with 27 elements. In index notation they are expressed as $(P_s)_{\alpha\beta}$, $(\mu_s)_{\alpha\beta}$, and $(Q_s)_{\alpha\beta\gamma}$, with α, β, and γ varying from 1 to 3.

If a summation is taken of the diagonal elements in the pressure tensor (3.17), one obtains

$$\sum_{\alpha=1}^{3}(P_s)_{\alpha\alpha} = m_s \int \mathrm{d}^3 v_s f_s (\mathbf{v}_s - \mathbf{u}_s)^2 = 3p_s, \tag{3.20}$$

where the second expression follows from Equation (3.15) and where $p_s = n_s k T_s$ is the *partial pressure* of the gas. When collisions are important, the diagonal elements of the pressure tensor are the most important elements and they are generally equal. As a consequence, it is convenient to remove these diagonal elements from the pressure tensor and consider them separately. This is accomplished by defining a new tensor, the *stress tensor*, $\boldsymbol{\tau}_s$.

$$\boldsymbol{\tau}_s = \mathbf{P}_s - p_s \mathbf{I}, \tag{3.21}$$

where $\mathbf{I}$ is a *unit dyadic* (diagonal elements equal to unity). In index notation, it is $\delta_{\alpha\beta}$. The stress tensor is a measure of the extent to which the gas deviates from an isotropic character. As collisions become more important, the gas becomes more isotropic and the stress tensor becomes negligible.

3.3   General transport equations

Transport equations that describe the spatial and temporal evolution of the physically significant velocity moments $(n_s, \mathbf{u}_s, T_s, \mathbf{P}_s, \mathbf{q}_s)$ can be obtained by multiplying the Boltzmann equation (3.7) with an appropriate function of velocity and then integrating over velocity space. However, before this procedure is applied, it is convenient to express the Boltzmann equation in a slightly different form. Given that $\mathbf{r}$, $\mathbf{v}$, and t are independent variables

$$\nabla \cdot (f_s \mathbf{v}_s) = \mathbf{v}_s \cdot \nabla f_s + f_s (\nabla \cdot \mathbf{v}_s) = \mathbf{v}_s \cdot \nabla f_s \tag{3.22}$$

and

$$\nabla_v \cdot (f_s \mathbf{a}_s) = \mathbf{a}_s \cdot \nabla_v f_s + f_s (\nabla_v \cdot \mathbf{a}_s) = \mathbf{a}_s \cdot \nabla_v f_s \tag{3.23}$$

because $\nabla_v \cdot \mathbf{a}_s = 0$ for the acceleration processes relevant to planetary atmospheres and ionospheres (Equation 3.8). Therefore, the Boltzmann equation (3.7) can also be written as

$$\frac{\partial f_s}{\partial t} + \nabla \cdot (f_s \mathbf{v}_s) + \nabla_v \cdot (f_s \mathbf{a}_s) = \frac{\delta f_s}{\delta t}. \tag{3.24}$$

In what follows, the transport equations are obtained from Equation (3.24), which is in terms of $\mathbf{v}_s$. The resulting transport equations are commonly referred to as being in the *conservative* form. Alternatively, the Boltzmann equation can be transformed into an equation for $\mathbf{c}_s$ before the velocity moments are taken. The two approaches are equivalent, but the use of Equation (3.24) is more straightforward for the calculation of the lower-order velocity moments (density, drift velocity, and energy). In either case, a general moment equation can be derived, which is called the *Maxwell transfer equation* (Appendix F).

An equation describing the evolution of the species density is obtained simply by integrating Equation (3.24) over all velocities

$$\int d^3 v_s \left[\frac{\partial f_s}{\partial t} + \nabla \cdot (f_s \mathbf{v}_s) + \nabla_v \cdot (f_s \mathbf{a}_s) \right] = \int d^3 v_s \frac{\delta f_s}{\delta t}, \tag{3.25}$$

where

$$\int d^3 v_s \frac{\partial f_s}{\partial t} = \frac{\partial}{\partial t} \int d^3 v_s f_s = \frac{\partial n_s}{\partial t}, \tag{3.26}$$

$$\int d^3 v_s \nabla \cdot (f_s \mathbf{v}_s) = \nabla \cdot \int d^3 v_s f_s \mathbf{v}_s = \nabla \cdot (n_s \mathbf{u}_s), \tag{3.27}$$

$$\int d^3 v_s \nabla_v \cdot (f_s \mathbf{a}_s) = \int_S dA_v (f_s \mathbf{a}_s) \cdot \hat{\mathbf{n}}_v = 0, \tag{3.28}$$

$$\int d^3 v_s \frac{\delta f_s}{\delta t} \equiv \frac{\delta n_s}{\delta t}. \tag{3.29}$$

In Equation (3.28), the divergence theorem is applied so that the velocity–space volume integral can be transformed into a velocity–space surface integral at infinity, where dA_v is the surface area element and $\hat{\mathbf{n}}_v$ is an outwardly directed unit normal. Since there are no particles with infinite velocities, f_s and the surface integral in Equation (3.28) approaches zero as v_s goes to infinity. Substituting Equations (3.26–3.29) into Equation (3.25) yields the *continuity equation*

$$\frac{\partial n_s}{\partial t} + \nabla \cdot (n_s \mathbf{u}_s) = \frac{\delta n_s}{\delta t}. \tag{3.30}$$

The equation describing the evolution of the species drift velocity is obtained by multiplying the Boltzmann equation (3.24) by $m_s \mathbf{c}_s$ and then integrating over all velocities

$$m_s \int d^3 v_s \left[\mathbf{c}_s \frac{\partial f_s}{\partial t} + \mathbf{c}_s \nabla \cdot (f_s \mathbf{v}_s) + \mathbf{c}_s \nabla_v \cdot (f_s \mathbf{a}_s) \right] = m_s \int d^3 v_s \mathbf{c}_s \frac{\delta f_s}{\delta t}, \tag{3.31}$$

where the terms can be integrated to obtain the following results:

$$m_s \int d^3 v_s (\mathbf{v}_s - \mathbf{u}_s) \frac{\partial f_s}{\partial t} = n_s m_s \frac{\partial \mathbf{u}_s}{\partial t} \tag{3.32}$$

$$m_s \int d^3 v_s (\mathbf{v}_s - \mathbf{u}_s) \nabla \cdot (f_s \mathbf{v}_s) = \nabla \cdot \mathbf{P}_s + n_s m_s (\mathbf{u}_s \cdot \nabla) \mathbf{u}_s \tag{3.33}$$

$$m_s \int d^3 v_s (\mathbf{v}_s - \mathbf{u}_s) \nabla_v \cdot (f_s \mathbf{a}_s) = -n_s m_s \langle \mathbf{a}_s \rangle \tag{3.34}$$

$$m_s \int d^3 v_s \mathbf{c}_s \frac{\delta f_s}{\delta t} \equiv \frac{\delta \mathbf{M}_s}{\delta t}. \tag{3.35}$$

In evaluating the integrals in Equations (3.32–3.35), use was made of the vector identity involving the divergence of a scalar multiplied by a vector, the divergence theorem which converts volume integrals into surface integrals, and the definitions of the transport properties (Equations 3.10–3.11, and 3.17). Finally, the substitution of Equations (3.8, 3.32–3.35) into Equation (3.31) yields the *momentum equation*

$$n_s m_s \frac{D_s \mathbf{u}_s}{Dt} + \nabla \cdot \mathbf{P}_s - n_s m_s \mathbf{G} - n_s e_s (\mathbf{E} + \mathbf{u}_s \times \mathbf{B}) = \frac{\delta \mathbf{M}_s}{\delta t}, \tag{3.36}$$

where D_s/Dt is the convective derivative

$$\frac{D_s}{Dt} = \frac{\partial}{\partial t} + \mathbf{u}_s \cdot \nabla. \tag{3.37}$$

In a similar manner, the energy, pressure tensor, and heat flow equations can be derived by multiplying the Boltzmann Equation (3.24) by $\frac{1}{2} m_s c_s^2$, $m_s \mathbf{c}_s \mathbf{c}_s$, and $\frac{1}{2} m_s c_s^2 \mathbf{c}_s$, respectively, and then integrating over velocity space. After a considerable amount of algebra, these equations can be expressed as follows:

Energy equation:

$$\frac{D_s}{Dt} \left(\frac{3}{2} p_s \right) + \frac{3}{2} p_s (\nabla \cdot \mathbf{u}_s) + \nabla \cdot \mathbf{q}_s + \mathbf{P}_s : \nabla \mathbf{u}_s = \frac{\delta E_s}{\delta t}, \tag{3.38}$$

Pressure tensor equation:

$$\frac{D_s \mathbf{P}_s}{Dt} + \nabla \cdot \mathbf{Q}_s + \mathbf{P}_s (\nabla \cdot \mathbf{u}_s) + \frac{e_s}{m_s} (\mathbf{B} \times \mathbf{P}_s - \mathbf{P}_s \times \mathbf{B})$$
$$+ \mathbf{P}_s \cdot \nabla \mathbf{u}_s + (\mathbf{P}_s \cdot \nabla \mathbf{u}_s)^T = \frac{\delta \mathbf{P}_s}{\delta t}. \tag{3.39}$$

Heat flow equation:

$$\frac{D_s \mathbf{q}_s}{Dt} + \mathbf{q}_s \cdot \nabla \mathbf{u}_s + \mathbf{q}_s (\nabla \cdot \mathbf{u}_s) + \mathbf{Q}_s : \nabla \mathbf{u}_s + \nabla \cdot \boldsymbol{\mu}_s$$

$$+ \left[\frac{D_s \mathbf{u}_s}{Dt} - \mathbf{G} - \frac{e_s}{m_s} (\mathbf{E} + \mathbf{u}_s \times \mathbf{B}) \right]$$

$$\cdot \left(\boldsymbol{\tau}_s + \frac{5}{2} p_s \mathbf{I} \right) - \frac{e_s}{m_s} \mathbf{q}_s \times \mathbf{B} = \frac{\delta \mathbf{q}_s}{\delta t}, \tag{3.40}$$

where

$$\frac{\delta E_s}{\delta t} \equiv \frac{m_s}{2} \int d^3 v_s c_s^2 \frac{\delta f_s}{\delta t}, \tag{3.41}$$

$$\frac{\delta \mathbf{P}_s}{\delta t} \equiv m_s \int d^3 v_s \mathbf{c}_s \mathbf{c}_s \frac{\delta f_s}{\delta t}, \tag{3.42}$$

$$\frac{\delta \mathbf{q}_s}{\delta t} \equiv \frac{m_s}{2} \int d^3 v_s c_s^2 \mathbf{c}_s \frac{\delta f_s}{\delta t}. \tag{3.43}$$

In Equations (3.38–3.40), the transpose of a tensor $\mathbf{A} = A_{\alpha\beta}$ is denoted by $\mathbf{A}^T = A_{\beta\alpha}$ and the operation $\mathbf{Q}_s : \nabla \mathbf{u}_s = \sum_\beta \sum_\gamma (Q_s)_{\alpha\beta\gamma} (\partial u_{s\beta} / \partial x_\gamma)$ corresponds to the double dot product of the two tensors $\mathbf{Q}_s$ and $\nabla \mathbf{u}_s$.

A few points should be noted about the general transport equations. First, the set of equations can be increased to an arbitrary size merely by taking additional velocity moments of the Boltzmann equation. For example, if the Boltzmann equation is multiplied by $m_s \mathbf{c}_s \mathbf{c}_s \mathbf{c}_s$ and integrated over velocity space, an equation describing the spatial and temporal evolution of the heat flow tensor $\mathbf{Q}_s$ will be obtained. Further, the general transport equations do not constitute a closed system because the equation governing the moment of order ℓ contains the moment of order $\ell + 1$. That is, the continuity equation describes the evolution of the density, but it also contains the drift velocity, and so on. Finally, it should be noted that the collision terms appearing on the right-hand sides of the general transport equations can be evaluated rigorously only for a unique interaction potential between the colliding particles, which will be presented later. For general interaction potentials, it is necessary to know the distribution functions of the colliding particles in order to evaluate the collision terms. Therefore, to obtain a useable system of transport equations, an approximate expression for the velocity distribution function is needed so that the system of equations can be closed and the collision terms can be evaluated.

3.4   Maxwellian velocity distribution

A relatively simple distribution function prevails when collisions dominate. As will be discussed later, in this case the species distribution function is driven toward a Maxwellian distribution function. If the different species in the gas mixture have relative drifts, but collisions between similar particles are significant, then f_s is driven toward a *local drifting Maxwellian*

$$f_s^M (\mathbf{r}, \mathbf{v}_s, t) = n_s(\mathbf{r}, t) \left[\frac{m_s}{2\pi k T_s(\mathbf{r}, t)} \right]^{3/2}$$
$$\cdot \exp\left\{ -m_s [\mathbf{v}_s - \mathbf{u}_s(\mathbf{r}, t)]^2 / 2k T_s(\mathbf{r}, t) \right\}. \tag{3.44}$$

When collisions dominate, f_s takes this form at all positions in space and at all times, which is why it is called a *local* drifting Maxwellian. Note that the drifting Maxwellian depends only on the density, drift velocity, and temperature moments.

It is easy to verify that the drifting Maxwellian is consistent with the general definitions for the density, drift velocity, and temperature (Equations 3.10, 3.11, and 3.15). For example, the density is obtained by integrating the distribution function over all velocities, and if Equation (3.44) is used the density definition (3.10) becomes

$$n_s = n_s \left(\frac{m_s}{2\pi k T_s} \right)^{3/2} \int d^3 v_s \exp\left[\frac{-m_s (\mathbf{v}_s - \mathbf{u}_s)^2}{2k T_s} \right] \tag{3.45}$$

The integral can be calculated by introducing the random velocity, $\mathbf{c}_s = \mathbf{v}_s - \mathbf{u}_s$, and by using the fact that $d^3 c_s = d^3 v_s$ (the introduction of $\mathbf{c}_s$ merely changes the origin of the coordinate system, but the integral is still over all of velocity space). Since the resulting integrand depends only on the magnitude of c_s, a spherical coordinate system can be used, with $d^3 c_s = 4\pi c_s^2 dc_s$, and then Equation (3.45) becomes

$$n_s = n_s \left(\frac{m_s}{2\pi k T_s} \right)^{3/2} \int_0^\infty (4\pi c_s^2 dc_s) \exp(-m_s c_s^2 / 2k T_s). \tag{3.46}$$

The integral, according to Appendix C, is $(2\pi k T_s / m_s)^{3/2}$ and, hence, Equation (3.46) reduces to $n_s = n_s$. Likewise, if the drifting Maxwellian is used in the general definitions for the drift velocity (3.11) and temperature (3.15), it can be shown that it is consistent with these definitions.

A schematic diagram of a drifting Maxwellian distribution function is shown in Figure 3.2. The peak of the distribution occurs at $\mathbf{v}_s = \mathbf{u}_s$. The distribution is symmetric about the peak and falls off exponentially from the peak in all directions. The distribution decreases by a factor of "e" when $|\mathbf{v}_s - \mathbf{u}_s| = (2k T_s / m_s)^{1/2}$ and, hence, the width of the Maxwellian is determined by the temperature and the mass. In three dimensions, the contours of constant f_s^M are concentric spheres with the centers at $\mathbf{v}_s = \mathbf{u}_s$. A two-dimensional cut through the distribution yields concentric circles, and a line through the Maxwellian yields the classic bell-shaped curve.

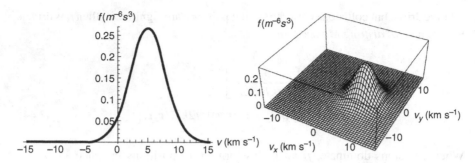

Figure 3.2 The Maxwellian velocity distribution. The left panel is a one-dimensional cut through the Maxwellian along the v_x-axis, and the right panel is a two-dimensional slice in the principal v_x–v_y plane. The Maxwellian shown is for a density of 10^5 cm^{-3}, $u_x = 5$ km s^{-1}, $u_y = 0$, $u_z = 0$, and $T = 1000$ K.

3.5   Closing the system of transport equations

As noted earlier, it is necessary to have an expression for f_s in order to close the system of general transport equations. A standard mathematical technique for obtaining approximate expressions for the species distribution function is to expand $f_s(\mathbf{r}, \mathbf{c}_s, t)$ in a complete orthogonal series of the form

$$f_s(\mathbf{r}, \mathbf{c}_s, t) = f_{so}(\mathbf{r}, \mathbf{c}_s, t) \sum_\alpha a_\alpha(\mathbf{r}, t) M_\alpha(\mathbf{c}_s), \tag{3.47}$$

where f_{so} is an "appropriate" zeroth-order velocity distribution function, M_α represents a complete set of orthogonal polynomials, a_α represents the unknown expansion coefficients, and the subscript α is used to indicate that the summation is generally over more than one coordinate index.[2-4] The zeroth-order distribution function and the set of orthogonal polynomials are generally chosen so that the series converges rapidly, and therefore, only a few terms in the series expansion are needed. If collisions are important, one would expect that the actual species distribution function is approximately Maxwellian at all locations and times. Consequently, it is logical to adopt a local Maxwellian as the zeroth-order distribution function[2]

$$f_{so} = f_s^M = n_s \left(\frac{m_s}{2\pi k T_s} \right)^{3/2} \exp(-m_s c_s^2 / 2k T_s), \tag{3.48}$$

where $\mathbf{c}_s = \mathbf{v}_s - \mathbf{u}_s$ and $n_s, \mathbf{u}_s, T_s$ depend on $\mathbf{r}$ and t. With a local Maxwellian as the zeroth-order distribution function and with a Cartesian coordinate system in velocity space, the associated orthogonal polynomials are the Hermite tensors. The unknown expansion coefficients are also tensors of all orders. For convenience, however, the expansion coefficients can be expressed in terms of the physically significant (and unknown) moments of the distribution function ($n_s, \mathbf{u}_s, T_s, \mathbf{P}_s, \mathbf{q}_s$, etc.) simply by taking the appropriate velocity moments of the series expansion (3.47).

To close the system of transport equations, the series expansion is first truncated at some level by setting all higher-order expansion coefficients (velocity moments) to zero. Only the transport equations that pertain to the velocity moments in the truncated series expansion are retained. However, as noted earlier, the transport equation for the moment of order ℓ contains the moment of order $\ell+1$. These higher-order velocity moments in the transport equations are not set to zero, but instead are expressed in terms of the lower-order moments with the aid of the "truncated" series expansion, which then yields a closed system of transport equations. For planetary ionospheres and atmospheres, a truncated series expansion that includes the stress tensor and heat flow vector is particularly useful. In this so-called *13-moment approximation*, the truncated series expansion for f_s takes the form

$$f_s = f_{so}\left[1 + \frac{m_s}{2kT_s p_s}\boldsymbol{\tau}_s : \mathbf{c}_s\mathbf{c}_s - \left(1 - \frac{m_s c_s^2}{5kT_s}\right)\frac{m_s}{kT_s p_s}\mathbf{q}_s \cdot \mathbf{c}_s\right], \tag{3.49}$$

where f_{so} is given by Equation (3.48). Note that in the 13-moment approximation the stress tensor and heat flow vector are put on an equal footing with the density, drift velocity, and temperature. The name 13-moment approximation stems from the fact that each species in the gas mixture is described by 13 parameters ($n_s = 1$, $\mathbf{u}_s = 3$, $T_s = 1$, $\mathbf{q}_s = 3$, $\boldsymbol{\tau}_s = 5$), where only five of the nine elements in the stress tensor are unknown, because it is defined to be symmetric ($\tau_{\alpha\beta} = \tau_{\beta\alpha}$) and traceless ($\sum_\alpha \tau_{\alpha\alpha} = 0$). As noted before, the double dot product is $\boldsymbol{\tau} : \mathbf{cc} = \sum_\alpha \sum_\beta \tau_{\alpha\beta}c_\beta c_\alpha$.

It is easy to show that by multiplying the 13-moment expression for f_s (Equation 3.49), respectively, with 1, $\mathbf{c}_s$, $\frac{1}{2}m_s c_s^2$, $m_s\mathbf{c}_s\mathbf{c}_s$, and $\frac{1}{2}m_s c_s^2\mathbf{c}_s$ and integrating over velocity space, the distribution function properly accounts for the density, drift velocity, temperature, stress tensor, and heat flow vector. However, the general transport equations (3.39, 3.40) have velocity moments (μ_s, $\mathbf{Q}_s$) that are of a higher order than what is available at the 13-moment level. These higher-order moments can now be expressed in terms of the 13 lower-order moments with the aid of the truncated series expansion (3.49). Specifically, by multiplying Equation (3.49) with $\frac{1}{2}m_s c_s^2\mathbf{c}_s\mathbf{c}_s$ and $m_s\mathbf{c}_s\mathbf{c}_s\mathbf{c}_s$, respectively, and integrating over all velocities, one obtains

$$\mu_s = \frac{5}{2}\frac{kT_s}{m_s}\left(p_s\mathbf{I} + \frac{7}{5}\boldsymbol{\tau}_s\right), \tag{3.50}$$

$$(\mathbf{Q}_s)_{\alpha\beta\gamma} = \frac{2}{5}\left[(q_s)_\alpha\delta_{\beta\gamma} + (q_s)_\gamma\delta_{\alpha\beta} + (q_s)_\beta\delta_{\alpha\gamma}\right], \tag{3.51}$$

where index notation is used in Equation (3.51). Using Equations (3.50) and (3.51), it is now possible to calculate the terms needed to close the system of general transport equations:

$$\nabla \cdot \mathbf{Q}_s = \frac{2}{5}\left[\nabla\mathbf{q}_s + (\nabla\mathbf{q}_s)^T + (\nabla \cdot \mathbf{q}_s)\mathbf{I}\right], \tag{3.52}$$

$$Q_s : \nabla \mathbf{u}_s = \frac{2}{5}[\mathbf{q}_s(\nabla \cdot \mathbf{u}_s) + (\nabla \mathbf{u}_s) \cdot \mathbf{q}_s + \mathbf{q}_s \cdot \nabla \mathbf{u}_s], \tag{3.53}$$

$$\nabla \cdot \boldsymbol{\mu}_s = \frac{5}{2} \frac{k}{m_s}\left[\nabla(T_s p_s) + \frac{7}{5}\nabla \cdot (T_s \boldsymbol{\tau}_s)\right]. \tag{3.54}$$

Only the pressure tensor (3.39) and heat flow (3.40) equations are affected by the closure, and these become:

Pressure tensor equation:

$$\frac{D_s \mathbf{P}_s}{Dt} + \frac{2}{5}[\nabla \mathbf{q}_s + (\nabla \mathbf{q}_s)^T + (\nabla \cdot \mathbf{q}_s)\mathbf{I}]$$

$$+ \mathbf{P}_s(\nabla \cdot \mathbf{u}_s) + \frac{e_s}{m_s}(\mathbf{B} \times \mathbf{P}_s - \mathbf{P}_s \times \mathbf{B})$$

$$+ \mathbf{P}_s \cdot \nabla \mathbf{u}_s + (\mathbf{P}_s \cdot \nabla \mathbf{u}_s)^T = \frac{\delta \mathbf{P}_s}{\delta t}, \tag{3.55}$$

Heat flow equation:

$$\frac{D_s \mathbf{q}_s}{Dt} + \frac{7}{5}\mathbf{q}_s \cdot \nabla \mathbf{u}_s + \frac{7}{5}\mathbf{q}_s(\nabla \cdot \mathbf{u}_s) + \frac{2}{5}(\nabla \mathbf{u}_s) \cdot \mathbf{q}_s$$

$$+ \frac{5}{2}\frac{k}{m_s}\left[\nabla(T_s p_s) + \frac{7}{5}\nabla \cdot (T_s \boldsymbol{\tau}_s)\right] + \left[\frac{D_s \mathbf{u}_s}{Dt} - \mathbf{G} - \frac{e_s}{m_s}(\mathbf{E} + \mathbf{u}_s \times \mathbf{B})\right]$$

$$\cdot \left(\boldsymbol{\tau}_s + \frac{5}{2}p_s\mathbf{I}\right) - \frac{e_s}{m_s}\mathbf{q}_s \times \mathbf{B} = \frac{\delta \mathbf{q}_s}{\delta t}. \tag{3.56}$$

3.6   13-moment transport equations

The closed system of transport equations at the 13-moment level of approximation is given by Equations (3.30), (3.36), (3.38), (3.55), and (3.56). For future reference, it is convenient to list these equations in one place. However, an equation describing the evolution of the stress tensor is generally more useful than the equation for the pressure tensor. The stress tensor equation can be obtained by subtracting $\frac{2}{3}\mathbf{I}$ times the energy equation (3.38) from the pressure tensor equation (3.55). Likewise, it is also convenient to simplify the heat flow equation (3.56) with the aid of the momentum equation (3.36). With these changes, the closed system of 13-moment transport equations becomes

$$\frac{\partial n_s}{\partial t} + \nabla \cdot (n_s \mathbf{u}_s) = \frac{\delta n_s}{\delta t}, \tag{3.57}$$

$$n_s m_s \frac{D_s \mathbf{u}_s}{Dt} + \nabla p_s + \nabla \cdot \boldsymbol{\tau}_s - n_s e_s(\mathbf{E} + \mathbf{u}_s \times \mathbf{B}) - n_s m_s \mathbf{G} = \frac{\delta \mathbf{M}_s}{\delta t}, \tag{3.58}$$

$$\frac{D_s}{Dt}\left(\frac{3}{2}p_s\right) + \frac{5}{2}p_s(\nabla \cdot \mathbf{u}_s) + \nabla \cdot \mathbf{q}_s + \boldsymbol{\tau}_s : \nabla \mathbf{u}_s = \frac{\delta E_s}{\delta t}, \tag{3.59}$$

$$\frac{D_s \boldsymbol{\tau}_s}{Dt} + \boldsymbol{\tau}_s(\nabla \cdot \mathbf{u}_s) + \frac{e_s}{m_s}\left[\mathbf{B} \times \boldsymbol{\tau}_s - \boldsymbol{\tau}_s \times \mathbf{B}\right]$$

$$+ p_s\left[\nabla \mathbf{u}_s + (\nabla \mathbf{u}_s)^T - \frac{2}{3}(\nabla \cdot \mathbf{u}_s)\mathbf{I}\right]$$

$$+ \frac{2}{5}\left[\nabla \mathbf{q}_s + (\nabla \mathbf{q}_s)^T - \frac{2}{3}(\nabla \cdot \mathbf{q}_s)\mathbf{I}\right]$$

$$+ \left[\boldsymbol{\tau}_s \cdot \nabla \mathbf{u}_s + (\boldsymbol{\tau}_s \cdot \nabla \mathbf{u}_s)^T - \frac{2}{3}(\boldsymbol{\tau}_s : \nabla \mathbf{u}_s)\mathbf{I}\right] = \frac{\delta \boldsymbol{\tau}_s}{\delta t}, \tag{3.60}$$

$$\frac{D_s \mathbf{q}_s}{Dt} + \frac{7}{5}\mathbf{q}_s \cdot \nabla \mathbf{u}_s + \frac{7}{5}\mathbf{q}_s(\nabla \cdot \mathbf{u}_s) + \frac{2}{5}(\nabla \mathbf{u}_s) \cdot \mathbf{q}_s$$

$$+ \frac{5}{2}\frac{kp_s}{m_s}\nabla T_s + \frac{1}{\rho_s}(\nabla \cdot \boldsymbol{\tau}_s) \cdot (p_s\mathbf{I} - \boldsymbol{\tau}_s)$$

$$+ \left(\frac{7}{2}\frac{k}{m_s}\nabla T_s - \frac{1}{\rho_s}\nabla p_s\right) \cdot \boldsymbol{\tau}_s - \frac{e_s}{m_s}\mathbf{q}_s \times \mathbf{B} = \frac{\delta \mathbf{q}'_s}{\delta t}, \tag{3.61}$$

where

$$\frac{\delta \boldsymbol{\tau}_s}{\delta t} = \frac{\delta \mathbf{P}_s}{\delta t} - \frac{2}{3}\frac{\delta E_s}{\delta t}\mathbf{I}, \tag{3.62}$$

$$\frac{\delta \mathbf{q}'_s}{\delta t} = \frac{\delta \mathbf{q}_s}{\delta t} - \frac{1}{\rho_s}\frac{\delta \mathbf{M}_s}{\delta t} \cdot \left(\boldsymbol{\tau}_s + \frac{5}{2}p_s\mathbf{I}\right), \tag{3.63}$$

and where $\rho_s = n_s m_s$ is the *mass density*. Note also that the relation (3.21), $\mathbf{P}_s = \boldsymbol{\tau}_s + p_s\mathbf{I}$, has been used.

The 13-moment system of equations is very powerful and can be used to describe a wide range of plasma and neutral gas flows, provided the species velocity distributions are not too far from Maxwellians. It can be applied to collision-dominated, transitional, and collisionless flows and provides for a continuous transition between these regimes. It can also be applied to subsonic, transonic, and supersonic flows as well as chemically reactive flows. As will be shown in the chapters that follow, in the collision-dominated limit, the 13-moment system of equations reduces to the Euler and Navier–Stokes equations depending on whether terms proportional to the zeroth or first power of the collisional mean-free-path are retained (Chapters 5 and 10). At the Navier–Stokes level, transport processes such as ordinary diffusion, thermal diffusion, thermal conduction, diffusion-thermal heat flow, thermoelectric heat flow, and viscosity are included at a level that corresponds to either the first or second approximation of Chapman and Cowling,[1] depending on the particular transport coefficient. In the collisionless limit, the 13-moment system of equations reduces to the Chew–Goldberger–Low (CGL) and extended CGL equations depending on whether terms proportional to the zeroth or first power of the Larmor radius are retained (Chapter 7). The 13-moment equations also account for collisionless heat flow and temperature anisotropies (Chapter 5).

Temperature anisotropies typically occur in a plasma when collisions are infrequent and there is a preferred direction, which can result from the presence of a strong magnetic field, a strong electric field, or a strong pressure gradient. In this case, the thermal spread of particles along the preferred direction can be different from that perpendicular to the preferred direction, which then yields different species temperatures parallel and perpendicular to the preferred direction. The definitions of the parallel and perpendicular temperatures that are consistent with the isotropic temperature definition (3.15) are

$$T_{s\parallel} = \frac{m_s}{k} \langle c_{s\parallel}^2 \rangle = \frac{m_s}{kn_s} \int d^3 v_s f_s (\mathbf{v}_s - \mathbf{u}_s)_\parallel^2, \tag{3.64}$$

$$T_{s\perp} = \frac{m_s}{2k} \langle c_{s\perp}^2 \rangle = \frac{m_s}{2kn_s} \int d^3 v_s f_s (\mathbf{v}_s - \mathbf{u}_s)_\perp^2. \tag{3.65}$$

By comparing Equations (3.15), (3.64), and (3.65), it is apparent that

$$T_s = \frac{1}{3} \left[T_{s\parallel} + 2T_{s\perp} \right]. \tag{3.66}$$

However, when there are different temperatures parallel and perpendicular to a preferred direction, there are also different heat flows because a heat flow is simply a flow of thermal energy. The definitions of the flow of parallel and perpendicular thermal energies that are consistent with the usual heat flow definition (3.16) are

$$\mathbf{q}_s^\parallel = n_s m_s \langle c_{s\parallel}^2 \mathbf{c}_s \rangle = m_s \int d^3 v_s f_s (\mathbf{v}_s - \mathbf{u}_s)_\parallel^2 (\mathbf{v}_s - \mathbf{u}_s), \tag{3.67}$$

$$\mathbf{q}_s^\perp = \frac{1}{2} n_s m_s \langle c_{s\perp}^2 \mathbf{c}_s \rangle = \frac{m_s}{2} \int d^3 v_s f_s (\mathbf{v}_s - \mathbf{u}_s)_\perp^2 (\mathbf{v}_s - \mathbf{u}_s), \tag{3.68}$$

where a comparison of definitions (3.16), (3.67), and (3.68) indicates that

$$\mathbf{q}_s = \frac{1}{2} \left[\mathbf{q}_s^\parallel + 2\mathbf{q}_s^\perp \right]. \tag{3.69}$$

In the 13-moment approximation, the fundamental velocity moments are n_s, $\mathbf{u}_s$, T_s, $\boldsymbol{\tau}_s$, and $\mathbf{q}_s$, and all other moments can be expressed in terms of these fundamental moments. As before, this can be accomplished by substituting the 13-moment expression for f_s (3.49) into the definitions for $T_{s\parallel}$, $T_{s\perp}$, $\mathbf{q}_s^\parallel$, and $\mathbf{q}_s^\perp$ and performing the integrals, which yields

$$T_{s\parallel} = T_s + \tau_s : \mathbf{e}_3\mathbf{e}_3/(n_s k), \tag{3.70}$$

$$T_{s\perp} = T_s + \tau_s : (\mathbf{I} - \mathbf{e}_3\mathbf{e}_3)/(2n_s k), \tag{3.71}$$

$$\mathbf{q}_s^{\parallel} = \frac{2}{5}(\mathbf{I} + 2\mathbf{e}_3\mathbf{e}_3) \cdot \mathbf{q}_s, \tag{3.72}$$

$$\mathbf{q}_s^{\perp} = \frac{2}{5}(2\mathbf{I} - \mathbf{e}_3\mathbf{e}_3) \cdot \mathbf{q}_s, \tag{3.73}$$

where $(\mathbf{e}_1, \mathbf{e}_2, \mathbf{e}_3)$ are unit vectors of an orthogonal coordinate system and where the preferred direction is along the $\mathbf{e}_3$ axis. Note that the diagonal elements of the stress tensor are responsible for the temperature anisotropy. Also note that $\mathbf{q}_s^{\parallel}$ and $\mathbf{q}_s^{\perp}$ are not independent but are related to $\mathbf{q}_s$ in specific ways. Finally, it should be noted that in the 13-moment approximation the temperature is assumed to be isotropic to the lowest order (i.e., T_s appears in the zeroth-order distribution f_{so}). The deviations from isotropy therefore appear via the correction terms in the series expansion (3.49). For the series to converge, the terms in the expansion must be small compared to unity and, hence, the temperature anisotropy and heat flow must be "small."

3.7 Generalized transport systems

In some plasma flows, the species velocity distributions may depart sufficiently from a Maxwellian such that the 13-moment approximation is not adequate. Provided that the departures are not too large, one can simply add more terms in the series expansion and then truncate the series at a higher level. The next appropriate level is the *20-moment approximation*, and at this level the species distribution function takes the following form[5]

$$f_s = f_{so}\left(1 + \frac{m_s}{2kT_s p_s}\tau_s : \mathbf{c}_s\mathbf{c}_s + \frac{m_s^2}{6k^2 T_s^2 p_s}\mathbf{Q}_s : \mathbf{c}_s\mathbf{c}_s\mathbf{c}_s - \frac{m_s}{kT_s p_s}\mathbf{q}_s \cdot \mathbf{c}_s\right), \tag{3.74}$$

where $\mathbf{Q}_s : \mathbf{c}_s\mathbf{c}_s\mathbf{c}_s = \sum_{\alpha,\beta,\gamma}(Q_s)_{\alpha\beta\gamma}(c_s)_\alpha(c_s)_\beta(c_s)_\gamma$. In the 20-moment approximation, the heat flow tensor is put on an equal footing with the density, drift velocity, temperature, and stress tensor. $\mathbf{Q}_s$ is symmetric with respect to a change in any two coordinate indices, and hence, there are 10 unknown elements in this tensor. This means there is a total of 20 parameters that describe each species in the gas mixture at this level of approximation. Therefore, the system of transport equations must be expanded to include flow equations for the 10 heat flow elements. Generally, however, the 20-moment system of transport equations is too complicated to be of practical use.

If the flow conditions are such that the departures of the species distribution functions cannot be adequately described by the 13-moment approximation, it is better to derive an entirely new set of transport equations that is based on a series expansion (3.47) about a "nonMaxwellian" zeroth-order distribution function f_{so}. The specific form of f_{so} depends on the specific problem that is to be solved. This zeroth-order distribution function may be obtained by solving a simple but related problem, it may be obtained from simple physical arguments, or it may be deduced from measurements. In practice, one considers only a limited number of terms in the series expansion, therefore the zeroth-order distribution should be selected with care. A well-chosen zeroth-order distribution function yields expansion coefficients that decrease rapidly as the order of the coefficients increases. However, for every zeroth-order distribution function there is an associated set of transport equations that describes the spatial and temporal evolution of the expansion coefficients, and consequently, if a complex zeroth-order distribution function is selected in order to get close to the "expected" form of f_s, it may be difficult or impossible to solve the resulting set of transport equations. Therefore, for highly nonMaxwellian flows, the zeroth-order distribution function must be reasonably close to the expected form of f_s so that the series expansion (3.47) can be truncated at a fairly low order, yet it must be simple enough to yield reasonable transport equations for the expansion coefficients. In applications involving plasma flows in planetary ionospheres, generalized transport equations have been derived for series expansions about several "nonMaxwellian" zeroth-order distribution functions, including bi-Maxwellian (two temperature), tri-Maxwellian (three temperature) and toroidal distribution functions.[6-8]

The transport equations based on a zeroth-order bi-Maxwellian velocity distribution are particularly useful for describing collisionless plasmas subjected to strong magnetic fields. In this case the zeroth-order velocity distribution takes the form

$$f_{so} = f_s^{BM}$$

$$= n_s \left(\frac{m_s}{2\pi k T_{s\parallel}} \right)^{1/2} \left(\frac{m_s}{2\pi k T_{s\perp}} \right) \exp\left(-\frac{m_s c_{s\parallel}^2}{2k T_{s\parallel}} - \frac{m_s c_{s\perp}^2}{2k T_{s\perp}} \right). \tag{3.75}$$

Note that with a bi-Maxwellian-based series expansion the anisotropic character of the distribution, as expressed by $T_{s\parallel}$ and $T_{s\perp}$, is accounted for in the weight factor, f_{so}, of the series expansion for f_s. In the Maxwellian-based series expansion, on the other hand, the temperature anisotropy enters through the stress terms in the series (3.49), which must be small for the series to converge. Therefore, a bi-Maxwellian-based series expansion can describe plasmas with much larger temperature anisotropies than a Maxwellian-based expansion with the same number of terms.[8]

3.8   Kinetic, Monte Carlo, and particle-in-cell methods

In planetary ionospheres at high altitudes, the plasma and neutral gases eventually become collisionless. The altitude where this occurs is called the exobase, which is defined to be the altitude where the collision mean-free-path (mfp) of the particles is equal to the density scale height. Below the exobase, the mfp is smaller than the density scale height and above the exobase it is larger than the density scale height. Because of the long-range nature of Coulomb collisions, plasma exobases are typically at higher altitudes than neutral exobases. For example, in the Earth's polar region, the neutral exobase is located at an altitude between 500 and 600 km, while the plasma exobase is between 1500 and 3000 km.

At the altitudes where either the neutral or plasma gas is collisionless, a kinetic model can be adopted, as was done for the solar and terrestrial polar winds. In this case, the Vlasov Equation (3.6) can be integrated in altitude for steady-state conditions, assuming that the velocity distribution function at the exobase is known. The integration yields the species velocity distribution function at all altitudes above the exobase, from which the various velocity moments (n_s, $\mathbf{u}_s$, T_s, etc.) can be calculated for each species. However, the result obtained depends on the assumed velocity distribution function at the exobase, which is generally not known in a collisionless (or almost collisionless) regime. Over the years, many different expressions for the velocity distribution function at the exobase have been adopted, including monoenergetic distributions, a truncated Maxwellian, a drifting or displaced Maxwellian, a truncated bi-Maxwellian, a drifting or displaced bi-Maxwellian, a Lorentzian, and a bi-Lorentzian.[9]

Other mathematical methods have also been used to describe collisionless (or almost collisionless) gases and plasmas. In the transition region between the collision-dominated and the collisionless polar wind flow, the Boltzmann equation with Coulomb collisions has been integrated numerically. However, the accuracy obtained for the velocity distribution function depends on the numerical grid spacing in velocity space, which extends to infinity.[10] Additional methods include Monte Carlo (Appendix P) and macroscopic particle-in-cell (PIC) techniques, whereby the motions of individual particles are followed as they are subjected to various forces and collision processes. However, the accuracy obtained depends on the number of particles followed. In the latest three-dimensional PIC simulation of the polar wind, one billion particles were followed during an idealized geomagnetic storm.[11] The simulation included the effects of gravity, polarization electrostatic field, magnetic mirror force, centripetal acceleration, ion self-collisions, low-altitude auroral ion energization, wave–particle interactions (WPI), and the $\mathbf{E} \times \mathbf{B}$ drift of the plasma flux tubes (see Chapter 12 for a discussion of the polar wind). These and other results are described in more detail in a series of review papers.[12–16]

3.9 Maxwell equations

The 13-moment and generalized systems of transport equations are only complete if the electric and magnetic fields that exist in the plasma are known. However, this is typically not the case because currents that flow in the plasma generate magnetic fields and differing ion and electron densities create electric fields. Therefore, in general, the Maxwell equations of electricity and magnetism must be solved along with the plasma transport equations. In a vacuum, these equations are given by

$$\nabla \cdot \mathbf{E} = \rho_c/\varepsilon_0, \tag{3.76a}$$

$$\nabla \times \mathbf{E} = -\frac{\partial \mathbf{B}}{\partial t}, \tag{3.76b}$$

$$\nabla \cdot \mathbf{B} = 0, \tag{3.76c}$$

$$\nabla \times \mathbf{B} = \mu_0 \mathbf{J} + \mu_0\varepsilon_0 \frac{\partial \mathbf{E}}{\partial t}, \tag{3.76d}$$

where the *charge density*, ρ_c, and the *current density*, $\mathbf{J}$, are given by

$$\rho_c = \sum_s n_s e_s, \tag{3.77}$$

$$\mathbf{J} = \sum_s n_s e_s \mathbf{u}_s, \tag{3.78}$$

and where ε_0 is the *permittivity* and μ_0 the *permeability* of free space.

3.10   Specific references

1. Chapman, S., and T. G. Cowling, *The Mathematical Theory of Non-Uniform Gases*, New York: Cambridge University Press, 1970.
2. Grad, H., On the kinetic theory of rarefied gases, *Comm. Pure Appl. Math.*, **2**, 331, 1949.
3. Burgers, J. M., *Flow Equations for Composite Gases*, New York: Academic, 1969.
4. Schunk, R. W., Mathematical structure of transport equations for multispecies flows, *Rev. Geophys. Space Phys.*, **15**, 429, 1977.
5. Grad, H., Principles of the kinetic theory of gases, *Handbook of Phys.*, **XII**, 205, New York: Springer, 1958.
6. Oraevskii, V., R. Chodura, and W. Fenberg, Hydrodynamic equations for plasmas in strong magnetic field – I Collisionless approximation, *Plasma Phys.*, **10**, 819, 1968.
7. St-Maurice, J.-P., and R. W. Schunk, Ion velocity distributions in the high-latitude ionosphere, *Rev. Geophys. Space Phys.*, **17**, 99, 1979.
8. Barakat, A. R., and R. W. Schunk, Transport equations for multicomponent anisotropic space plasma: A review, *Plasma Phys.*, **24**, 389, 1982.
9. Lemaire, J. F., W. K. Peterson, T. Chang, *et al.*, History of kinetic polar wind models and early observations, *J. Atmos. Solar-Terr. Phys.*, **69**, 1901, 2007.
10. Pierrard, V., Fonctions de distribution des vitesses des particles s'echappant de L'ionosphere, Ph.D. thesis, UCL, 1997.
11. Barakat, A. R., and R. W. Schunk, A three-dimensional model of the generalized polar wind, *J. Geophys. Res.*, **111**, A12314, doi: 10.1029/2006JA011662, 2006.
12. Tam, S. W. Y., T. Chang, and V. Pierrard, Kinetic modeling of the polar wind, *J. Atmos. Solar-Terr. Phys.*, **69**, 1984, 2007.
13. Yau, A. W., T. Abe, and W. K. Peterson, The polar wind: recent observations, *J. Atmos. Solar-Terr. Phys.*, **69**, 1936, 2007.

14. Schunk, R. W., Time-dependent simulations of the global polar wind, *J. Atmos. Solar-Terr. Phys.*, **69**, 2028, 2007.

15. Banerjee, S., and V. V. Gavrishchaka, Multimoment convecting flux tube model of the polar wind system with return current and microprocesses, *J. Atmos. Solar-Terr. Phys.*, **69**, 2071, 2007.

16. Gavrishchaka, V. V., S. Banerjee, and P. N. Guzdar, Large-scale oscillations and transport processes generated by multiscale inhomogeneities in the ionospheric field-aligned flows: a 3-D simulation with a dipole magnetic field, *J. Atmos. Solar-Terr. Phys.*, **69**, 2058, 2007.

3.11 Problems

Problem 3.1 Show that $\nabla_v \cdot \mathbf{a} = 0$ for the acceleration given in Equation (3.8).

Problem 3.2 Show that the right-hand sides of Equations (3.32) to (3.35) are correct.

Problem 3.3 Derive the energy transport equation by multiplying the Boltzmann equation (3.24) by $(1/2)m_s c_s^2$ and then integrating over velocity space.

Problem 3.4 Show that the local drifting Maxwellian distribution (3.44) is consistent with the general definitions for the drift velocity (3.11) and temperature (3.15) moments.

Problem 3.5 Show that the 13-moment distribution function (3.49) is consistent with the general definitions for the density (3.10), drift velocity (3.11), and temperature (3.15) moments.

Problem 3.6 Show that the 13-moment distribution function (3.49) is consistent with the general definition for the heat flow vector (3.16).

Problem 3.7 Subtract $\frac{2}{3}\mathbf{I}$ times the energy equation (3.38) from the pressure tensor equation (3.55) and thereby derive the stress tensor equation (3.60).

Problem 3.8 Using the definitions of T_s (3.15), $T_{s\parallel}$ (3.64), and $T_{s\perp}$ (3.65), show that they are related via Equation (3.66).

Problem 3.9 Using the definitions for $\mathbf{q}_s$ (3.16), $\mathbf{q}_s^{\parallel}$ (3.67), and $\mathbf{q}_s^{\perp}$ (3.68), show that they are related via Equation (3.69).

Problem 3.10 Substitute the 13-moment expression for f_s (3.49) into the definition for $T_{s\parallel}$ (3.64) and show that Equation (3.70) is correct.

Problem 3.11 Substitute the bi-Maxwellian distribution function (3.75) into the definition for the heat flow $\mathbf{q}_s$ (3.16) and obtain an expression that relates $\mathbf{q}_s$ to $T_{s\parallel}$ and $T_{s\perp}$.

Problem 3.12 Substitute the bi-Maxwellian distribution function (3.75) into the definitions for $\mathbf{q}_s^{\parallel}$ (3.67) and $\mathbf{q}_s^{\perp}$ (3.68) and obtain expressions that relate these vectors to $T_{s\parallel}$ and $T_{s\perp}$.

Problem 3.13 Consider nondrifting Maxwellian and bi-Maxwellian velocity distributions, where the parallel temperature for the bi-Maxwellian distribution is associated with the z-axis. Calculate the flux of particles across the $z = 0$ plane for the particles that move from the negative to the positive z-direction for both the Maxwellian and bi-Maxwellian distributions.

Problem 3.14 Consider the following expression for the distribution of a given plasma species:

$$f(\mathbf{r}, \mathbf{c}, t) = f_o \left[1 - \left(1 - \frac{mc^2}{5kT} \right) \frac{m}{kTp} \mathbf{q} \cdot \mathbf{c} \right],$$

$$f_o = n \left(\frac{m}{2\pi kT} \right)^{3/2} \exp \left(-\frac{mc^2}{2kT} \right),$$

where f is the 8-moment approximation in the Maxwellian-based expansion of the distribution function and n, $\mathbf{u}$, T, p, and $\mathbf{q}$ have the usual definitions. Note that $\mathbf{c}$ is the usual "random" velocity. Calculate the random flux crossing an imaginary plane from one side to the other for the 8-moment expression given above (see Appendix H). Take the plane perpendicular to the principal v_x-axis.

Problem 3.15 The velocity distribution for a nonequilibrium gas is given by

$$f(v) = \frac{n}{4\pi v_o^2} \delta(v - v_0),$$

where $\delta(v)$ is the Dirac delta function, v_0 is a constant, and v is the magnitude of the velocity. Calculate the density, drift velocity, and temperature.

Problem 3.16 The escape flux from a gravitationally bound planetary atmosphere is calculated by assuming that above a given critical altitude there are no more collisions, and particles having energies greater than what is necessary to overcome the gravitational pull of the planet will escape. In obtaining this expression for the "particle" flux it is assumed that at that critical level (called the exobase) the distribution function is a nondrifting Maxwellian. The speed which is necessary to overcome the gravitational pull at a given altitude, called the escape speed, is given by

$$v_{\text{esc}} = \left(\frac{2GM}{r} \right)^{1/2},$$

where G is the gravitational constant, M is the mass of the planet in question, and r is the geocentric distance to the exobase altitude. Calculate the escape flux in terms of the density, the most probable speed at this exobase, and the escape velocity (Appendix H).

Problem 3.17 The Lorentzian velocity distribution function is given by

$$f(\mathbf{r}, \mathbf{v}, t) = n(\mathbf{r}, t) \left[1 + \frac{mv^2}{2\alpha kT(\mathbf{r}, t)} \right]^{-(\alpha+1)},$$

where n is the density, T the temperature, m the mass, v the velocity, k is the Boltzmann constant, and α is a constant number. Calculate the density, drift velocity, and temperature moments of the Lorentzian distribution function.

Problem 3.18 Calculate the heat flow moment of the Lorentzian distribution function given in Problem 3.17.

Chapter 4

Collisions

Collisions play a fundamental role in the dynamics and energetics of ionospheres. They are responsible for the production of ionization, the diffusion of plasma from high to low density regions, the conduction of heat from hot to cold regions, the exchange of energy between different species, and other processes. The collisional processes can be either elastic or inelastic. The interactions leading to chemical reactions are discussed in Chapter 8. In an *elastic collision*, the momentum and kinetic energy of the colliding particles are conserved, while this is not the case in an *inelastic collision*. The exact nature of the collision process depends both on the relative kinetic energy of the colliding particles and on the type of particles. In general, for low energies, elastic collisions dominate, but as the relative kinetic energy increases, inelastic collisions become progressively more important. The order of importance is from elastic to rotational, vibrational, and electronic excitation, and then to ionization as the relative kinetic energy increases. However, the different collision processes may affect the continuity, momentum, and energy equations in different ways. For example, ionization of neutral gases by solar radiation and particle impact are the main sources of plasma in the ionospheres and these processes must be included in the continuity equation. On the other hand, ionization collisions are very infrequent compared with binary elastic collisions under most circumstances, and therefore, the momentum perturbation associated with the ionization process is generally not important and can be neglected in the momentum equation.

The various ionospheres correspond to partially ionized gases, and therefore, several collisional processes need to be considered, including Coulomb collisions, resonant and nonresonant ion–neutral interactions, electron–neutral interactions, and collisions between different neutral species. In the material that follows, the focus is on deriving the collision terms that appear on the right-hand side of the transport equations in order to elucidate the intrinsic limitations associated with the various simplified expressions. Another goal is to present collision terms that can be used in the applications of the transport equations to ionospheric problems.

4.1 Simple collision parameters

Some important collision parameters can be calculated by considering the simple scenario depicted in Figure 4.1. In this scenario, a large particle of radius r_0 (e.g., a neutral molecule) is surrounded by a gas of small particles (e.g., electrons) that has a constant density n. If the thermal motion is neglected and if the particles collide as hard spheres, then as the neutral molecule moves through the gas with a relative speed v, it produces a wake that, in time Δt has a volume $\sigma(v\Delta t)$, where $\sigma = \pi r_0^2$ is the cross section of the neutral molecule and $v\Delta t$ is the distance traveled by the neutral. The number of electrons in this volume, which corresponds to the total number of collisions between the neutral and electron gas, is $(\sigma v\Delta t)n$. Therefore, the number of collisions per unit time, the *collision frequency*, is given by

$$\nu = v\sigma n. \tag{4.1}$$

The *collision time* (i.e., the mean time between collisions) is simply the reciprocal of the collision frequency

$$\tau = \frac{1}{v\sigma n} \tag{4.2}$$

and the *mean-free-path* is just the speed multiplied by the mean time between collisions ($v\tau$)

$$\lambda_{\mathrm{mfp}} = \frac{1}{\sigma n}. \tag{4.3}$$

These results exhibit some intuitively obvious features of hard sphere collisions. That is, larger relative velocities, collision cross sections, and gas densities lead to greater collision frequencies and reduced mean-free-paths. Although these results are

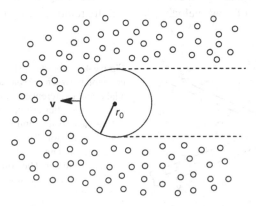

Figure 4.1 Schematic diagram showing a large particle of radius r_0 moving with velocity **v** through a background gas of small particles that is stationary and has a constant density n.

also true in general, a gas typically exhibits thermal motion, and hence, the relative neutral–electron speed will be different for different electrons, $g = |\mathbf{v}_n - \mathbf{v}_e|$. Also, the collision cross section is typically a function of the relative speed, $\sigma = \sigma(g)$, which means

$$\nu = g\sigma(g)n \tag{4.4}$$

$$\lambda_{mfp} = \frac{1}{\sigma(g)n}. \tag{4.5}$$

Hence, different particles in the gas will have different collision frequencies and mean-free-paths. To arrive at "average" quantities, it is necessary to take an average over the particle distribution functions. Typically, Maxwellians are used when this is done and the results then depend on the temperatures of the colliding species.

The simple collision parameters discussed above are useful for elucidating some basic collision features, but in practice there are different ways to define the collision frequency. The most useful is the collision frequency for momentum transfer, which is introduced later. At that time, the temperature-dependent average collision frequencies are presented.

4.2   Binary elastic collisions

To pursue a more rigorous determination of the various transport coefficients, it is necessary to study the dynamics of particle collisions. For now, the focus is on binary elastic collisions. In an *elastic collision*, the mass, momentum, and energy of the colliding particles are conserved in the collision process. That means ionization, chemical reactions, and electronic excitation do not occur. Figure 4.2 provides a schematic of a binary collision in a laboratory reference frame. The particle velocities before the collision are $\mathbf{v}_s$ and $\mathbf{v}_t$, while those after the collision are $\mathbf{v}'_s$ and $\mathbf{v}'_t$. The angle θ is the *scattering angle* and the *impact parameter b* is the distance of closest approach if the particles do not collide.

In dealing with binary collisions and in evaluating collision integrals, it is convenient to introduce the center-of-mass velocity, $\mathbf{V}_c$, and the relative velocity, $\mathbf{g}_{st}$,

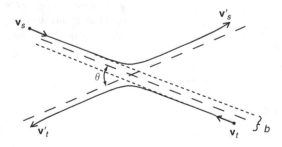

Figure 4.2 Binary elastic collision between two particles in a laboratory reference frame. The particle velocities are $\mathbf{v}_s$ and $\mathbf{v}_t$ before the collision and $\mathbf{v}'_s$ and $\mathbf{v}'_t$ after the collision. The angle θ is the scattering angle and b is the impact parameter. The collision depicted is for a repulsion.

of the colliding particles,

$$V_c = \frac{m_s v_s + m_t v_t}{m_s + m_t},$$ (4.6)

$$g_{st} = v_s - v_t,$$ (4.7)

where these expressions correspond to the velocities before the collision. These equations can also be inverted to give v_s and v_t in terms of V_c and g_{st}:

$$v_s = V_c + \frac{m_t}{m_s + m_t} g_{st},$$ (4.8)

$$v_t = V_c - \frac{m_s}{m_s + m_t} g_{st}.$$ (4.9)

After the collision, similar expressions hold, but now all of the velocities are primed:

$$V'_c = \frac{m_s v'_s + m_t v'_t}{m_s + m_t},$$ (4.10)

$$g'_{st} = v'_s - v'_t,$$ (4.11)

$$v'_s = V'_c + \frac{m_t}{m_s + m_t} g'_{st},$$ (4.12)

$$v'_t = V'_c - \frac{m_s}{m_s + m_t} g'_{st},$$ (4.13)

where use has already been made of the fact that the particle masses do not change in a collision. Conservation of momentum and kinetic energy in the collision yield additional relations,

$$m_s v_s + m_t v_t = m_s v'_s + m_t v'_t,$$ (4.14)

$$\frac{1}{2} m_s v_s^2 + \frac{1}{2} m_t v_t^2 = \frac{1}{2} m_s v'^2_s + \frac{1}{2} m_t v'^2_t,$$ (4.15)

and these can be used to relate V_c to V'_c and g_{st} to g'_{st}. The comparison of Equations (4.6) for V_c, (4.10) for V'_c, and the momentum conservation equation (4.14) indicates that

$$V'_c = V_c,$$ (4.16)

which means the center-of-mass velocity does not change in a collision. Substituting the velocities v_s (4.8), v_t (4.9), v'_s (4.12), and v'_t (4.13) into the energy Equation (4.15) yields, after cancellation of terms,

$$g_{st}^2 = g'^2_{st},$$ (4.17)

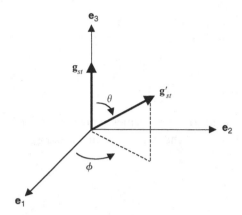

which indicates that the magnitude of the relative velocity does not change in a collision. The relative velocity merely changes direction, as shown in Figure 4.3.

The advantage of a center-of-mass reference frame in describing binary elastic collisions is now obvious; the center-of-mass velocity, $\mathbf{V}_c$, does not change and the magnitude of the relative velocity g_{st} is also constant. Therefore, if the initial velocities of the colliding particles and the scattering angle are known, the velocities after the collision can be calculated. In a gas, many particles can collide with a given particle and, hence, there is a distribution of initial velocities. This aspect of collisional dynamics is discussed later. The scattering angle, on the other hand, depends on the nature of the collision process. For interparticle force laws that vary inversely as the distance between the particles, r^{-a}, the scattering angle depends on the power of the force law, a, the magnitude of the relative velocity, and the impact parameter. Therefore, the trajectories of the colliding particles are governed by classical mechanics.[1]

Ultimately, the goal is either to calculate or measure the differential scattering cross section that appears in the Boltzmann collision integral (3.9) so that the integral can be evaluated. When a calculation of this cross section is possible, it is first necessary to calculate the trajectories of the colliding particles, and then the differential scattering cross section can be obtained. As a simple example of how these calculations are done, it is instructive to consider a Coulomb collision between an electron and a heavy ion (Figure 4.4). These particles are chosen because the ion becomes the center-of-mass and the relative velocity $\mathbf{g}_{st}$ is approximately equal to the electron velocity. In this simple collision scenario, the electron approaches the ion with an initial velocity v_0 and impact parameter b_0. For a *central force*, such as the Coulomb one, the force, $\mathbf{F}$, is directed along the line joining the two particles and it is associated with a potential energy, V

$$\mathbf{F} = -\frac{1}{4\pi\varepsilon_0}\frac{e^2}{r^2}\mathbf{e}_r, \tag{4.18}$$

$$V(r) = -\frac{1}{4\pi\varepsilon_0}\frac{e^2}{r}, \tag{4.19}$$

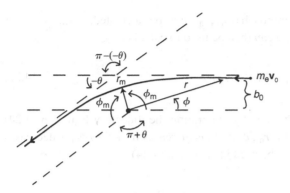

Figure 4.4 Trajectory of an electron during a collision with a heavy ion. The ion corresponds to the center-of-mass of the colliding particles and the electron velocity is the relative velocity. The (r, ϕ) are the polar coordinates and (r_m, ϕ_m) is the point where the particles are at a minimum distance. The scattering angle θ is defined to be positive for a repulsion.

where e is the magnitude of the electron charge, r is the distance between the charges, and $\mathbf{e}_r$ is a unit vector along r. $\mathbf{F}$ is the force on the electron.

For a central force, the collision trajectory lies in a plane and it is governed by the conservation of energy and momentum.[1] Using the *polar coordinates* (r, ϕ) shown in Figure 4.4, the conservation of energy and angular momentum yields

$$\frac{1}{2} m_e \left[\left(\frac{dr}{dt} \right)^2 + r^2 \left(\frac{d\phi}{dt} \right)^2 \right] + V(r) = \frac{1}{2} m_e v_0^2, \tag{4.20}$$

$$m_e r^2 \frac{d\phi}{dt} = m_e v_0 b_0, \tag{4.21}$$

where dr/dt and $r(d\phi/dt)$ are the radial and angular velocities at location (r, ϕ), respectively. The terms on the right-hand sides of Equations (4.20) and (4.21) are the initial energy and angular momentum, respectively. The equation for the trajectory, $r(\phi)$, can be obtained from

$$\frac{dr}{dt} = \frac{dr}{d\phi} \frac{d\phi}{dt}. \tag{4.22}$$

Using Equations (4.20) and (4.21) for dr/dt and $d\phi/dt$, respectively, one obtains

$$\left(\frac{dr}{d\phi} \right)^2 = \frac{r^4}{b_0^2} \left[1 - \frac{b_0^2}{r^2} - \frac{2V(r)}{m_e v_0^2} \right], \tag{4.23}$$

or

$$d\phi = \pm \frac{b_0}{r^2} \left[1 - \frac{b_0^2}{r^2} - \frac{2V(r)}{m_e v_0^2} \right]^{-1/2} dr, \tag{4.24}$$

where the $\pm$ signs arise from taking the square root. The choice of sign depends on which side of the point of minimum distance (r_m, ϕ_m) is being considered.

Because the trajectory is symmetric about the point of minimum distance, the *scattering angle* θ is related to θ_m by (Figure 4.4)

$$\theta = \pi - 2\phi_m. \tag{4.25}$$

Therefore, the point of closest approach (r_m, ϕ_m) must be calculated. At this location, $dr/d\phi = 0$, and Equation (4.23) can then be used to calculate r_m

$$r_m^2 - \frac{2V(r_m)}{m_e v_0^2} r_m^2 - b_0^2 = 0. \tag{4.26}$$

The other coordinate ϕ_m is obtained by integrating the trajectory Equation (4.24) from infinity $(r = \infty, \phi = 0)$ to (r_m, ϕ_m). As ϕ increases from 0 to ϕ_m, r decreases and, hence, the minus sign must be used in Equation (4.24)

$$\phi_m = \int_0^{\phi_m} d\phi = -\int_\infty^{r_m} dr \frac{b_0}{r^2} \left[1 - \frac{b_0^2}{r^2} - \frac{2V(r)}{m_e v_0^2} \right]^{-1/2}. \tag{4.27}$$

The expression for θ therefore becomes

$$\theta = \pi - 2b_0 \int_{r_m}^\infty \frac{dr}{r^2} \left[1 - \frac{b_0^2}{r^2} - \frac{2V(r)}{m_e v_0^2} \right]^{-1/2}. \tag{4.28}$$

Equation (4.28) for the scattering angle applies to any central force.

For the case of an electron and ion, $V(r)$ is given by Equation (4.19), and Equation (4.28) then becomes

$$\theta = \pi - 2b_0 \int_{r_m}^\infty \frac{dr}{r^2} \left[1 - \frac{b_0^2}{r^2} - \frac{2\alpha_0}{r} \right]^{-1/2}, \tag{4.29}$$

where

$$\alpha_0 = -\frac{1}{4\pi \varepsilon_0} \frac{e^2}{m_e v_0^2}. \tag{4.30}$$

Before evaluating the integral, it is necessary to first calculate r_m from Equation (4.26), which becomes

$$r_m^2 - 2\alpha_0 r_m - b_0^2 = 0. \tag{4.31}$$

The solution of this equation is $r_m = \alpha_0 + (\alpha_0^2 + b_0^2)^{1/2}$, where the $(+)$ sign in the quadratic formula is required to obtain a positive value for r_m. For what follows, it is useful to multiply and divide this solution by $-\alpha_0 + (\alpha_0^2 + b_0^2)^{1/2}$, so that r_m is cast in a more convenient form

$$r_m = \frac{b_0^2}{-\alpha_0 + (\alpha_0^2 + b_0^2)^{1/2}}. \tag{4.32}$$

The integral in Equation (4.29) can be evaluated by introducing a change of variables. Letting $x = 1/r$ in Equation (4.29), one obtains

$$\theta = \pi - 2b_0 \int\limits_{0}^{(1/r_m)} dx \left(1 - 2\alpha_0 x - b_0^2 x^2\right)^{-1/2}. \tag{4.33}$$

This integral can be evaluated using a standard table of integrals

$$\int dx(c_0 + c_1 x + c_2 x^2)^{-1/2} = \frac{1}{\sqrt{-c_2}} \sin^{-1}\left[\frac{-2c_2 x - c_1}{(c_1^2 - 4c_0 c_2)^{1/2}}\right], \tag{4.34}$$

which, for the coefficients in Equation (4.33), yields

$$\theta = \pi - 2b_0 \left\{\frac{1}{b_0} \sin^{-1}\left[\frac{b_0^2 x + \alpha_0}{(\alpha_0^2 + b_0^2)^{1/2}}\right]\right\}_0^{1/r_m}$$

$$= \pi - 2b_0\left\{\frac{1}{b_0} \sin^{-1}(1) - \frac{1}{b_0}\sin^{-1}\left[\frac{\alpha_0}{(\alpha_0^2 + b_0^2)^{1/2}}\right]\right\}$$

$$= 2\sin^{-1}\left[\frac{\alpha_0}{(\alpha_0^2 + b_0^2)^{1/2}}\right], \tag{4.35}$$

where $\sin^{-1}(1) = \pi/2$.

The case just discussed, where an electron collides with an ion, can be generalized to arbitrary Coulomb collisions by letting $-e^2 \to q_s q_t$ and $m_e v_0^2 \to \mu_{st} g_{st}^2$, where $\mu_{st} = m_s m_t/(m_s + m_t)$ is the reduced mass. Therefore, α_0 (Equation 4.30) becomes

$$\alpha_0 = \frac{1}{4\pi\varepsilon_0} \frac{q_s q_t}{\mu_{st} g_{st}^2}. \tag{4.36}$$

Also, an alternative form of Equation (4.35) is

$$\tan\left(\frac{\theta}{2}\right) = \frac{\alpha_0}{b_0} = \frac{1}{4\pi\varepsilon_0} \frac{q_s q_t}{\mu_{st} g_{st}^2 b_0}. \tag{4.37}$$

Note that if the charges have the same sign, θ is positive (repulsion), while if the charges have opposite signs, θ is negative (attractive). Also note that for $b_0 = \infty$, $\theta = 0$; for $b_0 = \alpha_0$, $\theta = \pi/2$; and for $b_0 = 0$, $\theta = \pi$. Hence, scattering occurs for all impact parameters, according to Equation (4.37). As will be discussed later, shielding by oppositely charged particles provides a cut-off for the maximum impact parameter applicable to Coulomb collisions.

In the case of Coulomb collisions, the variation of the interaction potential with the particle separation ($V \sim 1/r$) is well-known, but for collisions between neutral particles or between ions and neutrals, the interaction potential is not that easy to obtain. In principle, the forces between particles can be calculated using quantum mechanics, but in practice only very simple systems can be calculated that way.

Instead, most of the information on interparticle forces is obtained from experimental data. The procedure that is usually adopted is to use theory as a guide to the form of the force law and then to measure the diffusion (or mobility) of one species as it drifts through another species (Section 5.1). This transport property is also calculated using the Boltzmann collision integral and the assumed form for the interparticle force law, and the parameters in the force law are then adjusted until the measured and calculated transport properties agree. By conducting experiments with different species temperatures, the force laws can be deduced for different relative velocities, g_{st}, between the colliding particles.

Over the years, several different forms for inverse-power interaction potentials have been used.[2] For purely repulsive or purely attractive potentials, the experimental data can frequently be fitted with either inverse-power or exponential potentials

$$V = \pm \frac{K(\alpha)}{r^\alpha} \tag{4.38}$$

or

$$V = \pm V_0 e^{-r/r_0}, \tag{4.39}$$

where K is a function of α and $K(\alpha)$, α, V_0, and r_0 are positive constants. The $(+)$ sign corresponds to repulsion and the $(-)$ sign to attraction. If the potential energy of interaction has both attractive and repulsive components, so that it exhibits a potential well, it may be possible to represent it as a sum of two or more terms like those in Equations (4.38) and (4.39). The simplest combination is the so-called Lennard-Jones $(\alpha - \beta)$ interaction potential

$$V(r) = \frac{K(\alpha)}{r^\alpha} - \frac{K(\beta)}{r^\beta}; \quad \alpha > \beta, \tag{4.40}$$

where α and β are positive whole numbers and $K(\alpha)$ and $K(\beta)$ are constants. The first term is used to describe a short-range repulsive force and the second term a long-range attractive force. In particular, the Lennard-Jones (12–6) interaction potential has been very successful in describing elastic ion–neutral interactions. However, other interaction potentials have been used, including multiple inverse-power terms and combinations of exponential and inverse-power terms. This subject is discussed again when the various collision terms are calculated using the Boltzmann collision integral.

4.3   Collision cross sections

Up to this point, the focus has been on binary collisions. However, laboratory measurements usually involve a beam or flux of particles that is scattered off target particles, and the resulting scattering cross section is measured. Consider the

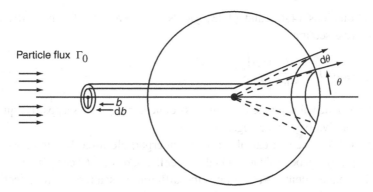

Figure 4.5 Scattering of particles in a symmetric center of force. A repulsive force is depicted.

scenario depicted in Figure 4.5. A homogeneous, monoenergetic, flux of identical particles, Γ_0, is incident on a single fixed-target molecule that acts like a center of force. For a repulsive force, the incident particles are scattered away from the molecule, with those having a smaller impact parameter b being scattered through a larger angle. An important cross section is the differential scattering cross section, which characterizes the angular distribution of the scattered particles. Because of the symmetry of the central force, the pattern of scattered particles is symmetric about the axis through the target particle (Figure 4.5). Hence, the angular distribution depends only on the polar scattering angle θ. Specifically, the *differential scattering cross section*, $\sigma_{st}(g_{st}, \theta)$, is defined as the number of particles scattered per solid angle $d\Omega$, per unit time, divided by the incident intensity. The relative velocity, g_{st}, is between the colliding particles, which, for the case shown in Figure 4.5, is simply the velocity of the incident flux. Note that g_{st} is included as a parameter in $\sigma_{st}(g_{st}, \theta)$ because different incident velocities yield different scattering patterns.

Given this definition of $\sigma_{st}(g_{st}, \theta)$, the number of particles scattered into a solid angle $d\Omega$ per unit time is

$$dN = \sigma_{st}(g_{st}, \theta)d\Omega\,\Gamma_0. \tag{4.41}$$

Again, because of the symmetry of the scattering process, $d\Omega = 2\pi \sin\theta\, d\theta$ (Figure 4.5), which yields

$$dN = 2\pi \sin\theta\, d\theta\, \sigma_{st}(g_{st}, \theta)\Gamma_0. \tag{4.42}$$

The number of particles scattered can also be related to the impact parameter, b

$$dN = \Gamma_0 2\pi b\, db. \tag{4.43}$$

Equating Equations (4.42) and (4.43) yields an expression for the differential scattering cross section

$$\sigma_{st}(g_{st}, \theta) = \frac{b}{\sin \theta} \left| \frac{db}{d\theta} \right|, \tag{4.44}$$

where the absolute value is taken because, as defined, $\sigma_{st}(g_{st}, \theta)$ is a positive quantity whereas the derivative can be negative.

Note that $db/d\theta$ can be calculated if the interparticle force law is known, and hence, $\sigma_{st}(g_{st}, \theta)$ can be evaluated. On the other hand, $\sigma_{st}(g_{st}, \theta)$ can also be directly measured in experiments. At any rate, the differential scattering cross section is needed to evaluate the Boltzmann collision integral.

There are additional collision cross sections that are important and they involve integrals over solid angle. For example, the *total scattering cross section* is defined as the number of particles scattered per unit time divided by the incident flux, which is all the particles scattered regardless of their direction

$$Q_T(g_{st}) = \int d\Omega \, \sigma_{st}(g_{st}, \theta). \tag{4.45}$$

The *momentum transfer cross section* is the total momentum transferred per unit time to the target molecule divided by the incident flux. For the case shown in Figure 4.5, where the target molecule is fixed, the momentum of an incident particle is $m_0 v_0$ and the incident momentum flux is $m_0 v_0 \Gamma_0$. After the particles interact with the target molecule, they are scattered at an angle θ. Therefore, the new momentum, in the incident direction, of a given particle is $m_0 v_0 \cos \theta$, and the momentum flux after scattering becomes $m_0 v_0 \Gamma_0 \cos \theta$. The change of momentum in the incident direction, which is the momentum transferred to the target molecule, is $m_0 v_0 \Gamma_0 (1 - \cos \theta)$. The total momentum transferred to the target molecule per unit time is obtained by integrating this quantity over $\sigma(g_{st}, \theta) d\Omega$. After dividing by the incident momentum flux, $m_0 v_0 \Gamma_0$, one obtains the momentum transfer cross section

$$Q_{st}^{(1)}(g_{st}) = \int d\Omega \, \sigma_{st}(g_{st}, \theta)(1 - \cos \theta). \tag{4.46}$$

As will be seen throughout this chapter, the momentum transfer cross section plays a prominent role in diffusion theory. However, other cross sections are important, and these arise when the collision terms for the higher velocity moments (stress, heat flow, etc.) are evaluated. The general form of the collision cross sections is

$$Q_{st}^{(l)}(g_{st}) = \int d\Omega \, \sigma_{st}(g_{st}, \theta)(1 - \cos^l \theta), \tag{4.47}$$

where the superscript l is an integer.

Because Coulomb collisions play an important role in ionospheres and correspond to long-range interactions, it is instructive to study this process in more detail. The first step is to evaluate the differential scattering cross section (4.44). The connection between the scattering angle θ and the impact parameter b for Coulomb collisions is given by Equation (4.37). Taking the derivative of Equation (4.37) yields an equation for $|db/d\theta|$:

$$\left|\frac{db}{d\theta}\right| = \frac{4\pi\varepsilon_0\mu_{st}g_{st}^2}{q_sq_t}\frac{b^2}{2\cos^2\left(\dfrac{\theta}{2}\right)},\tag{4.48}$$

and, therefore, the differential scattering cross section (4.44) is given by

$$\sigma_{st}(g_{st},\theta) = \frac{4\pi\varepsilon_0\mu_{st}g_{st}^2}{q_sq_t}\frac{b^3}{2\sin\theta\cos^2\left(\dfrac{\theta}{2}\right)}.\tag{4.49}$$

Eliminating b^3 with the aid of Equation (4.37), setting $\tan(\theta/2) = \sin(\theta/2)/\cos(\theta/2)$, and using the trigonometric identity $\sin\theta = 2\sin(\theta/2)\cos(\theta/2)$, yields the classical form of the *Rutherford scattering cross section* for Coulomb collisions

$$\sigma_{st}(g_{st},\theta) = \left(\frac{q_sq_t}{4\pi\varepsilon_0\mu_{st}g_{st}^2}\right)^2\frac{1}{4\sin^4\left(\dfrac{\theta}{2}\right)}.\tag{4.50}$$

Given that $2\sin^2(\theta/2) = 1 - \cos\theta$, this can also be expressed in the form

$$\sigma_{st}(g_{st},\theta) = \left(\frac{q_sq_t}{4\pi\varepsilon_0\mu_{st}g_{st}^2}\right)^2\frac{1}{(1-\cos\theta)^2}.\tag{4.51}$$

The momentum transfer cross section (4.46) for Coulomb collisions can now be calculated by using Equation (4.51)

$$Q_{st}^{(1)}(g_{st}) = 2\pi\alpha_0^2\int_0^\pi\frac{\sin\theta d\theta}{1-\cos\theta} = 2\pi\alpha_0^2\int_{-1}^1\frac{dx}{1-x} = -2\pi\alpha_0^2[\ln(1-x)]_{-1}^1,\tag{4.52}$$

where α_0 is defined in Equation (4.36) and where the integral was transformed by letting $x = \cos\theta$. Note that when the $x = 1$ ($\theta = 0$, $b = \infty$) limit is taken, the integral becomes infinite. That means the particles with infinitely small scattering angles ($b \to \infty$, $\theta \to 0$) contribute to make an infinite momentum transfer cross section. As it turns out, all of the cross sections (4.47) are infinite for Coulomb collisions. However, the situation can be remedied by putting a limit on the collision impact parameter b, which is equivalent to putting a limit on the scattering angle θ. The limit is justified because of *Debye shielding*. Specifically, when a charge is placed in a plasma, it is surrounded by charges of the opposite sign, and its potential is therefore confined to a spherical domain that has a radius approximately equal to the Debye length, $\lambda_D = (\varepsilon_0 kT/n_e e^2)^{1/2}$, where for simplicity, it is assumed that all the species have a common temperature. As a consequence, the potential field of an individual charged particle in a plasma does not effectively extend beyond a distance of about λ_D, which means the maximum impact parameter, b_{max}, should be set equal to λ_D. Associated with b_{max} is a minimum scattering angle, θ_{min}, and with this restriction, Equation (4.52) becomes

$$Q_{st}^{(1)}(g_{st}) = -2\pi\alpha_0^2\left[\ln(1-x)\right]_{-1}^{\cos\theta_{min}} = 2\pi\alpha_0^2\ln\left[\frac{2}{1-\cos\theta_{min}}\right]$$

$$= 2\pi\alpha_0^2\ln\left[\frac{1}{\sin^2\left(\frac{\theta_{min}}{2}\right)}\right]. \tag{4.53}$$

From Equation (4.35),

$$\sin^2\left(\frac{\theta_{min}}{2}\right) = \frac{\alpha_0^2}{\alpha_0^2 + b_{max}^2} = \frac{1}{1+\left(\dfrac{\lambda_D}{\alpha_0}\right)^2}, \tag{4.54}$$

where b_{max} has been replaced by λ_D in Equation (4.54). Substituting (4.54) into (4.53) yields

$$Q_{st}^{(1)}(g_{st}) = 2\pi\alpha_0^2\ln\left[1+\left(\frac{\lambda_D}{\alpha_0}\right)^2\right]. \tag{4.55}$$

The quantity λ_D/α_0 has significance and is usually denoted by Λ

$$\Lambda = \frac{\lambda_D}{\alpha_0} = \frac{4\pi\varepsilon_0\mu_{st}g_{st}^2}{q_s q_t}\lambda_D. \tag{4.56}$$

In general, different interacting species have different Λs, but the differences are small and neglected in multi-component gas mixtures. Also, an average of g_{st}^2 over a Maxwellian velocity distribution, $\mu_{st}\langle g_{st}^2\rangle = 3kT$, is usually inserted in Λ, where the differences in species temperatures are ignored. Therefore, Λ can be written as

$$\Lambda = 9\left(\frac{4\pi}{3}n_e\lambda_D^3\right) = 9N_{\lambda_D}, \tag{4.57}$$

where N_{λ_D} is the number of particles in a Debye sphere (Equation 2.5). Typically, N_{λ_D} is very large and, hence, Λ is very large.

Setting $\lambda_D/\alpha_0 = \Lambda$ in Equation (4.55), neglecting 1 compared to Λ^2, and using Equation (4.36) for α_0, yields the momentum transfer cross section for Coulomb collisions

$$Q_{st}^{(1)}(g_{st}) = 4\pi\left(\frac{q_s q_t}{4\pi\varepsilon_0\mu_{st}g_{st}^2}\right)^2\ln\Lambda, \tag{4.58}$$

where $\ln \Lambda$ is the *Coulomb logarithm*, which is typically between 10 and 25 for space plasmas.

4.4   Transfer collision integrals

The transfer integrals arise when velocity moments of the Boltzmann equation are taken, and they are just moments of the Boltzmann collision integral (3.9). If $\xi_s(\mathbf{c}_s)$ is a general velocity moment, the corresponding moment of the Boltzmann collision integral is

$$\int d^3c_s\, \xi_s(\mathbf{c}_s)\frac{\delta f_s}{\delta t} = \iiint d^3c_s\, d^3c_t\, d\Omega\, g_{st}\sigma_{st}(g_{st},\theta)\left(f_s'f_t' - f_sf_t\right)\xi_s. \quad (4.59)$$

For $\xi_s = 1$, $m_s\mathbf{c}_s$, $\frac{1}{2}m_sc_s^2$, $m_s\mathbf{c}_s\mathbf{c}_s$, and $\frac{1}{2}m_sc_s^2\mathbf{c}_s$, the moments of the Boltzmann collision integral are symbolically written as $\delta n_s/\delta t$, $\delta \mathbf{M}_s/\delta t$, $\delta E_s/\delta t$, $\delta \mathbf{P}_s/\delta t$, and $\delta \mathbf{q}_s/\delta t$, respectively (Equations 3.29, 3.35, 3.41, 3.42, and 3.43). Additional collision moments are $\delta \boldsymbol{\tau}_s/\delta t$ and $\delta \mathbf{q}_s'/\delta t$ (Equations 3.62 and 3.63). Although the different collision moments can be calculated by using Equation (4.59), it is mathematically more convenient to use the following equivalent form (Appendix G):

$$\int d^3c_s\, \xi_s(\mathbf{c}_s)\frac{\delta f_s}{\delta t} = \iiint d^3c_s\, d^3c_t\, d\Omega\, g_{st}\sigma_{st}(g_{st},\theta)f_sf_t(\xi_s' - \xi_s), \quad (4.60)$$

where $\xi_s' = \xi_s(\mathbf{c}_s')$ is the moment evaluated with the velocity found after the collision. The integrals of the type shown in Equation (4.60) are called *transfer integrals* because they represent the change in a transport property (momentum, energy, etc.) as a result of collisions. The multiple integrals in Equation (4.60) are easier to calculate than those in Equation (4.59) because they do not require the distribution functions after the collision, $f_s'f_t'$.

The calculation of the multiple integrals in Equation (4.60) has to be done in two steps. First, because $d\Omega$ is the solid angle in the colliding particles' center-of-mass reference frame, it is necessary to transform to this frame before integrating over the solid angle. Subsequently, it is necessary to transform back to the $(\mathbf{c}_s, \mathbf{c}_t)$ reference frame so that the integrals over d^3c_s and d^3c_t can be performed. The first step is common to all collision processes and will be done here, while the second step requires a knowledge of the specific velocity dependence of $\sigma_{st}(g_{st},\theta)$ and, as will be shown, this leads to additional complications.

To evaluate the integrals over $d\Omega$, the necessary transformation is from $(\mathbf{c}_s, \mathbf{c}_t)$ to $(\mathbf{V}_c, \mathbf{g}_{st})$, where the center-of-mass velocity, $\mathbf{V}_c$, and the relative velocity, $\mathbf{g}_{st}$, are given in Equations (4.6) and (4.7), respectively. The inversion of these equations yields $\mathbf{v}_s$ and $\mathbf{v}_t$ in terms of $\mathbf{V}_c$ and $\mathbf{g}_{st}$, and these are given in Equations (4.8) and (4.9), respectively. Because $\mathbf{v}_s = \mathbf{c}_s + \mathbf{u}_s$ and $\mathbf{v}_t = \mathbf{c}_t + \mathbf{u}_t$, Equations (4.6) to (4.9) can be easily modified to provide the transformations that are needed here

$$\mathbf{c}_s = \hat{\mathbf{V}}_c + \frac{m_t}{m_s + m_t} \mathbf{g}_{st}, \tag{4.61}$$

$$\hat{\mathbf{V}}_c = \frac{1}{m_s + m_t} [m_s \mathbf{c}_s + m_t \mathbf{c}_t - m_t (\mathbf{u}_s - \mathbf{u}_t)], \tag{4.62}$$

$$\mathbf{g}_{st} = \mathbf{c}_s - \mathbf{c}_t + \mathbf{u}_s - \mathbf{u}_t, \tag{4.63}$$

where

$$\hat{\mathbf{V}}_c = \mathbf{V}_c - \mathbf{u}_s, \tag{4.64}$$

and where an expression for $\mathbf{c}_t$ is not needed because the moments are for $\xi_s(\mathbf{c}_s)$. Note that it has already been shown that $\mathbf{V}_c$ and $|\mathbf{g}_{st}|$ do not change in a collision. Likewise, because the average drift of the gas as a whole does not change in an individual collision, the velocity $\hat{\mathbf{V}}_c$ (4.64) does not change in a collision.

In what follows, the integrals over solid angles are performed for $\xi_s = 1$, $m_s \mathbf{c}_s$, and $\frac{1}{2} m_s c_s^2$, and for the others only the final answers are given in order to avoid excessive algebra. The integrals over solid angles are of the form

$$\int d\Omega \, \sigma_{st}(g_{st}, \theta)(\xi'_s - \xi_s). \tag{4.65}$$

For $\xi_s = 1$, the integral is zero because the particle's mass does not change in an elastic collision. Therefore, the corresponding transfer integral (4.60) immediately yields

$$\frac{\delta n_s}{\delta t} = 0 \tag{4.66}$$

for all elastic collision processes.

For $\xi_s = m_s \mathbf{c}_s$, the velocity difference that is needed in the integral over $d\Omega$ is $(\mathbf{c}'_s - \mathbf{c}_s)$. However, $\mathbf{c}_s$ is given in Equation (4.61) and $\mathbf{c}'_s$ is the same equation with a prime on $\mathbf{g}'_{st}$ ($\mathbf{V}_c$ does not change in a collision), which yields

$$m_s(\mathbf{c}'_s - \mathbf{c}_s) = \mu_{st}(\mathbf{g}'_{st} - \mathbf{g}_{st}), \tag{4.67}$$

where $\mu_{st} = m_s m_t/(m_s + m_t)$ is the reduced mass. Using Equation (4.67), the solid angle integral (4.65) becomes

$$m_s \int d\Omega \, \sigma_{st}(g_{st}, \theta)(\mathbf{c}'_s - \mathbf{c}_s) = \mu_{st} \int d\Omega \, \sigma_{st}(g_{st}, \theta)(\mathbf{g}'_{st} - \mathbf{g}_{st}). \tag{4.68}$$

This integral can be evaluated by using the coordinate system shown in Figure 4.3, where $\mathbf{g}_{st}$ is taken along the z-axis. The relative velocity, $\mathbf{g}'_{st}$, after an elastic collision is rotated through a scattering angle θ, but its magnitude does not change. Therefore,

$$\mathbf{g}'_{st} = g_{st}(\sin\theta\,\cos\phi\,\mathbf{e}_1 + \sin\theta\,\sin\phi\,\mathbf{e}_2 + \cos\theta\mathbf{e}_3), \tag{4.69}$$

$$\mathbf{g}_{st} = g_{st}\mathbf{e}_3. \tag{4.70}$$

Using Equations (4.69) and (4.70), the solid angle integral can be expressed as

$$m_s \int d\Omega\,\sigma_{st}(g_{st},\theta)(\mathbf{c}'_s - \mathbf{c}_s)$$

$$= \mu_{st} \int_0^{2\pi} d\phi \int_0^{\pi} \sin\theta\,d\theta\,\sigma_{st}(g_{st},\theta)\big[\sin\theta\,\cos\phi\,\mathbf{e}_1$$

$$+ \sin\theta\,\sin\phi\,\mathbf{e}_2 + (\cos\theta - 1)\mathbf{e}_3\big]g_{st}$$

$$= -2\pi\mu_{st} \int_0^{\pi} \sin\theta\,d\theta\,\sigma_{st}(g_{st},\theta)(1 - \cos\theta)(g_{st}\mathbf{e}_3)$$

$$= -\mu_{st}\mathbf{g}_{st}Q_{st}^{(1)}, \tag{4.71}$$

where

$$Q_{st}^{(1)}(g_{st}) = \int d\Omega\,\sigma_{st}(g_{st},\theta)(1 - \cos\theta) \tag{4.72}$$

is the momentum transfer cross section that was deduced earlier using physical arguments (Equation (4.46)).

For $\xi_s = \frac{1}{2}m_s c_s^2$, the velocity difference that is relevant is $c_s'^2 - c_s^2$. To get c_s^2 you merely take $\mathbf{c}_s \cdot \mathbf{c}_s$ using Equation (4.61), which yields

$$c_s^2 = \hat{V}_c^2 + \frac{2m_t}{m_s + m_t}\hat{\mathbf{V}}_c \cdot \mathbf{g}_{st} + \frac{m_t^2}{(m_s + m_t)^2}g_{st}^2. \tag{4.73}$$

The quantity $c_s'^2$ is obtained from the same formula by evaluating all velocities after the collision. However, because $\hat{\mathbf{V}}_c$ and $|\mathbf{g}_{st}|$ do not change in a collision,

$$\frac{1}{2}m_s(c_s'^2 - c_s^2) = \mu_{st}\hat{\mathbf{V}}_c \cdot (\mathbf{g}'_{st} - \mathbf{g}_{st}), \tag{4.74}$$

and the integral over solid angle (4.65) then becomes

$$\frac{m_s}{2} \int d\Omega\,\sigma_{st}(g_{st},\theta)(c_s'^2 - c_s^2) = \hat{\mathbf{V}}_c \cdot \mu_{st} \int d\Omega\,\sigma_{st}(g_{st},\theta)(\mathbf{g}'_{st} - \mathbf{g}_{st})$$

$$= -\mu_{st}(\hat{\mathbf{V}}_c \cdot \mathbf{g}_{st})Q_{st}^{(1)}. \tag{4.75}$$

Note that the first integral on the right-hand side of Equation (4.75) is the same one that appeared in Equation (4.68) and this leads to the result given in Equation (4.71).

When ξ_s is set equal to $m_s \mathbf{c}_s \mathbf{c}_s$ and $\frac{1}{2} m_s c_s^2 \mathbf{c}_s$, the integrals over solid angles (4.65) involve a second-order tensor with respect to $\mathbf{g}_{st}$. The tensor involved is

$$
\begin{aligned}
\mathbf{g}'_{st}\mathbf{g}'_{st} - \mathbf{g}_{st}\mathbf{g}_{st} = g_{st}^2[&\sin^2\theta\cos^2\phi\,\mathbf{e}_1\mathbf{e}_1 + \sin^2\theta\cos\phi\sin\phi\,\mathbf{e}_1\mathbf{e}_2 \\
&+ \sin\theta\cos\theta\cos\phi\,\mathbf{e}_1\mathbf{e}_3 + \sin^2\theta\cos\phi\sin\phi\,\mathbf{e}_2\mathbf{e}_1 \\
&+ \sin^2\theta\sin^2\phi\,\mathbf{e}_2\mathbf{e}_2 + \sin\theta\cos\theta\sin\phi\,\mathbf{e}_2\mathbf{e}_3 \\
&+ \sin\theta\cos\theta\cos\phi\,\mathbf{e}_3\mathbf{e}_1 + \sin\theta\cos\theta\sin\phi\,\mathbf{e}_3\mathbf{e}_2 \\
&+ (\cos^2\theta - 1)\mathbf{e}_3\mathbf{e}_3]
\end{aligned}
\tag{4.76}
$$

for the coordinate system shown in Figure 4.3. The quantities such as $\mathbf{e}_1\mathbf{e}_2$ are unit tensors that define the nine tensor locations (like unit vectors defining three orthogonal directions). When Equation (4.76) is integrated over solid angle, many terms drop out because of the ϕ integration and the expression reduces to

$$
\int d\Omega\, \sigma_{st}(g_{st}, \theta)(\mathbf{g}'_{st}\mathbf{g}'_{st} - \mathbf{g}_{st}\mathbf{g}_{st}) = \frac{1}{2}(g_{st}^2\mathbf{I} - 3\mathbf{g}_{st}\mathbf{g}_{st})Q_{st}^{(2)},
\tag{4.77}
$$

where $\mathbf{I} = \mathbf{e}_1\mathbf{e}_1 + \mathbf{e}_2\mathbf{e}_2 + \mathbf{e}_3\mathbf{e}_3$ is the unit dyadic and

$$
Q_{st}^{(2)}(g_{st}) = \int d\Omega\, \sigma_{st}(g_{st}, \theta)(1 - \cos^2\theta)
\tag{4.78}
$$

is a higher-order collision cross section.

The rest of the details concerning the evaluation of the integrals over solid angle for the moments $\xi_s = m_s \mathbf{c}_s \mathbf{c}_s$ and $\frac{1}{2} m_s c_s^2 \mathbf{c}_s$ are not discussed here.[3] However, for future reference, it is useful to summarize these and the two moments derived above in one place:

$$
m_s \int d\Omega\, \sigma_{st}(g_{st}, \theta)(\mathbf{c}'_s - \mathbf{c}_s) = -\mu_{st}\mathbf{g}_{st}Q_{st}^{(1)}
\tag{4.79a}
$$

$$
\frac{m_s}{2} \int d\Omega\, \sigma_{st}(g_{st}, \theta)(c'^2_s - c^2_s) = -\mu_{st}(\hat{\mathbf{V}}_c \cdot \mathbf{g}_{st})Q_{st}^{(1)}
\tag{4.79b}
$$

$$
\begin{aligned}
m_s \int d\Omega\, \sigma_{st}(g_{st}, \theta)(\mathbf{c}'_s\mathbf{c}'_s - \mathbf{c}_s\mathbf{c}_s) = &-\mu_{st}(\hat{\mathbf{V}}_c\mathbf{g}_{st} + \mathbf{g}_{st}\hat{\mathbf{V}}_c)Q_{st}^{(1)} \\
&+ \frac{1}{2}\frac{\mu_{st}^2}{m_s}(g_{st}^2\mathbf{I} - 3\mathbf{g}_{st}\mathbf{g}_{st})Q_{st}^{(2)}
\end{aligned}
\tag{4.79c}
$$

$$
\begin{aligned}
\frac{m_s}{2} \int d\Omega\, \sigma_{st}(g_{st}, \theta)(c'^2_s\mathbf{c}'_s - c^2_s\mathbf{c}_s) = &-\frac{1}{2}\mu_{st}\left[\left(\hat{V}_c^2 + \frac{\mu_{st}^2 g_{st}^2}{m_s^2}\right)\mathbf{g}_{st} \right. \\
&\left. + 2\hat{\mathbf{V}}_c(\hat{\mathbf{V}}_c \cdot \mathbf{g}_{st})\right]Q_{st}^{(1)} + \frac{1}{2}\frac{\mu_{st}^2}{m_s}\hat{\mathbf{V}}_c \\
&\cdot \left(g_{st}^2\mathbf{I} - 3\mathbf{g}_{st}\mathbf{g}_{st}\right)Q_{st}^{(2)}.
\end{aligned}
\tag{4.79d}
$$

With the above expressions for the integrations over solid angles in the center-of-mass reference frame, the transfer integrals (4.60) become

$$\int d^3c_s \, \xi_s \frac{\delta f_s}{\delta t} = \iint d^3c_s \, d^3c_t \, f_s f_t g_{st} \left[\int d\Omega \, \sigma_{st}(g_{st}, \theta)(\xi'_s - \xi_s) \right]. \qquad (4.80)$$

The next step is to transform the results of the solid angle integrations (4.79a–d) from $(\hat{\mathbf{V}}_c, \mathbf{g}_{st})$ back to $(\mathbf{c}_s, \mathbf{c}_t)$ and then perform the integrals over d^3c_s and d^3c_t. However, these remaining integrals can be evaluated rigorously only for the so-called *Maxwell molecule interactions*, where $\sigma_{st} \sim 1/g_{st}$. In this case $g_{st}\sigma_{st}$ is a constant, which means that $g_{st}Q_{st}^{(1)}$ and $g_{st}Q_{st}^{(2)}$ are also constants and can be removed from the integrals. At that point, no integrations actually have to be performed because the integrals become recognizable velocity moments, such as n_s, n_t, $\mathbf{u}_s$, etc. On the other hand, for all other interactions, it is necessary to adopt approximate expressions for f_s and f_t in order to evaluate the transfer integrals.

4.5   Maxwell molecule collisions

Maxwell molecule collisions correspond to an interaction potential of $V \sim 1/r^4$ and $\sigma_{st} \sim 1/g_{st}$. In this case, the momentum transfer integral can be obtained from Equation (4.80) by setting $\xi_s = m_s \mathbf{c}_s$ and by using Equations (4.79a) and (3.35)

$$\frac{\delta \mathbf{M}_s}{\delta t} = -\mu_{st} \left[g_{st}Q_{st}^{(1)} \right] \iint d^3c_s \, d^3c_t \, f_s f_t \mathbf{g}_{st}. \qquad (4.81)$$

Expressing $\mathbf{g}_{st}$ in terms of $\mathbf{c}_s$ and $\mathbf{c}_t$ with the aid of Equation (4.63) and noting that $\langle \mathbf{c}_s \rangle = \langle \mathbf{c}_t \rangle = 0$, Equation (4.81) becomes

$$\frac{\delta \mathbf{M}_s}{\delta t} = n_s m_s \nu_{st}(\mathbf{u}_t - \mathbf{u}_s), \qquad (4.82)$$

where the *momentum transfer collision frequency* is defined as

$$\nu_{st} = \frac{n_t m_t}{m_s + m_t} \left[g_{st}Q_{st}^{(1)} \right]. \qquad (4.83)$$

The energy transfer integral is obtained from Equation (4.80) by setting $\xi_s = \frac{1}{2}m_s c_s^2$ and by using Equations (4.79b) and (3.41)

$$\frac{\delta E_s}{\delta t} = -\mu_{st} \left[g_{st}Q_{st}^{(1)} \right] \iint d^3c_s \, d^3c_t \, f_s f_t (\hat{\mathbf{V}}_c \cdot \mathbf{g}_{st}). \qquad (4.84)$$

Table 4.1 Neutral gas polarizabilities.[4,5]

Species	$\gamma_n(10^{-24}\ cm^3)$	Species	$\gamma_n(10^{-24}\ cm^3)$
CH_4	2.59	N_2	1.76
CO	1.97	N_2O	3.00
CO_2	2.63	Na	2.70
H	0.67	NH_3	2.22
H_2	0.82	NO	1.74
H_2O	1.48	O	0.77
He	0.21	O_2	1.60
N	1.13	SO_2	3.89

The term $\hat{\mathbf{V}}_c \cdot \mathbf{g}_{st}$ can be obtained from Equations (4.62) and (4.63)

$$\hat{\mathbf{V}}_c \cdot \mathbf{g}_{st} = \frac{1}{m_s + m_t}[m_s c_s^2 - m_t c_t^2 + (m_t - m_s)\mathbf{c}_t \cdot \mathbf{c}_s$$

$$- (m_t - m_s)\mathbf{c}_s \cdot (\mathbf{u}_s - \mathbf{u}_t) + 2m_t \mathbf{c}_t \cdot (\mathbf{u}_s - \mathbf{u}_t) - m_t(\mathbf{u}_s - \mathbf{u}_t)^2]. \tag{4.85}$$

When this term is substituted into Equation (4.84) and use is made of the relations $\langle c_s^2 \rangle = 3kT_s/m_s$ and $\langle c_t^2 \rangle = 3kT_t/m_t$ (see Chapter 3), the result is

$$\frac{\delta E_s}{\delta t} = \frac{n_s m_s v_{st}}{m_s + m_t}\left[3k(T_t - T_s) + m_t(\mathbf{u}_s - \mathbf{u}_t)^2\right]. \tag{4.86}$$

The transfer integrals for $\xi_s = m_s \mathbf{c}_s \mathbf{c}_s$ and $\frac{1}{2}m_s c_s^2 \mathbf{c}_s$ can be calculated in a manner similar to that for $\delta \mathbf{M}_s/\delta t$ and $\delta E_s/\delta t$, but the algebra is considerably more involved. For easy reference, these collision terms, as well as those derived above, are listed below. First, however, it should be noted that Maxwell molecule collisions are a reasonable approximation for elastic (nonresonant) ion–neutral interactions. As the ion approaches the neutral, the neutral becomes polarized and the interaction is between the ion and an induced dipole, for which the interaction potential is

$$V = -\frac{1}{4\pi \varepsilon_0} \frac{\gamma_n e^2}{2r^4}. \tag{4.87}$$

In Equation (4.87), γ_n is the neutral polarizability; values are given in Table 4.1 for the neutral gases relevant to ionospheres. For this interaction potential, it has been shown that the collision frequency (4.83) can be expressed as[6]

$$v_{in} = 2.21\pi \frac{n_n m_n}{m_i + m_n}\left(\frac{\gamma_n e^2}{\mu_{in}}\right)^{1/2}, \tag{4.88}$$

where subscripts i and n are used to emphasize that Maxwell molecule collisions only apply to elastic ion–neutral interactions.

The transfer integrals, including the two previously derived terms (4.82) and (4.86), for Maxwell molecule collisions are summarized as follows:

$$\frac{\delta n_i}{\delta t} = 0 \tag{4.89a}$$

$$\frac{\delta \mathbf{M}_i}{\delta t} = -\sum_n n_i m_i \nu_{in}(\mathbf{u}_i - \mathbf{u}_n) \tag{4.89b}$$

$$\frac{\delta E_i}{\delta t} = -\sum_n \frac{n_i m_i \nu_{in}}{m_i + m_n}\left[3k(T_i - T_n) - m_n(\mathbf{u}_i - \mathbf{u}_n)^2\right] \tag{4.89c}$$

$$\frac{\delta \mathbf{P}_i}{\delta t} = -\sum_n \frac{2m_i \nu_{in}}{m_i + m_n}\left\{\mathbf{P}_i - \frac{n_i}{n_n}\mathbf{P}_n - n_i m_n(\mathbf{u}_i - \mathbf{u}_n)(\mathbf{u}_i - \mathbf{u}_n) + \frac{3}{4}\frac{m_n}{m_i}\frac{Q_{in}^{(2)}}{Q_{in}^{(1)}}\right.$$

$$\left. \times \left[\boldsymbol{\tau}_i + \frac{\rho_i}{\rho_n}\boldsymbol{\tau}_n + \rho_i(\mathbf{u}_i - \mathbf{u}_n)(\mathbf{u}_i - \mathbf{u}_n) - \frac{\rho_i}{3}(\mathbf{u}_i - \mathbf{u}_n)^2\mathbf{I}\right]\right\} \tag{4.89d}$$

$$\frac{\delta \boldsymbol{\tau}_i}{\delta t} = -\sum_n \frac{2m_i \nu_{in}}{m_i + m_n}\left\{\boldsymbol{\tau}_i\left[1 + \frac{3m_n}{4m_i}\frac{Q_{in}^{(2)}}{Q_{in}^{(1)}}\right] - n_i m_n\right.$$

$$\left. \times \left[\frac{1}{\rho_n}\boldsymbol{\tau}_n + (\mathbf{u}_i - \mathbf{u}_n)(\mathbf{u}_i - \mathbf{u}_n) - \frac{1}{3}(\mathbf{u}_i - \mathbf{u}_n)^2\mathbf{I}\right]\left[1 - \frac{3}{4}\frac{Q_{in}^{(2)}}{Q_{in}^{(1)}}\right]\right\}. \tag{4.89e}$$

$$\frac{\delta \mathbf{q}_i}{\delta t} = -\sum_n \nu_{in}\left\{A_{in}^{(1)}\mathbf{q}_i + \frac{1}{2}A_{in}^{(2)}\mathbf{P}_i \cdot (\mathbf{u}_i - \mathbf{u}_n)\right.$$

$$+ \frac{\rho_i}{\rho_n}A_{in}^{(4)}\left[\mathbf{P}_n \cdot (\mathbf{u}_i - \mathbf{u}_n) - \mathbf{q}_n\right] + (\mathbf{u}_i - \mathbf{u}_n)$$

$$\left. \times \left[\frac{3}{2}p_n\frac{\rho_i}{\rho_n}A_{in}^{(4)} + \frac{3}{2}p_i A_{in}^{(3)} + \frac{1}{2}\rho_i(\mathbf{u}_i - \mathbf{u}_n)^2 A_{in}^{(4)}\right]\right\} \tag{4.89f}$$

where

$$A_{in}^{(1)} = \frac{1}{(m_i + m_n)^2}\left[3m_i^2 + m_n^2 + 2m_i m_n\frac{Q_{in}^{(2)}}{Q_{in}^{(1)}}\right] \tag{4.90a}$$

$$A_{in}^{(2)} = \frac{1}{(m_i + m_n)^2}\left[2(m_i - m_n)^2 + m_n(m_i - 3m_n)\frac{Q_{in}^{(2)}}{Q_{in}^{(1)}}\right] \tag{4.90b}$$

$$A_{in}^{(3)} = \frac{1}{(m_i + m_n)^2}\left[(m_i - m_n)^2 + m_n(m_n + 3m_i)\frac{Q_{in}^{(2)}}{Q_{in}^{(1)}}\right] \tag{4.90c}$$

$$A_{\text{in}}^{(4)} = \frac{2m_n^2}{(m_i + m_n)^2} \left[2 - \frac{Q_{\text{in}}^{(2)}}{Q_{\text{in}}^{(1)}} \right]. \tag{4.90d}$$

The ratio of collision cross sections, $Q_{\text{in}}^{(2)}/Q_{\text{in}}^{(1)}$, varies between 0.7 and 1.0 for an ion–neutral interaction dominated by an induced dipole attraction, with a short-range hard core repulsion.[7] A reasonable value to use is 0.8.

The transfer integrals for Maxwell molecule collisions are valid for arbitrary drift velocity differences and arbitrary temperature differences between the ion and neutral gases.

4.6   Collision terms for Maxwellian velocity distributions

As noted earlier, for general collision processes it is necessary to have an approximate expression for the species distribution functions, in order to evaluate the transfer integrals (4.80). The simplest situation is when each species in the gas can be described by drifting Maxwellian distributions (Equation 3.44). This case is known as the *five-moment approximation* because each species in the gas is characterized by five parameters ($n_s = 1$, $\mathbf{u}_s = 3$, $T_s = 1$). With regard to the transfer integrals, the integrations over solid angles have already been done and the results are given in Equations (4.79a–d). When Maxwell molecule collisions were considered, the next step was to transform from the ($\hat{\mathbf{V}}_c$, $\mathbf{g}_{st}$) system back to the ($\mathbf{c}_s$, $\mathbf{c}_t$) system and then to integrate over $d^3c_s\, d^3c_t$. However, for general collision processes, this latter transformation of velocities is not useful and, as will be shown, it is more convenient to perform the velocity integrations in terms of the relative velocity, g_{st}.

For Maxwellian velocity distributions and for particle collisions that are governed by inverse-power interaction potentials, it is possible to derive collision terms that are valid for arbitrary drift velocity differences and arbitrary temperature differences between the interacting species in a gas mixture. However, even in this simple five-moment approximation, the calculations are laborious. Therefore, in what follows, only the momentum transfer collision term is calculated and only in the limit when the drift velocity differences between the various species are much smaller than typical thermal speeds. Although only this case is considered, it is still more than adequate to establish clearly how the collision terms are evaluated. Then, for future reference, the general collision terms are listed for important special cases.

The momentum transfer collision integral is obtained from Equation (4.80) using $\xi_s = m_s \mathbf{c}_s$

$$\frac{\delta \mathbf{M}_s}{\delta t} = \iint d^3c_s\, d^3c_t\, f_s f_t g_{st} \left[m_s \int d\Omega \sigma_{st}(g_{st}, \theta)(\mathbf{c}_s' - \mathbf{c}_s) \right]. \tag{4.91}$$

The integration over a solid angle is given in Equation (4.79a), so that $\delta \mathbf{M}_s / \delta t$ becomes

$$\frac{\delta \mathbf{M}_s}{\delta t} = -\mu_{st} \iint d^3 c_s \, d^3 c_t f_s f_t g_{st} Q_{st}^{(1)} \mathbf{g}_{st}. \tag{4.92}$$

When the colliding gases are described by Maxwellian distribution functions, the term $f_s f_t$ can be expressed as

$$f_s f_t = n_s n_t \left(\frac{m_s}{2\pi k T_s} \right)^{3/2} \left(\frac{m_t}{2\pi k T_t} \right)^{3/2} \exp \left(-\frac{m_s c_s^2}{2k T_s} - \frac{m_t c_t^2}{2k T_t} \right). \tag{4.93}$$

The integrations over $d^3 c_s$ and $d^3 c_t$ can be performed by introducing the following velocities[3]

$$\mathbf{c}_* = \mathbf{V}_c - \mathbf{u}_c + \beta \Delta \mathbf{u} + \beta \mathbf{g}, \tag{4.94}$$

$$\mathbf{g}_* = -\mathbf{g} - \Delta \mathbf{u}, \tag{4.95}$$

where $\mathbf{V}_c$ is the center-of-mass velocity (4.6), $\mathbf{g}$ is the relative velocity (4.7) and

$$\mathbf{u}_c = \frac{m_s \mathbf{u}_s + m_t \mathbf{u}_t}{m_s + m_t}, \quad \Delta \mathbf{u} = \mathbf{u}_t - \mathbf{u}_s. \tag{4.96}$$

The parameter β and other temperature-dependent factors to be used later are

$$a^2 = \frac{2k T_s T_t}{m_t T_s + m_s T_t}, \quad \alpha^2 = \frac{2k T_{st}}{\mu_{st}}, \quad \beta = \frac{\mu_{st}}{m_s + m_t} \frac{T_t - T_s}{T_{st}}. \tag{4.97}$$

Note that the subscripts s and t have been temporarily left off $\mathbf{g}_{st}$. In Equation (4.97), μ_{st} and T_{st} are the reduced mass and reduced temperature, respectively;

$$\mu_{st} = \frac{m_s m_t}{m_s + m_t}, \tag{4.98}$$

$$T_{st} = \frac{m_s T_t + m_t T_s}{m_s + m_t}. \tag{4.99}$$

The velocity transformation is from $(\mathbf{c}_s, \mathbf{c}_t)$ to $(\mathbf{c}_*, \mathbf{g}_*)$, and with the latter velocities defined in Equations (4.94) and (4.95), the connection between the two sets of velocities is

$$\mathbf{c}_s = \mathbf{c}_* - \psi \mathbf{g}_*, \tag{4.100}$$

$$\mathbf{c}_t = \mathbf{c}_* + (1 - \psi) \mathbf{g}_*, \tag{4.101}$$

where

$$\psi = \frac{m_t T_s}{m_t T_s + m_s T_t}, \quad (1 - \psi) = \frac{m_s T_t}{m_t T_s + m_s T_t} \tag{4.102}$$

and where the transformation of $d^3c_s\, d^3c_t$ is accomplished with the aid of a Jacobian (Appendix C)

$$d^3c_s\, d^3c_t = d^3c_*\, d^3g_*. \tag{4.103}$$

Using Equations (4.100), (4.101), and (4.97), the term $f_s f_t$ (4.93) can be written as

$$f_s f_t = \frac{n_s n_t}{\pi^3 a^3 \alpha^3} \exp\left(-\frac{c_*^2}{a^2} - \frac{g_*^2}{\alpha^2}\right). \tag{4.104}$$

Substituting Equations (4.103) and (4.104) into the expression for $\delta \mathbf{M}_s / \delta t$ (4.92) yields

$$\frac{\delta \mathbf{M}_s}{\delta t} = -\frac{\mu_{st} n_s n_t}{\pi^3 a^3 \alpha^3} \int d^3 c_* \, e^{-c_*^2/a^2} \int d^3 g_* \, e^{-g_*^2/\alpha^2} g Q_{st}^{(1)}(g)\mathbf{g}. \tag{4.105}$$

The first integral can be easily evaluated using a spherical coordinate system in velocity space because the integrand depends only on the magnitude of c_*

$$\int d^3 c_* \, e^{-c_*^2/a^2} = 4\pi \int_0^\infty dc_* c_*^2 e^{-c_*^2/a^2} = \pi^{3/2} a^3. \tag{4.106}$$

In the second integral, Equation (4.95) is used to express g_*^2 in terms of $\mathbf{g}$ and $\Delta \mathbf{u}$

$$g_*^2 = g^2 + 2\mathbf{g} \cdot \Delta \mathbf{u} + (\Delta u)^2 \tag{4.107}$$

and use is made of the fact that $d^3 g_* = d^3 g$ (the only difference being a displacement of the origin of velocity space). With these changes, the second integral in Equation (4.105) becomes

$$\int d^3 g \, g Q_{st}^{(1)}(g)\mathbf{g} \exp\left[-\frac{g^2 + 2\mathbf{g} \cdot \Delta \mathbf{u} + (\Delta u)^2}{\alpha^2}\right]. \tag{4.108}$$

This integral can be evaluated by using a spherical coordinate system with $\Delta \mathbf{u}$ taken along the polar axis, as shown in Figure 4.6. In this coordinate system, Equation (4.108) becomes

$$\int_0^\infty dg \, g^4 Q_{st}^{(1)}(g) \int_0^{2\pi} d\phi' \int_0^\pi \sin\theta'\, d\theta'$$

$$\times \exp\left[-\frac{g^2 + 2g(\Delta u)\cos\theta' + (\Delta u)^2}{\alpha^2}\right]$$

$$\times [\sin\theta' \cos\phi' \mathbf{e}_1 + \sin\theta' \sin\phi' \mathbf{e}_2 + \cos\theta' \mathbf{e}_3], \tag{4.109}$$

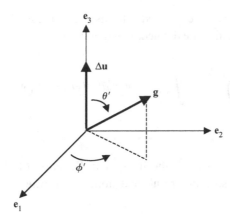

Figure 4.6 Coordinate system in velocity space used to evaluate the general transfer collision integrals. $\Delta \mathbf{u} = \mathbf{u}_t - \mathbf{u}_s$.

where $d^3g = g^2 \sin\theta' \, d\theta' \, d\phi'$. After integrating over $d\phi'$ and letting $x = \cos\theta'$, Equation (4.109) can be expressed as

$$2\pi \mathbf{e}_3 \int\limits_0^\infty dg \, g^4 Q_{st}^{(1)}(g) \int\limits_{-1}^1 dx \, x \exp\left[-\frac{g^2 + 2g(\Delta u)x + (\Delta u)^2}{\alpha^2}\right]. \qquad (4.110)$$

The integrals in Equation (4.110) can be evaluated for inverse-power interaction potentials, and the resulting momentum transfer collision terms are valid for arbitrary $(\mathbf{u}_t - \mathbf{u}_s)$ and $(T_t - T_s)$. This is discussed in more detail later. However, it is instructive to consider the limit of small relative drift velocities, i.e., when the drift velocity differences are much smaller than thermal speeds. In this limit, the exponential in Equation (4.110) can be expanded as follows:

$$\exp\left[-\frac{g^2 + 2g(\Delta u)x + (\Delta u)^2}{\alpha^2}\right] = e^{-g^2/\alpha^2} e^{-[2g(\Delta u)x + (\Delta u)^2]/\alpha^2}$$

$$= e^{-g^2/\alpha^2}\left[1 - \frac{2g(\Delta u)x}{\alpha^2}\right], \qquad (4.111)$$

where the second exponential on the right-hand side is expanded in a series because the argument is small (Appendix D) and where the $(\Delta u/\alpha)^2$ term is second-order in Δu and can be neglected [α is a thermal speed; see Equation (4.97)]. Substituting Equation (4.111) into Equation (4.110) yields

$$2\pi \mathbf{e}_3 \int\limits_0^\infty dg \, g^4 Q_{st}^{(1)}(g) e^{-g^2/\alpha^2} \int\limits_{-1}^1 dx \left[x - \frac{2g(\Delta u)}{\alpha^2}x^2\right]. \qquad (4.112)$$

The integration over dx is $-(4/3)g(\Delta u)/\alpha^2$ and Equation (4.112) then becomes

$$-\frac{8\pi}{3}\frac{(\Delta u)}{\alpha^2}\mathbf{e}_3 \int\limits_0^\infty dg \, e^{-g^2/\alpha^2} g^5 Q_{st}^{(1)}(g). \qquad (4.113)$$

This remaining integral over g can be expressed in terms of the so-called *Chapman–Cowling collision integrals*,[8] which are defined, in general, as

$$\Omega_{st}^{(l,j)} = \frac{1}{\sqrt{4\pi}} \left(\frac{\mu_{st}}{2kT_{st}} \right)^{(2j+3)/2} \int\limits_0^\infty dg_{st}\, e^{-\mu_{st}g_{st}^2/(2kT_{st})} g_{st}^{(2j+3)} Q_{st}^{(l)}(g_{st}).$$

(4.114)

Remembering that $\alpha = (2kT_{st}/\mu_{st})^{1/2}$, the integral in Equation (4.113) can be expressed as $\alpha^5 \sqrt{4\pi} \Omega_{st}^{(1,1)}$. Using this result and noting that $\Delta u \mathbf{e}_3 = \mathbf{u}_t - \mathbf{u}_s$, Equation (4.113) becomes

$$-\frac{16\pi^{3/2}}{3} \alpha^3 \Omega_{st}^{(1,1)} (\mathbf{u}_t - \mathbf{u}_s).$$

(4.115)

The term given by Equation (4.115) is the result for the second integral in Equation (4.105) in the limit of small $(\mathbf{u}_t - \mathbf{u}_s)$, while the result for the first integral is given in Equation (4.106). The substitution of these results into Equation (4.105) yields the final expression for $\delta \mathbf{M}_s / \delta t$

$$\frac{\delta \mathbf{M}_s}{\delta t} = n_s m_s \nu_{st} (\mathbf{u}_t - \mathbf{u}_s),$$

(4.116)

where the *momentum transfer collision frequency* is defined as

$$\nu_{st} = \frac{16}{3} \frac{n_t m_t}{m_s + m_t} \Omega_{st}^{(1,1)}.$$

(4.117)

The advantage of introducing the Chapman–Cowling collision integrals (4.114) is that they have been evaluated for many collision processes. They also appear in the higher-order transport equations, as will be shown in Section 4.7.

As noted earlier, the integral in Equation (4.110) can be evaluated without approximation for some specific collision processes. Hard-sphere interactions are an example of such a case, and this case is outlined here as an illustration. The collision cross section for hard spheres, $Q_{st} = \pi \sigma^2$, is a constant (σ is the sum of the radii of the colliding particles). For this case, Equation (4.110) becomes

$$2\pi \mathbf{e}_3 Q_{st}^{(1)} \int\limits_0^\infty dg\, g^4 \int\limits_{-1}^1 dx\, x \exp\left[-\frac{g^2 + 2g(\Delta u)x + (\Delta u)^2}{\alpha^2} \right]$$

$$= \mathbf{e}_3 Q_{st}^{(1)} \left[-\frac{8\pi}{3} \alpha^4 (\Delta u) \Phi_{st} \right],$$

(4.118)

where

$$\Phi_{st} = \frac{3\sqrt{\pi}}{8} \left(\varepsilon_{st} + \frac{1}{\varepsilon_{st}} - \frac{1}{4\varepsilon_{st}^3} \right) \mathrm{erf}(\varepsilon_{st}) + \frac{3}{8} \left(1 + \frac{1}{2\varepsilon_{st}^2} \right) e^{-\varepsilon_{st}^2}, \quad (4.119)$$

$$\varepsilon_{st} = \frac{|\mathbf{u}_t - \mathbf{u}_s|}{(2kT_{st}/\mu_{st})^{1/2}}, \quad (4.120)$$

and erf is the error function (Appendix D). The substitution of (4.118) and (4.106) into (4.105) yields an expression for $\delta\mathbf{M}_s/\delta t$ that is completely general for hard-sphere collisions

$$\frac{\delta\mathbf{M}_s}{\delta t} = n_s m_s \left[\frac{16}{3} \frac{n_t m_t}{m_s + m_t} \frac{\alpha}{\sqrt{4\pi}} Q_{st}^{(1)} \right] (\mathbf{u}_t - \mathbf{u}_s) \Phi_{st}. \quad (4.121)$$

Using the definitions of α (4.97) and the Chapman–Cowling collision integral (4.114), one obtains

$$\Omega_{st}^{(1,1)} = \frac{\alpha}{\sqrt{4\pi}} Q_{st}^{(1)} \quad (4.122)$$

for hard-sphere interactions. This result, in combination with the collision frequency definition (4.117), yields the final expression for $\delta\mathbf{M}_s/\delta t$,

$$\frac{\delta\mathbf{M}_s}{\delta t} = n_s m_s \nu_{st} (\mathbf{u}_t - \mathbf{u}_s) \Phi_{st}. \quad (4.123)$$

Equation (4.123) is valid for arbitrary drift velocity differences and arbitrary temperature differences between the interacting gases. Although it was derived for hard-sphere interactions, the general form is valid for all central force interactions; only ν_{st} and Φ_{st} change. If species s collides with several other species, a sum over species t should appear in Equation (4.123). Also, an equation for the energy transfer collision term, $\delta E_s/\delta t$, can be derived in a manner similar to that described above for $\delta\mathbf{M}_s/\delta t$. For future reference, the general expressions for the five-moment collision terms are summarized as follows:

$$\frac{\delta n_s}{\delta t} = 0, \quad (4.124a)$$

$$\frac{\delta\mathbf{M}_s}{\delta t} = \sum_t n_s m_s \nu_{st} (\mathbf{u}_t - \mathbf{u}_s) \Phi_{st}, \quad (4.124b)$$

$$\frac{\delta E_s}{\delta t} = \sum_t \frac{n_s m_s \nu_{st}}{m_s + m_t} \left[3k(T_t - T_s) \Psi_{st} + m_t (\mathbf{u}_s - \mathbf{u}_t)^2 \Phi_{st} \right], \quad (4.124c)$$

where ν_{st} is given in Equation (4.117) and where Φ_{st} and Ψ_{st} are velocity-dependent correction factors that are different for different collision processes.

For the ionospheres, the important velocity-dependent correction factors pertain to Coulomb, hard-sphere, and elastic ion–neutral (Maxwell molecule) collisions. These are summarized as follows:

Coulomb:

$$\Phi_{st} = \frac{3\sqrt{\pi}}{4} \frac{\operatorname{erf}(\varepsilon_{st})}{\varepsilon_{st}^3} - \frac{3}{2} \frac{1}{\varepsilon_{st}^2} e^{-\varepsilon_{st}^2}, \tag{4.125a}$$

$$\Psi_{st} = e^{-\varepsilon_{st}^2}, \tag{4.125b}$$

Hard sphere:

$$\Phi_{st} = \frac{3\sqrt{\pi}}{8} \left(\varepsilon_{st} + \frac{1}{\varepsilon_{st}} - \frac{1}{4\varepsilon_{st}^3} \right) \operatorname{erf}(\varepsilon_{st}) + \frac{3}{8} \left(1 + \frac{1}{2\varepsilon_{st}^2} \right) e^{-\varepsilon_{st}^2}, \tag{4.126a}$$

$$\Psi_{st} = \frac{\sqrt{\pi}}{2} \left(\varepsilon_{st} + \frac{1}{2\varepsilon_{st}} \right) \operatorname{erf}(\varepsilon_{st}) + \frac{1}{2} e^{-\varepsilon_{st}^2}, \tag{4.126b}$$

Maxwell molecule:

$$\Phi_{st} = 1, \tag{4.127a}$$

$$\Psi_{st} = 1, \tag{4.127b}$$

where ε_{st} is given by Equation (4.120) and $\operatorname{erf}(x)$ is the error function (Appendix D). Note that the hard-sphere result for Φ_{st} in Equation (4.126a) is the same as that derived above (4.119). It is repeated in this summary to provide an easy reference.

In the limit of very small relative drifts between the interacting species ($\varepsilon_{st} \ll 1$), $\Phi_{st} = \Psi_{st} = 1$ for all inverse-power interaction potentials, including those listed above. In the opposite limit of very large relative drifts ($\varepsilon_{st} \gg 1$), Φ_{st} and $\Psi_{st} \to 0$ for Coulomb collisions, while $\Phi_{st} \to 3\pi^{1/2}\varepsilon_{st}/8$ and $\Psi_{st} \to \pi^{1/2}\varepsilon_{st}/2$ for hard-sphere interactions.

4.7 Collision terms for 13-moment velocity distributions

The assumption that each species in the gas is described by separate drifting Maxwellians is not adequate for most of the ionospheres because this level of approximation does not take into account stress and heat flow processes. To include these and other effects, it is necessary to assume that each species in the gas can be represented by a 13-moment distribution function (Equation 3.49). Unfortunately, general collision terms have not been derived for this expression because of the associated mathematical difficulties involved. Collision terms for the 13-moment approximation have been derived in the limit of small drift velocity differences between the interacting species, but arbitrary temperature differences. These collision terms are known as *Burgers semilinear collision terms*. If it is further assumed that the species temperature differences are small compared to the individual species temperatures, the 13-moment collision terms are known as *Burgers linear collision terms*.

In the linear approximation, the gas is assumed to be sufficiently collision-dominated that the distributions are approximately Maxwellian. That is, the heat flow and stress terms in the 13-moment expansions for f_s and f_t can be treated as small quantities. In addition, as noted before, $(\mathbf{u}_t - \mathbf{u}_s)$ and $(T_t - T_s)$ are also treated as small quantities. Therefore, products or powers of $\mathbf{q}_s$, $\mathbf{q}_t$, $\boldsymbol{\tau}_s$, $\boldsymbol{\tau}_t$, $(\mathbf{u}_t - \mathbf{u}_s)$, and $(T_t - T_s)$ are neglected. Starting from the 13-moment expressions for f_s and f_t, (3.49), the term $f_s f_t$ that appears in the transfer collision integral (4.80) can be simplified when the small quantities are neglected and it reduces to

$$
f_s f_t = n_s n_t \left(\frac{m_s}{2\pi k T_s} \right)^{3/2} \left(\frac{m_t}{2\pi k T_t} \right)^{3/2} \exp\left(-\frac{m_s c_s^2}{2kT_s} - \frac{m_t c_t^2}{2kT_t} \right)
$$

$$
\times \left[1 + \frac{m_s}{2kT_s p_s} \boldsymbol{\tau}_s : \mathbf{c}_s \mathbf{c}_s + \frac{m_t}{2kT_t p_t} \boldsymbol{\tau}_t : \mathbf{c}_t \mathbf{c}_t - \left(1 - \frac{m_s c_s^2}{5kT_s} \right) \frac{m_s}{kT_s p_s} \mathbf{q}_s \cdot \mathbf{c}_s \right.
$$

$$
\left. - \left(1 - \frac{m_t c_t^2}{5kT_t} \right) \frac{m_t}{kT_t p_t} \mathbf{q}_t \cdot \mathbf{c}_t \right]. \tag{4.128}
$$

The procedure used to calculate the transfer collision integrals with $f_s f_t$ given by Equation (4.128) is similar to that described in the previous subsection for the five-moment approximation and small relative drifts, except for the additional assumption of small species temperature differences. For example, $\delta \mathbf{M}_s / \delta t$ is still given by Equation (4.92), and the change in velocity integration variables given in Equations (4.94) to (4.103) is still needed. Now, however, the product of drifting Maxwellians in Equation (4.104) must be replaced with the velocity transformation of Equation (4.128). Therefore, the double integral in Equation (4.105) will contain a series of terms, and they have to be integrated in a manner similar to the procedure outlined via Equations (4.106) to (4.116).

The linear collision terms for the 13-moment approximation are not derived here because of the extensive algebra involved.[3] However, they are summarized below because of their wide applicability in aeronomy and space physics. A convenient form for the 13-moment *linear* collision terms is[9]

$$
\frac{\delta n_s}{\delta t} = 0, \tag{4.129a}
$$

$$
\frac{\delta \mathbf{M}_s}{\delta t} = -\sum_t n_s m_s \nu_{st} (\mathbf{u}_s - \mathbf{u}_t) + \sum_t \nu_{st} \frac{z_{st} \mu_{st}}{kT_{st}} \left(\mathbf{q}_s - \frac{\rho_s}{\rho_t} \mathbf{q}_t \right), \tag{4.129b}
$$

$$
\frac{\delta E_s}{\delta t} = -\sum_t \frac{n_s m_s \nu_{st}}{m_s + m_t} 3k (T_s - T_t), \tag{4.129c}
$$

$$\frac{\delta \mathbf{P}_s}{\delta t} = -\sum_t \frac{2m_s \nu_{st}}{m_s + m_t} \left[\mathbf{P}_s - \frac{n_s}{n_t} \mathbf{P}_t + \frac{3}{10} z''_{st} \frac{m_t}{m_s} \left(\boldsymbol{\tau}_s + \frac{\rho_s}{\rho_t} \boldsymbol{\tau}_t \right) \right] - \frac{3}{5} z''_{ss} \nu_{ss} \boldsymbol{\tau}_s,$$

$$\text{(4.129d)}$$

$$\frac{\delta \mathbf{q}_s}{\delta t} = -\sum_t \nu_{st} \left[D^{(1)}_{st} \mathbf{q}_s - D^{(4)}_{st} \frac{\rho_s}{\rho_t} \mathbf{q}_t \right.$$

$$\left. + \frac{5}{2} p_s (\mathbf{u}_s - \mathbf{u}_t) \left(1 - \frac{m_t z_{st}}{m_s + m_t} \right) \right] - \frac{2}{5} z''_{ss} \nu_{ss} \mathbf{q}_s, \qquad \text{(4.129e)}$$

and $\delta \boldsymbol{\tau}_s / \delta t$ and $\delta \mathbf{q}'_s / \delta t$ can be obtained from the definitions given in Equations (3.62) and (3.63), respectively;

$$\frac{\delta \boldsymbol{\tau}_s}{\delta t} = -\sum_t \frac{2m_s \nu_{st}}{m_s + m_t} \left[\boldsymbol{\tau}_s \left(1 + \frac{3}{10} z''_{st} \frac{m_t}{m_s} \right) \right.$$

$$\left. - \frac{n_s}{n_t} \boldsymbol{\tau}_t \left(1 - \frac{3}{10} z''_{st} \right) \right] - \frac{3}{5} z''_{ss} \nu_{ss} \boldsymbol{\tau}_s, \qquad \text{(4.129f)}$$

$$\frac{\delta \mathbf{q}'_s}{\delta t} = -\frac{2}{5} z''_{ss} \nu_{ss} \mathbf{q}_s - \sum_t \nu_{st} \left\{ \mathbf{q}_s \left[D^{(1)}_{st} + \frac{5}{2} z_{st} \frac{\mu_{st}}{m_s} \frac{T_s}{T_{st}} \right] \right.$$

$$\left. - \frac{\rho_s}{\rho_t} \mathbf{q}_t \left[D^{(4)}_{st} + \frac{5}{2} z_{st} \frac{\mu_{st}}{m_s} \frac{T_s}{T_{st}} \right] - \frac{5}{2} p_s \frac{m_t z_{st}}{m_s + m_t} (\mathbf{u}_s - \mathbf{u}_t) \right\}, \qquad \text{(4.129g)}$$

and where

$$D^{(1)}_{st} = \frac{1}{(m_s + m_t)^2} \left[3m_s^2 - \frac{5}{2} m_t (m_s + m_t) z_{st} + m_t^2 z'_{st} + \frac{4}{5} m_s m_t z''_{st} \right],$$

$$\text{(4.130a)}$$

$$D^{(4)}_{st} = \frac{1}{(m_s + m_t)^2} \left[3m_t^2 - \frac{5}{2} m_t (m_s + m_t) z_{st} + m_t^2 z'_{st} - \frac{4}{5} m_t^2 z''_{st} \right],$$

$$\text{(4.130b)}$$

$$z_{st} = 1 - \frac{2}{5} \frac{\Omega^{(1,2)}_{st}}{\Omega^{(1,1)}_{st}}, \qquad \text{(4.131a)}$$

$$z'_{st} = \frac{5}{2} + \frac{2}{5} \frac{\Omega^{(1,3)}_{st} - 5\Omega^{(1,2)}_{st}}{\Omega^{(1,1)}_{st}}, \qquad \text{(4.131b)}$$

$$z''_{st} = \frac{\Omega^{(2,2)}_{st}}{\Omega^{(1,1)}_{st}}, \qquad \text{(4.131c)}$$

$$z_{st}''' = \frac{\Omega_{st}^{(2,3)}}{\Omega_{st}^{(1,1)}}. \tag{4.131d}$$

In these equations, $\rho = nm$ is the mass density, μ_{st} is the reduced mass (4.98), T_{st} is the reduced temperature (4.99), ν_{st} is the momentum transfer collision frequency (4.117) and $\Omega_{st}^{(l,j)}$ is the Chapman–Cowling collision integral (4.114). Note that the parameters z_{st}, z_{st}', z_{st}'', z_{st}''', $D_{st}^{(1)}$, and $D_{st}^{(4)}$ become pure numbers once the identity of the colliding particles is known. Values for these parameters and the associated momentum transfer collision frequency are given in Section 4.8 for the collision processes relevant to the ionospheres.

As will be shown in Chapter 5, the heat flow terms that appear in the momentum collision term (4.129b) account for thermal diffusion effects, and they also provide corrections to ordinary diffusion. The drift velocity terms in the heat flow collision term (4.129e) account for thermoelectric and diffusion thermal effects.

The *semilinear* collision terms are valid for small drift velocity differences and arbitrary temperature differences between the interacting species. For this case, the continuity, momentum, and energy collision terms (4.129a–c) are unchanged, but the pressure tensor (4.129d) and heat flow (4.129e) collision terms (and consequently $\delta\boldsymbol{\tau}_s/\delta t$ and $\delta\mathbf{q}_s'/\delta t$) are modified;

$$\frac{\delta\mathbf{P}_s}{\delta t} = \sum_t \frac{n_s m_s \nu_{st}}{m_s + m_t} 2k(T_t - T_s)\mathbf{I} - \sum_t \frac{2m_s \nu_{st}}{m_s + m_t} \frac{T_t}{T_{st}}\left(\boldsymbol{\tau}_s - \frac{n_s T_s}{n_t T_t}\boldsymbol{\tau}_t\right)$$
$$- \sum_t{}' \frac{\nu_{st}}{m_s + m_t}\left[\frac{3}{5}m_t z_{st}'' - 2\mu_{st}(1 - z_{st})\frac{(T_t - T_s)}{T_{st}}\right]\left(\boldsymbol{\tau}_s + \frac{\rho_s}{\rho_t}\boldsymbol{\tau}_t\right), \tag{4.132a}$$

$$\frac{\delta\mathbf{q}_s}{\delta t} = \sum_t \frac{n_s m_s \nu_{st}}{m_s + m_t}(\mathbf{u}_t - \mathbf{u}_s)\left\{\frac{5}{2}kT_s\left[\frac{T_t}{T_{st}} + \frac{m_t}{m_s}\frac{T_{st}}{T_s}(1 - z_{st})\right] - k(T_t - T_s)y_{st}\right\}$$
$$- \sum_t{}' \nu_{st}\mathbf{q}_s\left[\frac{3m_s^2}{(m_s + m_t)^2}\frac{T_t^2}{T_{st}^2} + B_{st}^{(3)}\left(z_{st}' - \frac{5}{2}z_{st}\right) - B_{st}^{(1)}\right.$$
$$\left. + \frac{m_s m_t}{(m_s + m_t)^2}\frac{T_t}{T_{st}}\left(\frac{4}{5}z_{st}'' - \frac{5}{2}\frac{T_s}{T_{st}}z_{st}\right)\right]$$
$$+ \sum_t{}' \nu_{st}\frac{\rho_s}{\rho_t}\mathbf{q}_t\left[\frac{3m_t^2}{(m_s + m_t)^2}\frac{T_s^2}{T_{st}^2} + B_{st}^{(3)}\left(z_{st}' - \frac{5}{2}z_{st}\right) + B_{st}^{(2)}\right.$$
$$\left. - \frac{m_s m_t}{(m_s + m_t)^2}\frac{T_s}{T_{st}}\left(\frac{4}{5}\frac{m_t}{m_s}z_{st}'' + \frac{5}{2}\frac{T_t}{T_{st}}z_{st}\right)\right], \tag{4.132b}$$

where

$$y_{st} = \frac{m_t}{m_s + m_t} \left[2z''_{st} - 5\frac{T_s}{T_{st}} - \frac{15}{2}\frac{m_s(1 - z_{st})}{m_s + m_t}\frac{(T_t - T_s)}{T_{st}} \right], \qquad (4.133a)$$

$$B_{st}^{(1)} = \frac{m_t\mu_{st}}{(m_s + m_t)^2}\frac{(T_t - T_s)}{T_{st}} \left[\frac{4}{5}z'''_{st} - 2z''_{st} + \frac{m_s}{m_t}\frac{T_t}{T_{st}}(6 - 11z_{st}) \right], \quad (4.133b)$$

$$B_{st}^{(2)} = \frac{m_t\mu_{st}}{(m_s + m_t)^2}\frac{(T_t - T_s)}{T_{st}} \left[-\frac{4}{5}z'''_{st} + 2z''_{st} + \frac{T_s}{T_{st}}(6 - 11z_{st}) \right], \quad (4.133c)$$

$$B_{st}^{(3)} = \frac{m_t^2}{(m_s + m_t)^2} \left[1 + \frac{3m_s^2}{(m_s + m_t)^2}\frac{(T_t - T_s)^2}{T_{st}^2} \right]. \qquad (4.133d)$$

The prime on the summations in Equations (4.132a,b) means that the case $t = s$ is included.

4.8   Momentum transfer collision frequencies

The relevant collision processes for ionospheres, which are partially ionized gases, include Coulomb interactions, nonresonant ion–neutral interactions, resonant charge exchange, and electron–neutral interactions. In what follows, expressions for the appropriate Chapman–Cowling collision integrals and momentum transfer collision frequencies are presented.

The transport cross sections and collision integrals that are needed in the evaluation of the collision terms have been calculated for a general *inverse-power interparticle force law* of the form[8]

$$F = \frac{K_{st}}{r^a}, \qquad (4.134)$$

where F is the magnitude of the force, K_{st} and a are constants, and r is the distance between the particles. For such a force law,

$$Q_{st}^{(l)} = 2\pi A_l(a)\left(\frac{K_{st}}{\mu_{st}g_{st}^2} \right)^{2/(a-1)}, \qquad (4.135)$$

$$\Omega_{st}^{(l,j)} = \frac{\sqrt{\pi}}{2}A_l(a)\Gamma\left(j + 2 - \frac{2}{a - 1} \right)\left(\frac{K_{st}}{\mu_{st}} \right)^{2/(a-1)}\left(\frac{2kT_{st}}{\mu_{st}} \right)^{(a-5)/[2(a-1)]}, \qquad (4.136)$$

where

$$A_l(a) = \int_0^\infty (1 - \cos^l\theta)\hat{b}_0\,d\hat{b}_0. \qquad (4.137)$$

Table 4.2 Values of $A_1(a)$ and $A_2(a)$ for selected values of a.[8]

a	$A_1(a)$	$A_2(a)$
5	0.422	0.436
7	0.385	0.357
9	0.382	0.332
11	0.383	0.319
15	0.393	0.309
∞	0.5	0.333

In Equations (4.135) to (4.137), $\Gamma(x)$ is a gamma function (Appendix D), θ is the scattering angle, and $\hat{b}_0 = b(\mu_{st} g_{st}^2 / K_{st})^{1/(a-1)}$ is a nondimensional impact parameter. The integral in Equation (4.137) can be evaluated numerically by quadrature, and the resulting values of $A_l(a)$ are pure numbers that depend only on l and a. Values are given in Table 4.2 for various combinations of these parameters. The momentum transfer collision frequency (4.117) and the various ratios of the Chapman–Cowling integrals (4.131a–c) for inverse-power force laws can be expressed as

$$\nu_{st} = \frac{8\sqrt{\pi}}{3} A_1(a) \Gamma\left(3 - \frac{2}{a-1}\right) \frac{n_t m_t}{m_s + m_t} \left(\frac{K_{st}}{\mu_{st}}\right)^{2/(a-1)}$$

$$\times \left(\frac{2kT_{st}}{\mu_{st}}\right)^{(a-5)/[2(a-1)]}, \tag{4.138}$$

$$z_{st} = -\frac{1}{5}\frac{a-5}{a-1}, \tag{4.139a}$$

$$z'_{st} = \frac{5}{2} - \frac{2}{5}\frac{(a+1)(3a-5)}{(a-1)^2}, \tag{4.139b}$$

$$z''_{st} = \frac{3a-5}{a-1}\frac{A_2(a)}{A_1(a)}. \tag{4.139c}$$

For *Coulomb interactions*, the quantities that are needed in the 13-moment collision terms (4.129a–g) are (in cgs units)

$$\nu_{st} = \frac{16\sqrt{\pi}}{3}\frac{n_t m_t}{m_s + m_t}\left(\frac{2kT_{st}}{\mu_{st}}\right)^{-3/2}\frac{e_s^2 e_t^2}{\mu_{st}^2}\ln\Lambda, \tag{4.140}$$

$$z_{st} = \frac{3}{5}, \qquad z'_{st} = \frac{13}{10}, \qquad z''_{st} = 2, \tag{4.141a}$$

Table 4.3 The collision frequency coefficients B_{st} for ion–ion interactions.

| | | | | | t | | | | | |
s	H^+	He^+	C^+	N^+	O^+	CO^+	N_2^+	NO^+	O_2^+	CO_2^+
H^+	0.90	1.14	1.22	1.23	1.23	1.25	1.25	1.25	1.25	1.26
He^+	0.28	0.45	0.55	0.56	0.57	0.59	0.59	0.60	0.60	0.61
C^+	0.102	0.18	0.26	0.27	0.28	0.31	0.31	0.31	0.31	0.32
N^+	0.088	0.16	0.23	0.24	0.25	0.28	0.28	0.28	0.28	0.30
O^+	0.077	0.14	0.21	0.22	0.22	0.25	0.25	0.26	0.26	0.27
CO^+	0.045	0.085	0.13	0.14	0.15	0.17	0.17	0.17	0.18	0.19
N_2^+	0.045	0.085	0.13	0.14	0.15	0.17	0.17	0.17	0.18	0.19
NO^+	0.042	0.080	0.12	0.13	0.14	0.16	0.16	0.16	0.17	0.18
O_2^+	0.039	0.075	0.12	0.12	0.13	0.15	0.15	0.16	0.16	0.17
CO_2^+	0.029	0.055	0.09	0.09	0.10	0.12	0.12	0.12	0.12	0.14

$$D_{st}^{(1)} = \left(3m_s^2 + \frac{1}{10}m_s m_t - \frac{1}{5}m_t^2\right)\Big/(m_s + m_t)^2, \tag{4.141b}$$

$$D_{st}^{(4)} = \left(\frac{6}{5}m_t^2 - \frac{3}{2}m_s m_t\right)\Big/(m_s + m_t)^2, \tag{4.141c}$$

where $\ln \Lambda$ is the Coulomb logarithm (4.57). For the ionospheres, $\ln \Lambda \sim 15$, and the Coulomb collision frequency can be approximated numerically by

$$\nu_{st} = 1.27\frac{Z_s^2 Z_t^2 M_{st}^{1/2}}{M_s}\frac{n_t}{T_{st}^{3/2}}, \tag{4.142}$$

where M_s is the particle mass in atomic mass units, M_{st} is the reduced mass in atomic mass units, Z_s and Z_t are the particle charge numbers, n_t is in cm^{-3}, and T_{st} is in kelvins. For ion–ion interactions this reduces further to

$$\nu_{st} = B_{st}\frac{n_t}{T_t^{3/2}}, \tag{4.143}$$

where B_{st} is a numerical coefficient; values are given in Table 4.3 for the ion species found in the ionospheres. Equation (4.142) also reduces further for electron–electron and electron–ion interactions;

$$\nu_{ei} = 54.5\frac{n_i Z_i^2}{T_e^{3/2}}, \tag{4.144}$$

$$\nu_{ee} = \frac{54.5}{\sqrt{2}}\frac{n_e}{T_e^{3/2}}, \tag{4.145}$$

where subscript e denotes electrons and subscript i denotes ions.

Table 4.4 The collision frequency coefficients $C_{in} \times 10^{10}$ for nonresonant ion–neutral interactions.

Ion	Neutral							
	H	He	N	O	CO	N_2	O_2	CO_2
H^+	R^a	10.6	26.1	R	35.6	33.6	32.0	41.4
He^+	4.71	R	11.9	10.1	16.9	16.0	15.3	20.0
C^+	1.69	1.71	5.73	4.94	8.74	8.26	8.01	10.7
N^+	1.45	1.49	R	4.42	7.90	7.47	7.25	9.73
O^+	R	1.32	4.62	R	7.22	6.82	6.64	8.95
CO^+	0.74	0.79	2.95	2.58	R	4.24	4.49	6.18
N_2^+	0.74	0.79	2.95	2.58	4.84	R	4.49	6.18
NO^+	0.69	0.74	2.79	2.44	4.59	4.34	4.27	5.89
O_2^+	0.65	0.70	2.64	2.31	4.37	4.13	R	5.63
CO_2^+	0.47	0.51	2.00	1.76	3.40	3.22	3.18	R

^{a}R means that the collisional interaction is resonant.

Ion–neutral interactions can be either resonant or nonresonant. *Nonresonant ion–neutral interactions* occur between unlike ions and neutrals, and they correspond to a long-range polarization attraction coupled with a short-range repulsion. As noted in Section 4.5, such an interaction can be approximated by a Maxwell molecule interaction, with the momentum transfer collision frequency given in Equation (4.88). For a given ion–neutral pair, this nonresonant collision frequency takes a particularly simple form

$$\nu_{in} = C_{in} n_n, \tag{4.146}$$

where n_n is in cm^{-3} and C_{in} is a numerical coefficient; values are given in Table 4.4 for some of the different ion–neutral combinations found in the ionospheres. The other quantities that are needed in the 13-moment collision terms (4.129a–g) are

$$z_{st} = 0, \qquad z'_{st} = 1, \qquad z''_{st} = 2, \tag{4.147a}$$

$$D^{(1)}_{st} = \left(3m_s^2 + m_t^2 + \frac{8}{5} m_s m_t \right) \Big/ (m_s + m_t)^2, \tag{4.147b}$$

$$D^{(4)}_{st} = \frac{12}{5} m_t^2 / (m_s + m_t)^2. \tag{4.147c}$$

When these quantities are substituted into the 13-moment collision terms (4.129a–g), they are in agreement with the linearized version of the general Maxwell molecule collision terms given in Equations (4.89a–f).

At elevated temperatures ($T > 300$ K), the interaction between an ion and its parent neutral is dominated by a *resonant charge exchange*. That is, as the ion and neutral approach each other, an electron jumps across from the neutral to the ion, thereby changing identities. In this way, a fast ion can become a fast neutral after the collision, which results in a large transfer of momentum and energy between the colliding particles. Although a resonant charge exchange is technically not an elastic collision, it is pseudo-elastic in the sense that very little energy is lost in the collision and, therefore, the Boltzmann collision integral can be used to calculate the relevant transport properties. However, for collisions between an ion and its parent neutral, the different collision cross sections, $Q_{in}^{(l)}$, are dominated by different processes. Specifically, the collision integrals with $l = 1$ are governed by the charge exchange mechanism, while for collision integrals with $l = 2$ the charge exchange mechanism cancels and elastic scattering dominates.

For resonant charge exchange, it is the energy-dependent charge exchange cross section, Q_E, that is generally measured

$$Q_E = (A' - B' \log_{10} \varepsilon_{in})^2, \tag{4.148}$$

where $\varepsilon_{in} = \mu_{in} g_{in}^2/2$ (in eV) is the relative kinetic energy of the colliding particles, and A' and B' (in cm) are constants that are different for different gases. It can be shown that the connection between the charge exchange and the momentum transfer cross sections is $Q_{st}^{(1)} = 2Q_E$.

Using this result, the desired Chapman–Cowling collision integrals become[9]

$$\Omega_{in}^{(1,1)} = \left(\frac{kT_{in}}{2\pi \mu_{in}}\right)^{1/2} \Bigg[(39.84B^2 - 17.85AB + 2A^2)$$
$$+ (8.923B^2 - 2AB) \log_{10} \frac{T_{in}}{M} + \frac{B^2}{2}\left(\log_{10} \frac{T_{in}}{M}\right)^2 \Bigg], \tag{4.149a}$$

$$\Omega_{in}^{(1,2)} = 3\left(\frac{kT_{in}}{2\pi \mu_{in}}\right)^{1/2} \Bigg[(41.14B^2 - 18.13AB + 2A^2)$$
$$+ (9.067B^2 - 2AB) \log_{10} \frac{T_{in}}{M} + \frac{B^2}{2}\left(\log_{10} \frac{T_{in}}{M}\right)^2 \Bigg], \tag{4.149b}$$

$$\Omega_{in}^{(1,3)} = 12\left(\frac{kT_{in}}{2\pi \mu_{in}}\right)^{1/2} \Bigg[(42.12B^2 - 18.35AB + 2A^2)$$
$$+ (9.176B^2 - 2AB) \log_{10} \frac{T_{in}}{M} + \frac{B^2}{2}\left(\log_{10} \frac{T_{in}}{M}\right)^2 \Bigg] \tag{4.149c}$$

$$\Omega_{in}^{(2,2)} = 0.8\pi \left(\frac{\gamma_n e^2}{\mu_{in}}\right)^{1/2}, \tag{4.149d}$$

where

$$A = A' + B'[13.4 - \log_{10} M], \tag{4.150a}$$
$$B = 2B', \tag{4.150b}$$

Table 4.5 Momentum transfer collision frequencies for resonant ion–neutral interactions.[5,10] Densities are in cm^{-3}.

Species	T_r, K	ν_{in}, s^{-1}
H^+, H	> 50	$2.65 \times 10^{-10} n(H) T_r^{1/2} (1 - 0.083 \log_{10} T_r)^2$
He^+, He	> 50	$8.73 \times 10^{-11} n(He) T_r^{1/2} (1 - 0.093 \log_{10} T_r)^2$
N^+, N	> 275	$3.83 \times 10^{-11} n(N) T_r^{1/2} (1 - 0.063 \log_{10} T_r)^2$
O^+, O	> 235	$3.67 \times 10^{-11} n(O) T_r^{1/2} (1 - 0.064 \log_{10} T_r)^2$
N_2^+, N_2	> 170	$5.14 \times 10^{-11} n(N_2) T_r^{1/2} (1 - 0.069 \log_{10} T_r)^2$
O_2^+, O_2	> 800	$2.59 \times 10^{-11} n(O_2) T_r^{1/2} (1 - 0.073 \log_{10} T_r)^2$
H^+, O	> 300	$6.61 \times 10^{-11} n(O) T_i^{1/2} (1 - 0.047 \log_{10} T_i)^2$
O^+, H	> 300	$4.63 \times 10^{-12} n(H) (T_n + T_i/16)^{1/2}$
CO^+, CO	> 525	$3.42 \times 10^{-11} n(CO) T_r^{1/2} (1 - 0.085 \log_{10} T_r)^2$
CO_2^+, CO_2	> 850	$2.85 \times 10^{-11} n(CO_2) T_r^{1/2} (1 - 0.083 \log_{10} T_r)^2$

$T_r = (T_i + T_n)/2$. The CO^+ and CO_2^+ collision frequencies were calculated, not measured.

and where γ_n is the neutral polarizability (Table 4.1), $T_{in} = (T_i + T_n)/2$ is the reduced temperature, M is the ion or neutral mass in atomic mass units, and A' and B' are the constants that appear in the charge exchange cross section (4.148). Using these collision integrals, the 13-moment collision terms for resonant charge exchange can be readily obtained from Equations (4.129a–g), (4.130a,b), and (4.131a–d).

A less rigorous, but relatively simple, approach has been widely used with regard to resonant charge exchange.[5] In this approach, the energy-dependent charge exchange cross section (4.148) is replaced with a Maxwellian-averaged cross section, $\langle Q_E \rangle$, before the Chapman–Cowling collision integrals are evaluated. When this is done,[9] the resonant charge exchange collision terms reduce to the hard-sphere collision terms (discussed later), with the hard-sphere cross section, $\pi \sigma^2$, replaced by $2\langle Q_E \rangle$ and the hard-sphere value of $Q_{in}^{(2)}/Q_{in}^{(1)}$ replaced with the charge exchange value of $\frac{1}{3}$. The resulting momentum transfer collision frequency for resonant charge exchange using this less rigorous approach becomes

$$\nu_{in} = \frac{8}{3\sqrt{\pi}} n_n \left[\frac{2k(T_i + T_n)}{m_i} \right]^{1/2} \left[A' + 3.96B' - B' \log_{10}(T_i + T_n) \right]^2,$$

(4.151)

where A' and B' are the constants that appear in Equation (4.148) for Q_E. Values for ν_{in} are given in Table 4.5 for the collisions relevant to most ionospheres. These expressions for ν_{in} have been widely used in the momentum collision terms (4.129b), without the heat flow corrections, and in the energy collision term (4.129c) throughout aeronomy and space physics. The extension of this less rigorous approach to the stress and heat flow equations is discussed in Reference 9.

The parameter for *elastic electron–neutral interactions* that is generally measured is the velocity-dependent momentum transfer cross section $Q_{en}^{(1)}$. For low-energy electron collisions with the neutrals typically found in the ionospheres, this cross section can be expressed in the form

$$Q_{en}^{(1)} = R_1 + R_2 v_e + R_3 v_e^2 + R_4 v_e^3, \qquad (4.152)$$

where R_1, R_2, R_3, and R_4 are experimentally determined constants and where the electron velocity v_e is approximately equal to the electron–neutral relative velocity, g_{en}. Using this expression, the Chapman–Cowling collision integrals (4.114) for $l = 1$ become

$$\Omega_{en}^{(1,j)} = \frac{1}{4\sqrt{\pi}} \left(\frac{2kT_e}{m_e} \right)^{1/2} \left[R_1 \Gamma(j+2) + R_2 \left(\frac{2kT_e}{m_e} \right)^{1/2} \Gamma\left(j + \frac{5}{2} \right) \right.$$

$$\left. + R_3 \left(\frac{2kT_e}{m_e} \right) \Gamma(j+3) + R_4 \left(\frac{2kT_e}{m_e} \right)^{3/2} \Gamma\left(j + \frac{7}{2} \right) \right], \qquad (4.153)$$

where $\Gamma(x)$ is a gamma function. The other quantities that are required to evaluate the 13-moment collision terms are

$$\nu_{st} = \frac{16}{3} n_n \Omega_{en}^{(1,1)}, \qquad (4.154)$$

$$D_{en}^{(1)} = -\frac{5}{2} z_{en} + z'_{en}, \qquad (4.155a)$$

$$D_{en}^{(4)} = 3 - \frac{5}{2} z_{en} + z'_{en} - \frac{4}{5} z''_{en}, \qquad (4.155b)$$

where z_{en}, z'_{en}, and z''_{en} are defined in Equations (4.131a–c) and where terms of order m_e/m_n have been neglected compared to terms of order 1. Table 4.6 provides momentum transfer collision frequencies for the elastic electron–neutral interactions relevant to the ionospheres. With regard to z_{en} and z'_{en}, they can be calculated from Equations (4.131a) and (4.131b), respectively, for each electron–neutral collision pair. However, the calculation of z''_{en}, is problematic because it requires a knowledge of $Q_{en}^{(2)}$ and, therefore, a knowledge of the differential scattering cross section, $\sigma_{en}(v_e, \theta)$. Unfortunately, most experiments measure $Q_{en}^{(1)}$, not the differential scattering cross section. In some cases, this problem can be circumvented because for low-energy electron collisions with some neutrals, such as He and O, the momentum transfer cross section, $Q_{en}^{(1)}$, is approximately constant. Hence, for these neutrals, the thermal electrons collide with them as hard spheres, for which $z''_{en} = 2$ (Equation 4.157a).

Table 4.6 Momentum transfer collision frequencies for electron–neutral interactions.[11,12] Densities are in cm^{-3}.

Species	ν_{en}, s^{-1}
N_2	$2.33 \times 10^{-11} n(N_2)(1 - 1.21 \times 10^{-4} T_e) T_e$
O_2	$1.82 \times 10^{-10} n(O_2)(1 + 3.6 \times 10^{-2} T_e^{1/2}) T_e^{1/2}$
O	$8.9 \times 10^{-11} n(O)(1 + 5.7 \times 10^{-4} T_e) T_e^{1/2}$
He	$4.6 \times 10^{-10} n(He) T_e^{1/2}$
H	$4.5 \times 10^{-9} n(H)(1 - 1.35 \times 10^{-4} T_e) T_e^{1/2}$
CO	$2.34 \times 10^{-11} n(CO)(T_e + 165)$
CO_2	$3.68 \times 10^{-8} n(CO_2)(1 + 4.1 \times 10^{-11} |4500 - T_e|^{2.93})$

The quantities needed for *hard-sphere* interactions in the 13-moment collision terms are:

$$\nu_{st} = \frac{8}{3\sqrt{\pi}} \frac{n_t m_t}{m_s + m_t} \left(\frac{2kT_{st}}{\mu_{st}} \right)^{1/2} (\pi \sigma^2), \tag{4.156}$$

$$z_{st} = -\frac{1}{5}, \qquad z'_{st} = \frac{13}{10}, \qquad z''_{st} = 2, \tag{4.157a}$$

$$D_{st}^{(1)} = \left(3m_s^2 + \frac{21}{10} m_s m_t + \frac{9}{5} m_t^2 \right) \Big/ (m_s + m_t)^2, \tag{4.157b}$$

$$D_{st}^{(4)} = \left(\frac{16}{5} m_t^2 + \frac{1}{2} m_s m_t \right) \Big/ (m_s + m_t)^2, \tag{4.157c}$$

where σ is the sum of the radii of the colliding particles.

Finally, it should be noted that the momentum transfer collision frequencies are not symmetric with respect to a change of indices, but satisfy the relation

$$n_s m_s \nu_{st} = n_t m_t \nu_{ts}. \tag{4.158}$$

4.9 Specific references

1. Goldstein, H., *Classical Mechanics*, Second Edn., Reading, MA: Addison-Wesley, 1980.
2. Hirschfelder, J. O., C. F. Curtiss, and R. B. Bird, *Molecular Theory of Gases and Liquids*, New York: Wiley, 1964.
3. Burgers, *Flow Equations for Composite Gases*, New York: Academic, 1969.
4. Henry, R. J. W., Elastic scattering from atomic oxygen and photodetachment from O^-, *Phys. Rev.*, **162**, 56, 1967.
5. Banks, P. M., and G. Kockarts, *Aeronomy*, New York: Academic, 1973.
6. Dalgarno, A., M. R. C. McDowell, and A. Williams, The mobilities of ions in unlike gases, *Phil. Trans. Roy. Soc. London, Ser. A*, **250**, 411, 1958.

7. St.-Maurice, J.-P., and R. W. Schunk, Ion velocity distributions in the high-latitude ionosphere, *Rev. Geophys. Space Phys.*, **17**, 99, 1979.

8. Chapman, S., and T. G. Cowling, *The Mathematical Theory of Non-Uniform Gases*, New York: Cambridge University Press, 1970.

9. Schunk, R. W., Transport equations for aeronomy, *Planet. Space Sci.*, **23**, 437, 1975.

10. Butler, D. M., The ionosphere of Venus, Ph.D. dissertation, Houston, TX: Rice University, 1975.

11. Itikawa, Y., Momentum transfer cross sections for electron collisions with atoms and molecules, *At. Data Nucl. Data Tables*, **21**, 69, 1978.

12. Schunk, R. W., and A. F. Nagy, Ionospheres of the terrestrial planets, *Rev. Geophys. Space Phys.*, **18**, 813, 1980.

4.10   Problems

Problem 4.1 Show that the center-of-mass velocity does not change in a binary elastic collision.

Problem 4.2 Show that the magnitude of the relative velocity does not change in a binary elastic collision.

Problem 4.3 Starting from Equation (4.28), calculate the scattering angle θ for an inverse-power interaction potential of the form $V(r) = -K_0/r^2$. The parameter K_0 is a constant and r is the separation between the particles.

Problem 4.4 Given the definitions of $\mathbf{V}_c$ (4.6), $\hat{\mathbf{V}}_c$ (4.64), $\mathbf{g}_{st}$ (4.7), and $\mathbf{c}_s$ (3.14), show that Equations (4.61) to (4.63) are correct.

Problem 4.5 Use index notation and derive the nine elements that are associated with the second-order tensor in Equation (4.76).

Problem 4.6 Using Equations (4.100), (4.101), and (4.97), show that the product $f_s f_t$ for Maxwellian velocity distributions is given by the expression in (4.104).

Problem 4.7 Show that the velocity-dependent correction factors for Coulomb collisions (4.125a,b) approach unity in the limit of small relative drifts between interacting species.

Problem 4.8 Show that the velocity-dependent correction factors for hard-sphere interactions (4.126a,b) have the following limits when the relative drift between the interacting species is large: $\Phi_{st} \to 3\pi^{1/2}\varepsilon_{st}/8$ and $\Psi_{st} \to \pi^{1/2}\varepsilon_{st}/2$.

Problem 4.9 Show that when Equations (4.147a–c) are used for z_{st}, z'_{st}, z''_{st}, $D_{st}^{(1)}$, and $D_{st}^{(4)}$, the 13-moment collision terms (4.129a–g) are in agreement with the *linearized* version of the general Maxwell molecule collision terms (4.89a–f).

Problem 4.10 Consider the following Boltzmann equation:

$$\frac{q}{m}\mathbf{E}_0 \cdot \nabla_u f = -\nu_0[f - f^M],$$

where

$$f^M = n\left(\frac{m}{2\pi kT_0}\right)^{3/2} \exp\left(-\frac{mv^2}{2kT_0}\right),$$

and where f, v, n, m, and q are the species distribution function, velocity, density, mass, and charge, respectively, $\mathbf{E}_0$ is a constant electric field, T_0 is a constant temperature, v_0 is a constant collision frequency, and k is Boltzmann's constant. Derive the continuity, momentum, and energy equations associated with this Boltzmann equation.

Problem 4.11 The so-called Lorentz collision model is a differential collision operator that describes electron collisions with cold ions; it is given by

$$\frac{\delta f}{\delta t} = \frac{2\pi n_i e^4 \ln \Lambda}{m^2} \nabla_v \cdot \left(\frac{v^2 \mathbf{I} - \mathbf{v}\mathbf{v}}{v^3} \cdot \nabla_v f\right)$$

$$= \frac{2\pi n_i e^4 \ln \Lambda}{m^2} \frac{\partial}{\partial v_\alpha}\left(\frac{v^2 \delta_{\alpha\beta} - v_\alpha v_\beta}{v^3} \frac{\partial f}{\partial v_\beta}\right),$$

where the first expression is in dyadic notation and the second in index notation. Also, f, v, m, and e are the electron distribution function, velocity, mass, and charge, respectively, n_i is the ion density, and $\ln \Lambda$ is the Coulomb logarithm. Calculate the density, drift velocity, and temperature moments of this collision term.

Problem 4.12 Consider a collision between molecules 1 and 2 in which molecule 2 is initially at rest. The deflection angle in the center-of-mass coordinate system is denoted by χ_{cm}, as indicated in Figure 4.7. Show that the angle of deflection, χ_{1-lab}, is given by the following relation:

$$\tan \chi_{1-lab} = \frac{\sin \chi_{cm}}{\cos \chi_{cm} + (m_1/m_2)}.$$

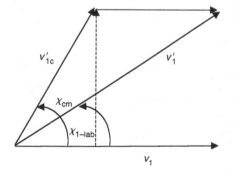

Figure 4.7 Diagram associated with Problem 4.12.

Problem 4.13 Consider the two-body collision shown in Figure 4.8. Show that the fractional energy loss between particles 1 and 2, having masses m_1 and m_2, respectively, is given by the following expression, if particle 2 is initially at rest. [Hint: Start out by writing the cosine law relation for $(\mathbf{v}'_1 - \mathbf{v}_2)^2$, expressing v'_{1c} and v_{2c} in terms of g]:

$$\frac{E_1 - E'_1}{E_1} = \frac{2(m_1 m_2)}{(m_1 + m_2)^2}(1 - \cos \chi).$$

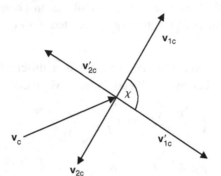

Figure 4.8 Diagram associated with Problem 4.13.

Chapter 5

Simplified transport equations

The 13-moment system of transport equations was introduced in Chapter 3 and several associated sets of collision terms were derived in Chapter 4. However, a rigorous application of the 13-moment system of equations for a multi-species plasma is rather difficult and it has been a common practice to use significantly simplified equation sets to study ionospheric behavior. The focus of this chapter is to describe, in some detail, the transport equations that are appropriate under different ionospheric conditions. The description includes a clear presentation of the major assumptions and approximations needed to derive the various simplified sets of equations so that potential users know the limited range of their applicability.

The equation sets discussed in this chapter are based on the assumption of *collision dominance*, for which the species velocity distribution functions are close to drifting Maxwellians. This assumption implies that the stress and heat flow terms in the 13-moment expression of the velocity distribution (3.49) are small. Simplified equations are derived for different levels of ionization, including weakly, partially, and fully ionized plasmas. A *weakly ionized plasma* is one in which Coulomb collisions can be neglected and only ion–neutral and electron–neutral collisions need to be considered. In a *partially ionized plasma*, collisions between ions, electrons, and neutrals have to be accounted for. Finally, in a *fully ionized plasma*, ion and electron collisions with neutrals are negligible. Note that in the last case, neutral particles can still be present, and in many fully ionized plasmas the neutrals are much more abundant than the charged particles. The plasma is fully ionized in the collisional sense because of the long-range nature of Coulomb interactions.

The topics in this chapter progress from very simple to more complex sets of transport equations. First, the well-known coefficients of diffusion, viscosity, and thermal conduction are derived using simple mean-free-path arguments. Next, completely general continuity, momentum, and energy equations are derived for the special case when all species in the plasma can be described by drifting Maxwellian velocity distributions (i.e., no stress or heat flow effects). This is followed by a discussion of transport effects in a weakly ionized plasma, for which simplifications are possible because Coulomb collisions are negligible. Then, for partially and fully ionized plasmas, the momentum equation is used to describe several important transport processes that can occur along a strong magnetic field, including multi-species ion diffusion, supersonic ion outflow, and time-dependent plasma expansion phenomena. Following these topics, the momentum equation is again used to describe first cross-**B** diffusion, and then electrical conduction, both along and across **B**. At this point, simplified equations are presented for the stress tensor and heat flow vector and their validity is discussed. This naturally leads into a discussion of higher-order diffusion effects, including heat flow corrections to ordinary diffusion, thermal diffusion, and thermoelectric effects. Finally, a summary is presented that indicates what sets of equations are to be used for different ionospheric applications.

Several different species have to be considered in this chapter, and therefore, it is useful to standardize the subscript convention. Throughout this and subsequent chapters, subscript e is for electrons, i for ions, n for neutrals, and j for any charged species (e.g., different ions or either ions or electrons).

5.1 Basic transport properties

Diffusion, viscosity, and thermal conduction are well-known transport processes, but before presenting a rigorous derivation of the associated transport coefficients, it is instructive to derive their general form using simple mean-free-path considerations. The analysis assumes that the mean-free-path, λ, is much smaller than the scale length for variation of any of the macroscopic gas properties (density, drift velocity, and temperature).

In the first example, the net flux of particles across a plane is calculated for a nondrifting isothermal gas with a density that decreases uniformly in the x-direction (Figure 5.1a). The plane at x is where the flux of particles is to be calculated and the planes at $x + \Delta x$ and $x - \Delta x$ are on the two sides, approximately a mean-free-path

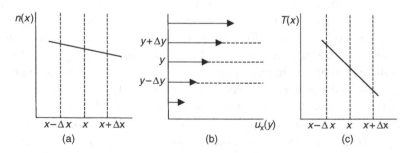

Figure 5.1 Simple density (a), flow velocity (b), and temperature (c) profiles used in the mean-free-path analysis of diffusion, viscosity, and thermal conduction, respectively.

away. If $\langle c \rangle$ is the average thermal speed (Equation H.21), then the *thermal particle flux* (average number of particles per unit area per unit time) crossing the plane at x is $n\langle c \rangle/4$ (Equation H.26). If the gas density is uniform, the net particle flux crossing the plane is zero because the thermal flux moving to the right cancels the thermal flux moving to the left. However, when the density varies with x, $n(x)$, the net particle flux crossing the plane is not zero. In this case, the particles that reach x from the left are associated with a density $n(x - \Delta x)$, because on the average that is where they had their last collision. Hence, their contribution to the particle flux at x is $\langle c \rangle n(x - \Delta x)/4$. The contribution to the particle flux at x from the particles on the right is $- \langle c \rangle n(x + \Delta x)/4$. Therefore, the *net particle flux*, Γ, crossing the plane at x is

$$\Gamma = \frac{\langle c \rangle}{4} n(x - \Delta x) - \frac{\langle c \rangle}{4} n(x + \Delta x). \tag{5.1}$$

Because Δx is small, the densities can be expanded in a Taylor series about x

$$\Gamma = \frac{\langle c \rangle}{4} \left[n(x) - \frac{dn}{dx} \Delta x - n(x) - \frac{dn}{dx} \Delta x \right] = -\frac{\langle c \rangle}{2} \frac{dn}{dx} \Delta x. \tag{5.2}$$

But $\Delta x \approx \lambda = \langle c \rangle / \nu$, and therefore

$$\Gamma = -\frac{\langle c \rangle^2}{2\nu} \frac{dn}{dx}. \tag{5.3}$$

For a Maxwellian velocity distribution $\langle c \rangle = (8kT/\pi m)^{1/2}$ (Equation H.21), which yields

$$\Gamma = -\frac{4}{\pi} \frac{kT}{m\nu} \frac{dn}{dx} = -D \frac{dn}{dx} \tag{5.4}$$

where

$$D = 1.3 \frac{kT}{m\nu}. \tag{5.5}$$

Equation (5.4) is *Fick's law* and it indicates that the particle flux is proportional to the density gradient. The proportionality factor, D, is the *diffusion coefficient*. Except for the numerical factor, the simple mean-free-path analysis produces the correct form for D. A more rigorous value for the numerical factor in Equation (5.5) is given in Section 5.14.

The substitution of Fick's law (5.4) for the particle flux into the continuity equation (3.57) leads to the classical diffusion equation

$$\frac{\partial n}{\partial t} = D \frac{\partial^2 n}{\partial x^2}, \tag{5.6}$$

where, for simplicity, D is taken to be constant and the production and loss of particles are neglected. Equation (5.6) is a parabolic partial differential equation. This equation contains a first-order time derivative and a second-order spatial derivative, therefore, its solution requires one initial condition and two boundary conditions on x. An indication of how diffusion works can be obtained by considering a simple example. Assume that there is a background gas with a uniform density. Then, at $t = 0$, N particles per unit area of another gas are created at $x = 0$ on the y–z plane. The sudden appearance of these new particles corresponds to the initial condition and the boundary conditions are that the density of these new particles goes to zero as x goes to $\pm\infty$. For $t > 0$, the newly created particles diffuse away from the $x = 0$ plane through the background gas and this process is described by the diffusion coefficient. For this simple scenario, the solution to Equation (5.6) is

$$n(x, t) = \frac{N}{2(\pi Dt)^{1/2}} e^{-x^2/4Dt}. \tag{5.7}$$

Figure 5.2 shows the temporal evolution of the density profiles. Each profile is a standard Gaussian curve with the peak at $x = 0$. As t increases, the density at the peak decreases and the curve broadens. As $t \to \infty$, $n(x, t) \to 0$ for all values of x. As $t \to 0$, $n(x, t) \to 0$ for all x, except for $x = 0$ where the solution is not defined.

Another important transport property is *viscosity*, which corresponds to the transport of momentum in a direction perpendicular to the flow direction when a perpendicular velocity gradient exists. This is illustrated in Figure 5.1b. In this simple example, the gas flow is in the x-direction, but the magnitude of the velocity varies with y, $u_x(y)$. Consider the plane shown by the dashed line at an arbitrary location y, $u_x(y)$. The planes at $y + \Delta y$ and $y - \Delta y$ are on the two sides, approximately a mean-free-path away. Viscosity arises because of the thermal motion of the particles in a direction perpendicular to the flow direction. For the simple case of an isothermal, constant density gas, the particles that cross the plane at y from below carry momentum $[n\langle c \rangle /4]mu_x(y - \Delta y)$, while the particles that cross the plane from above carry momentum $[n\langle c \rangle /4]mu_x(y + \Delta y)$. If the velocities above and below the plane at y are the same, there is no net transfer of x-momentum because what is carried up balances what is carried down. However, when there is a velocity gradient, there is a net transfer of x-momentum per unit area per unit time across the plane at y and this is the *viscous stress* τ_{yx}. The x-momentum carried upward minus that carried downward is

$$\tau_{yx} = \frac{nm\langle c \rangle}{4} \left[u_x(y - \Delta y) - u_x(y + \Delta y) \right]. \tag{5.8}$$

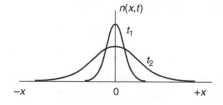

Figure 5.2 Density profiles versus x at selected times showing the effect of particle diffusion away from the origin. The profiles are obtained from Equation (5.7).

The quantity Δy is small, therefore the velocities can be expanded in a Taylor series about y:

$$\tau_{yx} = \frac{nm\langle c \rangle}{4}\left[u_x(y) - \frac{\partial u_x}{\partial y}\Delta y - u_x(y) - \frac{\partial u_x}{\partial y}\Delta y \right]$$

$$= -\frac{nm\langle c \rangle}{2}\frac{\partial u_x}{\partial y}\Delta y. \tag{5.9}$$

As before, $\Delta y \approx \lambda = \langle c \rangle / \nu$, and hence

$$\tau_{yx} = -nm\frac{\langle c \rangle^2}{2\nu}\frac{\partial u_x}{\partial y}. \tag{5.10}$$

As before, given a Maxwellian velocity distribution, $\langle c \rangle$ can be expressed in terms of the temperature T, to yield

$$\tau_{yx} = -\eta\frac{\partial u_x}{\partial y}, \tag{5.11}$$

where

$$\eta = 1.3\frac{nkT}{\nu} \tag{5.12}$$

is the *coefficient of viscosity*. As in the case of the diffusion coefficient, the simple mean-free-path analysis produces the correct form for η.

It is instructive to consider a simple scenario to see how viscosity affects a flowing gas. A classic problem is a one-dimensional flow between parallel plates. The gas flows in the x-direction and the parallel plates are at $y = 0$ and a. The plates are infinite in the x- and z-directions, and their velocities are $V_0 (y = a)$ and zero $(y = 0)$ in the x-direction. The layer of the gas near the upper plate will acquire the velocity V_0 because of friction between the upper plate and the gas, and this information will then be transmitted to the rest of the gas via viscosity. When viscosity dominates the flow, the steady state momentum equation (3.58) reduces to

$$\nabla \cdot \tau \approx 0, \tag{5.13}$$

which for this simple problem becomes (using Equation 5.11)

$$\frac{d}{dy}\left(\eta\frac{du_x}{dy} \right) = 0, \tag{5.14}$$

or

$$\frac{d^2 u_x}{dy^2} = 0 \tag{5.15}$$

when η is assumed to be constant. For the adopted boundary conditions, the solution
of Equation (5.15) is

$$u_x(y) = V_0 \frac{y}{a}. \tag{5.16}$$

The velocity displays a linear variation with y, and this is the smallest gradient that
is possible for this problem. Hence, viscosity acts to smooth velocity gradients.

The last transport coefficient that is instructional to consider is the thermal conductivity. In this case, one is interested in the flow of thermal energy per unit area
per unit time, which is the *heat flow*. The thermal energy is translational and thus
for a monatomic gas, $m\langle c^2\rangle/2 = 3kT/2$, and the flux of particles carrying this
energy is $n\langle c\rangle/4$ in the x-direction (Figure 5.1c). For a Maxwellian, this latter flux
is $n\langle c\rangle/4 = n(kT/2\pi m)^{1/2}$ (Equation H.26). Therefore, for a gas with a constant
density, the net flux of thermal energy crossing the plane at x is

$$q = \frac{3}{2\sqrt{2\pi m}} nk^{3/2} \left[T^{3/2}(x - \Delta x) - T^{3/2}(x + \Delta x) \right], \tag{5.17}$$

where the first term corresponds to those particles that come from $x - \Delta x$ and are
moving to the right, while the second term corresponds to the particles from $x + \Delta x$
that are moving to the left. As before, the temperature terms can be expanded in a
Taylor series about x

$$q = \frac{3}{2\sqrt{2\pi m}} nk^{3/2} \left(T^{3/2} - \frac{3}{2} T^{1/2} \frac{dT}{dx} \Delta x - T^{3/2} - \frac{3}{2} T^{1/2} \frac{dT}{dx} \Delta x \right)$$

$$= \frac{-9}{2\sqrt{2\pi m}} nk^{3/2} T^{1/2} \frac{dT}{dx} \Delta x. \tag{5.18}$$

With $\Delta x \approx \lambda = \langle c\rangle/\nu$, Equation (5.18) becomes

$$q = -\lambda \frac{dT}{dx}, \tag{5.19}$$

where

$$\lambda = 2.9 \frac{nk^2 T}{m\nu} \tag{5.20}$$

is the *thermal conductivity*. Again, a simple mean-free-path analysis is able to produce the correct form for the thermal conductivity. Equation (5.19) indicates that in
response to a temperature gradient, the heat is conducted from the hot to the cold
regions of the gas, which is intuitively obvious. Thermal conduction is very important in the energy balance of ionospheres, and several examples are given in later
chapters after this process has been treated more rigorously.

5.2 The five-moment approximation

In the five-moment approximation, the species velocity distribution is assumed to be adequately represented by a drifting Maxwellian (3.44). At this level of approximation, stress, heat flow, and all higher-order moments are neglected, and each species in the gas is expressed in terms of just the density, drift velocity, and temperature. The drift velocity has three components, therefore there are a total of five parameters describing each species. The spatial and temporal evolution of these five parameters is governed by the continuity, momentum, and energy equations (3.57–59). The truncation of this reduced system of transport equations is obtained by using the drifting Maxwellian velocity distribution to express the higher-order moments in terms of the lower-order moments $(n_s, \mathbf{u}_s, T_s)$. As shown in Appendix H, this procedure yields

$$\mathbf{q}_s = \tau_s = 0, \tag{5.21a}$$

$$\mathbf{P}_s = (n_s k T_s)\mathbf{I} = p_s\mathbf{I}, \tag{5.21b}$$

where $\mathbf{I}$ is the unit dyadic. Note that in the five-moment approximation, heat flow is not included and the pressure tensor is diagonal and isotropic (i.e., the three diagonal elements are the same).

As shown in Chapter 4, completely general collision terms have been derived for the five-moment approximation. These collision terms are valid for arbitrary inverse-power force laws, large temperature differences, and large relative drifts between the interacting species (4.124a–c). Using these collision terms in the continuity, momentum, and energy equations (3.57–59), and adopting the truncation (or closure) conditions (5.21a,b), the system of transport equations for the *five-moment approximation* becomes

$$\frac{\partial n_s}{\partial t} + \nabla \cdot (n_s\mathbf{u}_s) = 0, \tag{5.22a}$$

$$n_s m_s \frac{D_s\mathbf{u}_s}{Dt} + \nabla p_s - n_s m_s\mathbf{G} - n_s e_s[\mathbf{E} + \mathbf{u}_s \times \mathbf{B}]$$
$$= \sum_t n_s m_s \nu_{st}\Phi_{st}(\mathbf{u}_t - \mathbf{u}_s), \tag{5.22b}$$

$$\frac{D_s}{Dt}\left(\frac{3}{2}p_s\right) + \frac{5}{2}p_s(\nabla \cdot \mathbf{u}_s)$$
$$= \sum_t \frac{n_s m_s \nu_{st}}{m_s + m_t}\left[3k(T_t - T_s)\Psi_{st} + m_t(\mathbf{u}_s - \mathbf{u}_t)^2\Phi_{st}\right]. \tag{5.22c}$$

The five-moment approximation has significant limitations. Specifically, processes that yield anisotropic pressures, thermal diffusion, and thermal conduction are not included because heat flow and stress are not considered at this level of approximation.

5.3   Transport in a weakly ionized plasma

In many of the ionospheres, the low-altitude domain is generally characterized as a *weakly ionized gas*, in that Coulomb collisions are not important. The transport processes are dominated by electron and ion collisions with the neutral particles. In this case, the heat flow terms that appear on the right-hand side of the momentum equation (3.58; 4.129b) are absent for nonresonant ion–neutral collisions and are negligibly small for electron–neutral collisions. Under these circumstances, the momentum equation (3.58; 4.129b) for the charged particles reduces to

$$n_j m_j \left[\frac{\partial \mathbf{u}_j}{\partial t} + (\mathbf{u}_j \cdot \nabla) \mathbf{u}_j \right] + \nabla p_j + \nabla \cdot \boldsymbol{\tau}_j - n_j m_j \mathbf{G}$$

$$- e_j n_j \left[\mathbf{E} + \mathbf{u}_j \times \mathbf{B} \right] = n_j m_j \nu_{jn} (\mathbf{u}_n - \mathbf{u}_j), \tag{5.23}$$

where subscript n corresponds to neutrals and subscript j to any charged species.

In the so-called *diffusion approximation*, the inertial terms are neglected. The effect of this can be seen by comparing these terms to the pressure gradient term. Assuming that L is a characteristic scale length in the plasma, the ratio of the second and third terms in Equation (5.23) is

$$\frac{n_j m_j u_j^2 / L}{n_j k T_j / L} \sim \frac{u_j^2}{(kT_j/m_j)} \sim M_j^2, \tag{5.24}$$

where the single-species *Mach number*, M_j, is the drift speed, u_j, divided by a factor proportional to the thermal speed, $(kT_j/m_j)^{1/2}$, for species j. Therefore, the nonlinear inertial term can be neglected when $M_j^2 \ll 1$, or for *subsonic flow*. In a similar manner, the ratio of the first and third terms in Equation (5.23) is

$$\frac{n_j m_j u_j / \tau'}{n_j k T_j / L} \sim \frac{\left(\frac{L}{\tau'}\right) u_j}{(kT_j/m_j)} \sim M_j \frac{L/\tau'}{(kT_j/m_j)^{1/2}}, \tag{5.25}$$

where τ' is a characteristic time constant for the plasma. Equation (5.25) indicates that the $\partial \mathbf{u}_j / \partial t$ term can be neglected if the time constant for the plasma process is long. In practice, the neglect of the $\partial \mathbf{u}_j / \partial t$ term acts to eliminate plasma wave phenomena. Therefore, in summary, the *diffusion approximation* is valid for a slowly varying, subsonic flow.

At this point, it is instructive to consider a simple diffusion situation in which a constant electric field, $\mathbf{E}_0$, exists in a weakly ionized plasma, but $\mathbf{B}$, $\mathbf{G}$, $\boldsymbol{\tau}_j$, and $\mathbf{u}_n$ are negligible. In this case, the diffusion approximation of Equation (5.23) becomes

$$\nabla p_j - e_j n_j \mathbf{E}_0 = -n_j m_j \nu_{jn} \mathbf{u}_j. \tag{5.26}$$

For an isothermal plasma (T_j = constant), Equation (5.26) can be expressed as

$$\mathbf{\Gamma}_j = -D_j \nabla n_j \pm \bar{\mu}_j n_j \mathbf{E}_0, \tag{5.27}$$

where

$$D_j = \frac{kT_j}{m_j \nu_{jn}}, \tag{5.28}$$

$$\bar{\mu}_j = \frac{|e_j|}{m_j \nu_{jn}} \tag{5.29}$$

are the *diffusion* and *mobility coefficients*, respectively. In Equation (5.27), $\mathbf{\Gamma}_j = n_j \mathbf{u}_j$ is the particle flux and the $\pm$ signs correspond to ions and electrons, respectively. For $\mathbf{E}_0 = 0$, Equation (5.27) reduces to Fick's law,

$$\mathbf{\Gamma}_j = -D_j \nabla n_j, \tag{5.30}$$

which was derived earlier using mean-free-path considerations (5.4).

It is also instructive to consider the effects of stress and heat flow in a weakly ionized gas because they account for nonMaxwellian effects (3.49) and correspond to a higher level of approximation. Such effects are important, for example, in the terrestrial E and F regions at high latitudes, where convection electric fields induce relative ion–neutral drifts as large as several kilometers per second. The electric fields, which are directed perpendicular to the geomagnetic field, originate in the magnetosphere and are mapped down along the $\mathbf{B}$ field to the ionosphere (Section 2.3). In the E region, the dominant ion–neutral interactions are nonresonant, and therefore, the Maxwell molecule collision terms (4.89a–f) are appropriate. However, to simplify the collision terms, it is assumed that there is only one neutral species, that $m_i = m_n$, that $Q_{in}^{(2)} = Q_{in}^{(1)}$, and that the neutrals have a drifting Maxwellian velocity distribution ($\mathbf{q}_n = \tau_n = 0$). Note that these are reasonable assumptions at terrestrial E region altitudes for both NO^+ and O_2^+ ions. The momentum (3.58, 4.89b), energy (3.59, 4.89c), stress (3.60, 4.89e), and heat flow (3.61, 4.89f) equations for the simple case of a steady state, homogeneous plasma subjected to an imposed perpendicular electric field, $\mathbf{E}_\perp$, reduce to[1]

$$\frac{e_i}{m_i}(\mathbf{E}_\perp + \mathbf{u}_i \times \mathbf{B}) = \nu_{in}(\mathbf{u}_i - \mathbf{u}_n), \tag{5.31}$$

$$0 = 3k(T_n - T_i) + m_i(\mathbf{u}_i - \mathbf{u}_n)^2, \tag{5.32}$$

$$\mathbf{b} \times \tau_i - \tau_i \times \mathbf{b} + \frac{7}{4}\frac{\nu_{in}}{\omega_{c_i}}\tau_i = \frac{1}{4}\frac{\nu_{in}}{\omega_{c_i}}n_i m_i \left[(\mathbf{u}_i - \mathbf{u}_n)(\mathbf{u}_i - \mathbf{u}_n) \right.$$

$$\left. - \frac{1}{3}(\mathbf{u}_i - \mathbf{u}_n)^2 \mathbf{I} \right], \tag{5.33}$$

$$\mathbf{b} \times \mathbf{q}_i + \frac{3}{2}\frac{\nu_{in}}{\omega_{c_i}}\mathbf{q}_i = \frac{1}{2}\frac{\nu_{in}}{\omega_{c_i}}\left[\frac{5}{2}\boldsymbol{\tau}_i \cdot (\mathbf{u}_i - \mathbf{u}_n) + \frac{1}{3}n_i m_i(\mathbf{u}_i - \mathbf{u}_n)^2(\mathbf{u}_i - \mathbf{u}_n)\right],$$

$$(5.34)$$

where $\mathbf{b}$ is a unit vector directed along the geomagnetic field and $\omega_{c_i} = e_i B/m_i$ is the ion cyclotron frequency (Equation 2.7).

The momentum (5.31) and energy (5.32) equations can be readily solved and the solutions are

$$(\mathbf{u}_i - \mathbf{u}_n) = \frac{e_i}{m_i}\left(\frac{\nu_{in}}{\nu_{in}^2 + \omega_{c_i}^2}\mathbf{E}'_\perp + \frac{\omega_{c_i}}{\nu_{in}^2 + \omega_{c_i}^2}\mathbf{E}'_\perp \times \mathbf{b}\right),$$

$$(5.35)$$

$$T_i = T_n + \frac{m_i}{3k}(\mathbf{u}_i - \mathbf{u}_n)^2,$$

$$(5.36)$$

where

$$\mathbf{E}'_\perp = \mathbf{E}_\perp + \mathbf{u}_n \times \mathbf{B}.$$

$$(5.37)$$

The two terms on the right-hand side of Equation (5.35) correspond, respectively, to the *Pedersen* and *Hall* components of the relative ion–neutral drift. The parallel component of the relative drift is zero in this case because $\mathbf{E}'_\perp$ is directed perpendicular to $\mathbf{B}$ and gravity is ignored. The energy Equation (5.36) shows clearly that the ion temperature is greater than the neutral temperature, because of the frictional interactions associated with the relative ion–neutral drift. Note, however, that when the collision term dominates the energy equation and there is only one neutral species, the collision frequency drops out of Equation (5.32).

The stress (5.33) and heat flow (5.34) equations can also be readily solved by introducing a right-handed Cartesian coordinate system with unit vectors pointing in the $\mathbf{B}, \mathbf{E}'_\perp \times \mathbf{B}$, and $\mathbf{E}'_\perp$ directions, respectively.[1] Here, however, it is instructive to consider only the two limiting cases of strong ($\nu_{in}/\omega_{c_i} \to \infty$) and weak ($\nu_{in}/\omega_{c_i} \to 0$) collisions. The effect of the magnetic field is negligible for strong collisions, and the solutions of the stress tensor (5.33) and heat flow (5.34) equations are

$$\boldsymbol{\tau}_i = \frac{1}{7}n_i m_i\left[(\mathbf{u}_i - \mathbf{u}_n)(\mathbf{u}_i - \mathbf{u}_n) - \frac{1}{3}(\mathbf{u}_i - \mathbf{u}_n)^2\mathbf{I}\right],$$

$$(5.38)$$

$$\mathbf{q}_i = \frac{4}{21}n_i m_i(\mathbf{u}_i - \mathbf{u}_n)^2(\mathbf{u}_i - \mathbf{u}_n).$$

$$(5.39)$$

Note that both a stress and heat flow can develop in a *weakly ionized homogeneous plasma* due to a relative ion–neutral drift. The magnitude of the stress tensor is proportional to $|\mathbf{u}_i - \mathbf{u}_n|^2$, while the magnitude of the heat flow is proportional to $|\mathbf{u}_i - \mathbf{u}_n|^3$.

In the small collision frequency limit, all of the components of $\boldsymbol{\tau}_i$ and $\mathbf{q}_i$, except the parallel components, can be obtained from Equations (5.33) and (5.34) by setting

$\nu_{in}/\omega_{c_i} = 0$, which yields

$$\mathbf{b} \times \boldsymbol{\tau}_i - \boldsymbol{\tau}_i \times \mathbf{b} = 0, \tag{5.40}$$

$$\mathbf{b} \times \mathbf{q}_i = 0. \tag{5.41}$$

Equation (5.41) indicates that the heat flow perpendicular to $\mathbf{B}$ goes to zero as $\nu_{in}/\omega_{c_i} \to 0$. Equation (5.40) indicates that the stress tensor is diagonal, and using the fact that the sum of the diagonal elements is zero (3.21), the solution to Equation (5.40) can be expressed in the form

$$\boldsymbol{\tau}_i = \tau_{i\parallel}\mathbf{bb} + \tau_{i\perp}(\mathbf{I} - \mathbf{bb}), \tag{5.42}$$

where

$$\tau_{i\perp} = -\frac{1}{2}\tau_{i\parallel} \tag{5.43}$$

and where the subscripts $\parallel$ and $\perp$ denote components parallel and perpendicular to $\mathbf{B}$, respectively. Therefore, in the limit $\nu_{in}/\omega_{c_i} \to 0$, the stress tensor becomes isotropic in the plane perpendicular to $\mathbf{B}$. The parallel components of $\boldsymbol{\tau}_i$ and $\mathbf{q}_i$ are obtained by taking the parallel components of Equations (5.33) and (5.34), respectively, which yield

$$\tau_{i\parallel} = -\frac{1}{21}n_i m_i(\mathbf{u}_i - \mathbf{u}_n)^2, \tag{5.44}$$

$$q_{i\parallel} = 0. \tag{5.45}$$

Thus, in the collisionless limit, $\mathbf{q}_i = 0$ (Equations 5.41, 5.45) for the case considered.

The above analysis indicates that in general a relative ion–neutral drift in a weakly ionized plasma induces both a stress and heat flow, and these processes account for the deviations from the zeroth-order drifting Maxwellian distribution (3.49). The effect of heat flow is to cause an asymmetric velocity distribution. For example, if the ions drift in the x-direction with a bulk velocity, $\mathbf{u}_0$, and there is also a heat flow present in the x-direction due to a relative ion-neutral drift (5.39), the velocity distribution takes the asymmetric form shown in Figure 5.3a. Relative to the drifting Maxwellian, the effect of a positive x-directed heat flow is to remove particles from the tail in the minus v_x-direction and increase the number of particles in the $+v_x$ tail, which acts to produce an asymmetric velocity distribution along the v_x-axis.

The effect of the stress tensor is to distort the isotropic pressure distribution that is characteristic of a drifting Maxwellian (5.21b). For example, in the limit of $\nu_{in}/\omega_{c_i} \to 0$, the stress tensor (5.42) is diagonal, but anisotropic. Therefore, the pressure tensor, $\mathbf{P}_i = p_i\mathbf{I} + \boldsymbol{\tau}_i$, is also anisotropic, which means that there are different pressures (or temperatures) parallel and perpendicular to $\mathbf{B}$. Using Equations (3.70) and (3.71), which relate the parallel and perpendicular temperatures to the stress

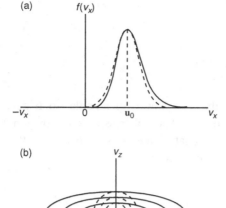

Figure 5.3 (a) Velocity distribution with a bulk drift, $\mathbf{u}_0$, and a heat flow in the x-direction (solid curve) and the corresponding unmodified drifting Maxwellian distribution (dashed curve). (b) Contours of an anisotropic velocity distribution with a bulk drift in the z-direction and an enhanced temperature in the x-direction (solid curves). The dashed curves are for a drifting Maxwellian distribution.

tensor, and the expressions for $\tau_{i\parallel}$ (5.44) and $\tau_{i\perp}$ (5.43), the temperatures can be expressed as

$$T_{i\parallel} = T_i - \frac{1}{21}\frac{m_i}{k}(\mathbf{u}_i - \mathbf{u}_n)^2, \tag{5.46}$$

$$T_{i\perp} = T_i + \frac{1}{42}\frac{m_i}{k}(\mathbf{u}_i - \mathbf{u}_n)^2. \tag{5.47}$$

Figure 5.3b shows the effect of an anisotropic stress tensor on the ion velocity distribution for the case when the ions drift along $\mathbf{B}$ with a velocity $\mathbf{u}_0$ and $T_{i\perp} > T_{i\parallel}$. Note that the thermal spread (width of the distribution) perpendicular to $\mathbf{B}$ is greater than that parallel to $\mathbf{B}$.

For simplicity, in the analysis presented above, it was assumed that the plasma was homogeneous, that steady-state conditions prevailed, that gravity was negligible, and that there was only one neutral species with $m_n = m_i$ and $Q_{in}^{(2)} = Q_{in}^{(1)}$. As it turns out, the plasma in the high-latitude terrestrial E region (Figure 2.16) is basically homogeneous in the direction perpendicular to $\mathbf{B}$, but spatial variations along $\mathbf{B}$ are present and in general are important. Also, at these lower altitudes, the *diffusion approximation* is valid (Chapter 12). Taking these facts into account and dropping the above simplifying assumptions, the transport equations that are appropriate for the high-latitude terrestrial E region are

$$\frac{\partial n_i}{\partial t} + \frac{\partial}{\partial r}(n_i u_{i\parallel}) = \frac{\delta n_i}{\delta t}, \tag{5.48a}$$

$$\frac{\partial p_i}{\partial r} + \frac{\partial \tau_{i\parallel}}{\partial r} + n_i m_i g_\parallel - n_i e_i E_\parallel = n_i m_i \sum_n \nu_{in}(u_n - u_i)_\parallel, \tag{5.48b}$$

$$e_i\left[\mathbf{E}_\perp + \mathbf{u}_{i\perp} \times \mathbf{B}\right] = m_i \sum_n \nu_{in}(\mathbf{u}_i - \mathbf{u}_n)_\perp, \tag{5.48c}$$

$$0 = \sum_n \frac{\nu_{in}}{m_i + m_n}\left[3k(T_n - T_i) + m_n(\mathbf{u}_i - \mathbf{u}_n)^2\right], \tag{5.48d}$$

where r is the spatial coordinate along $\mathbf{B}$, $E_\parallel$ and $\mathbf{E}_\perp$ are the components of the electric field parallel and perpendicular to $\mathbf{B}$, respectively, $g_\parallel$ is the component of gravity along $\mathbf{B}$, and $\delta n_i/\delta t$ accounts for the production and loss of ionization, which is discussed in Chapter 9.

The parallel component of the stress tensor, $\tau_{i\parallel}$, has a general form that is similar to the spatially homogeneous expression (5.38) because the main component of $(\mathbf{u}_i - \mathbf{u}_n)$ is primarily perpendicular to $\mathbf{B}$, which is the direction where the plasma is homogeneous. If the various neutral species have displaced Maxwellian velocity distributions with a common temperature and common drift velocity, the general expression for $\tau_{i\parallel}$ is given by[1]

$$\tau_{i\parallel} = \frac{R_i}{S_i} n_i m_i\left[(\mathbf{u}_i - \mathbf{u}_n)_\parallel^2 - \frac{1}{3}(\mathbf{u}_i - \mathbf{u}_n)^2\right], \tag{5.49a}$$

where

$$R_i = \sum_n \frac{m_n \nu_{in}}{m_i + m_n}\left[1 - \frac{3}{4}\frac{Q_{in}^{(2)}}{Q_{in}^{(1)}}\right], \tag{5.49b}$$

$$S_i = \sum_n \frac{m_i \nu_{in}}{m_i + m_n}\left[1 + \frac{3}{4}\frac{m_n}{m_i}\frac{Q_{in}^{(2)}}{Q_{in}^{(1)}}\right]. \tag{5.49c}$$

For most ionospheric applications, the contribution of $(\mathbf{u}_i - \mathbf{u}_n)_\parallel^2$ to $\tau_{i\parallel}$ can be neglected. For a mixture of NO^+ or O_2^+ with either N_2 or O_2, $R_i/S_i \approx \frac{1}{4}$. For a gas mixture composed of O^+ and either N_2 or O_2, $R_i/S_i \approx \frac{1}{3}$.

5.4 Transport in partially and fully ionized plasmas

In the previous section, the transport equations that are applicable to ionospheric regions where Coulomb collisions are negligible were discussed. In the rest of this chapter, the effect of Coulomb collisions on the transport processes will no longer be neglected, and their inclusion leads to some interesting new phenomena. Also, for some of the ionospheres, the rotation of the planet is sufficiently fast that centripetal acceleration and the Coriolis force become significant at the altitudes where Coulomb collisions are important. Under these circumstances, it is customary to adopt a coordinate system that is fixed to the rotating planet, which introduces Coriolis and centripetal acceleration terms in the momentum equation (Chapter 10). Although these latter processes are neglected in the derivations that follow, it is useful to list the general momentum equation here both for future reference and so that the reader can clearly see what processes are neglected in the various sets of simplified transport equations that will be presented. Therefore, in a *rotating reference frame*, the momentum equation (3.58, 4.129b) for the charged particles is given by

$$\rho_s \frac{D_s \mathbf{u}_s}{Dt} + \nabla p_s + \nabla \cdot \boldsymbol{\tau}_s - n_s e_s (\mathbf{E} + \mathbf{u}_s \times \mathbf{B})$$

$$+ \rho_s \left[-\mathbf{G} + 2\boldsymbol{\Omega}_r \times \mathbf{u}_s + \boldsymbol{\Omega}_r \times (\boldsymbol{\Omega}_r \times \mathbf{r}) \right]$$

$$= \sum_t n_s m_s \nu_{st} (\mathbf{u}_t - \mathbf{u}_s) + \sum_t \nu_{st} \frac{z_{st} \mu_{st}}{k T_{st}} \left(\mathbf{q}_s - \frac{\rho_s}{\rho_t} \mathbf{q}_t \right), \tag{5.50}$$

where the linear collision terms are adopted (4.129b) and where $\boldsymbol{\Omega}_r$ is the planet's angular velocity and $\mathbf{r}$ is the radius vector from the center of the planet.

Equation (5.50) is very general and can be used to describe a wide range of transport processes. However, at the altitudes where the ionospheres are partially ionized, the momentum equation can usually be simplified because the diffusion approximation is valid. To demonstrate this, it is convenient to consider the strongly magnetized planets at middle and high magnetic latitudes, where the **B** field is nearly vertical. Above some altitude, approximately 160 km for the Earth, the ion and electron collision frequencies are much smaller than the corresponding cyclotron frequencies and, as a consequence, the plasma is constrained to move along the **B** field like beads on a string. In certain regions, electric fields can cause the entire ionosphere to convect horizontally across the magnetic field, but this latter motion is distinct from the field-aligned motion and the two can simply be added vectorially. The field-aligned motion is influenced by gravity, as well as by density and temperature gradients. Owing to the small electron mass, gravity causes a slight charge separation, with the lighter electrons tending to settle on top of the heavier ions. This slight charge separation results in a *polarization electrostatic field*, which prevents a further charge separation. After this electrostatic field develops, the ions and electrons move together as a single gas under the influence of gravity and the density and temperature gradients. Such a motion is called *ambipolar diffusion*.

It is useful to distinguish between major and minor ions before deriving the ambipolar diffusion equation. A *major ion* is a species whose density is comparable to the electron density, and consequently, it is important in maintaining the overall charge neutrality in the plasma. A *minor ion*, on the other hand, is essentially a trace species whose density is much smaller than that of the electrons, and hence, its contribution to the charge neutrality is negligibly small. In what follows, ambipolar diffusion equations will be derived for both major and minor ions.

5.5   Major ion diffusion

In the diffusion approximation, wave phenomena are not considered ($\partial \mathbf{u}_s / \partial t \to 0$) and the flow is subsonic ($\mathbf{u}_s \cdot \nabla \mathbf{u}_s \to 0$). Also, because the ions and electrons move together, *charge neutrality* ($n_e = n_i$) and *zero current* ($n_e\mathbf{u}_e = n_i\mathbf{u}_i$) conditions prevail, where it is assumed that the plasma contains major ions, electrons, and, for convenience, one neutral species. In addition, for a partially ionized plasma, the heat flow terms in Equation (5.50) are small and will be ignored for now, as will the Coriolis and centripetal accleration terms. With these assumptions, the ion and electron momentum equation (5.50) along the magnetic field reduce to

$$\nabla_\parallel p_i + (\nabla \cdot \boldsymbol{\tau}_i)_\parallel - n_i e \mathbf{E}_\parallel - n_i m_i \mathbf{G}_\parallel$$

$$= n_i m_i \nu_{ie}(\mathbf{u}_e - \mathbf{u}_i)_\parallel + n_i m_i \nu_{in}(\mathbf{u}_n - \mathbf{u}_i)_\parallel, \tag{5.51}$$

$$\nabla_\parallel p_e + (\nabla \cdot \boldsymbol{\tau}_e)_\parallel + n_e e \mathbf{E}_\parallel - n_e m_e \mathbf{G}_\parallel$$

$$= n_e m_e \nu_{ei}(\mathbf{u}_i - \mathbf{u}_e)_\parallel + n_e m_e \nu_{en}(\mathbf{u}_n - \mathbf{u}_e)_\parallel, \tag{5.52}$$

where $\mathbf{E}_\parallel$ is the polarization electrostatic field that develops because of the very slight charge separation. Letting $n_e = n_i$, $\mathbf{u}_e = \mathbf{u}_i$, and using the fact that $n_i m_i \nu_{ie} = n_e m_e \nu_{ei}$, (4.158), the addition of Equations (5.51) and (5.52) yields

$$\nabla_\parallel (p_e + p_i) + (\nabla \cdot \boldsymbol{\tau}_i)_\parallel + (\nabla \cdot \boldsymbol{\tau}_e)_\parallel - n_i(m_i + m_e)\mathbf{G}_\parallel$$

$$= n_i(m_i \nu_{in} + m_e \nu_{en})(\mathbf{u}_n - \mathbf{u}_i)_\parallel. \tag{5.53}$$

In Equation (5.53), $m_e \nu_{en} \ll m_i \nu_{in}$ because of the small electron mass (see Section 4.8). Likewise, $\boldsymbol{\tau}_e$ is much smaller than $\boldsymbol{\tau}_i$ because the stress tensor is proportional to the particle mass (5.38). Neglecting terms that contain the electron mass, and setting $p_e = n_e k T_e$ and $p_i = n_i k T_i$, Equation (5.53) reduces to the *ambipolar diffusion equation*,

$$\mathbf{u}_{i\parallel} = \mathbf{u}_{n\parallel} - D_a\left[\frac{1}{n_i}\nabla_\parallel n_i + \frac{1}{T_p}\nabla_\parallel T_p - \frac{m_i \mathbf{G}_\parallel}{2kT_p} + \frac{(\nabla \cdot \boldsymbol{\tau}_i)_\parallel}{2n_i k T_p}\right], \tag{5.54}$$

where the *ambipolar diffusion coefficient* (D_a) and *plasma temperature* (T_p) are given by

$$D_a = \frac{2kT_p}{m_i \nu_{in}}, \tag{5.55}$$

$$T_p = \frac{T_e + T_i}{2}. \tag{5.56}$$

Equation (5.54) applies along the magnetic field for strongly magnetized ionospheres, and it also applies in the vertical direction for unmagnetized ionospheres. Letting r correspond to the spatial coordinate either along $\mathbf{B}$ or in the vertical direction for the unmagnetized case, Equation (5.54) can also be expressed in the form

$$\frac{1}{n_i}\frac{\partial n_i}{\partial r} = -\frac{m_i g}{2kT_p} - \frac{1}{T_p}\frac{\partial T_p}{\partial r} - \frac{\partial \tau_{i\parallel}/\partial r}{2n_i k T_p} + \frac{(u_n - u_i)}{D_a}, \tag{5.57}$$

where $\mathbf{G}_\parallel = -g\mathbf{e}_r$. Note that $\nu_{in} \propto n_n$ and thus $D_a \propto 1/n_n$. Therefore, D_a increases exponentially with altitude because the neutral density decreases exponentially with altitude (Figure 2.14). As a consequence, the last term in Equation (5.57) rapidly becomes unimportant as altitude increases. If the stress term is also neglected, Equation (5.57) reduces to the classical *diffusive equilibrium equation*

$$\frac{1}{n_i}\frac{\partial n_i}{\partial r} = -\frac{1}{H_p} - \frac{1}{T_p}\frac{\partial T_p}{\partial r}, \tag{5.58}$$

where H_p is the *plasma scale height*

$$H_p = \frac{2kT_p}{m_i g}. \tag{5.59}$$

Equation (5.58) can be easily integrated for an isothermal ionosphere ($T_p =$ constant), and if the variation of gravity with altitude is ignored, the integration yields

$$n_i = (n_i)_0 e^{-(r-r_0)/H_p}, \tag{5.60}$$

where the subscript 0 corresponds to some reference altitude. Therefore, in the diffusive equilibrium region, the major ion (or electron) density decreases exponentially with altitude at a rate governed by the plasma scale height (Figure 2.16).

5.6   Polarization electrostatic field

In the derivation of the ambipolar diffusion equation, the ion (5.51) and electron (5.52) momentum equations were added and the polarization electrostatic field dropped out. However, in many applications an explicit expression for this electric field is needed. Basically, it is the electron motion that leads to the creation of this field, and hence, it can be obtained from the electron momentum equation (5.52). Neglecting the terms that contain m_e, the polarization electrostatic field effectively becomes

$$e\mathbf{E}_\parallel = -\frac{1}{n_e}\mathbf{\nabla}_\parallel p_e. \tag{5.61}$$

This expression is valid regardless of the number of ion species in the plasma because all of the electron–ion collision terms drop out owing to the small electron mass.

Equation (5.61) can be expressed in a convenient, alternate form for the special case of an isothermal electron gas. Letting $\mathbf{E}_\parallel = -\mathbf{\nabla}_\parallel \Phi$, where Φ is the *electrostatic potential*, and assuming that T_e is constant, Equation (5.61) becomes

$$\frac{e}{kT_e}\frac{\partial \Phi}{\partial r} = \frac{1}{n_e}\frac{\partial n_e}{\partial r}, \tag{5.62}$$

where, as before, r is the spatial coordinate either along $\mathbf{B}$ or in the vertical direction, depending on whether the planet is magnetized or not. Equation (5.62) can be easily integrated to obtain the well-known *Boltzmann relation*

$$n_e = (n_e)_0 e^{e\Phi/kT_e}, \tag{5.63}$$

where $(n_e)_0$ is the equilibrium electron density that prevails when $\Phi = 0$. The Boltzmann relation is widely used in both plasma physics and space physics, but it must be remembered that it is derived from a simplified electron momentum Equation (5.52) that does not contain the inertial terms. Therefore, in addition to the isothermal restriction, it is also restricted to slowly varying phenomena and subsonic electron drifts.

As noted above, ambipolar diffusion occurs as a result of the polarization electrostatic field that develops in response to a slight electron–ion charge separation. This electric field is established very rapidly by the electrons, before the ions have time to move. An estimate of the distance over which charge separation occurs can be obtained with the aid of the Boltzmann relation. Consider a plasma that is initially neutral ($n_e = n_i = n_0$). Subsequently, the electrons move a small distance away and a polarization electric field, $\mathbf{E}^*$, is established, which is governed by Gauss' law (3.76a)

$$\nabla \cdot \mathbf{E}^* = e(n_i - n_e)/\varepsilon_0. \tag{5.64}$$

For this electrostatic field, $\mathbf{E}^* = -\nabla\Phi^*$, and hence, Gauss' law becomes the *Poisson equation*

$$\nabla^2\Phi^* = -e(n_i - n_e)/\varepsilon_0. \tag{5.65}$$

The ions are unperturbed because they do not have time to move, so $n_i = n_0$. The electron density, on the other hand, does change and it is described by the Boltzmann relation with $(n_e)_0 = n_0$. For a small charge separation, the potential energy, $e\Phi^*$, is much smaller than the electron thermal energy, kT_e, and therefore, the exponential in Equation (5.63) can be expanded for a small argument, which yields

$$n_e = n_0\left(1 + \frac{e\Phi^*}{kT_e}\right), \tag{5.66}$$

where only the first two terms in the series expansion are retained. Substituting $n_i = n_0$ and the electron density (5.66) into the Poisson equation (5.65) yields

$$\nabla^2\Phi^* = \Phi^*/\lambda_D^2, \tag{5.67}$$

where $\lambda_D = (\varepsilon_0 kT_e/n_0 e^2)^{1/2}$ is the *electron Debye length* that was introduced earlier (Equation 2.4). With one spatial dimension, say x, the solution to Equation (5.67) is

$$\Phi^* = c_0 e^{-|x|/\lambda_D}, \tag{5.68}$$

where c_0 is an integration constant. This solution indicates that the polarization electrostatic field is established over a distance of about λ_D, which for the ionospheres is of the order of a few centimeters (Table 2.6). Therefore, ambipolar diffusion applies over distances greater than a few centimeters.

In summary, the polarization electrostatic field (5.61) exists at all altitudes where the diffusion approximation is valid. At all altitudes, there is a slight electron–ion charge separation that occurs over a distance of about λ_D, which is a few centimeters in the ionospheres.

5.7   Minor ion diffusion

The diffusion equation for a minor ion species in a plasma composed primarily of major ions, electrons, and neutrals can be obtained from the general momentum equation (5.50). As with the major ion, the diffusion approximation implies that the inertial terms are negligibly small, and if the Coriolis force, centripetal acceleration, and heat flow terms are also neglected, the momentum equation for the minor ion (subscript ℓ) reduces to

$$\nabla_\parallel p_\ell + (\nabla \cdot \boldsymbol{\tau}_\ell)_\parallel - n_\ell e_\ell \mathbf{E}_\parallel - n_\ell m_\ell \mathbf{G}_\parallel$$

$$= n_\ell m_\ell \big[\nu_{\ell e}(\mathbf{u}_e - \mathbf{u}_\ell)_\parallel + \nu_{\ell n}(\mathbf{u}_n - \mathbf{u}_\ell)_\parallel + \nu_{\ell i}(\mathbf{u}_i - \mathbf{u}_\ell)_\parallel \big], \qquad (5.69)$$

where, as before, this equation applies either along $\mathbf{B}$ for strongly magnetized ionospheres or in the vertical direction for unmagnetized ionospheres. The momentum exchange between the minor ions and electrons is negligible because of the small electron mass. Also, collisions with the neutrals are usually negligible compared with collisions with the major ions because of the long-range nature of Coulomb collisions ($\nu_{\ell i} \gg \nu_{\ell n}$). Neglecting these collision terms and setting $p_\ell = n_\ell k T_\ell$, Equation (5.69) can be expressed in the form

$$\mathbf{u}_{\ell\parallel} = \mathbf{u}_{i\parallel} - D_\ell \left[\frac{1}{n_\ell}\nabla_\parallel n_\ell + \frac{1}{T_\ell}\nabla_\parallel T_\ell - \frac{m_\ell \mathbf{G}_\parallel}{kT_\ell} - \frac{e_\ell \mathbf{E}_\parallel}{kT_\ell} + \frac{(\nabla \cdot \boldsymbol{\tau}_\ell)_\parallel}{n_\ell kT_\ell} \right],$$

$$(5.70)$$

where the *minor ion diffusion coefficient*, D_ℓ, is given by

$$D_\ell = \frac{kT_\ell}{m_\ell \nu_{\ell i}}. \qquad (5.71)$$

Equation (5.70) indicates that the major ions affect the minor ions in three ways. First, as the major ions diffuse along $\mathbf{B}$, they tend to drag the minor ions with them. Also, when the minor ions try to diffuse in response to their density and temperature gradients, their motion is impeded by collisions with the major ions. Finally, the polarization electrostatic field that appears in Equation (5.70) is established by the charge separation between the major ions and electrons. Using Equation (5.61) for

$\mathbf{E}_\parallel$ and setting $p_e = n_e k T_e$, Equation (5.70) takes the classical form for the minor ion, ambipolar diffusion equation;

$$\mathbf{u}_{\ell\parallel} = \mathbf{u}_{i\parallel} - D_\ell \left[\frac{1}{n_\ell} \nabla_\parallel n_\ell + \frac{1}{T_\ell} \nabla_\parallel (T_\ell + T_e) - \frac{m_\ell \mathbf{G}_\parallel}{kT_\ell} \right.$$

$$\left. + \frac{T_e}{T_\ell n_e} \nabla_\parallel n_e + \frac{(\nabla \cdot \boldsymbol{\tau}_\ell)_\parallel}{n_\ell k T_\ell} \right]. \tag{5.72}$$

The characteristic solutions for a minor ion species can be illustrated by assuming that steady-state conditions prevail, the ionosphere is isothermal, the variation of gravity with altitude is negligible, and stress effects are unimportant. With these assumptions, the scalar version of Equation (5.72) can be written as

$$n_\ell u_\ell = n_\ell u_i - D_\ell \left[\frac{dn_\ell}{dr} + n_\ell \left(\frac{1}{H_\ell} + \frac{T_e}{T_\ell} \frac{1}{n_e} \frac{dn_e}{dr} \right) \right], \tag{5.73}$$

where r is the spatial coordinate, as before, and H_ℓ is the *minor ion scale height*, given by

$$H_\ell = \frac{kT_\ell}{m_\ell g}. \tag{5.74}$$

As altitude increases, the major ion velocity $u_i \to 0$, and its density distribution becomes a diffusive equilibrium distribution (Equation 5.58). Also, ionization and chemical reactions are not important for the minor ion at high altitudes, and therefore, its steady state continuity equation reduces to $d(n_\ell u_\ell)/dr = 0$ or $n_\ell u_\ell = F_\ell$, where F_ℓ is a constant. With this information, Equation (5.73) becomes

$$F_\ell = -D_\ell \left[\frac{dn_\ell}{dr} + n_\ell \left(\frac{1}{H_\ell} - \frac{T_e}{T_\ell H_p} \right) \right]. \tag{5.75}$$

Taking the derivative of Equation (5.75), bearing in mind that F_ℓ, H_ℓ, H_p, T_ℓ, and T_e are assumed to be constant, one obtains the following second-order differential equation for n_ℓ:

$$\frac{d^2 n_\ell}{dr^2} + \left[\frac{1}{H_p} + \left(\frac{1}{H_\ell} - \frac{T_e}{T_\ell H_p} \right) \right] \frac{dn_\ell}{dr} + \left(\frac{1}{H_\ell} - \frac{T_e}{T_\ell H_p} \right) \frac{1}{H_p} n_\ell = 0, \tag{5.76}$$

where use was made of the fact that

$$\frac{1}{D_\ell} \frac{dD_\ell}{dr} = \frac{1}{H_p}. \tag{5.77}$$

The latter result follows from Equation (5.71), which shows that $D_\ell \propto 1/\nu_{\ell i} \propto 1/n_i$. However, n_i decreases exponentially with altitude at a rate governed by H_p (5.60), and hence, D_ℓ increases exponentially with altitude at this rate.

The two linearly independent solutions of the minor ion equation (5.76) are

$$
n_\ell = (n_\ell)_0 \exp\left[\left(\frac{T_e}{T_\ell H_p} - \frac{1}{H_\ell}\right)(r - r_0)\right], \tag{5.78a}
$$

$$
n_\ell = (n_\ell)_0 \exp\left(-\frac{r - r_0}{H_p}\right), \tag{5.78b}
$$

where r_0 is a reference altitude and $(n_\ell)_0$ is the minor ion density at this altitude. The general solution for n_ℓ is a linear combination of solutions (5.78a) and (5.87b), with appropriate integration constants. However, the minor ion behavior can be better understood by separately examining the two linearly independent solutions.

The first solution (5.78a) corresponds to *diffusive equilibrium* for a minor ion in the presence of major ions and electrons. If $T_e \sim T_i \sim T_\ell = T$, then

$$
\left(\frac{T_e}{T_\ell H_p} - \frac{1}{H_\ell}\right) \sim \frac{g}{kT}\left(\frac{m_i}{2} - m_\ell\right). \tag{5.79}
$$

For heavy minor ions ($m_\ell > m_i/2$), this quantity is negative and the minor ion density (5.78a) decreases exponentially with altitude above the reference level. On the other hand, for light minor ions ($m_\ell < m_i/2$), the quantity in Equation (5.79) is positive and the minor ion density (5.78a) increases exponentially with altitude above the reference level. The solution is valid up to the altitude where species ℓ is no longer a minor ion. The physical reason for this behavior can be understood by recognizing the fact that $E_\parallel$ is controlled by the major ions and electrons. The magnitude of this field is such as to counterbalance the gravitational force on the major ions and keep them from separating from the much lighter electrons. This means that minor ion species that are lighter than $m_i/2$ will experience a net upward force.

The second solution (5.78b) indicates that the minor ion density decreases exponentially with altitude with the same scale height as the major ion. This solution corresponds to the *maximum upward flow* of the minor ion that the plasma will sustain. The upward flow velocity increases exponentially with altitude at the same rate that the density decreases with altitude, because $n_\ell u_\ell = F_\ell = $ constant. For this solution, the minor ion always remains minor. However, at some altitude the flow becomes supersonic, and hence, the neglect of the nonlinear inertial term in the momentum equation is no longer justified.

5.8   Supersonic ion outflow

The field lines near the magnetic poles of planets with intrinsic magnetic fields extend deep into space in an antisunward direction. Along these so-called open field lines, thermal ions and electrons can escape the topside ionosphere. The outflow begins at low altitudes, but as the ions diffuse upward their speed increases and eventually the flow becomes supersonic. The nonlinear inertial term in the momentum equation (5.50) must be retained for supersonic ion outflow and the situation becomes more complex. To illustrate this case, it is convenient to make the following simplifying assumptions: (a) there is only one ion species; (b) the flow is ambipolar ($n_i = n_e$, $u_i = u_e$); (c) the ionosphere is isothermal; (d) steady-state conditions prevail; (e) the neutrals are stationary; and (f) the stress, heat flow, Coriolis, and centripetal acceleration terms are not important.

The ion momentum equation along the **B** field reduces, with the above mentioned assumptions, to

$$n_i m_i u_i \frac{du_i}{dr} + k(T_e + T_i)\frac{dn_i}{dr} + n_i m_i g = -n_i m_i \nu_{in} u_i, \tag{5.80}$$

where Equation (5.61) was used for the polarization electrostatic field and where the ambipolar flow assumption was also employed. Equation (5.80) can be expressed in the following form:

$$u_i \frac{du_i}{dr} + \frac{V_S^2}{n_i}\frac{dn_i}{dr} + g = -\nu_{in} u_i, \tag{5.81}$$

where

$$V_S = \left[\frac{k(T_e + T_i)}{m_i}\right]^{1/2} \tag{5.82}$$

is the *ion-acoustic speed*. The density gradient in Equation (5.81) can be related to the velocity gradient with the aid of the continuity equation. In the steady state case, assuming no sources or sinks, this equation is simply given by

$$\mathbf{\nabla} \cdot (n_i \mathbf{u}_i) = \frac{1}{A}\frac{d}{dr}(An_i u_i) = 0, \tag{5.83}$$

where the divergence is taken in a curvilinear coordinate system and A is the cross-sectional area of the flux tube (see Section 11.1). For radial outflow in a spherical geometry (e.g., solar wind), $A \sim r^2$; whereas for ion outflow along dipolar field lines near the magnetic pole (e.g., polar wind), $A \sim r^3$ (Appendix B). Using Equation (5.83), the density gradient can be expressed as

$$\frac{1}{n_i}\frac{dn_i}{dr} = -\frac{1}{u_i}\frac{du_i}{dr} - \frac{1}{A}\frac{dA}{dr}, \tag{5.84}$$

and the substitution of this result into Equation (5.80) yields

$$(u_i^2 - V_S^2)\frac{1}{u_i}\frac{du_i}{dr} - \frac{V_S^2}{A}\frac{dA}{dr} + g = -\nu_{in} u_i. \tag{5.85}$$

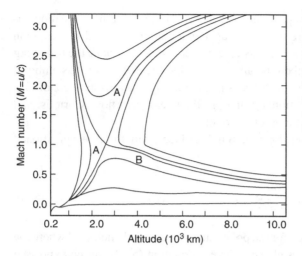

Figure 5.4 Schematic diagram showing the possible solutions to the Mach number equation for H^+ outflow in the terrestrial polar wind. Curve B corresponds to subsonic flow, and curve A corresponds to the solution that exhibits a transition from subsonic to supersonic outflow.[2]

With the introduction of the *ion-acoustic Mach number*,

$$M = \frac{u_i}{V_S},$$
(5.86)

Equation (5.85) can be cast in the following form:

$$\frac{dM}{dr} = \frac{M}{M^2 - 1}\left(\frac{1}{A}\frac{dA}{dr} - \frac{g}{V_S^2} - \frac{\nu_{in}}{V_S}M\right).$$
(5.87)

This equation corresponds to a first-order, nonlinear, ordinary differential equation for the Mach number. Note that the equation contains singularities at $M = \pm 1$, at the points of transition from subsonic to supersonic flow in the upward ($M = 1$) or downward ($M = -1$) directions.

Figure 5.4 shows schematically the different solutions that are possible for an outflow situation. The solutions are presented in a Mach number versus altitude format. All of the solutions that remain subsonic ($M < 1$) at all altitudes are possible physical solutions. The Mach number (flow velocity) is small at low altitudes for these solutions, increases to a peak value that is less than unity, and then decreases to a small value at high altitudes. On the other hand, for supersonic flow, only the critical solution (labeled A) is a physical solution. For this case, the ion flow is subsonic at low altitudes, passes through the singularity point $M = 1$, and then is supersonic at high altitudes. Which solution prevails is determined by the pressure difference between high and low altitudes.

Additional insight concerning the subsonic versus supersonic nature of the flow can be gained by examining the sign of the terms in Equation (5.87). At low altitudes in the terrestrial ionosphere the flow is upward and subsonic ($0 < M < 1$), and hence, $M/(M^2 - 1)$ is negative. Also, at low altitudes, gravity dominates and the sum of the terms in the curved brackets is negative. The net result is that $dM/dr > 0$ and the Mach number (flow velocity) increases with altitude. As altitude increases,

$v_{in} \to 0$ and gravity ($g \sim 1/r^2$) decreases more rapidly than the area term ($\frac{1}{A}\frac{dA}{dr} \sim \frac{3}{r}$), which means that at some altitude the terms in the curved brackets will change sign and become positive. If M is still less than unity at this altitude, dM/dr becomes negative and the Mach number (flow velocity) then decreases with altitude. This behavior corresponds to the subsonic solution labeled B in Figure 5.4. On the other hand, if M becomes greater than unity at the altitude where the sum of the terms in the curved brackets becomes positive, then $dM/dr > 0$, as it is at low altitudes, and the Mach number (flow velocity) continues to increase. This situation corresponds to the supersonic solution labeled A in Figure 5.4. As noted above, which solution prevails is determined by the pressure difference between high and low altitudes. In the terrestrial ionosphere, both types of flow occur.

The supersonic flow described above is similar to what occurs in a *Lavalle rocket nozzle*. In this case, an initially subsonic flow [$M/(M^2 - 1) < 0$] enters a converging nozzle ($dA/dr < 0$), which yields a positive dM/dr. When the flow just passes the sonic point [$M/(M^2 - 1) > 0$], the nozzle is designed to diverge ($dA/dr > 0$), and hence, dM/dr remains positive. The net result is a smooth transition from subsonic to supersonic flow. In the solar and terrestrial polar winds, gravity acts as the convergent nozzle in the subsonic flow regime, and the diverging magnetic field acts as the divergent nozzle in the supersonic regime. In the case of neutral gas outflow from comets, the gas–dust friction acts as the convergent nozzle and the spherical expansion acts as the divergent nozzle.

As a final issue concerning the transition from subsonic to supersonic flow, it should be noted that the singularity in Equation (5.87) arises only because the time derivative and stress terms in the momentum equation were neglected when Equation (5.87) was derived. When these terms are included, the singularity does not occur. Nevertheless, the above physical description is still an instructive and realistic account of what occurs in a transition from subsonic to supersonic flow.

5.9 Time-dependent plasma expansion

The previous discussion concerning the supersonic flow of an electrically neutral plasma was restricted to steady-state conditions. However, additional transport features occur during a time-dependent plasma expansion, and the results are relevant to a wide range of plasma flows in aeronomy and space physics.[3–5] It is instructive to consider a simple one-dimensional expansion scenario involving the collisionless expansion of an electrically neutral plasma into a vacuum. Figure 5.5 shows a schematic of the initial setup. At $t = 0$, the half-space $r < 0$ contains a single-ion electrically neutral plasma and the half-space $r > 0$ is a vacuum. For $t > 0$, the plasma is allowed to expand into the vacuum. At first, the electrons stream ahead of the ions into the vacuum because of their greater thermal speed, but after a short time a polarization electrostatic field develops that acts both to slow the electron expansion and accelerate the ion expansion. Once this polarization field develops, the expansion is ambipolar, and the ions and electrons move together as a single fluid.

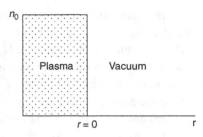

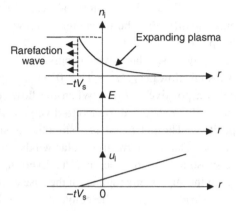

Figure 5.5 Self-similar solution for the expansion of a single-ion plasma into a vacuum.[4] The initial plasma–vacuum configuration is shown in the top panel and the plasma expansion features at time t are shown in the bottom panel.

For this simple expansion scenario, the plasma is assumed to be collisionless and isothermal, and the effects of gravity and stress are ignored. Therefore, in the ambipolar expansion phase ($n_e = n_i$, $u_e = u_i$), the continuity and momentum equations for the ions (or electrons) reduce to

$$\frac{\partial n_i}{\partial t} + \frac{\partial}{\partial r}(n_i u_i) = 0, \tag{5.88}$$

$$\frac{\partial u_i}{\partial t} + u_i \frac{\partial u_i}{\partial r} + \frac{V_S^2}{n_i} \frac{\partial n_i}{\partial r} = 0, \tag{5.89}$$

where Equation (5.61) was used for the polarization electrostatic field and where V_S is the ion-acoustic speed (5.82). Note that these equations are similar to those used to describe supersonic ion outflow (Equations 5.83 and 5.81).

Equations (5.88) and (5.89) yield *self-similar solutions*, which depend only on the ratio r/t of the independent variables r and t. With the introduction of the self-similar parameter ξ, which is defined to be

$$\xi = \frac{r}{tV_S}, \tag{5.90}$$

the derivatives with respect to r and t can be expressed as

$$\frac{\partial}{\partial r} = \frac{\partial \xi}{\partial r} \frac{d}{d\xi} = \frac{1}{tV_S} \frac{d}{d\xi}, \tag{5.91}$$

$$\frac{\partial}{\partial t} = \frac{\partial \xi}{\partial t}\frac{d}{d\xi} = -\frac{\xi}{t}\frac{d}{d\xi}. \tag{5.92}$$

With the aid of Equations (5.91) and (5.92), the continuity (5.88) and momentum (5.89) equations become, respectively

$$\frac{1}{n_i}\frac{dn_i}{d\xi}(u_i - \xi V_S) + \frac{du_i}{d\xi} = 0, \tag{5.93}$$

$$\frac{du_i}{d\xi}(u_i - \xi V_S) + \frac{V_S^2}{n_i}\frac{dn_i}{d\xi} = 0, \tag{5.94}$$

and the solution of these equations is

$$n_i = n_0 e^{-(\xi+1)}, \tag{5.95a}$$

$$u_i = V_S(\xi + 1). \tag{5.95b}$$

Note that the solution is only valid for $(\xi + 1) > 0$. For $(\xi + 1) < 0$, the plasma is unperturbed. This condition enters through the boundary condition for the solution of the continuity equation, which is that at $(\xi + 1) = 0$, $n_i = n_0$ (the unperturbed plasma density). The associated polarization electrostatic field can now be obtained from Equation (5.61), and the result is

$$E = \frac{(kT_e/e)}{tV_S}. \tag{5.95c}$$

The self-similar solution (5.95a–c) for the expansion of a single-ion plasma into a vacuum in shown in Figure 5.5. For $t > 0$, a rarefaction wave propagates into the plasma at the ion-acoustic speed. The density in the expansion region decreases exponentially with distance (5.95a) and the profile is concave at all times. The associated polarization electrostatic field does not vary with position, but its magnitude decreases inversely with time (5.95c). The ion drift velocity increases linearly with distance (5.95b) because of the ion acceleration associated with the electric field. However, at a given distance r, the ion drift velocity decreases as t^{-1} in parallel with the decrease in the electric field.

5.10   Diffusion across B

Up to this point, the focus has been on transport either along **B** for a planet with a strong magnetic field or in the vertical direction for an unmagnetized planet. However, plasma transport across a magnetic field can play an important role in certain ionospheric regions. To illustrate the effects of cross-field transport, it is convenient to consider a plasma that spans all levels of ionization, from weakly ionized at low altitudes to fully ionized at high altitudes. For simplicity, it is also convenient to consider a three-component plasma composed of ions, electrons, and neutrals. Assuming that the diffusion approximation is valid ($\partial \mathbf{u}_s / \partial t \to 0$, $\mathbf{u}_s \cdot \nabla \mathbf{u}_s \to 0$), stress and heat flow are not important ($\boldsymbol{\tau}_s = \mathbf{q}_s = 0$), and adopting an inertial reference frame (no Coriolis or centripetal acceleration terms), the momentum equation (5.50) perpendicular to $\mathbf{B}$ for the charged particles (subscript j) reduces to

$$\nabla p_j - n_j e_j (\mathbf{E}_\perp + \mathbf{u}_j \times \mathbf{B}) - n_j m_j \mathbf{G} = n_j m_j \nu_{jn} (\mathbf{u}_n - \mathbf{u}_j), \tag{5.96}$$

where $\mathbf{E}_\perp$ is an applied electric field that is perpendicular to $\mathbf{B}$ and where electron–ion collisions are neglected because the momentum transfer associated with them is small. It is convenient in solving Equation (5.96) first to transform to a reference frame moving with the neutral wind ($\mathbf{u}_j \to \mathbf{u}'_j + \mathbf{u}_n$), which introduces an effective electric field that is given by $\mathbf{E}'_\perp = \mathbf{E}_\perp + \mathbf{u}_n \times \mathbf{B}$. Therefore, Equation (5.96) becomes

$$\nabla p_j - n_j e_j (\mathbf{E}'_\perp + \mathbf{u}'_j \times \mathbf{B}) - n_j m_j \mathbf{G} = -n_j m_j \nu_{jn} \mathbf{u}'_j. \tag{5.97}$$

At high altitudes, collisions with the neutrals are negligible because the neutral densities decrease exponentially with altitude (see Figures 2.14, 2.19, 2.22, and 2.24). In this case, the transport across $\mathbf{B}$ can be easily obtained by taking the cross product of Equation (5.97) with $\mathbf{B}$, which yields

$$\mathbf{u}'_{j\perp} = \mathbf{u}_E + \mathbf{u}_D + \mathbf{u}_G, \tag{5.98}$$

where the *electromagnetic drift* ($\mathbf{u}_E$), *diamagnetic drift* ($\mathbf{u}_D$), and *gravitational drift* ($\mathbf{u}_G$) are given by

$$\mathbf{u}_E = \frac{\mathbf{E}'_\perp \times \mathbf{B}}{B^2}, \tag{5.99}$$

$$\mathbf{u}_D = -\frac{1}{n_j e_j} \frac{\nabla p_j \times \mathbf{B}}{B^2}, \tag{5.100}$$

$$\mathbf{u}_G = \frac{m_j}{e_j} \frac{\mathbf{G} \times \mathbf{B}}{B^2}, \tag{5.101}$$

and where $(\mathbf{u}'_j \times \mathbf{B}) \times \mathbf{B} = -B^2 \mathbf{u}'_{j\perp}$. Note that the electrons and ions drift across $\mathbf{B}$ together in the presence of a perpendicular electric field, but they drift in opposite directions in the presence of pressure gradients and gravity. It should also be noted that when collisions are unimportant, the resulting drifts are perpendicular to both $\mathbf{B}$ and the force causing the drift.

At the altitudes where collisions are important, it is possible to have perpendicular drifts both in the direction of the force, $\mathbf{F}_\perp$, and in the $\mathbf{F}_\perp \times \mathbf{B}$ direction. Assuming that the forces in Equation (5.97) have components perpendicular to $\mathbf{B}$, this equation can be expressed in the form

$$\mathbf{u}'_{j\perp} = -\frac{D_j}{p_j} \nabla_\perp p_j \pm \bar{\mu}_j \mathbf{E}'_\perp + \frac{1}{\nu_{jn}} \mathbf{G}_\perp \pm \frac{\omega_{cj}}{\nu_{jn}} (\mathbf{u}'_{j\perp} \times \mathbf{b}), \tag{5.102}$$

where $D_j = kT_j/(m_j\nu_{jn})$ is the diffusion coefficient (5.28), $\bar{\mu}_j = |e_j|/(m_j\nu_{jn})$ is the mobility coefficient (5.29), $\omega_{cj} = |e_j|B/m_j$ is the cyclotron frequency (2.7), $\mathbf{b} = \mathbf{B}/B$ is the unit vector, and the $\pm$ signs correspond to ions and electrons, respectively. Equation (5.102) can be readily solved by first expressing it in terms of the individual Cartesian velocity components. The resulting solution is given by

$$\mathbf{u}'_{j\perp} = -\frac{D_{j\perp}}{p_j}\mathbf{\nabla}_\perp p_j \pm \bar{\mu}_{j\perp}\mathbf{E}'_\perp + \frac{\nu_{jn}}{\nu^2_{jn} + \omega^2_{cj}}\mathbf{G}_\perp + \frac{\mathbf{u}_E + \mathbf{u}_D + \mathbf{u}_G}{1 + \nu^2_{jn}/\omega^2_{cj}}, \quad (5.103)$$

where

$$D_{j\perp} = \frac{D_j}{1 + \omega^2_{cj}/\nu^2_{jn}}, \quad (5.104)$$

$$\bar{\mu}_{j\perp} = \frac{\bar{\mu}_j}{1 + \omega^2_{cj}/\nu^2_{jn}}. \quad (5.105)$$

In the limit of $\nu_{jn}/\omega_{cj} \to 0$, $D_{j\perp} \to 0, \bar{\mu}_{j\perp} \to 0$, Equation (5.103) reduces to Equation (5.98). In the opposite limit of $\nu_{jn}/\omega_{cj} \to \infty$, $D_{j\perp} \to D_j$, $\bar{\mu}_{j\perp} \to \bar{\mu}_j$, Equation (5.103) reduces to the expression that prevails when $\mathbf{B} = 0$ (Equation 5.27).

5.11 Electrical conductivities

Electric currents play an important role in the dynamics and energetics of the ionospheres. For ionospheres that are not influenced by strong intrinsic magnetic fields, the electric currents can generate self-consistent magnetic fields that are sufficiently strong to affect the large-scale plasma motions. Under these circumstances, Maxwell's equations must be solved along with the plasma transport equations. Although such a procedure is straightforward, it is generally more convenient to use the so-called magnetohydrodynamics (MHD) approximation to the transport equations, which is discussed in Chapter 7. On the other hand, for the currents that flow in strongly magnetized ionospheres, the self-consistent magnetic fields generated by the currents are too small to affect the large-scale plasma dynamics, and hence, the intrinsic magnetic field can be taken as a known field. In this latter case, currents can flow both along and across $\mathbf{B}$ in response to imposed electric fields. The currents along $\mathbf{B}$ are carried by the electrons because their mobility is much greater than that of the ions (Equation 5.29). However, both ions and electrons contribute to the current that flows across $\mathbf{B}$. Typically, the current flows down along $\mathbf{B}$ from high to low altitudes, across $\mathbf{B}$ at low altitudes, and then back up along $\mathbf{B}$ to high altitudes, forming an electrical circuit that spans all levels of ionization.

It is convenient first to consider the cross-$\mathbf{B}$ current, which is typically driven by a perpendicular electric field, $\mathbf{E}_\perp$. Generally, for the electric field strengths found in the strongly magnetized ionospheres, the electric field dominates the perpendicular momentum equation (5.50), which reduces to

$$-\frac{e_j}{m_j}(\mathbf{E}_\perp + \mathbf{u}_j \times \mathbf{B}) = \sum_t \nu_{jt}(\mathbf{u}_t - \mathbf{u}_j),$$

(5.106)

where subscript j corresponds to electrons or one of the ion species and where the summation over t involves all of the other species.

For a given ion species (subscript i), the momentum transfer to the electrons is negligible because of the small electron mass. Also, the momentum exchange with other ion species is much smaller than that with the neutrals because the ion drifts are nearly equal and $n_i \ll n_n$. In addition, the different neutral species typically have the same drift velocity, $\mathbf{u}_n$. With this information, Equation (5.106), for ion species i, can be simplified and it becomes

$$\frac{e_i}{m_i}(\mathbf{E}_\perp + \mathbf{u}_i \times \mathbf{B}) = \nu_i(\mathbf{u}_i - \mathbf{u}_n),$$

(5.107)

where

$$\nu_i = \sum_n \nu_{in}.$$

(5.108)

To solve Equation (5.107), it is convenient first to transform the equation to a reference frame moving with the neutral wind ($\mathbf{u}_i \to \mathbf{u}_i' + \mathbf{u}_n$), which yields

$$\frac{e_i}{m_i \nu_i}\mathbf{E}_\perp' + \frac{\omega_{c_i}}{\nu_i}\mathbf{u}_i' \times \mathbf{b} = \mathbf{u}_i',$$

(5.109)

where $\mathbf{E}_\perp' = \mathbf{E}_\perp + \mathbf{u}_n \times \mathbf{B}$ is an effective electric field (5.37), $\omega_{c_i} = e_i B/m_i$, and $\mathbf{b} = \mathbf{B}/B$. The next step is to solve for the individual velocity components using a Cartesian coordinate system with $\mathbf{E}_\perp'$ along the x-axis and $\mathbf{b}$ along the z-axis. After the velocity components are obtained, they can be recast in terms of vectors, which yields

$$\mathbf{u}_{i\perp}' = \frac{e_i}{m_i}\left(\frac{\nu_i}{\nu_i^2 + \omega_{c_i}^2}\mathbf{E}_\perp' - \frac{\omega_{c_i}}{\nu_i^2 + \omega_{c_i}^2}\mathbf{b} \times \mathbf{E}_\perp'\right).$$

(5.110)

The final form for the result is obtained by transforming back to the original reference frame ($\mathbf{u}_i' \to \mathbf{u}_i - \mathbf{u}_n$) and then multiplying by $n_i e_i$:

$$\mathbf{J}_{i\perp} = n_i e_i \mathbf{u}_{n\perp} + \sigma_i\left(\frac{\nu_i^2}{\nu_i^2 + \omega_{c_i}^2}\mathbf{E}_\perp' - \frac{\nu_i \omega_{c_i}}{\nu_i^2 + \omega_{c_i}^2}\mathbf{b} \times \mathbf{E}_\perp'\right),$$

(5.111)

where $\mathbf{J}_{i\perp} = n_i e_i \mathbf{u}_{i\perp}$ is the perpendicular ion current and σ_i is the *ion conductivity*, given by

$$\sigma_i = \frac{n_i e_i^2}{m_i \nu_i}.$$

(5.112)

The momentum loss of the electrons to the ions is much smaller than that to the neutrals because the neutral density is typically much greater than the ion density. Therefore, when Equation (5.106) is applied to the electrons, it reduces to an equation similar to the ion equation (5.107), except for the sign of the charge. This electron momentum equation is solved in a manner similar to that discussed above for the ions, and the solution is

$$\mathbf{J}_{e\perp} = -n_e e \mathbf{u}_{n\perp} + \sigma_e \left(\frac{v_e^2}{v_e^2 + \omega_{c_e}^2} \mathbf{E}'_\perp + \frac{v_e \omega_{c_e}}{v_e^2 + \omega_{c_e}^2} \mathbf{b} \times \mathbf{E}'_\perp \right), \qquad (5.113)$$

where $\mathbf{J}_{e\perp} = -e n_e \mathbf{u}_{e\perp}$ is the perpendicular electron current, $\omega_{c_e} = |e|B/m_e$, and

$$v_e = \sum_n v_{en}, \qquad (5.114)$$

$$\sigma_e = \frac{n_e e^2}{m_e v_e}. \qquad (5.115)$$

The total perpendicular current is simply, $\mathbf{J}_\perp = \mathbf{J}_{e\perp} + \sum_i J_{i\perp}$, which can be obtained from Equations (5.111) and (5.113), and the result is

$$\mathbf{J}_\perp = \left(\sum_i n_i e_i - n_e e \right) \mathbf{u}_{n\perp} + \sigma_P(\mathbf{E}_\perp + \mathbf{u}_n \times \mathbf{B}) + \sigma_H \mathbf{b} \times (\mathbf{E}_\perp + \mathbf{u}_n \times \mathbf{B}),$$

$$(5.116)$$

where the *Pedersen*, σ_P, and *Hall*, σ_H, *conductivities* are given by

$$\sigma_P = \sum_i \sigma_i \frac{v_i^2}{v_i^2 + \omega_{c_i}^2} + \sigma_e \frac{v_e^2}{v_e^2 + \omega_{c_e}^2}, \qquad (5.117)$$

$$\sigma_H = -\sum_i \sigma_i \frac{v_i \omega_{c_i}}{v_i^2 + \omega_{c_i}^2} + \sigma_e \frac{v_e \omega_{c_e}}{v_e^2 + \omega_{c_e}^2} \qquad (5.118)$$

and where $\mathbf{E}'_\perp = \mathbf{E}_\perp + \mathbf{u}_n \times \mathbf{B}$ was used in Equation (5.116). Typically, there is a very small net charge in the ionospheres, and hence, the first term in Equation (5.116) can be neglected. Also, the electron contribution to the Pedersen and Hall conductivities can be simplified because $v_e \ll \omega_{c_e}$ in most cases. Keeping only the terms that are of order v_e/ω_{c_e} the Pedersen and Hall conductivities reduce to

$$\sigma_P = \sum_i \sigma_i \frac{v_i^2}{v_i^2 + \omega_{c_i}^2}, \qquad (5.119)$$

$$\sigma_H = -\sum_i \sigma_i \frac{v_i \omega_{c_i}}{v_i^2 + \omega_{c_i}^2} + \frac{v_e \sigma_e}{\omega_{c_e}}. \qquad (5.120)$$

These results indicate that, in general, the electrons contribute to the Hall current, but not to the Pedersen current.

The electron current along $\mathbf{B}$ can be obtained from the parallel component of Equation (5.50). Neglecting both the terms on the left-hand side of this equation that contain the electron mass, and the heat flow terms on the right-hand side, which will be discussed in the next section, the parallel component of Equation (5.50) becomes

$$\nabla_\parallel p_e + e n_e \mathbf{E}_\parallel = \sum_t n_e m_e \nu_{et}(\mathbf{u}_t - \mathbf{u}_e)_\parallel, \tag{5.121}$$

where $\mathbf{E}_\parallel$ is an applied electric field that is much larger than the polarization field. Typically, when an electric current is induced along $\mathbf{B}$, the electron drift velocity is much greater than the ion and neutral drift velocities and the latter velocities can be neglected. Using this fact, and setting $p_e = n_e k T_e$, Equation (5.121) becomes

$$k T_e \nabla_\parallel n_e + k n_e \nabla_\parallel T_e + e n_e \mathbf{E}_\parallel = -n_e m_e \nu'_e \mathbf{u}_{e\parallel}, \tag{5.122}$$

where

$$\nu'_e = \sum_i \nu_{ei} + \sum_n \nu_{en}. \tag{5.123}$$

With the introduction of the field-aligned current density $\mathbf{J}_\parallel = -e n_e \mathbf{u}_{e\parallel}$, Equation (5.122) can be expressed in the following form:

$$\mathbf{J}_\parallel = \sigma_e \left(\mathbf{E}_\parallel + \frac{k T_e}{e n_e} \nabla_\parallel n_e \right) + \bar{\varepsilon}_e \nabla_\parallel T_e, \tag{5.124}$$

where σ_e is the *parallel electrical conductivity* and $\bar{\varepsilon}_e$ is the *current flow conductivity due to thermal gradients*. These coefficients are given by

$$\sigma_e = \frac{n_e e^2}{m_e \nu'_e}, \tag{5.125a}$$

$$\bar{\varepsilon}_e = \frac{n_e e k}{m_e \nu'_e}. \tag{5.125b}$$

Note that the σ_e defined in Equation (5.125a) is similar to that defined previously in Equation (5.115), with the only difference being the electron collision frequency. It should also be noted that when the applied electric field dominates, Equation (5.124) reduces to *Ohm's law* for electron motion along $\mathbf{B}$, which is

$$\mathbf{J}_\parallel = \sigma_e \mathbf{E}_\parallel. \tag{5.126}$$

The conductivities given in Equations (5.125a) and (5.125b) have been widely used in ionospheric studies, but they correspond only to the first approximation to these coefficients. In the next section, these field-aligned conductivities will be derived again including the effect of electron heat flow on the momentum balance.

As will be seen, the heat flow provides an important correction to the electrical conductivity (5.125a).

5.12 Electron stress and heat flow

Simplified expressions for the electron stress tensor and heat flow vector can be obtained when the collision frequency is large, and hence, the electron velocity distribution is very close to a drifting Maxwellian (i.e., small $\boldsymbol{\tau}_e$ and $\mathbf{q}_e$). However, in the ionospheres, electron transport effects are generally important at all altitudes, and therefore, it is necessary to consider electron interactions with other electrons, ions, and neutrals. To simplify the electron collision terms, it is convenient to assume that the various ion and neutral species have displaced Maxwellian velocity distribution functions ($\boldsymbol{\tau}_i = \mathbf{q}_i = 0$; $\boldsymbol{\tau}_n = \mathbf{q}_n = 0$), and that terms of order m_e/m_i and m_e/m_n can be neglected compared with terms of order unity. With these assumptions, the linear collision terms (4.129) for the electrons become

$$\frac{\delta \mathbf{M}_e}{\delta t} = -\sum_i \rho_e \nu_{ei}(\mathbf{u}_e - \mathbf{u}_i)$$

$$-\sum_n \rho_e \nu_{en}(\mathbf{u}_e - \mathbf{u}_n) + \frac{m_e}{kT_e}\mathbf{q}_e\left(\frac{3}{5}\sum_i \nu_{ei} + \sum_n \nu_{en} z_{en}\right),$$

(5.127a)

$$\frac{\delta E_e}{\delta t} = -\sum_i \frac{\rho_e \nu_{ei}}{m_i} 3k(T_e - T_i) - \sum_n \frac{\rho_e \nu_{en}}{m_n} 3k(T_e - T_n),$$

(5.127b)

$$\frac{\delta \mathbf{P}_e}{\delta t} = -\sum_i \frac{\rho_e \nu_{ei}}{m_i} 2k(T_e - T_i)\mathbf{I} - \sum_n \frac{\rho_e \nu_{en}}{m_n} 2k(T_e - T_n)\mathbf{I} - \frac{6}{5}\nu_{ea}\boldsymbol{\tau}_e,$$

(5.127c)

$$\frac{\delta \mathbf{q}_e}{\delta t} = -p_e \sum_i \nu_{ei}(\mathbf{u}_e - \mathbf{u}_i) - \frac{5}{2}p_e \sum_n \nu_{en}(\mathbf{u}_e - \mathbf{u}_n)(1 - z_{en}) - \frac{4}{5}\nu_{eb}\mathbf{q}_e,$$

(5.127d)

where

$$\nu_{ea} = \nu_{ee} + \sum_i \nu_{ei} + \frac{1}{2}\sum_n \nu_{en} z_{en}'',$$

(5.128)

$$\nu_{eb} = \nu_{ee} - \frac{1}{4}\sum_i \nu_{ei} + \frac{5}{4}\sum_n \nu_{en}\left(z_{en}' - \frac{5}{2}z_{en}\right).$$

(5.129)

The collision-dominated transport equations are obtained from the 13-moment system of equations (3.57–61, 5.127a–d) by using a perturbation scheme in which

τ_e and $\mathbf{q}_e$ are treated as small quantities. To lowest order in the perturbation scheme, stress and heat flow effects are neglected and the resulting continuity, momentum, and energy equations correspond to the Euler equations. However, the Euler approximation is not useful for the electron gas because electron heat flow is almost always important. To the next order in the perturbation scheme, τ_e and $\mathbf{q}_e$ are expressed in terms of n_e, $\mathbf{u}_e$, and T_e with the aid of the stress tensor (3.60, 5.127c) and heat flow (3.61, 5.127d) equations. This is accomplished by assuming that terms containing $\nu\tau_e$, $\nu\mathbf{q}_e$, $\omega_{c_e}\tau_e$, and $\omega_{c_e}\mathbf{q}_e$ are the same order as terms that just contain the lower-order moments n_e, $\mathbf{u}_e$, and T_e, while all other terms containing τ_e and $\mathbf{q}_e$ are of order $1/\nu$ and, therefore, are negligible. Retaining only those terms of order 1, the electron stress tensor and heat flow equations become

$$\tau_e - \frac{5\omega_{c_e}}{6\nu_{ea}}(\mathbf{b} \times \tau_e - \tau_e \times \mathbf{b}) = -\eta_e\left[\nabla\mathbf{u}_e + (\nabla\mathbf{u}_e)^T - \frac{2}{3}(\nabla \cdot \mathbf{u}_e)\mathbf{I}\right],$$

(5.130)

$$\mathbf{q}_e + \frac{5\omega_{c_e}}{4\nu_{ec}}\mathbf{q}_e \times \mathbf{b} = -\lambda_e\nabla T_e + \frac{15}{8}\frac{p_e}{\nu_{ec}}\sum_i \nu_{ei}(\mathbf{u}_e - \mathbf{u}_i)$$

$$+ \frac{25}{8}\frac{p_e}{\nu_{ec}}\sum_n \nu_{en}z_{en}(\mathbf{u}_e - \mathbf{u}_n),$$

(5.131)

where the coefficients of viscosity and thermal conductivity are

$$\eta_e = \frac{5p_e}{6\nu_{ea}},$$

(5.132)

$$\lambda_e = \frac{25}{8}\frac{kp_e}{m_e\nu_{ec}},$$

(5.133)

and where

$$\nu_{ec} = \nu_{ee} + \frac{13}{8}\sum_i \nu_{ei} + \frac{5}{4}\sum_n \nu_{en}z'_{en}.$$

(5.134)

Therefore, the *closed system of Navier–Stokes equations* for the electron gas is composed of the stress tensor (5.130) and heat flow (5.131) equations and the following continuity, momentum, and energy equations:

$$\frac{\partial n_e}{\partial t} + \nabla \cdot (n_e\mathbf{u}_e) = \frac{\delta n_e}{\delta t},$$

(5.135a)

$$\rho_e\frac{D_e\mathbf{u}_e}{Dt} + \nabla p_e - \rho_e\mathbf{G} + n_e e(\mathbf{E} + \mathbf{u}_e \times \mathbf{B}) + \nabla \cdot \tau_e$$

$$= -\sum_i \rho_e\nu_{ei}(\mathbf{u}_e - \mathbf{u}_i) - \sum_n \rho_e\nu_{en}(\mathbf{u}_e - \mathbf{u}_n)$$

$$+ \frac{m_e}{kT_e}\mathbf{q}_e\left(\frac{3}{5}\sum_i \nu_{ei} + \sum_n \nu_{en}z_{en}\right),$$

(5.135b)

$$\frac{D_e}{Dt}\left(\frac{3}{2}p_e\right) + \frac{5}{2}p_e(\nabla \cdot \mathbf{u}_e) + \nabla \cdot \mathbf{q}_e + \tau_e : \nabla\mathbf{u}_e$$

$$= -\sum_i \frac{\rho_e \nu_{ei}}{m_i} 3k(T_e - T_i) - \sum_n \frac{\rho_e \nu_{en}}{m_n} 3k(T_e - T_n). \tag{5.135c}$$

Several factors should be noted about the electron transport equations. First, stresses arise as a result of velocity gradients, and when $\mathbf{B} = 0$, the stress tensor takes the classical Navier–Stokes form (see Section 10.3). Also, for a heat flow along $\mathbf{B}$, Equation (5.131) indicates that $\mathbf{q}_e \sim -\lambda_e \nabla T_e$, as expected, but there are additional terms proportional to $(\mathbf{u}_e - \mathbf{u}_i)$ and $(\mathbf{u}_e - \mathbf{u}_n)$. This indicates that an electron heat flow is induced by a relative drift between the electrons and other species, which is called a *thermoelectric effect*.

Additional insight about the collision-dominated electron equations can be gained by considering a fully ionized plasma composed of electrons and one singly ionized ion species. For such mixtures, relatively simple expressions for η_e and λ_e can be obtained by using Equation (4.140) for the Coulomb collision frequencies, and these expressions are

$$\eta_e = \frac{5}{8\sqrt{\pi}(1 + \sqrt{2})} \frac{m_e^{1/2}(kT_e)^{5/2}}{e^4 \ln \Lambda}, \tag{5.136}$$

$$\lambda_e = \frac{75}{4\sqrt{\pi}(8 + 13\sqrt{2})} \frac{k(kT_e)^{5/2}}{m_e^{1/2}e^4 \ln \Lambda}. \tag{5.137}$$

The terms containing the $\sqrt{2}$ account for electron–ion collisions. Note that both η_e and λ_e are proportional to $T_e^{5/2}$ for a fully ionized plasma. Electron heat flow is known to be important in all of the ionospheres, but the effects of viscous stress have not been rigorously evaluated. Nevertheless, viscous stress is expected to be negligible because the electron drift velocity and its gradient are typically small, and hence, the $\nabla \cdot \tau_e$ term cannot compete with the other terms in the electron momentum equation.

For the fully ionized gas under consideration, the electron heat flow parallel to $\mathbf{B}$ (5.131) can be expressed in the form

$$\mathbf{q}_{e\parallel} = -\lambda_e \nabla_\parallel T_e - \frac{15}{8} \frac{\nu_{ei}}{\nu_{ec}} \frac{kT_e}{e} \mathbf{J}_\parallel, \tag{5.138}$$

where $\mathbf{J}_\parallel = n_e e(\mathbf{u}_i - \mathbf{u}_e)_\parallel$ is the current density. Neglecting the τ_e term and the terms containing m_e, the electron momentum equation (5.135b) parallel to $\mathbf{B}$ can also be expressed in the form

$$\frac{n_e e^2}{m_e \nu_{ei}}\left(\mathbf{E}_\parallel + \frac{kT_e}{n_e e}\nabla_\parallel n_e\right) + \frac{e n_e k}{m_e \nu_{ei}}\nabla_\parallel T_e = \mathbf{J}_\parallel + \frac{3}{5}\frac{e}{kT_e}\mathbf{q}_{e\parallel}. \tag{5.139}$$

Note that Equation (5.139) is essentially the fully ionized limit of Equation (5.124), except for the heat flow term, which was neglected in the derivation of Equation (5.124). As will be seen, the heat flow affects the momentum balance and provides corrections to the electrical conductivity and the current flow conductivity due to thermal gradients. This can be shown by eliminating $\mathbf{q}_{e\|}$ in Equation (5.139) with the aid of Equation (5.138). After doing this, the final forms for the electron momentum (5.139) and heat flow (5.138) equations are given by

$$\mathbf{J}_{\|} = \sigma'_{e}\left(\mathbf{E}_{\|} + \frac{kT_e}{en_e}\mathbf{\nabla}_{\|}n_e\right) + \bar{\varepsilon}'_{e}\mathbf{\nabla}_{\|}T_e, \tag{5.140}$$

$$\mathbf{q}_{e\|} = -\lambda_e\mathbf{\nabla}_{\|}T_e - \beta_e\mathbf{J}_{\|}, \tag{5.141}$$

where the conductivities can be expressed as

$$\sigma'_e = \frac{n_e e^2}{m_e \nu_{ei}}\frac{1}{g_{\sigma 0}}, \tag{5.142}$$

$$\bar{\varepsilon}'_e = \frac{n_e k e}{m_e \nu_{ei}}\frac{1}{g_{\varepsilon 0}}, \tag{5.143}$$

$$\lambda_e = \frac{5 n_e k^2 T_e}{m_e \nu_{ei}}\frac{1}{g_{\lambda 0}}, \tag{5.144}$$

$$\beta_e = \frac{15}{8}\frac{\nu_{ei}}{\nu_{ec}}\frac{kT_e}{e}. \tag{5.145}$$

In Equations (5.142–145), the parameters $g_{\sigma 0}, g_{\varepsilon 0}$, and $g_{\lambda 0}$ are pure numbers to be discussed below. The thermal conductivity (5.144) is the same as that given previously in Equation (5.137), but expressed in a different form. The coefficient β_e is the *thermoelectric coefficient* and it accounts for the electron heat flow associated with a current. The equation for $\mathbf{J}_{\|}$ (5.140) is similar to Equation (5.124), but the conductivities are modified because of the electron heat flow. The conductivities in Equations (5.142) and (5.143) correspond to the second approximation, whereas those in Equations (5.125a,b) correspond to the first approximation to these conductivities.[6]

The different levels of approximation can be traced to the expression for the species velocity distribution function, which in general is an infinite series about some zeroth-order weight factor. For the 13-moment approximation, only a few terms are retained in the series expansion for f_s (Equation 3.49), and consequently, only a few terms appear in the 13-moment expressions for the linear collision terms (see Equations 4.129a–g). On the other hand, for the general case of an infinite series for f_s, *each* of the linear collision terms (4.129a–g) would contain an infinite series of progressively higher-order velocity moments. These terms would describe higher-order distortions (beyond $\mathbf{\tau}_s$ and $\mathbf{q}_s$) of the species velocity distribution. Naturally,

Table 5.1 Comparison of electron transport parameters.[1]

Parameter	13-moment	Exact values
g_{σ_0}	0.518	0.506
g_{ε_0}	0.287	0.297
g_{λ_0}	3.731	1.562
$e\beta_e/kT_e$	0.804	0.779

the more terms retained in the expansion for f_s, and hence in the collison terms, the more accurate are the associated conductivites.

An exact numerical solution of the electron Boltzmann equation has been obtained for a fully ionized gas,[7] and this solution is equivalent to keeping all of the terms in the infinite series for f_s. The resulting conductivities have been expressed in the same forms as those given in Equations (5.142–145), and the 13-moment results can be compared to the exact conductivities simply by comparing the corresponding correction factors g_{σ_0}, g_{ε_0}, and g_{λ_0} and the coefficient of β_e. This comparison is shown in Table 5.1, where the 13-moment values were calculated with the aid of the Coulomb collision frequencies (4.144) and (4.145). Except for g_{λ_0}, which is in error by more than a factor of two, all of the 13-moment conductivities are in excellent agreement with the exact values.[7]

The electron conductivities (5.142–145) can be generalized to include electron interactions with several ion and neutral species simply by replacing ν_{ei} with $\nu_e' = \sum_i \nu_{ei} + \sum_n \nu_{en}$ (5.123), but in this case the exact values of the g-correction factors depend on both the degree of ionization and the specific neutral species under consideration.[8,9] Typically, these g-correction factors vary by factors of two to three, as the degree of ionization is varied from the weakly to fully ionized states.

Perhaps the most widely used conductivity is the electron thermal conductivity because electron heat flow is an important process in all of the ionospheres. In many ionospheric applications, the following relatively simple, electron thermal conductivity has been used[10]

$$\lambda_e = \frac{7.7 \times 10^5 T_e^{5/2}}{1 + 3.22 \times 10^4 \dfrac{T_e^2}{n_e} \displaystyle\sum_n n_n \langle Q_{en}^{(1)} \rangle}, \tag{5.146}$$

where the units are eV cm^{-1} s^{-1} K^{-1} and where $\langle Q_{en}^{(1)} \rangle$ is a Maxwellian average of the momentum transfer cross section (4.46). This expression was derived using mean-free-path considerations, but in the derivation a slight algebraic error was made. The number in the denominator should be 2.16, not 3.22.[11] However, as it turns out, the number 3.22 yields slightly better results when values calculated from Equation (5.146) are compared with the more rigorous values obtained from the generalization of Equation (5.144). Typically, the errors associated with the approximate λ_e (5.146) are less than 5%, and reach a maximum of 18%. Such errors are acceptable in ionospheric studies, because the uncertainties associated with the electron–neutral momentum transfer cross sections are generally larger.

5.13   Ion stress and heat flow

Collision-dominated expressions for the ion stress tensor and heat flow vector can be derived using a perturbation scheme similar to that used for the electrons,[1] and the resulting equations are also similar. For example, in the limit of a fully ionized plasma with a single ion component, the collision-dominated stress tensor is given by the following equation

$$\boldsymbol{\tau}_i + \frac{5\omega_{c_i}}{6\nu_{ii} + 10\nu_{ie}} (\mathbf{b} \times \boldsymbol{\tau}_i - \boldsymbol{\tau}_i \times \mathbf{b}) = -\eta_i \left[\nabla \mathbf{u}_i + (\nabla \mathbf{u}_i)^{\mathrm{T}} - \frac{2}{3}(\nabla \cdot \mathbf{u}_i)I \right],$$

(5.147)

where

$$\eta_i = \frac{5p_i}{6\nu_{ii} + 10\nu_{ie}}.$$

(5.148)

In the terrestrial case, viscous stress is not important for the ions. For the neutrals, on the other hand, it is important, as will be shown in Section 10.3. The reason for this difference relates to the magnitude of the main flow and the direction of the velocity gradient. For the neutrals, the main flows are horizontal, at speeds of from $100–800 \text{ m s}^{-1}$, while the velocity gradients are in the vertical direction. These conditions yield large viscous stress effects. For the ions, the velocity gradients are also primarily in the vertical direction, but the flows can be either in the horizontal or vertical directions. The vertical ion drifts are usually of the order of $10–50 \text{ m s}^{-1}$, and because of the low speeds, the viscous effects associated with them are small. Large horizontal ion flows can occur, with speeds up to several km s^{-1}, but they are $\mathbf{E} \times \mathbf{B}$ drifts, and they exhibit little variation with altitude. The lack of a velocity gradient implies that viscous stress is not important even for these large drifts.

Although the viscous stress effects associated with velocity gradients (5.147) are typically not important, large ion–neutral relative drifts do result in important stress effects, as discussed previously in Section 5.3 for a weakly ionized plasma (Equation 5.49a). The extension of this result to a single-ion partially ionized plasma is straightforward, and the modified expression is given by

$$\tau_{i\|} = \frac{R_i}{S_i + 0.6\nu_{ii}} n_i m_i \left[(\mathbf{u}_i - \mathbf{u}_n)_\|^2 - \frac{1}{3}(\mathbf{u}_i - \mathbf{u}_n)^2 \right],$$

(5.149)

where R_i and S_i are still given by Equations (5.49b, c). The appropriate ion momentum equation is (5.48b), with the polarization electrostatic field, $E_\|$, given by Equation (5.61).

The derivation of a collision-dominated expression for the ion heat flow is more involved than that for the electrons. In the electron derivation (Section 5.12), the ions and neutrals were assumed to have drifting Maxwellian velocity distributions, which simplified the analysis. This simplification is reasonable for the electrons because the small electron mass acts to decouple the electrons from the other species. On the other hand, when the ion equations are derived, it is not appropriate to assume simplified forms for the electron and neutral velocity distributions; the full 13-moment expression must be used. This more general procedure leads to additional transport effects, and these are discussed in Section 5.14.

5.14   Higher-order diffusion processes

The 13-moment system of transport equations can describe ordinary diffusion, thermal diffusion, and thermoelectric transport processes at a level of approximation that is equivalent to Chapman and Cowling's so-called first and second approximations,[12] depending on the process. However, the 13-moment approach has an advantage over the Chapman–Cowling method in that the different components of the gas mixture can have separate temperatures. The classical forms for the diffusion and heat flow equations are obtained from the 13-moment momentum (3.58) and heat flow (3.61) equations by making several simplifying assumptions. First, the linear collision terms (4.129) are adopted. Also, the inertial and stress terms in the momentum equation are neglected. In the heat flow equation (3.61), all terms proportional to $\mathbf{q}_s$ and $\boldsymbol{\tau}_s$ are neglected, except the $\mathbf{q}_s$ terms multiplied by a collision frequency (collision-dominated conditions). Finally, only diffusion and heat flows either along a strong magnetic field or in the vertical direction are considered, and density and temperature gradients perpendicular to this direction are assumed to be small.

With the above assumptions, the momentum (3.58, 4.129b) and heat flow (3.61, 4.129g) equations for a fully ionized plasma reduce to

$$\nabla p_s - n_s m_s \mathbf{G} - n_s e_s \mathbf{E} = n_s m_s \sum_t \nu_{st}(\mathbf{u}_t - \mathbf{u}_s)$$

$$+ \frac{3}{5}\sum_t \nu_{st}\frac{\mu_{st}}{kT_{st}}\left(\mathbf{q}_s - \frac{\rho_s}{\rho_t}\mathbf{q}_t\right), \tag{5.150}$$

$$-\frac{5}{2}\frac{kp_s}{m_s}\nabla T_s = \mathbf{q}_s\left[\frac{4}{5}\nu_{ss} + \sum_{t\neq s}\nu_{st}\left(D_{st}^{(1)} + \frac{3}{2}\frac{\mu_{st}}{m_s}\frac{T_s}{T_{st}}\right)\right]$$

$$-\sum_{t\neq s}\nu_{st}\left(D_{st}^{(4)} + \frac{3}{2}\frac{\mu_{st}}{m_s}\frac{T_s}{T_{st}}\right)\frac{\rho_s}{\rho_t}\mathbf{q}_t$$

$$-\frac{3}{2}p_s\sum_{t\neq s}\frac{m_t\nu_{st}}{m_s + m_t}(\mathbf{u}_s - \mathbf{u}_t), \tag{5.151}$$

where ν_{ss}, ν_{st}, $D_{st}^{(1)}$, and $D_{st}^{(4)}$ are those relevant to Coulomb collisions (4.140, 4.141a–c). In addition to these equations, the plasma is governed by charge neutrality and charge conservation with no current (ambipolar diffusion), and for a three-component plasma, these conditions are

$$n_e = n_i Z_i + n_j Z_j,$$ (5.152)

$$n_e \mathbf{u}_e = n_i Z_i \mathbf{u}_i + n_j Z_j \mathbf{u}_j.$$ (5.153)

In Equations (5.152) and (5.153), subscript e is for the electrons and subscripts i and j are for the two ion species.

The application of the heat flow equation (5.151) to the two ion species and the electrons yields three coupled equations because $\mathbf{q}_e$, $\mathbf{q}_i$, and $\mathbf{q}_j$ appear in each equation. The simultaneous solution of the three equations for the individual heat flows yields equations of the form

$$\mathbf{q}_e = -\lambda_e \nabla T_e + \delta_{ei}(\mathbf{u}_e - \mathbf{u}_i) + \delta_{ej}(\mathbf{u}_e - \mathbf{u}_j),$$ (5.154)

$$\mathbf{q}_i = -K'_{ji} \nabla T_i - K_{ij} \nabla T_j + R_{ij}(\mathbf{u}_i - \mathbf{u}_j),$$ (5.155)

$$\mathbf{q}_j = -K_{ji} \nabla T_i - K'_{ij} \nabla T_j - R_{ji}(\mathbf{u}_i - \mathbf{u}_j),$$ (5.156)

where λ_e, K_{ij}, K_{ji}, K'_{ij}, and K'_{ji} are *thermal conductivities* and δ_{ei}, δ_{ej}, R_{ij}, and R_{ji} are *diffusion thermal coefficients*.[13] These expressions are given in Appendix I. Note that a flow of heat is induced in *both* ion gases as a result of a temperature gradient in either gas or as a result of a relative drift between the ion gases. The latter process is known as a *diffusion thermal effect*. When this process operates in the electron gas, it is called a *thermoelectric effect*, as discussed previously in Section 5.12 (Equation 5.141). It should also be noted that ∇T_e terms do not appear in the ion heat flow equations and that ∇T_i and ∇T_j terms do not appear in the electron heat flow equation. This occurs because in deriving Equations (5.154–156), terms of the order of $(m_e/m_i)^{1/2}$ and $(m_e/m_j)^{1/2}$ were neglected. With regard to the underlying physics, the fact that a temperature gradient in one ion gas can induce a heat flow in another ion gas can be traced to the collision process. If heat flows in ion gas i due to a ∇T_i, then the ion gas has a nonMaxwellian velocity distribution (Equation 3.49). The nonMaxwellian feature in gas i is communicated to gas j via collisions, and they induce a similar nonMaxwellian feature in gas j.

Turning to the momentum equation (5.150), the primary function of the electrons is to establish the polarization electrostatic field that produces ambipolar diffusion. Taking into account the small electron mass and using equation (5.154) for $\mathbf{q}_e$, the electron momentum equation (5.150) can be expressed in the form[13,14]

$$e\mathbf{E} = -\frac{1}{n_e} \nabla p_e - \frac{(15\sqrt{2}/8)(n_i Z_i^2 + n_j Z_j^2)k \nabla T_e}{n_i Z_i + n_j Z_j + (13\sqrt{2}/8)(n_i Z_i^2 + n_j Z_j^2)}.$$ (5.157)

The second term on the right-hand side of Equation (5.157) is a *thermal diffusion* process, and it describes the effect of heat flow on the electron momentum balance. Again, ion temperature gradient terms do not appear in Equation (5.157) because terms of the order of $(m_e/m_i)^{1/2}$ and $(m_e/m_j)^{1/2}$ are neglected.

Ion diffusion equations of the classical form can now be obtained from the momentum equation (5.150) by explicitly writing the momentum equations for ion species i and j, by eliminating the polarization electrostatic field with the aid of Equation (5.157), by using Equations (5.154–156) for the ion and electron heat flows, and by taking into account the small electron mass. When the resulting equations are solved for the ion drift velocities, the following diffusion equations are obtained:

$$\mathbf{u}_i = \mathbf{u}_j - D_i \left[\frac{1}{n_i} \nabla n_i - \frac{m_i \mathbf{G}}{kT_i} + \frac{1}{T_i} \nabla T_i + Z_i \frac{T_e/T_i}{n_e} \nabla n_e \right.$$

$$\left. + \frac{(Z_i - \gamma_i)}{T_i} \nabla T_e + \frac{n_j}{n_i + n_j} \left(\frac{\alpha_{ij}}{T_i} \nabla T_i - \frac{\alpha_{ij}^*}{T_i} \nabla T_j \right) \right], \tag{5.158}$$

$$\mathbf{u}_j = \mathbf{u}_i - D_j \left[\frac{1}{n_j} \nabla n_j - \frac{m_j \mathbf{G}}{kT_j} + \frac{1}{T_j} \nabla T_j + Z_j \frac{T_e/T_j}{n_e} \nabla n_e \right.$$

$$\left. + \frac{(Z_j + \gamma_j)}{T_j} \nabla T_e - \frac{n_i}{n_i + n_j} \left(\frac{\alpha_{ij}}{T_j} \nabla T_i - \frac{\alpha_{ij}^*}{T_j} \nabla T_j \right) \right], \tag{5.159}$$

where

$$D_i = \frac{kT_i}{m_i \nu_{ij}} \frac{1}{1 - \Delta_{ij}}, \tag{5.160}$$

$$D_j = \frac{kT_j}{m_j \nu_{ji}} \frac{1}{1 - \Delta_{ij}}. \tag{5.161}$$

In the above equations, α_{ij}, α_{ij}^*, γ_i, and γ_j are *thermal diffusion coefficients* and Δ_{ij} is a correction factor for *ordinary diffusion*.[13] All of these coefficients arise as a result of the effect that heat flow has on the momentum balance. The complete expressions are given in Appendix I. It should be noted that the correction factor, Δ_{ij}, is less than one, so the effect of heat flow is to enhance ordinary diffusion. Heat flow also induces an additional diffusion via temperature gradients, which is called *thermal diffusion*. In particular, a temperature gradient in either of the ion gases or in the electron gas causes thermal diffusion in both ion gases. The effect of thermal diffusion is to drive the heavy ions toward the hotter regions, which usually means toward higher altitudes.

Diffusion and heat flow equations have also been derived for a three-component partially ionized plasma composed of electrons (subscript e), ions (subscript i), and neutrals (subscript n). The technique used to derive these equations is similar to that described above for the fully ionized plasma, and the resulting equations are

given by

$$\mathbf{q}_i = -K'_{ni}\nabla T_i - K_{in}\nabla T_n + R_{in}(\mathbf{u}_i - \mathbf{u}_n), \tag{5.162}$$

$$\mathbf{q}_n = -K_{ni}\nabla T_i - K'_{in}\nabla T_n - R_{ni}(\mathbf{u}_i - \mathbf{u}_n) \tag{5.163}$$

$$\mathbf{u}_i = \mathbf{u}_n - D_a\left[\frac{1}{n_i}\nabla n_i - \frac{m_i\mathbf{G}}{k(T_e + T_i)} + \frac{\nabla(T_e + T_i)}{T_e + T_i}\right.$$
$$\left. + \frac{2\omega}{T_e + T_i}\nabla T_n + \frac{2\omega^*}{T_e + T_i}\nabla T_i\right], \tag{5.164}$$

where

$$D_a = \frac{k(T_e + T_i)}{m_i\nu_{in}}\frac{1}{1 - \Delta_{in}}, \tag{5.165}$$

and where ω and ω^* are thermal diffusion coefficients, the Ks are thermal conductivities, the Rs are diffusion thermal coefficients, and Δ_{in} is a correction factor for the ambipolar diffusion coefficient; the corresponding expressions are given in Appendix I. Typically, these processes are not as important in a partially ionized gas as they are in a fully ionized gas.

The accuracy of the various ion and neutral transport coefficients can be determined only in the limit of equal species temperatures. In this limit the 13-moment system of transport equations yields ordinary diffusion coefficients that correspond to the second approximation to these coefficients, while the resulting thermal diffusion coefficients and thermal conductivities correspond to the first approximation. The accuracies of these levels of approximation have been studied,[12,15–16] and it appears that for a fully ionized plasma the various ion transport coefficients are accurate to within 20–30%. For a partially ionized plasma, the transport coefficients are accurate to within 5%.

Sometimes it is useful to have an expression for the ion thermal conductivity that is not as complicated as those given by Equations (5.155), (5.156), and (5.162). A simplified expression can be derived with the aid of a few assumptions. The starting point for the derivation is the heat flow equation (5.151). Assuming that the $\mathbf{q}_t$ and $(\mathbf{u}_s - \mathbf{u}_t)$ terms are negligible and that $T_s \approx T_{st}$, Equation (5.151) can be simplified and written in the form

$$\mathbf{q}_s = -\lambda_s\nabla T_s, \tag{5.166}$$

where the thermal conductivity (λ_s) is given by

$$\lambda_s = \frac{25}{8}\frac{n_s k^2 T_s}{m_s\nu_{ss}}\left[1 + \frac{5}{4}\sum_{t\neq s}\frac{\nu_{st}}{\nu_{ss}}\left(D_{st}^{(1)} + \frac{3}{2}\frac{\mu_{st}}{m_s}\right)\right]^{-1}, \tag{5.167}$$

and where $p_s = n_s k T_s$ was used in arriving at Equation (5.167). Equations (5.166) and (5.167) can be applied to any species and the general expression for $D_{st}^{(1)}$ is given by Equation (4.130a). The expression in Equation (5.167) is what Chapman and Cowling call the first approximation to the thermal conductivity.[12]

For ions, a convenient form can be obtained by using Equation (4.142) for v_{ss}, and the thermal conductivity (5.167) becomes

$$\lambda_i = 3.1 \times 10^4 \frac{T_i^{5/2}}{M_i^{1/2}Z_i^4}\left[1 + \frac{5}{4}\sum_{t\neq i}\frac{v_{it}}{v_{ii}}\left(D_{it}^{(1)} + \frac{3}{2}\frac{\mu_{it}}{m_i}\right)\right]^{-1}, \qquad (5.168)$$

where the units are eV cm^{-1} s^{-1} K^{-1} and where M_i is the ion mass in atomic mass units and Z_i is the ion charge number. Note that subscript i is used in Equation (5.168) to emphasize that the expression only applies to ions. The summation over the subscript t pertains to neutrals and other ion species. Also, as noted above, Equation (5.167) corresponds to the first approximation to the thermal conductivity. For ions, this first approximation conductivity has to be corrected to achieve agreement with the more rigorous values obtained from a numerical solution of the Boltzmann equation.[7] The corrections are made by multiplying v_{ii} by 0.8, which was already done in arriving at the numerical factor 3.1×10^4 in Equation (5.168).

A more explicit numerical expression for λ_i can be obtained for a fully ionized plasma, and the result is

$$\lambda_i = 3.1 \times 10^4 \frac{T_i^{5/2}}{M_i^{1/2}Z_i^4}\left[1 + 1.75\sum_{j\neq i}\frac{Z_j^2\,n_j}{Z_i^2\,n_i}\left(\frac{M_j}{M_i + M_j}\right)^{1/2}\right.$$

$$\left. \times \frac{3M_i^2 + \frac{8}{5}M_iM_j + \frac{13}{10}M_j^2}{(M_i + M_j)^2}\right]^{-1}, \qquad (5.169)$$

where subscripts i and j are used to emphasize that the expression only applies to a plasma in which ion–ion collisions are dominant.

5.15 Summary of appropriate use of transport equations

The topics in this chapter progressed from very simple to more complex sets of transport equations. This progression had the advantage of clearly showing the reader, in a step-by-step fashion, how the various transport processes affect a plasma. However, for practical applications, it is usually the final, more complex, equations that are needed. Hence, for the practitioner, the following summary indicates what equations are needed for different applications:

1. In the *five-moment approximation*, stress and heat flow effects are assumed to be negligible, and the properties of the plasma are described by only five parameters (n_s, $\mathbf{u}_s$, T_s). The appropriate continuity, momentum, and

energy equations are given by Equations (5.22a–c). The collision terms that appear in these equations are valid for arbitrary temperature differences and arbitrary relative drifts between the interacting species. However, these equations cannot describe anisotropic pressure distributions, thermal diffusion, and thermal conduction effects because stress and heat flow are not considered.

2.   In a *weakly ionized plasma*, Coulomb collisions are negligible, and the drift speeds are typically subsonic, so that the diffusion approximation is valid. Under these circumstances, the appropriate continuity, momentum, and energy equations are given by Equations (5.48a–d). The continuity and momentum equations apply either along **B** for a planet with a strong intrinsic magnetic field or in the vertical direction for an unmagnetized planet. The stress tensor component $\tau_{i\parallel}$ is given by Equations (5.49a–c), but it is only important if there is a large relative ion–neutral drift in the horizontal direction.

3.   For *partially and fully ionized plasmas*, the general continuity, momentum, and energy equations are given by Equations (3.57–59). However, for a coordinate system that is fixed to a rotating planet, Coriolis and centripetal acceleration terms may need to be added to the momentum equation, and this equation is given by Equation (5.50). The general momentum equation for motion either along **B** for magnetized planets or in the vertical direction for unmagnetized planets can be simplified in the diffusion approximation (ambipolar, subsonic flow). For a three-component fully ionized plasma (electrons and two ion species), the ion diffusion equations are given by Equations (5.158) and (5.159). The associated collision-dominated expressions for the ion and electron heat flows are given by Equations (5.154–156). The ion conductivities are given in Appendix I and Equation (5.146) can be used for the electron thermal conductivity. For a three-component partially ionized plasma (electrons, one ion, and one neutral species), the appropriate ambipolar diffusion and heat flow equations are given by Equations (5.162–164), where again the conductivities are given in Appendix I. Simplified, albeit less rigorous, ion thermal conductivities are given by Equations (5.168) and (5.169). When thermal diffusion and diffusion thermal heat flow are not important, the ambipolar diffusion Equation (5.164) reduces to Equation (5.54), except for the $\tau_{i\parallel}$ term. This latter term is important if there are large relative ion–neutral drifts in the horizontal direction. In this case, the $\tau_{i\parallel}$ appropriate for a partially ionized plasma is given by Equation (5.149). Note that if thermal diffusion and stress are both important, the ambipolar diffusion Equation (5.164) must be augmented with the stress term that appears in Equation (5.54). For supersonic flow either along a magnetic field for a magnetized planet or in the vertical direction for an unmagnetized planet, the momentum equation must include the inertial terms. For steady state and simple time-dependent expansions, the appropriate momentum equations are, respectively, Equations (5.87) and (5.89).

4.   For *diffusion across* **B**, all levels of ionization generally need to be considered. The equation describing the diffusion of charged particles across **B** is Equation (5.103), which depends on the collision-to-cyclotron frequency ratio of the charged particle under consideration. In the small collision frequency limit, only the **E** × **B** (5.99), diamagnetic (5.100), and gravitational (5.101) drifts survive, which are perpendicular to both **B** and the force causing the drift. In the high collision-frequency limit, the drift is in the direction of the forces causing the drift (5.27).

5.   *Electrical currents* typically flow along **B** from high to low altitudes, across **B** at low altitudes, and then back up along **B** to high altitudes, forming an electrical circuit that spans all levels of ionization. The current flow across **B** is given by Equation (5.116), and the associated Pedersen and Hall conductivities are given by Equations (5.119) and (5.120), respectively. The current flow along **B** is given by Equation (5.140). The associated electrical conductivity and the current flow conductivity due to thermal gradients are given by Equations (5.142) and (5.143), respectively, for a fully ionized plasma. The g-correction factors are given in Table 5.1 for a fully ionized plasma. For a partially ionized plasma, the quantity ν_{ei} in the expressions for the conductivities must be replaced with the total electron collision frequency (5.123), and the g-correction factors become dependent on the degree of ionization. The variation of the g-correction factors with the ratio ν_{en}/ν_{ei} is given in Reference 9.

5.16   Specific references

1. Schunk, R. W., Transport equations for aeronomy, *Planet. Space Sci.*, **23**, 437, 1975.
2. Banks, P. M., and T. E. Holzer, High-latitude plasma transport: the polar wind, *J. Geophys. Res.*, **74**, 6317, 1969.
3. Gurevich, A. V., L. V. Pariiskaya, and L. P. Pitaevskii, Self-similar motion of a rarefied plasma, *Sov. Phys. JETP Engl. Transl.* **22**, 449, 1966.
4. Singh, N., and R. W. Schunk, Numerical calculations relevant to the initial expansion of the polar wind, *J. Geophys. Res.*, **87**, 9154, 1982.
5. Schunk, R. W., and E. P. Szuszczewicz, Plasma expansion characteristics of ionized clouds in the ionosphere: macroscopic formulation, *J. Geophys. Res.*, **96**, 1337, 1991.
6. Schunk, R. W., Mathematical structure of transport equations for multispecies flows, *Rev. Geophys. Space Phys.*, **15**, 429, 1977.
7. Spitzer, L., and R. Härm, Transport phenomena in a completely ionized gas, *Phys. Rev.*, **89**, 977, 1953.
8. Shkarofsky, I. P., Values of transport coefficients in a plasma for any degree of ionization based on a Maxwellian distribution, *Can. J. Phys.*, **39**, 1619, 1961.
9. Schunk, R. W., and J. C. G. Walker, Transport properties of the ionospheric electron gas, *Planet. Space Sci.*, **18**, 1535, 1970.

10. Banks, P. M., Charged particle temperatures and electron thermal conductivity in the upper atmosphere, *Annls. Geophys.*, **22**, 577, 1966.

11. Nagy, A. F., and T. E. Cravens, Ionosphere: energetics, in *Venus II*, University of Arizona Press, 1997.

12. Chapman, S., and T. G. Cowling, *The Mathematical Theory of Non-Uniform Gases*, New York: Cambridge University Press, New York: 1970.

13. Conrad, J. R., and R. W. Schunk, Diffusion and heat flow equations with allowance for large temperature differences between interacting species, *J. Geophys. Res.*, **84**, 811, 1979.

14. Schunk, R. W., and J. C. G. Walker, Thermal diffusion in the topside ionosphere for mixtures which include multiply-charged ions, *Planet. Space Sci.*, **17**, 853, 1969.

15. Meador, W. E., and L. D. Staton, Electrical and thermal properties of plasmas, *Phys. Fluids*, **8**, 1694, 1965.

16. Devoto, R. S., Transport properties of ionized monoatomic gases, *Phys. Fluids*, **9**, 1230, 1966.

5.17   General references

Burgers, J. M., *Flow Equations for Composite Gases*, New York: Academic, 1969.

Chapman, S., and T. G. Cowling, *The Mathematical Theory of Non-Uniform Gases*, New York: Cambridge University Press, 1970.

Grad, H., Principles of the kinetic theory of gases, *Handb. Phys.*, **XII**, 205, 1958.

Hirschfelder, J. O., C. F. Curtiss, and R. G. Bird, *Molecular Theory of Gases and Liquids*, New York: Wiley, 1964.

Kennard, E. H., *Kinetic Theory of Gases*, New York: McGraw-Hill, 1938.

Present, R. D., *Kinetic Theory of Gases*, New York: McGraw-Hill, 1958.

Schunk, R. W., Transport equations for aeronomy, *Planet. Space Sci.*, **23**, 437, 1975.

Schunk, R. W., Mathematical structure of transport equations for multispecies flows, *Rev. Geophys. Space Phys.*, **15**, 429, 1977.

Spitzer, L., *Physics of Fully Ionized Gases*, New York: Wiley, 1967.

Tanenbaum, B. S., *Plasma Physics*, New York: McGraw-Hill, 1967.

5.18   Problems

Problem 5.1 Show that the density expression given in Equation (5.7) is a solution to Equation (5.6).

Problem 5.2 Consider the simple scenario where a gas with a constant density and temperature has a flow velocity in the x-direction. If the velocity varies with x, $u_x(x)$, calculate the viscous stress component τ_{xx} using a mean-free-path approach.

Problem 5.3 Consider the simple scenario where a gas flows between infinite parallel plates. The gas, which has a constant density and temperature, flows in the

x-direction and the parallel plates are at $y = 0$ and a. The plate velocities are V_0 at $y = 0$ and V_1 at $y = a$. Calculate the velocity component $u_x(y)$.

Problem 5.4 Consider a stationary gas with a constant density. The gas is confined between infinite parallel plates, which are located at $y = 0$ and a. The plate temperatures are T_0 at $y = 0$ and T_1 at $y = a$. Assume thermal conduction dominates the energy balance ($\nabla \cdot \mathbf{q} \approx 0$) and that the thermal conductivity of the gas is given by Equation (5.20), with ν constant. Calculate the temperature as a function of y.

Problem 5.5 Show that Equations (5.35) and (5.36) are solutions to the momentum (5.31) and energy (5.32) equations.

Problem 5.6 Show that the expression for the isotropic stress tensor (5.42) is the solution to Equation (5.40). Note that in index notation the isotropic stress tensor is given by $(\tau_i)_{\alpha\beta} = \tau_{i\parallel} b_\alpha b_\beta + \tau_{i\perp}(\delta_{\alpha\beta} - b_\alpha b_\beta)$, where α and β are the coordinate indices.

Problem 5.7 Calculate plasma scale heights (Equation 5.59) for Venus, Earth, and Mars at an altitude of 400 km. Assume that $T_p = 1000$ K.

Problem 5.8 In deriving the ambipolar diffusion equation for a minor ion (5.72), it was implicitly assumed that the minor ion is singly charged. Derive the ambipolar diffusion equation for the case when the minor ion is multiply charged, and then obtain the two linearly independent solutions for multiply charged minor ions that are equivalent to Equations (5.78a,b).

Problem 5.9 Derive a Mach number equation for a minor ion species that is similar to Equation (5.87). Assume that the major ions and electrons are in diffusive equilibrium and that $n_e = n_i$, where subscript i corresponds to the major ion. Adopt the same assumptions used in the derivation of Equation (5.87).

Problem 5.10 Show that the solution of Equation (5.102) for $\mathbf{u}'_{j\perp}$ is given by Equation (5.103).

Problem 5.11 Show that the solution of Equation (5.109) for $\mathbf{u}'_i$ is given by Equation (5.110).

Problem 5.12 Show that the weakly ionized expression for the stress tensor (5.49a) is modified for the case of a single-ion, partially ionized plasma, and that the result is given by Equation (5.149).

Problem 5.13 Show that the fully ionized expression for the ion thermal conductivity (5.169) follows from Equation (5.168) and then calculate the ion conductivities for a plasma composed of H^+, He^+, and O^+.

Problem 5.14 Consider a partially ionized, electrically neutral, four-component plasma composed of hot electrons ($n_h, \mathbf{u}_h, T_h$), cold electrons ($n_c, \mathbf{u}_c, T_c$), ions

$(n_i, \mathbf{u}_i, T_i)$, and one neutral species $(n_n, \mathbf{u}_n = 0, T_n)$. Derive an ambipolar diffusion equation for the plasma. Include gravity and temperature gradients, but ignore $\mathbf{B}$.

Problem 5.15 For the four-component plasma described in Problem 5.14, derive a Mach number equation for steady state, ambipolar, supersonic plasma flow along a strong diverging magnetic field.

Chapter 6

Wave phenomena

Plasma waves are prevalent throughout the ionospheres. The waves can just have fluctuating electric fields or they can have both fluctuating electric and magnetic fields. Also, the wave amplitudes can be either small or large, depending on the circumstances. Small amplitude waves do not appreciably affect the plasma, and in many situations they can be used as a diagnostic of physical processes that are operating in the plasma. Large amplitude waves, on the other hand, can have a significant effect on the plasma dynamics and energetics. In general, there is a myriad of waves that can propagate in a plasma, and it is not possible, or warranted, to give a detailed discussion here. Instead, the focus in this chapter is on just the fundamental wave modes that can propagate in both magnetized and unmagnetized plasmas. First, the general characteristics of waves are presented. This is followed by a discussion of small amplitude waves in both unmagnetized and magnetized plasmas, including high frequency (electron) waves and low frequency (ion) waves. Next, the effect that collisions have on the waves is illustrated, and this is followed by a presentation of wave excitation mechanisms (plasma instabilities). Finally, large amplitude shock waves and double layers are discussed.

6.1 General wave properties

Many types of waves can exist in the plasma environments that characterize the ionospheres. Hence, it is useful to first introduce some common wave nomenclature before discussing the various wave types. It is also useful to distinguish between background plasma properties and wave induced properties. In what follows, subscript 0 designates background plasma properties, and subscript 1 designates both the wave and the perturbed plasma properties associated with the wave. The waves can be *electrostatic*, for which there is only a fluctuating electric field, $\mathbf{E}_1$, or

159

Table 6.1 Wave characteristics.

Electrostatic wave	$\mathbf{E}_1 \neq 0, \mathbf{B}_1 = 0$
Electromagnetic wave	$\mathbf{E}_1 \neq 0, \mathbf{B}_1 \neq 0$
Longitudinal mode	$\mathbf{E}_1 \parallel \mathbf{K}$
Transverse mode	$\mathbf{E}_1 \perp \mathbf{K}$
Parallel propagation	$\mathbf{K} \parallel \mathbf{B}_0$
Perpendicular propagation	$\mathbf{K} \perp \mathbf{B}_0$
Cut-off	$K \to 0$
Resonance	$K \to \infty$

electromagnetic, for which there are both fluctuating electric, $\mathbf{E}_1$, and magnetic, $\mathbf{B}_1$, fields. In the case of a *longitudinal mode*, the *propagation constant*, $\mathbf{K}$, which defines the direction of propagation of the wave, and the fluctuating electric field, $\mathbf{E}_1$, are parallel, whereas for a *transverse mode* they are perpendicular. Also, in a magnetized plasma with a background magnetic field, $\mathbf{B}_0$, the waves can propagate along the magnetic field ($\mathbf{K} \parallel \mathbf{B}_0$), perpendicular to it ($\mathbf{K} \perp \mathbf{B}_0$), or at an arbitrary angle. For easy reference, the nomenclature is summarized in Table 6.1.

The starting point for a discussion of wave phenomena is the set of Maxwell equations (3.76a–d). For electrostatic waves ($\mathbf{B}_1 = 0$), only two of the four Maxwell equations are relevant, and these are

$$\nabla \cdot \mathbf{E}_1 = \rho_{1c}/\varepsilon_0, \tag{6.1}$$

$$\nabla \times \mathbf{E}_1 = 0, \tag{6.2}$$

where $\rho_{1c} = \sum_s e_s n_{s1}$ is the perturbed charge density. The curl equation (6.2) can be satisfied by introducing a scalar potential, Φ_1, such that

$$\mathbf{E}_1 = -\nabla \Phi_1 \tag{6.3}$$

because $\nabla \times (\nabla \Phi_1) = 0$ (Appendix B). The substitution of Equation (6.3) into Equation (6.1) then yields a second-order, partial differential equation for the potential, which is known as the *Poisson equation*, and is given by

$$\nabla^2 \Phi_1 = -\rho_{1c}/\varepsilon_0. \tag{6.4}$$

For electrostatic waves, the effect of the plasma enters through the perturbed charge density. Given a knowledge of $\rho_{1c}(\mathbf{r}, t)$, the perturbed potential can be obtained from a solution of Equation (6.4), and then $\mathbf{E}_1$ can be obtained from Equation (6.3). However, as will be discussed later, for small amplitude, sinusoidal waves, the electrostatic waves are longitudinal ($\mathbf{K} \parallel \mathbf{E}_1$). In this case, Equation (6.2) is automatically satisfied and only Equation (6.1) needs to be considered. For some electrostatic waves, charge neutrality is maintained not only in the background plasma, but also in the plasma wave perturbation. For these waves, equation (6.1) can be replaced by the charge neutrality condition $\sum_s e_s n_{s1} = 0$.

The full set of Maxwell equations is needed for electromagnetic waves ($\mathbf{E}_1 \neq 0$, $\mathbf{B}_1 \neq 0$). However, a more useful form of these equations can be obtained by taking the curl of the curl equations (3.76b,d). For example, the curl of Faraday's law (3.76b) yields

$$\nabla \times (\nabla \times \mathbf{E}_1) = -\frac{\partial}{\partial t}(\nabla \times \mathbf{B}_1), \tag{6.5}$$

where the spatial and temporal derivatives can be interchanged for coordinate systems that are fixed in space. Now, substituting Ampère's law (3.76d) into Equation (6.5) and using the vector relation $\nabla \times (\nabla \times \mathbf{E}_1) = \nabla(\nabla \cdot \mathbf{E}_1) - \nabla^2 \mathbf{E}_1$ (Appendix B) yields the following equation:

$$\nabla^2 \mathbf{E}_1 - \mu_0 \varepsilon_0 \frac{\partial^2 \mathbf{E}_1}{\partial t^2} - \nabla(\nabla \cdot \mathbf{E}_1) = \mu_0 \frac{\partial \mathbf{J}_1}{\partial t}, \tag{6.6}$$

where $\mathbf{J}_1 = \sum_s e_s (n_s \mathbf{u}_s)_1$ is the perturbed current density.

A similar equation for $\mathbf{B}_1$ can be obtained by first taking the curl of Ampère's law (3.76d) and then performing manipulations similar to those that led to Equation (6.6). However, in practice, the electric field, $\mathbf{E}_1$, is typically obtained first from Equation (6.6), and then the associated magnetic field, $\mathbf{B}_1$, is obtained from Faraday's law, which is $\nabla \times \mathbf{E}_1 = -\partial \mathbf{B}_1/\partial t$ (3.76b). Note that for electromagnetic waves the effect of the plasma can enter through both the perturbed current density, $\mathbf{J}_1$, and the perturbed charge density, ρ_{1c}, via $\nabla \cdot \mathbf{E}_1$ (3.76a).

In a vacuum ($\rho_{1c} = 0$, $\mathbf{J}_1 = 0$), Equation (6.6) reduces to the classical wave equation, which is

$$\nabla^2 \mathbf{E}_1 - \frac{1}{c^2} \frac{\partial^2 \mathbf{E}_1}{\partial t^2} = 0 \tag{6.7}$$

where $\nabla \cdot \mathbf{E}_1 = 0$ (6.2) and $c = 1/\sqrt{\mu_0 \varepsilon_0}$ is the *speed of light*.

For the special case of small amplitude, sinusoidal perturbations, the fluctuating electric field can be expressed in the form

$$\mathbf{E}_1(\mathbf{r}, t) = \mathbf{E}_{10} \cos(\mathbf{K} \cdot \mathbf{r} - \omega t) \tag{6.8}$$

where $\mathbf{E}_{10}$ is a constant vector, $\mathbf{K}$ is the *propagation vector* of the wave, and ω is the *wave frequency*. The magnitude of $\mathbf{K}$ is the *wave number* and it is related to the *wavelength*, λ, by $K = 2\pi/\lambda$. Waves of the type given in Equation (6.8) are known as *plane waves*. Note that for plane waves, the spatial, $\mathbf{r}$, and temporal, t, variations appear in the cosine function, and that they are characterized by a single frequency, ω, and propagation vector, $\mathbf{K}$. Also, for plane waves, it is mathematically convenient to introduce complex functions, so that Equation (6.8) can be expressed in the form

$$E_1(\mathbf{r}, t) = \mathbf{E}_{10} e^{i(\mathbf{K} \cdot \mathbf{r} - \omega t)} \tag{6.9}$$

where $e^{i\alpha} = \cos \alpha + i \sin \alpha$, and i is the square root of minus one. Therefore, the original form given in Equation (6.8) can be recovered simply by taking the real part of the expression in Equation (6.9). The advantage of using Equation (6.9) is that when ∇ and $\partial/\partial t$ operate on this exponential form of a plane wave, they become

$$\nabla \to i\mathbf{K}, \quad \frac{\partial}{\partial t} \to -i\omega. \tag{6.10}$$

There are several other important wave properties that should be noted. The first concerns the phase of the cosine function. This function has a constant phase when $(\mathbf{K} \cdot \mathbf{r} - \omega t)$ is constant. The velocity at which a constant phase propagates is called the *phase velocity*, V_{ph}, and it is given by

$$V_{\text{ph}} = \frac{\omega}{K}. \tag{6.11}$$

The velocity at which the energy or information propagates is called the *group velocity*, V_{g}, and it is given by

$$V_{\text{g}} = \frac{d\omega}{dK}. \tag{6.12}$$

The flow of energy (energy per unit area and per unit time) for an electromagnetic wave is in the direction of $\mathbf{K}$ and it is given by the *Poynting vector*, which is

$$\mathbf{S} = \mathbf{E}_1 \times \mathbf{H}_1. \tag{6.13}$$

For sinusoidal waves (6.9), the Poynting vector is time dependent. However, what is generally of interest is the time-averaged flow of energy, which can be calculated from the expression[1]

$$\langle \mathbf{S} \rangle = \frac{1}{2} \text{Re} \, \mathbf{E}_1 \times \mathbf{H}_1^* \tag{6.14}$$

where $\mathbf{H}_1^*$ is the complex conjugate of $\mathbf{H}_1$ and "Re" means that the real part of the expression should be used.

The substitution of the plane wave solution (6.9) into the vacuum wave equation (6.7) leads to a relation between K and ω, which is called the *dispersion relation*, and the result is

$$\omega^2 = c^2 K^2. \tag{6.15}$$

Note that in this case, the phase and group velocities are the same

$$\frac{\omega}{K} = \frac{d\omega}{dK} = \pm c \tag{6.16}$$

where the $\pm$ sign indicates that the waves can propagate in opposite directions. Also, although Equation (6.15) is called a dispersion relation, there is no dispersion in this case, because the phase velocity (6.16) does not depend on frequency.

Equation (6.15) indicates that if a plane wave solution is assumed, the wave number, K, and the wave frequency, ω, are related. However, this is not the only restriction on the wave parameters $(\mathbf{E}_1, \mathbf{B}_1, \mathbf{K}, \omega)$. From Faraday's law (3.76b), it is clear that if $\mathbf{E}_1$ has a plane wave form, then $\mathbf{B}_1$ must also have this form, otherwise Faraday's law cannot be satisfied. In fact, all four of the Maxwell equations (3.76a–d) must be satisfied if the plane wave solution is correct. When the plane wave solution (6.9) is substituted into Maxwell's vacuum equations ($\rho_{1c} = 0$, $\mathbf{J}_1 = 0$), the following additional constraints on the wave parameters are obtained:

$$\mathbf{K} \cdot \mathbf{E}_1 = 0, \tag{6.17a}$$

$$\mathbf{K} \times \mathbf{E}_1 = \omega \mathbf{B}_1, \tag{6.17b}$$

$$\mathbf{K} \cdot \mathbf{B}_1 = 0, \tag{6.17c}$$

$$\mathbf{K} \times \mathbf{B}_1 = -\frac{\omega}{c^2} \mathbf{E}_1. \tag{6.17d}$$

These additional constraints indicate that $\mathbf{K}$, $\mathbf{E}_1$, and $\mathbf{B}_1$ are perpendicular to each other and that $\mathbf{E}_1 \times \mathbf{B}_1$ points in the direction of $\mathbf{K}$, as shown in Figure 6.1. Such an electromagnetic wave is called a *transverse wave*, because $\mathbf{E}_1$ and $\mathbf{B}_1$ are perpendicular to the direction of propagation of the wave. Equations (6.17b) and (6.17d)

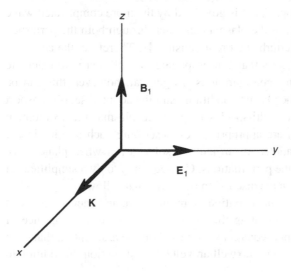

Figure 6.1 Directions of the wave parameters for a transverse electromagnetic wave propagating in a vacuum. The dispersion relation is $\omega^2 = c^2 K^2$.

also indicate that the magnitudes of the fluctuating electric and magnetic fields are related, and the relation is $E_1 = cB_1$.

In a plasma, the perturbed charge density, ρ_{1c}, and current density, $\mathbf{J}_1$, must also have a plane wave form similar to Equation (6.9), when the electric, $\mathbf{E}_1$, and magnetic, $\mathbf{B}_1$, field perturbations have this form. For *electrostatic waves*, either Gauss' law (6.1) or the Poisson equation (6.4) is appropriate, and when a plane wave solution is assumed, these equations become, respectively,

$$i\mathbf{K} \cdot \mathbf{E}_1 = \rho_{1c}/\varepsilon_0 \tag{6.18}$$

and

$$K^2 \Phi_1 = \rho_{1c}/\varepsilon_0. \tag{6.19}$$

For *electromagnetic waves*, the general wave equation (6.6) is appropriate. When a plane wave solution is assumed, this equation becomes

$$\left(\frac{\omega^2}{c^2} - K^2\right)\mathbf{E}_1 + \mathbf{K}\left(\mathbf{K} \cdot \mathbf{E}_1\right) = -i\omega\mu_0\mathbf{J}_1. \tag{6.20}$$

Note that for either electrostatic or electromagnetic waves, the wave equations, which are partial differential equations, become algebraic equations when plane wave solutions are assumed.

6.2   Plasma dynamics

As noted in the previous section, the propagation of electrostatic waves in a plasma is governed by either Gauss' law (6.18) or the Poisson equation (6.19), and the effect of plasma enters through the perturbed charge density, ρ_{1c}. On the other hand, for electromagnetic waves the propagation is governed by the more complicated wave equation (6.20), and the effect of the plasma can enter through both the perturbed charge density, ρ_{1c}, and the perturbed current density, $\mathbf{J}_1$. Therefore, the next step in determining the types of waves that can propagate in a plasma is to calculate ρ_{1c} and $\mathbf{J}_1$ for different plasma configurations. In general, however, this can be difficult, depending on both the plasma conditions and the adopted set of transport equations. For example, for a multi-species magnetized plasma, the 13-moment transport equations (3.57–61) are appropriate for describing each species in the plasma. However, these equations are difficult to solve, even when plane wave solutions can be assumed for the perturbations. Consequently, only a simplified set of transport equations is used in the discussion of waves that follows.

It is assumed that the five-moment continuity, momentum, and energy equations (5.22a–c) are adequate for representing the plasma dynamics in the presence of waves. These simplified transport equations are based on the assumption that each species in the plasma has a drifting Maxwellian velocity distribution. In addition to this limitation, gravity and collisions are ignored. In a plasma, the electrodynamic forces are much more important than gravity, and hence, the neglect of gravity in calculating the normal wave modes is not restrictive. However, gravity is important for wave phenomena in neutral atmospheres, and this is discussed in Chapter 10. The effect that collisions have on waves is discussed in Section 6.12.

With the above simplifications, the continuity, momentum, and energy equations (5.22a–c) become

$$\frac{\partial n_s}{\partial t} + \nabla \cdot (n_s \mathbf{u}_s) = 0, \tag{6.21}$$

$$n_s m_s \left[\frac{\partial \mathbf{u}_s}{\partial t} + (\mathbf{u}_s \cdot \nabla)\mathbf{u}_s \right] + \nabla p_s - n_s e_s(\mathbf{E} + \mathbf{u}_s \times \mathbf{B}) = 0, \tag{6.22}$$

$$\frac{D_s p_s}{Dt} + \gamma p_s(\nabla \cdot \mathbf{u}_s) = 0, \tag{6.23}$$

where $\gamma = 5/3$ is the ratio of specific heats, which follows from the Euler equations.

The energy equation (6.23) can be cast in a more convenient form with the aid of the continuity equation (6.21), which indicates that

$$\nabla \cdot \mathbf{u}_s = -\frac{1}{n_s}\frac{D_s n_s}{Dt}. \tag{6.24}$$

When this expression is substituted into the energy equation (6.23), the result is

$$\frac{D_s p_s}{Dt} - \frac{\gamma p_s}{n_s}\frac{D_s n_s}{Dt} = 0, \tag{6.25}$$

which can also be written as

$$\frac{D_s}{Dt}\left(\frac{p_s}{\rho_s^\gamma}\right) = 0 \tag{6.26a}$$

or

$$\frac{p_s}{\rho_s^\gamma} = \text{constant.} \tag{6.26b}$$

Equation (6.26b) is known as the *equation of state* for a plasma. Although Equation (6.26b) was derived from the Euler energy equation (6.23), for which $\gamma = 5/3$, the equation of state (6.26b) is frequently used with other values of γ. Note that $\gamma = 5/3$ corresponds to an adiabatic flow and $\gamma = 1$ corresponds to an isothermal flow.

In the momentum equation (6.22), ∇p_s is needed, and using Equation (6.26b), this term can be expressed in the form

$$\nabla p_s = \frac{\gamma p_s}{\rho_s}\nabla \rho_s = \frac{\gamma k T_s}{m_s}\nabla \rho_s, \tag{6.27}$$

where $p_s = n_s k T_s$ has been used to obtain the second expression in Equation (6.27). At this point, it is useful to generalize the expression for γ by letting it be different for different species in the plasma ($\gamma \rightarrow \gamma_s$). With the use of a generalized γ_s in Equation (6.27), the momentum equation (6.22) becomes

$$n_s m_s \left[\frac{\partial \mathbf{u}_s}{\partial t} + (\mathbf{u}_s \cdot \nabla) \mathbf{u}_s \right] + \gamma_s k T_s \nabla n_s - n_s e_s (\mathbf{E} + \mathbf{u}_s \times \mathbf{B}) = 0. \qquad (6.28)$$

The transport equations that describe the plasma dynamics in response to waves are the continuity equation (6.21) and the momentum equation (6.28). The normal wave modes that can propagate in a plasma governed by these transport equations (or any other set of transport equations) are obtained as follows. First, the equilibrium state of the plasma has to be calculated. Then, the equilibrium state is disturbed by adding small perturbations to the plasma and electromagnetic parameters (n_s, $\mathbf{u}_s$, $\mathbf{E}$, and $\mathbf{B}$). Next, the perturbed parameters are substituted into the transport equations, and the transport equations are linearized because the perturbations are small. The perturbed parameters are also assumed to be described by plane waves, and with this assumption, the partial differential equations for the perturbed parameters are converted into a set of linear algebraic equations. When the algebraic equations are solved, the result is the dispersion relation, which relates K and ω.

The above normal mode analysis will be applied to a simple plasma situation. Initially, the plasma is assumed to be electrically neutral, uniform, and steady. The initial density, n_{s0}, is therefore constant. The plasma is also assumed to have a constant drift ($\mathbf{u}_{s0}$) and to be subjected to perpendicular electric, $\mathbf{E}_0$, and magnetic, $\mathbf{B}_0$, fields. For this initial equilibrium state, the continuity equation (6.21) is automatically satisfied because n_{s0} and $\mathbf{u}_{s0}$ are constant. The momentum equation (6.28) indicates that the constant plasma drift that exists in the equilibrium state can have both parallel and perpendicular components, relative to $\mathbf{B}_0$. The perpendicular drift component is governed by

$$\mathbf{E}_0 + \mathbf{u}_{s0} \times \mathbf{B}_0 = 0. \qquad (6.29a)$$

The solution of this equation for $\mathbf{u}_{s0}$ leads to the well-known electrodynamics drift (Equation 5.99), which is

$$(\mathbf{u}_{s0})_\perp = \frac{\mathbf{E}_0 \times \mathbf{B}_0}{B_0^2}. \qquad (6.29b)$$

However, in the equilibrium state, the plasma can also have a constant drift, $(\mathbf{u}_{s0})_\parallel$, parallel to the magnetic field because such a drift satisfies the parallel component of the momentum equation (6.28). Therefore, in the equilibrium state, the total plasma drift is given by

$$\mathbf{u}_{s0} = (\mathbf{u}_{s0})_\parallel + (\mathbf{u}_{s0})_\perp. \qquad (6.30)$$

Now that the equilibrium state of the plasma is established, the characteristic waves (normal modes) that can propagate in the plasma can be calculated by perturbing this equilibrium state. This is accomplished by perturbing the plasma parameters and the electric and magnetic fields, as follows:

$$n_s(\mathbf{r}, t) = n_{s0} + n_{s1}(\mathbf{r}, t), \tag{6.31a}$$

$$\mathbf{u}_s(\mathbf{r}, t) = \mathbf{u}_{s0} + \mathbf{u}_{s1}(\mathbf{r}, t), \tag{6.31b}$$

$$\mathbf{E}(\mathbf{r}, t) = \mathbf{E}_0 + \mathbf{E}_1(\mathbf{r}, t), \tag{6.31c}$$

$$\mathbf{B}(\mathbf{r}, t) = \mathbf{B}_0 + \mathbf{B}_1(\mathbf{r}, t), \tag{6.31d}$$

where the subscript 1 is used to denote a small perturbation. These perturbed quantities are then substituted into the continuity (6.21) and momentum (6.28) equations, and this yields equations that govern the behavior of n_{s1} and $\mathbf{u}_{s1}$.

The substitution of Equations (6.31a–d) into the continuity equation (6.21) yields

$$\frac{\partial}{\partial t}(n_{s0} + n_{s1}) + \nabla \cdot \left[(n_{s0} + n_{s1})(\mathbf{u}_{s0} + \mathbf{u}_{s1})\right] = 0. \tag{6.32}$$

The perturbations are assumed to be small, and hence, nonlinear terms like $n_{s1}\mathbf{u}_{s1}$ are negligible compared with linear terms. Also, other terms drop out because n_{s0} and $\mathbf{u}_{s0}$ are constant, and therefore, Equation (6.32) reduces to

$$\frac{\partial n_{s1}}{\partial t} + n_{s0}\nabla \cdot \mathbf{u}_{s1} + \mathbf{u}_{s0} \cdot \nabla n_{s1} = 0. \tag{6.33}$$

Note that Equation (6.33) is linear in the perturbed quantities, as it should be.

The perturbed momentum equation is obtained in a similar fashion. The substitution of the perturbed quantities (6.31a–d) into the momentum equation (6.28) yields

$$(n_{s0} + n_{s1})m_s\left[\frac{\partial}{\partial t}(\mathbf{u}_{s0} + \mathbf{u}_{s1}) + (\mathbf{u}_{s0} + \mathbf{u}_{s1}) \cdot \nabla(\mathbf{u}_{s0} + \mathbf{u}_{s1})\right]$$

$$+ \gamma_s kT_s \nabla(n_{s0} + n_{s1})$$

$$- (n_{s0} + n_{s1})e_s\left[\mathbf{E}_0 + \mathbf{E}_1 + (\mathbf{u}_{s0} + \mathbf{u}_{s1}) \times (\mathbf{B}_0 + \mathbf{B}_1)\right] = 0. \tag{6.34}$$

Neglecting the nonlinear terms, taking account of the fact that the equilibrium parameters (n_{s0}, $\mathbf{u}_{s0}$, $\mathbf{E}_0$, $\mathbf{B}_0$) are constant, and using the equilibrium momentum equation (6.29a), leads to the following momentum equation for the perturbed quantities:

$$n_{s0}m_s\left[\frac{\partial \mathbf{u}_{s1}}{\partial t} + (\mathbf{u}_{s0} \cdot \nabla)\mathbf{u}_{s1}\right] + \gamma_s kT_s \nabla n_{s1}$$

$$- n_{s0}e_s(\mathbf{E}_1 + \mathbf{u}_{s1} \times \mathbf{B}_0 + \mathbf{u}_{s0} \times \mathbf{B}_1) = 0. \tag{6.35}$$

For small perturbations, the perturbed quantities can be described by plane waves;

$$n_{s1}, \mathbf{u}_{s1}, \mathbf{E}_1, \mathbf{B}_1 \propto e^{i(\mathbf{K}\cdot\mathbf{r}-\omega t)}. \tag{6.36}$$

Substituting the plane-wave solution (6.36) into the perturbed continuity (6.33) and momentum (6.35) equations, and remembering that $\nabla \to i\mathbf{K}$ and $\partial/\partial t \to -i\omega$ (Equation 6.10), yields

$$(\omega - \mathbf{K}\cdot\mathbf{u}_{s0})n_{s1} = n_{s0}\mathbf{K}\cdot\mathbf{u}_{s1} \tag{6.37}$$

$$i(\omega - \mathbf{K}\cdot\mathbf{u}_{s0})\mathbf{u}_{s1} - i\mathbf{K}\frac{\gamma_s kT_s}{n_{s0}m_s}n_{s1} + \frac{e_s}{m_s}(\mathbf{E}_1 + \mathbf{u}_{s1}\times\mathbf{B}_0 + \mathbf{u}_{s0}\times\mathbf{B}_1) = 0. \tag{6.38}$$

Note that by assuming a plane wave solution, the partial differential equations (6.33) and (6.35) for the perturbed density, n_{s1}, and drift velocity, $\mathbf{u}_{s1}$, are converted into algebraic equations.

For electrostatic waves ($\mathbf{B}_1 = 0$), the perturbed parameters ($n_{s1}, \mathbf{u}_{s1}, \mathbf{E}_1$) are governed by the continuity equation (6.37), the momentum equation (6.38), and Gauss' law (6.18). For electromagnetic waves, the same continuity (6.37) and momentum (6.38) equations govern the behavior of n_{s1} and $\mathbf{u}_{s1}$, respectively, but $\mathbf{E}_1$ is described by the perturbed wave equation (6.20). Also, $\mathbf{B}_1$ is obtained from the plane wave form of Faraday's law (6.17b).

Depending on what wave modes are of interest, the relevant equations can be solved for the unknown parameters ($n_{s1}, \mathbf{u}_{s1}, \mathbf{E}_1, \mathbf{B}_1$), and the result is a dispersion relation that describes the characteristics of the waves. This can be done for the general case, but the resulting dispersion relation is complex. The alternative approach is to study the different wave modes separately, which is more instructive, and this is what is done in the sections that follow.

First, the dispersion relations for *electrostatic waves* in an *unmagnetized plasma* are derived, including high frequency (electron plasma) and low frequency (ion-acoustic) waves. Then, *electrostatic waves* in a *magnetized plasma* are discussed, and this again includes both high frequency (upper hybrid) and low frequency (lower hybrid and ion-cyclotron) waves. Next, the dispersion relation for *electromagnetic waves* in an *unmagnetized plasma* is presented. This is followed by a discussion of *electromagnetic waves* in a *magnetized plasma*, including both high frequency (ordinary, extraordinary, L, and R) and low frequency (Alfvén and magnetosonic) waves.

6.3   Electron plasma waves

Electron plasma waves are high frequency electrostatic waves that can propagate in any direction in an unmagnetized plasma and along the magnetic field in a magnetized plasma. The basic characteristics of these waves can be elucidated by considering a two-component, fully ionized plasma that is electrically neutral ($n_{e0} = n_{i0}$), stationary ($\mathbf{u}_{e0} = \mathbf{u}_{i0} = 0$), uniform, and steady. Also, there are no imposed electric or magnetic fields ($\mathbf{E}_0 = \mathbf{B}_0 = 0$). The fact that the waves are *high frequency* means that the ions do not participate in the wave motion. Physically, the ion inertia is too large and the ions cannot respond to the rapidly fluctuating waves. Therefore, the ion equations of motion can be ignored, and the ions merely provide a stationary background of positive charge.

For these electrostatic waves, the relevant equations are the electron continuity (6.37) and momentum (6.38) equations and Gauss' law (6.18). With the above simplifications, these equations become

$$\omega n_{e1} = n_{e0}\mathbf{K} \cdot \mathbf{u}_{e1}, \tag{6.39a}$$

$$i\omega \mathbf{u}_{e1} - i\mathbf{K}\frac{\gamma_e k T_e}{n_{e0} m_e}n_{e1} - \frac{e}{m_e}\mathbf{E}_1 = 0, \tag{6.39b}$$

$$i\mathbf{K} \cdot \mathbf{E}_1 = -e n_{e1}/\varepsilon_0, \tag{6.39c}$$

where subscript $s = $ e for electrons and where $\rho_{1c} = e(n_{i1} - n_{e1}) = -e n_{e1}$, because the ions cannot respond to the high-frequency waves. The dispersion relation is obtained by solving Equations (6.39a–c) for the unknown perturbations (n_{e1}, $\mathbf{u}_{e1}$, $\mathbf{E}_1$). The same dispersion relation is obtained regardless of which parameter is solved for. The easiest solution is obtained by first taking the scalar product of $\mathbf{K}$ with the momentum equation (6.39b), which yields

$$i\omega(\mathbf{K} \cdot \mathbf{u}_{e1}) - i K^2 \frac{\gamma_e k T_e}{n_{e0} m_e}n_{e1} - \frac{e}{m_e}(\mathbf{K} \cdot \mathbf{E}_1) = 0. \tag{6.40}$$

When $\mathbf{K} \cdot \mathbf{u}_{e1}$ from the continuity equation (6.39a) and $\mathbf{K} \cdot \mathbf{E}_1$ from Gauss' law (6.39c) are substituted into (6.40), the result is

$$n_{e1}\left(-\omega^2 + \frac{\gamma_e k T_e}{m_e}K^2 + \frac{n_{e0}e^2}{m_e\varepsilon_0}\right) = 0. \tag{6.41}$$

Now, $n_{e1} \neq 0$, because the plasma was disturbed and, therefore, the solution to Equation (6.41) yields the dispersion relation for *electron plasma waves*, which is

$$\omega^2 = \omega_{p_e}^2 + 3V_e^2 K^2. \tag{6.42}$$

In Equation (6.42), ω_{p_e} is the *electron plasma frequency* and V_e is the *electron thermal speed*, and these are given by

$$\omega_{p_e} = \left(\frac{n_{e0}e^2}{\varepsilon_0 m_e}\right)^{1/2}, \tag{6.43}$$

$$V_e = \left(\frac{k T_e}{m_e}\right)^{1/2}. \tag{6.44}$$

Also, in Equation (6.42), $\gamma_e = 3$ was used because the density compressions are one-dimensional.[2] Note that ω_{pe} was introduced previously in Equation (2.6).

The so-called *cold plasma* ($T_e = 0$) approximation is assumed for many applications. In this case, the dispersion relation for electron plasma waves (6.42) becomes

$$\omega^2 = \omega_{pe}^2, \tag{6.45}$$

which describes *plasma oscillations*. Note that Equation (6.45) does not describe waves because K does not appear in this expression. In a cold plasma, a disturbance created locally does not propagate to other parts of the plasma, but remains a local disturbance.

6.4   Ion-acoustic waves

Ion-acoustic waves are the low frequency version of electron plasma waves. That is, they are low frequency electrostatic waves that can propagate in any direction in an unmagnetized plasma and along the magnetic field in a magnetized plasma. However, for these waves, the ion equations of motion must be considered in addition to the electron equations of motion. As with electron plasma waves, the basic characteristics of the ion-acoustic waves can be elucidated by considering a two-component, electrically neutral ($n_{e0} = n_{i0}$) plasma that is stationary ($\mathbf{u}_{e0} = \mathbf{u}_{i0} = 0$), uniform, and steady. It is also not subjected to either electric or magnetic fields ($\mathbf{E}_0 = \mathbf{B}_0 = 0$) in the analysis that follows.

The relevant equations for ion-acoustic waves are the electron and ion continuity (6.37) and momentum (6.38) equations and Gauss' law (6.18). The electron continuity and momentum equations, for the plasma under consideration, are the same as those previously given in Equations (6.39a,b). The ion continuity and momentum equations are similar, and are given by

$$\omega n_{i1} = n_{i0} \mathbf{K} \cdot \mathbf{u}_{i1}, \tag{6.46a}$$

$$i\omega \mathbf{u}_{i1} - i\mathbf{K} \frac{\gamma_i k T_i}{n_{i0} m_i} n_{i1} + \frac{e}{m_i} \mathbf{E}_1 = 0, \tag{6.46b}$$

where subscript i denotes ions. For this wave, Gauss' law (6.18) must take account of both the electron and ion density perturbations, and the correct form is given by

$$i\mathbf{K} \cdot \mathbf{E}_1 = e(n_{i1} - n_{e1})/\varepsilon_0. \tag{6.46c}$$

As was done for electron plasma waves, it is convenient to take the scalar product of $\mathbf{K}$ with the electron momentum equation (6.39b), which resulted in Equation (6.40). The electron continuity equation indicates that $\mathbf{K} \cdot \mathbf{u}_{e1} = \omega n_{e1}/n_{e0}$, and when this result is substituted into Equation (6.40), the result is

$$i\left(\omega^2 - K^2 \frac{\gamma_e k T_e}{m_e}\right)\frac{n_{e1}}{n_{e0}} - \frac{e}{m_e}\mathbf{K} \cdot \mathbf{E}_1 = 0. \tag{6.47}$$

When Equation (6.47) is multiplied by $m_e n_{e0}/i$, the equation becomes

$$\left[m_e\omega^2 - K^2(\gamma_e k T_e)\right]n_{e1} - \frac{en_{e0}}{i}\mathbf{K} \cdot \mathbf{E}_1 = 0. \tag{6.48}$$

Likewise, when the same algebraic manipulations are performed on the ion momentum equation (6.45b), a similar equation is obtained, and it is given by

$$\left[m_i\omega^2 - K^2(\gamma_i k T_i)\right]n_{i1} + \frac{en_{i0}}{i}\mathbf{K} \cdot \mathbf{E}_1 = 0. \tag{6.49}$$

The term containing m_e in Equation (6.48) can be neglected compared with the other terms in this momentum equation. This is equivalent to neglecting the electron inertial term, which is valid for low frequency waves. Neglecting the m_e term, adding Equations (6.48) and (6.49), and then dividing by m_i, leads to the following result:

$$\left(\omega^2 - K^2 \frac{\gamma_i k T_i}{m_i}\right)n_{i1} - K^2 \frac{\gamma_e k T_e}{m_i}n_{e1} = 0, \tag{6.50}$$

where the $\mathbf{K} \cdot \mathbf{E}_1$ terms cancel because $n_{e0} = n_{i0}$.

At this point it is necessary to obtain a relationship between n_{e1} and n_{i1}. This can be obtained by substituting $\mathbf{K} \cdot \mathbf{E}_1$ from Gauss' law (6.46c) into the electron momentum equation (6.48), and the result is

$$-K^2(\gamma_e k T_e)n_{e1} + \frac{e^2 n_{e0}}{\varepsilon_0}(n_{i1} - n_{e1}) = 0. \tag{6.51}$$

Multiplying Equation (6.51) by $\varepsilon_0/(e^2 n_{e0})$ and then solving for n_{e1} leads to the following equation:

$$n_{e1} = \frac{n_{i1}}{1 + \gamma_e K^2 \lambda_D^2} \tag{6.52}$$

where $\lambda_D = (\varepsilon_0 k T_e/e^2 n_{e0})^{1/2}$ is the Debye length (Equation 2.4).

The substitution of Equation (6.52) into Equation (6.50) leads to one equation for one unknown and, hence, to the dispersion relation for *ion plasma waves*, which is

$$\omega^2 = K^2\left(\frac{\gamma_i k T_i}{m_i} + \frac{\gamma_e k T_e}{m_i(1 + \gamma_e K^2 \lambda_D^2)}\right). \tag{6.53}$$

It is instructive to express the term $K^2\lambda_D^2$ in terms of the wavelength, λ, and the result is

$$K^2\lambda_D^2 = \frac{4\pi^2\lambda_D^2}{\lambda^2} \tag{6.54}$$

where $K = 2\pi/\lambda$. For long wavelength waves ($\lambda \gg \lambda_D$), $K^2\lambda_D^2 \ll 1$, and in this limit Equation (6.53) becomes the dispersion relation for *ion-acoustic (sound)* waves,

$$\omega^2 = K^2 V_S^2, \tag{6.55}$$

where V_S is the *ion-acoustic speed*,

$$V_S = \left(\frac{\gamma_i k T_i + \gamma_e k T_e}{m_i}\right)^{1/2}. \tag{6.56}$$

Note that V_S agrees with the expression introduced earlier if $\gamma_e = \gamma_i = 1$ (Equation 5.82). Also note that the dispersion relation for sound waves in a plasma (Equation 6.55) is similar to the dispersion relation for sound waves in a neutral gas (Equation 10.32). Finally, when $K^2\lambda_D^2 \ll 1$, $n_{e1} = n_{i1}$ (Equation 6.52) and, therefore, charge neutrality is maintained not only in the background plasma, but in the perturbation as well.

6.5   Upper hybrid oscillations

Upper hybrid oscillations are high frequency electrostatic oscillations that are directed perpendicular to a magnetic field. The fact that the oscillations are high frequency means that only the electron equations of motion are needed. As in the previous cases, the dispersion relation for upper hybrid oscillations is derived by considering a two-component, electrically neutral ($n_{e0} = n_{i0}$), stationary ($\mathbf{u}_{e0} = \mathbf{u}_{i0} = 0$), uniform, and steady plasma. The plasma is not subjected to an electric field ($\mathbf{E}_0 = 0$), but there is an imposed magnetic field, $\mathbf{B}_0$. It is also assumed that the plasma is cold ($T_e = 0$). This means that the pressure gradient term in the momentum equation is not considered. Without thermal motion, it is not possible to have a wave and, hence, the dispersion relation to be derived will actually describe localized oscillations.

The relevant equations for upper hybrid oscillations are the electron continuity (6.37) and momentum (6.38) equations and Gauss' law (6.18). With the above assumptions, these equations become

$$\omega n_{e1} = n_{e0}\mathbf{K} \cdot \mathbf{u}_{e1}, \tag{6.57a}$$

$$i\omega \mathbf{u}_{e1} - \frac{e}{m_e}(\mathbf{E}_1 + \mathbf{u}_{e1} \times \mathbf{B}_0) = 0, \tag{6.57b}$$

$$i\mathbf{K} \cdot \mathbf{E}_1 = -e n_{e1}/\varepsilon_0. \tag{6.57c}$$

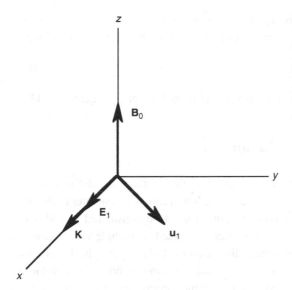

Figure 6.2 Directions of the wave and plasma parameters for electrostatic waves that propagate in a direction perpendicular to an imposed magnetic field $\mathbf{B}_0$. The velocity $\mathbf{u}_1$ is in the x–y plane. These directions are relevant to both upper hybrid and lower hybrid waves.

For simplicity, a Cartesian coordinate system is adopted with the magnetic field taken along the z-axis and $\mathbf{K}$ taken along the x-axis, as shown in Figure 6.2. The fluctuating electric field, $\mathbf{E}_1$, is also in the x-direction because $\mathbf{K} \parallel \mathbf{E}_1$ for electrostatic waves. However, $\mathbf{u}_{e1}$ has both x and y components. For this coordinate system, the continuity equation (6.57a) becomes

$$n_{e1} = \frac{n_{e0}K(u_{e1})_x}{\omega}. \tag{6.58}$$

An expression for $(u_{e1})_x$ can be obtained from the x and y components of the momentum equation (6.57b), which are given by

$$i\omega(u_{e1})_x - \frac{eE_1}{m_e} - \omega_{c_e}(u_{e1})_y = 0, \tag{6.59a}$$

$$i\omega(u_{e1})_y + \omega_{c_e}(u_{e1})_x = 0, \tag{6.59b}$$

where $\omega_{c_e} = eB_0/m_e$ is the electron cyclotron frequency (Equation 2.7). These two equations can be readily solved to obtain $(u_{e1})_x$, which is given by

$$(u_{e1})_x = \frac{-i\omega}{\omega^2 - \omega_{c_e}^2} \frac{eE_1}{m_e}. \tag{6.60}$$

The substitution of Equation (6.60) for $(u_{e1})_x$ into the equation for n_{e1} (6.58) leads to the following result:

$$n_{e1} = \frac{-iK}{\omega^2 - \omega_{c_e}^2} \frac{n_{e0}eE_1}{m_e}. \tag{6.61}$$

The final substitution of n_{e1} (6.61) into Gauss' law (6.57c) leads to an equation for E_1, from which the dispersion relation for *upper hybrid oscillations* is obtained

$$\omega^2 = \omega_{pe}^2 + \omega_{ce}^2, \tag{6.62}$$

where $\omega_{pe} = (n_{e0}e^2/\varepsilon_0 m_e)^{1/2}$ is the electron plasma frequency (Equation 6.43).

6.6   Lower hybrid oscillations

Lower hybrid oscillations are low frequency electrostatic oscillations that are directed perpendicular to a magnetic field. The low frequency character of the oscillations means that the ion equations of motion must be considered in addition to the electron equations. However, other than the need to include the ion motion, the plasma configuration is the same as that used to study upper hybrid oscillations (Figure 6.2). The relevant electron equations are the same as those given previously in Equations (6.57a,b). The ion continuity and momentum equations are similar to the electron equations, and for a cold plasma ($T_i = 0$) are given by

$$\omega n_{i1} = n_{i0}\mathbf{K} \cdot \mathbf{u}_{i1}, \tag{6.63a}$$

$$i\omega\mathbf{u}_{i1} + \frac{e}{m_i}(\mathbf{E}_1 + \mathbf{u}_{i1} \times \mathbf{B}_0) = 0. \tag{6.63b}$$

As was the case for ion plasma waves, charge neutrality can be assumed for low frequency waves, provided that the wavelengths are longer than the plasma Debye length (Equation 6.52). Therefore, instead of using Gauss' law (6.18), the fluctuating plasma is assumed to remain neutral, which means that

$$n_{e1} = n_{i1}. \tag{6.63c}$$

Previously, in the derivation of the dispersion relation for upper hybrid oscillations, the electron continuity and momentum equations were solved for the coordinate system shown in Figure 6.2, and the resulting expression for n_{e1} is given by Equation (6.61). Using similar mathematical manipulations, the ion continuity (6.63a) and momentum (6.63b) equations can be solved to yield an expression for n_{i1}, which is given by

$$n_{i1} = \frac{iK}{\omega^2 - \omega_{ci}^2} \frac{n_{i0}eE_1}{m_i}, \tag{6.64}$$

where $\omega_{ci} = eB_0/m_i$ is the *ion-cyclotron frequency* (Equation 2.7).

The charge neutrality condition (6.63c) indicates that n_{e1} (6.61) and n_{i1} (6.64) can be equated, and this yields the following equation:

$$\frac{-1}{m_e(\omega^2 - \omega_{ce}^2)} = \frac{1}{m_i(\omega^2 - \omega_{ci}^2)}, \tag{6.65}$$

where $n_{e0} = n_{i0}$. Equation (6.65) can be cast in the form

$$m_i(\omega^2 - \omega_{c_i}^2) = -m_e(\omega^2 - \omega_{c_e}^2),$$ (6.66)

which can also be written as

$$\omega^2(m_e + m_i) = m_e\omega_{c_e}^2 + m_i\omega_{c_i}^2 = e^2 B_0^2 \frac{(m_e + m_i)}{m_e m_i}$$

$$= \omega_{c_e}\omega_{c_i}(m_e + m_i).$$ (6.67)

Therefore, the final expression for *lower hybrid oscillations* is given by

$$\omega^2 = \omega_{c_e}\omega_{c_i}.$$ (6.68)

6.7 Ion-cyclotron waves

Ion-cyclotron waves are low frequency electrostatic waves that propagate in a direction that is *almost* perpendicular to a magnetic field. The difference between these waves and lower hybrid oscillations can be traced to how charge neutrality is maintained in the perturbed plasma. For lower hybrid oscillations, the propagation is *exactly* perpendicular to B_0. In this case, the electron (6.58) and ion (6.63a) continuity equations indicate that if $n_{e1} = n_{i1}$, then $(u_{e1})_x = (u_{i1})_x$. In other words, when the propagation is exactly perpendicular to B_0, the electrons must move across B_0 to maintain charge neutrality. However, the electrons can more easily move along B_0 than across B_0. Consequently, when K has a small parallel component, the ion motion across B_0 can be neutralized by an electron flow along B_0. This difference in charge neutralization leads to ion-cyclotron waves.

The plasma configuration considered here is the same as that used in the discussion of lower hybrid oscillations, except that here K is almost, but not exactly, perpendicular to B_0. Figure 6.3 shows the directions of the wave and plasma parameters for this case. Now, $K = K_x e_1 + K_z e_3$, where (e_1, e_2, e_3) are unit vectors for the Cartesian coordinate system shown in this figure. However, $K_x \gg K_z$ because the direction of propagation is almost perpendicular to B_0. The ion motion is predominantly across B_0 and is the same as that calculated for lower hybrid oscillations. Therefore, the expression (6.64) for n_{i1} is the same, except that K and E_1 in Equation (6.64) now pertain to the x-components of these parameters

$$n_{i1} = \frac{iK_x}{\omega^2 - \omega_{c_i}^2} \frac{n_{i0}eE_{1x}}{m_i}.$$ (6.69)

For electron flow along B_0, the governing equation is the parallel component of the momentum equation (6.38), which becomes

$$i\omega m_e(u_{e1})_z - iK_z \frac{\gamma_e kT_e}{n_{e0}} n_{e1} - eE_{1z} = 0,$$ (6.70)

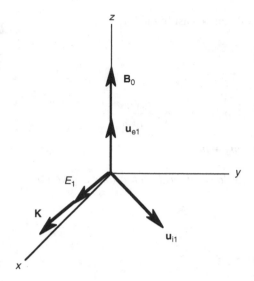

Figure 6.3 Directions of the wave and plasma parameters for electrostatic waves that propagate at almost 90° to a constant magnetic field $\mathbf{B}_0$. The velocity $\mathbf{u}_{i1}$ is in the x–y plane, $\mathbf{u}_{e1}$ is along $\mathbf{B}_0$, and $\mathbf{K}$ is in the x–z plane. These directions are relevant to electrostatic ion-cyclotron waves.

where the electrons are no longer assumed to be cold (i.e., $T_e \neq 0$). For low frequency waves, the inertial term, which contains m_e, can be neglected. Therefore, Equation (6.70) can be easily solved to obtain an expression for n_{e1}, which is

$$n_{e1} = -\frac{e n_{e0} E_{1z}}{iK_z(\gamma_e k T_e)}. \tag{6.71}$$

Equations (6.69) and (6.71) can be equated because of charge neutrality ($n_{e1} = n_{i1}$), and the result is

$$E_{1z} = \frac{(\gamma_e k T_e) K_x (K_z E_{1x})}{m_i(\omega^2 - \omega_{c_i}^2)}. \tag{6.72}$$

For electrostatic waves, the components of the electric field are related because $\nabla \times \mathbf{E}_1 = 0$. This curl equation becomes $\mathbf{K} \times \mathbf{E}_1 = 0$ for plane waves, which means $K_z E_{1x} = K_x E_{1z}$. Substituting this result into Equation (6.72) yields the following equation:

$$\omega^2 = \omega_{c_i}^2 + K_x^2 \frac{\gamma_e k T_e}{m_i}. \tag{6.73}$$

Now, $K^2 = K_x^2 + K_z^2 \approx K_x^2$, and $V_S = (\gamma_e k T_e/m_i)^{1/2}$ in this application, because the ions were assumed to be cold (Equation 5.56). Therefore, Equation (6.73) can also be written as

$$\omega^2 = \omega_{c_i}^2 + K^2 V_S^2, \tag{6.74}$$

which is the dispersion relation for *electrostatic ion-cyclotron waves.*

6.8 Electromagnetic waves in a plasma

The propagation of electromagnetic waves in a vacuum was discussed in Section 6.1, and it was shown that the wave is transverse ($\mathbf{E}_1 \perp \mathbf{K}$) and that $\omega^2 = c^2 K^2$. The focus here is on the propagation of electromagnetic waves in a plasma. The full set of Maxwell equations is needed for electromagnetic waves, and when a plane wave solution is assumed, the resulting general wave equation (6.6) takes the algebraic form given in equation (6.20). The effect of the plasma can enter through both the perturbed current density, $\mathbf{J}_1$, and the perturbed charge density, ρ_{1c}. However, for purely transverse waves $\mathbf{K} \cdot \mathbf{E}_1 = 0$, and in this case, the general wave equation reduces to

$$\left(\frac{\omega^2}{c^2} - K^2 \right) \mathbf{E}_1 = -i\omega\mu_0 \mathbf{J}_1. \tag{6.75}$$

The perturbed current density is obtained by a linearization of the total current density, $\mathbf{J}$, which for a two-component plasma is given by

$$\mathbf{J} = n_i e \mathbf{u}_i - n_e e \mathbf{u}_e. \tag{6.76}$$

The linearization is accomplished by first perturbing the densities and drift velocities in the usual manner (Equations 6.31a–d)

$$n_e = n_{e0} + n_{e1}, \tag{6.77a}$$

$$n_i = n_{i0} + n_{i1}, \tag{6.77b}$$

$$\mathbf{u}_e = \mathbf{u}_{e0} + \mathbf{u}_{e1}, \tag{6.77c}$$

$$\mathbf{u}_i = \mathbf{u}_{i0} + \mathbf{u}_{i1}. \tag{6.77d}$$

Substituting Equations (6.77a–d) into Equation (6.76) and neglecting the nonlinear terms yields the following expression for the current density:

$$\mathbf{J} = \mathbf{J}_0 + \mathbf{J}_1, \tag{6.78}$$

where

$$\mathbf{J}_0 = n_{e0} e (\mathbf{u}_{i0} - \mathbf{u}_{e0}), \tag{6.79}$$

$$\mathbf{J}_1 = n_{e0} e (\mathbf{u}_{i1} - \mathbf{u}_{e1}) + n_{i1} e \mathbf{u}_{i0} - n_{e1} e \mathbf{u}_{e0}, \tag{6.80}$$

and where it is assumed that charge neutrality prevails in the undisturbed plasma ($n_{i0} = n_{e0}$). The current $\mathbf{J}_0$ is the current that flows in the undisturbed plasma, and $\mathbf{J}_1$ is the perturbed current associated with the electromagnetic wave.

It is instructive to consider first the propagation of electromagnetic waves in a plasma that is not subjected to either electric or magnetic fields ($\mathbf{E}_0 = \mathbf{B}_0 = 0$). For simplicity, the plasma is also assumed to be electrically neutral ($n_{e0} = n_{i0}$), stationary

($\mathbf{u}_{e0} = \mathbf{u}_{i0} = 0$), cold ($T_e = T_i = 0$), uniform, and steady. When an electromagnetic wave propagates through such a plasma, a current is induced and the disturbed plasma then affects the electromagnetic wave. When light waves or microwaves propagate through a plasma, only the electrons can respond because the wave frequencies are high. For these waves, the relevant equations are the electron continuity (6.37) and momentum (6.38) equations, the electromagnetic wave equation (6.75), and the expression for the perturbed current density (6.80).

With these simplifying assumptions, the perturbed current density (6.80) and the electron momentum equation (6.38) reduce to

$$\mathbf{J}_1 = -n_{e0}e\mathbf{u}_{e1}, \tag{6.81}$$

$$i\omega\mathbf{u}_{e1} - \frac{e}{m_e}\mathbf{E}_1 = 0. \tag{6.82}$$

Substituting $\mathbf{u}_{e1}$ from Equation (6.82) into the equation for $\mathbf{J}_1$ (6.81) and then substituting that result into the wave equation (6.75), yields an equation for $\mathbf{E}_1$, which is given by

$$\mathbf{E}_1\left(\frac{\omega^2}{c^2} - K^2 - \mu_0\varepsilon_0\omega_{pe}^2\right) = 0, \tag{6.83}$$

where ω_{pe} is the electron plasma frequency (Equation 6.43). The fluctuating electric field is not zero and, therefore, the quantity in the brackets must be zero, which yields

$$\omega^2 = \omega_{pe}^2 + c^2K^2, \tag{6.84}$$

where, as before, $\mu_0\varepsilon_0 = 1/c^2$. Equation (6.84) is the dispersion relation for high frequency electromagnetic waves propagating in an unmagnetized plasma. For these waves, the phase velocity is greater than the speed of light ($V_{ph} > c$), but the group velocity, V_g, is less than c. Specifically, from Equation (6.84),

$$\frac{\omega^2}{K^2} = c^2 + \frac{\omega_{pe}^2}{K^2} > c^2, \tag{6.85}$$

$$\frac{d\omega}{dK} = \frac{c}{(\omega/K)}c < c. \tag{6.86}$$

It is of interest to determine the frequencies of electromagnetic waves that can propagate through an unmagnetized plasma. These can be obtained from the dispersion relation (6.84) by solving for K, and the result is

$$K = \frac{\sqrt{\omega^2 - \omega_{pe}^2}}{c}. \tag{6.87}$$

When $\omega > \omega_{p_e}$, K is real, and the wave propagates through the plasma. When $\omega = \omega_{p_e}$, $K = 0$, and this is called the *cut-off frequency*. Finally, when $\omega < \omega_{p_e}$, $K = i|K|$ is imaginary, and the wave is damped. The damping distance can be obtained from the plane wave expression for $\mathbf{E}_1$, which is given by Equation (6.9). When $K = i|K|$, this expression becomes

$$\mathbf{E}_1(\mathbf{r}, t) = \mathbf{E}_{10} e^{-|K|x} \cos(\omega t), \tag{6.88}$$

where, for simplicity, a one-dimensional situation was assumed and where the real part of the plane wave expression was taken. Equation (6.88) indicates that as an electromagnetic wave, with frequency $\omega < \omega_{p_e}$, tries to propagate through the plasma, it is damped exponentially with distance. The *penetration depth* or *skin depth*, δ, is given by

$$\delta = \frac{1}{|K|} = \frac{c}{\sqrt{\omega_{p_e}^2 - \omega^2}}. \tag{6.89}$$

Physically, the results on wave damping can be understood as follows. The plasma frequency defines the electrons' ability to adjust to an imposed oscillating electric field. When $\omega < \omega_{p_e}$, the electrons can easily adjust to the imposed electric field and establish an oppositely directed, polarization electric field that cancels the imposed electric field. The decay of the electric field then leads to a decay of the associated magnetic field because Faraday's law indicates that $\mathbf{B}_1 = \mathbf{K} \times \mathbf{E}_1/\omega$ (Equation 6.17b). On the other hand, when $\omega > \omega_{p_e}$, the electrons cannot fully adjust to the imposed, oscillating electric field, and the electromagnetic wave is modified as it passes through the plasma, but it does not decay.

6.9   Ordinary and extraordinary waves

Ordinary and extraordinary waves are high frequency electromagnetic waves that propagate in a direction perpendicular to a magnetic field, $\mathbf{B}_0$. For the ordinary wave (O mode), the wave electric field is parallel to the background magnetic field ($\mathbf{E}_1 \parallel \mathbf{B}_0$), whereas for the extraordinary wave (X mode), it is perpendicular to the background magnetic field ($\mathbf{E}_1 \perp \mathbf{B}_0$). As noted before, high frequency means that only the electron motion needs to be considered. Also, as in the previous wave analyses, the dispersion relation is derived by considering a two-component, electrically neutral ($n_{e0} = n_{i0}$), stationary ($\mathbf{u}_{e0} = \mathbf{u}_{i0} = 0$), uniform, and steady plasma. The plasma is magnetized ($\mathbf{B}_0 \neq 0$), but there is no imposed electric field ($\mathbf{E}_0 = 0$). In addition, the plasma is assumed to be cold ($T_e = 0$).

The relevant plasma transport equations are the electron continuity (6.37) and momentum (6.38) equations, and the expression for the perturbed current density (6.80). For the above equilibrium plasma configuration, these equations reduce to

$$\omega n_{e1} = n_{e0}\mathbf{K} \cdot \mathbf{u}_{e1}, \tag{6.90a}$$

$$i\omega\mathbf{u}_{e1} - \frac{e}{m_e}(\mathbf{E}_1 + \mathbf{u}_{e1} \times \mathbf{B}_0) = 0, \tag{6.90b}$$

$$\mathbf{J}_1 = -n_{e0}e\mathbf{u}_{e1}. \tag{6.90c}$$

The wave equation is also needed in addition to these transport equations. However, as it turns out, the extraordinary wave is not a purely transverse wave. Therefore, Equation (6.75) cannot be used because $\mathbf{K} \cdot \mathbf{E}_1 \neq 0$. The complete wave equation (6.20) is needed to describe the extraordinary wave.

The orientations of the wave vectors for the ordinary wave are shown in Figure 6.4. This wave is purely transverse ($\mathbf{K} \cdot \mathbf{E}_1 = 0$) and, hence, the reduced wave equation (6.75) is applicable. In addition, for this wave, the fluctuating electric field is parallel to the magnetic field ($\mathbf{E}_1 \parallel \mathbf{B}_0$), and this electric field induces a velocity that is also parallel to the magnetic field ($\mathbf{u}_{e1} \parallel \mathbf{B}_0$). Under these circumstances, the $\mathbf{u}_{e1} \times \mathbf{B}_0$ term vanishes and the resulting system of equations (6.90a–c, 6.20) reduces to the equivalent equations for an unmagnetized plasma (6.81, 6.82). Therefore, the dispersion relation for the *ordinary wave* is the same as that obtained for an electromagnetic wave in an unmagnetized plasma (Equation 6.84).

The ordinary wave, and all of the other waves considered up to this point, are *linearly polarized*, which means that the electric field, $\mathbf{E}_1$, always lies along one axis. For the extraordinary wave, on the other hand, a component of $\mathbf{E}_1$ along $\mathbf{K}$ develops as the wave propagates in the plasma, and therefore, the wave becomes partly longitudinal ($\mathbf{E}_{1x} \parallel \mathbf{K}$) and partly transverse ($\mathbf{E}_{1y} \perp \mathbf{K}$), as shown in Figure 6.4. The two components of the electric field are out of phase by 90° and their magnitudes are

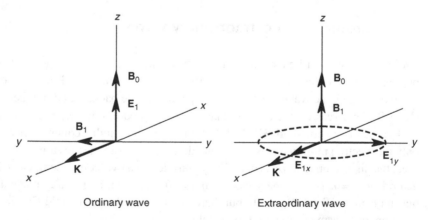

Figure 6.4 Directions of the wave parameters for the ordinary and extraordinary electromagnetic waves. Both waves propagate in a direction perpendicular to a magnetic field. The ordinary wave is linearly polarized and the extraordinary wave is elliptically polarized.

not equal. Consequently, as the wave propagates, the tip of the electric field vector traces out an ellipse every wave period, and this is called *elliptic polarization*.

It is necessary to allow $\mathbf{E}_1$ and $\mathbf{u}_{e1}$ to have both x and y components in order to describe the extraordinary wave. In this case, the x and y components of the momentum equation (6.90b) become

$$i\omega(u_{e1})_x - \frac{e}{m_e}\left[E_{1x} + (u_{e1})_y B_0\right] = 0, \tag{6.91a}$$

$$i\omega(u_{e1})_y - \frac{e}{m_e}\left[E_{1y} - (u_{e1})_x B_0\right] = 0. \tag{6.91b}$$

These two equations can be easily solved to yield the individual velocity components, which are given by

$$(u_{e1})_x = \frac{-(e/m_e\omega)}{1 - \omega_{c_e}^2/\omega^2}\left(iE_{1x} + \frac{\omega_{c_e}}{\omega}E_{1y}\right) \tag{6.92a}$$

$$(u_{e1})_y = \frac{(e/m_e\omega)}{1 - \omega_{c_e}^2/\omega^2}\left(\frac{\omega_{c_e}}{\omega}E_{1x} - iE_{1y}\right). \tag{6.92b}$$

Now, substituting the expression for $\mathbf{J}_1$ (6.90c) into the wave equation (6.20), leads to the following result:

$$(\omega^2 - K^2 c^2)\mathbf{E}_1 + c^2\mathbf{K}(KE_{1x}) = i\omega\frac{n_{e0}e}{\varepsilon_0}\mathbf{u}_{e1}, \tag{6.93}$$

where $c^2 = 1/\mu_0\varepsilon_0$. After taking the x and y components of Equation (6.93), and using Equations (6.92a) and (6.92b) for the velocity components, the wave equation (6.93) becomes

$$E_{1x}(\omega^2 - \omega_{c_e}^2 - \omega_{pe}^2) + E_{1y}\left(i\omega_{pe}^2\frac{\omega_{c_e}}{\omega}\right) = 0, \tag{6.94a}$$

$$E_{1x}(-i\omega_{pe}^2\omega_{c_e}\omega) + E_{1y}\left[(\omega^2 - K^2 c^2)(\omega^2 - \omega_{c_e}^2) - \omega_{pe}^2\omega^2\right] = 0. \tag{6.94b}$$

The two equations can be solved for either E_{1x} or E_{1y} and then the dispersion relation is obtained

$$(\omega^2 - \omega_{c_e}^2 - \omega_{pe}^2)(\omega^2 - K^2 c^2)(\omega^2 - \omega_{c_e}^2)$$
$$- (\omega^2 - \omega_{c_e}^2 - \omega_{pe}^2)\omega_{pe}^2\omega^2 - \omega_{pe}^4\omega_{c_e}^2 = 0. \tag{6.95}$$

Equation (6.95) can be simplified via several algebraic manipulations,[2] and the result is the classical form for the dispersion relation that describes the *extraordinary wave*

$$\omega^2 = K^2 c^2 + \omega_{pe}^2\frac{\omega^2 - \omega_{pe}^2}{\omega^2 - (\omega_{pe}^2 + \omega_{c_e}^2)}. \tag{6.96}$$

Typically, not all frequencies can propagate in a plasma, and this is true for both the ordinary (O-mode) and extraordinary (X-mode) waves. There are generally both cut-offs and resonances. A *cut-off* is the frequency at which the wave number $K \to 0$, whereas a resonance is the frequency at which $K \to \infty$. A wave is usually reflected at a cut-off and absorbed at a resonance. The cut-offs and resonances divide the frequency domain into propagation and nonpropagation bands.

The dispersion relation for the ordinary wave, which is $\omega^2 = \omega_{pe}^2 + c^2 K^2$ (Equation 6.84), has one cut-off and no resonances. The cut-off frequency is $\omega = \omega_{pe}$. Therefore, the ordinary wave can propagate in a plasma only for frequencies $\omega > \omega_{pe}$.

The extraordinary wave (Equation 6.96) has one resonance and two cut-offs. The resonance occurs when $K \to \infty$, and an inspection of the dispersion relation (6.96) indicates that the nonzero frequency at which $K \to \infty$ is $\omega^2 = \omega_{pe}^2 + \omega_{ce}^2$, which is the upper hybrid frequency (Equation 6.62). Therefore, as an extraordinary wave, which is partly electrostatic and partly electromagnetic, approaches a resonance, both ω/K and $d\omega/dK \to 0$, and the wave energy is converted into electrostatic upper hybrid oscillations.

The cut-offs for the extraordinary wave are obtained from the dispersion relation (6.96) by setting $K = 0$, which yields

$$\omega^2 = \omega_{pe}^2 \frac{\omega^2 - \omega_{pe}^2}{\omega^2 - (\omega_{pe}^2 + \omega_{ce}^2)}. \tag{6.97}$$

This equation can be rearranged as follows:

$$1 = \frac{\omega_{pe}^2}{\omega^2} \frac{1 - \dfrac{\omega_{pe}^2}{\omega^2}}{1 - \dfrac{\omega_{pe}^2}{\omega^2} - \dfrac{\omega_{ce}^2}{\omega^2}}$$

$$\left(1 - \frac{\omega_{pe}^2}{\omega^2}\right) - \frac{\omega_{ce}^2}{\omega^2} = \frac{\omega_{pe}^2}{\omega^2}\left(1 - \frac{\omega_{pe}^2}{\omega^2}\right)$$

$$\left(1 - \frac{\omega_{pe}^2}{\omega^2}\right)^2 = \frac{\omega_{ce}^2}{\omega^2}$$

$$\omega^2 \mp \omega\omega_{ce} - \omega_{pe}^2 = 0 \tag{6.98}$$

where the $\mp$ signs appear when the square root is taken of Equation (6.98). For both the minus sign and the plus sign in Equation (6.98), two roots appear when the quadratic formula is applied. However, for each case, only the positive frequency is considered. Negative frequencies are associated with negative K and correspond to waves propagating in the opposite direction. With this caveat in mind, the solution of Equation (6.98) is

$$\omega_R = \frac{1}{2}\left[\omega_{c_e} + (\omega_{c_e}^2 + 4\omega_{p_e}^2)^{1/2}\right], \tag{6.99a}$$

$$\omega_L = \frac{1}{2}\left[-\omega_{c_e} + (\omega_{c_e}^2 + 4\omega_{p_e}^2)^{1/2}\right]. \tag{6.99b}$$

The frequencies ω_R and ω_L are called the right-hand and left-hand cut-offs of the extraordinary wave.

The ordering of the resonance (at the upper hybrid frequency) and the cut-offs with respect to frequency magnitude is $\omega_L^2 < (\omega_{p_e}^2 + \omega_{c_e}^2) < \omega_R^2$. The propagation characteristics for the extraordinary wave are as follows:

$\omega < \omega_L$ no propagation,

$\omega_L < \omega < \omega_h$ propagation,

$\omega_h < \omega < \omega_R$ no propagation,

$\omega_R < \omega$ propagation,

where $\omega_h^2 = \omega_{p_e}^2 + \omega_{c_e}^2$.

6.10 L and R waves

The L and R waves are high frequency, transverse, electromagnetic waves that propagate along a magnetic field. The wave electric field, which is perpendicular to **K**, has two orthogonal components that have equal amplitudes, but are out of phase by 90°. Consequently, as the wave propagates along the magnetic field, **B**$_0$, the electric field vector, **E**$_1$, rotates about **B**$_0$ and its tip traces out a circle every wave period (Figure 6.5). Hence, the L and R waves are *circularly polarized*.

The relevant plasma transport equations are the same as those used to describe the ordinary and extraordinary waves (Equations 6.90a–c), and the appropriate wave

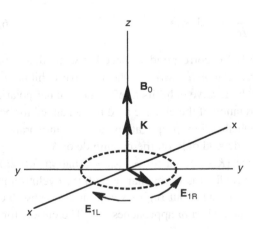

Figure 6.5 Directions of the wave parameters for the L and R electromagnetic waves. Both waves propagate along the magnetic field and are circularly polarized, but the electric field vectors of the waves rotate in opposite directions.

equation is the transverse wave equation (6.75). Both $\mathbf{E}_1$ and $\mathbf{u}_{e1}$ have x and y components, as was the case for the extraordinary wave, and the solution for $(u_{e1})_x$ and $(u_{e1})_y$ in terms of E_{1x} and E_{1y} is the same as obtained previously (Equations 6.92a,b). The difference results from the fact that here the propagation is along $\mathbf{B}_0$, not perpendicular to $\mathbf{B}_0$. Using the fact that $\mathbf{K} \parallel \mathbf{B}_0$, and using the expressions for the perturbed current (6.90c) and the perturbed velocities (6.92a,b), the x and y components of the transverse wave equation (6.75) become

$$E_{1x}\left(\omega^2 - K^2 c^2 - \frac{\omega_{pe}^2}{1 - \omega_{ce}^2/\omega^2}\right) + E_{1y}\left(i\frac{\omega_{ce}}{\omega}\frac{\omega_{pe}^2}{1 - \omega_{ce}^2/\omega^2}\right) = 0,$$

$$\tag{6.100a}$$

$$E_{1x}\left(-i\frac{\omega_{ce}}{\omega}\frac{\omega_{pe}^2}{1 - \omega_{ce}^2/\omega^2}\right) + E_{1y}\left(\omega^2 - K^2 c^2 - \frac{\omega_{pe}^2}{1 - \omega_{ce}^2/\omega^2}\right) = 0.$$

$$\tag{6.100b}$$

The solution of Equations (6.100a) and (6.100b) for either E_{1x} or E_{1y} yields the following relation:

$$\omega^2 - K^2 c^2 - \frac{\omega_{pe}^2}{1 - \omega_{ce}^2/\omega^2} = \pm\frac{\omega_{ce}}{\omega}\frac{\omega_{pe}^2}{1 - \omega_{ce}^2/\omega^2}$$

or

$$\omega^2 - K^2 c^2 = \frac{\omega_{pe}^2 \omega}{\omega^2 - \omega_{ce}^2}(\omega \pm \omega_{ce}). \tag{6.101}$$

There are two waves that can propagate along $\mathbf{B}_0$, corresponding to the $\pm$ signs, and these are given by

$$\omega^2 = K^2 c^2 + \frac{\omega_{pe}^2}{1 - \omega_{ce}/\omega} \qquad \text{(R wave)}, \tag{6.102a}$$

$$\omega^2 = K^2 c^2 + \frac{\omega_{pe}^2}{1 + \omega_{ce}/\omega} \qquad \text{(L wave)}. \tag{6.102b}$$

Equations (6.102a) and (6.102b) correspond, respectively, to the dispersion relations for the *R and L electromagnetic waves*. The R wave exhibits a right-hand circular polarization and the L wave exhibits a left-hand circular polarization (Figure 6.5). The direction of rotation of the electric field is unchanged for both the R and L waves regardless of whether they propagate parallel or antiparallel to $\mathbf{B}_0$ because the dispersion relations depend only on the magnitude of K.

The R wave has a resonance $(K \to \infty)$ at $\omega = \omega_{ce}$ (Equation 6.102a). The direction of rotation of the electric field is in resonance with the cyclotron motion of the electrons. As the electrons gyrate about $\mathbf{B}_0$, they continuously absorb energy from the R wave, and it is damped when ω approaches ω_{ce}. The cut-off for the R wave is obtained by setting $K = 0$ in Equation (6.102a), and the cut-off occurs at $\omega = \omega_R$ (Equation 6.99a). The propagation features of the R wave are as follows:

$\omega < \omega_{c_e}$ propagation (whistler wave),

$\omega_{c_e} < \omega < \omega_R$ no propagation,

$\omega > \omega_R$ propagation.

The L wave does not have a resonance ($K \to \infty$) because the electric field rotates in the opposite direction to the gyration motion of the electrons. However, the L wave does have a cut-off ($K = 0$), and Equation (6.102b) indicates that this occurs at the frequency $\omega = \omega_L$ (Equation 6.99b). The propagation characteristics of the L wave are given by:

$\omega < \omega_L$ no propagation,

$\omega > \omega_L$ propagation.

6.11   Alfvén and magnetosonic waves

Alfvén and magnetosonic waves are low frequency, transverse, electromagnetic waves that propagate in a magnetized plasma. The Alfvén wave propagates along the magnetic field and the magnetosonic wave propagates across the magnetic field. Both waves are linearly polarized. The low-frequency nature of the waves means that both the electron and ion motion must be considered. The dispersion relations for these waves can be derived in a manner similar to that used to derive the dispersion relations for the high frequency electromagnetic waves (O-mode, X-mode, L and R waves). In this case, the appropriate equations are the electron continuity (6.37) and momentum (6.38) equations, similar equations for the ions, an expression for $\mathbf{J}_1$ that includes the ion motion [$\mathbf{J}_1 = n_{e0}e(\mathbf{u}_{i1} - \mathbf{u}_{e1})$; Equation (6.80)], and the transverse wave equation (6.75). For the Alfvén wave, the plasma is assumed to be cold ($T_e = T_i = 0$), but that is not the case for the magnetosonic wave.

Although the dispersion relations for the Alfvén and magnetosonic waves can be derived using the equations mentioned above, they can be derived more easily starting from the so-called ideal magnetohydrodynamic (MHD) equations, and this is done in Chapter 7. However, in the MHD approximation, the displacement current, $\partial \mathbf{E}/\partial t$, in Faraday's law (3.76d) is ignored because all MHD phenomena are assumed to be low frequency. The neglect of the displacement current does not affect the dispersion relation for the Alfvén wave, but it does modify the one for the magnetosonic wave.

When the plasma transport and wave equations mentioned above are used to derive the dispersion relations for the Alfvén and magnetosonic waves, these dispersion relations are, respectively, given by

$$\omega^2 = K^2 V_A^2,$$ (6.103)

$$\omega^2 = K^2 \frac{V_S^2 + V_A^2}{1 + V_A^2/c^2},$$ (6.104)

where $V_A = B_0/(\mu_0 n_{i0} m_i)^{1/2}$ is the Alfvén speed (Equation 7.88) and V_S is the ion-acoustic or sound speed (Equation 6.56). As noted above, the dispersion relation for the Alfvén wave (6.103) is the same as that obtained using the ideal MHD equations (7.90). A comparison of (6.104) with the corresponding MHD dispersion relation for magnetosonic waves (7.91) indicates that the effect of including the displacement current in Faraday's law is to add the V_A^2/c^2 term in the denominator of Equation (6.104). In the limit when $V_A \ll c$, which is typically the case, the two dispersion relations become equivalent.

6.12   Effect of collisions

Collisions have an important effect on many plasma processes that occur in the ionospheres, and it is natural to ask whether they can affect wave phenomena. The effect of collisions on waves can be determined simply by rederiving the wave dispersion relations including the collision terms in the electron and ion momentum equations. As an example, the dispersion relation for electrostatic *electron plasma waves* is rederived with allowance for electron–ion collisions. As before (Section 6.3), the plasma is assumed to be unmagnetized, electrically neutral ($n_{e0} = n_{i0}$), stationary ($\mathbf{u}_{e0} = \mathbf{u}_{i0} = 0$), uniform, and steady. The ions do not participate in the wave motion because the wave frequency is high and they have a large inertia.

The normal modes of the plasma are obtained by first linearizing the electron continuity and momentum equations and Gauss' law, and then assuming plane wave solutions. Equations (6.39a–c) are the result of this procedure for electron plasma waves when collisions are not considered. These equations are applicable here, except that an electron–ion collision term must be added to the right-hand side of the momentum equation (6.39b). For electron–ion collisions, the appropriate collision term for a Maxwellian plasma is given by Equation (4.124b), and in the limit of small relative drifts between the electrons and ions, this collision term reduces to

$$\frac{\delta \mathbf{M}_e}{\delta t} = n_e m_e \nu_{ei}(\mathbf{u}_i - \mathbf{u}_e),$$ (6.105)

where ν_{ei} is the electron–ion collision frequency (4.144). For the case considered here, the ion density and electron temperature are constant, and therefore, ν_{ei} is constant. In the derivation leading to Equation (6.39b), the electron momentum equation was divided by $-n_{e0} m_e$ and, therefore, the collision term that is consistent with Equation (6.39b) is

$$-\frac{1}{n_{e0}m_e}\frac{\delta \mathbf{M}_e}{\delta t} = -\nu_{ei}(\mathbf{u}_i - \mathbf{u}_e)$$

$$= \nu_{ei}\mathbf{u}_{e1}, \tag{6.106}$$

where the second expression is the linearized form of the collision term ($\mathbf{u}_i = \mathbf{u}_{i0}$; $\mathbf{u}_e = \mathbf{u}_{e0} + \mathbf{u}_{e1}$).

With the addition of the linearized electron–ion collision term on the right-hand side of Equation (6.39b), this momentum equation becomes

$$i\omega\mathbf{u}_{e1} - i\mathbf{K}\frac{3V_e^2}{n_{e0}}n_{e1} - \frac{e}{m_e}\mathbf{E}_1 = \nu_{ei}\mathbf{u}_{e1}, \tag{6.107}$$

where $\gamma = 3$ for a one-dimensional compression and $V_e^2 = kT_e/m_e$ (Equation 6.44). The scalar product of $\mathbf{K}$ with Equation (6.107) yields

$$i\omega(\mathbf{K} \cdot \mathbf{u}_{e1}) - iK^2\frac{3V_e^2}{n_{e0}}n_{e1} - \frac{e}{m_c}(\mathbf{K} \cdot \mathbf{E}_1) = \nu_{ei}(\mathbf{K} \cdot \mathbf{u}_{e1}). \tag{6.108}$$

Now, $\mathbf{K} \cdot \mathbf{u}_{e1} = \omega n_{e1}/n_{e0}$ (Equation 6.39a) and $\mathbf{K} \cdot \mathbf{E}_1 = -en_{e1}/i\varepsilon_o$ (Equation 6.39c). When these expressions are substituted into Equation (6.108), the following relation is obtained:

$$\omega^2 + i\nu_{ei}\omega = \omega_{pe}^2 + 3K^2V_e^2. \tag{6.109}$$

Equation (6.109) is the dispersion relation for *electron plasma waves* with allowance for electron–ion collisions.

The effect of collisions can be easily seen by considering the limit $\nu_{ei} \to \infty$. In this limit, Equation (6.109) becomes $\omega^2 + i\nu_{ei}\omega \approx 0$, and the nontrivial root is $\omega = -i\nu_{ei}$. The substitution of this result into the plane wave solution (6.9) indicates that the wave perturbation is damped, exponentially with time, as follows

$$\mathbf{E}_1(\mathbf{r}, t) = \mathbf{E}_{10}e^{-\nu_{ei}t}e^{i(\mathbf{K}\cdot\mathbf{r})}. \tag{6.110}$$

The damping rate is ν_{ei}^{-1}. Although the above analysis is for electron plasma waves, this result has general validity. That is, the effect of collisions is to damp waves, and in general, the damping is effective when the collision frequency is greater than the wave frequency (6.109). Physically, waves correspond to a coherent motion, and collisions act to scatter the particles and destroy the coherent wave motion.

6.13   Two-stream instability

In Section 6.8, it was shown that when an electromagnetic wave, with a frequency ω less than ω_{pe}, tries to propagate in an unmagnetized plasma, it is damped exponentially with distance (Equation 6.88). In Section 6.12, it was shown that waves can be damped exponentially with time when the collision frequency is greater than the wave frequency (Equation 6.110). However, wave amplitudes can also grow exponentially, both with time and distance, when there is an energy source. In this case, the plasma becomes *unstable*. A plasma can become unstable when there is a relative drift between different species (*streaming instabilities*), when a heavy fluid lies on top of a light fluid (*Rayleigh–Taylor instability*), and when the species velocity distributions are nonMaxwellian (*velocity-space instabilities*).

It is instructive to consider the so-called *two-stream instability*, as one example of an unstable plasma. In this case, it is assumed that there is a relative drift between the electrons and ions in a two-component, fully ionized plasma, and this relative drift is a possible energy source for waves. For simplicity, the plasma is also assumed to be electrically neutral ($n_{e0} = n_{i0}$), cold ($T_e = T_i = 0$), unmagnetized ($\mathbf{B}_0 = 0$), uniform, and steady. There is no external electric field ($\mathbf{E}_0 = 0$), and the ions are stationary ($\mathbf{u}_{i0} = 0$), but the electrons have an initial drift relative to the ions ($\mathbf{u}_{e0} \neq 0$).

The procedure for studying the stability of the plasma is the same as that used to calculate the normal modes of a plasma (Sections 6.1 and 6.2). That is, the ion and electron continuity and momentum equations are perturbed and linearized, and then plane wave solutions are assumed (Equations 6.37 and 6.38). At that point, the possible excitation of either electrostatic or electromagnetic waves can be considered. For electrostatic waves, Gauss' law (6.18) is used, while for electromagnetic waves the general wave equation (6.20) is applicable.

Typically, in either unmagnetized or strongly magnetized plasmas, electrostatic waves are more easily excited than electromagnetic waves and, therefore, they are considered here.[2] For this case, the electron and ion continuity (6.37) and momentum (6.38) equations and Gauss' law (6.18) become

$$(\omega - \mathbf{K} \cdot \mathbf{u}_{e0})n_{e1} = n_{e0}\mathbf{K} \cdot \mathbf{u}_{e1}, \tag{6.111a}$$

$$\omega n_{i1} = n_{i0}\mathbf{K} \cdot \mathbf{u}_{i1}, \tag{6.111b}$$

$$i(\omega - \mathbf{K} \cdot \mathbf{u}_{e0})\mathbf{u}_{e1} - \frac{e}{m_e}\mathbf{E}_1 = 0, \tag{6.111c}$$

$$i\omega\mathbf{u}_{i1} + \frac{e}{m_i}\mathbf{E}_1 = 0, \tag{6.111d}$$

$$i\mathbf{K} \cdot \mathbf{E}_1 = e(n_{i1} - n_{e1})/\varepsilon_0. \tag{6.111e}$$

The scalar product of $\mathbf{K}$ with the electron and ion momentum equations (6.111c,d) yields, respectively,

$$\mathbf{K} \cdot \mathbf{u}_{e1} = \frac{e}{im_e}\frac{\mathbf{K} \cdot \mathbf{E}_1}{(\omega - \mathbf{K} \cdot \mathbf{u}_{e0})}, \tag{6.112}$$

$$\mathbf{K} \cdot \mathbf{u}_{i1} = -\frac{e}{i\omega m_i} \mathbf{K} \cdot \mathbf{E}_1. \tag{6.113}$$

Substituting these results into the appropriate continuity equations (6.111a,b) yields expressions for n_{e1} and n_{i1} in terms of the electric field

$$n_{e1} = \frac{n_{e0}e}{im_e} \frac{\mathbf{K} \cdot \mathbf{E}_1}{(\omega - \mathbf{K} \cdot \mathbf{u}_{e0})^2}, \tag{6.114}$$

$$n_{i1} = \frac{-n_{i0}e}{im_i} \frac{\mathbf{K} \cdot \mathbf{E}_1}{\omega^2}. \tag{6.115}$$

Finally, substituting Equations (6.114) and (6.115) into Coulomb's law (6.111e) leads to the dispersion relation for the *electrostatic two-stream instability*, which is

$$1 = \frac{\omega_{pi}^2}{\omega^2} + \frac{\omega_{pe}^2}{(\omega - \mathbf{K} \cdot \mathbf{u}_{e0})^2}, \tag{6.116}$$

where $\omega_{ps} = (n_{s0}e^2/m_s\varepsilon_0)^{1/2}$ is the plasma frequency for species s.

In the limit of $m_i \to \infty$, $\omega_{pi} \to 0$ and the dispersion relation (6.116) reduces to

$$(\omega - \mathbf{K} \cdot \mathbf{u}_{e0})^2 = \omega_{pe}^2. \tag{6.117}$$

This relation is equivalent to the expression derived earlier for a cold, stationary plasma (Equation 6.45), except there is a *Doppler shift* of the frequency by the amount $\mathbf{K} \cdot \mathbf{u}_{e0}$.

In the general case of a finite m_i, the dispersion relation (6.116) is a fourth-order equation for ω. If all four roots are real, the plasma is stable. If any of the roots are complex, then the plasma is unstable because complex roots always occur in complex conjugate pairs. One of the complex roots corresponds to a damped wave and the other to a growing wave. Typically, the solution for the growing wave dominates and the plasma becomes unstable.

The plasma stability for the dispersion relation (6.116) can be determined by graphical means by introducing the following function:

$$y(K, \omega) = \frac{\omega_{pi}^2}{\omega^2} + \frac{\omega_{pe}^2}{(\omega - \mathbf{K} \cdot \mathbf{u}_{e0})^2}, \tag{6.118}$$

where $y(K, \omega) = 1$ yields the dispersion relation (6.116). A sketch of y versus ω for a fixed $\mathbf{K}$ is shown in Figure 6.6. Note that when ω approaches either $+\infty$ or $-\infty$, $y \to 0$. Also, note that $y \to \infty$ when ω approaches 0 and $\mathbf{K} \cdot \mathbf{u}_{e0}$. In the sketch of y versus ω, two general cases are possible in the central portion of the curve, as shown. When the line at unity intersects $y(K, \omega)$ at four distinct points, there are four real roots and the plasma is stable for the adopted value of $\mathbf{K}$. On the other hand, when the line at unity intersects $y(K, \omega)$ at two points, there are two complex roots and the plasma is unstable. Therefore, it is necessary to determine whether the minimum value of $y(K, \omega)$ shown in Figure 6.6 lies above or below the line at unity.

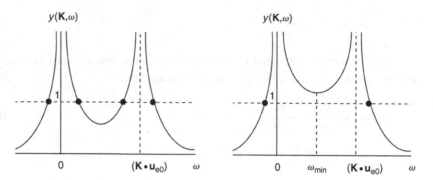

Figure 6.6 Graphical solution of the two-stream dispersion relation (6.116) for cases when the plasma is stable (left plot) and when it is unstable (right plot). The dispersion relation is $y(K, \omega) = 1^2$.

The minimum of the function is obtained from $\partial y / \partial \omega = 0$, which yields

$$\omega_{p_i}^2 (\omega - \mathbf{K} \cdot \mathbf{u}_{e0})^3 + \omega_{p_e}^2 \omega^3 = 0, \tag{6.119a}$$

or

$$\frac{m_e}{m_i} (\omega - \mathbf{K} \cdot \mathbf{u}_{e0})^3 + \omega^3 = 0. \tag{6.119b}$$

The minimum of $y(K, \omega)$ occurs at a frequency (ω_{min}) that is in the range $0 < \omega_{min} < \mathbf{K} \cdot \mathbf{u}_{e0}$. Therefore, an approximate value for ω_{min} can be obtained by assuming $\omega_{min} \ll \mathbf{K} \cdot \mathbf{u}_{e0}$ and, hence, by neglecting ω in comparison with $\mathbf{K} \cdot \mathbf{u}_{e0}$ in (6.119b). This approximate value is

$$\omega_{min} \approx \left(\frac{m_e}{m_i} \right)^{1/3} \mathbf{K} \cdot \mathbf{u}_{e0}. \tag{6.120}$$

Note that this solution of (6.119b) for ω_{min} is consistent with the assumption that $\omega_{min} \ll \mathbf{K} \cdot \mathbf{u}_{e0}$. Now, the substitution of ω_{min} (6.120) into the expression for y (6.118) yields

$$y_{min} \approx \frac{\omega_{p_i}^2}{\left(\dfrac{m_e}{m_i} \right)^{2/3} (\mathbf{K} \cdot \mathbf{u}_{e0})^2} + \frac{\omega_{p_e}^2}{(\mathbf{K} \cdot \mathbf{u}_{e0})^2}. \tag{6.121}$$

When $y_{min} > 1$, the plasma is unstable and this occurs for

$$(\mathbf{K} \cdot \mathbf{u}_{e0})^2 \lesssim \omega_{p_e}^2. \tag{6.122}$$

Finally, it should be noted that relative drifts between interacting species are common in the ionospheres and, therefore, streaming instabilities can play an important role in the plasma dynamics and energetics.

6.14   Shock waves

The focus up to this point has been on the types of waves that can propagate in ionized gases. Only small disturbances were considered and, hence, the continuity, momentum, and energy equations could be linearized and plane wave solutions could be assumed. Under these circumstances, the set of partial differential equations for the perturbed parameters could be converted into a set of linear algebraic equations from which the dispersion relation was obtained.

A set of algebraic equations for the perturbed parameters can also be obtained in the opposite limit of *shock waves*, which are sharp discontinuities that occur in supersonic flows in response to either an obstacle or changing conditions in the region ahead of the flow. However, for shock waves, the resulting algebraic equations are nonlinear.

The occurrence of shock waves can be traced to the fact that waves propagate with a finite speed in a neutral or ionized gas. When a subsonic flow approaches an obstacle, the waves created by the obstacle propagate back into the gas, and they carry the information that the gas is approaching an obstacle. The gas then gradually adjusts its flow properties to accommodate the obstacle. For a gas flow that is almost sonic, the flow and wave speeds are nearly equal. Therefore, the waves cannot propagate very far from the obstacle before they are overtaken by the flow. Hence, the adjustment of the flow to the obstacle is distributed over a smaller spatial region than for a low-speed flow. For a supersonic flow, the drift speed is greater than the wave speed and, consequently, the waves created by the obstacle cannot propagate back into the gas. In this case, the adjustment of the flow to the obstacle is abrupt and occurs in a narrow spatial region that is either at or close to the obstacle. The shock thickness is of the order of a few mean-free-paths for a collision-dominated gas. The net effect of the shock is either to stop the flow or to slow down and deflect the flow around the obstacle, which leads to both density and temperature enhancements on the side of the shock that is closer to the obstacle. Physically, these conditions result because of the enhanced collision frequency in the shock, which acts to convert flow energy into random (thermal) energy.

The classical treatment of shock waves starts with the Euler equations (5.22a–c). These equations, for a single-component neutral gas and with gravity neglected, become

$$\frac{\partial \rho}{\partial t} + \nabla \cdot (\rho \mathbf{u}) = 0, \tag{6.123}$$

$$\rho \left[\frac{\partial \mathbf{u}}{\partial t} + (\mathbf{u} \cdot \nabla)\mathbf{u} \right] + \nabla p = 0, \tag{6.124}$$

$$\frac{\partial p}{\partial t} + \mathbf{u} \cdot \nabla p + \gamma p (\nabla \cdot \mathbf{u}) = 0, \tag{6.125}$$

where $\rho = nm$ is the mass density and $\gamma = 5/3$ is the ratio of specific heats. As it turns out, Equations (6.123) to (6.125), and the forthcoming analysis of shocks, also apply to a single-component ionized gas under certain conditions, but this will be discussed later.

The momentum (6.124) and energy (6.125) equations are not in their most convenient form for shock studies. The momentum equation can be modified by multiplying the continuity equation (6.123) by $\mathbf{u}$ and then adding it to the momentum equation (6.124), which yields

$$\frac{\partial}{\partial t}(\rho \mathbf{u}) + \nabla \cdot (\rho \mathbf{uu}) + \nabla p = 0. \tag{6.126}$$

A modified equation for the flow of energy can be obtained by first taking the scalar product of $\mathbf{u}$ with the momentum equation (6.124), which yields

$$\rho \left[\frac{\partial}{\partial t}\left(\frac{u^2}{2}\right) + (\mathbf{u} \cdot \nabla)\frac{u^2}{2} \right] + \mathbf{u} \cdot \nabla p = 0. \tag{6.127}$$

This equation can also be written in the form

$$\frac{\partial}{\partial t}\left(\frac{1}{2}\rho u^2\right) + \mathbf{u} \cdot \nabla\left(\frac{1}{2}\rho u^2\right) - \frac{u^2}{2}\left(\frac{\partial \rho}{\partial t} + \mathbf{u} \cdot \nabla\rho\right) + \mathbf{u} \cdot \nabla p = 0. \tag{6.128}$$

Now, the continuity equation (6.123) indicates that $\partial \rho/\partial t + \mathbf{u} \cdot \nabla\rho = -\rho(\nabla \cdot \mathbf{u})$. When this expression is substituted into Equation (6.128), and the second and third terms are combined, the result is

$$\frac{\partial}{\partial t}\left(\frac{1}{2}\rho u^2\right) + \nabla \cdot \left(\frac{1}{2}\rho u^2 \mathbf{u}\right) + \mathbf{u} \cdot \nabla p = 0. \tag{6.129}$$

An expression for $\mathbf{u} \cdot \nabla p$ can be obtained from the energy equation (6.125), which can be written in the form

$$\frac{\partial p}{\partial t} + \mathbf{u} \cdot \nabla p + \nabla \cdot (\gamma p \mathbf{u}) - \mathbf{u} \cdot \nabla(\gamma p) = 0. \tag{6.130}$$

Equation (6.130) then yields

$$\mathbf{u} \cdot \nabla p = \frac{1}{\gamma - 1}\frac{\partial p}{\partial t} + \frac{\gamma}{\gamma - 1}\nabla \cdot (p\mathbf{u}). \tag{6.131}$$

Substituting Equation (6.131) into Equation (6.129) yields the final form for the energy flow equation, which is

$$\frac{\partial}{\partial t}\left(\frac{1}{2}\rho u^2 + \frac{p}{\gamma - 1}\right) + \nabla \cdot \left[\left(\frac{1}{2}\rho u^2 + \frac{\gamma p}{\gamma - 1}\right)\mathbf{u}\right] = 0. \tag{6.132}$$

Equations (6.123), (6.126), and (6.132) correspond, respectively, to the continuity, momentum, and energy equations that are typically used in shock studies. If the

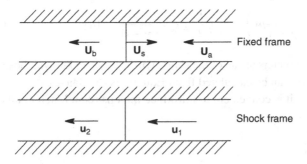

Figure 6.7 Shock dynamics in a one-dimensional system. The top schematic shows a shock propagating through a gas in a fixed reference frame. The gas velocities ahead and behind the shock are $\mathbf{U_a}$ and $\mathbf{U_b}$, respectively, and $\mathbf{U_s}$ is the shock velocity. The bottom schematic shows the same flow in the shock reference frame.

conditions ahead of the shock are known, these equations are sufficient to determine the conditions behind the shock. To illustrate this point, consider the simple one-dimensional situation shown in Figure 6.7. The top schematic shows a shock propagating through a gas with velocity $\mathbf{U_s}$. The velocities $\mathbf{U_a}$ and $\mathbf{U_b}$ are the gas velocities ahead and behind the shock, respectively. The bottom schematic shows the flow dynamics in a reference frame fixed to the shock. The velocities $\mathbf{u_1}$ and $\mathbf{u_2}$ are the velocities ahead and behind the shock, respectively. Note that $\mathbf{u_1} = \mathbf{U_a} - \mathbf{U_s}$. The situation depicted in Figure 6.7 is for a *normal shock*, for which the fluid velocity is perpendicular to the shock structure.

In the shock reference frame, the flow is assumed to be steady and the flow is also assumed to be homogeneous on both sides of the shock. The shock is therefore treated as a discontinuity and the goal is to calculate the "jump" in the gas properties as the shock is crossed. For this one-dimensional, steady, constant-area flow, the continuity (6.123), momentum (6.126), and energy (6.132) equations become

$$\frac{d}{dx}[\rho u] = 0, \tag{6.133a}$$

$$\frac{d}{dx}[\rho u^2 + p] = 0, \tag{6.133b}$$

$$\frac{d}{dx}\left[\left(\frac{1}{2}\rho u^2 + \frac{\gamma p}{\gamma - 1}\right)u\right] = 0. \tag{6.133c}$$

Equations (6.133a–c) indicate that the quantities in the square brackets are conserved as the shock is crossed, and therefore, the parameters on the two sides of the shock are connected by the following relations:

$$\rho_1 u_1 = \rho_2 u_2, \tag{6.134a}$$

$$\rho_1 u_1^2 + p_1 = \rho_2 u_2^2 + p_2, \tag{6.134b}$$

$$\left(\frac{1}{2}\rho_1 u_1^2 + \frac{\gamma}{\gamma-1}p_1\right)u_1 = \left(\frac{1}{2}\rho_2 u_2^2 + \frac{\gamma}{\gamma-1}p_2\right)u_2. \tag{6.134c}$$

Hence, if the gas parameters ahead of the shock (ρ_1, u_1, p_1) are known, those behind the shock (ρ_2, u_2, p_2) can be calculated from Equations (6.134a–c).

For what follows, it is convenient to introduce the upstream Mach number

$$M_1^2 = \frac{\rho_1 u_1^2}{\gamma p_1}, \tag{6.135}$$

where $\gamma p_1/\rho_1$ is the square of the thermal speed (Equation 10.33). Equation (6.135) indicates that $p_1 = \rho_1 u_1^2/(\gamma M_1^2)$, and this result should be substituted into Equations (6.134b) and (6.134c) before the equations are solved. Now, an equation for u_2 can be obtained by substituting ρ_2 (from Equation 6.134a) and p_2 (from Equation 6.134c) into Equation (6.134b), which yields

$$u_2^2 - \frac{2\gamma}{\gamma+1}\left(1 + \frac{1}{\gamma M_1^2}\right)u_1 u_2 + \left(\frac{\gamma-1}{\gamma+1} + \frac{2}{(\gamma+1)M_1^2}\right)u_1^2 = 0, \tag{6.136}$$

or

$$(u_2 - u_1)\left[u_2 - \left(\frac{\gamma-1}{\gamma+1} + \frac{2}{(\gamma+1)M_1^2}\right)u_1\right] = 0. \tag{6.137}$$

There are two solutions to this quadratic equation. One solution is simply that $u_2 = u_1$, which is the solution when a shock does not exist. The other solution provides the change in velocity across a shock, which is

$$\frac{u_2}{u_1} = \frac{\gamma-1}{\gamma+1} + \frac{2}{(\gamma+1)M_1^2}. \tag{6.138a}$$

With this expression for u_2/u_1, it is now possible to obtain expressions for ρ_2/ρ_1 and p_2/p_1 from Equations (6.134a) and (6.134c), respectively, and these expressions are given by

$$\frac{\rho_2}{\rho_1} = \frac{(\gamma+1)M_1^2}{2+(\gamma-1)M_1^2}, \tag{6.138b}$$

$$\frac{p_2}{p_1} = \frac{2\gamma M_1^2 - \gamma + 1}{\gamma+1}. \tag{6.138c}$$

Equations (6.138a–c) are the *Rankine–Hugoniot relations* for the jump conditions across a shock.

In addition to the three jump conditions (6.138a–c), it is useful to have an expression for the Mach number behind the shock, M_2. This expression can be easily obtained by starting with the ratio

$$\frac{M_2^2}{M_1^2} = \frac{\rho_2}{\rho_1}\frac{u_2^2}{u_1^2}\frac{p_1}{p_2}. \tag{6.139}$$

The substitution of Equations (6.138a–c) into Equation (6.139) then yields

$$M_2^2 = \frac{2 + (\gamma - 1)M_1^2}{2\gamma M_1^2 - \gamma + 1}. \tag{6.140}$$

For *weak shocks* ($M_1 \to 1$), the density, velocity, and pressure are continuous and $M_2 \to 1$. For *strong shocks* ($M_1 \gg 1$), Equations (6.138a–c) and (6.140) take the following forms:

$$\frac{u_2}{u_1} \to \frac{\gamma - 1}{\gamma + 1} = \frac{1}{4}, \tag{6.141a}$$

$$\frac{\rho_2}{\rho_1} \to \frac{\gamma + 1}{\gamma - 1} = 4, \tag{6.141b}$$

$$\frac{p_2}{p_1} \to \frac{2\gamma}{\gamma + 1}M_1^2 = \frac{5}{4}M_1^2, \tag{6.141c}$$

and

$$M_2^2 \to \frac{\gamma - 1}{2\gamma} = \frac{1}{5}, \tag{6.142}$$

where the numerical factors are for $\gamma = 5/3$. Therefore, for strong hydrodynamic shocks, the maximum density compression and velocity decrease behind the shock are a factor of four. Likewise, there is a maximum limit to the decrease in the Mach number behind the shock, which is $M_2 = 0.45$. However, there is no limit to the pressure (i.e., temperature) increase behind the shock, according to Equation (6.141c).

The discussion of shock waves presented above was based on the Euler equations (6.123–125), which in turn are based on the assumption of a Maxwellian velocity distribution (3.44). Therefore, the Rankine–Hugoniot relations (6.138a–c) are valid provided the fluid is Maxwellian on both sides of the shock. However, for high Mach number flows, this may not be the case, and then the limiting values for the jump conditions (Equations 6.141a–c, 6.142) are not appropriate.

Also, when a plasma is treated as a single-component, electrically neutral gas, the Euler equations (6.123–125) are valid to lowest order under certain conditions (see Equations 7.45a,c,e). Specifically, they are valid both for an unmagnetized plasma flow and for a plasma flow along a strong magnetic field. Under these circumstances, the Rankine–Hugoniot relations are valid, and they properly describe the jump conditions across shocks in both the solar and terrestrial polar winds.[3,4]

6.15   Double layers

A current is induced when an electric field is applied to either an unmagnetized plasma or along the magnetic field of a magnetized plasma. In collision-dominated plasmas, the relationship between the induced current and the applied electric field is given by Ohm's law (Equation 5.124), which reduces to $\mathbf{J} = \sigma\mathbf{E}$ (Equation 5.126) when density and temperature gradients are negligible. When the plasma is not collision-dominated, an electron–ion two-stream instability may be triggered, depending on the strength of the current. In this case, the plasma can become turbulent, and an *anomalous resistivity* can arise as a result of electron "collisions" with the oscillating, wave electric fields. As a consequence, the classical collision-dominated conductivity is not valid and an anomalous conductivity must be calculated to obtain a relationship between the applied electric field and the induced current. Also, when a large electric field is applied to a dilute plasma, an electrostatic double layer can form.

An *electrostatic double layer* is a narrow region that contains a large electric field relative to the electric fields that exist in the plasma surrounding the double layer (Figure 6.8). The potential drop across the double layer, Φ_0, is generally larger than the equivalent *thermal potential*, kT_e/e, of the plasma. The potential varies monotonically across the double layer, and the potential drop is supported by two distinct layers of oppositely charged particles. The electron and ion layers, which are separated by some tens of Debye lengths, have approximately the same number of particles. Therefore, the double layer is electrically neutral when it is viewed as a single structure. This feature is consistent with the fact that the electric fields are small outside the double layer. Not all double layers are associated with currents. Double layers can form at the boundaries between plasmas with different temperatures or densities, and under these circumstances, they prevent a current flow from one plasma to the other. Current-carrying double layers have been suggested to exist both on auroral field lines in the terrestrial magnetosphere and in the Jovian magnetosphere.[5,6] Current-free double layers have been deduced to occur at high altitudes in the terrestrial polar cap as a result of the interaction of cold ionospheric and hot magnetospheric plasmas.[7,8]

The particles both inside and outside of a double layer, DL, are shown in Figure 6.9. For this case, the DL electric field points to the right. All of the electrons in the plasma on the right that move toward the DL penetrate the DL and are accelerated as they pass through it. On the other hand, the bulk of the ions in the plasma on the right that move toward the DL are reflected by the DL electric field and only the very energetic ions can penetrate the DL. For the plasma on the left, the reverse occurs. The ions penetrate the DL and are accelerated, while the bulk of the electrons are reflected. For the case shown, there is a net current flow from left to right.

Double layers have been studied for many years and there is an extensive literature on these nonlinear potential structures.[9–11] The first self-consistent theory of a double layer was developed in 1929 by Langmuir, who studied a strong double layer in a one-dimensional geometry for steady-state conditions.[9] Langmuir also assumed

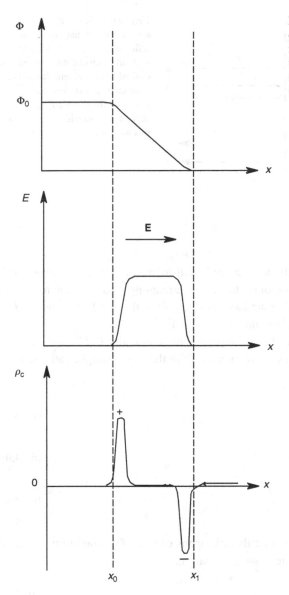

Figure 6.8 Schematic diagram of an electrostatic double layer, including the spatial variations of the potential (top), electric field (middle), and charge density (bottom).

that the plasmas on the two sides of the double layer were cold, unmagnetized, and collisionless. Although Langmuir's theory is very simple, it can explain some important DL features and, therefore, it is instructive to consider it here. The situation studied by Langmuir is shown schematically in Figure 6.8. The region $x < x_0$ contains a cold plasma, and at the edge of the double layer ($x = x_0$), there is an ion flux ($n_{i0}u_{i0} = \Gamma_i$) that enters the double layer (Figure 6.9). All of the electrons in the region $x < x_0$ that approach the double layer are reflected because the double layer is assumed to be strong. Likewise, the region $x > x_1$ contains a cold plasma, and there is an electron flux ($n_{e1}u_{e1} = \Gamma_e$) that enters the double layer at $x = x_1$.

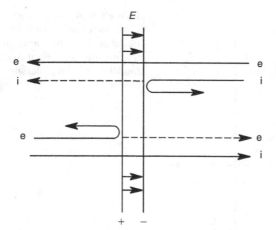

Figure 6.9 Schematic diagram showing the transmitted and reflected particles associated with an electric double layer. The dashed lines indicate that some high-energy particles can penetrate the double layer even though the electric field opposes this motion.

In the region $x > x_1$, all of the ions are reflected by the double layer because it is strong. In this simple scenario, only the counterstreaming ions and electrons exist in the double layer. At the left boundary ($x = x_0$), $\Phi = \Phi_0$ and $E = 0$, where E is the electric field. At the right boundary ($x = x_1$), $\Phi = E = 0$.

The equations that govern this scenario are the continuity (3.57) and momentum (3.58) equations and Gauss' law (3.76a). With the above simplifications, these equations reduce to

$$\frac{d}{dx}(n_s u_s) = 0, \tag{6.143a}$$

$$m_s u_s \frac{du_s}{dx} - e_s E = 0, \tag{6.143b}$$

$$\frac{dE}{dx} = \sum_s n_s e_s / \varepsilon_0, \tag{6.143c}$$

where subscript s corresponds to either electrons or ions. The continuity equation (6.143a) can be easily integrated, and the result is

$$n_s u_s = \text{constant.} \tag{6.144}$$

Likewise, after setting $E = -d\Phi/dx$, integrating the momentum equation (6.143b) results in

$$\frac{1}{2}m_s u_s^2 + e_s \Phi = \text{constant.} \tag{6.145}$$

For the ions, the constants of integration are determined at the left boundary ($x = x_0$), where $n_i = n_{i0}$, $u_i = u_{i0}$, and $\Phi = \Phi_0$. For the electrons, the constants are determined at the right boundary ($x = x_1$), where $n_e = n_{e1}$, $u_e = u_{e1}$, and $\Phi = 0$. When the various constants of integration are evaluated, the continuity (6.144) and momentum (6.145) equations for the ions and electrons become

$$n_i u_i = n_{i0} u_{i0} = \Gamma_i, \tag{6.146}$$

$$n_e u_e = n_{e1} u_{e1} = \Gamma_e, \tag{6.147}$$

$$\frac{1}{2} m_i u_i^2 + e\Phi = \frac{1}{2} m_i u_{i0}^2 + e\Phi_0, \tag{6.148}$$

$$\frac{1}{2} m_e u_e^2 - e\Phi = \frac{1}{2} m_e u_{e1}^2. \tag{6.149}$$

The drift energy of the ions at $x = x_0$ is $m_i u_{i0}^2/2$, and the drift energy of the electrons at $x = x_1$ is $m_e u_{e1}^2/2$; both are negligible compared to the energy these charged particles gain as they are accelerated by the strong double layer. Neglecting these terms and then solving the momentum equations (6.148) and (6.149) for the ion and electron velocities leads to the following results:

$$u_i = \left(\frac{2e}{m_i}\right)^{1/2} (\Phi_0 - \Phi)^{1/2}, \tag{6.150}$$

$$u_e = \left(\frac{2e}{m_e}\right)^{1/2} \Phi^{1/2}. \tag{6.151}$$

When these expressions are substituted into the corresponding continuity equations, (6.146) and (6.147), the result is

$$n_i = \Gamma_i \left(\frac{2e}{m_i}\right)^{-1/2} (\Phi_0 - \Phi)^{-1/2}, \tag{6.152}$$

$$n_e = \Gamma_e \left(\frac{2e}{m_e}\right)^{-1/2} \Phi^{-1/2}. \tag{6.153}$$

Using Equations (6.152) and (6.153), along with $E = -d\Phi/dx$, Gauss' law (6.143c) can be expressed in the form

$$\frac{d}{dx}\left(\frac{d\Phi}{dx}\right) = -\frac{\Gamma_i}{\varepsilon_0}\left(\frac{m_i e}{2}\right)^{1/2} (\Phi_0 - \Phi)^{-1/2} + \frac{\Gamma_e}{\varepsilon_0}\left(\frac{m_e e}{2}\right)^{1/2} \Phi^{-1/2}. \tag{6.154}$$

When Equation (6.154) is multiplied by $(d\Phi/dx)$, it becomes

$$\frac{d}{dx}\left[\frac{1}{2}\left(\frac{d\Phi}{dx}\right)^2\right] = \frac{\Gamma_i}{\varepsilon_0}\left(\frac{m_i e}{2}\right)^{1/2} (\Phi_0 - \Phi)^{-1/2}\left(-\frac{d\Phi}{dx}\right)$$
$$+ \frac{\Gamma_e}{\varepsilon_0}\left(\frac{m_e e}{2}\right)^{1/2} \Phi^{-1/2}\left(\frac{d\Phi}{dx}\right). \tag{6.155}$$

Equation (6.155) can now be integrated from some position x to the DL boundary at $x = x_1$, and the result is

$$\frac{1}{2}(E_1^2 - E^2) = \frac{2\Gamma_i}{\varepsilon_0}\left(\frac{m_i e}{2}\right)^{1/2}\{(\Phi_0 - \Phi)^{1/2}\}_x^{x_1}$$

$$+ \frac{2\Gamma_e}{\varepsilon_0}\left(\frac{m_e e}{2}\right)^{1/2}\{\Phi^{1/2}\}_x^{x_1}. \tag{6.156}$$

However, at $x = x_1$, $E = \Phi = 0$, and therefore, Equation (6.156) becomes

$$E^2 = \frac{2\Gamma_i}{\varepsilon_0}(2m_i e)^{1/2}\left[(\Phi_0 - \Phi)^{1/2} - \Phi_0^{1/2}\right]$$

$$+ \frac{2\Gamma_e}{\varepsilon_0}(2m_e e)^{1/2}\Phi^{1/2}. \tag{6.157}$$

Now, at the left double layer boundary ($x = x_0$), $\Phi = \Phi_0$ and $E = 0$. When these conditions are substituted into Equation (6.157), the *Langmuir condition* is obtained

$$\Gamma_e = \left(\frac{m_i}{m_e}\right)^{1/2}\Gamma_i, \tag{6.158}$$

which indicates that the electron flux entering the double layer must be much greater than the ion flux for a stationary and steady double layer. The substitution of Γ_i from Equation (6.158) into Equation (6.157) yields the following expression for E^2:

$$E^2 = \frac{2\Gamma_e}{\varepsilon_0}(2m_e e)^{1/2}\left[(\Phi_0 - \Phi)^{1/2} - \Phi_0^{1/2} + \Phi^{1/2}\right]. \tag{6.159}$$

It is instructive to examine the behavior of Equation (6.159) close to the left boundary at $x = x_0$. Near this boundary, $\Phi \approx \Phi_0$, and the term in the square brackets is approximately given by $(\Phi_0 - \Phi)^{1/2}$. Using this approximation and $E = -d\Phi/dx$ in Equation (6.159), yields an equation of the form

$$\left(\frac{d\Phi}{dx}\right)^2 = \frac{2\Gamma_e}{\varepsilon_0}(2m_e e)^{1/2}(\Phi_0 - \Phi)^{1/2}. \tag{6.160}$$

When the square root of Equation (6.160) is taken, the result is

$$\frac{d\Phi}{dx} = -\left(\frac{2\Gamma_e}{\varepsilon_0}\right)^{1/2}(2m_e e)^{1/4}(\Phi_0 - \Phi)^{1/4}, \tag{6.161a}$$

or

$$\frac{-d\Phi}{(\Phi_0 - \Phi)^{1/4}} = \left(\frac{2\Gamma_e}{\varepsilon_0}\right)^{1/2}(2m_e e)^{1/4}\,dx, \tag{6.161b}$$

where the minus sign is used when the square root is taken because $d\Phi/dx$ is negative. When Equation (6.161b) is integrated from x_0 to x, the result is

$$\frac{4}{3}(\Phi_0 - \Phi)^{3/4} = \left(\frac{2\Gamma_e}{\varepsilon_0}\right)^{1/2} (2m_e e)^{1/4}(x - x_0). \tag{6.162}$$

Equation (6.162) can also be written as

$$J_e(x - x_0)^2 = \frac{4\varepsilon_0}{9}\left(\frac{2e}{m_e}\right)^{1/2}(\Phi_0 - \Phi)^{3/2}, \tag{6.163}$$

where $J_e = e\Gamma_e$ is the magnitude of the electron current density that enters the double layer.

Although Equation (6.163) is only valid near $x = x_0$, it displays the correct functional form, in general. Specifically, when Equation (6.159) is numerically integrated from $x = x_0$ to $x = x_1$, Equation (6.163) is obtained, except that the right-hand side should be multiplied by 1.865 and $(x - x_0)^2$ should be replaced with d^2, where d is the width of the double layer. With these changes, the general result for strong double layers is given by

$$J_e d^2 = 1.17\varepsilon_0\left(\frac{e}{m_e}\right)^{1/2}\Phi_0^{3/2}, \tag{6.164a}$$

or

$$\Phi_0 = 0.9\left(\frac{m_e}{e\varepsilon_0^2}\right)^{1/3} d^{4/3}J_e^{2/3}. \tag{6.164b}$$

6.16 Summary of important formulas

Electrostatic waves:

$\omega^2 = \omega_{pe}^2 + 3V_e^2 K^2$	Electron plasma waves	$\mathbf{B}_0 = 0$ or
		$\mathbf{K} \parallel \mathbf{B}_0$
$\omega^2 = \omega_{pe}^2$	Electron plasma oscillations	$T_e = 0$
$\omega^2 = \omega_{pe}^2 + \omega_{ce}^2$	Upper hybrid oscillations	$\mathbf{K} \perp \mathbf{B}_0$
$\omega^2 = K^2 V_S^2$	Ion-acoustic waves	$\mathbf{B}_0 = 0$ or
		$\mathbf{K} \parallel \mathbf{B}_0$
$\omega^2 = \omega_{ci}^2 + K^2 V_S^2$	Ion-cyclotron waves	$\mathbf{K} \perp \mathbf{B}_0$
$\omega^2 = \omega_{ce}\omega_{ci}$	Lower hybrid oscillations	$\mathbf{K} \perp \mathbf{B}_0$
		$T_e = T_i = 0$

Electromagnetic waves:

$\omega^2 = K^2c^2 + \omega_{p_e}^2$ 　　　　Ordinary wave (O-mode)　　$\mathbf{B}_0 = 0$ or

$\mathbf{K} \perp \mathbf{B}_0; \mathbf{E}_1 \parallel \mathbf{B}_0$

$T_e = 0$

$\omega^2 = K^2c^2$

$+ \omega_{p_e}^2 \dfrac{\omega^2 - \omega_{p_e}^2}{\omega^2 - (\omega_{p_e}^2 + \omega_{c_e}^2)}$ 　　Extraordinary wave (X-mode)　$\mathbf{K} \perp \mathbf{B}_0$

$\mathbf{E}_1 \perp \mathbf{B}_0$

Elliptic polarization

$T_e = 0$

$\omega^2 = K^2c^2 + \dfrac{\omega_{p_e}^2}{1 - \omega_{c_e}/\omega}$ 　　R Wave (whistler wave)　　$\mathbf{K} \parallel \mathbf{B}_0$

$\mathbf{E}_1 \perp \mathbf{B}_0$

Circular polarization

$T_e = 0$

$\omega^2 = K^2c^2 + \dfrac{\omega_{p_e}^2}{1 + \omega_{c_e}/\omega}$ 　　L wave　　　　　　　　$\mathbf{K} \parallel \mathbf{B}_0$

$\mathbf{E}_1 \perp \mathbf{B}_0$

Circular Polarization

$T_e = 0$

$\omega^2 = K^2V_A^2$ 　　　　　　　Alfvén wave　　　　　　　$\mathbf{K} \parallel \mathbf{B}_0$

$\mathbf{E}_1 \perp \mathbf{B}_0$

Linear polarization

$\omega^2 = K^2 \dfrac{V_S^2 + V_A^2}{1 + V_A^2/c^2}$ 　　Magnetosonic wave　　　　$\mathbf{K} \perp \mathbf{B}_0$

$\mathbf{E}_1 \perp \mathbf{B}_0$

Linear polarization

where $\omega_{p_s} = (n_s e^2/\varepsilon_0 m_s)^{1/2}$ is the plasma frequency of species s, $\omega_{c_s} = |e_s|B/m_s$ is the cyclotron frequency, $V_A = B/(\mu_0 n_i m_i)^{1/2}$ is the Alfvén speed, $V_S = \left[(\gamma_i kT_i + \gamma_e kT_e)/m_i\right]^{1/2}$ is the ion-acoustic speed, and $V_e = (kT_e/m_e)^{1/2}$ is the electron thermal speed.

Hydrodynamic shocks:

$$\frac{u_2}{u_1} = \frac{\gamma - 1}{\gamma + 1} + \frac{2}{(\gamma + 1)M_1^2},$$

$$\frac{\rho_2}{\rho_1} = \frac{(\gamma + 1)M_1^2}{2 + (\gamma - 1)M_1^2},$$

$$\frac{p_2}{p_1} = \frac{2\gamma M_1^2 - \gamma + 1}{\gamma + 1},$$

$$M_2^2 = \frac{2 + (\gamma - 1)M_1^2}{2\gamma M_1^2 - \gamma + 1},$$

where the jump conditions are relevant to the shock reference frame, and where subscript 1 corresponds to the flow conditions ahead of the shock and subscript 2 to the flow conditions behind the shock.

Strong double layers:

$$E^2 = \frac{2J_e}{\varepsilon_0}\left(\frac{2m_e}{e}\right)^{1/2}\left[(\Phi_0 - \Phi)^{1/2} - \Phi_0^{1/2} + \Phi^{1/2}\right],$$

$$J_e = \left(\frac{m_i}{m_e}\right)^{1/2}J_i,$$

$$\Phi_0 = 0.9\left(\frac{m_e}{e\varepsilon_0^2}\right)^{1/3}d^{4/3}J_e^{2/3},$$

where J_e and J_i are the current densities that enter the double layer, d is the thickness of the double layer, and Φ_0 is the potential jump across the double layer.

6.17   Specific references

1. Jackson, J. D., *Classical Electrodynamics*, New York: Wiley, 1998.
2. Chen, F. F., *Introduction to Plasma Physics and Controlled Fusion*, New York: Plenum, 1985.
3. Sonett, C. P., and D. S. Colburn, The $SI^+ - SI^-$ pair and interplanetary forward-reverse shock ensembles, *Planet. Space Sci.*, **13**, 675, 1965.
4. Singh, N., and R. W. Schunk, Temporal behavior of density perturbations in the polar wind, *J. Geophys. Res.*, **90**, 6487, 1985.
5. Mozer, F. S., C. W. Carlson, M. K. Hudson, *et al.*, Observations of paired electrostatic shocks in the polar magnetosphere, *Phys. Rev. Lett.*, **38**, 292, 1977.
6. Shawhan, S. D., C.-G. Fälthammar, and L. P. Block, On the nature of large auroral zone electric fields at $1 - R_E$ altitude, *J. Geophys. Res.*, **83**, 1049, 1978.
7. Winningham, J. D., and C. Gurgiolo, DE-2 photoelectron measurements consistent with a large scale parallel electric field over the polar cap, *Geophys. Res. Lett.*, **9**, 977, 1982.
8. Barakat, A. R., and R. W. Schunk, Effect of hot electrons on the polar wind, *J. Geophys. Res.*, **89**, 9771, 1984.
9. Langmuir, I., The interaction of electron and positive ion space charges in cathode sheaths, *Phys. Rev.*, **33**, 954, 1929.
10. Block, L. P., Potential double layers in the ionosphere, in *Cosmic Electrodynamics*, **3**, 349, 1972.
11. Carlqvist, P., Some theoretical aspects of electrostatic double layers, in *Wave Instabilities in Space Plasmas*, ed. by P. J. Palmadesso and K. Papadopoulos, 83, Boston: D. Reidel, 1979.

6.18   General references

Bittencourt, J. A., *Fundamentals of Plasma Physics*, Brazil, 1995.

Cap, F. F., *Handbook of Plasma Instabilities*, vol. 1 and 2, New York: Academic Press, 1976.

Emanuel, G., *Gasdynamics: Theory and Applications*, New York: American Institute of Aeronautics and Astronautics, 1986.

Gary, S. P., *Theory of Space Plasma Microinstabilities*, Cambridge, UK: Cambridge University Press, 1993.

Ichimaru, S., *Plasma Physics: An Introduction to Statistical Physics of Charged Particles*, Menlo Park, CA: Benjamin/Cummings, 1986.

Nicholson, D. R., *Introduction to Plasma Physics*, New York: Wiley, 1983.

Stix, T. H., *Waves in Plasmas*, New York: American Institute of Physics, 1992.

Tidman, D. A., and N. A. Krall, *Shock Waves in Collisionless Plasmas*, New York: Wiley, 1971.

6.19   Problems

Problem 6.1 Consider a collisionless, spatially uniform, electrically neutral, unmagnetized, electron–ion plasma. If the plasma drifts with a constant velocity $u_{e0} = u_{i0} = u_0$, derive the dispersion relation for electron plasma waves that propagate both parallel and perpendicular to u_0.

Problem 6.2 Consider a collisionless, spatially uniform, electrically neutral, unmagnetized plasma. The plasma is composed of electrons (subscript e) and two ion species (subscripts i and j). All three species can be described by an equation of state (6.27). Derive the dispersion relation for ion-acoustic waves.

Problem 6.3 For the plasma described in Problem 6.1, derive the dispersion relation for ion-acoustic waves that propagate in the direction of u_0.

Problem 6.4 A collisionless, spatially uniform, electrically neutral, electron–ion plasma is subjected to a constant magnetic field B_0. Derive the dispersion relation for upper hybrid oscillations assuming that T_e is constant, but not zero.

Problem 6.5 A collisionless, spatially uniform, electrically neutral, electron–ion plasma drifts with a velocity u_0 in the direction of a constant magnetic field B_0. Derive the dispersion relation for high frequency, electrostatic oscillations both parallel and perpendicular to B_0. Assume the plasma is cold.

Problem 6.6 Consider a collisionless, spatially uniform, electrically neutral, stationary plasma. The plasma is composed of electrons (subscript e) and two ion species (subscripts i and j). The plasma is also subjected to a magnetic field B_0. Derive the dispersion relation for electrostatic ion-cyclotron waves.

Problem 6.7 A constant magnetic field B_0 permeates a collisionless, fully ionized, three-species plasma. The plasma is composed of hot electrons (n_h, T_h), cold electrons (n_c, T_c), and ions (n_i, T_i). Initially, the plasma is electrically neutral, homogeneous, and stationary. Derive the dispersion relation for electrostatic ion-cyclotron waves.

Problem 6.8 Consider the plasma described in Problem 6.1. Derive the dispersion relation for high frequency "light waves" that propagate in the direction of $\mathbf{u}_0$. Assume that T_e is constant, but not zero.

Problem 6.9 Consider the plasma described in Problem 6.6 and derive the dispersion relation for Alfvén waves.

Problem 6.10 Consider the plasma described in Problem 6.7 and derive the dispersion relation for Alfvén waves.

Problem 6.11 An electron–ion plasma is homogeneous, electrically neutral, stationary and unmagnetized. If the electron–ion collision term is given by Equation (6.105), and if ν_{ei} is assumed to be constant, derive the dispersion relation for high frequency "light waves."

Problem 6.12 Add the collision term ($-n_s m_s \nu_s \mathbf{u}_s$) to the right-hand side of Equation (6.28) and then derive the dispersion relation for ion-acoustic waves for the case of an unmagnetized plasma that is stationary, homogeneous, and electrically neutral. Assume that ν_s is constant for both electrons and ions.

Problem 6.13 A collisionless, homogeneous, electrically neutral, electron–ion plasma drifts under the influence of perpendicular electric, $\mathbf{E}_0$, and magnetic, $\mathbf{B}_0$, fields (see Equation 6.29b). Consider the stability of the plasma with regard to high frequency, electrostatic oscillations both parallel and perpendicular to $\mathbf{B}_0$. Assume that the plasma is cold.

Problem 6.14 An electron beam of density n_b and velocity $\mathbf{u}_b$ propagates through a background electron plasma of density n_p. The background plasma is cold, stationary, and uniform, and there are no background electric or magnetic fields. Assume that the ions are immobile and that they provide the charge neutrality of the system. Discuss the stability of the system for $\omega > \omega_p$ and for $\omega < \omega_p$, where ω is the wave frequency and ω_p is the plasma frequency of the background plasma.

Problem 6.15 Derive the Langmuir condition (6.158) for the case when there are two ion species in the presence of a strong double layer.

Chapter 7

Magnetohydrodynamic formulation

The 13-moment system of transport equations was presented in Chapter 3 and several associated sets of collision terms were derived in Chapter 4. These 13-moment transport equations, in combination with the Maxwell equations for the electric and magnetic fields, are very general and can be applied to describe a wide range of plasma flows in the ionospheres. However, the complete system of equations for a multi-species plasma is difficult to solve under most circumstances, and therefore, simplified sets of transport equations have been used over the years. The simplified sets of equations that are based on the assumption of collision dominance were presented in Chapter 5. In this chapter, certain simplified transport equations are derived in which the plasma is treated as a single conducting fluid, rather than a mixture of individual plasma species. These single-fluid transport equations, along with the Maxwell equations, are known as the *single-fluid magnetohydrodynamic (MHD) equations*.

The outline of this chapter is as follows. First, the single-fluid transport equations are derived from the 13-moment system of equations. Subsequently, a generalized Ohm's law is derived for a fully ionized plasma. This naturally leads to simplifications that yield the *classical* set of MHD equations. The classical MHD equations are then applied to important specific cases, including a discussion of pressure balance, the diffusion of a magnetic field into a plasma, the concept of a **B** field frozen in a plasma, the derivation of the spiral magnetic field associated with rotating magnetized bodies, and the derivation of the double-adiabatic energy equations for a collisionless anisotropic plasma. These topics are followed by a derivation of the MHD waves and shocks that can exist in a plasma.

7.1   General MHD equations

To treat a gas mixture as a single conducting fluid, it is necessary to add the contributions of the individual species and obtain both total and average parameters for the gas mixture. Some of the fundamental parameters are the total mass density, ρ, the charge density, ρ_c, the average drift velocity, $\mathbf{u}$, and the total current density, $\mathbf{J}$, which are defined as

$$\rho = \sum_s n_s m_s, \tag{7.1}$$

$$\rho_c = \sum_s n_s e_s, \tag{7.2}$$

$$\mathbf{u} = \sum_s n_s m_s \mathbf{u}_s \Big/ \sum_s n_s m_s, \tag{7.3}$$

$$\mathbf{J} = \sum_s n_s e_s \mathbf{u}_s. \tag{7.4}$$

Note that the average drift velocity (7.3) is the same as that used in the early classical work on transport theory and was introduced previously in Equation (3.13). The quantities ρ_c and $\mathbf{J}$ have also been defined before in Equations (3.77) and (3.78), but it is convenient to list them again for easy reference.

All of the higher-order transport properties, such as the temperature, pressure tensor, and heat flow vector, are defined for this single-fluid treatment relative to the average drift velocity of the gas mixture (Equation 7.3) and not the individual species drift velocities (Equation 3.14). Therefore, in this case, the random or thermal velocity is defined as

$$\mathbf{c}_s^* = \mathbf{v}_s - \mathbf{u} \tag{7.5}$$

and the important transport properties become[1,2]

$$\frac{3}{2}kT_s^* = \frac{1}{2}m_s \langle c_s^{*2} \rangle, \tag{7.6}$$

$$\mathbf{q}_s^* = \frac{1}{2}n_s m_s \langle c_s^{*2} \mathbf{c}_s^* \rangle, \tag{7.7}$$

$$\mathbf{P}_s^* = n_s m_s \langle \mathbf{c}_s^* \mathbf{c}_s^* \rangle, \tag{7.8}$$

$$\tau_s^* = \mathbf{P}_s^* - p_s^* \mathbf{I}, \tag{7.9}$$

where $p_s^* = n_s k T_s^*$ is the partial pressure of species s.

When $\mathbf{u}$ is used to define the transport properties, it is customary to introduce a species *diffusion velocity*, $\mathbf{w}_s$, to describe the mean flow of a given species relative to the average drift velocity of the gas mixture

$$\mathbf{w}_s = \mathbf{u}_s - \mathbf{u}. \tag{7.10}$$

The difference in the definitions of the transport properties, defined in Equations (3.15–17, 3.21) and in Equations (7.6–9), can be expressed in terms of the diffusion velocities $\mathbf{w}_s$ by noting that

$$\mathbf{c}_s^* = \mathbf{c}_s + \mathbf{w}_s, \tag{7.11}$$

which follows from Equations (3.14), (7.5), and (7.10). Substituting Equation (7.11) into Equations (7.6–9) and taking account of the definitions in Equations (3.15–17, 3.21), the following expressions, connecting the different definitions of the transport properties, are obtained:

$$T_s^* = T_s + m_s w_s^2 / 3k, \tag{7.12}$$

$$\mathbf{q}_s^* = \mathbf{q}_s + \frac{5}{2} p_s \mathbf{w}_s + \mathbf{w}_s \cdot \boldsymbol{\tau}_s + \frac{1}{2} n_s m_s w_s^2 \mathbf{w}_s, \tag{7.13}$$

$$\mathbf{P}_s^* = \mathbf{P}_s + n_s m_s \mathbf{w}_s \mathbf{w}_s, \tag{7.14}$$

$$\boldsymbol{\tau}_s^* = \boldsymbol{\tau}_s + n_s m_s [\mathbf{w}_s \mathbf{w}_s - (w_s^2/3)\mathbf{I}]. \tag{7.15}$$

Therefore, the total transport properties for a single-fluid description are simply given by

$$p = \sum_s n_s k T_s^*, \tag{7.16}$$

$$\mathbf{q} = \sum_s \mathbf{q}_s^*, \tag{7.17}$$

$$\mathbf{P} = \sum_s \mathbf{P}_s^*, \tag{7.18}$$

$$\boldsymbol{\tau} = \sum_s \boldsymbol{\tau}_s^*. \tag{7.19}$$

Now that the transport properties have been redefined in terms of $\mathbf{u}$, it is possible to derive the single-fluid continuity, momentum, and energy equations starting from the 13-moment system of equations (3.57–61). The equation describing the flow of the total mass density, ρ, is obtained by multiplying the continuity equation (3.57) by m_s and summing over all species in the gas mixture, which yields

$$\frac{\partial}{\partial t} \left(\sum_s n_s m_s \right) + \nabla \cdot \left(\sum_s n_s m_s \mathbf{u}_s \right) = \frac{\delta}{\delta t} \left(\sum_s n_s m_s \right), \tag{7.20}$$

or

$$\frac{\partial \rho}{\partial t} + \nabla \cdot (\rho \mathbf{u}) = 0, \tag{7.21}$$

where ρ and $\mathbf{u}$ are defined in Equations (7.1) and (7.3), respectively, and where it is assumed that there is no net production or loss of particles in the gas mixture ($\delta \rho / \delta t = 0$). In a similar manner, an equation describing the evolution of the charge density, ρ_c, is obtained by multiplying the continuity equation (3.57) by e_s and summing over all of the species, which yields

$$\frac{\partial \rho_c}{\partial t} + \nabla \cdot \mathbf{J} = 0, \tag{7.22}$$

where it is assumed that $\delta \rho_c / \delta t = 0$.

The momentum equation for the gas mixture is obtained by summing the individual momentum equations (3.58), which yields

$$\sum_s \rho_s \frac{D_s \mathbf{u}_s}{Dt} + \sum_s \nabla \cdot \mathbf{P}_s - \rho \mathbf{G} - \rho_c \mathbf{E} - \mathbf{J} \times \mathbf{B} = 0, \tag{7.23}$$

where ρ, ρ_c, and $\mathbf{J}$ are defined by Equations (7.1), (7.2), and (7.4), respectively. Note that the collision terms cancel when the individual momentum equations are summed. The first term in Equation (7.23) can be expressed in an alternate form by using both the individual continuity equation (3.57) and the continuity equation for the gas mixture (7.21), as follows:

$$\begin{aligned}
\sum_s \rho_s \frac{D_s \mathbf{u}_s}{Dt} &= \sum_s \rho_s \frac{\partial \mathbf{u}_s}{\partial t} + \sum_s \rho_s (\mathbf{u}_s \cdot \nabla) \mathbf{u}_s \\
&= \sum_s \left[\frac{\partial}{\partial t} (\rho_s \mathbf{u}_s) - \mathbf{u}_s \frac{\partial \rho_s}{\partial t} \right] + \sum_s \rho_s (\mathbf{u}_s \cdot \nabla) \mathbf{u}_s \\
&= \frac{\partial}{\partial t} (\rho \mathbf{u}) + \sum_s \mathbf{u}_s \nabla \cdot (\rho_s \mathbf{u}_s) + \sum_s \rho_s (\mathbf{u}_s \cdot \nabla) \mathbf{u}_s \\
&= \rho \frac{\partial \mathbf{u}}{\partial t} + \mathbf{u} \frac{\partial \rho}{\partial t} + \sum_s \nabla \cdot (\rho_s \mathbf{u}_s \mathbf{u}_s).
\end{aligned} \tag{7.24}$$

Likewise, the pressure tensor term in Equation (7.23) can be cast in a more convenient form

$$\begin{aligned}
\sum_s \nabla \cdot \mathbf{P}_s &= \nabla \cdot \left[\sum_s \mathbf{P}_s^* - \sum_s \rho_s (\mathbf{u}_s - \mathbf{u})(\mathbf{u}_s - \mathbf{u}) \right] \\
&= \nabla \cdot \mathbf{P} - \sum_s \nabla \cdot (\rho_s \mathbf{u}_s \mathbf{u}_s - \rho_s \mathbf{u}_s \mathbf{u} - \rho_s \mathbf{u} \mathbf{u}_s + \rho_s \mathbf{u} \mathbf{u}) \\
&= \nabla \cdot \mathbf{P} - \sum_s \nabla \cdot (\rho_s \mathbf{u}_s \mathbf{u}_s) + \nabla \cdot (\rho \mathbf{u} \mathbf{u}),
\end{aligned} \tag{7.25}$$

where use has been made of Equations (7.10), (7.14), and (7.18). Substituting Equations (7.24) and (7.25) into Equation (7.23), yields the momentum equation for the gas mixture

$$\rho \frac{D\mathbf{u}}{Dt} + \nabla \cdot \mathbf{P} - \rho \mathbf{G} - \rho_c \mathbf{E} - \mathbf{J} \times \mathbf{B} = 0, \tag{7.26}$$

where the convective derivative for the composite gas is given by

$$\frac{D}{Dt} = \frac{\partial}{\partial t} + \mathbf{u} \cdot \nabla, \tag{7.27}$$

and where use has been made of both the continuity equation (7.21) and the tensor relation $\nabla \cdot (\rho \mathbf{uu}) = \mathbf{u} \nabla \cdot (\rho \mathbf{u}) + \rho \mathbf{u} \cdot \nabla \mathbf{u}$.

The species momentum equations can also be used to derive an equation for $\mathbf{J}$ in a manner similar to that used to derive the momentum equation (7.26). Multiplying Equation (3.58) by e_s/m_s and summing over all species yields

$$\sum_s n_s e_s \frac{D_s \mathbf{u}_s}{Dt} + \sum_s \nabla \cdot \left(\frac{e_s}{m_s} \mathbf{P}_s \right) - \rho_c \mathbf{G}$$

$$- \sum_s \frac{n_s e_s^2}{m_s} (\mathbf{E} + \mathbf{u}_s \times \mathbf{B}) = \sum_s \frac{e_s}{m_s} \frac{\delta \mathbf{M}_s}{\delta t}, \tag{7.28}$$

where now the collision terms do not cancel. The inertial and pressure tensor terms can be manipulated in a manner similar to that which led to Equations (7.24) and (7.25), and the result is

$$\sum_s n_s e_s \frac{D_s \mathbf{u}_s}{Dt} = \frac{\partial \mathbf{J}}{\partial t} + \sum_s \nabla \cdot (n_s e_s \mathbf{u}_s \mathbf{u}_s), \tag{7.29}$$

$$\sum_s \nabla \cdot \left(\frac{e_s}{m_s} \mathbf{P}_s \right) = \sum_s \nabla \cdot \left(\frac{e_s}{m_s} \mathbf{P}_s^* \right) - \sum_s \nabla \cdot (n_s e_s \mathbf{u}_s \mathbf{u}_s)$$

$$+ \nabla \cdot (\mathbf{u} \mathbf{J} + \mathbf{J} \mathbf{u} - \rho_c \mathbf{uu}). \tag{7.30}$$

The substitution of Equations (7.29) and (7.30) into Equation (7.28) leads to an equation governing the spatial and temporal evolution of the current density, and this equation is given by

$$\frac{\partial \mathbf{J}}{\partial t} + \nabla \cdot (\mathbf{u} \mathbf{J} + \mathbf{J} \mathbf{u} - \rho_c \mathbf{uu}) + \nabla \cdot \left(\sum_s \frac{e_s}{m_s} \mathbf{P}_s^* \right)$$

$$- \rho_c \mathbf{G} - \sum_s \frac{n_s e_s^2}{m_s} (\mathbf{E} + \mathbf{u}_s \times \mathbf{B}) = \sum_s \frac{e_s}{m_s} \frac{\delta \mathbf{M}_s}{\delta t}. \tag{7.31}$$

A single-fluid energy equation can be derived simply by summing the individual energy equations (3.59) over all the species in the gas mixture. Using algebraic manipulations similar to those used to derive the equations for $\mathbf{u}$ and $\mathbf{J}$, the energy equation can be cast in the following form:

$$\frac{D}{Dt} \left(\frac{3}{2} p \right) + \frac{5}{2} p \nabla \cdot \mathbf{u} + \boldsymbol{\tau} : \nabla \mathbf{u} + \nabla \cdot \mathbf{q} - \mathbf{J} \cdot (\mathbf{E} + \mathbf{u} \times \mathbf{B}) + \rho_c \mathbf{u} \cdot \mathbf{E} = 0$$

$$\tag{7.32}$$

where the continuity (7.21) and momentum (7.26) equations must be used to get the energy equation in the form given in (7.32).

In summary, the general MHD equations for a single-fluid conducting gas are composed of the mass continuity equation (7.21), the charge continuity equation (7.22), the momentum equation (7.26), the equation for the current density (7.31), the energy equation (7.32), and the complete set of Maxwell equations (3.76a–d). Although these general MHD equations are not as complicated as the complete 13-moment set of transport equations, they are still difficult to solve, and typically, additional simplifications are made before they are used.[2-4] The most frequently used additional simplifications are discussed in Section 7.3.

7.2   Generalized Ohm's law

The equation for the current density (7.31) is not in its classical form. Typically, when this equation is used, the following additional assumptions are made: (1) the gas consists only of electrons and one singly ionized ion species; (2) charge neutrality prevails ($n_e = n_i = n$); (3) the linear collision terms (4.129b) are appropriate and the heat flow contribution to these collision terms can be neglected; and (4) terms of order m_e/m_i can be neglected compared with terms of order one.

Equation (7.31), for such a two-component plasma that is electrically neutral ($\rho_c = 0$), becomes

$$\frac{\partial \mathbf{J}}{\partial t} + \nabla \cdot (\mathbf{uJ} + \mathbf{Ju}) + e\nabla \cdot \left(\frac{\mathbf{P}_i^*}{m_i} - \frac{\mathbf{P}_e^*}{m_e}\right) - \frac{ne^2}{m_e}(\mathbf{E} + \mathbf{u}_e \times \mathbf{B})$$

$$-\frac{ne^2}{m_i}(\mathbf{E} + \mathbf{u}_i \times \mathbf{B}) = \frac{e}{m_i}\frac{\delta \mathbf{M}_i}{\delta t} - \frac{e}{m_e}\frac{\delta \mathbf{M}_e}{\delta t}. \tag{7.33}$$

Also, using the charge neutrality condition and taking account of the small electron mass, the expressions for $\mathbf{u}$ (Equation 7.3) and $\mathbf{J}$ (Equation 7.4) become

$$\mathbf{u} \approx \mathbf{u}_i, \tag{7.34}$$

$$\mathbf{J} = ne(\mathbf{u}_i - \mathbf{u}_e). \tag{7.35}$$

An expression for $\mathbf{u}_e$ in terms of $\mathbf{J}$ and $\mathbf{u}$ can now be obtained from Equations (7.34) and (7.35), and the result is

$$\mathbf{u}_e \approx \mathbf{u} - \mathbf{J}/ne. \tag{7.36}$$

Consider the third term on the left-hand side of Equation (7.33). Only the $\mathbf{P}_e^*$ term will survive because of the small electron mass. With regard to the fourth and fifth terms, the electric field term divided by m_i can be neglected compared with the electric field term divided by m_e. The magnetic field terms can be expressed in the form

$$-\frac{ne^2}{m_e}\left(\mathbf{u}_e \times \mathbf{B} + \frac{m_e}{m_i}\mathbf{u}_i \times \mathbf{B}\right) = -\frac{ne^2}{m_e}\left[\left(\mathbf{u} - \frac{\mathbf{J}}{ne}\right) \times \mathbf{B} + \frac{m_e}{m_i}\mathbf{u} \times \mathbf{B}\right]$$

$$= -\frac{ne^2}{m_e}(\mathbf{u} \times \mathbf{B} - \mathbf{J} \times \mathbf{B}/ne), \qquad (7.37)$$

where Equations (7.34) and (7.36) were used for $\mathbf{u}_i$ and $\mathbf{u}_e$, respectively, and where the term containing m_e/m_i was neglected.

The linear collision term for the electrons is obtained from Equation (4.129b), and it becomes

$$\frac{e}{m_e}\frac{\delta \mathbf{M}_e}{\delta t} = en\nu_{ei}(\mathbf{u}_i - \mathbf{u}_e) = \nu_{ei}\mathbf{J}. \qquad (7.38)$$

Likewise, the ion collision term can be expressed in the form

$$\frac{e}{m_i}\frac{\delta \mathbf{M}_i}{\delta t} = \frac{e}{m_i}n_im_i\nu_{ie}(\mathbf{u}_e - \mathbf{u}_i) = \frac{e}{m_i}n_em_e\nu_{ei}(\mathbf{u}_e - \mathbf{u}_i) = -\left(\frac{m_e}{m_i}\right)\nu_{ei}\mathbf{J},$$

$$(7.39)$$

where $n_im_i\nu_{ie} = n_em_e\nu_{ei}$ (Equation 4.158). Therefore, a comparison of Equations (7.39) and (7.38) indicates that the ion collision term can be neglected compared with the electron collision term.

Using Equations (7.37) and (7.38), and neglecting the terms discussed above that are small, Equation (7.33) reduces to

$$\frac{\partial \mathbf{J}}{\partial t} + \nabla \cdot (\mathbf{u}\mathbf{J} + \mathbf{J}\mathbf{u}) - \frac{e}{m_e}\nabla \cdot \mathbf{P}_e^* - \frac{ne^2}{m_e}(\mathbf{E} + \mathbf{u} \times \mathbf{B}) + \frac{e}{m_e}\mathbf{J} \times \mathbf{B} = -\nu_{ei}\mathbf{J},$$

$$(7.40)$$

Multiplying (7.40) by $-m_e/ne^2$, this equation can be written as

$$-\frac{m_e}{ne^2}\left[\frac{\partial \mathbf{J}}{\partial t} + \nabla \cdot (\mathbf{u}\mathbf{J} + \mathbf{J}\mathbf{u})\right] + \frac{1}{ne}\nabla \cdot \mathbf{P}_e^* - \frac{1}{ne}\mathbf{J} \times \mathbf{B} + \mathbf{E} + \mathbf{u} \times \mathbf{B} = \mathbf{J}/\sigma_e,$$

$$(7.41)$$

where

$$\sigma_e = \frac{ne^2}{m_e\nu_{ei}}. \qquad (7.42)$$

Note that σ_e is the so-called first approximation to the parallel conductivity of a fully ionized plasma (see Equations 5.125a and 5.142). Finally, the neglect of the terms in the square brackets that are multiplied by m_e leads to the *generalized Ohm's law* for an MHD plasma, which is

$$\frac{1}{ne}(\nabla \cdot \mathbf{P}_e^* - \mathbf{J} \times \mathbf{B}) + \mathbf{E} + \mathbf{u} \times \mathbf{B} = \mathbf{J}/\sigma_e. \tag{7.43}$$

The $\mathbf{J} \times \mathbf{B}$ term contains the Hall current effect. The ratio of this term to the conductivity term is simply ω_{c_e}/ν_{ei}. Therefore, when the collision frequency is much greater than the cyclotron frequency, the Hall current effect is negligible. Even when the collision frequency is not large, it is often possible to neglect both the Hall current and pressure tensor terms. Under these conditions, the generalized Ohm's law reduces to

$$\mathbf{J} = \sigma_e(\mathbf{E} + \mathbf{u} \times \mathbf{B}). \tag{7.44}$$

7.3 Simplified MHD equations

As noted earlier, the general set of MHD equations is complicated and rarely used. Instead, a simplified set of MHD equations is used that is based on several additional assumptions. First, charge neutrality is assumed ($\rho_c = 0$). Next, in the momentum equation (7.26), the pressure tensor is typically assumed to be diagonal and isotropic, $\mathbf{P} = p\mathbf{I}$, so that $\nabla \cdot \mathbf{P} = \nabla p$. This means that the stress tensor is negligible and only the scalar pressure is important. Two additional assumptions generally made are that the simplified form of Ohm's law (Equation 7.44) can be used and that the energy equation (7.32) can be replaced by an equation of state. Both of these assumptions are difficult to justify a priori, but in many applications they can be, at least, partially justified after the solutions are obtained. Finally, it is assumed that the phenomena under consideration vary slowly in time, being governed by ion time scales. Under these circumstances, the displacement current, $\varepsilon_0 \partial \mathbf{E}/\partial t$, in the Maxwell $\nabla \times \mathbf{B}$ equation (3.76d) can be neglected.

It is convenient to list the set of simplified MHD equations in one place because of its wide use by the scientific community. With these assumptions, the equations for mass continuity (7.21), current continuity (7.22), momentum (7.26), the current density (7.31), and energy (7.32) reduce, respectively, to the following set of equations:

$$\frac{\partial \rho}{\partial t} + \nabla \cdot (\rho \mathbf{u}) = 0, \tag{7.45a}$$

$$\nabla \cdot \mathbf{J} = 0, \tag{7.45b}$$

$$\rho \frac{D\mathbf{u}}{Dt} + \nabla p - \rho \mathbf{G} - \mathbf{J} \times \mathbf{B} = 0, \tag{7.45c}$$

$$\mathbf{J} = \sigma_e(\mathbf{E} + \mathbf{u} \times \mathbf{B}), \tag{7.45d}$$

$$p = C\rho^\gamma, \tag{7.45e}$$

where the equation of state was introduced previously in Equation (6.26b) and C is
a constant. The associated set of simplified Maxwell equations is

$$\nabla \times \mathbf{E} = -\frac{\partial \mathbf{B}}{\partial t}, \tag{7.45f}$$

$$\nabla \times \mathbf{B} = \mu_0 \mathbf{J}. \tag{7.45g}$$

Note that Equation (7.45b) is redundant because it can be obtained by taking the
divergence of Equation (7.45g). Also, the $\nabla \cdot \mathbf{B}$ and $\nabla \cdot \mathbf{E}$ equations do not have the
same status as the two curl equations. From Faraday's law (7.45f), $\partial/\partial t(\nabla \cdot \mathbf{B}) = 0$,
and hence, the requirement that $\nabla \cdot \mathbf{B} = 0$ can be specified as an initial condition.
Likewise, $\nabla \cdot \mathbf{E} = 0$ is not imposed as an additional constraint because charge neutrality has already been assumed. The electric field is completely determined by the
two curl equations and Ohm's law.[5] In reality, however, when numerical computa-
tions are performed, it is important to verify that the solutions to the simplified set of
MHD equations are, at least, consistent with $\nabla \cdot \mathbf{E}$ and $\nabla \cdot \mathbf{B}$ being very small. This
is necessary because the numerical techniques employed tend to introduce errors
that eventually cause these conditions to be violated.

7.4   Pressure balance

It is instructive to consider the balance of pressure for the special case of a steady state
($\partial/\partial t = 0$), incompressible ($\rho = $ constant; $\nabla \cdot \mathbf{u} = 0$), and irrotational ($\nabla \times \mathbf{u} = 0$)
MHD flow. In this case, the momentum equation (7.45c) becomes

$$\rho(\mathbf{u} \cdot \nabla)\mathbf{u} + \nabla p - \mathbf{J} \times \mathbf{B} = 0, \tag{7.46}$$

where gravity is neglected. When $\mathbf{J}$ is eliminated with the aid of Ampère's
law (7.45g), Equation (7.46) becomes

$$\rho(\mathbf{u} \cdot \nabla)\mathbf{u} + \nabla p - \frac{1}{\mu_0}(\nabla \times \mathbf{B}) \times \mathbf{B} = 0. \tag{7.47}$$

The third term in Equation (7.47) can be cast in a more convenient form by using
one of the vector relations given in Appendix B, which is

$$\frac{1}{2}\nabla(\mathbf{B} \cdot \mathbf{B}) = (\mathbf{B} \cdot \nabla)\mathbf{B} - (\nabla \times \mathbf{B}) \times \mathbf{B} \approx -(\nabla \times \mathbf{B}) \times \mathbf{B}. \tag{7.48}$$

The second result in Equation (7.48) is true provided that $\mathbf{B}$ does not vary appreciably
along its direction, which is a situation that frequently occurs. The first term in
Equation (7.47) can also be cast in a more convenient form by using the same vector
relation

$$\frac{1}{2}\nabla(\mathbf{u} \cdot \mathbf{u}) = (\mathbf{u} \cdot \nabla)\mathbf{u} + \mathbf{u} \times (\nabla \times \mathbf{u}) = (\mathbf{u} \cdot \nabla)\mathbf{u}, \tag{7.49}$$

where for irrotational flow $\nabla \times \mathbf{u} = 0$.

Substituting Equations (7.48) and (7.49) into Equation (7.47) yields an equation of the form

$$\nabla \left(\frac{1}{2}\rho u^2 + p + \frac{B^2}{2\mu_0} \right) = 0, \tag{7.50}$$

where use has been made of the fact that ρ is constant for an incompressible fluid. Equation (7.50) indicates that the quantity in the brackets is a constant. For the special case of a one-dimensional flow where $\mathbf{B}$ is perpendicular to $\mathbf{u}$ and all quantities vary only in the $\mathbf{u}$ direction, the assumption of incompressible flow ($\nabla \cdot \mathbf{u} = 0$) implies that both ρ and $\mathbf{u}$ are constants. Hence, for this case, $\rho u^2/2$ can be added to the quantity in the brackets of Equation (7.50) and the result is

$$\rho u^2 + p + \frac{B^2}{2\mu_0} = \text{constant}. \tag{7.51}$$

As it turns out, the one-dimensional result (7.51) is also valid for a compressible flow (see Section 7.9 and Problem 7.11). In Equation (7.51), p is the *kinetic pressure*, ρu^2 is the *dynamic pressure*, and $B^2/2\mu_0$ is the *magnetic pressure*. Therefore, Equation (7.51) indicates that, for the conditions assumed, the total pressure is a constant. Note that such a pressure balance has been observed, for example, at the day side Venus ionopause region (Section 13.2).

Consider the special case of a stationary plasma ($\mathbf{u} = 0$). If the plasma has a pressure gradient, then Equation (7.51) implies that the magnetic pressure must vary in an opposite sense so that the total pressure remains constant. Therefore, the magnetic field must be weak in the regions where the density is high, and vice versa. The reduction of the magnetic field in the high-density regions is caused by a *diamagnetic current*, which can be obtained from the momentum equation (7.46). Taking the cross product of this equation with $\mathbf{B}$ yields

$$\nabla p \times \mathbf{B} = (\mathbf{J} \times \mathbf{B}) \times \mathbf{B} = -B^2 \mathbf{J}_\perp, \tag{7.52}$$

or

$$\mathbf{J}_\perp = -\frac{\nabla p \times \mathbf{B}}{B^2}, \tag{7.53}$$

where $\mathbf{J}_\perp$ is the current perpendicular to $\mathbf{B}$. An indication of the magnitude of the diamagnetic effect is given by the ratio of the kinetic and magnetic pressures, and this ratio is called the β *of the plasma*

$$\beta = \frac{p}{B^2/2\mu_0}. \tag{7.54}$$

When β is small, the magnetic field is strong, and it has a dominating effect on the plasma dynamics. On the other hand, when β is large, the magnetic field is weak, and it does not appreciably affect the plasma dynamics.

7.5   Magnetic diffusion

Magnetized and unmagnetized plasmas frequently come into contact. This occurs, for example, when the magnetized solar wind impacts the unmagnetized ionosphere of Venus (Section 13.2). When such plasmas come into contact, it is important to know whether or not the magnetic field in the one plasma can penetrate the other plasma. As will be shown, the extent to which a magnetic field can penetrate a plasma depends on the conductivity of the plasma.

An equation that describes the diffusion of a magnetic field in a plasma can be obtained from the simplified set of MHD equations (7.45a–g). For simplicity, assume that the conductivity of the plasma is constant. Substituting the electric field obtained from Ohm's law (7.45d) into Faraday's law (7.45f) results in the following equation:

$$\frac{\partial \mathbf{B}}{\partial t} = -\frac{1}{\sigma_e} \nabla \times \mathbf{J} + \nabla \times (\mathbf{u} \times \mathbf{B}). \tag{7.55}$$

Now, the substitution of $\mathbf{J}$, obtained from Ampère's law (7.45g), into Equation (7.55) yields

$$\frac{\partial \mathbf{B}}{\partial t} = -\frac{1}{\mu_0 \sigma_e} \nabla \times (\nabla \times \mathbf{B}) + \nabla \times (\mathbf{u} \times \mathbf{B})$$

$$= -\frac{1}{\mu_0 \sigma_e} \left[\nabla (\nabla \cdot \mathbf{B}) - \nabla^2 \mathbf{B} \right] + \nabla \times (\mathbf{u} \times \mathbf{B}), \tag{7.56}$$

where a vector identity, given in Appendix B, was used for the $\nabla \times (\nabla \times \mathbf{B})$ expression. Given that $\nabla \cdot \mathbf{B} = 0$ (3.76c), Equation (7.56) reduces to

$$\frac{\partial \mathbf{B}}{\partial t} = \frac{1}{\mu_0 \sigma_e} \nabla^2 \mathbf{B} + \nabla \times (\mathbf{u} \times \mathbf{B}). \tag{7.57}$$

When $\mathbf{u} = 0$, Equation (7.57) takes the classical form of a diffusion equation, and hence, the first term on the right-hand side of (7.57) accounts for *magnetic diffusion*. The second term is the flow term. Clearly, when σ_e is large, the flow term dominates, while when it is small, magnetic diffusion dominates. An estimate of the relative importance of the two processes can be obtained by taking the ratio of the flow and diffusion terms, which is called the *magnetic Reynolds number*. Letting L correspond to the characteristic scale length of the gradients, the ratio of the two terms is given by

$$\frac{\left| \nabla \times (\mathbf{u} \times \mathbf{B}) \right|}{\left| \nabla^2 \mathbf{B} / \mu_0 \sigma_e \right|} = \frac{uB/L}{B/(\mu_0 \sigma_e L^2)} = uL\mu_0 \sigma_e. \tag{7.58}$$

In the ionospheres, the plasma conductivities are typically very large, and therefore, the flow term dominates. In this case, it can be shown that the magnetic field is effectively *frozen in* the plasma.[5] Hence, both the magnetic field and plasma have the same velocity. This velocity can be obtained from Ohm's law (7.45d) which, in the limit of $\sigma_e \to \infty$, becomes

$$\mathbf{E} + \mathbf{u} \times \mathbf{B} = 0. \tag{7.59}$$

Taking the cross product of Equation (7.59) with $\mathbf{B}$ results in the well-known expression for the $\mathbf{E} \times \mathbf{B}$ drift velocity

$$\mathbf{u}_\perp = \frac{\mathbf{E} \times \mathbf{B}}{B^2}. \tag{7.60}$$

Therefore, when the plasma conductivity is large, both the plasma and magnetic field move with the $\mathbf{E} \times \mathbf{B}$ drift velocity (also see Equation 5.99).

7.6   Spiral magnetic field

The solar wind is a classic example of a plasma with a magnetic field that is frozen into the flow.[6] Beyond about ten solar radii, R_s, the conductivity and magnetic Reynolds number are extremely large. Consequently, as the solar wind moves radially away from the Sun, the magnetic field cannot diffuse through the plasma and it is carried with the plasma into interplanetary space. If the Sun did not rotate, the magnetic field would extend radially outward in all directions; but because the Sun rotates, the magnetic field lines are twisted into *Archimedes' spirals* (Figure 2.6).

The basic configuration of the Sun's magnetic field can be obtained by considering the simple case of a spherically symmetric, purely radial solar wind. Beyond about $10 R_s$, the solar wind velocity does not vary appreciably, and for this simple analysis, it is assumed to be constant. Now, consider a spherical coordinate system (r, θ, ϕ) fixed to the rotating Sun, with the polar axis aligned with the Sun's rotation axis and ϕ positive in the direction of rotation. In the inertial (nonrotating) reference frame, the solar wind velocity components are $(u_r, 0, 0)$, and in the rotating frame they are given by (Equation 10.1)

$$\mathbf{U} = \mathbf{u} - \mathbf{\Omega}_s \times \mathbf{r} \tag{7.61}$$

where $\mathbf{U}$ is used to designate the velocity in the rotating frame and $\mathbf{\Omega}_s$ is the Sun's rotation rate (2.7×10^{-6} rad s^{-1}). From Equation (7.61), the velocity components in the equatorial plane of the rotating reference frame are

$$U_r = u_r, \tag{7.62a}$$

$$U_\phi = -\Omega_s r. \tag{7.62b}$$

The difference in the plasma motion seen in the inertial and rotating reference frames is shown in Figure 7.1. In the inertial frame, the plasma element expands radially outward and the Sun's counterclockwise rotation causes the magnetic field to be twisted. In the rotating frame, on the other hand, the plasma elements appear to be moving both outward and in a clockwise direction. The magnetic field moves with

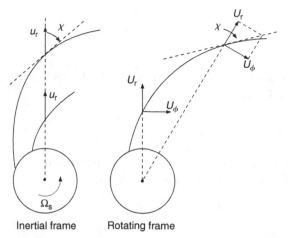

Figure 7.1 The magnetic field configuration and radial expansion of the solar wind as seen in inertial and rotating reference frames. The solar wind velocity components in the inertial frame are $(u_r, 0, 0)$ and those in the rotating frame are $(U_r, 0, U_\phi)$. Ω_s is the Sun's rotation rate and χ is the angle between the magnetic field and the radial direction.

Inertial frame   Rotating frame

these plasma elements because it is frozen into the flow. Therefore, the trajectory of a magnetic field line is the same as the trajectory of a plasma element in the rotating reference frame, which is defined by

$$\frac{U_r}{U_\phi} = -\frac{u_r}{\Omega_s r}. \tag{7.63}$$

Using $U_r = dr/dt$ and $U_\phi = r d\phi/dt$, Equation (7.63) becomes

$$\frac{dr}{d\phi} = -\frac{u_r}{\Omega_s}. \tag{7.64}$$

This equation can be easily integrated to obtain an equation for the trajectory, and the result is

$$r - r_0 = -\frac{u_r}{\Omega_s}(\phi - \phi_0), \tag{7.65}$$

where r_0 and ϕ_0 are reference positions (at $10\,R_s$).

The angle that the magnetic field (i.e., trajectory) makes with the radial direction is called the *spiral angle* χ. From Figure 7.1, this angle is given by

$$\tan \chi = \frac{|U_\phi|}{U_r} = \frac{\Omega_s r}{u_r}. \tag{7.66}$$

At the Earth's orbit, $u_r = 400$ km s^{-1} and $r = 1$ AU $\approx 1.5 \times 10^{11}$ meters. Therefore, $\tan \chi \approx 1$ and $\chi \approx 45°$. This value, which was obtained from a simple analysis, is in remarkable agreement with measurements of the spiral angle shown in Figure 2.6.

The magnetic field can be obtained by starting with the equation $\mathbf{\nabla} \cdot \mathbf{B} = 0$, which simply yields

$$B_r = B_0 \left(\frac{r_0}{r} \right)^2 \tag{7.67a}$$

because the flow is spherically symmetric. From Figure 7.1, the azimuthal component, B_ϕ, is related to B_r by

$$B_\phi = -B_r \tan \chi = -B_0 \frac{r_0^2 \Omega_s}{r u_r}. \tag{7.67b}$$

Combining Equations (7.67a) and (7.67b), the magnitude of the magnetic field in the equatorial plane can be written as

$$|\mathbf{B}| = B_0 \left(\frac{r_0}{r} \right)^2 \left(1 + \frac{r^2 \Omega_s^2}{u_r^2} \right)^{1/2}. \tag{7.67c}$$

7.7 Double-adiabatic energy equations

When a magnetized plasma becomes collisionless, it is unlikely that the plasma pressure will remain isotropic. Instead, there will generally be different pressures parallel, $p_\parallel$, and perpendicular, $p_\perp$, to $\mathbf{B}$. This situation arises because the charged particles cannot effectively move across $\mathbf{B}$, and hence, the thermal spread along $\mathbf{B}$ tends to be different from that across $\mathbf{B}$. In the MHD approximation, the pressure tensor equation is similar in form to the 13-moment pressure tensor equation (3.55). Also, for a strongly magnetized plasma ($\beta \ll 1$), the dominant terms in the pressure tensor equation are those containing $\mathbf{B}$. Consequently, to lowest order, the MHD pressure tensor equation reduces to

$$\mathbf{B} \times \mathbf{P} - \mathbf{P} \times \mathbf{B} = 0, \tag{7.68}$$

which has as its solution

$$\mathbf{P} = p_\parallel \mathbf{bb} + p_\perp (\mathbf{I} - \mathbf{bb}), \tag{7.69}$$

where $\mathbf{I}$ is the unit dyadic and $\mathbf{b} = \mathbf{B}/B$ is a unit vector in the direction of the local magnetic field. Note that a diagonal pressure tensor of the form in Equation (7.69) is consistent with a bi-Maxwellian velocity distribution (Equation 3.75). In this case, the relations connecting the pressures and temperatures are $p_\parallel = nkT_\parallel$ and $p_\perp = nkT_\perp$.

Equations describing the temporal and spatial evolution of the parallel and perpendicular pressures can be obtained by taking the scalar products of $\mathbf{bb}$ and $(\mathbf{I} - \mathbf{bb})$, respectively, with the MHD pressure tensor equation, which is similar to Equation (3.55). The result of these operations is

$$\frac{1}{p_\parallel}\frac{Dp_\parallel}{Dt} = -\mathbf{\nabla}\cdot\mathbf{u} - 2\mathbf{bb}:\mathbf{\nabla}\mathbf{u}, \tag{7.70}$$

$$\frac{1}{p_\perp}\frac{Dp_\perp}{Dt} = -2\mathbf{\nabla}\cdot\mathbf{u} + \mathbf{bb}:\mathbf{\nabla}\mathbf{u}. \tag{7.71}$$

For a highly conducting plasma, Ohm's law reduces to $\mathbf{E}+\mathbf{u}\times\mathbf{B}=0$ (Equation 7.59). Taking the curl of this equation and using $\mathbf{\nabla}\times\mathbf{E}=-\partial\mathbf{B}/\partial t$ and $\mathbf{\nabla}\cdot\mathbf{B}=0$, Ohm's law can be written in the form

$$\frac{D\mathbf{B}}{Dt} + \mathbf{B}(\mathbf{\nabla}\cdot\mathbf{u}) - \mathbf{B}\cdot\mathbf{\nabla}\mathbf{u} = 0. \tag{7.72}$$

The parallel component of Equation (7.72) can be obtained by taking the scalar product of $\mathbf{b}$ with this equation, which then yields an equation for $\mathbf{bb}:\mathbf{\nabla}\mathbf{u}$ that is given by

$$\mathbf{bb}:\mathbf{\nabla}\mathbf{u} = \frac{1}{B}\frac{DB}{Dt} + \mathbf{\nabla}\cdot\mathbf{u}. \tag{7.73}$$

An expression for $\mathbf{\nabla}\cdot\mathbf{u}$ can be obtained from the continuity equation (7.45a)

$$\mathbf{\nabla}\cdot\mathbf{u} = -\frac{1}{\rho}\frac{D\rho}{Dt}. \tag{7.74}$$

When Equations (7.73) and (7.74) are substituted into the equations for the parallel (7.70) and perpendicular (7.71) pressures, these equations become the well-known *double-adiabatic energy equations*, which are given by[7]

$$\frac{D}{Dt}\left(\frac{p_\parallel B^2}{\rho^3}\right) = 0, \tag{7.75a}$$

$$\frac{D}{Dt}\left(\frac{p_\perp}{B\rho}\right) = 0. \tag{7.75b}$$

With an anisotropic pressure distribution (7.69), the momentum equation (7.26) is modified because of the $\mathbf{\nabla}\cdot\mathbf{P}$ term, which can be expressed as

$$\mathbf{\nabla}\cdot\mathbf{P} = \mathbf{\nabla}\cdot\left[p_\parallel\mathbf{bb} + p_\perp(\mathbf{I}-\mathbf{bb})\right] = \mathbf{\nabla}\cdot\left[p_\perp\mathbf{I} + (p_\parallel-p_\perp)\frac{\mathbf{BB}}{B^2}\right]. \tag{7.76}$$

The first term becomes

$$\mathbf{\nabla}\cdot(p_\perp\mathbf{I}) = \mathbf{\nabla}p_\perp \tag{7.77}$$

and the second term can be expanded using a tensor identity, as follows:

$$\nabla \cdot \left[(p_\parallel - p_\perp) \frac{\mathbf{B}}{B^2} \mathbf{B} \right] = (p_\parallel - p_\perp) \frac{\mathbf{B}}{B^2} (\nabla \cdot \mathbf{B}) + \mathbf{B} \cdot \nabla \left[(p_\parallel - p_\perp) \frac{\mathbf{B}}{B^2} \right],$$

(7.78)

where $\nabla \cdot \mathbf{B} = 0$. Therefore, using Equations (7.77) and (7.78), the divergence of the pressure tensor (7.76) becomes

$$\nabla \cdot \mathbf{P} = \nabla p_\perp + \mathbf{B} \cdot \nabla \left[(p_\parallel - p_\perp) \frac{\mathbf{B}}{B^2} \right].$$

(7.79)

In summary, the closed system of transport equations in the double-adiabatic limit is the simplified MHD equations (7.45a–g), but with ∇p in the momentum equation (7.45c) replaced with $\nabla \cdot \mathbf{P}$ in (7.79) and with the equation of state (7.45e) replaced with the double-adiabatic energy equations (7.75a and b). However, note that these equations are applicable only if heat flow is negligible in the plasma under consideration.

7.8 Alfvén and magnetosonic waves

The discussion in Chapter 6 of the characteristic waves that can propagate in both magnetized and unmagnetized plasmas did not consider the low frequency waves that can propagate in a highly conducting, magnetized plasma. Although these waves could have been treated with the standard transport equations presented in that chapter, the waves are more easily derived from the simplified MHD equations (7.45a–g).

For the wave analysis, gravity is neglected, and the plasma conductivity is assumed to be infinite so that Ohm's law (7.45d) reduces to $\mathbf{E} = -\mathbf{u} \times \mathbf{B}$. Also, the equation of state (7.45e) can be expressed as $\nabla p = V_S^2 \nabla \rho$, where $V_S = (\gamma p / \rho)^{1/2}$ is the sound speed, and $\mathbf{J} = (\nabla \times \mathbf{B})/\mu_0$ from Ampère's law (7.45g). With this information, the simplified set of MHD equations becomes

$$\frac{\partial \rho}{\partial t} + \nabla \cdot (\rho \mathbf{u}) = 0,$$

(7.80a)

$$\rho \left[\frac{\partial \mathbf{u}}{\partial t} + (\mathbf{u} \cdot \nabla)\mathbf{u} \right] + V_S^2 \nabla \rho - \frac{1}{\mu_0} (\nabla \times \mathbf{B}) \times \mathbf{B} = 0,$$

(7.80b)

$$\nabla \times (\mathbf{u} \times \mathbf{B}) = \frac{\partial \mathbf{B}}{\partial t}.$$

(7.80c)

In calculating the characteristic waves that can propagate in the plasma, it is assumed that the plasma is initially uniform and that the mass density, ρ_0, pressure, p_0, and magnetic field, $\mathbf{B}_0$, are constant. Also, there are no imposed electric fields

($E_0 = 0$) and the plasma is stationary ($u_0 = 0$). Then, the plasma is perturbed as follows:

$$\rho = \rho_0 + \rho_1(\mathbf{r}, t), \tag{7.81a}$$

$$\mathbf{u} = \mathbf{u}_1(\mathbf{r}, t), \tag{7.81b}$$

$$p = p_0 + p_1(\mathbf{r}, t), \tag{7.81c}$$

$$\mathbf{E} = \mathbf{E}_1(\mathbf{r}, t), \tag{7.81d}$$

$$\mathbf{B} = \mathbf{B}_0 + \mathbf{B}_1(\mathbf{r}, t), \tag{7.81e}$$

where subscript 1 denotes a small perturbation. Substituting the perturbed parameters (7.81a–e) into Equations (7.80a–c) and retaining only those terms that are linear in the perturbed parameters leads to the following equations:

$$\frac{\partial \rho_1}{\partial t} + \rho_0(\nabla \cdot \mathbf{u}_1) = 0, \tag{7.82a}$$

$$\rho_0 \frac{\partial \mathbf{u}_1}{\partial t} + V_S^2 \nabla \rho_1 - \frac{1}{\mu_0}(\nabla \times \mathbf{B}_1) \times \mathbf{B}_0 = 0, \tag{7.82b}$$

$$\nabla \times (\mathbf{u}_1 \times \mathbf{B}_0) = \frac{\partial \mathbf{B}_1}{\partial t}, \tag{7.82c}$$

where $V_S = (\gamma p_0 / \rho_0)^{1/2}$ in Equation (7.82b).

The perturbations can be assumed to be sinusoidal,

$$\rho_1, \mathbf{u}_1, p_1, \mathbf{E}_1, \mathbf{B}_1 \propto e^{i(\mathbf{K} \cdot \mathbf{r} - \omega t)}, \tag{7.83}$$

because the perturbations are small. Therefore, when ∇ and $\partial/\partial t$ operate on perturbed quantities, they can simply be replaced by $\nabla \to i\mathbf{K}$ and $\partial/\partial t \to -i\omega$. In this case, the partial differential equations (7.82a–c) reduce to the algebraic equations:

$$-\omega \rho_1 + \rho_0 \mathbf{K} \cdot \mathbf{u}_1 = 0, \tag{7.84a}$$

$$-\omega \rho_0 \mathbf{u}_1 + V_S^2 \rho_1 \mathbf{K} - \frac{1}{\mu_0}(\mathbf{K} \times \mathbf{B}_1) \times \mathbf{B}_0 = 0, \tag{7.84b}$$

$$\mathbf{K} \times (\mathbf{u}_1 \times \mathbf{B}_0) = -\omega \mathbf{B}_1. \tag{7.84c}$$

There are three equations for the three unknowns ($\rho_1, \mathbf{u}_1, \mathbf{B}_1$), and the goal is to obtain one equation for one unknown, which then leads to the dispersion relation for the possible wave modes. An expression for ρ_1 in terms of $\mathbf{u}_1$ can be obtained from Equation (7.84a), and the result is

$$\rho_1 = \frac{\rho_0}{\omega} \mathbf{K} \cdot \mathbf{u}_1. \tag{7.85}$$

Likewise, $\mathbf{B}_1$ can be expressed in terms of $\mathbf{u}_1$ with the aid of Equation (7.84c). When the double cross product in Equation (7.84c) is expanded, using the vector relation $\mathbf{A} \times (\mathbf{B} \times \mathbf{C}) = (\mathbf{A} \cdot \mathbf{C})\mathbf{B} - (\mathbf{A} \cdot \mathbf{B})\mathbf{C}$ given in Appendix B, the equation for $\mathbf{B}_1$ becomes

$$\mathbf{B}_1 = \frac{1}{\omega}\big[(\mathbf{K} \cdot \mathbf{u}_1)\mathbf{B}_0 - (\mathbf{K} \cdot \mathbf{B}_0)\mathbf{u}_1\big]. \tag{7.86}$$

Now, an equation for $\mathbf{u}_1$ can be obtained by expanding the double cross product in Equation (7.84b) with the vector relation given above and by substituting Equations (7.85) and (7.86) into Equation (7.84b), and the result is

$$\mathbf{u}_1\big[-\omega^2 + V_A^2(\mathbf{K} \cdot \mathbf{b})^2\big] - \mathbf{b}V_A^2(\mathbf{K} \cdot \mathbf{b})(\mathbf{K} \cdot \mathbf{u}_1)$$
$$+ \mathbf{K}\big[(V_A^2 + V_S^2)(\mathbf{K} \cdot \mathbf{u}_1) - V_A^2(\mathbf{K} \cdot \mathbf{b})(\mathbf{b} \cdot \mathbf{u}_1)\big] = 0, \tag{7.87}$$

where $\mathbf{b} = \mathbf{B}_0/B_0$ is a unit vector and V_A is the *Alfvén velocity*

$$V_A = \frac{B_0}{\sqrt{\mu_0 \rho_0}}. \tag{7.88}$$

Equation (7.87) is the dispersion relation for the characteristic waves that can propagate in a single-component, highly conducting plasma. As it turns out, three distinct waves are possible. Two of the waves propagate along $\mathbf{B}_0$ and one propagates in a direction perpendicular to $\mathbf{B}_0$. For one of the parallel propagating waves, the perturbed velocity, $\mathbf{u}_1$, is also parallel to $\mathbf{B}_0$ (Figure 7.2). In this case, the dispersion relation (7.87) simply reduces to

$$\omega^2 = K^2 V_S^2. \tag{7.89}$$

For this wave, $\mathbf{B}_1 = 0$ (Equation 7.86), $\mathbf{E}_1 = 0$ ($\mathbf{E}_1 = -\mathbf{u}_1 \times \mathbf{B}_0$), and $\rho_1 = \rho_0 K u_1/\omega$ (Equation 7.85). Therefore, Equation (7.89) is the dispersion relation for ordinary acoustic waves (Equation 6.56).

For the other parallel propagating wave, the perturbed velocity, $\mathbf{u}_1$, is perpendicular to $\mathbf{B}_0$ (Figure 7.2). When $\mathbf{u}_1 \perp \mathbf{K}$ and $\mathbf{K} \parallel \mathbf{B}_0$, the dispersion relation (7.87) reduces to

$$\omega^2 = K^2 V_A^2, \tag{7.90}$$

which is known as the *Alfvén wave*. For this wave, $\rho_1 = 0$ (Equation 7.85), $\mathbf{B}_1 = -(B_0 K/\omega)\mathbf{u}_1$ (Equation 7.86), and $\mathbf{E}_1 = -\mathbf{u}_1 \times \mathbf{B}_0$. Therefore, the Alfvén wave is an electromagnetic wave that propagates along $\mathbf{B}_0$.

The third wave propagates in a direction that is perpendicular to $\mathbf{B}_0$. If $\mathbf{K}$ is also perpendicular to $\mathbf{u}_1$, then Equation (7.87) reduces to $\omega = 0$, which is a trivial solution. For $\mathbf{K} \parallel \mathbf{u}_1$ (Figure 7.2), the dispersion relation (7.87) can be easily solved

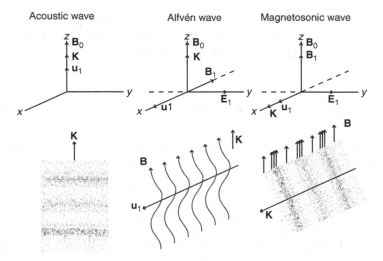

Figure 7.2 Characteristics of the three primary waves that can propagate in a single-component, compressible, conducting fluid. For the acoustic mode, there are no electric or magnetic field fluctuations and the density perturbations are along $\mathbf{B}_0$. For the Alfvén wave, there are no density perturbations and the magnetic field fluctuations are in the x-direction, producing the kinks in $\mathbf{B}$. There are density perturbations associated with the magnetosonic wave and they are in the x-direction. The associated $\mathbf{B}$ field always points along the z-axis, but because the field is frozen in the plasma, there are compressions and rarefactions of $\mathbf{B}$ similar to those in ρ_1.

by first taking the scalar product of Equation (7.87) with $\mathbf{K}$, which yields

$$\omega^2 = K^2(V_S^2 + V_A^2). \tag{7.91}$$

This is the *magnetosonic wave*. For this wave, there is a density compression and expansion because $\rho_1 = \rho_0 K u_1/\omega$ (Equation 7.85), and this is why "sonic" appears in the name. However, the wave also has electric and magnetic perturbations associated with it, where $\mathbf{B}_1 = (K u_1/\omega)\mathbf{B}_0$ (Equation 7.86) and $\mathbf{E}_1 = -\mathbf{u}_1 \times \mathbf{B}_0$. Note that the wave is electromagnetic in nature because $\mathbf{E}_1$, $\mathbf{B}_1$, and $\mathbf{K}$ are orthogonal and $\mathbf{E}_1 \times \mathbf{B}_1$ points in the $\mathbf{K}$ direction.

Note that the three waves given by Equations (7.89), (7.90), and (7.91) are actually nondispersive because ω/K does not depend on the frequency.

The above analysis only considered those waves that propagate either along or perpendicular to $\mathbf{B}_0$. However, waves can also propagate at an angle to $\mathbf{B}_0$. Again, there are three modes that are possible. If α is the angle between the wave vector $\mathbf{K}$ and $\mathbf{B}_0$, the three waves are given by the following dispersion relations[4,8]

$$\frac{\omega}{K} = V_A \cos\alpha, \tag{7.92}$$

$$\left(\frac{\omega}{K}\right)^2 = \frac{1}{2}(V_A^2 + V_S^2) \pm \frac{1}{2}\left[(V_A^2 + V_S^2)^2 - 4V_A^2 V_S^2 \cos^2\alpha\right]^{1/2}. \tag{7.93}$$

The first mode (7.92) is known as the *oblique Alfvén wave*. In Equation (7.93), the plus sign yields the *fast MHD wave*, while the minus sign yields the *slow MHD wave*. Note that the additional plus and minus signs that result from taking the square root of Equation (7.93) merely indicate that both the fast and slow MHD waves can propagate in opposite directions in the plasma.

7.9   Shocks and discontinuities

In Section 6.14, the Rankine–Hugoniot relations, which describe the jump conditions across a shock, were derived for the case of ordinary hydrodynamic shocks. Here, shocks and discontinuities are discussed for the more general case of a collisionless, magnetized plasma. Compared with ordinary hydrodynamics, the passage of a shock through a collisionless plasma is more complex. The main reason is that energy and momentum can be transferred from the plasma particles to the electric and magnetic fields, and these fields must be taken into account when the conservation equations are applied to both the pre-shock and post-shock plasmas. Also, plasma instabilities and turbulence can be excited as a result of the processes associated with the shock. Therefore, in general, shocks can be laminar, turbulent, or a mixture of both features.[9]

Only MHD shocks and discontinuities are discussed in this section, and the starting point is the simplified (or ideal) MHD equations (7.45a–g). These equations, like the Euler equations (6.123–125) used to describe ordinary shocks, are not in a convenient form to obtain the jump conditions across a shock. Therefore, they must be converted to a conservative form, as was done with the Euler equations (6.123–125). In comparing the MHD continuity (7.45a), momentum (7.45c), and energy (7.45e) equations with the corresponding Euler equations (6.123–125), it is apparent that the continuity and energy equations are the same because (7.45e) can also be written as $D/Dt(p/\rho^\gamma) = 0$. The only difference between the momentum equations is the appearance of the $\mathbf{J} \times \mathbf{B}$ term in the MHD momentum equation (7.45c) because gravity is ignored. Hence, most of the work needed to convert the MHD momentum and energy equations to a conservative form has already been done in connection with the conversion of the Euler equations, and only the $\mathbf{J} \times \mathbf{B}$ term needs to be considered here.

The $\mathbf{J} \times \mathbf{B}$ term in the momentum equation (7.45c) can be converted to a conservative form by first expressing $\mathbf{J}$ in terms of $\nabla \times \mathbf{B}$ via Equation (7.45g), which yields

$$-\mathbf{J} \times \mathbf{B} = -\frac{1}{\mu_0}(\nabla \times \mathbf{B}) \times \mathbf{B} = -\frac{1}{\mu_0}\left[(\mathbf{B} \cdot \nabla)\mathbf{B} - \nabla\left(\frac{B^2}{2}\right)\right]$$

$$= -\frac{1}{\mu_0}\left[\nabla \cdot (\mathbf{BB}) - \mathbf{B}(\nabla \cdot \mathbf{B}) - \nabla\left(\frac{B^2}{2}\right)\right]$$

$$= -\frac{1}{\mu_0}\nabla \cdot \left(\mathbf{BB} - \frac{B^2}{2}\mathbf{I}\right), \tag{7.94}$$

where $\mathbf{I}$ is the unit dyadic, $\nabla \cdot \mathbf{B} = 0$, and where the second and third expressions result from the use of the vector relations given in Appendix B. Adding the $-\mathbf{J} \times \mathbf{B}$ expression (7.94) to the Euler momentum equation (6.126) yields the conservative form of the MHD momentum equation (7.45c), which is

$$\frac{\partial}{\partial t}(\rho \mathbf{u}) + \nabla \cdot \left[\rho \mathbf{u}\mathbf{u} + p\mathbf{I} - \frac{1}{\mu_0}\left(\mathbf{B}\mathbf{B} - \frac{B^2}{2}\mathbf{I} \right) \right] = 0. \qquad (7.95)$$

The conservative form of the energy equation (7.45e) is obtained by first taking the scalar product of $\mathbf{u}$ with the momentum equation (7.45c) and then substituting the continuity (7.45a) and energy (7.45e) equations into this modified momentum equation. However, all of the algebraic manipulations have already been done in connection with the derivation of the Euler energy equation (6.132), except for the additional term $-(\mathbf{J} \times \mathbf{B}) \cdot \mathbf{u}$. This term can be converted to a conservative form as follows:

$$-\mathbf{u} \cdot (\mathbf{J} \times \mathbf{B}) = -\frac{1}{\mu_0}\mathbf{u} \cdot (\nabla \times \mathbf{B}) \times \mathbf{B} = \frac{1}{\mu_0}(\mathbf{u} \times \mathbf{B}) \cdot (\nabla \times \mathbf{B}), \qquad (7.96)$$

where the first expression follows from equation (7.45g) and the second from a vector relation given in Appendix B. For a highly conducting plasma ($\sigma_e \to \infty$), Equation (7.45d) indicates that $\mathbf{E} = -\mathbf{u} \times \mathbf{B}$ and, therefore, Equation (7.96) can be written as

$$-\mathbf{u} \cdot (\mathbf{J} \times \mathbf{B}) = -\frac{1}{\mu_0}\mathbf{E} \cdot (\nabla \times \mathbf{B}) = \frac{1}{\mu_0}\left[\nabla \cdot (\mathbf{E} \times \mathbf{B}) - \mathbf{B} \cdot (\nabla \times \mathbf{E}) \right]$$

$$= \frac{1}{\mu_0}\left[\nabla \cdot (\mathbf{E} \times \mathbf{B}) + \frac{\partial}{\partial t}\left(\frac{B^2}{2} \right) \right], \qquad (7.97)$$

where the second expression follows from the vector relation $\nabla \cdot (\mathbf{E} \times \mathbf{B}) = \mathbf{B} \cdot (\nabla \times \mathbf{E}) - \mathbf{E} \cdot (\nabla \times \mathbf{B})$ and the third from Equation (7.45f). Adding the term given in Equation (7.97) to the Euler energy equation (6.132) yields the conservative form of the MHD energy equation, which is

$$\frac{\partial}{\partial t}\left(\frac{1}{2}\rho u^2 + \frac{p}{\gamma - 1} + \frac{B^2}{2\mu_0} \right)$$

$$+ \nabla \cdot \left[\left(\frac{1}{2}\rho u^2 + \frac{\gamma p}{\gamma - 1} \right)\mathbf{u} + \frac{1}{\mu_0}\mathbf{E} \times \mathbf{B} \right] = 0 \qquad (7.98)$$

where $\mathbf{E} \times \mathbf{B}/\mu_0 = \mathbf{E} \times \mathbf{H}$ is the *Poynting vector*.

The equations that are appropriate for MHD shocks and discontinuities are the continuity (7.45a), momentum (7.95), and energy (7.98) equations, coupled with $\nabla \times \mathbf{E} = -\partial \mathbf{B}/\partial t$ (7.45f), and $\mathbf{E} = -\mathbf{u} \times \mathbf{B}$ for $\sigma_e \to \infty$ (7.45d). As with hydrodynamic shocks, the equations are applied in the reference frame of the shock or discontinuity, and steady-state conditions are assumed. It is also assumed that the plasma is homogeneous on both sides of the discontinuity over at least a short distance. In the shock reference frame and for steady state ($\partial/\partial t \to 0$) conditions, the MHD equations become

$$\nabla \cdot (\rho \mathbf{u}) = 0, \tag{7.99a}$$

$$\nabla \cdot \left[\rho \mathbf{u}\mathbf{u} + p\mathbf{I} - \frac{1}{\mu_0}\left(\mathbf{B}\mathbf{B} - \frac{B^2}{2}\mathbf{I} \right) \right] = 0, \tag{7.99b}$$

$$\nabla \cdot \left[\left(\frac{1}{2}\rho u^2 + \frac{\gamma p}{\gamma - 1} \right)\mathbf{u} + \frac{1}{\mu_0}\mathbf{E} \times \mathbf{B} \right] = 0, \tag{7.99c}$$

$$\nabla \times (\mathbf{u} \times \mathbf{B}) = \mathbf{0}. \tag{7.99d}$$

The procedure for obtaining the jump conditions across a shock or discontinuity is the standard procedure used in electromagnetic theory to obtain boundary conditions on the electric and magnetic fields.[5] Specifically, for the divergence equations (for example, $\nabla \cdot \mathbf{A} = 0$), a so-called *Gaussian pillbox* (a cylinder) is created so that its axis is normal to the discontinuity and the discontinuity cuts the cylinder in half, with the top and bottom of the cylinder on opposite sides of the discontinuity. The divergence expression ($\nabla \cdot \mathbf{A} = 0$) is then integrated over the volume of the cylinder, and because of the divergence theorem (Appendix C), the volume integral can be converted into an integral over the surface of the cylinder. As the sides of the cylinder go to zero, a condition is obtained that relates the normal component of the quantity under the divergence operator (i.e., $\mathbf{A}$) on the two sides of the discontinuity ($A_{1n} = A_{2n}$), where subscript n indicates a normal component. For a curl equation, a similar procedure is employed, but Stokes theorem (Appendix C) is applied to a loop. The net effect is that a *divergence* implies that the *normal component* of the quantity is continuous, and a *curl* implies that the *tangential component* is continuous.

At this point, it is useful to introduce the normal (subscript n) and tangential (subscript t) components of the vectors, relative to the surface of the discontinuity. Also, as with hydrodynamic shocks, subscripts 1 and 2 denote the quantities on the opposite sides of the discontinuity. However, before the momentum equation (7.99b) is converted into a jump condition, the normal and tangential components of this vector equation should be taken. With these conditions in mind, the jump relations associated with Equations (7.99a–d) become

$$(\rho u_n)_1 = (\rho u_n)_2, \tag{7.100a}$$

$$\left[\rho u_n^2 + p - \frac{1}{\mu_0}\left(B_n^2 - \frac{B^2}{2} \right) \right]_1 = \left[\rho u_n^2 + p - \frac{1}{\mu_0}\left(B_n^2 - \frac{B^2}{2} \right) \right]_2, \tag{7.100b}$$

$$\left[\rho u_n u_t - \frac{B_n B_t}{\mu_0} \right]_1 = \left[\rho u_n u_t - \frac{B_n B_t}{\mu_0} \right]_2, \tag{7.100c}$$

Table 7.1 Classification scheme for MHD discontinuities.[4]

Contact discontinuity	$u_n = 0, B_n \neq 0$
Tangential discontinuity	$u_n = 0, B_n = 0$
Parallel shock	$u_n \neq 0, B_t = 0$
Perpendicular shock	$u_n \neq 0, B_n = 0$
Oblique shock	$u_n \neq 0, B_t \neq 0, B_n \neq 0$

$$\left[\left(\frac{1}{2}\rho u^2 + \frac{\gamma p}{\gamma - 1} \right) u_n + \frac{B_t}{\mu_0}(u_n B_t - u_t B_n) \right]_1$$

$$= \left[\left(\frac{1}{2}\rho u^2 + \frac{\gamma p}{\gamma - 1} \right) u_n + \frac{B_t}{\mu_0}(u_n B_t - u_t B_n) \right]_2, \qquad (7.100d)$$

$$(u_n B_t - u_t B_n)_1 = (u_n B_t - u_t B_n)_2, \qquad (7.100e)$$

where $(\mathbf{E} \times \mathbf{B}) \cdot \mathbf{n} = B_t(u_n B_t - u_t B_n)$ and the tangential component of $\mathbf{E}$ is $(u_n B_t - u_t B_n)$. Given the five parameters on the one side of the MHD discontinuity (ρ_1, u_{1n}, u_{1t}, B_{1n}, B_{1t}), Equations (7.100a–e) are sufficient to determine these parameters on the other side of the discontinuity.

For ordinary (nonMHD) shocks ($\mathbf{B} = 0$) and for the case of a normal shock ($u_t = 0$), Equations (7.100a–e) reduce to the jump conditions given previously (Equations 6.134a–c), which led to the Rankine–Hugoniot relations (6.138a–c). In general, various situations can occur in a magnetized plasma and a classification scheme for MHD discontinuities has been established, depending on whether the plasma or magnetic field penetrate the discontinuity. This classification scheme is given in Table 7.1.

7.10   Specific references

1. Chapman, S., and T. G. Cowling, *The Mathematical Theory of Non-Uniform Gases*, New York: Cambridge University Press, 1970.
2. Schunk, R. W., Mathematical structure of transport equations for multispecies flows, *Rev. Geophys. Space Phys.*, **15**, 429, 1977.
3. St-Maurice, J.-P., and R. W. Schunk, Ion-neutral momentum coupling near discrete high-latitude ionospheric features, *J. Geophys. Res.*, **87**, 1711, 1982.
4. Siscoe, G. L., Solar system magnetohydrodynamics, in *Solar-Terrestrial Physics*, ed. by R. L. Carovillano and J. M. Forbes, **11**, Dordrecht, Netherlands: D. Reidel, 1983.
5. Jackson, J. D., *Classical Electrodynamics*, New York: Wiley, 1998.
6. Parker, E. N., *Interplanetary Dynamical Processes*, New York: Interscience, 1963.
7. Chew, G. F., M. L. Goldberger, and F. E. Low, The Boltzmann equation and the one-fluid hydromagnetic equations in the absence of particle collisions, *Proc. Roy. Soc., Ser. A.*, **236**, 112, 1956.

8. Bittencourt, J. A., *Fundamentals of Plasma Physics*, Brazil, 1995.

9. Tidman, D. A., and N. A. Krall, *Shock Waves in Collisionless Plasmas*, New York: Wiley Interscience, 1971.

7.11   General references

Chen, F. F., *Introduction to Plasma Physics and Controlled Fusion*, New York: Plenum Press, 1984.

Cravens, T. E., *Physics of Solar System Plasmas*, Cambridge, UK: Cambridge University Press, 1997.

Hargreaves, J. K., *The Solar-Terrestrial Environment*, Cambridge, UK: Cambridge University Press, 1992.

Hones, E. W., *Magnetic Reconnection in Space and Laboratory Plasmas*, Geophysical Monograph, **30**, Washington, DC: American Geophysical Union, 1984.

Krall, H. A., and A. W. Trivelpiece, *Principles of Plasma Physics*, McGraw-Hill Company, 1973.

Lui, A. T. Y., *Magnetotail Physics*, Baltimore, MD: The Johns Hopkins University Press, 1987.

Moore, T. E., and J. H. Waite, Jr., *Modeling Magnetospheric Plasma*, Geophysical Monograph, **44**, Washington, DC: American Geophysical Union, 1988.

7.12   Problems

Problem 7.1 Show that the different definitions for the stress tensor, τ_s (Equation 3.21) and τ_s^* (Equation 7.9), are related via Equation (7.15).

Problem 7.2 The heat flow tensor in the single-fluid treatment is defined as $\mathbf{Q}_s^* = n_s m_s \langle \mathbf{c}_s^* \mathbf{c}_s^* \mathbf{c}_s^* \rangle$. Express this tensor in terms of $\mathbf{Q}_s$ (Equation 3.19) and the diffusion velocity $\mathbf{w}_s$ (Equation 7.10).

Problem 7.3 Show that the momentum equation (7.23) can be expressed in the form given by Equation (7.24) when the continuity equations (3.57) and (7.21) are used.

Problem 7.4 Derive the single-fluid MHD energy equation (7.32) from the individual species energy equations (3.59).

Problem 7.5 Using the parameters given in Tables 2.4 and 2.6, calculate β for the Earth at 300 km and Jupiter at 3000 km.

Problem 7.6 The trajectory of the Sun's magnetic field in the equatorial plane is given by Equation (7.65). Calculate the Sun's **B**-field trajectory at an arbitrary solar latitude Θ using the same assumptions adopted in the derivation of Equation (7.65).

Problem 7.7 Ignore the heat flow and collision terms in Equation (3.55) and assume that the pressure tensor is given by Equation (7.69). Show that the parallel and perpendicular pressures are then governed by Equations (7.70) and (7.71), respectively.

Problem 7.8 The MHD equations in the double-adiabatic limit are given by (Section 7.7):

$$\frac{\partial \rho}{\partial t} + \nabla \cdot (\rho \mathbf{u}) = 0,$$

$$\rho \frac{D\mathbf{u}}{Dt} + \nabla p_\perp + \mathbf{B} \cdot \nabla \left[(p_\parallel - p_\perp) \frac{\mathbf{B}}{B^2} \right] - \frac{1}{\mu_0} (\nabla \times \mathbf{B}) \times \mathbf{B} = 0,$$

$$\nabla \times (\mathbf{u} \times \mathbf{B}) = \frac{\partial \mathbf{B}}{\partial t},$$

$$\frac{p_\parallel B^2}{\rho^3} = \text{constant},$$

$$\frac{p_\perp}{B\rho} = \text{constant},$$

where it is assumed that $\mathbf{G} = 0$ and $\sigma_e \rightarrow \infty$. Linearize this system of equations by assuming $\rho = \rho_0 + \rho_1$, $\mathbf{u} = \mathbf{u}_1$, $p_\perp = p_{\perp 0} + p_{\perp 1}$, $p_\parallel = p_{\parallel 0} + p_{\parallel 1}$, and $\mathbf{B} = \mathbf{B}_0 + \mathbf{B}_1$, where ρ_0, $p_{\perp 0}$, $p_{\parallel 0}$, and $\mathbf{B}_0$ are constants, and then derive the dispersion relation for plane waves that propagate along $\mathbf{B}_0$.

Problem 7.9 Derive the dispersion relations (7.92) and (7.93), which describe the oblique Alfvén wave and the fast and slow MHD waves.

Problem 7.10 Show that the jump conditions across a shock or discontinuity that are associated with Equations (7.99a–d) are given by Equations (7.100a–e).

Problem 7.11 Starting from Equation (7.95) show that the pressure balance Equation (7.51) is valid for a compressible, one-dimensional, plasma flow with $\mathbf{B}$ perpendicular to $\mathbf{u}$ and all spatial variations only in the $\mathbf{u}$ direction.

Chapter 8

Chemical processes

Chemical processes are of major importance in determining the equilibrium distribution of ions in planetary ionospheres, even though photoionization and, in some cases, impact ionization are responsible for the initial creation of the electron–ion pairs. This is particularly apparent for the ionospheres of Venus and Mars because they determine the dominant ion species (Sections 13.2 and 13.3). The major neutral constituent in the thermosphere of both Venus and Mars is CO_2, and yet the major ion is O_2^+, as a result of ion–neutral chemistry. Therefore, a thorough knowledge of the controlling chemical processes is necessary for a proper understanding of ionospheric structure and behavior. The dividing line between chemical and physical processes is somewhat artificial and often determined by semantics. In this chapter the discussion centers on reactions involving ions, electrons, and neutral constituents; photoionization and impact ionization are discussed in Chapter 9.

8.1   Chemical kinetics

The area of science concerned with the study of chemical reactions is known as *chemical kinetics*. This branch of science examines the reaction processes from various points of view. A chemical reaction in which the phase of the reactant does not change is called a *homogeneous reaction*, whereas a chemical process in which different phases are involved is referred to as a *heterogeneous reaction*. In the context of atmospheric chemistry, heterogeneous reactions involve surfaces and are significant in some of the lower atmospheric chemical processes (e.g., the Antarctic ozone hole), but do not play an important role in ionospheric chemistry. The chemical change that takes place in a chemical reaction is generally represented by the following, so-called *stoichiometric*, equation:

$$aA + bB \rightarrow cC + dD, \tag{8.1}$$

where A and B denote the reactants, C and D represent the product molecules, and a, b, c, and d indicate the number of molecules of the various species involved in the reaction. The *dissociative recombination* of O_2^+ with an electron is an example of such a reaction

$$O_2^+ + e^- \rightarrow O + O. \tag{8.2}$$

Reactions that proceed in both directions are called *reversible*; the *accidentally resonant charge exchange reaction*, shown below, is an example of such a reversible reaction

$$O^+ + H \leftrightarrow H^+ + O. \tag{8.3}$$

These are called *elementary reactions* because the products are formed directly from the reactants. In the terrestrial ionosphere, O^+ can directly recombine with an electron, but this process is very slow. In most cases, O^+ recombines through a multi-step process involving *intermediate species*;

$$O^+ + N_2 \rightarrow NO^+ + N, \tag{8.4}$$

$$NO^+ + e \rightarrow N + O. \tag{8.5}$$

It is common practice in ionospheric and atmospheric work to denote the number density of a given species A as [A], $n(A)$, or n_A. In the rest of the book, the choice between these symbols will be based on simplicity. The SI unit for concentration is moles per cubic decimeter (note that a cubic decimeter is a liter). However, in ionospheric work the common unit for number density is molecules per cubic centimeter (cm^{-3}).

The relatively low densities present in upper atmospheres imply that the most common reactions of importance in ionospheric chemistry are the two-body or *bimolecular* reactions represented by Equation (8.1). In the lower thermospheres, three-body or *termolecular* reactions may become important. An example of such a reaction is the *three-body recombination* of atomic oxygen

$$O + O + M \rightarrow O_2 + M, \tag{8.6}$$

where M denotes a third body.

The reaction rate of these chemical processes is a function of the concentration of the reactant species and in the next section we show how to obtain a general expression for this rate, using kinetic collision theory. At this point, just assume that the rate is proportional to the densities of the reactants and write the rate of reaction,

R, for the bimolecular reaction between species A and B as

$$R = k_{AB}[A]^i[B]^j, \tag{8.7}$$

where k_{AB} is the reaction rate constant and i and j are the *orders* of the reaction with respect to constituents A and B, respectively. The overall order of the reaction is given by the sum of i and j. To avoid the use of awkward symbols, numerical subscripts, e.g., k_1, will be used to distinguish among the reaction rates in the rest of the book.

It is generally advisable to evaluate the various time constants associated with the different processes involved in any complex problem, in order to assess which are the controlling ones. For example, in the ionospheres both chemical and transport processes are potentially important. However, if the time constant for chemistry is much shorter than that for transport, one may be able to neglect the latter.

The only first-order reaction of importance that needs to be considered in an ionosphere is the *spontaneous de-excitation* of a molecule, atom, or ion. A good example of such a process is the transition of an excited oxygen atom from its ^1D state to the ground ^3P state (Figure 8.1a)

$$O(^1D) \rightarrow O(^3P) + h\nu(630/636\,\text{nm}). \tag{8.8}$$

where the de-excitation results in a photon, $h\nu$, that has a wavelength of either 630 nm or 636 nm (the oxygen red line; Section 8.7). In certain altitude regions this transition time is fast compared to transport processes, and in this case, the relevant continuity equation is

$$\frac{d[O(^1D)]}{dt} = -k_1[O(^1D)]. \tag{8.9}$$

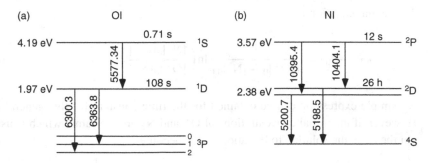

Figure 8.1 (a) Simple diagram showing the low-lying energy levels of oxygen atoms (OI). (b) Simple diagram showing the low-lying energy levels of nitrogen atoms (NI). Note that the wavelengths of the various transitions are in Å.

The solution of Equation (8.9) for the time variation of $[O(^1D)]$, in terms of the initial density, $[O(^1D)]_0$, is

$$[O(^1D)] = [O(^1D)]_0 e^{-k_1(t-t_0)}. \tag{8.10}$$

If the characteristic time constant, τ_1, is defined as the time during which the initial concentration drops to $1/e$ of its initial value, it is related to k_1, by

$$\tau_1 = 1/k_1. \tag{8.11}$$

For the sake of consistency, the rate at which spontaneous de-excitation takes place was written as k_1. However, the rate is usually denoted as A, and referred to as the *Einstein A factor or coefficient*; it has a value of $8.6 \times 10^{-3} \text{ s}^{-1}$.[1]

Two types of second-order reactions are possible. The first of these, in which two identical species are involved, is not significant in ionospheric chemistry. The second type, in which two different reactants are involved, is important and is discussed in what follows. A representative example of such a reaction is the ion–atom interchange reaction, indicated by Equation (8.4). The continuity equation, without the transport terms, for this reaction is

$$\frac{d[O^+]}{dt} = \frac{d[N_2]}{dt} = -k_2[O^+][N_2]. \tag{8.12}$$

In solving Equation (8.12), it is common to define a variable $[X]$, which is equal to the number of O^+ and N_2 ions or molecules that have reacted in a unit volume during time, t. Thus, Equation (8.12) can be rewritten as

$$\frac{d[X]}{dt} = k_2([O^+]_0 - [X])([N_2]_0 - [X]), \tag{8.13}$$

where $[O^+]_0$ and $[N_2]_0$ are the initial densities at $t = t_0$. The solution to this differential equation is

$$k_2(t - t_0) = \frac{1}{([O^+]_0 - [N_2]_0)} \ln\left(\frac{[N_2]_0[O^+]}{[O^+]_0[N_2]}\right). \tag{8.14}$$

No simple expression can be obtained for the time constant for this general case. However, if the initial concentrations of O^+ and N_2 are the same, which is usually not the case, the solution to Equation (8.13) becomes

$$k_2(t - t_0) = \frac{([O^+]_0 - [O^+])}{[O^+]_0[O^+]} = \frac{([N_2]_0 - [N_2])}{[N_2]_0[N_2]}. \tag{8.15}$$

The time taken for the initial densities to decrease by a factor of two, in this special case, is

$$\tau_2 = \frac{1}{k_2[O^+]_0} = \frac{1}{k_2[N_2]_0}. \tag{8.16}$$

In many ionospheric applications, one of the species participating in a bimolecular reaction may remain approximately constant. In the terrestrial ionosphere, the charge exchange represented by Equation (8.4) is a good example of such a process. The densities of N_2 are orders of magnitude larger than that of O^+ and thus are not affected by this reaction. Therefore, this charge exchange behaves like a first-order reaction.

Finally, we look at third-order reactions of the type represented by Equation (8.6). In that specific case the relevant differential equation is

$$\frac{d[O]}{dt} = -2k_3[O]^2[M]. \tag{8.17}$$

A number of different cases are possible for termolecular reactions. In the case of atomic oxygen recombination, the two cases that are appropriate to consider are (1) the situation where the third body is O and (2) the case when $[M] \gg [O]$, so that this reaction does not cause the concentration of M to change. The differential equation corresponding to the first case, when all three reactants are the same, can be written as

$$\frac{d[O]}{dt} = -3k_3[O]^3 \tag{8.18}$$

and the solution becomes

$$\frac{1}{[O]^2} = \frac{1}{[O]_0^2} + 6k_3(t - t_0). \tag{8.19}$$

In this case the time necessary for the initial concentration to drop to half of its initial value is

$$\tau_3 = \frac{1}{2k_3[O]_0^2}. \tag{8.20}$$

The solution to the differential equation in the case when the third species M is a constant is

$$\frac{1}{[O]} = \frac{1}{[O]_0} + 2k_3^*(t - t_0), \tag{8.21}$$

where $k_3^* = k_3[M]$. The time it takes for the initial concentration to drop to half of its initial value is

$$\tau_3^* = \frac{1}{2k_3^*[O]_0}. \tag{8.22}$$

A solution to the continuity equation without the transport term for a termolecular reaction for species A is easy to obtain if both [B] and [C] are much larger than [A]. In that case, the solution takes the form of Equation (8.10). If only one of the constituents can be assumed to be time independent, the differential equation takes the form of Equation (8.12). Finally, if all three reactants are varying with time, one needs to use the method of partial fractions to obtain a solution. However, such reactions are not likely to be of importance in most upper atmospheres.

8.2   Reaction rates

A person interested in studying the ionosphere needs to have quantitative information on the reaction rates and an understanding of the various factors influencing these rates. A physically intuitive way to begin a discussion of reaction rates is to use collision theory to calculate the rate of bimolecular reactions. The collision rate between two groups of molecules can be calculated in a straightforward manner (Section 4.3). Consider two groups of molecules with densities and velocities n_s, $\mathbf{v}_s$, and n_t, $\mathbf{v}_t$, respectively. The rate at which a single molecule of group t collides with molecules of group s can be expressed in terms of a stationary t molecule and s molecules moving with relative velocity, $\mathbf{g}_{st} = \mathbf{v}_s - \mathbf{v}_t$. The flux of s molecules, Γ_s, encountering molecule t with velocity increment $d^3 v_s$, is

$$d\Gamma_s = g_{st} f_s(\mathbf{v}_s) d^3 v_s, \tag{8.23}$$

where $f_s(\mathbf{v}_s)$ is the velocity distribution function of the s molecules. The differential cross section $\sigma_{st}(g_{st}, \theta)$ gives the fraction of particles scattered into a solid angle $d\Omega = \sin\theta \, d\theta \, d\phi$, where θ and ϕ are the spatial polar and azimuthal angles, respectively (Figure 4.5). If one integrates over all scattering angles and uses the definition of the total scattering cross section, Q_T, given by Equation (4.45), then the total number of collisions, dN_{st}, between molecules s and t within the velocity increments $d^3 v_s \, d^3 v_t$ is (G.4)

$$dN_{st} = \xi_{st} g_{st} Q_T(g_{st}) f_s(\mathbf{v}_s) f_t(\mathbf{v}_t) d^3 v_s \, d^3 v_t, \tag{8.24}$$

where ξ_{st} is $\frac{1}{2}$ if s and t are identical and unity otherwise. Equation (8.24) can be written in terms of the center-of-mass and relative velocities. If both particle populations are characterized by Maxwellian velocity distributions, then the integral over the center-of-mass velocities can be easily carried out. Writing the relative velocity in spherical coordinates and integrating over the solid angle associated with the relative velocity leads to the following expression for the collision rate, N_{st}:

$$N_{st} = 4\pi \xi_{st} n_s n_t \left(\frac{\mu_{st}}{2\pi k T_{st}} \right)^{3/2} \int_0^\infty dg_{st} g_{st}^3 Q_T(g_{st}) \exp\left(-\frac{\mu_{st} g_{st}^2}{2k T_{st}} \right), \tag{8.25}$$

where μ_{st} is the reduced mass (4.98) and T_{st} is defined by Equation (4.99).

If the total cross section, Q_T, is independent of the relative speed, g_{st}, the following simple expression for N_{st} is obtained:

$$N_{st} = \xi_{st} Q_T n_s n_t \sqrt{\frac{8kT_{st}}{\pi \mu_{st}}}. \tag{8.26}$$

Equation (8.26) allows for the situation when the two gases are characterized by different temperatures, T_s and T_t.

An approach very similar to the one used for collision rates has also been used to calculate simple, bimolecular chemical reaction rates. A variety of approximations have been used in making these simple calculations, which involve assumptions concerning the appropriate cross sections, minimum approach distances, or velocities necessary for a reaction to take place. One approach, presented here,[2] assumes that when the two particles approach within a critical distance, d_c, they stick together to form an intermediate complex that eventually breaks up into the final products. If there are no forces between the two molecules, the cross section is simply πd_c^2. However, when the particles are close together, strong repulsive forces are present and only those molecules that have sufficient energy to overcome this potential barrier can approach to within the critical distance. The magnitude of this potential barrier corresponds to the minimum energy necessary to form the complex, and this energy is commonly referred to as the *activation energy*, E_a. Therefore, the cross section for a bimolecular chemical reaction, in the presence of repelling potentials and using these assumptions, can be written as[2]

$$\begin{aligned} Q_T(E) &= \pi d_c^2 (1 - E_a/E_{st}) & E_{st} &> E_a \\ &= 0 & E_{st} &< E_a. \end{aligned} \tag{8.27}$$

Writing the relative kinetic energy, E_{st}, in terms of the magnitude of the relative velocity, $E_{st} = 1/2(\mu_{st} g_{st}^2)$, the energy-dependent cross section (8.27) can be substituted into Equation (8.25) to give

$$N_{st} = 4\pi \xi_{st} \pi d_c^2 n_s n_t \left(\frac{\mu_{st}}{2\pi kT_{st}}\right)^{3/2} \int\limits_{g_c}^{\infty} dg_{st} g_{st}^3 \left(1 - \frac{2E_a}{\mu_{st} g_{st}^2}\right) \exp\left(-\frac{\mu_{st} g_{st}^2}{2kT_{st}}\right), \tag{8.28}$$

where

$$g_c = \left(\frac{2E_a}{\mu_{st}}\right)^{1/2}. \tag{8.29}$$

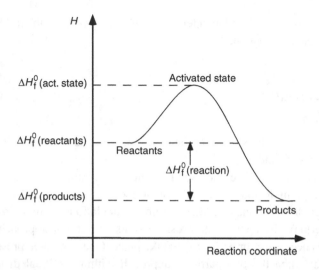

Figure 8.2 A schematic diagram indicating the variation of enthalpies of formation in a reaction.

After performing the integral in Equation (8.28), the following relation for the rate at which bimolecular chemical reactions take place is obtained:

$$N_{st} = 2\xi_{st}d_c^2 n_s n_t \left(\frac{2\pi kT_{st}}{\mu_{st}}\right)^{1/2} \exp\left(-\frac{E_a}{kT_{st}}\right). \tag{8.30}$$

Expressing this rate in terms of the conventional bimolecular chemical reaction rate, k_{st}, one gets the well known *Arrhenius equation*

$$k_{st} = 2d_c^2 \sqrt{\frac{2\pi k}{\mu_{st}}} \sqrt{T_{st}} \exp\left(-\frac{E_a}{kT_{st}}\right). \tag{8.31}$$

A way to think of bimolecular reactions is illustrated in Figure 8.2, which is plotted in terms of enthalpy change; it is assumed that an intermediate, activated state is formed during the reaction. This figure simply indicates that if the total enthalpy change of the reactants is greater than that of the products, the reaction results in an energy release; such a reaction is said to be *exothermic*. If the sum of the enthalpy changes of the products is greater than that of the reactants, the reaction is *endothermic*. Given typical ionospheric temperatures of less than a few thousand degrees, exothermic reactions are the dominant ones. At this point it is appropriate to remind the reader of the definition of enthalpy, H

$$H = E + pV \tag{8.32}$$

where E, p, and V denote energy, pressure, and volume, respectively. However, in general, pressure remains constant for ionospheric reactions, and therefore, ΔH is in effect ΔE. Given the *enthalpies of formation* (ΔH_f^0) of the reactants and products,

Table 8.1 Standard enthalpies of formation at
298.15 K.[3, 4]

Species (gaseous)	ΔH_f^0 (kJ mole^{-1})
C	715
CH_3	147
CH_4	−74.8
C_2H_2	228
CO	−111
CO_2	−394
H	218
H_2	0.0
H_2O	−242
H_3^+	1107
H_3O^+	581
He	0.0
K	89.2
N	473
N_2	0.0
$N_2^+(^2\Sigma)$	1503
NO	90.3
$NO^+(^1\Sigma)$	990
Na	108
$O(^3P)$	249
$O(^1D)$	439
$O(^1S)$	653
$O^+(^4S)$	1563
$O^-(^2P)$	105
O_2	0.0
$O_2(^1\Delta_g)$	94.2
$O_2^+(^2\Pi)$	1171
$O_2^-(^2\Pi)$	−48.6
O_3	143
OH	38.9
S	277
SO	5.01
SO_2	−297

it is possible to calculate whether a reaction is exothermic or endothermic and the excess energy available from the exothermic reactions. The important ionospheric reaction between O^+ and N_2 (8.4) provides a good example (note that 1 eV particle^{-1} is equal to 9.649×10^4 joules mole^{-1} or 2.305×10^4 calories mole^{-1}):

$$O^+(1563) + N_2(0) \rightarrow NO^+(990) + N(473), \tag{8.33}$$

where the numbers in the brackets are the enthalpies of formation in kilo-joules mole^{-1} obtained from Table 8.1, and all reactants and products are assumed to be in their ground state. The excess of the total enthalpies of formation of the reactants over that of the products is 100 kJ mole^{-1}. Thus, the reaction is exothermic and the excess energy that is available as kinetic energy of the products is 1.04 eV. Table 8.1 gives the enthalpies of formation at the standard temperature of 298.15 K; this is usually denoted by a superscript, thus the standard notation is $\Delta H_{\mathrm{f}}^{0}$. Note that all species are assumed to be in their ground state, unless indicated otherwise in this table. Also, for the sake of simplicity, the $\Delta H_{\mathrm{f}}^{0}$ values for the ions are not given in Table 8.1, except in a few cases, because they are simply equal to the value of the neutral gas $\Delta H_{\mathrm{f}}^{0}$ plus the ionization potential, given in Table 9.1.

8.3   Charge exchange processes

The two most important chemical processes of direct ionospheric relevance are the charge exchange and recombination reactions. Charge exchange reactions can be important with respect to momentum transfer (Section 4.8 and Appendix G), energy balance, hot atom formation, and ion chemistry (Chapter 10). In simple charge exchange reactions between ions and their parent atom or molecule, or in accidentally resonant charge exchange, the reactants tend to maintain their kinetic energy after the transfer of the electric charge. Therefore, this process can provide a rapid means for energetic ions to become energetic neutral particles or vice versa. In a more general charge exchange process, as shown in Equation (8.4), the reaction may be an important source of a given ion species.

It has been shown[5] that the energy dependence of both symmetric and accidentally resonant charge exchange reactions have a very similar behavior and that their cross sections can be expressed as given by Equation (4.148). The corresponding momentum transfer collision frequencies are given in Table 4.5.

The accidentally resonant charge exchange reaction between hydrogen and oxygen ions, shown by Equation (8.3), is of great importance in a number of ionospheres. For example in the terrestrial upper ionosphere it is the main source of H$^+$. For convenience the reaction is rewritten in more detail as

$$O^{+}(^{4}S) + H(^{2}S) \leftrightarrow H^{+} + O(^{3}P_{J}) + \Delta E. \tag{8.34}$$

The energy differences for the different J values are shown in Table 8.2. The reaction rate for the reverse of the reaction indicated by (8.34) was measured to be 3.75×10^{-10} cm^3 s^{-1} at 300 K.[6] The reaction rates for the forward and reverse reactions are related by the products of the *partition functions*, Θ, of the reactants involved multiplied by $\exp(-\Delta E/kT)$. The partition function is defined as

$$\Theta = \sum_{i} g_i \exp(-E_i/kT) \tag{8.35}$$

where E_i is the energy of the atom or ion in the ith level and g_i is the degeneracy or statistical weight of the energy level. The partition functions for O$^+$, H, and H$^+$

Table 8.2 Energy differences for the different J values of the ground state of atomic oxygen.

ΔE (eV)	0.00000	0.01965	0.02808
J	2	1	0

are given by the ground state degeneracies, $(2L+1)(2S+1)$, which are 4, 2, and 1, respectively. The general expression for the partition function of O is

$$\Theta(\mathrm{O}) = \sum_{J=0}^{2} (2J+1)\exp(-E_J/kT). \qquad (8.36)$$

where $J = 2$ is the lowest energy level, with the $J = 1$ level 0.01965 eV above $J = 2$ and the $J = 0$ level 0.02808 eV above $J = 2$. In the terrestrial upper ionosphere, the temperatures are sufficiently high that in most cases the exponential factor in Equation (8.36) can be neglected, leading to a partition function value of 9 for atomic oxygen. However, in other applications, for example in the ionosphere of Venus where the neutral gas temperatures are less than 300 K, the exponential factor cannot be neglected. For these applications, the following expression can be used to approximately evaluate the oxygen partition function:

$$\Theta(\mathrm{O}) = 5 + 3\exp(-223/kT) + \exp(-325/kT). \qquad (8.37)$$

Charge exchange reactions are also important in auroral studies, in which the interaction of energetic precipitating ions and neutrals are investigated. For example, in the ionosphere and upper atmosphere of the Earth, Jupiter, and Saturn, precipitating ions of keV energy undergo numerous charge exchange reactions leading to important ionization and heating effects. These interactions are not commonly considered to be chemical reactions, but this is only an artificial and semantic distinction; they will be discussed further in Section 9.5.

There are hundreds of bimolecular ion–molecule reactions of importance for ionospheric studies. It is impossible to list them all in this book, but a small subset of these reactions is given in Table 8.3.

No laboratory or other direct information exists for some ion–neutral reactions. In the case of nonpolar molecules, an approximate upper limit for the reaction rate can be obtained by using the *Langevin model*, which assumes that the reaction is controlled by the ion-induced dipole potential,[8] as discussed in Section 4.5. The reaction rate resulting from this approximation, expressed in $\mathrm{cm}^3\,\mathrm{s}^{-1}$, is

$$k \approx 3 \times 10^{-9} \left(\frac{\gamma_n}{\mu_{st}}\right)^{1/2}, \qquad (8.38)$$

where γ_n is the polarizability of the neutral reactant and μ_{st} is the reduced mass. The polarizability of a typical small polyatomic molecule is of the order of $10^{-24}\,\mathrm{cm}^3$;

Table 8.3 Room temperature bimolecular ion–molecule reaction rates.[7]

Reaction number	Reaction	Rate constant ($cm^3 s^{-1}$)
R1	$C^+ + CO_2 \rightarrow CO^+ + CO$	9.9×10^{-10}
R2	$CH_3^+ + CH_4 \rightarrow C_2H_5^+ + H_2$	1.1×10^{-9}
R3	$CO^+ + O \rightarrow O^+ + CO$	1.4×10^{-10}
R4	$CO^+ + CO_2 \rightarrow CO_2^+ + CO$	1.1×10^{-9}
R5	$CO_2^+ + O \rightarrow O^+ + CO_2$	9.6×10^{-11}
R6	$CO_2^+ + O \rightarrow O_2^+ + CO$	1.64×10^{-10}
R7	$CO_2^+ + NO \rightarrow NO^+ + CO_2$	1.23×10^{-10}
R8	$CO_2^+ + H \rightarrow HCO^+ + O$	2.7×10^{-10}
R9	$H^+ + H_2(v \geq 4) \rightarrow H_2^+ + H^a$	see Figure 8.3
R10	$H^+ + O \rightarrow O^+ + H$	3.75×10^{-10}
R11	$H^+ + CO_2 \rightarrow HCO^+ + O$	3.8×10^{-9}
R12	$H_2^+ + H_2 \rightarrow H_3^+ + H$	2.0×10^{-9}
R13	$H_2^+ + CH_4 \rightarrow CH_3^+ + H_2 + H$	2.3×10^{-9}
R14	$H_2^+ + H \rightarrow H^+ + H_2$	6.4×10^{-10}
R15	$HNC^+ + CH_4 \rightarrow HCNH^+ + CH_3$	1.1×10^{-9}
R16	$H_2O^+ + H_2O \rightarrow H_3O^+ + OH$	1.85×10^{-9}
R17	$H_2O^+ + H_2 \rightarrow H_3O^+ + H$	7.6×10^{-10}
R18	$H_2O^+ + CH_4 \rightarrow H_3O^+ + CH_3$	1.12×10^{-9}
R19	$H_2O^+ + NH_3 \rightarrow NH_3^+ + H_2O$	2.21×10^{-9}
R20	$H_2O^+ + NH_3 \rightarrow NH_4^+ + OH$	9.45×10^{-10}
R21	$H_3^+ + CH_4 \rightarrow CH_5^+ + H_2$	2.4×10^{-9}
R22	$H_3^+ + NH_3 \rightarrow NH_4^+ + H_2$	4.4×10^{-9}
R23	$H_3^+ + H_2O \rightarrow H_3O^+ + H_2$	5.3×10^{-9}
R24	$H_3O^+ + NH_3 \rightarrow NH_4^+ + H_2O$	2.23×10^{-9}
R25	$He^+ + CH_4 \rightarrow CH_2^+ + H_2 + He$	8.5×10^{-10}
R26	$He^+ + CH_4 \rightarrow H^+ + CH_3^+ + He$	4.4×10^{-10}
R27	$He^+ + CO_2 \rightarrow CO^+ + O + He$	7.8×10^{-10}
R28	$He^+ + CO_2 \rightarrow O^+ + CO + He$	1.4×10^{-10}
R29	$He^+ + N_2 \rightarrow N^+ + N + He$	7.8×10^{-10}
R30	$He^+ + N_2 \rightarrow N_2^+ + He$	5.2×10^{-10}
R31	$He^+ + O_2 \rightarrow O^+ + O + He$	9.7×10^{-10}
R32	$N^+ + CH_4 \rightarrow CH_3^+ + NH$	5.75×10^{-10}
R33	$N^+ + CH_4 \rightarrow HCNH^+ + H_2$	4.14×10^{-10}
R34	$N^+ + CH_4 \rightarrow HCN^+ + H_2 + H$	1.15×10^{-10}
R35	$N^+ + CO_2 \rightarrow CO_2^+ + N$	9.2×10^{-10}
R36	$N^+ + CO_2 \rightarrow CO^+ + NO$	2.0×10^{-10}
R37	$N^+ + O_2 \rightarrow O_2^+ + N$	3.07×10^{-10}
R38	$N^+ + O_2 \rightarrow NO^+ + O$	2.32×10^{-10}
R39	$N^+ + O_2 \rightarrow O^+ + NO$	4.6×10^{-11}
R40	$N_2^+ + CH_4 \rightarrow CH_3^+ + N_2 + H$	1.04×10^{-9}
R41	$N_2^+ + CH_4 \rightarrow CH_2^+ + N_2 + H_2$	1.0×10^{-10}
R42	$N_2^+ + CO_2 \rightarrow CO_2^+ + N_2$	8.0×10^{-10}
R43	$N_2^+ + NO \rightarrow NO^+ + N_2$	4.1×10^{-10}

Table 8.3 *(Cont.)*

Reaction number	Reaction	Rate constant ($cm^3 s^{-1}$)
R44	$N_2^+ + O \rightarrow NO^+ + N$	1.3×10^{-10}
R45	$N_2^+ + O \rightarrow O^+ + N_2$	9.8×10^{-12}
R46	$N_2^+ + O_2 \rightarrow O_2^+ + N_2$	5.0×10^{-11}
R47	$O^+ + N_2 \rightarrow NO^+ + N$	1.2×10^{-12}
R48	$O^+ + O_2 \rightarrow O_2^+ + O$	2.1×10^{-11}
R49	$O^+ + NO \rightarrow NO^+ + O$	8.0×10^{-13}
R50	$O^+ + CO_2 \rightarrow O_2^+ + CO$	1.1×10^{-9}
R51	$O^+ + H \rightarrow H^+ + O$	6.4×10^{-10}
R52	$O_2^+ + NO \rightarrow NO^+ + O_2$	4.6×10^{-10}
R53	$O_2^+ + N \rightarrow NO^+ + O$	1.5×10^{-10}

^{a} v corresponds to the vibrational state.

values of polarizability for the most important neutral constituents are given in Table 4.1. The reaction rate for polar molecules is expected to be considerably larger than the Langevin value.

8.4   Recombination reactions

The most direct recombination process, called *radiative recombination*, is the inverse of photoionization

$$X^+ + e^- \rightarrow X^* + h\nu, \tag{8.39}$$

where X^* indicates that the product atom or molecule may be in an excited state. However, the radiative recombination rate is small (Table 8.4) and in most cases this is a negligibly slow process.

The chemical loss process that most frequently dominates ionospheric abundances is dissociative recombination, an example of which was given by Equation (8.5). The product atoms may be in an excited state and the excess energy goes to the kinetic energy of the products. A very important dissociative recombination reaction in the terrestrial, Venus, and Mars ionospheres is that of the ground electronic state of O_2^+, for which the different, energetically permitted branches are

$$O_2^+(X^2\Pi_g) + e^- \rightarrow O(^3P) + O(^3P) + [6.99\,eV] \qquad (0.22)$$
$$\rightarrow O(^3P) + O(^1D) + [5.02\,eV] \qquad (0.42)$$
$$\rightarrow O(^1D) + O(^1D) + [3.06\,eV] \qquad (0.31)$$
$$\rightarrow O(^3P) + O(^1S) + [2.80\,eV] \qquad (<0.01)$$
$$\rightarrow O(^1D) + O(^1S) + [0.84\,eV] \qquad (0.05). \tag{8.40}$$

Table 8.4 Radiative recombination rates.[9]

Reaction	Rate constant (cm^3 s^{-1})
C^+	$4.2 \times 10^{-12}(250/T_e)^{0.7}$
H^+	$4.8 \times 10^{-12}(250/T_e)^{0.7}$
He^+	$4.8 \times 10^{-12}(250/T_e)^{0.7}$
N^+	$3.6 \times 10^{-12}(250/T_e)^{0.7}$
Na^+	$3.2 \times 10^{-12}(250/T_e)^{0.7}$
O^+	$3.7 \times 10^{-12}(250/T_e)^{0.7}$

Table 8.5 Dissociative recombination rates (J. L. Fox, private communication).[11-15]

Reaction	Rate (cm^3s^{-1})
CH_3^+	$3.5 \times 10^{-7}(300/T_e)^{0.5}$
CH_4^+	$3.5 \times 10^{-7}(300/T_e)^{0.5}$
CO^+	$2.75 \times 10^{-7}(300/T_e)^{0.55}$
CO_2^+	$4.2 \times 10^{-7}(300/T_e)^{0.75}$
$HCNH^+$	$3.5 \times 10^{-7}(300/T_e)^{0.5}$
H_2^+	$1.6 \times 10^{-8}(300/T_e)^{0.43}$ for $v = 0^a$ $2.3 \times 10^{-7}(300/T_e)^{0.4}$ for $v \neq 0$
H_3^+	$4.6 \times 10^{-6}(T_e)^{-0.65}$
H_2O^+ and H_3O^+	$1.57 \times 10^{-5}(T_e)^{-0.569}$ for $T_e < 800$ K $4.73 \times 10^{-5}T_e^{-0.74}$ for 800 K $< T_e < 4000$ K $1.03 \times 10^{-3}T_e^{-1.111}$ for $T_e > 4000$ K
NH_3^+	$3.3 \times 10^{-7}(300/T_e)^{0.5}$
N_2^+	$2.2 \times 10^{-7}(300/T_e)^{0.39}$
NO^+	$4.0 \times 10^{-7}(300/T_e)^{0.5}$
O_2^+	$2.4 \times 10^{-7}(300/T_e)^{0.70}$
OH^+	$3.75 \times 10^{-8}(300/T_e)^{0.5}$

^{a} v corresponds to the vibrational level.

For such a reaction, the value of the total *recombination rate*, namely the rate at which the sum of all branches takes place, and the *branching ratios*, which indicate the fraction going to each branch, must be specified. The excess energy for a given branch is shown inside the square brackets and the measured branching ratios[10] are given in the curved brackets. The dissociative recombination of H_2^+, H_2O^+, H_3^+, and H_3O^+ are of great importance in the ionospheres of the outer planets and comets, especially the last two;

$$H_2^+ + e^- \rightarrow H + H, \tag{8.41}$$

$$H_2O^+ + e^- \rightarrow H + OH$$
$$\rightarrow H_2 + O, \tag{8.42}$$

$$H_3^+ + e^- \rightarrow H + H_2$$
$$\rightarrow H + H + H, \tag{8.43}$$

$$H_3O^+ + e^- \rightarrow H + H_2O$$
$$\rightarrow H_2 + OH$$
$$\rightarrow H_3 + O. \tag{8.44}$$

The total dissociative recombination rate and the branching ratios may depend on the vibrational state of the ion and the energy or temperature of the electrons. The present understanding of dissociative recombination rates comes from a combination of laboratory, space-based measurements and theoretical calculations. There are still many uncertainties associated with these values; the information presented in Table 8.5 represents the best accepted values at this time.

8.5   Negative ion chemistry

In the lower ionosphere, where the neutral gas density is relatively large, negative ions may be formed. Such negative ions are believed to be important in the terrestrial D region (Section 11.4) and have been observed at Titan (Section 13.6). However, because of the difficulties associated with their measurements, only minimal information is available about these ions. The formation of negative ions is believed to start by the collision between an electron and neutral particles, in which the electron becomes attached to a neutral particle. The most important of these *attachment reactions* at Earth is the one involving two oxygen molecules:

$$O_2 + O_2 + e^- \rightarrow O_2^- + O_2 + 0.5 \text{ eV}, \tag{8.45}$$

although the following reaction may also be significant:

$$O_2 + N_2 + e^- \rightarrow O_2^- + N_2 + 0.5 \text{ eV}. \tag{8.46}$$

These negative ions may be lost by a variety of mechanisms. The most likely of these processes are *photodetachment*,

$$O_2^- + h\nu(<2.44 \text{ } \mu\text{m}) \rightarrow O_2 + e^-, \tag{8.47}$$

Table 8.6 Photodetachment and photodissociation
rates at 1 AU.[16]

Reaction	Rate (s^{-1})
$O^- + h\nu \rightarrow O + e^-$	1.4
$O_2^- + h\nu \rightarrow O_2 + e^-$	3.8×10^{-1}
$O_3^- + h\nu \rightarrow O_3 + e^-$	4.7×10^{-2}
$OH^- + h\nu \rightarrow OH + e^-$	1.1
$CO_3^- + h\nu \rightarrow CO_3 + e^-$	2.2×10^{-2}
$NO_2^- + h\nu \rightarrow NO_2 + e^-$	8.0×10^{-4}
$NO_3^- + h\nu \rightarrow NO_3 + e^-$	5.2×10^{-2}
$O_3^- + h\nu \rightarrow O^- + O_2$	0.47
$O_4^- + h\nu \rightarrow O_2^- + O_2$	0.24
$CO_3^- + h\nu \rightarrow O^- + CO_2$	0.15
$CO_4^- + h\nu \rightarrow O_2^- + CO_2$	6.2×10^{-3}

associative detachment,

$$O_2^- + O \rightarrow O_3 + e^- + 0.6 \text{ eV}, \tag{8.48}$$

two-body ion–ion recombination,

$$O_2^- + O_2^+ \rightarrow 2O_2 + 11.6 \text{ eV}, \tag{8.49}$$

or *collisional detachment* involving an excited atom or molecule

$$O_2^- + O_2(^1\Delta_g) \rightarrow 2O_2 + e^- + 0.5 \text{ eV}. \tag{8.50}$$

Photodissociation can also change the negative ion species in the following way:

$$O_3^- + h\nu \rightarrow O^- + O_2. \tag{8.51}$$

In reality, the formation of the initial negative ions O_2^-, as well as O^-, is just the beginning of a long chain of chemical reactions leading to more and more complex negative ions. A more detailed discussion of these steps is given in Section 11.4, which includes a short discussion of the terrestrial D region. Here, in Table 8.6, the rate coefficients for some of the more important negative ion reactions are presented.[16]

8.6   Excited state chemistry

The presence of a significant population of excited neutral or ionized species can have a major impact on ionospheric chemistry. Neutral or ion species in an excited state, corresponding to forbidden electronic transitions (Section 8.7), have relatively long lifetimes, which in turn can result in significant populations. Vibrationally excited molecules can also have a significant impact on reaction rates. In general, electron-ically or vibrationally excited species have reaction rates that are different, often higher, than the corresponding ground state ones. Also, a certain reaction that is endothermic when the reactants are in a ground state may become exothermic if one of the reactants is in an excited state. In general, the various potentially important excited states need to be considered as separate species, adding a potentially major complexity to ionospheric calculations. Nevertheless, such details are often neces-sary to insure that the resulting models provide a realistic description of the true nature of the ionosphere.

A good example to demonstrate the importance of metastable species is the reac-tions involving the ^2D state of atomic nitrogen (see the energy level diagram shown in Figure 8.1b). One of the sources of excited atomic nitrogen in the terrestrial high-latitude upper atmosphere is *electron impact dissociative excitation* of N_2

$$N_2 + e^- \rightarrow N(^2D) + N + e. \tag{8.52}$$

Dissociative recombination of N_2^+ and NO^+ may also produce $N(^2D)$

$$N_2^+ + e^- \rightarrow N(^2D) + N, \tag{8.53}$$

$$NO^+ + e^- \rightarrow N(^2D) + O. \tag{8.54}$$

The reaction between $N(^2D)$ and molecular oxygen is the main source of NO in the lower thermosphere,

$$N(^2D) + O_2 \rightarrow NO + O. \tag{8.55}$$

The rate of formation of NO by the reaction between a ground state atomic nitrogen and O_2 is highly temperature dependent,[16] $4.4 \times 10^{-12} \exp(-3220/T)$ cm^3 s^{-1}, because of the relatively large activation energy of the reaction. At the temperatures found in the lower terrestrial thermosphere ($\sim$350 K), the ground state reaction is negligible, and reaction (8.56), involving the metastable ^2D state of atomic nitrogen, with a rate coefficient of 6×10^{-12} cm^{3}s^{-1}, is the dominant one.[17]

Another example of the potential importance of excited state chemistry is asso-ciated with the loss of H^+ in the upper ionospheres of the giant planets. Radiative recombination is very slow, so charge exchange with the major neutral background constituent, H_2, needs to be considered. However, the charge exchange reaction

$$H^+ + H_2 \rightarrow H_2^+ + H \tag{8.56}$$

is endothermic, unless H_2 is in a vibrationally excited state $v \geq 4$. As indicated in Figure 8.3, the reaction is reasonably fast for $v \geq 4$. Therefore, this reaction may be a potentially important loss of H^+, if the vibrational temperature of H_2 is sufficiently elevated. This is discussed in more detail in Section 13.4.

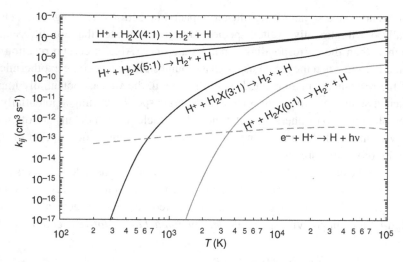

Figure 8.3 The kinetic temperature and vibrational level dependent reaction rate of $H^+ + H_2$. The rates plotted are for various different vibrational levels and rotational levels, $J = 1$. (Courtesy of Dr. D. Shemansky).

8.7　Optical emissions; airglow and aurora

Excited atmospheric species eventually drop to a lower level of excitation by either *spontaneous emission* of a photon or by losing energy via a collision. *Collisional de-excitation* is commonly referred to as *quenching*. The value of the Einstein *transition* probability of spontaneous de-excitation (Section 8.1) depends on the electric and magnetic dipole and electric quadrupole contributions. The electric dipole component is about 10^5 times that of the magnetic dipole one and about 10^8 times the electric quadrupole value. Therefore, if the electric dipole term is zero, because of symmetry properties, the transition probability becomes very small, and the corresponding transition is referred to as a *forbidden* one. Atoms and molecules in an excited state from which the de-excitation transition is forbidden are said to be in a *metastable* state.

　　The optical emissions from excited atmospheric species are referred to as *aurora* if particle impact excitation, other than that due to photoelectrons, is the original source of the excitation energy. The emissions are referred to as *airglow* if solar radiation is the initial source of energy causing the excitation. This book does not discuss in any detail the processes leading to these optical emissions; there are a number of good references available on this topic (see the General References). However, one emission line of the dayglow is discussed in this section as a representative example. Otherwise, further discussion of this topic will only come up in the context of ionospheric relevance.

　　The oxygen atom has two low-lying electronic states at 1.97 and 4.19 eV, the ^1D and ^1S states, respectively, as shown in Figure 8.1a. The spontaneous transition from the ^1D state to the ground, $^3P_2$, state results in the emission of a photon at

630 nm. This, so-called *oxygen red line* is an important emission in the terrestrial night and dayglow, as well as the aurora. Here, we concentrate on this specific dayglow emission line. The processes which are plausible production sources for $O(^1D)$ are[18]

$$e^* + O(^3P) \rightarrow e^* + O(^1D), \tag{8.57}$$

$$O_2 + h\nu \rightarrow O + O(^1D), \tag{8.58}$$

$$O_2^+ + e^- \rightarrow O + O(^1D), \tag{8.59}$$

$$O(^1S) \rightarrow h\nu(\lambda = 558\,\text{nm}) + O(^1D), \tag{8.60}$$

$$N(^2D) + O_2 \rightarrow NO + O(^1D), \tag{8.61}$$

$$N(^2D) + O \rightarrow N(^4S) + O(^1D), \tag{8.62}$$

$$N(^2P) + O_2 \rightarrow NO + O(^1D), \tag{8.63}$$

$$N(^2P) + O \rightarrow N(^4S) + O(^1D), \tag{8.64}$$

$$N^+ + O_2 \rightarrow NO^+ + O(^1D), \tag{8.65}$$

$$O^+(^2D) + O(^3P) \rightarrow O^+(^4S) + O(^1D), \tag{8.66}$$

$$e^* + O_2 \rightarrow e^* + O + O(^1D), \tag{8.67}$$

$$e_{th} + O \rightarrow e_{th} + O(^1D), \tag{8.68}$$

where e^* and e_{th} represent energetic (photoelectron) and thermal electrons, respectively. If no excited state is indicated, it means that depending on the circumstances, any of the energetically permitted states is possible. The possible loss reactions are

$$O(^1D) + N_2 \rightarrow O(^3P) + N_2, \tag{8.69}$$

$$O(^1D) + O_2 \rightarrow O(^3P) + O_2, \tag{8.70}$$

$$O(^1D) + O(^3P) \rightarrow O(^3P) + O(^3P), \tag{8.71}$$

$$O(^1D) + e^- \rightarrow O(^3P) + e^-, \tag{8.72}$$

$$O(^1D) \rightarrow O(^3P) + h\nu(\lambda = 630, \text{or } 636\,\text{nm}). \tag{8.73}$$

In the case of steady-state conditions and negligible transport effects, the continuity equation simplifies to a balance between production and loss rates. Writing the continuity equation for an excited species X^* and then solving that algebraic equation for the number density of X^*, $n(X^*)$, and multiplying it by the Einstein coefficient (equal to 1/lifetime; Section 8.1) for the transition of interest, A_λ, the

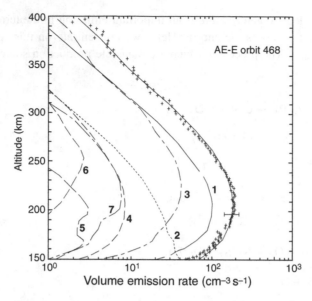

Figure 8.4 A comparison of calculated and measured 6300 Å airglow intensities. Symbols 1, 2, and 3 indicate the contributions from photoelectron impact (8.58), photodissociation (8.59), and dissociative recombination (8.60), respectively.[18]

following equation for the steady state emission rate, $R(\lambda)$, is obtained,

$$R(\lambda) = A_\lambda n(X^*) = \frac{\sum_p p(X^*)}{\dfrac{\sum_\lambda A_\lambda(X^*)}{A_\lambda} + \left(\dfrac{\sum_Q k_Q n(Q)}{A_\lambda}\right)}, \tag{8.74}$$

where $p(X^*)$ represents the various production rates of X^* by the different source mechanisms, the k_Q are the quenching rate coefficients and the summations are over all production, quenching, and radiative de-excitation processes. Figure 8.4 shows a comparison of the calculated 630 nm emission rate and that measured[18] by the visible airglow photometer (VAE) carried aboard the *AE-E* satellite.[19] The agreement is very good between the modeled and observed intensities and shows that photoelectron impact excitation, photodissociation, and dissociative recombination are the three main source processes.

8.8   Specific references

1. Froese-Fischer, C. and G. Tachier, Breit Pauli energy levels, lifetimes and transition probabilities for the beryllium-like to neon-like sequences, *Atomic Data and Nuclear Tables*, **87**, 1, 2004.

2. Present, R. D., *Kinetic Theory of Gases*, New York: McGraw-Hill, 1958.

3. DeMore, W. B., S. P. Sander, D. M. Golden *et al.*, *Chemical Kinetics and Photochemical Data for Use in Stratospheric Modeling*, JPL Pub. 97-4, 1997.

4. Rosenstock, H. M., K. Draxl, B. W. Steiner, and J. T. Herron, Energetics of gaseous ions, *J. Phys. Chem. Ref. Data*, **6**, Suppl. No. 1, 1977.

5. Dalgarno, A., The mobilities of ions in their parent gases, *Phil. Trans. Roy. Soc. London, Ser. A*, **250**, 426, 1958.

6. Fehsenfeld, F. C., and E. E. Ferguson, Thermal energy reaction rate constants for H^+ and CO^+ with O and NO, *J. Chem. Phys.*, **56**, 3066, 1972.

7. Anicich, V. G., Evaluated bimolecular ion-molecule gas phase kinetics of positive ions for use in modeling the chemistry of planetary atmospheres, cometary comae and interstellar clouds, *J. Phys. Chem. Ref. Data*, **22**, 1469, 1994; An index of the literature for bimolecular gas cation–molecule reaction kinetics, *JPL* Publ-03-19, 2003.

8. Steinfeld, J. I., J. S. Francisco, and W. L. Hase, *Chemical Kinetics and Dynamics*, Englewood Cliffs, NJ: Prentice Hall, 1989.

9. Dalgarno, A., and D. R. Bates, Electronic recombination, in *Atomic and Molecular Processes*, ed. by D. R. Bates, **245**, New York: Academic Press, 1962.

10. Kella, D., L. Vejby-Christensen, P. J. Johnson, H. B. Pedersen, and L. H. Andersen, The source of green light emission determined from a heavy-ion storage ring experiment, *Science*, **276**, 1530, 1997.

11. Fox, J. L., Hydrocarbon ions in the ionospheres of Titan and Jupiter, *Dissociative Recombination: Theory, Experiment and Applications*, **40**, City, NJ: World Scientific, 1996.

12. Haberli, R. M., M. R. Combi, T. I. Gombosi, D. L. de Zeeww, and K. E. Powell, Quantitative analysis of H_2O^+ coma images using a multiscale MHD model with detailed ion chemistry, *Icarus*, **130**, 373, 1997.

13. Sundström, G., J. R. Mowatt, H. Danared *et al.*, Destruction rate of H_3^+ by low energy electrons measured in a storage-ring experiment, *Science*, **263**, 785, 1994.

14. Peverall, R., S. Rosén, J. R. Peterson *et al.*, Dissociative recombinaton and excitation of O_2^+: Cross sections, product yields and implications for studies of ionospheric airglows, *J. Chem. Phys.*, **114**, 6679, 2001.

15. Viggiano, A. A., A. Ehlerding, F. Hellberg *et al.*, Rate constants and branching ratios for the dissociative recombination of CO_2^+, *J. Chem. Phys.*, **122**, 226101, 2005.

16. Turunen, E., H. Matveinen, J. Tolvanen, and H. Ranta, D-region ion chemistry model, *STEP Handbook of Ionospheric Models*, ed. by R. W. Schunk, pp. 1–25, Logan: Solar-Terrestrial Energy Program, Utah State University Press, 1996.

17. Cleary, D. D., Daytime high-latitude rocket observations of the NO γ, δ, and ε bands, *J. Geophys. Res.*, **91**, 11337, 1986.

18. Solomon, S. C., and V. J. Abreu, The 630 nm dayglow, *J. Geophys. Res.*, **94**, 6817, 1989.

19. Hays, P. B., G. Carignan, B. C. Kennedy, G. G. Shepherd, and J. C. G. Walker, The visible-airglow experiment on Atmosphere Explorer, *Radio Sci.*, **8**, 369, 1973.

8.9   General references

Banks, P. M., and G. Kockarts, *Aeronomy*, New York: Academic Press, 1973.

Brekke, A., *Physics of the Upper Polar Atmosphere*, New York: John Wiley & Sons, 1997.

Gardiner, W. C., *Rates and Mechanisms of Chemical Reactions*, Menlo Park, CA: Benjamin/Cummings Pub. Co., 1969.

Gombosi, T. I., *Gaskinetic Theory*, Cambridge, UK: Cambridge University Press, 1994.

Huestis, D. L., S. W. Bougher, J. L. Fox *et al.*, Cross sections and reaction rates for comparative planetary aeronomy, *Space Sci. Rev.*, **139**, 63, doi 10.1007/s11214-008-9383-7, 2008.

Jones, A. V., *Aurora*, Dordrecht, Holland: D. Reidel Publishing Co., 1974.

Laidler, K. J., and J. H. Meiser, *Physical Chemistry*, Boston: Houghton Mifflin Co., 1995.

Rees, M. H., *Physics and Chemistry of the Upper Atmosphere*, Cambridge, UK: Cambridge University Press, 1989.

Weston, R. E., and H. A. Schwarz, *Chemical Kinetics*, Englewood Cliffs, NJ: Prentice-Hall Inc., 1972.

8.10   Problems

Problem 8.1 The half-life of a first-order reaction is 15 minutes. What is the rate constant of this reaction? What fraction of the reactant remains after 45 minutes?

Problem 8.2

(a)   Show that if the initial concentrations of the species reacting in a second-order reaction are the same, then Equation (8.15) follows from Equation (8.13).

(b)   A gas species A is removed via a second-order reaction with B. If the rate constant for this loss reaction is 1×10^{-9} cm^3 s^{-1} and the initial densities of A and B are the same, what is the half-life of constituent A if its initial concentration was 1×10^6 cm^{-3}?

Problem 8.3 Obtain Equation (8.25) from Equation (8.24) assuming that both particle populations are characterized by Maxwellian velocity distributions. For the sake of simplicity, you may assume that both gases have the same temperature, T. Also note that $d^3 v_s \, d^3 v_t = d^3 V_c \, d^3 g_{st}$.

Problem 8.4 Show that Equation (8.31) reduces to Equation (8.26) when the activation energy, E_a, is zero.

Problem 8.5 Measurements show that the rate of a given reaction doubles when the temperature is raised from 300 to 310 K. What is the activation energy of this reaction expressed in eV particle^{-1}?

Problem 8.6 The activation energy of the reaction

$$H + CH_4 \rightarrow H_2 + CH_3$$

is $49.8\,kJ\,mole^{-1}$. Given the enthalpies of formation in Table 8.1, estimate the activation energy of the reverse reaction. (Look at Figure 8.2 for guidance.)

Problem 8.7 Calculate the excess energy resulting from the following reaction:

$$O^+({}^4S) + CO_2 \rightarrow O_2^+({}^2\Pi) + CO$$

Express your results in terms of $eV\,particle^{-1}$.

Problem 8.8 On the planet Imaginus, at a given altitude, the O^+, O_2^+, and electron densities are 5×10^5, 5×10^5, and $10^6\,cm^{-3}$, respectively, and the neutral gas, electron, and ion temperatures are all 1000 K. The O_2 density at this altitude is $10^8\,cm^{-3}$.

(a)   Calculate the radiative recombination rate of O^+.

(b)   Compare the rate from (a) with the charge exchange rate between O^+ and O_2 and the dissociative recombination rate of O_2^+.

(c)   If the loss of O^+ is controlled by the two-step process of charge exchange followed by dissociative recombination, as calculated in (b) above, which of these two processes is the rate limiting (the slow) one?

(d)   If the time constant for transport at this altitude is 10^5 seconds and the loss of O^+ is determined by charge exchange with O_2, will transport or chemistry dominate at this altitude? (Compare time constants!)

Problem 8.9 Use the steady state continuity equation for the excited $O({}^1D)$ atom to obtain Equation (8.75).

Chapter 9

Ionization and energy exchange processes

Solar extreme ultraviolet (EUV) radiation and particle, mostly electron, precipitation are the two major sources of energy input into the thermospheres and ionospheres in the solar system. A schematic diagram showing the energy flow in a thermosphere–ionosphere system caused by solar EUV radiation is shown in Figure 9.1. Relatively long wavelength photons ($\gtrsim 900$ Å) generally cause dissociation, while shorter wavelengths cause ionization; the exact distribution of these different outcomes depends on the relevant cross sections and the atmospheric species. The only true sinks of energy, as far as the ionospheres are concerned, are airglow and neutral heating of the thermosphere. Even the escaping photoelectron flux can be reflected or become the incoming flux for a conjugate ionosphere. The specific distribution of the way that energy flows through the system is very important in determining the composition and thermal structure of the ionospheric plasmas. This chapter begins with a discussion of the absorption of the ionizing and dissociating solar radiation and the presentation of information needed to calculate ionization and energy deposition rates. This material is followed by a description of particle transport processes. The chapter ends with a presentation of electron and ion heating and cooling rates that can be used in practical applications.

9.1   Absorption of solar radiation

Radiative transfer calculations of the solar EUV energy deposition into the thermosphere are relatively simple because absorption is the only dominant process. To illustrate the basic physical principles, it is convenient first to make the following simplifying assumptions: (*a*) the radiation is monochromatic (single wavelength), (*b*) the atmosphere consists of a single absorbing species, which decreases exponentially with altitude with a constant characteristic length, H, and (*c*) the atmosphere

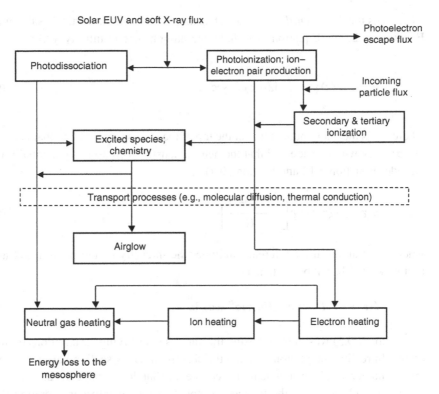

Figure 9.1 Block diagram of the energy flow in the upper atmosphere.

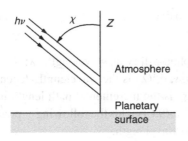

Figure 9.2 Schematic diagram showing monochromatic radiation penetrating a plane and horizontally stratified atmosphere.

is plane and horizontally stratified in the manner shown by Figure 9.2. Let σ^a be the *absorption cross section*, $I(s_\lambda)$ the *photon flux*, and $n(z)$ the *neutral density*, where s_λ is the distance along the path of the photons and z is altitude. As the photon flux penetrates the atmosphere, it is attenuated by absorption. The decrease in intensity of the flux after it travels an incremental distance, ds_λ, is

$$dI(s_\lambda) = -I(s_\lambda)n(z)\sigma^a ds_\lambda. \tag{9.1}$$

With the assumption of a plane stratified atmosphere

$$ds_\lambda = -dz \sec \chi \tag{9.2}$$

where χ is the *solar zenith angle*, which is measured from the vertical. Substituting Equation (9.2) into Equation (9.1) and integrating from z to infinity, yields

$$\ln[I_\infty/I(z)] = \int_z^\infty dz\, n(z)\sigma^a \sec\chi,$$ (9.3)

where I_∞ is the unattenuated flux at the top of the atmosphere. Given that neither χ nor σ^a vary with altitude and that the neutral density decreases exponentially with altitude (Equation 9.12 and Section 10.7);

$$n(z) = n(z_0) \exp\left[-\frac{(z-z_0)}{H}\right]$$ (9.4)

where z_0 is an arbitrary reference altitude, the intensity of the photon flux at an arbitrary altitude can be written as

$$I(z, \chi) = I_\infty \exp[-Hn(z)\sigma^a \sec\chi].$$ (9.5)

Equation (9.5) was obtained using the various simplifying assumptions outlined above. In reality, the photon flux and the absorption cross section vary with wavelength, numerous absorbing neutral species exist that do not have the same altitude variation, and of course the planets are not flat. Taking into consideration these factors, Equation (9.1) has to be modified to the general form

$$dI(z, \lambda, \chi) = -\sum_s n_s(z)\sigma_s^a(\lambda)I(z, \lambda)\, ds_\lambda$$ (9.6)

where $I(z, \lambda)$ is the intensity of the solar photon flux at wavelength, λ; $n_s(z)$ is the number density of the absorbing species, s; $\sigma_s^a(\lambda)$ is the wavelength-dependent absorption cross section of species s; and ds_λ is the incremental path length in the direction of the flux. The integration of Equation (9.6) yields the following expression for the solar flux as a function of altitude and wavelength:

$$I(z, \lambda, \chi) = I_\infty(\lambda) \exp\left[-\int_\infty^z \sum_s n_s(z)\sigma_s^a(\lambda)\, ds_\lambda\right],$$ (9.7)

where $I_\infty(\lambda)$ is the flux at the top of the atmosphere and the integration is to be carried out along the optical path. The argument of the exponential in Equation (9.7) is defined as the *optical depth* or *optical thickness*, τ, thus

$$\tau(z, \lambda, \chi) \equiv \int_\infty^z \sum_s n_s(z)\sigma_s^a(\lambda)\, ds_\lambda$$ (9.8)

and

$$I(z, \lambda, \chi) = I_\infty(\lambda) \exp\left[-\tau(z, \lambda, \chi)\right]. \tag{9.9}$$

The evaluation of the optical depth requires a detailed knowledge of the atmospheric densities and all the relevant absorption cross sections. If the atmosphere is assumed to be plane stratified and some simplifying assumptions are made regarding the gas temperatures, simple expressions can be obtained for the optical depth. The vertical distribution of a given neutral atmospheric species can be simply written as (Section 10.7)

$$n_s(z) = n_s(z_0)\frac{T_s(z_0)}{T_s(z)} \exp\left[-\int_{z_0}^{z} \frac{dz'}{H_s(z')}\right] \tag{9.10}$$

where the *neutral gas scale height*, H_s, is defined as

$$H_s(z) \equiv \frac{kT_s(z)}{m_s g(z)}. \tag{9.11}$$

If one assumes that both the temperature and the scale height are altitude independent, the following well-known exponential relation for the density is obtained:

$$n_s(z) = n_s(z_0) \exp\left[-\frac{(z - z_0)}{H_s}\right]. \tag{9.12}$$

The vertical *column density* is easily obtained from Equation (9.12) and is given by

$$\int_{z_0}^{\infty} n_s(z)\, dz = n_s(z_0)H_s(z_0). \tag{9.13}$$

The relationship given by Equation (9.13) is true even if T_s is not independent of altitude. To show this, a new dimensionless parameter, h, known as *reduced height*, has to be introduced. The reduced height is defined by the following relation:

$$dh \equiv dz/H_s, \tag{9.14}$$

where $h = 0$ corresponds to an altitude z_0. Using this new parameter, the following expression for the pressure variation is obtained by substituting Equation (9.14) into Equation (9.10) and setting $p_s = n_s k T_s$:

$$p_s(h) = p_s(0) \exp(-h). \tag{9.15}$$

The vertical column density is then given by

$$\int_{z_0}^{\infty} n_s(z') \, dz' = \int_{0}^{\infty} n_s(h') H_s(h') \, dh'$$

$$= \int_{0}^{\infty} \frac{p_s(h')}{kT_s} \frac{kT_s}{mg} \, dh'$$

$$= \frac{p_s(0)}{mg} \int_{0}^{\infty} \exp(-h') \, dh'$$

$$= n_s(z_0) H_s(z_0), \tag{9.16}$$

where the only assumption made was that g is not a function of altitude. Next, if the atmosphere is assumed to be plane and horizontally stratified (Equation 9.2 and Figure 9.2), then the expression for the optical depth can be simply written as

$$\tau(z, \lambda, \chi) = \sec\chi \sum_{s} n_s(z) \sigma_s^a(\lambda) H_s. \tag{9.17}$$

The plane stratified assumption (that is, the $-\sec\chi \, dz$ approximation for ds_λ) is good for χ less than about 75°, but at larger zenith angles the curvature of the planetary surface and changing densities with solar zenith angle makes the atmospheric column content a much more complicated function of χ. A so-called *Chapman function*, $Ch(z_0, \chi_0)$, has been used in the past,[1] which is defined by the following relation:

$$\int_{z_0}^{\infty} n_s(z) \, ds_\lambda \equiv n_s(z_0) H_s(z_0) Ch(z_0, \chi_0). \tag{9.18}$$

A great deal of effort used to be devoted to obtaining good analytic expressions for this Chapman function.[1] However, with the availability of high speed computers, an exact evaluation of the optical depth is relatively easy, as long as the necessary information on the wavelength-dependent absorption cross sections and the densities, as a function of altitude and solar zenith angle, are available.

9.2 Solar EUV intensities and absorption cross sections

Solar radiation in the EUV and X-ray range of wavelengths excites, dissociates, and ionizes the neutral constituents in the upper atmosphere. These emissions come from different regions of the solar atmosphere (chromosphere, transition region, and corona), and therefore, both the short- and long-term variabilities are wavelength dependent. Although measurements of solar ultraviolet radiation began in 1946,[2] it was not until the 1970s that quantitative information for the wavelength region of thermospheric and ionospheric interest, 5–185 nm, became available.

The currently existing database is limited, consisting of a few rocket measurements,[3, 4] and two extended satellite observations. Measurements by a spectrophotometer[5] on board the *Atmosphere Explorer* (AE) satellites between 1974 and 1981 and a solar EUV experiment[6] (SEE) currently on the Thermosphere, Ionosphere, Mesosphere Energetics, and Dynamics (*TIMED*) satellite are the current base of our information.

A detailed knowledge of the behavior of this important wavelength region of the solar flux is necessary for quantitative studies of the thermosphere and ionosphere. Two important, but conflicting, criteria need to be considered in the creation and dissemination of this spectral information. One of these is the desire to keep the data as compact as possible, while the other is the need to have sufficient spectral details to make its use meaningful in potential applications (e.g., theoretical calculations).

A number of different solar EUV models have been introduced over the years. One widely used model is the so-called EUVAC Solar Flux Model.[7] This EUVAC model uses only 37 wavelength intervals, covering a range of 5 to 105 nm, and its basic parameters are given in Table J.1 in Appendix J.

In the EUVAC model, the following simple factor accounts for solar activity variations:

$$P = (F10.7 + \langle F10.7\rangle)/2 \tag{9.19}$$

which is used to scale each wavelength bin of solar photon flux, I_i, for different levels of solar activity via the expression

$$I_i = F74113_i[1 + A_i(P - 80)]. \tag{9.20}$$

In these equations, $F74113_i$ is the modified reference flux, as given in Table J.1; A_i is the scaling factor for each interval, also given in Table J.1; $F10.7$ is the 10.7 cm solar radio flux in $WHz^{-1}m^{-2}$, multiplied by 10^{22}; and $\langle F10.7\rangle$ is F10.7 flux averaged over 81 days. Thus, this model provides an estimate of the unattenuated solar flux for any period.

The EUVAC model is appropriate for calculating parameters such as ionization rates, but its relatively coarse wavelength resolution is not good enough to calculate certain parameters of aeronomic interest adequately, such as detailed photoelectron fluxes, for example. A high resolution model which has flexible wavelength binning was developed recently to overcome this shortcoming of the simpler EUVAC, if necessary. This high resolution model is called HEUVAC[8] and its wavelength bins can range from 0.1 to 100 nm. There are other solar flux models also available (e.g., NRLEUV[9] and SOLAR2000,[10]) but they are all based on the same general database, and while there are some differences among them, they lead to very similar results.

It is clear from Equation (9.7) that to calculate the optical depth and the solar flux as a function of altitude, the absorption cross sections for the various atmospheric species are necessary. Because the solar flux is given in discrete intervals, one needs to average the corresponding absorption cross sections over the same wavelength intervals. Table J.2 gives wavelength averaged absorption cross sections, corresponding to the EUVAC solar flux intervals, for N_2, O_2, O, N, CO_2, CO, CH_4, H_2O, He, and SO_2.

9.3   Photoionization

Photoionization of the neutral gas constituents in planetary atmospheres produces free electron–ion pairs, and this is the major source of ionization in most ionospheres. The energy of the ionizing photons exceeds, in general, the threshold ionization energy (see Table 9.1 for the ground-state ionization energy of some common neutral gas species), with the excess going either into electron kinetic energy or excitation of the resulting ion. The reason that the electrons pick up the bulk of the kinetic energy is that the ions are much more massive than the electrons, and therefore, the ions acquire very little recoil energy during the photoionization process.

Before presenting a rigorous expression for the energy-dependent photoelectron production rate, it is instructive to derive a simple expression for this rate using the same three simplifying assumptions that led to the simple expression for $I(z, \chi)$ given by Equation (9.5). If the probability of a photon absorption, resulting in the production of an ion–electron pair, is denoted by η, then this rate of production, sometimes called the *Chapman production function*, P_c, can be written as

$$P_c(z, \chi) = I(z, \chi)\eta\sigma^a n(z) = I_\infty \exp[-Hn(z)\sigma^a \sec\chi]\eta\sigma^a n(z). \qquad (9.21)$$

With the advent of high-speed computers, this highly simplified equation (9.21) is not of much practical use. However, it is extremely useful for gaining physical insight. This equation clearly indicates that the production rate is proportional to the product of the intensity of solar ionizing radiation, which increases with altitude, and the neutral gas density, which decreases with altitude. The altitude of the peak production rate can be obtained by differentiating Equation (9.21) and setting it equal to zero; this gives

$$z_{max} = z_0 + H \ln[n(z_0)H\sigma^a \sec\chi], \qquad (9.22)$$

where z_0 is an arbitrary reference altitude. This result shows that z_{max} increases with increasing solar zenith angle, just as one would expect from intuition. Substituting Equation (9.22) into Equation (9.21) gives the following expression for the maximum production rate:

$$P_c(z_{max}, \chi) = \frac{I_\infty \eta \cos\chi}{H \exp(+1)}. \qquad (9.23)$$

Note that the peak production rate increases with decreasing solar zenith angle and is a maximum for an overhead Sun. This can be seen in Figure 9.3, which is a plot of

Table 9.1 Ionization threshold potentials.

Neutral	eV	nm
C	11.26	110.1
CH_4	12.55	98.79
CO	14.01	88.49
CO_2	13.77	90.04
H	13.60	91.16
H_2	15.43	80.35
H_2O	12.62	98.24
He	24.59	50.42
Mg	7.646	162.2
N	14.55	85.33
N_2	15.58	79.58
NH_3	10.16	121.9
NO	9.264	133.8
Na	5.139	241.3
O	13.62	91.03
O_2	12.06	102.8
OH	13.18	94.07
S	10.36	119.7
SO	10.0	124.0
SO_2	12.34	100.5

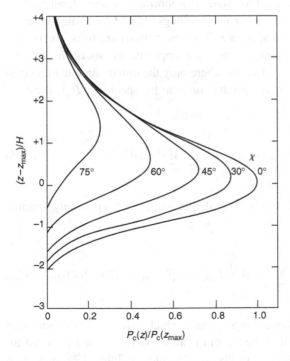

Figure 9.3 Plot of the normalized Chapman production function (9.21).

Equation (9.21) versus z/H, normalized to the maximum production rate, $P_c(z_{max})$, for the case where z_0 is taken to be z_{max} for an overhead Sun ($\chi = 0°$). This altitude is also the same as the altitude for unit optical depth ($\tau = 1$) for an overhead Sun.

Realistic detailed calculations of the electron production rate as a function of altitude, energy, and solar zenith angle are more complicated than that for the Chapman production function given by Equation (9.21). The initial photoelectron energy depends on the final state of the ion, as well as the energy of the ionizing photon. Therefore, to calculate the energy distribution of the newly created photoelectrons, one needs to know not only the total *ionization cross sections* listed in Table J.2, but also the ionization cross sections for each excited ionic state (or the equivalent information through the *branching ratios* of the final ion states). The branching ratio multiplied by the total ionization cross section gives the ionization cross section of a given state.

The expression for the altitude, energy, and solar zenith angle dependent *photoelectron production rate*, $P_e(E, \chi, z)$, can be written as

$$P_e(E, \chi, z) = \sum_l \sum_s n_s(z) \int_0^{\lambda_{si}} I_\infty(\lambda) \exp[-\tau(\lambda, \chi, z)] \sigma_s^i(\lambda) p_s(\lambda, E_l) \, d\lambda,$$

$$(9.24)$$

where $\sigma_s^i(\lambda)$ is the wavelength-dependent total ionization cross section, $p_s(\lambda, E_l)$ is the branching ratio for a given final ion state with ionization energy level E_l, $E = E_\lambda - E_l$, E_λ is the energy corresponding to wavelength λ, and λ_{si} is the ionization threshold wavelength for neutral species s. The summations are to be carried out over all species, s, and ion states, l. There are applications where the detailed photoelectron spectrum is not needed, but where only the ionization rate of a given ion species is needed. This total *ion production rate* for species s, $P_{ts}(z)$, can be written as

$$P_{ts}(z, \chi) = n_s(z) \int_0^{\lambda_{si}} I_\infty(\lambda) \exp\left[-\tau(z, \chi, \lambda)\right] \sigma_s^i(\lambda) \, d\lambda. \qquad (9.25)$$

If the total ionization rate for all species P_t is all that is needed, P_{ts} is simply summed over all s, giving

$$P_t(z, \chi) = \sum_s P_{ts} = \sum_s n_s(z) \int_0^{\lambda_{si}} I_\infty(\lambda) \exp\left[-\tau(z, \chi, \lambda)\right] \sigma_s^i(\lambda) \, d\lambda. \quad (9.26)$$

A very useful parameter is the *ionization frequency*, which is defined as the ionization rate per unit neutral gas particle at the top of the atmosphere. It was calculated for a number of important ions and the results are presented in Table 9.2.[9] Ionization

Table 9.2 Total ionization frequencies[a] at 1 AU (J. A. Fennelly, Private communication, 1998).

Species	Solar minimum[b]	Solar maximum[c]
CH_4	5.753×10^{-7}	1.708×10^{-6}
CO	4.245×10^{-7}	1.127×10^{-6}
CO_2	6.696×10^{-7}	1.695×10^{-6}
H_2	7.46×10^{-8}	1.407×10^{-7}
H_2O	4.286×10^{-7}	1.184×10^{-6}
He	5.283×10^{-8}	1.276×10^{-7}
N_2	3.35×10^{-7}	9.476×10^{-7}
O	2.44×10^{-7}	6.346×10^{-7}
O_2	4.90×10^{-7}	1.594×10^{-6}
SO_2	1.147×10^{-6}	3.278×10^{-6}

[a]These ionization frequencies are for ions that are the same as the parent neutrals. Units are s^{-1}. Calculated using EUVAC.[7]
[b]Corresponds to solar flux F74133.
[c]Corresponds to solar flux F79050N.

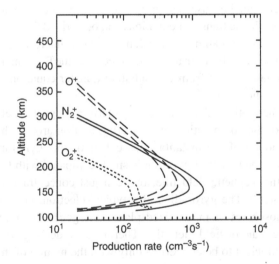

Figure 9.4 Calculated photoionization and total (photon plus photoelectron) production rates for the terrestrial upper atmosphere, corresponding to solar minimum conditions and a solar zenith angle of 65°. In each pair of curves the one with the smaller ionization rate is for photoionization only.[11]

frequencies are useful because the production rate of a given ion species, at altitudes well above the peak production rate, can be calculated simply by multiplying the appropriate neutral density with the corresponding ionization frequency.

Representative examples of calculated photoionization rates for Earth and Venus are shown in Figures 9.4 and 9.5, respectively. The rate of other analogous processes (e.g., molecular dissociation) can also be calculated, using Equation (9.25), if the ionization cross sections are replaced by the appropriate cross sections.

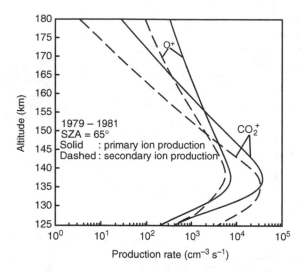

Figure 9.5 Calculated photon and photoelectron (denoted as secondary) ionization rates for the upper atmosphere of Venus,[12] corresponding to solar maximum conditions and a solar zenith angle of 65°.

9.4   Superthermal electron transport

The transport calculations for electrons in the atmosphere are significantly more difficult than those for EUV radiation because scattering and local sources play an important role. Electron and ion kinetic transport equations can be derived from the Boltzmann equation (3.7), and numerous authors have done so.[13] It is interesting to point out that the radiative transport equation can also be derived from the Boltzmann equation,[14] and some authors have started from the radiative transfer equation to obtain the electron transport equations.[15]

In a collisionless plasma, the motion of a charged particle in a magnetic field can be considered to consist of the combination of a gyrating motion around the magnetic field line and the motion of the instantaneous center of this gyration, called the *guiding center*. When the radius of gyration is small compared with the characteristic dimensions of the magnetic field line, one can just concentrate on the motion of the guiding center.[16] The gyroradius of a typical electron, created by photoionization in the ionospheres of magnetized planets, is generally small compared with an ionospheric scale or field "length." Therefore, in dealing with photoelectron transport, it is sufficient to be concerned only with the motion of the guiding centers. With this approach, one expresses the distribution function in terms of the guiding center parameters and then averages over a gyration period. If one further neglects drift motions perpendicular to the magnetic field and gravitational acceleration and assumes that any externally imposed electric field is parallel to the magnetic field line, the following equation is obtained:

$$\frac{\partial f}{\partial t} + \mu v \frac{\partial f}{\partial r} + \frac{e}{m}\varepsilon_\parallel \mu \frac{\partial f}{\partial v} + \left(\frac{e}{m}\varepsilon_\parallel - \frac{v^2}{2B}\frac{dB}{dr}\right)\frac{1-\mu^2}{v}\frac{\partial f}{\partial \mu} = \frac{\delta f}{\delta t}, \quad (9.27)$$

where $\mu = \cos \alpha$, $\alpha =$ pitch angle, $r =$ distance along the magnetic field line, $\varepsilon_\parallel =$ externally imposed parallel electric field, e is the electron charge, v is the magnitude of the velocity, and all other symbols have been defined earlier. [Note that in the rest of the book, the electric field is denoted by E, consistent with conventional notation. However, E is also the commonly used symbol for energy. Therefore, to distinguish between energy and electric field, ε is used for the latter in Equations (9.27–29).] It is often convenient to write Equation (9.27) in terms of the flux of particles and change from the variable v to the kinetic energy, E. Most of the direct measurements of these fluxes are in terms of flux versus energy. Assuming that the particle velocity changes slowly along the field lines, this modified form of the transport equation is

$$\sqrt{\frac{m}{2E}}\frac{\partial \Phi}{\partial t} + \mu \frac{\partial \Phi}{\partial r} + e\varepsilon_\parallel E \mu \frac{\partial}{\partial E}\left(\frac{\Phi}{E}\right)$$

$$+ \left(\frac{e\varepsilon_\parallel}{E} - \frac{1}{B}\frac{dB}{dr}\right)\frac{1-\mu^2}{2}\frac{\partial \Phi}{\partial \mu} = \sqrt{\frac{m}{2E}}\frac{\delta \Phi}{\delta t}, \qquad (9.28)$$

where Φ is the flux

$$\Phi \equiv \frac{2E}{m^2}f \qquad (9.29)$$

and E is in eV. This definition of Φ leads to flux units such as $\text{cm}^{-2}\,\text{s}^{-1}\,\text{eV}^{-1}\,\text{ster}^{-1}$, which are the normal units in spherical coordinates. Electron transport equations of this form have been solved for the case of the terrestrial photoelectron fluxes.[17] These types of relatively detailed, time-dependent calculations are only necessary for special circumstances, such as the refilling of empty plasmaspheric field tubes. A much simpler, steady state formulation of this equation has been used for the calculation of ionospheric photoelectron fluxes.

For most ionospheric applications, except possibly right at sunrise or sunset, steady-state conditions can be assumed. Furthermore, it is also appropriate to neglect the presence of external electric fields and the divergence of the magnetic field. In that case, Equation (9.28) simplifies to the following:

$$\mu \frac{\partial \Phi}{\partial r} = \sqrt{\frac{m}{2E}}\frac{\delta \Phi}{\delta t}. \qquad (9.30)$$

In solving this equation, the photoelectron flux is typically divided into a number of equal angular components or streams.[18, 19] It has been demonstrated, using Monte Carlo calculations, that given all the uncertainties associated with the differential scattering cross sections, it is generally sufficient to consider only two streams in the ionosphere.[20] The two-stream equations can be written as follows:

$$\langle \mu \rangle \frac{\partial \Phi^+}{\partial r} = -n_s\sigma_s^t \Phi^+ + \frac{n_s\sigma_s^e}{2}(\Phi^+ + \Phi^-) + \frac{Q_0}{2}, \qquad (9.31)$$

$$-\langle\mu\rangle\frac{\partial\Phi^-}{\partial r} = -n_s\sigma_s^t\Phi^- + \frac{n_s\sigma_s^e}{2}(\Phi^+ + \Phi^-) + \frac{Q_0}{2} \tag{9.32}$$

where n_s is number density of the scattering background species, and σ_s^t and σ_s^e are the total and elastic scattering cross sections for species s, respectively. Also, the following definitions for the upward and downward fluxes have been introduced:

$$\Phi^+(r) \equiv \int_0^{2\pi} d\phi \int_0^1 d\mu \, \Phi(\phi,\mu,r), \tag{9.33}$$

$$\Phi^-(r) \equiv \int_0^{2\pi} d\phi \int_{-1}^0 d\mu \, \Phi(\phi,\mu,r), \tag{9.34}$$

$$Q_0(r) = \int_0^{2\pi} d\phi \int_{-1}^1 d\mu \, P_e(\phi,\mu,r). \tag{9.35}$$

Furthermore, in arriving at Equations (9.31) and (9.32) it was assumed that the electron production rate, P_e, is isotropic, that the average of the cosine of the pitch angle, $\langle\mu\rangle$, is altitude independent, and that the elastic forward and backward scattering probabilities are equal to 1/2. Equations (9.31) and (9.32) are written, for the sake of simplicity, assuming the presence of only one scattering or absorbing species. However, for a multi-constituent atmosphere one only needs to sum over the various species, s, to arrive at the appropriate equations. These equations give the flux at one energy. Energy-dependent calculations can be carried out by assuming that the flux is zero above some energy, E_{ub}, and then solving the equations for monotonically decreasing energies taking into account the particles that cascade from higher energies by adding an effective production term, Q_{casc}, to the right-hand side of Equations (9.31) and (9.32). This Q_{casc} corresponds to all particles that cascade to energy, E, from energies between E and E_{ub}. The above discussions of electron transport assumed the presence of a magnetic field and used the guiding center approximation. However, these equations can and have been used to calculate electron transport in the ionospheres of nonmagnetic planets. In general these were done for vertical, one-dimensional calculations.

Figure 9.6 shows a comparison of measured photoelectron fluxes with those calculated by the two-stream method. The noticeable peaks in the 20 to 30 eV energy range are due to the photoionization of the neutral species into various excited states by the very strong HeII 304 Å solar line. The steep drop above about 60 eV is caused by the corresponding decrease in the relevant solar flux. The increase in photoelectron flux at the low energies is the result of both the increase in the solar flux and electrons cascading downward in energy via inelastic collisions.

At low altitudes, where collisions are sufficiently frequent, photoelectrons are created and lost essentially at the same location, therefore transport is negligible. In

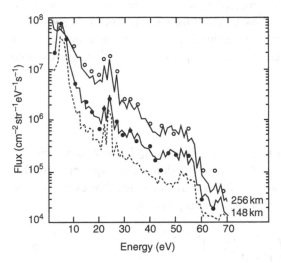

Figure 9.6 Calculated and measured photoelectron fluxes at two different altitudes in the terrestrial upper atmosphere, corresponding to solar minimum conditions. The solid curves are the values calculated with the standard model and the open and filled circles are the *Atmosphere Explorer-E* measurements. The dashed curve shows calculated results obtained by changing the EUV flux and cross section models.[11]

that case, so-called local calculations are appropriate, in which one simply equates the source and loss terms. In this high collision region, it is reasonable to assume that the distribution function, $f(E)$, is isotropic, and it is written in terms of the kinetic energy. The rate at which particles are lost, L, from an energy increment dE can be written as

$$L = \sum_s \sum_l n_s \sigma_{sl}(E) v(E) f(E) \, dE, \tag{9.36}$$

where σ_{sl} is the inelastic collision cross section, the subscript s denotes the different atmospheric species, the subscript l denotes inelastic loss processes and $f(E)$ is the un-normalized particle distribution function. The rate at which new particles are created within this energy increment dE is denoted simply as $Q(E) \, dE$. The rate at which particles are scattered into this energy increment dE from higher energies can be expressed as

$$S = \sum_s \sum_l n_s \sigma_{sl}(E + \Delta E_{sl}) v(E + \Delta E_{sl}) f(E + \Delta E_{sl}) \, dE, \tag{9.37}$$

where ΔE_{sl} is the energy loss suffered by the particle colliding with species s in an inelastic collision l. Equating the source and loss terms and then solving for $f(E)$ gives

$$f(E) = \frac{Q(E) + \sum_s \sum_l n_s \sigma_{sl}(E + \Delta E_{sl}) v(E + \Delta E_{sl}) f(E + \Delta E_{sl})}{\sum_s \sum_l n_s \sigma_{sl}(E) v(E)}. \tag{9.38}$$

Here again note that the expression for $f(E)$ contains $f(E + \Delta E_{sl})$. Thus, this equation is normally solved by assuming that $f(E)$ is zero above some upper boundary value, $E > E_{ub}$, and then "work down" in energy. Note that both in this local approximation and in the two- or multiple-stream approach the energy loss processes are considered to be discrete. This is an appropriate assumption for all interactions except for the electron–electron one, which is basically a continuous loss process. For the latter interaction, an effective collision cross section, σ_{eff}, has been used[18] to approximate this continuous loss process

$$\sigma_{\text{eff}} = \frac{1}{\Delta E} \frac{1}{n_e} \frac{dE}{dz},$$ (9.39)

where n_e is the thermal electron density and $(dE/dz)/n_e$ is the *loss function* or *stopping cross section* for electron–electron interactions. The expression for this electron–electron stopping cross section, Σ_e, is rather complex; however, a simple and quite good approximation was obtained that is given by[21]

$$\Sigma_e(E) = \frac{3.37 \times 10^{-12}}{E^{0.94} n_e^{0.03}} \left(\frac{E - E_e}{E - 0.53 E_e} \right)^{2.36},$$ (9.40)

where $E_e = 8.618 \times 10^5 T_e$ and T_e is the thermal electron temperature.

An analogous expression to Equation (9.38) can be obtained from Equations (9.31) and (9.32) by neglecting the $\partial \Phi / \partial r$ transport terms. The expression thus obtained for the upward flux, under no transport (local) conditions is

$$\Phi^+(E) = \frac{Q_0(E)/2 + Q_{\text{casc}}^+(E)}{\sum_s \sum_l n_s \sigma_{sl}},$$ (9.41)

where Q_{casc}^+ corresponds to the rate at which particles cascade down to energy, E, from higher energies and have upward-directed velocities; the expression for Φ^- has the same form.

At this point the so-called continuous loss approximation should be mentioned; this can also be used when transport is negligible. If ΔE_{sl} is small, $f(E)$ is a relatively smooth function, and one can use a Taylor series expansion about E in Equation (9.38) and arrive at the following expression:

$$\sum_s \sum_l n_s \sigma_{sl}(E) v(E) f(E)\, dE = Q(E)\, dE + \sum_s \sum_l n_s \sigma_{sl}(E) v(E) f(E)\, dE$$

$$+ \sum_s \sum_l n_s \left\{ \frac{\partial}{\partial E} \left[\sigma_{sl}(E) v(E) f(E) \right]_E \Delta E_{sl} \right\} dE.$$ (9.42)

Canceling terms on both sides and integrating from E to ∞, one gets

$$f(E) = \frac{\displaystyle\int_E^\infty Q(E)\, dE}{\displaystyle\sum_s \sum_l n_s \sigma_{sl}(E) v(E) \Delta E_{sl}} = \frac{\displaystyle\int_E^\infty Q(E)\, dE}{dE/dt},$$ (9.43)

where dE/dt is the energy loss rate (note that $n\sigma v$ is the collision frequency). Equation (9.43) is intuitively very clear: the number of particles at a given energy, E, is directly proportional to the production rate at all energies above E and inversely proportional to the loss rate at E (consider the analogy with flow in a pipeline).

Particle impact ionization rates can be calculated in a way analogous to photoionization. That is, once the particle flux, $\Phi_p(z,E)$, is determined as a function of altitude and energy the ionization rate of ion species, s, in a given state, l, with energy E_{sl}, is:

$$P_{sl}(z,E_{sl}) = n_s(z) \int_{E_{sl}}^{\infty} \Phi_p(z,E)\sigma_s^i(E)p_s(E,E_{sl})\, dE \qquad (9.44)$$

where the ionization cross sections and branching ratios refer to the relevant impact processes. If the total ionization rate of a given ion is needed, it is obtained by summing over all l, and if the total ionization rate over all species is desired one sums over all l and s. Figure 9.7 shows calculated electron–ion pair production rates for monoenergetic electron fluxes precipitating into the terrestrial atmosphere. As expected, the higher the energy of the electron flux the deeper into the atmosphere it penetrates. Also the column integral of the ionization rate increases with increasing electron energy. It is well established that on average it takes about 35 eV to produce an electron–ion pair. (This value does depend on atmospheric species and electron energy, but it is a good first approximation for most atmospheric species and electron energies above about 100 eV.) This value can be used to get estimates of column ionization rates and is also useful as a first-order check in complex calculations.

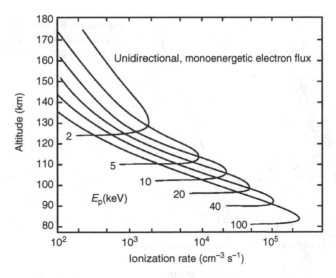

Figure 9.7 Calculated electron–ion pair production rates for monoenergetic electron fluxes of 10^8 electrons cm^{-2} s^{-1} precipitating into the terrestrial atmosphere.[22]

9.5   Superthermal ion and neutral particle transport

The transport of superthermal ions and neutral gas particles is somewhat more complicated than that of electrons because additional processes, such as charge exchange and ionization, are involved. These processes require the solution of simultaneous coupled transport equations for both the ions and the neutrals. For example, protons can capture an electron from an atmospheric neutral species, M (Section 8.3);

$$H^+ + M \rightarrow H + M^+. \tag{9.45}$$

The reverse process can turn a neutral hydrogen into a proton;

$$H + M^+ \rightarrow H^+ + M, \tag{9.46}$$

or a neutral hydrogen can also become a proton via *ionization-stripping*;

$$H + M \rightarrow H^+ + M + e^-. \tag{9.47}$$

A further complication is that while the ion motion is confined to a helical path along the field line, the neutral particles move in a straight line in the direction of the velocity they acquired at their creation (neglecting gravity). This means that an initially narrow precipitating beam can spread out significantly as it penetrates the atmosphere. Figure 9.8 shows a sketch of this phenomenon. The extent of this spreading is determined by the fraction of time that the incident particle spends in its neutral versus charged state.

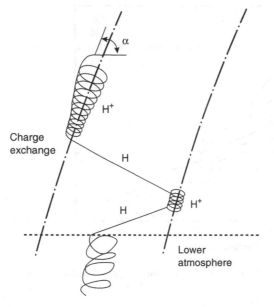

Figure 9.8 Representative path of a charged ion entering a magnetized atmosphere.[23]

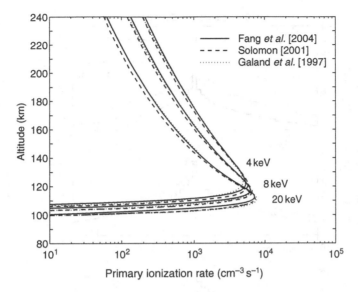

Figure 9.9 Comparison of calculated primary ionization rates versus altitude obtained by different methods. The Fang et al.[27] results are shown by solid lines and those of Solomon[26] by dashed lines; the results of Galand et al.[25] fall right on top of the Fang et al. results.

Self-consistent calculations of ion–neutral precipitation have been carried out using a variety of approaches. Among these model calculations during the last couple of decades are relatively simple one-dimensional studies employing the two-stream approach,[24] the relevant multi-stream transport equations,[25] or Monte Carlo methods.[26] However, very recently multi-dimensional studies have also been published.[27] These latter studies used the so-called direct simulation Monte Carlo (DSMC) method.[28] Figure 9.9 shows the results of three different one-dimensional calculations of the primary ionization rates for incident 1 erg cm^{-2} sec^{-1} Maxwellian proton fluxes with characteristic energy of $E_0 = 4, 8$, and 20 keV. The agreement among these results is very good and the calculations show that, just as was the case for electrons, and as expected, the flux penetrates deeper into the atmosphere as the incident energy increases. The importance of beam spreading has been established by the 3D, Monte Carlo calculations.[27] Figure 9.10a shows how a 1 erg cm^{-2} s^{-1}, monoenergetic, 10 keV proton beam injected at the top of the atmosphere spreads for an assumed vertical magnetic field; Figures 9.10b and 9.10c indicate the spreading of the associated neutral hydrogen beam and ionization rate. It should be noted that ion or neutral precipitation in general has a spatial spread. Studies, using the Monte Carlo method, of such extended precipitation events have now been published.[29]

The charge exchange process, Equation (9.45), creates energetic neutral atoms (ENAs) that, neglecting gravity, move in a straight line. These ENAs provide an opportunity to "image" the energetic ion population. This measurement approach

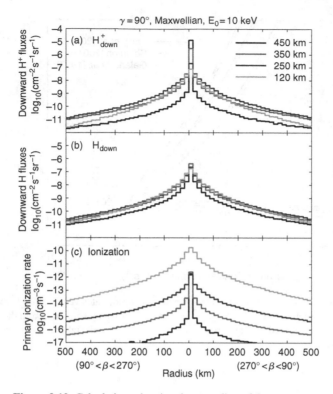

Figure 9.10 Calculations showing the spreading of the protons, neutral hydrogen, and ionization rates for a precipitating 10 keV proton beam in a vertical magnetic field at different altitudes (β is the azimuthal angle).[27]

has become an important tool for remotely studying three-dimensional plasma populations in our solar system.[30, 31]

9.6   Electron and ion heating rates

The energy absorbed in the thermosphere from either solar radiation or particle precipitation is partitioned among a number of different channels. The block diagram in Figure 9.1 shows the various major routes that the absorbed energy takes; the main processes are ionization, excitation, heating, and transport. Some of the excited species undergo spontaneous de-excitation, which leads to airglow or auroral emissions, as briefly mentioned in Section 8.7. A fraction of the absorbed energy goes to the neutrals, ions, and electrons as kinetic energy. The calculation of the associated heating rates is complex because one needs to understand, in detail, all the energy sharing processes. The total amount of incident solar energy absorbed in a unit volume per unit time, $Q_{\text{total}}(z)$, is simply equal to

$$Q_{\text{total}}(z) = \sum_s n_s(z) \int E(\lambda) I(z, \lambda) \sigma_s^a(\lambda) \, d\lambda, \tag{9.48}$$

where $E(\lambda)$ is the energy of the absorbed photon in eV. However, the calculation of the fractions going to the different processes is very difficult and the concept of a *heating efficiency* has been widely used. The heating efficiency for a given constituent is defined as the fraction of the absorbed energy that goes locally to heating that constituent.

The question that needs to be discussed in this section is what fraction of the absorbed energy goes to heating the electrons and the ions. In general, the solar energy first goes to the electrons, which in turn transfer some of that energy, via Coulomb collisions, to the ion gas. For this reason, our discussion concentrates on the energy input calculations to the electron gas. The electron energy distribution function typically has been, somewhat arbitrarily, divided into thermal and non-thermal (superthermal) components. The electron–electron collision cross section is inversely proportional to energy, and therefore, at low energies the large number of collisions among the electrons result in a Maxwell–Boltzmann distribution. Consequently, these low-energy electrons can be characterized by a temperature. At higher energies, the electron–electron collisions are less frequent and inelastic collisions with the neutral background species become more important. Therefore, the distribution function becomes highly nonthermal and controlled mostly by the source processes and inelastic collisions. A distribution function measured at high latitude in the terrestrial ionosphere[32] clearly demonstrates this behavior, as shown in Figure 9.11.

The transport equations discussed in Chapter 3 are derived in terms of the total particle population. Therefore, the term $\delta E/\delta t$ in the energy equation (3.38) refers to the energy gained by the whole population. One could try to work with two

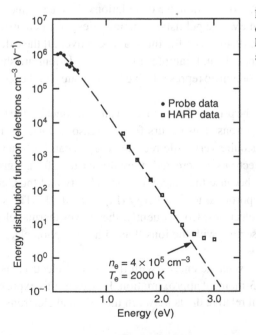

Figure 9.11 Electron energy distribution observed in the high-latitude terrestrial upper atmosphere.[32]

sets of fluid equations, one for the thermal and one for the superthermal electrons, but this would be very difficult to do for the latter, given its highly nonthermal character. So, it has been the accepted approach to use the fluid equations for the thermal electrons, which involve the bulk of the population, and then calculate the corresponding heating and cooling rates taking into account all elastic and inelastic collisional processes that the thermal electrons undergo. These heating, Q_e, and cooling, L_e, rates are discussed separately in this and the following section.

The transition energy, E_T, between the thermal and nonthermal population has, in general, been taken to be the energy where the distribution deviates detectably from a Maxwell–Boltzmann one. It has been shown that the thermal electron heating rate consists of three terms: one due to collisions between the thermal and superthermal electrons, one due to newly created electrons with energy less than E_T, and a term evaluated on the energy surface at E_T.[33] It has been the general practice to consider only the first contribution to the heating rate. In that case the electron heating rate, $Q_e(z)$, can be calculated from the following relation:

$$Q_e(z) = \int_{E_T}^{\infty} \Phi_e(z, E) \left(\frac{dE}{dz} \right)_e dE, \tag{9.49}$$

where Φ_e is the electron flux, $(dE/dz)_e = \Sigma_e$, as given by Equation (9.40), and is the rate at which an electron of energy E loses energy to the ambient thermal electrons in traveling a unit distance. It was shown that the term given by Equation (9.49) is the dominant term[33] (within a factor of two), and given the uncertainties associated with these calculations, no significant new effort has gone into improving them.

Most of the published photoelectron heating calculations were based on multi-stream, generally two-stream, models. In these calculations the energy increments are discrete, of the order of an eV. The published heating rates are calculated in the manner indicated by Equation (9.49), with the integral taken over all the calculated fluxes, except for the lowest increment. Examples of such calculations are shown in Figures 9.12 and 9.13, corresponding to representative heating rates for the terrestrial and Venus ionospheres.

With regard to ion heating, the primary heat source in an ionosphere is the thermal electrons and not the photoelectrons. This occurs for two reasons. First, during the ionization process, the ions acquire very little recoil energy because of their large mass. Also, after the photoelectrons are created, they do not transfer a significant amount of energy to the ions because they have a large velocity and the Coulomb cross section is inversely proportional to the energy (Equation 4.51). The same is true for precipitating auroral electrons. Consequently, the slower thermal electrons have a larger collision cross section with the ions than either the photoelectrons or the auroral electrons.

The rate at which the ions exchange energy with the thermal electrons is given by Equation (4.129c) in the 13-moment approximation. However, this expression is valid only in the limit of small relative drifts between the ions and electrons. When

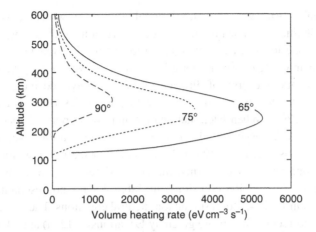

Figure 9.12 Calculated electron heating rates for the terrestrial ionosphere for the three indicated solar zenith angles.[34]

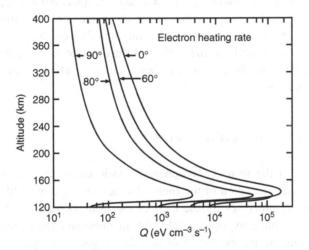

Figure 9.13 Calculated electron heating rates for the Venus ionosphere. The heating rates were calculated assuming that the photoelectron transport is not inhibited by an induced magnetic field.[35]

the relative ion–electron drift is large, Equation (4.129c) should be replaced with the five-moment nonlinear collision term (4.124c), which is a better approximation. Note that the two collision terms agree in the limit of small relative ion–electron drifts. In both expressions, the appropriate collision frequency is given by Equation (4.140). The velocity-dependent correction factors, Φ_{st} and Ψ_{st}, can be set equal to unity for ion–electron collisions because ε_{st} is small owing to the large electron thermal speed (Equation 4.120).

In addition to the heat gained from the thermal electrons, the ions can be heated as a result of exothermic chemical reactions and frictional interactions with other species. The heating from exothermic chemical reactions is typically small and the heating rates are generally not well known. Frictional interactions with other species can result in very significant ion heating rates in certain regions of the ionospheres. For example, when one ion species drifts relative to another ion species, such as in the terrestrial polar wind, or when ions drift through slower moving neutrals, frictional heating occurs as energy of directed motion is converted into random thermal energy. Note that this type of frictional heating is not described by the linear collision term (4.129c). When frictional heating may be important, the linear collision term should be replaced with the nonlinear collision term (4.124c). Likewise, the nonlinear collision term (4.124b) should then be used in the momentum equation. For the nonlinear collision terms, the appropriate collision frequencies are given in Table 4.3 for ion–ion collisions, in Table 4.4 for nonresonant ion–neutral collisions and in Table 4.5 for resonant ion–neutral collisions. The associated velocity-dependent correction factors are given by Equations (4.125a) and (4.125b) for ion–ion collisions, and they are equal to one for nonresonant ion–neutral collisions (Equations 4.127a,b). For resonant ion–neutral interactions, the hard sphere expressions for the correction factors Φ_{st} (4.126a) and Ψ_{st} (4.126b) are approximately valid. However, under most circumstances, the values obtained from the hard sphere correction factors are close enough to unity that they can be set equal to one with little error. This is especially true in view of the fact that there is generally a large uncertainty associated with the resonant charge exchange cross sections and, hence, collision frequencies.

9.7   Electron and ion cooling rates

In the lower altitudes of the various ionospheres, elastic collisions, along with rotational and vibrational excitation of the molecular neutrals, are most likely to be the dominant cooling processes for the thermal electron population. The fine structure excitation of atomic oxygen can also be an important mechanism. At high electron temperatures, the excitation of atomic oxygen to its lowest electronic state, ^1D (Figure 8.1), may also need to be considered. At high altitudes, where the ionospheric plasma approaches a fully ionized condition, Coulomb collisions with the ambient ions become the dominant energy loss mechanism for the electrons.

The calculation of thermal electron cooling rates for inelastic collisional processes requires a knowledge of the excitation cross sections. These cross sections are either calculated or measured, as a function of electron energy. Then, average (i.e., temperature-dependent) cooling rates for the thermal electrons are obtained by integrating the energy-dependent excitation cross sections over Maxwellian electron and neutral velocity distributions. In some cases, the cooling rates are fitted with convenient analytic expressions. Unfortunately, many of the inelastic electron cooling rates were calculated, not measured, and the calculated rates were generally based on important simplifying assumptions. Also, many of the cooling rates that are in use today are more than 30 years old. Nevertheless, electron cooling rates are required for ionospheric energy balance calculations, and therefore, the cooling rates that are currently available are given in what follows.

Letting $L_e(X)$ represent the cooling rate in eV cm^{-3}s^{-1} due to an inelastic collision with neutral species X (densities in cm^{-3} and temperatures in K), these cooling rates are given by the following expressions:

N_2 rotation[36]

$$L_e(N_2) = 3.5 \times 10^{-14} n_e n(N_2)(T_e - T_n)/T_e^{1/2}. \tag{9.50}$$

O_2 rotation[37]

$$L_e(O_2) = 5.2 \times 10^{-15} n_e n(O_2)(T_e - T_n)/T_e^{1/2}. \tag{9.51}$$

H_2 rotation[38]

The following expression was fit for a neutral temperature of 1000 K, appropriate for the outer planetary ionospheres; for significantly different neutral temperatures one needs to use the complex expressions given in the original reference.[38]

$$L_e(H_2) = 2.278 \times 10^{-11} n_e n(H_2)\{\exp[2.093 \times 10^{-4}(T_e - T_n)^{1.078} - 1]\}. \tag{9.52}$$

CO_2 rotation[39]

$$L_e(CO_2) = 5.8 \times 10^{-14} n_e n(CO_2)(T_e - T_n)/T_e^{1/2}. \tag{9.53}$$

CO rotation[40]

$$L_e(CO) = -\langle Q \rangle \exp\left\{\left[J_{max}\left(1 - \frac{2T_n}{T_e}\right) - \frac{T_n}{T_e} + 1\right]\frac{B_0}{kT_n}\right\}$$

$$\cdot \left(\frac{kT_n}{B_0}\right)^{15/8}\left[\Gamma\left(\frac{15}{8}\right) + \frac{1}{2}\left(\frac{kT_n}{B_0}\right)^{1/2}\Gamma\left(\frac{19}{8}\right)a\right.$$

$$\left. + \frac{1}{6}\frac{kT_n}{B_0}\Gamma\left(\frac{23}{8}\right)a^2\right]a, \tag{9.54}$$

where

$$J_{max} = (1.5kT_n/B_0)^{1/2}, \tag{9.55a}$$

$$a = \frac{-2B_0(T_e - T_n)}{kT_e T_n}, \tag{9.55b}$$

$$\langle Q \rangle = 2.36 \times 10^7 n_e n(CO)\frac{B_0^{7/4}2^{9/4}}{(kT_e)^{1/4}}R\sigma_r\Gamma\left(\frac{5}{4}\right), \tag{9.55c}$$

and where $\sigma_r = 4.90 \times 10^{-19}$ cm^2, R is the Rydberg energy, $B_0 = 2.4 \times 10^{-4}$ eV is the CO rotational constant, and $\Gamma(Y)$ is the gamma function of quantity Y.

H$_2$O rotation[41]

$$L_e(H_2O) = n_e n(H_2O)\left[a + b \ln(T_e/T_n)\right]\left[(T_e - T_n)/T_e^{5/4}\right], \tag{9.56}$$

where

$$a = 1.052 \times 10^{-8} + 6.043 \times 10^{-10} \ln(T_n), \tag{9.57a}$$

$$b = 4.18 \times 10^{-9} + 2.026 \times 10^{-10} \ln(T_n). \tag{9.57b}$$

CH$_4$ rotation[42]

No analytic expression has yet been presented for this loss function. Calculated values for different T_n, as a function of T_e, are presented in Figure 9.14.

N$_2$ vibration[36]

$$L_e(N_2) = n_e n(N_2)\left\{1 - \exp(-E_1/T_{vib})\right\}$$

$$\times \sum_{v=1}^{10} Q_{0v}\left\{1 - \exp\left[v E_1(T_e^{-1} - T_{vib}^{-1})\right]\right\}$$

$$+ n_e n(N_2)\left\{1 - \exp(-E_1/T_{vib})\right\} \exp(-E_1/T_{vib})$$

$$\sum_{v=2}^{9} Q_{1v}\left\{1 - \exp[(v-1)E_1(T_e^{-1} - T_{vib}^{-1})]\right\}, \tag{9.58}$$

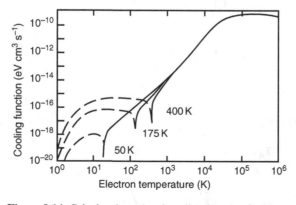

Figure 9.14 Calculated rotational cooling function for Maxwellian electrons in CH$_4$ versus electron temperature for the rotational temperatures shown.[42]

Table 9.3 Coefficients for calculations of Q_{0v} for $1500 \leq T_e \leq 6000$ K.[36]

v	A_{0v}	B_{0v}, K^{-1}	C_{0v}, K^{-2}	D_{0v}, K^{-3}	F_{0v}, K^{-4}	δ_{0v}
1	2.025	$8.782 \cdot 10^{-4}$	$2.954 \cdot 10^{-7}$	$-9.562 \cdot 10^{-11}$	$7.252 \cdot 10^{-15}$	0.06
2	-7.066	$1.001 \cdot 10^{-2}$	$-3.066 \cdot 10^{-6}$	$4.436 \cdot 10^{-10}$	$-2.449 \cdot 10^{-14}$	0.08
3	-8.211	$1.092 \cdot 10^{-2}$	$-3.369 \cdot 10^{-6}$	$4.891 \cdot 10^{-10}$	$-2.706 \cdot 10^{-14}$	0.10
4	-9.713	$1.204 \cdot 10^{-2}$	$-3.732 \cdot 10^{-6}$	$5.431 \cdot 10^{-10}$	$-3.008 \cdot 10^{-14}$	0.10
5	-10.353	$1.243 \cdot 10^{-2}$	$-3.850 \cdot 10^{-6}$	$5.600 \cdot 10^{-10}$	$-3.100 \cdot 10^{-14}$	0.13
6	-10.819	$1.244 \cdot 10^{-2}$	$-3.771 \cdot 10^{-6}$	$5.385 \cdot 10^{-10}$	$-2.936 \cdot 10^{-14}$	0.15
7	-10.183	$1.185 \cdot 10^{-2}$	$-3.570 \cdot 10^{-6}$	$5.086 \cdot 10^{-10}$	$-2.769 \cdot 10^{-14}$	0.15
8	-12.698	$1.309 \cdot 10^{-2}$	$-3.952 \cdot 10^{-6}$	$5.636 \cdot 10^{-10}$	$-3.071 \cdot 10^{-14}$	0.15
9	-14.710	$1.409 \cdot 10^{-2}$	$-4.249 \cdot 10^{-6}$	$6.058 \cdot 10^{-10}$	$-3.300 \cdot 10^{-14}$	0.15
10	-17.538	$1.600 \cdot 10^{-2}$	$-4.916 \cdot 10^{-6}$	$7.128 \cdot 10^{-10}$	$-3.941 \cdot 10^{-14}$	0.15

Table 9.4 Coefficients for calculations of Q_{0v} for $300 \leq T_e \leq 1500$ K.[36]

v	A_{0v}	B_{0v}, K^{-1}	C_{0v}, K^{-2}	D_{0v}, K^{-3}	F_{0v}, K^{-4}	δ_{0v}
1	-6.462	$3.151 \cdot 10^{-2}$	$-4.075 \cdot 10^{-5}$	$2.439 \cdot 10^{-8}$	$-5.479 \cdot 10^{-12}$	0.14

Table 9.5 Coefficients for calculations of Q_{1v} for $1500 \leq T_e \leq 6000$ K.[36]

v	A_{1v}	B_{1v}, K^{-1}	C_{1v}, K^{-2}	D_{1v}, K^{-3}	F_{1v}, K^{-4}	δ_{1v}
2	-3.413	$7.326 \cdot 10^{-3}$	$-2.200 \cdot 10^{-6}$	$3.128 \cdot 10^{-10}$	$-1.702 \cdot 10^{-14}$	0.11
3	-4.160	$7.803 \cdot 10^{-3}$	$-2.352 \cdot 10^{-6}$	$3.352 \cdot 10^{-10}$	$-1.828 \cdot 10^{-14}$	0.11
4	-5.193	$8.360 \cdot 10^{-3}$	$-2.526 \cdot 10^{-6}$	$3.606 \cdot 10^{-10}$	$-1.968 \cdot 10^{-14}$	0.12
5	-5.939	$8.807 \cdot 10^{-3}$	$-2.669 \cdot 10^{-6}$	$3.806 \cdot 10^{-10}$	$-2.073 \cdot 10^{-14}$	0.08
6	-8.261	$1.010 \cdot 10^{-2}$	$-3.039 \cdot 10^{-6}$	$4.318 \cdot 10^{-10}$	$-2.347 \cdot 10^{-14}$	0.10
7	-8.185	$1.010 \cdot 10^{-2}$	$-3.039 \cdot 10^{-6}$	$4.318 \cdot 10^{-10}$	$-2.347 \cdot 10^{-14}$	0.12
8	-10.823	$1.199 \cdot 10^{-2}$	$-3.620 \cdot 10^{-6}$	$5.159 \cdot 10^{-10}$	$-2.810 \cdot 10^{-14}$	0.09
9	-11.273	$1.283 \cdot 10^{-2}$	$-3.879 \cdot 10^{-6}$	$5.534 \cdot 10^{-10}$	$-3.016 \cdot 10^{-14}$	0.09

where $E_1 = 3353$ K $(0.2889$ eV$)$, $T_{\text{vib}} = T_n$ and

$$\log Q_{0v} = A_{0v} + B_{0v} T_e + C_{0v} T_e^2 + D_{0v} T_e^3 + F_{0v} T_e^4 - 16 , \tag{9.59a}$$

$$\log Q_{1v} = A_{1v} + B_{1v} T_e + C_{1v} T_e^2 + D_{1v} T_e^3 + F_{1v} T_e^4 - 16. \tag{9.59b}$$

The A, B, C, D, and F coefficients are given in Tables 9.3 to 9.5.

O$_2$ vibration[43]

$$L_e(O_2) = n_e n(O_2) Q(T_e) \left\{ 1 - \exp\left[2239(T_e^{-1} - T_n^{-1}) \right] \right\}, \tag{9.60}$$

where

$$\log_{10}[Q(T_e)] = -19.9171 + 0.0267 T_e - 3.9960 \times 10^{-5} T_e^2$$
$$+ 3.5187 \times 10^{-8} T_e^3 - 1.9228 \times 10^{-11} T_e^4$$
$$+ 6.6865 \times 10^{-15} T_e^5 - 1.4791 \times 10^{-18} T_e^6$$
$$+ 2.0127 \times 10^{-22} T_e^7 - 1.5346 \times 10^{-26} T_e^8$$
$$+ 5.0148 \times 10^{-31} T_e^9. \tag{9.61}$$

H_2 vibration[44]

$$L_e(H_2) = 1.17 \times 10^{-6} k n_e n(H_2) \exp\left[-5253.7/(T_e - T_n)\right], \tag{9.62a}$$
$$\text{for } (T_e - T_n) \leq 1870 \, \text{K}$$
$$L_e(H_2) = 1.00 \times 10^{-10} k n_e n(H_2)\left[-7663.1 + 4.4485(T_e - T_n)\right], \tag{9.62b}$$
$$\text{for } 1870 \, \text{K} < (T_e - T_n) \leq 2700 \, \text{K}.$$

A slightly more accurate, but much more complex expression is given in a more recent reference.[38]

CO_2 and CO vibration[40]

$$L_e(X) = -2.36 \times 10^7 n_e n(X) \left(\frac{2}{kT_e}\right)^{3/2} \sum_{j=1}^{M} w_j$$
$$\cdot \left\{\exp\left[-\frac{w_j(T_e - T_n)}{kT_e T_n}\right] - 1\right\} S_j(X), \tag{9.63}$$

where

$$S_j(X) = D_j + R_j, \tag{9.64a}$$

$$D_j = A_j w_j^2 e^{-C_j} \sum_{i=1}^{2} \left\{\frac{\Gamma(\delta_j + i)}{C_j^{\delta_j + i}} - \frac{\Gamma(\delta_j + i)}{\left[(1/\beta_j) + C_j\right]^{\delta_j + i}}\right.$$
$$+ \frac{\Gamma(\nu_j + i)}{\left[(1/\gamma_j) + C_j\right]^{\nu_j + i}} - \frac{1}{\beta_j} \cdot \frac{\Gamma(\delta_j + i + 1)}{\left[(1/\gamma_j) + C_j\right]^{\delta_j + i + 1}}\right\}, \tag{9.64b}$$

$$C_j = w_j/kT_e, \tag{9.64c}$$

$$R_j = \sum_{n=1}^{L} R_{jn}, \tag{9.64d}$$

Table 9.6 Vibrational cross section fitting parameters.[40]

Gas	Transition	W	A	δ	β	ν	γ
CO_2	010	0.083	3.07E-16	−6.72E-1	5.44E-1	3.19E-1	2.29E-1
CO_2	020 + 100	0.167	3.87E-17	−6.02E-1	1.08E3	4.98E-2	1.74E0
CO_2	001	0.291	3.92E-16	−7.75E-1	7.32E-1	3.25E-1	6.11E0
CO	$\nu' = 1$	0.266	4.32E-17	−6.65E-1	1.24E1	6.21E-1	7.19E-1

		Resonance terms			
Gas	Transition	n	σ_n	λ_n	$E_n{}^0$
CO_2	010	1	1.47E-16	1.08E0	3.78E0
CO_2	010	2	3.44E-16	2.25E0	7.23E0
CO_2	020 + 100	1	1.76E-16	8.51E-1	3.63E0
	020 + 100	2	2.95E-17	3.90E0	7.42E0
CO_2	001	1	2.41E-17	9.41E-1	7.42E0
CO	$\nu' = 1$	1	2.17E-16	8.57E-1	2.49E0

$$R_{jn} = \sigma_n \exp\left[\frac{1}{kT_e}\left(\frac{\lambda_n^2}{4kT_e} - E_n^0\right)\right]\left\{\frac{\lambda_n^2}{2}\exp(-\eta_t^2)\right.$$
$$\left. + \lambda_n\left(E_n^0 - \frac{\lambda_n^2}{2kT_e}\right)\pi^{1/2}[1 - \text{erf}(\eta_t)]/2\right\}, \tag{9.64e}$$

$$\eta_t = \frac{1}{\lambda_n}\left[w_j + \left(\frac{\lambda_n^2}{2kT_e} - E_n^0\right)\right], \tag{9.64f}$$

and where $n(X)$ is either the CO_2 or the CO number density, M is the number of vibrational modes considered for excitation from the ground vibrational state for species X, w_j is the threshold for process j, erf is the error function, L is the number of resonances, and the remaining quantities A_j, δ_j, β_j, ν_j, γ_j, σ_n, λ_n, and E_n^0 are given in Table 9.6.

H_2O vibration[41]
The analytic expression for the cooling function is extremely complex, but can be found in Reference [41]. This cooling function must be multiplied by $n_e n(H_2O)$ to obtain the cooling rate L_e. Calculated values for different T_n, as a function of T_e, are presented in Figure 9.15.

CH_4 vibration[42]
No analytic expression has yet been presented for this cooling function, which must be multiplied by $n_e n(CH_4)$ to obtain the cooling rate L_e. Calculated values for different T_n, as a function of T_e, are presented in Figure 9.16.

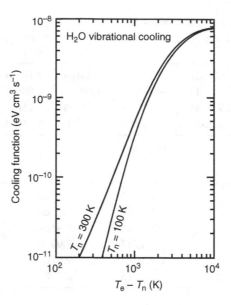

Figure 9.15 Calculated vibrational cooling function for H_2O versus the difference between the electron and neutral temperatures for the neutral temperatures shown.[41]

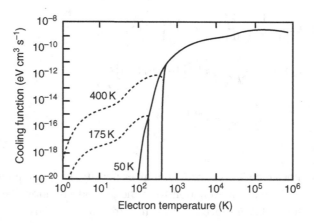

Figure 9.16 Calculated vibrational cooling function for Maxwellian electrons in CH_4 versus electron temperature for the neutral temperatures shown.[42]

O fine structure[45]

$$L_e(O) = n_e n(O) D^{-1} \Big(S_{10} \big\{ 1 - \exp[98.9(T_e^{-1} - T_n^{-1})] \big\}$$

$$+ S_{20} \big\{ 1 - \exp[326.6(T_e^{-1} - T_n^{-1})] \big\}$$

$$+ S_{21} \big\{ 1 - \exp[227.7(T_e^{-1} - T_n^{-1})] \big\} \Big), \tag{9.65}$$

where

$$D = 5 + \exp(-326.6\, T_n^{-1}) + 3\exp(-227.7\, T_n^{-1}), \tag{9.66a}$$

$$S_{21} = 1.863 \cdot 10^{-11}, \tag{9.66b}$$

$$S_{20} = 1.191 \cdot 10^{-11}, \tag{9.66c}$$

$$S_{10} = 8.249 \cdot 10^{-16}\, T_e^{0.6} \exp(-227.7\, T_n^{-1}). \tag{9.66d}$$

O(^{1}D) excitation[46]

$$L_e(O(^1D)) = 1.57 \times 10^{-12} n_e n(O) \exp\left(d\,\frac{T_e - 3000}{3000 T_e}\right)$$

$$\cdot \left[\exp\left(-22713\,\frac{T_e - T_n}{T_e T_n}\right) - 1\right], \tag{9.67}$$

where

$$d = 2.4 \times 10^4 + 0.3(T_e - 1500) - 1.947 \times 10^{-5}(T_e - 1500)(T_e - 4000). \tag{9.68}$$

With regard to elastic collisions, both the electron–ion and electron–neutral cooling rates are given by the five-moment energy exchange term (4.124c). For Coulomb collisions, the velocity-dependent correction factors, Φ_{st} and Ψ_{st} are given by Equations (4.125a) and (4.125b), respectively, and the associated Coulomb collision frequency is given by either Equation (4.140) or Equation (4.144). For elastic electron–neutral interactions, the velocity-dependent correction factors can be set equal to unity because ε_{st} is small owing to the large electron thermal speed (Equation 4.120). A number of the appropriate electron–neutral collision frequencies are given in Table 4.6.

The relative importance of the various electron cooling rates depends on the ionospheric and atmospheric conditions, and hence, on latitude, longitude, altitude, local time, season, solar cycle, and geomagnetic activity. Figure 9.17 shows a comparison of the electron cooling rates from at recent simulation of the northern, high-latitude, terrestrial ionosphere.[47] The simulation conditions were for winter (day 357), medium solar activity ($F_{10.7} = 150$), and high geomagnetic activity ($K_p = 6$). The figure shows altitude profiles of the cooling rates at 50° magnetic latitude, 130° magnetic longitude, and 5 UT. Also shown are the total cooling, L_e, and heating, Q_e, rates. The difference between L_e and Q_e is due to thermal conduction. Note that Figure 9.17 corresponds to a snapshot at one time and one location in an evolving ionosphere, and the comparison of the various cooling rates can be significantly different at other times and locations.

As far as the ions are concerned, the main cooling of the ion gases in the ionospheres results from collisions with the neutrals. This cooling is automatically included if the five-moment energy exchange term (4.124c) is used to describe

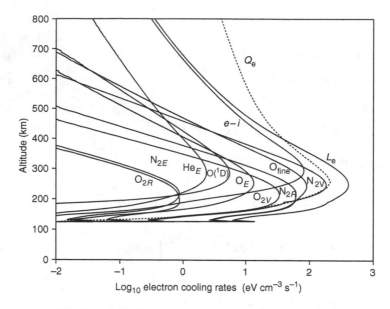

Figure 9.17 Electron cooling rates as a function of altitude in the northern terrestrial ionosphere.[47] L_e and Q_e denote the total electron cooling and heating rates, respectively. Subscripts R, V, and E represent the cooling rates associated with rotational, vibrational, and elastic collisions, respectively. The curves labeled $O(^1D)$, O_{fine}, and $e - i$ are the electron cooling rates associated with excitation of O to the ^1D state, excitation of the fine structure levels of O, and Coulomb collisions with ions, respectively. (Courtesy of M. David).

the ion–neutral interactions. The associated velocity-dependent correction factors and collision frequencies are the same as those discussed above with regard to ion heating rates.

9.8   Specific references

1. Smith, F. L., and C. Smith, Numerical evaluation of Chapman's grazing incidence integral Ch(χ, x), *J. Geophys. Res.*, **77**, 3592, 1972.
2. Baum, W. A., F. S. Johnson, J. J. Oberly, *et al.*, Solar ultraviolet spectrum to 88 kilometers, *Phys. Rev.*, **70**, 781, 1946.
3. Heroux, L., and J. E. Higgins, Summary of full disk solar fluxes between 250 and 1940 Å, *J. Geophys. Res.*, **82**, 3307, 1977.
4. Woods, T. N., and G. J. Rottman, Solar EUV irradiance derived from a sounding rocket experiment on 10 November 1988, *J. Geophys. Res.*, **95**, 6227, 1990.
5. Hinteregger, H. E., D. E. Bedo, and J. E. Manson, The EUV spectrophotometer on Atmosphere Explorer, *Radio Sci.*, **8**, 349, 1973.
6. Woods, T. N., F. G. Eparvier, S. M. Bailey, *et al.*, Solar EUV experiment (SEE): mission overview and first results, *J. Geophys. Res.*, **110**, A01312, doi:10.1029/2004JA010765, 2005.

7. Richards, P. G., J. A. Fennelly, and D. G. Torr, EUVAC: a solar EUV flux model for aeronomic calculations, *J. Geophys. Res.*, **99**, 8981, 1994.

8. Richards, P. G., T. N. Woods, and W. K. Peterson, HEUVAC: a new high resolution EUV proxy model, *Adv. Space Res.*, **37**, 315, 2006.

9. Lean, J. L., H. P. Warren, J. T. Mariska, and J. Bishop, A new model of solar EUV irradiance variability: 2. Comparison with empirical models and observations and implications for space weather, *J. Geophys. Res.*, **108 A(2)**, 1059, doi:10.1029/2001JA009238, 2003.

10. Tobiska, W. K., T. Woods, F. Eparvier *et al.*, The SOLAR2000 empirical solar irradiance model and forecast tool, *J. Atmos. Sol.-Terr. Phys.*, **62**, 1233, 2000.

11. Richards, P. G., and D. G. Torr, Ratios of photoelectron to EUV ionization rates for aeronomic studies, *J. Geophys. Res.*, **93**, 4060, 1988.

12. Kim, J., A. F. Nagy, T. E. Cravens, and A. J. Kliore, Solar cycle variations of the electron densities near the ionospheric peak of Venus, *J. Geophys. Res.*, **94**, 11997, 1989.

13. Khazanov, G. V., T. Neubert, and G. D. Gefan, A unified theory of ionosphere-plasmasphere transport of suprathermal electrons, *IEEE Trans. Plasma Sci.*, **22**, 187, 1994.

14. Oxenius, J., *Kinetic Theory of Particles and Photons*, Berlin: Springer Verlag, 1986.

15. Stolarski, R. S., Analytic approach to photoelectron transport, *J. Geophys. Res.*, **77**, 2862, 1972.

16. Northrop, T. G., *The Adiabatic Motion of Charged Particles*, New York: Wiley Interscience, 1963.

17. Khazanov, G. V., and M. W. Liemohn, Nonsteady state ionosphere-plasmasphere coupling of superthermal electrons, *J. Geophys. Res.*, **100**, 9669, 1995.

18. Nagy, A. F., and P. M. Banks, Photoelectron fluxes in the ionosphere, *J. Geophys. Res.*, **75**, 6260, 1970.

19. Strickland, D. J., D. L. Book, T. P. Coffey, and J. A. Fedder, Transport equation techniques for the deposition of auroral electrons, *J. Geophys. Res.*, **81**, 2755, 1976.

20. Solomon, S. C., Auroral electron transport using the Monte Carlo method, *Geophys. Res. Let.*, **20**, 185, 1993.

21. Swartz, W. E., J. S. Nisbet, and A. E. S. Green, Analytic expression for the energy transfer rate from photoelectrons to thermal electrons, *J. Geophys. Res.*, **76**, 8425, 1971.

22. Rees, M. H., *Physics and Chemistry of the Upper Atmosphere*, Cambridge, UK: Cambridge University Press, 1989.

23. Davidson, G. T., Expected spatial distribution of low-energy protons precipitated in the auroral zones, *J. Geophys. Res.*, **70**, 1061, 1965.

24. Kozyra, J. U., T. E. Cravens, and A. F. Nagy, Energetic O^+ precipitation, *J. Geophys. Res.*, **87**, 2481, 1982.

25. Galand, M., J. Lilensten, W. Kofman, and R. B. Sidje, Proton transport model in the ionosphere: 1. Multistream approach of the transport equations, *J. Geophys. Res.*, **102**, 22261, 1997.

26. Solomon, S. C., Auroral particle transport using Monte Carlo and hybrid methods, *J. Geophys. Res.*, **106**, 107, 2001.

27. Fang, X., M. W. Liemohn, and J. U. Kozyra, Quantification of the spreading effect of auroral proton precipitation, *J. Geophys. Res.*, **109**, A04309, doi:10.1029/2003JA010119, 2004.

28. Bird, G. A., *Molecular Gas Dynamics and the Direct Simulation of Gas Flows*, Oxford: Clarendon Press, 1994.

29. Fang, X., M. W. Liehmohn, J. U. Kozyra, and S. C. Solomon, Study of proton arc spreading effect on primary ionization rates, *J. Geophys. Res.*, **110**, A07302, doi:10.1029/2004JA010915, 2005.

30. Roelof, E. C. and A. J. Skinner, Extraction of ion distribuion from magnetospheric ENA and EUV images, *Space Sci. Rev.*, **91**, 437, 2000.

31. Paranicas, C., D. G. Mitchell, E. C. Roelof, *et al.*, Periodic intensity variation in global ENA images of Saturn, *Geophys. Res. Lett.*, **32**, L21101, doi:10.1029/2005GL023656, 2005.

32. Hays, P. B., and A. F. Nagy, Thermal electron energy distribution measurements in the ionosphere, *Planet. Space Sci.*, **21**, 1301, 1973.

33. Hoegy, W. R., Thermal electron heating rate: a derivation, *J. Geophys. Res.*, **89**, 977, 1984.

34. Rasmussen, C. E., J. J. Sojka, R. W. Schunk, V. B. Wickwar, and O. de la Beaujardiere, Comparison of simultaneous Chatanika and Millstone Hill temperature measurements with ionospheric model predictions, *J. Geophys. Res.*, **93**, 1922, 1988.

35. Cravens, T. E., T. I. Gombosi, J. Kozyra, *et al.* Model calculations of the dayside ionosphere of Venus: Energetics, *J. Geophys. Res.*, **85**, 7778, 1980.

36. Pavlov, A. V., New electron energy transfer rates for vibrational excitation of N_2, *Ann. Geophysicae*, **16**, 176, 1998a.

37. Pavlov, A. V., New electron energy transfer and cooling rates by excitation of O_2, *Ann. Geophysicae*, **16**, 1007, 1998b.

38. Waite, J. H. and T. E. Cravens, Vibrational and rotational cooling of electrons by moleculear hydrogen, *Planet. Space Sci.*, **29**, 1333, 1981.

39. Dalgarno, A., Inelastic collisions at low energies, *Can. J. Chem.*, **47**, 1723, 1969.

40. Porter, H. S., and H. G. Mayr, CO_2 and CO electron vibrational cooling rates, *J. Geophys. Rev.*, **84**, 6705, 1979.

41. Cravens, T. E., and A. Korosmezey, Vibrational and rotational cooling of electrons by water vapor, *Planet. Space Sci.*, **34**, 961, 1986.

42. Gan, L., and T. E. Cravens, Electron impact cross sections and cooling rates for methane, *Planet. Space Sci.*, **40**, 1535, 1992.

43. Jones, D. B., L. Campbell, M. J. Bottema, and M. J. Brunger, New electron-energy transfer rates for vibrational excitation of O_2, *New Journal of Physics*, **5**, 114.1, 2003.

44. Henry, R. J. W., and M. B. McElroy, The absorption of extreme ultraviolet solar radiation by Jupiter's upper atmosphere, *J. Atmos. Sci.*, **26**, 912, 1969.

45. Pavlov, A. V. and K. A. Berrington, Cooling rate of thermal electrons by electron impact excitation of fine structure levels of atomic oxygen, *Ann. Geophysicae*, **17**, 919, 1999.

46. Henry, R. J. W., P. G. Burke, and A. L. Sinfailam, Scattering of electrons by C, N, O, N^+, O^+, and O^{++}, *Phys. Rev.*, **178**, 218, 1969.

47. David, M., R. W. Schunk, and J. J. Sojka, The effect of downward electron heat flow and electron cooling processes in the high-latitude ionosphere, *J. Atmos, Solar-Terr. Phys.*, 2008.

9.9 General references

Banks, P. M. and G. Kockarts, *Aeronomy*, New York: Academic Press, 1973.

Bauer, S. J., *Physics of Planetary Ionospheres*, Berlin: Springer Verlag, 1973.

Chamberlain, J. W. and D. M. Hunten, *Theory of Planetary Atmospheres*, New York: Academic Press, 1987.

Chandrasekhar, S., *Radiative Transfer*, New York: Dover Publications Inc., 1960.

Fox, J. L., Aeronomy, in *Atomic, Molecular and Optical Physics Handbook*, ed. by G. W. F. Drake **940**, Woodbury, NY: American Institute of Physics Press, 1996.

Fox, J. L., M. I. Galand, and R. E. Johnson, Energy deposition in planetary atmospheres by charged particles and solar photons, in *Comparative Aeronomy*, Space Sciences Series of ISSI, New York: Springer, 2008.

Rees, M. H., *Physics and Chemistry of the Upper Atmosphere*, Cambridge, UK: Cambridge University Press, 1989.

Schunk, R. W. and A. F. Nagy, Electron temperatures in the F region of the ionosphere: theory and observations, *Rev. Geophys. Space Phys.*, **16**, 355, 1978.

Schunk, R. W. and A. F. Nagy, Ionospheres of the terrestrial planets, *Rev. Geophys. Space Phys.*, **18**, 813, 1980.

9.10 Problems

Problem 9.1 The atmosphere of the planet Imaginus consists of only molecular nitrogen, and the surface density is 5×10^{15} cm^{-3}. You can assume that the gas temperature is 1000 K and is constant with altitude; the acceleration due to gravity is 1000 cm s^{-2} and is also constant with altitude. This planet is twice as far from the Sun as is Earth. Calculate the intensity of the solar radiation of the 977.02 and the 303.78 Å solar lines for an overhead Sun, as a function of altitude, assuming that F10.7 and $\langle$F10.7$\rangle$ are both 150. Use the EUVAC model for these calculations. Also, calculate the total ionization rate due to these lines as a function of altitude. (Specifically calculate for 400, 375, 350, 325, 300, and 200 km.)

Problem 9.2 Calculate the approximate volume absorption rate of solar energy (eV cm^{-3} s^{-1}) at an altitude of 120 km for the planet Imaginus. In carrying out these calculations, you can assume the following values for solar radiation at the top of the atmosphere:

λ (nm)	I (photons cm^{-2} s^{-1} nm^{-1})
100–150	10^{11}
0–100	10^{8}

You should assume that the mean energy of the photons in each of these intervals is that at the midpoint. The only atmospheric species that needs to be considered is the molecule X_2, which is in diffusive equilibrium throughout the atmosphere. The wavelength thresholds for dissociation and ionization are 150 and 100 nm, respectively. The cross sections for photodissociation and photoionization are 1×10^{-21} and 5×10^{-18} cm^2, respectively. The number density at the surface of the planet is 5×10^{15} cm^{-3} and the scale height of X_2 is 3×10^6 cm, independent of altitude.

Problem 9.3 On the planet Imaginus, described in Problem 9.2, the photodissociation process, in the wavelength region 100 nm $< \lambda <$ 150 nm, results in two ground state X atoms, with the excess energy going into kinetic energy

$$X_2 + h\nu(100\,\text{nm} < \lambda < 150\,\text{nm}) \rightarrow X + X + \text{KE}.$$

What fraction of the total absorbed energy goes toward heating the ambient neutral gas in this wavelength region?

Problem 9.4 On the planet Imaginus, the neutral atmosphere consists only of atomic oxygen. The radiation impacting on this planet is monochromatic at a wavelength of 304 Å (30.4 nm). Half the oxygen ions are created in their ground state (^{4}S) and half in the first excited state (^{2}D), which is 3.31 eV above the ground state. The ionization threshold for oxygen is 13.62 eV. Sketch the energy distribution function of the newly created photoelectrons, showing the appropriate energies.

Problem 9.5 Starting from Equation (9.21) show that the altitude of the maximum production rate and the corresponding rate are given by Equations (9.22) and (9.23), respectively.

Problem 9.6 Show that Equation (9.41) does follow from Equation (9.32) in the case of low altitude, no transport conditions, with multiple neutral species and a cascading term, Q_{casc}^+.

Problem 9.7 On planet Imaginus, where the atmosphere consists of atomic X only, the energy distribution function of photoelectrons with energies greater than 50 eV is

$$f(E > 50\,\text{eV}) = f_0/E^2$$

and the photoionization source term for all energies is

$$Q(E) = Q_0 \exp\left(-\frac{E}{E_\text{p}}\right),$$

where f_0 is the value of the distribution function at 1 eV and E_p is a characteristic energy of the source function. Using Equation (9.38), write down the expression for the electron distribution function at 50 eV, given that the constituent X has only two excited states at 3 and 5 eV, respectively, and only one ionization level at 12 eV. Furthermore, you can assume that the excitation and ionization cross sections are the same, energy independent, and denoted by σ. Finally, assume that all secondary electrons are produced with an energy of 20 eV and tertiary ionization is negligible.

Chapter 10

Neutral atmospheres

Neutral atmospheres play a crucial role with regard to the formation, dynamics, and energetics of ionospheres, and therefore, an understanding of ionospheric behavior requires a knowledge of atmospheric behavior. A general description of the atmospheres that give rise to the ionospheres was given in Chapter 2. In this chapter, the processes that operate in upper atmospheres are described, and the equations presented have general applicability. However, the discussion of specifics is mainly directed toward the terrestrial upper atmosphere (see Chapter 2 for a limited description of other solar system neutral atmospheres) because our knowledge of this atmosphere is much more extensive than that for all of the other atmospheres (i.e., other planets, moons, and comets).

Typically, the lower domain of an upper atmosphere is turbulent, and the various atomic and molecular species are thoroughly mixed. However, as altitude increases, molecular diffusion rapidly becomes important and a diffusive separation of the various neutral species occurs. For Earth, this diffusive separation region extends from about 110 to 500 km, and most of the ionosphere and atmosphere interactions occur in this region. At higher altitudes the collisional mean-free-path becomes very long and the neutral particles basically follow ballistic trajectories. For the case of light neutrals, such as hydrogen and helium, and more energetic heavier gases, some of the ballistic trajectories can lead to the escape of particles from the atmosphere.

The topics in this chapter progress from the main processes that operate in the diffusive separation region of an upper atmosphere to the escape of atoms from the top of the atmosphere. First, atmospheric rotation is discussed because it has a significant effect on the horizontal flow of an atmosphere. Next, the Euler and Navier–Stokes equations are derived because they provide the framework for studies of the dynamics and energetics of upper atmospheres. This is followed by a discussion of atmospheric waves, including gravity waves and tides. Then, the discussion progresses from the neutral density structure, to the escape of terrestrial hydrogen, and to atmospheric energetics. Finally, topics relevant to an exosphere are discussed, including the escape of particles and the distribution of hot neutrals.

10.1   Rotating atmospheres

The 13-moment transport equations given in Equations (3.57) to (3.61) are relevant to an inertial reference frame. However, in most cases, atmospheric behavior is studied from a reference frame that is fixed to a rotating body. Hence, it is necessary to transform the transport equations from an inertial to a rotating reference frame. The main result of such a transformation is the appearance of Coriolis and centripetal acceleration terms in the Momentum Equation (3.58).

In a rotating reference frame, the velocity of interest is that relative to the rotating body, $\mathbf{u}_{\text{rot}}$. If $\mathbf{\Omega}_r$ is the angular velocity of the body, the connection between the velocity in an *inertial (nonrotating)* reference frame, $\mathbf{u}_{\text{int}}$, and that in the rotating reference frame is

$$\mathbf{u}_{\text{int}} = \mathbf{u}_{\text{rot}} + \mathbf{\Omega}_r \times \mathbf{r}, \tag{10.1}$$

where $\mathbf{r}$ is the radius vector from the center of the planet. Equation (10.1) is the well-known result from classical mechanics that links velocities in inertial and rotating reference frames.

In addition to a difference in velocities, as seen in the inertial and rotating frames, there is also a difference in total time derivatives if they operate on vectors. However, there is no difference if they operate on scalars. This can be shown by considering the simple situation in which a rotating planet is embedded in a constant vector field $\mathbf{W}$, where $\mathbf{W}$ is assumed to be perpendicular to the planet's angular velocity $\mathbf{\Omega}_r$. To an observer on the planet, the vector $\mathbf{W}$ appears to continuously change its direction, making a complete rotation in a time $2\pi/\mathbf{\Omega}_r$. In the rotating reference frame, the time rate of change of the *constant* vector $\mathbf{W}$ is

$$\left(\frac{d\mathbf{W}}{dt}\right)_{\text{rot}} = -\mathbf{\Omega}_r \times \mathbf{W}, \tag{10.2}$$

while in the inertial reference frame it is

$$\left(\frac{d\mathbf{W}}{dt}\right)_{\text{int}} = 0. \tag{10.3}$$

If $\mathbf{W}$ is now allowed to vary with position and time, the time rate of change of $\mathbf{W}$ in the inertial frame will not necessarily be zero. In this case, the connection between time derivatives in the inertial and rotating reference frames is

$$\left(\frac{d\mathbf{W}}{dt}\right)_{\text{rot}} = \left(\frac{d\mathbf{W}}{dt}\right)_{\text{int}} - \mathbf{\Omega}_r \times \mathbf{W}, \tag{10.4}$$

which follows from Equation (10.2).

Up to this point, the observers were fixed in the inertial and rotating reference frames. If they are now allowed to move with velocities $\mathbf{u}_{int}$ and $\mathbf{u}_{rot}$, respectively, the time derivatives d/dt in Equation (10.4) become the *convective* time derivatives

$$\left(\frac{D\mathbf{W}}{Dt}\right)_{int} = \left(\frac{\partial}{\partial t} + \mathbf{u}_{int} \cdot \nabla\right)\mathbf{W}, \tag{10.5}$$

$$\left(\frac{D\mathbf{W}}{Dt}\right)_{rot} = \left(\frac{\partial}{\partial t} + \mathbf{u}_{rot} \cdot \nabla\right)\mathbf{W}, \tag{10.6}$$

and Equation (10.4) becomes

$$\left(\frac{D\mathbf{W}}{Dt}\right)_{int} = \left(\frac{D\mathbf{W}}{Dt}\right)_{rot} + \mathbf{\Omega}_r \times \mathbf{W}. \tag{10.7}$$

When $\mathbf{W} = \mathbf{r}$, Equation (10.7) yields the well-known result (10.1). When $\mathbf{W} = \mathbf{u}_{int}$, Equation (10.7) becomes

$$\left(\frac{D\mathbf{u}_{int}}{Dt}\right)_{int} = \left(\frac{D\mathbf{u}_{int}}{Dt}\right)_{rot} + \mathbf{\Omega}_r \times \mathbf{u}_{int}. \tag{10.8}$$

Eliminating $\mathbf{u}_{int}$ on the right-hand side of Equation (10.8) with the aid of Equation (10.1) yields the following result:

$$\left(\frac{D\mathbf{u}_{int}}{Dt}\right)_{int} = \left[\frac{D}{Dt}(\mathbf{u}_{rot} + \mathbf{\Omega}_r \times \mathbf{r})\right]_{rot} + \mathbf{\Omega}_r \times (\mathbf{u}_{rot} + \mathbf{\Omega}_r \times \mathbf{r})$$

$$= \left(\frac{D\mathbf{u}_{rot}}{Dt}\right)_{rot} + 2\mathbf{\Omega}_r \times \mathbf{u}_{rot} + \mathbf{\Omega}_r \times (\mathbf{\Omega}_r \times \mathbf{r}), \tag{10.9}$$

where $\mathbf{\Omega}_r$ is assumed to be constant. The second and third terms on the right-hand side of (10.9) represent *Coriolis* and *centripetal acceleration*, respectively. Therefore, if the momentum equation (3.58) is applied in a rotating reference frame, Coriolis and centripetal acceleration terms must be added to the equation.

10.2   Euler equations

The Euler and Navier–Stokes equations of hydrodynamics can be derived from the 13-moment system of transport equations (3.57–61; 4.129a–g) by using a simple perturbation scheme. However, in the derivation of these equations, it is convenient to consider a single-component neutral gas because this corresponds to the classical case in which the equations apply. In the perturbation scheme, the collision frequency is assumed to be sufficiently large that the neutral velocity distribution is very nearly Maxwellian. In this collision-dominated limit, τ_n and $\mathbf{q}_n$ are small and of order $1/\nu_{nn}$ compared to n_n, $\mathbf{u}_n$, and T_n, which are of order one. To lowest order in the perturbation scheme, stress and heat flow effects are neglected, and the

13-moment equations of continuity, momentum, and energy, in a *rotating reference frame*, reduce to

$$\frac{\partial n_n}{\partial t} + \nabla \cdot (n_n \mathbf{u}_n) = 0, \tag{10.10a}$$

$$\rho_n \frac{D_n \mathbf{u}_n}{Dt} + \nabla p_n + \rho_n [2\mathbf{\Omega}_r \times \mathbf{u}_n + \mathbf{\Omega}_r \times (\mathbf{\Omega}_r \times \mathbf{r}) - \mathbf{G}] = 0, \tag{10.10b}$$

$$\frac{D_n}{Dt}\left(\frac{3}{2}p_n\right) + \frac{5}{2}p_n(\nabla \cdot \mathbf{u}_n) = 0, \tag{10.10c}$$

where $\rho_n = n_n m_n$ is the mass density. Equations (10.10a–c) correspond to the *Euler hydrodynamic equations* for a neutral gas. However, the energy equation (10.10c) can be cast in a more familiar form by eliminating the $(\nabla \cdot \mathbf{u}_n)$ term with the aid of the continuity equation (10.10a). When this is done, the energy equation reduces to the simple adiabatic energy equation with the ratio of specific heats equal to $\frac{5}{3}$

$$\frac{D_n}{Dt}\left(p_n \rho_n^{-5/3}\right) = 0. \tag{10.11}$$

Also, it should be noted that the Euler hydrodynamic equations (10.10a–c) are equivalent to the five-moment equations (5.22a–c) if the collision terms are neglected in the latter system of equations. Therefore, the Euler equations pertain to the case when the neutral velocity distribution is a drifting Maxwellian.

10.3   Navier–Stokes equations

To next order in the perturbation scheme, the stress tensor (3.60, 4.129f) and heat flow (3.61, 4.129g) equations are used to express $\boldsymbol{\tau}_n$ and $\mathbf{q}_n$ in terms of n_n, $\mathbf{u}_n$, and T_n. This is accomplished by noting that terms containing $\nu_{nn}\boldsymbol{\tau}_n$ and $\nu_{nn}\mathbf{q}_n$ are of order one, while all other terms containing $\boldsymbol{\tau}_n$ and $\mathbf{q}_n$ are of order $1/\nu_{nn}$. Retaining only those terms that are of order one, the stress tensor equation reduces to

$$p_n\left[\nabla \mathbf{u}_n + (\nabla \mathbf{u}_n)^T - \frac{2}{3}(\nabla \cdot \mathbf{u}_n)\mathbf{I}\right] = -\frac{6}{5}\nu_{nn}\boldsymbol{\tau}_n, \tag{10.12}$$

or

$$\boldsymbol{\tau}_n = -\eta_n\left[\nabla \mathbf{u}_n + (\nabla \mathbf{u}_n)^T - \frac{2}{3}(\nabla \cdot \mathbf{u}_n)\mathbf{I}\right], \tag{10.13}$$

where η_n is the *coefficient of viscosity*

$$\eta_n = \frac{5p_n}{6\nu_{nn}}. \tag{10.14}$$

Likewise, retaining only those terms of order one in the heat flow equation (3.61, 4.129g), it reduces to

$$\frac{5}{2}\frac{kp_n}{m_n}\nabla T_n = -\frac{4}{5}\nu_{nn}\mathbf{q}_n,$$ (10.15)

or

$$\mathbf{q}_n = -\lambda_n \nabla T_n,$$ (10.16)

where λ_n is the *thermal conductivity*

$$\lambda_n = \frac{25}{8}\frac{kp_n}{m_n\nu_{nn}}.$$ (10.17)

Note that both η_n and λ_n are proportional to p_n/ν_{nn}. This is consistent with the initial assumption that stress and heat flow are of order $1/\nu_{nn}$ in comparison with n_n, $\mathbf{u}_n$, and T_n. Also note that as the collision frequency decreases, the importance of stress and heat flow increases because η_n and λ_n become large. However, the expressions for τ_n (10.13) and $\mathbf{q}_n$ (10.16) were derived assuming that the collision frequency is large, and when the gas starts to become collisionless these equations are no longer valid.

A comparison of the viscosity (10.14) and thermal conductivity (10.17) coefficients with the corresponding coefficients (5.12) and (5.20) derived using mean-free-path considerations indicates that they are of the same form, except for the numerical factors. The comparison of Equation (10.14) and Equation (10.17) with those in the original work of Chapman and Cowling[1] can be made by setting ν_{nn} equal to the hard sphere collision frequency (4.156), which yields

$$\eta_n = \frac{5\sqrt{\pi}}{16}\frac{(m_n kT_n)^{1/2}}{\pi\sigma^2},$$ (10.18)

$$\lambda_n = \frac{75\sqrt{\pi}}{64}\frac{k}{m_n}\frac{(m_n kT_n)^{1/2}}{\pi\sigma^2}.$$ (10.19)

These expressions correspond to what Chapman and Cowling[1] call the first approximation to these coefficients. Both η_n and λ_n are directly proportional to $T_n^{1/2}$ for hard-sphere interactions and inversely proportional to the collision cross section, $\pi\sigma^2$. The latter proportionality indicates that viscosity and thermal conduction are more important for atomic species, such as H, He, and O, than for molecular species, such as N_2, O_2, and CO_2, because of the smaller collision cross section.

The *Navier–Stokes system of equations* consists of the continuity (3.57), momentum (3.58), and energy (3.59) equations coupled with the collision-dominated expressions for the stress tensor (10.13) and heat flow vector (10.16). This system can be expressed in the following form for a *rotating reference frame*:

$$\frac{\partial n_n}{\partial t} + \nabla \cdot (n_n \mathbf{u}_n) = 0, \tag{10.20a}$$

$$\frac{D_n \mathbf{u}_n}{Dt} + 2\boldsymbol{\Omega}_r \times \mathbf{u}_n + \boldsymbol{\Omega}_r \times (\boldsymbol{\Omega}_r \times \mathbf{r}) - \mathbf{G} + \frac{1}{\rho_n} \nabla p_n$$

$$- \bar{\eta}_n \left[\nabla^2 \mathbf{u}_n + \frac{1}{3} \nabla (\nabla \cdot \mathbf{u}_n) \right]$$

$$- \frac{\bar{\eta}_n}{2T_n} \nabla T_n \cdot \left[\nabla \mathbf{u}_n + (\nabla u_n)^T - \frac{2}{3} (\nabla \cdot \mathbf{u}_n) \mathbf{I} \right] = 0, \tag{10.20b}$$

$$c_v n_n \frac{D_n T_n}{Dt} + p_n (\nabla \cdot \mathbf{u}_n) - \nabla \cdot (\lambda_n \nabla T_n)$$

$$- \eta_n \left[\nabla \mathbf{u}_n + (\nabla \mathbf{u}_n)^T - \frac{2}{3} (\nabla \cdot \mathbf{u}_n) \mathbf{I} \right] : \nabla \mathbf{u}_n = 0, \tag{10.20c}$$

where $\bar{\eta}_n = \eta_n / \rho_n$ is the *kinetic viscosity* and $c_v = 3k/2$ is the *specific heat at constant volume*. The distinction between the Navier–Stokes system of equations and the complete 13-moment system of equations is that in the 13-moment approximation, $\boldsymbol{\tau}_n$ and $\mathbf{q}_n$ are put on an equal footing with n_n, $\mathbf{u}_n$, and T_n, while in the Navier–Stokes approximation, $\boldsymbol{\tau}_n$ and $\mathbf{q}_n$ are not independent moments, but instead are expressed in terms of the fundamental moments n_n, $\mathbf{u}_n$, and T_n and their first derivatives. Therefore, the Navier–Stokes equations are valid only for very small deviations from a drifting Maxwellian velocity distribution.

The perturbation scheme based on an expansion in powers of $1/\nu_{nn}$ can be continued to higher levels of approximation. However, the continuation of this scheme to higher levels leads to expressions for $\boldsymbol{\tau}_n$ and $\mathbf{q}_n$, which contain space and time derivatives of increasing order. When these expressions are then substituted into the momentum and energy equations, they yield partial differential equations of an even higher order. Consequently, to obtain solutions to the resulting set of flow equations, it is necessary to specify boundary conditions not only for n_n, $\mathbf{u}_n$, and T_n, but also for several derivatives of these quantities. The latter requirement precludes the usefulness of these higher-order Navier–Stokes equations.

10.4   Atmospheric waves

Wave phenomena are prevalent in planetary atmospheres and arise as a result of perturbations induced by both external and internal sources. In general, the waves can be classified into three main groups, with the primary designation being the spatial scale length of the wave. On the largest scale are *planetary waves* and *tides*, which are global in nature and exhibit coherent patterns in both latitude and longitude. In the terrestrial lower thermosphere, the planetary waves have periods of about 2, 5, and 16 days, while the tidal modes have periods of about 8, 12, and 24 hours. These large-scale waves contain both *migrating modes*, which are fixed in local time and may be driven, for example, by heating at the subsolar point, and *stationary modes*, which are fixed with respect to a rotating planet.

On a smaller spatial scale are *atmospheric gravity waves* (AGWs), which arise because of the buoyancy forces in the atmosphere. These waves are not global and, therefore, the curvature of the planet is not relevant. The waves typically have a localized source and propagate with a limited range of wavelengths. For the Earth, gravity waves can be generated in the stratosphere and mesosphere and then propagate to thermospheric heights or they can be generated *in situ* in the thermosphere. In the lower atmosphere, AGWs can be generated by perturbations in the jetstream, the flow of air over mountains, thunderstorms, volcanoes, and earthquakes. In the upper atmosphere, they can be generated by variations in the Joule and particle heating rates, the Lorentz forcing at high latitudes, the breaking of upward propagating tides, the movement of the solar terminator, and solar eclipses. Typically, gravity waves are divided into large-scale and medium-scale waves. The large-scale AGWs have horizontal wavelengths of about 1000 km, wave periods of more than an hour, and horizontal velocities of 500–1000 m s^{-1}. The medium-scale AGWs have horizontal wavelengths of several hundred kilometers, wave periods of about 5–60 minutes, and horizontal velocities of 100–300 m s^{-1}.

The smallest spatial scales pertain to acoustic waves. However, these waves, which are ordinary sound waves, do not play a prominent role in the dynamics or energetics of upper atmospheres.

The general treatment of atmospheric waves is very complicated and detailed descriptions can be found in classic books.[2, 3] This chapter focuses on simple descriptions of gravity waves and tides, with the goal being the elucidation of the basic physics. With this approach, the reader should have a sufficient knowledge of gravity waves and tides to understand their effects on the ionospheres that will be discussed in later chapters.

10.5   Gravity waves

For the analysis of gravity waves, consider only the characteristic modes that can exist in the atmosphere and ignore the source and dissipation mechanisms. In this case, viscous effects and thermal conduction can be neglected. In general, the Coriolis and centripetal acceleration terms can also be neglected, because the wave periods are typically much less than planetary rotation periods. Under these circumstances, the continuity, momentum, and energy equations for a single-component neutral gas (10.10a–c) reduce to

$$\frac{\partial \rho}{\partial t} + \nabla \cdot (\rho \mathbf{u}) = 0, \tag{10.21a}$$

$$\rho \left(\frac{\partial}{\partial t} + \mathbf{u} \cdot \nabla \right) \mathbf{u} + \nabla p - \rho \mathbf{G} = 0, \tag{10.21b}$$

$$\left(\frac{\partial}{\partial t} + \mathbf{u} \cdot \nabla\right)p + \gamma p(\nabla \cdot \mathbf{u}) = 0, \tag{10.21c}$$

where $\gamma = 5/3$ is the ratio of specific heats and where the subscript n has been omitted.

The characteristic waves that can propagate in an atmosphere are obtained by perturbing a given atmospheric state. For simplicity, the initial atmosphere is assumed to be isothermal ($T_0 = $ constant), stationary ($\mathbf{u}_0 = 0$), horizontally stratified, and in hydrostatic equilibrium (Equation 10.58)

$$\nabla p_0 = \rho_0 \mathbf{G}, \tag{10.22}$$

where subscript 0 is used to designate the initial unperturbed state. A Cartesian coordinate system can be introduced because the wavelengths of AGWs are typically much smaller than planetary radii. Letting the coordinates (x, y, z) correspond to (eastward, northward, upward), Equation (10.22) can be expressed in the form

$$\frac{1}{n_0}\frac{dn_0}{dz} = -\frac{1}{H_0}, \tag{10.23}$$

where $p_0 = n_0 k T_0$ and $H_0 = k T_0/mg$ is the atmospheric scale height (Equations 9.11 and 10.56). Equation (10.23) indicates that the initial atmospheric state varies only with z, and in an exponential manner

$$p_0, \rho_0 \propto e^{-z/H_0}. \tag{10.24}$$

The perturbation of the initial atmospheric state is accomplished by setting $\rho = \rho_0 + \rho_1, p = p_0 + p_1$, and $\mathbf{u} = \mathbf{u}_1$, where subscript 1 denotes the perturbations, which are assumed to be small. Substituting these quantities into Equations (10.21a–c), and retaining only those terms that are linear in the perturbations, yields the following equations:

$$\frac{\partial \rho_1}{\partial t} + \mathbf{u}_1 \cdot \nabla \rho_0 + \rho_0(\nabla \cdot \mathbf{u}_1) = 0, \tag{10.25a}$$

$$\rho_0 \frac{\partial \mathbf{u}_1}{\partial t} + \nabla p_1 - \rho_1 \mathbf{G} = 0, \tag{10.25b}$$

$$\frac{\partial p_1}{\partial t} + \mathbf{u}_1 \cdot \nabla p_0 + \gamma p_0(\nabla \cdot \mathbf{u}_1) = 0, \tag{10.25c}$$

where in the derivation use has been made of Equation (10.22). For what follows, it is convenient to express the equations in terms of ρ_1/ρ_0 and p_1/p_0 instead of ρ_1 and p_1. Also, the terms containing $\nabla \rho_0$ and ∇p_0 can be expressed in terms of the atmospheric scale height, H_0, by using Equation (10.24)

$$\mathbf{u}_1 \cdot \nabla \rho_0 = -\frac{\rho_0}{H_0} u_{1z}, \tag{10.26}$$

$$\mathbf{u}_1 \cdot \nabla p_0 = -\frac{p_0}{H_0} u_{1z}, \tag{10.27}$$

where u_{1z} is the vertical component of the perturbed velocity. With these changes, Equations (10.25a–c) become

$$\frac{\partial}{\partial t}\left(\frac{\rho_1}{\rho_0}\right) - \frac{1}{H_0} u_{1z} + (\nabla \cdot \mathbf{u}_1) = 0, \tag{10.28a}$$

$$\frac{\partial \mathbf{u}_1}{\partial t} + \frac{p_1}{\rho_0 p_0}\nabla p_0 + \frac{p_0}{\rho_0}\nabla\left(\frac{p_1}{p_0}\right) - \frac{\rho_1}{\rho_0}\mathbf{G} = 0, \tag{10.28b}$$

$$\frac{\partial}{\partial t}\left(\frac{p_1}{p_0}\right) - \frac{1}{H_0} u_{1z} + \gamma(\nabla \cdot \mathbf{u}_1) = 0. \tag{10.28c}$$

For small perturbations, the perturbed quantities can be described by plane waves

$$\left(\frac{\rho_1}{\rho_0}\right), \left(\frac{p_1}{p_0}\right), \mathbf{u}_1 \propto e^{i(\mathbf{K}\cdot\mathbf{r}-\omega t)}, \tag{10.29}$$

where $\mathbf{K}$ is the wave vector and ω is the wave frequency. For plane waves, the space and time derivatives of perturbed quantities can be easily obtained and Equations (10.28a–c) become

$$-i\omega\left(\frac{\rho_1}{\rho_0}\right) - \frac{1}{H_0} u_{1z} + i(\mathbf{K} \cdot \mathbf{u}_1) = 0, \tag{10.30a}$$

$$-i\omega\mathbf{u}_1 + \frac{p_1}{\rho_0 p_0}\nabla p_0 + \frac{p_0}{\rho_0}i\mathbf{K}\left(\frac{p_1}{p_0}\right) - \frac{\rho_1}{\rho_0}\mathbf{G} = 0, \tag{10.30b}$$

$$-i\omega\left(\frac{p_1}{p_0}\right) - \frac{1}{H_0} u_{1z} + i\gamma(\mathbf{K} \cdot \mathbf{u}_1) = 0. \tag{10.30c}$$

Equations (10.30a–c) can be solved to obtain the perturbed quantities and the result is a general dispersion relation for the wave modes that can propagate in the atmosphere. However, first consider two important limiting cases. If gravity is ignored $(\mathbf{G} \to 0, H_0 \to \infty)$, the initial atmosphere is homogeneous and Equations (10.30a–c) for the perturbed quantities reduce to

$$-\omega\rho_1 + \rho_0(\mathbf{K} \cdot \mathbf{u}_1) = 0, \tag{10.31a}$$

$$K^2 p_1 - \omega\rho_0(\mathbf{K} \cdot \mathbf{u}_1) = 0, \tag{10.31b}$$

$$-\omega p_1 + \gamma p_0(\mathbf{K} \cdot \mathbf{u}_1) = 0, \tag{10.31c}$$

where the momentum equation (10.30b) was dotted with $\mathbf{K}$. The solution for the perturbed quantities $(\rho_1, p_1, \mathbf{K} \cdot \mathbf{u}_1)$ can be obtained by substituting the equations into each other until there is one equation with an unknown. Regardless of what parameter is solved for, the same dispersion relation is obtained. Alternatively, Equations (10.31a–c) can be solved by the matrix method of linear algebra. In the present situation, the dispersion relation can be easily obtained by solving Equations (10.31b) and (10.31c) for p_1, which yields

$$\omega^2 = c_0^2 K^2, \tag{10.32}$$

where c_0 is the *sound speed* in the neutral gas,

$$c_0^2 = \frac{\gamma p_0}{\rho_0} = \gamma g H_0 = \frac{\gamma k T_0}{m}. \tag{10.33}$$

For sound waves, both K and ω are real, which means the waves propagate without growth or attenuation. Also, $\omega/K = d\omega/dK = \pm c_0$, where the $\pm$ signs correspond to sound waves that propagate in opposite directions. Hence, for sound waves there is no dispersion, because ω/K is constant.

The second limiting case that is worth discussing is for a negligible pressure disturbance ($p_1 = 0$). The resulting waves are then produced as a result of a balance between gravity and acceleration effects. For this situation, Equations (10.30a–c) reduce to

$$-i\omega\rho_1 + \rho_0 \left(iK_z - \frac{1}{H_0} \right) u_{1z} = 0, \tag{10.34a}$$

$$-i\omega\rho_0 u_{1z} + g\rho_1 = 0, \tag{10.34b}$$

$$\left(i\gamma K_z - \frac{1}{H_0} \right) u_{1z} = 0, \tag{10.34c}$$

where $\mathbf{K}$ and $\mathbf{u}_1$ are assumed to be only in the vertical direction and where $\mathbf{G} = -g\mathbf{e}_z$. Equation (10.34b) yields $\rho_1 = i\omega\rho_0 u_{1z}/g$, while Equation (10.34c) yields $iK_z = 1/\gamma H_0$. Substituting these results into Equation (10.34a) leads to the following expression:

$$\omega_b^2 = (\gamma - 1)\frac{g^2}{c_0^2}. \tag{10.35}$$

Note that the wave vector $\mathbf{K}$ does not appear in Equation (10.35) and, hence, the disturbance is not a propagating wave, but is a local buoyancy oscillation. The frequency given by (10.35) is called the *buoyancy frequency* or the *Brunt–Väisälä frequency*, and it is the natural frequency at which a local parcel of air oscillates if it is disturbed from its equilibrium.

Gravity waves basically propagate in the horizontal direction, but they usually have a small vertical component. For simplicity, the propagation is assumed to be in the x–z plane so that $\mathbf{K}$ only has x and z components. Note that this assumption does not lead to a new restriction because the horizontal directions are not coupled owing to the neglect of viscosity and Coriolis effects. With propagation in the x–z plane, the momentum equation (10.30b) becomes two equations, one for u_{1x} and one for u_{1z}. Therefore, Equations (10.30a–c) become four equations for the four unknown perturbations (ρ_1/ρ_0, p_1/p_0, u_{1x}, u_{1z}). These equations can then be solved by the matrix technique for linear equations, and the solution is nontrivial only when the following condition is satisfied:

$$
\begin{vmatrix}
-i\omega & 0 & iK_x & \left(iK_z - \dfrac{1}{H_0}\right) \\[2ex]
0 & iK_x\dfrac{c_0^2}{\gamma} & -i\omega & 0 \\[2ex]
g & \left(iK_z - \dfrac{1}{H_0}\right)\dfrac{c_0^2}{\gamma} & 0 & -i\omega \\[2ex]
0 & -i\omega & i\gamma K_x & \left(i\gamma K_z - \dfrac{1}{H_0}\right)
\end{vmatrix} = 0,
\qquad (10.36)
$$

where the columns correspond to the coefficients of ρ_1/ρ_0, p_1/p_0, u_{1x}, and u_{1z}, respectively, and the rows correspond to the continuity, x-momentum, z-momentum, and energy equations, respectively. The expansion of the determinant leads to the following dispersion relation that relates K and ω:

$$
\omega^4 - \omega^2 c_0^2(K_x^2 + K_z^2) + (\gamma - 1)g^2 K_x^2 - i\gamma g\omega^2 K_z = 0.
\qquad (10.37)
$$

When $g = 0$, the dispersion relation (10.37) reduces to that derived earlier for sound waves (10.32), for which both K and ω are real. When gravity is included, K_x, K_z, and ω cannot all be real. For a horizontally propagating wave, ω and K_x must be real, and therefore, it is necessary to assume that K_z is complex; $K_z = K_{zr} + iK_{zi}$, where K_{zr} and K_{zi} are the real and imaginary parts, respectively. With allowance for a complex K_z, Equation (10.37) becomes[4]

$$
\omega^4 - \omega^2 c_0^2(K_x^2 + K_{zr}^2 - K_{zi}^2) + \gamma g K_{zi}\omega^2 + (\gamma - 1)g^2 K_x^2
$$
$$
- i\omega^2 K_{zr}(\gamma g + 2c_0^2 K_{zi}) = 0,
\qquad (10.38)
$$

and from the imaginary part of this equation, one obtains

$$
K_{zi} = -\frac{\gamma g}{2c_0^2} = -\frac{1}{2H_0}.
\qquad (10.39)
$$

With a complex K_z, the velocity perturbation given in Equation (10.29) becomes

$$
\mathbf{u}_1 \propto e^{z/2H_0} e^{i(K_x x + K_{zr} z - \omega t)}
\qquad (10.40)
$$

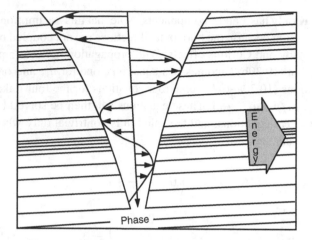

Figure 10.1 Schematic diagram of the characteristics of a large-scale gravity wave. The arrows show the neutral velocity variation with height, while the nearly horizontal lines show the density variation. Also shown are the directions of the group (energy) and phase velocities of the gravity wave.[4]

and waves that propagate in this manner are called *internal gravity waves*. These waves have the property that the wave perturbation energy, $\frac{1}{2}\rho_0 u_{1z}^2$, is constant because $\rho_0 \propto e^{-z/H_0}$ (Equation 10.24) and $u_{1z} \propto e^{z/2H_0}$ (Equation 10.40). Note that the wave amplitude grows as the wave propagates toward higher altitudes.

For large-scale gravity waves, the horizontal wavelength is much greater than the vertical wavelength ($\lambda_x \gg \lambda_z$), and the wave frequency is much smaller than the Brunt–Väisälä frequency ($\omega \ll \omega_b$). For such waves, the wavefronts are nearly horizontal, as shown in Figure 10.1. As a consequence, the group velocity has a slightly upward tilt, while the phase velocity has a sharply downward tilt.

10.6   Tides

Tides are global-scale atmospheric oscillations that arise primarily as a result of solar or lunar influences. The tides can be either gravitationally or thermally induced, and in both cases they are called *migrating tides* because the tidal perturbations move westward relative to a fixed location on the rotating Earth. A *diurnal tide* has a 24-hour period and a wavelength equal to the Earth's circumference, while a *semi-diurnal tide* has a 12-hour period and a wavelength equal to one-half of the Earth's circumference. In the terrestrial lower thermosphere (100 km), the main tidal source is the heating associated with the absorption of solar radiation by water vapor and ozone; and at mesospheric altitudes, the semi-diurnal tide dominates. However, this tide can propagate upward to thermospheric heights, and as it does, its amplitude grows. Tides can also be excited *in situ* in the thermosphere by the absorption of UV and EUV solar radiation. Above about 250 km, the solar-driven diurnal tide dominates, while between 100 and 250 km both diurnal and semi-diurnal components are present.

The mathematical treatment of atmospheric tides is more complicated than that of gravity waves because tides have long wavelengths and low frequencies. Consequently, both the curvature of the Earth and the Coriolis force must be taken into account. The classical theory of tides neglects dissipation processes, such as collisions, viscosity, and thermal conduction, and focuses on determining the normal modes of the atmosphere. Also, only a single species neutral gas is considered. Therefore, the starting place for the derivation of the tidal theory is the Euler equations (10.21a–c), but with the Coriolis term $2\rho\mathbf{\Omega}_r \times \mathbf{u}$ added to the momentum equation (10.21b). As with gravity waves, the atmosphere is assumed to be at rest ($\mathbf{u}_0 = 0$) and in hydrostatic equilibrium (10.22) prior to the tidal wave perturbation. The small perturbations are introduced in the usual manner by letting $\rho = \rho_0 + \rho_1$, $p = p_0 + p_1$, and $\mathbf{u} = \mathbf{u}_1$. However, an additional gravitational perturbation is introduced through the use of a potential, Ψ_1, such that $\mathbf{G} \rightarrow \mathbf{G} - \nabla\Psi_1$, where Ψ_1 is assumed to be small. The perturbation on gravity may arise, for example, as a result of the moon's influence on the Earth's upper atmosphere. When these perturbations are inserted into the Euler equations and only the linear terms are retained, the continuity, momentum, and energy equations are similar to the gravity wave equations (10.25a–c), except for the momentum equation which contains two additional terms. This modified momentum equation is given by

$$\rho_0 \frac{\partial \mathbf{u}_1}{\partial t} + \nabla p_1 - \rho_1 \mathbf{G} + \rho_0 \nabla \Psi_1 + 2\rho_0 \mathbf{\Omega}_r \times \mathbf{u}_1 = 0. \tag{10.41}$$

At this point, the tidal wave theory departs significantly from the gravity wave theory. First, the tidal waves are global, not local, and, therefore, a spherical coordinate system is needed. It is convenient to align the polar axis with the rotation vector, $\mathbf{\Omega}_r$, and to let r be the geocentric distance, θ be co-latitude, and ϕ be the azimuthal angle (positive toward the east). Next, several assumptions are introduced in the classical theory to simplify the mathematics. These assumptions are as follows: (1) the Coriolis and acceleration terms in the radial momentum equation are ignored; (2) the Coriolis terms involving the radial velocity are ignored; (3) the atmosphere is assumed to be a thin shell, so that $r = R_E + z \approx R_E$, where z is altitude; (4) the variation of g and r with altitude is ignored in the thin shell and $\partial/\partial r = \partial/\partial z$; and (5) ρ_0, p_0, and T_0 vary with z, but not with θ, ϕ, or t.

With the above assumptions, the continuity (10.25a), momentum (10.41), and energy (10.25c) equations become[3, 5]

$$\frac{\partial \rho_1}{\partial t} + u_{1z}\frac{\partial \rho_0}{\partial z} + \rho_0 \chi_1 = 0, \tag{10.42a}$$

$$\frac{\partial p_1}{\partial z} + g\rho_1 = -\rho_0 \frac{\partial \Psi_1}{\partial z}, \tag{10.42b}$$

$$\frac{\partial u_{1\theta}}{\partial t} - 2\Omega_r \cos\theta\, u_{1\phi} + \frac{1}{\rho_0 R_E}\frac{\partial p_1}{\partial \theta} = -\frac{1}{R_E}\frac{\partial \Psi_1}{\partial \theta}, \tag{10.42c}$$

$$\frac{\partial u_{1\phi}}{\partial t} + 2\Omega_r \cos\theta u_{1\theta} + \frac{1}{\rho_0 R_E \sin\theta}\frac{\partial p_1}{\partial\phi} = \frac{-1}{R_E \sin\theta}\frac{\partial\Psi_1}{\partial\phi}, \tag{10.42d}$$

$$\frac{\partial p_1}{\partial t} + u_{1z}\frac{\partial p_0}{\partial z} + \gamma p_0\chi_1 = 0, \tag{10.42e}$$

where

$$\chi_1 = (\nabla\cdot\mathbf{u}_1) = \frac{1}{R_E \sin\theta}\left[\frac{\partial}{\partial\theta}(\sin\theta u_{1\theta}) + \frac{\partial u_{1\phi}}{\partial\phi}\right] + \frac{\partial u_{1z}}{\partial z}. \tag{10.43}$$

The normal modes of the system are obtained by ignoring the gravitational forcing function, Ψ_1, and by substituting the equations into each other to obtain one second-order differential equation for one of the five unknowns (ρ_1, u_{1z}, $u_{1\theta}$, $u_{1\phi}$, p_1). However, it is best to solve for χ_1 (10.43) as the unknown. Also, in solving for χ_1, the time and longitudinal dependencies of the perturbed quantities are assumed to be periodic and of the form $\exp[im(\phi + 2\pi t/T_d)]$, where T_d is the length of the solar day. The index m must be an integer in order to obtain single-valued results, and positive values of m correspond to westward traveling oscillations that keep pace with the subsolar point. The values $m = 1$ and 2 correspond to diurnal and semi-diurnal oscillations, respectively.

The resulting partial differential equation for χ_1 is given by

$$H_0\frac{\partial^2\chi_1}{\partial z^2} + \left(\frac{dH_0}{dz} - 1\right)\frac{\partial\chi_1}{\partial z} - \frac{g}{4R_E^2\Omega_r^2}F\left[\left(\frac{\gamma-1}{\gamma} + \frac{dH_0}{dz}\right)\chi_1\right] = 0, \tag{10.44}$$

where F is the operator

$$F = \frac{1}{\sin\theta}\frac{\partial}{\partial\theta}\left(\frac{\sin\theta}{a^2 - \cos^2\theta}\frac{\partial}{\partial\theta}\right) - \frac{m}{a^2 - \cos^2\theta}\left(\frac{m}{\sin^2\theta} + \frac{1}{a}\frac{a^2 + \cos^2\theta}{a^2 - \cos^2\theta}\right), \tag{10.45}$$

and where $a = \pi m/(T_d\Omega_r)$. Equation (10.44) can be solved by the separation of variables technique because it is a linear equation. Letting $\chi_1(z,\theta) = Z(z)\Theta(\theta)$ and $1/h$ be the separation constant, Equation (10.44) separates into the following ordinary differential equations:

$$F\left[\Theta(\theta)\right] + \frac{4R_E^2\Omega_r^2}{gh}\Theta(\theta) = 0, \tag{10.46}$$

$$H_0\frac{d^2Z}{dz^2} + \left(\frac{dH_0}{dz} - 1\right)\frac{dZ}{dz} + \left(\frac{\gamma-1}{\gamma} + \frac{dH_0}{dz}\right)\frac{Z}{h} = 0. \tag{10.47}$$

Equation (10.46) for Θ is called the *Laplace tidal equation*. The separation constant h has dimensions of length and is called the *equivalent depth*. For each value of m, there is a series of eigenfunctions, $\Theta_{mn}(\theta)$, with associated eigenvalues, h_{mn}, that satisfy Equation (10.46), where $n = 1, 2, 3$, etc. The quantity $e^{im\phi}\Theta_{mn}(\theta)$ is called

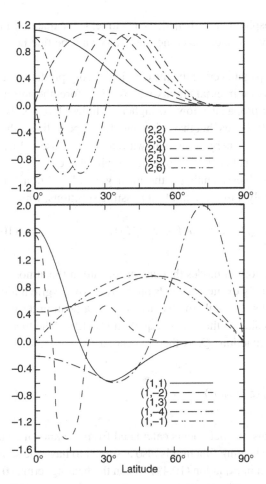

Figure 10.2 Semi-diurnal (top) and diurnal (bottom) Hough functions for latitudes between 0 and 90°.[8]

a *Hough function*.[6, 7] The Hough functions are denoted by (m, n), with m specifying the longitudinal dependence and n the latitudinal dependence. Figure 10.2 shows the latitudinal variation of some of the Hough functions for both the diurnal ($m = 1$) and semi-diurnal ($m = 2$) tidal components. When n is even, the Hough functions are symmetric about the equator, and when n is odd, they are anti-symmetric. Negative values of n correspond to nonpropagating modes.

For each tidal mode (m, n), there is a wavenumber, K_{mn}, that describes the vertical propagation characteristics of the wave, which is given by

$$K_{mn}^2 = \frac{1}{H_0 h_{mn}} \left[\frac{(\gamma - 1)}{\gamma} + \frac{dH_0}{dz} \right] - \frac{1}{4H_0^2}. \tag{10.48}$$

At a given altitude, the solution to the radial equation (10.47) varies as $Z_{mn}(z) \propto \exp(iK_{mn}z)$. Therefore, if K_{mn} is real, the tide propagates upward, whereas if K_{mn} is imaginary, the tide is *evanescent* (the wave decays as it tries to propagate vertically). Given the earlier definition of H_0, if the atmospheric temperature decreases with altitude, as it does in the mesosphere, then dH_0/dz is negative. If the quantity in the square brackets is also negative, K_{mn} can become imaginary and the tide becomes evanescent.

For the tidal modes that propagate vertically, the classical theory predicts that the wave energy is constant, as it is for internal gravity waves. Therefore, the amplitude of the tide grows as the wave propagates toward higher altitudes. Eventually, the tide may *break*, forming gravity waves, but that is beyond the scope of linear theory. Also, in the vertical direction, the energy flow is upward and the phase velocity is downward, which is similar to what occurs for internal gravity waves.

With the separation of variables technique, the most general solution for the perturbed parameter χ_1 is simply a linear sum of all possible solutions, which is

$$\chi_1 = \sum_m \sum_n Z_{mn}(z)\Theta_{mn}(\theta) \exp[im(\phi + 2\pi t/T_d)]. \qquad (10.49)$$

This solution describes the normal modes that can propagate in the atmosphere, but which tidal modes are actually excited depends on the form of the gravitational, Ψ_1, and solar heating functions. For a thermally driven tide, a heating term must be included on the right-hand side of the energy equation (10.42e). The measured (or prescribed) solar heating function, Q, is then expanded in a series of Hough functions

$$Q(z,\theta) = \sum_m \sum_n q_{mn}(z)\Theta_{mn}(\theta), \qquad (10.50)$$

where the $q_{mn}(z)$ are the expansion coefficients calculated from the known heating function $Q(z,\theta)$. Note that the heating function is also assumed to have the same ϕ and time dependencies as given in Equation (10.49). When the heating term (10.50) is added to the energy equation (10.42e) and a new equation for χ_1 is obtained, the resulting radial equation for $Z(z)$ is very complicated. Nevertheless, such a procedure was used to explain why the semi-diurnal tide dominates in the lower atmosphere.[3]

10.7 Density structure and controlling processes

At the lower boundary of the terrestrial thermosphere, which is called the mesopause, the major neutral gas components, molecular nitrogen and oxygen, are well mixed. Numerous important minor species, such as atomic oxygen, hydrogen, nitric oxide, ozone, and hydroxyl, are also present and chemically very active in the mesosphere and lower thermosphere. Density and composition vary significantly with latitude and longitude, as well as time, but vertical variations are, in general, much larger than horizontal ones. This is also generally true for most other upper atmospheres, with only few exceptions, such as Io, where the atmosphere is believed to be supplied, to a large degree, by volcanic sources. Therefore, the discussion is limited to altitude variations in this section. To establish the altitude variation of a given species, one needs to solve the vertical component of the corresponding continuity equation. This sounds simple, but of course it is not. The controlling chemistry has to be established if the species are not inert, and the relevant velocities and temperatures have to be determined.

This discussion begins by using the momentum equation (10.20b) to examine the altitude distribution of chemically inert neutral gas species. In the vertical direction, the neutral velocities are generally small and vary slowly with time, so that the diffusion approximation is valid. This means that the inertial, viscous stress, and Coriolis terms in Equation (10.20b) can be neglected. Centripetal acceleration for planets, where it is important, is usually combined with gravity, and the resulting expression is referred to as *effective gravity*. Also, when several neutral species are present in the gas, collision terms appear on the right-hand side of (10.20b) (see Equation 4.129b). The heat flow collision terms account for corrections to ordinary diffusion and thermal diffusion effects, which are negligible under most circumstances for neutral gases.

The vertical component of the momentum equation (10.20b) for the neutral gas, with allowance for the above outlined simplifications, can be written as

$$\nabla p_s - n_s m_s \mathbf{G} = -\sum_{t \neq s} m_s n_s \nu_{st} (\mathbf{u}_s - \mathbf{u}_t), \tag{10.51}$$

where subscripts s and t distinguish different species. After defining the diffusive flux, Γ_s, as $n_s \mathbf{u}_s$, one can write

$$\Gamma_s = -\frac{1}{\sum_{t \neq s} m_s \nu_{st}} \left(\nabla p_s - n_s m_s \mathbf{G} - \sum_{t \neq s} m_s n_s \nu_{st} \mathbf{u}_t \right). \tag{10.52}$$

Considering the vertical component of Equation (10.52) and defining *the diffusion coefficient*, D_s, as (see Section 5.3)

$$D_s = \frac{kT_s}{\sum_{t \neq s} m_s \nu_{st}}, \tag{10.53}$$

the expression for the vertical flux can be written as

$$\Gamma_{sz} = -\frac{D_s}{kT_s} \left(\frac{\partial p_s}{\partial z} + n_s m_s g - \sum_{t \neq s} m_s n_s \nu_{st} u_{tz} \right)$$

$$= -D_s \left(\frac{\partial n_s}{\partial z} + \frac{n_s}{T_s} \frac{\partial T_s}{\partial z} + n_s \frac{m_s g}{kT_s} - \frac{1}{kT_s} \sum_{t \neq s} m_s n_s \nu_{st} u_{tz} \right). \tag{10.54}$$

In most typical upper atmospheric situations, the last term in Equation (10.54) can be set to zero because the collision frequency is small. Under these circumstances the *diffusive equilibrium* solution, $\Gamma_{sz} = 0$, for $n_s(z)$ is

$$n_s(z) = n_s(z_0) \frac{T_s(z_0)}{T_s(z)} \exp\left(-\int_{z_0}^{z} \frac{dz'}{H_s}\right),$$
(10.55)

where H_s is the scale height of the neutral species, s (Equation 9.11)

$$H_s = \frac{kT_s}{m_s g}.$$
(10.56)

Equation (10.54) describes vertical transport due to molecular diffusion; note that this process tends to separate the atmospheric constituents according to their mass. Turbulence, on the other hand, mixes the atmospheric constituents and thus works against this tendency to separate. At lower altitudes ($z \lesssim 100$ km in the terrestrial atmosphere) this mixing process dominates and the region is called the *homosphere*. At higher altitudes ($z \gtrsim 125$ km in the terrestrial atmosphere) molecular diffusion prevails and the region is called the *heterosphere*. The concept of a *homopause* or *turbopause*, a sharp boundary between these two regions, has been used in the past. It is commonly taken as the altitude where $K_z = D_s$, where K_z is the vertical eddy diffusion coefficient, which is introduced in Equation (10.57). However, in reality there is a region of transition where both processes are significant. Over the years a number of measurements have been carried out to establish the location of the terrestrial turbopause. These measurements have been based on two different general methods. One has used rocket released gas trails to observe the transition from turbulent to diffusive regions; Figure 10.3 shows the observed trail from such a release.[9] The other approach used the measured altitude distribution of inert species to infer the altitude where diffusive separation begins. The results from such measurements and theoretical fits are shown in Figure 10.4.[10]

Figure 10.3 Photograph of a sodium vapor trail released from a sounding rocket showing the transition from a turbulent to a diffusive region in the atmosphere.[9]

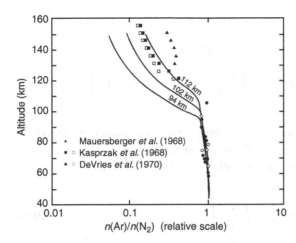

Figure 10.4 Calculated and measured height variations of the ratio of argon to molecular nitrogen densities. The calculations assumed three different turbopause heights, as indicated.[10]

The vertical velocity, u_{Esz}, due to eddy diffusion (mixing), can be written as

$$u_{Esz} = -K_z \frac{1}{n_s/n} \frac{\partial (n_s/n)}{\partial z} = -K_z \left(\frac{1}{n_s} \frac{\partial n_s}{\partial z} - \frac{1}{n} \frac{\partial n}{\partial z} \right), \tag{10.57}$$

where K_z is the vertical eddy diffusion coefficient and n is the total density. The simplified form of the steady state diffusion equation for the total density (often referred to as the *hydrostatic relation*, namely the balance between the gravitational force and the pressure gradient, see Equation (10.22)) is

$$\frac{dp}{dz} + n\langle m \rangle g = 0, \tag{10.58}$$

where

$$\langle m \rangle = \frac{\sum_s n_s m_s}{n}. \tag{10.59}$$

Substituting Equation (10.58) into Equation (10.57) the expression for the vertical flux due to eddy diffusion, Γ_{Esz}, becomes

$$\Gamma_{Esz} = n_s u_{Esz} = -K_z \left(\frac{\partial n_s}{\partial z} + \frac{n_s}{T} \frac{\partial T}{\partial z} + \frac{n_s}{H} \right), \tag{10.60}$$

where H and T are the scale height and the temperature of the mixed gas

$$H = \frac{kT}{\langle m \rangle g}. \tag{10.61}$$

Total vertical flux, Γ_{sz}, can be obtained by using Equations (10.54) and (10.57),

$$\Gamma_{sz}(z) = -D_s \left(\frac{\partial n_s}{\partial z} + \frac{n_s}{T_s} \frac{\partial T_s}{\partial z} + \frac{n_s}{H_s} \right) - K_z n \frac{\partial}{\partial z} \left(\frac{n_s}{n} \right). \tag{10.62}$$

Setting this flux equal to zero, one can obtain the following expression for the diffusive equilibrium number density:

$$n_s(z) = n_s(z_0) \frac{T_s(z_0)}{T_s(z)} \exp\left[-\int_{z_0}^{z} \left(\frac{1}{H_s} + \frac{\Lambda}{H} \right)(1 + \Lambda)^{-1} dz' \right],$$ (10.63)

where $\Lambda = K_z/D_s$. If $\Lambda = 0$, there is no turbulence and Equation (10.63) reduces to Equation (10.55), the diffusive equilibrium solution. On the other hand, when $\Lambda = \infty$, Equation (10.63) reduces to the fully mixed solution, which is

$$n_s(z) = n_s(z_0) \frac{T_s(z_0)}{T_s(z)} \exp\left(-\int_{z_0}^{z} \frac{dz'}{H} \right).$$ (10.64)

In case the flux is not zero a solution for $n_s(z)$ is possible,[11] if one assumes that the temperature is constant with altitude and so is the eddy diffusion coefficient, K_z. The reference altitude is taken to be at the homopause, so that $D_s = K_z \exp(h)$, where h is measured in units of the scale height of the mixed gas. The solution for n_s is

$$n_s(z) = A \exp(-h)[1 + \exp(h)]^{1-\Psi_s} + \frac{\Gamma_{sz} H}{K_z(1 - \Psi_s)} \exp(-h),$$

$$n_s(z) = B \exp(-h)[1 + \exp(h)]^{1-\Psi_s}$$

$$- \frac{\Gamma_{sz} H}{K_z(1 - \Psi_s)} \exp(-h)\{[1 + \exp(h)]^{1-\Psi_s} - 1\},$$ (10.65)

where Ψ_s = mass ratio = $m_s/\langle m \rangle$, Γ_{sz} is the vertical flux, and A and B are integration constants. The two forms of the relation for n_s are equivalent, with the first being useful for an upward flux and the second for a downward flux. For the case of zero flux, the asymptotes for large negative and positive values of h become, as expected,

$$n_s(z) \rightarrow A \exp(-h) \qquad \text{for } h \ll 0,$$
$$n_s(z) \rightarrow A \exp(-\Psi_s h) \quad \text{for } h \gg 0.$$ (10.66)

So far, only the steady state situation has been discussed. For a time-dependent problem the full continuity equation needs to be solved. In cases when transport, either molecular diffusion or eddy mixing, is dominant the time constants (approximated as being of the order of a scale height divided by the velocity) are, respectively

$$\tau_{\text{DM}} \approx H_s^2/D_s \quad \text{and} \quad \tau_{\text{DE}} \approx H^2/K_z.$$ (10.67)

A number of different ways show that the time constants are approximately correct. One way is to look at the continuity equation (10.20a), and noting that for the case

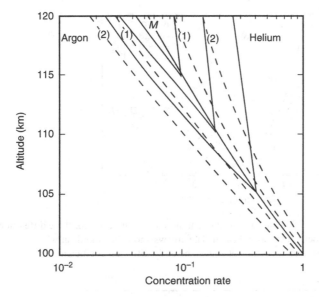

Figure 10.5 Calculations showing the time evolution of argon and helium distributions. The solid lines correspond to a steady state distribution of complete mixing and a sharp transition to diffusive separation at three different altitudes. The dashed curves marked by (1) and (2) show the calculated distributions 10 and 30 days from a completely mixed initial state.[12]

of molecular diffusion, $n_n u_n$ is of the order of $D_s \partial n_s / \partial z$ (10.54), therefore,

$$\frac{\partial n_s}{\partial t} \sim \frac{\partial}{\partial z}\left(D_s \frac{\partial n_s}{\partial z}\right). \tag{10.68}$$

Using scaling arguments and denoting the molecular diffusion time as τ_{DM}, Equation (10.68) can be written as

$$\frac{n_s}{\tau_{DM}} \sim \frac{D_s n_s}{H_s^2} \rightarrow \tau_{DM} \sim \frac{H_s^2}{D_s}. \tag{10.69}$$

(For a more rigorous derivation see Problem 10.7.) Figure 10.5 shows the results of calculations on the evolution of argon and helium densities from completely mixed initial conditions, after 10 and 30 days, respectively.[12] Note that argon tends more rapidly toward a diffusive altitude distribution than helium because the difference between the mixed and diffusive case is less for argon. Figure 10.6 shows steady state density profiles for a variety of eddy diffusion coefficients.[13]

The above discussions are appropriate for inert species, such as the noble gases He and Ar, as well as N_2, which is not affected by chemical processes to any significant degree. On the other hand, chemistry plays an important role in establishing the altitude distribution of oxygen and hydrogen in the terrestrial upper atmosphere. Molecular oxygen undergoes photodissociation at altitudes above about 100 km. In the lower thermosphere and upper mesosphere, the major loss mechanism for

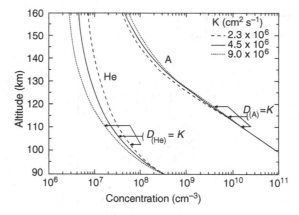

Figure 10.6 Calculated argon and helium density variations for three different eddy diffusion coefficients, as indicated. The arrows show the altitude at which the molecular diffusion coefficient equals the eddy diffusion coefficient.[13]

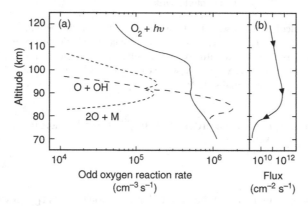

Figure 10.7 (a) Calculated molecular oxygen dissociation rate and atomic oxygen recombination rates. (b) Calculated vertical atomic oxygen flux.[14]

atomic oxygen is the three-body recombination reaction, $O + O + M \rightarrow O_2 + M$, introduced in Section 8.1 as Equation (8.6). At even lower altitudes, recombination with OH dominates. Rate calculations show that recombination is much slower than dissociation above about 90 km.[14] Therefore, atomic oxygen, newly created by dissociation, is transported downward to lower altitudes, where it recombines, and the freshly formed oxygen molecules are transported upward to replace those lost by dissociation at the higher altitudes. Figure 10.7 shows calculated photolysis and recombination rates as well as the downward atomic oxygen flux, which is equal to twice the upward flux of O_2.[14] Both eddy and molecular transport are involved, although eddy diffusion is the most important transport process at the lower altitudes.

To establish the altitude distribution of atomic oxygen, a continuity equation has to be solved that takes into account both eddy and molecular diffusion, as well

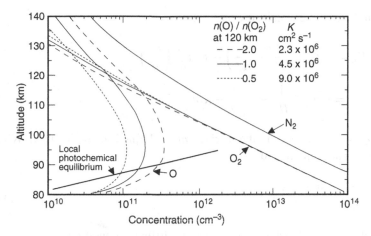

Figure 10.8 Calculated atomic and molecular oxygen and molecular nitrogen densities for three different assumed constant eddy diffusion coefficients.[13]

as the relevant photochemical processes. This cannot be done analytically, but has to be done numerically. The results of a one-dimensional numerical model[13] are shown in Figure 10.8. In this model the eddy diffusion coefficient was assumed to be independent of altitude and was varied to arrive at different O/O_2 ratios at 120 km. Chemical equilibrium solutions are also shown for comparison.

10.8 Escape of terrestrial hydrogen

The idea of the escape of light atmospheric gases, such as hydrogen, from the atmosphere was first proposed about 150 years ago. However, the processes controlling the actual escape of hydrogen from the terrestrial atmosphere have only been clarified about 30 years ago. Most of the attention in the past centered around the escape from the top of the atmosphere. This topic is discussed in Section 10.10. This section outlines the processes controlling the hydrogen distribution and flow velocities in the mesosphere and thermosphere, which play a very significant role in establishing the actual *escape flux* from the higher altitudes. It is not surprising that the escape flux can depend strongly on the "available" upward flux at lower altitudes. An expression for this *limiting flux* is developed here.[15] Hydrogen is a minor species, which makes the derivation of this limiting flux easier, but this is not a necessary condition.

The total vertical flux of a minor species in a stationary background gas is given by Equation (10.62). Now, if the temperature gradient terms are neglected and the logarithmic derivative of n_s for any arbitrary altitude distribution is denoted as

$$\frac{1}{n_s}\frac{\partial n_s}{\partial z}\bigg|_{\text{arbitrary}} = -\frac{1}{H_s^*}, \tag{10.70}$$

then Equation (10.62) can be rewritten as

$$\Gamma_{sz} = -D_s n_s \left(\frac{1}{H_s} - \frac{1}{H_s^*} \right) - K_z n \frac{\partial}{\partial z} \left(\frac{n_s}{n} \right)$$

$$= -b_s \xi_s \left(\frac{1}{H_s} - \frac{1}{H_s^*} \right) - K_z n \frac{\partial \xi_s}{\partial z},$$ (10.71)

where $b_s = D_s n$ and ξ_s is the *mixing ratio*, defined as $\xi_s = n_s/n$. When the constitutents are fully mixed

$$\frac{\partial \xi_s}{\partial z} = \frac{\partial}{\partial z} \left(\frac{n_s}{n} \right) = 0.$$ (10.72)

Next, if it is assumed that $m_s < \langle m \rangle$, then for diffusive conditions

$$\frac{\partial \xi_s}{\partial z} > 0.$$ (10.73)

Given the above assumption, the bracketed term on the right-hand side of Equation (10.71) is negative for mixed conditions, and negative approaching zero as it moves to diffusive equilibrium conditions. Given that $m_s < \langle m \rangle$, the first term is always positive and the maximum upward flux corresponds to complete mixing, which makes the second term zero. The expression for this limiting flux, Γ_ℓ, is

$$\Gamma_{sz} \bigg|_{\text{limiting}} = \frac{b_s \xi_s}{H} \left(1 - \frac{m_s}{\langle m \rangle} \right) = \Gamma_\ell.$$ (10.74)

Finally, for light gases, such as hydrogen, Equation (10.74) simplifies to

$$\Gamma_\ell \approx \frac{b_s \xi_s}{H}.$$ (10.75)

Substituting this result back into Equation (10.71), and after some rearranging, one obtains

$$\Gamma_{sz} = \Gamma_\ell - (K_z + D_s) n \frac{\partial \xi_s}{\partial z}.$$ (10.76)

This equation states that if the flow is not limiting there is a gradient in the mixing ratio (note that if Γ_{sz} is greater than Γ_ℓ the mixing ratio decreases with altitude, choking off the flow). In general, if escape from the top of the atmosphere is not a limiting factor, the upward flow is very close to the limiting flux.

The source of hydrogen in the homosphere, which supplies the upper atmosphere and the escaping flux, is discussed next. The main sources of hydrogen, which eventually escapes from the atmospheres of Earth, Venus, and Mars, are believed to be H_2O, H_2, and CH_4. An important question pertaining to all three planets is how quickly H_2 is converted to H at higher altitudes. Photolysis is very slow, and

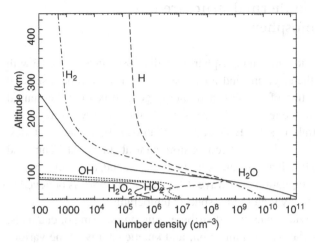

Figure 10.9 Plot of the calculated density distributions of hydrogen-carrying gases in the terrestrial atmosphere.[16]

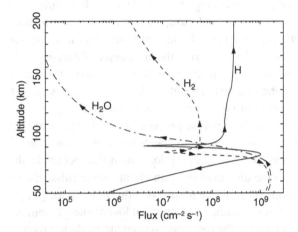

Figure 10.10 Plot of the calculated vertical fluxes of the major hydrogen-carrying species.[16]

on Earth the following two-step process in which H_2 is split by the formation of an O_2 molecule is believed to be the important one;

$$H_2 + O \rightarrow OH + H,$$

$$OH + O \rightarrow O_2 + H. \tag{10.77}$$

The results of model calculations of the density distributions of the major hydrogen-carrying gases and their fluxes for the terrestrial atmosphere are shown in Figures 10.9 and 10.10.[16] These calculations confirm that for the Earth, in general, the mixing ratio of the total hydrogen is about the same at the homopause as in the lower stratosphere. Furthermore, the escape flux is approximately equal to the limiting flux and all H_2 is converted into H below the critical level or exobase (Section 10.10).

10.9   Energetics and thermal structure of the Earth's thermosphere

The topic of energy deposition into an ionosphere was discussed in Section 9.6 with respect to the heating of the electrons and ions. One of the issues to be discussed in this section is what fraction of the absorbed energy goes to heating the neutral gas in the thermosphere. In the terrestrial thermosphere, the primary source of the energy going to the neutrals is solar EUV radiation, but at the higher latitudes, particle precipitation and Joule heating become dominant at times. At Venus and Mars solar heating is also dominant, whereas at the outer planets a combination of solar, energetic particle precipitation, and wave dissipation processes is believed to be responsible for the relatively high thermospheric temperatures.

The absorbed solar ultraviolet energy is distributed among three major channels: radiation or airglow, dissociation and ionization, and kinetic energy of the various thermospheric or ionospheric constituents, as outlined in Section 9.1 and indicated in Figure 9.1. An evaluation of the neutral gas heating efficiency, which is the fraction of absorbed energy that goes to heating (Section 9.6), needs accurate and comprehensive calculations of the chemical and transport processes, involving both neutral and ionized constituents. Such detailed calculations have been carried out for the thermospheres of Earth, Venus, and Mars during the last couple of decades.[17, 18]

In this section the discussion is limited to the terrestrial case and, because of the many processes involved in the establishment of the heating efficiency, only a single process, the photodissociation of O_2, is briefly discussed as a simple representative example. The major direct dissociation process for O_2 in the terrestrial thermosphere is photodissociation by radiation in the *Schumann–Runge contin-uum* ($\sim$125–175 nm). In addition, practically every ionization that occurs in the thermosphere ultimately leads to the dissociation of an O_2 molecule (dissociative recombination and other chemical processes are involved in this indirect route; e.g., $O_2^+ + e^- \rightarrow O + O$). These processes constitute an important loss of energy from the middle and upper thermosphere because the resulting oxygen atoms do not recombine locally, but are transported down to below about 100 km, where they recombine (Section 10.7). Therefore, each dissociation removes 5.12 eV from the thermosphere above about 100 km. Calculations indicate that about 33% of the total absorbed solar energy goes to O_2 dissociation. The average energy of a photon in the Schumann–Runge continuum is about 7.6 eV and the probability is high that one of the resulting oxygen atoms is in the ^1D state, taking a further 1.97 eV. Therefore, on average, there is about 0.5 eV left over, which goes into the kinetic energy of the oxygen atoms created by the dissociation, and thus, about 6.6% of the absorbed Schumann–Runge radiation goes directly into heating. Some of the energy that goes into exciting the ^1D state is recovered as heating via the quenching (collisional de-excitation) process (Section 8.7), which occurs mainly at lower altitudes.

Figure 10.11 shows the local energy loss rates for the various major channels, calculated for a typical, terrestrial solar minimum case.[17] Clearly, the total local losses do not balance the total energy deposition rate at the higher altitudes because

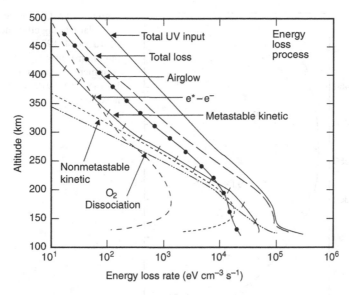

Figure 10.11 A comparison of calculated solar UV and EUV energy deposition rates and major energy partitions.[17]

of the downward transport of long lived species. The calculated heating efficiencies for a number of different cases are shown in Figure 10.12. These results indicate that the heating efficiency is not a constant, but varies with altitude, local time, season, and solar cycle. The calculated neutral gas heating efficiency for Venus is shown in Figure 10.13.[18] Neutral gas heating efficiency calculations for precipitating particles have also been carried out and the processes involved are similar.[19] The results of calculations for auroral electron fluxes with Maxwellian energy spectra of several characteristic energies and a range of energy deposition rates, for terrestrial conditions, are shown in Figure 10.14.

The simplest form of the energy equation, appropriate for the terrestrial thermosphere, is Equation (3.59), with the second and fourth terms on the left-hand side neglected. Considering the steady state vertical component of this energy equation, one can write for the "global mean"

$$-\frac{\partial}{\partial z}\left(\lambda_n \frac{\partial T}{\partial z}\right) = \frac{\delta E}{\delta t} = \bar{Q}_{heat}, \qquad (10.78)$$

where $\bar{Q}_{heat}$ is the globally averaged net heating rate and T is the globally averaged neutral gas temperature. This equation states that the only process by which heat is transported vertically is thermal conduction. Furthermore, no heat flows into the terrestrial thermosphere from the top. Therefore, the integral of the net heating above a given altitude has to be transported down through that level by heat conduction, thus setting the temperature gradient at that altitude. Considering the situation at 150 km of the terrestrial thermosphere and assuming that the integrated net heating

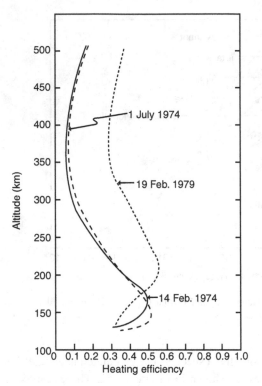

Figure 10.12 Calculated neutral gas solar heating efficiencies for February 14, 1974 (winter, solar minimum), July 1, 1974 (summer, solar minimum) and February 19, 1979 (winter, solar maximum).[17]

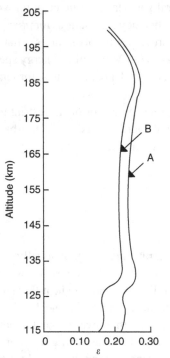

Figure 10.13 Calculated neutral gas solar heating efficiencies ε for Venus. The curve marked A is the standard value and that marked B is the minimum value.[18]

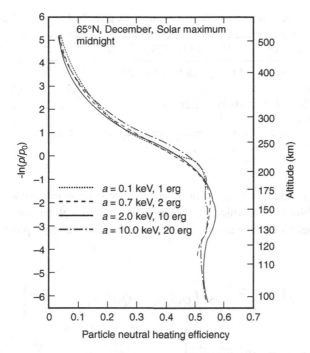

Figure 10.14 Calculated neutral gas heating efficiency for precipitating auroral electron fluxes with Maxwellian energy spectra of several characteristic energies and a range of energy deposition rates. The altitude scale is in terms of normalized pressure for a specific atmospheric condition.[19]

above 150 km is $0.3 \, \text{erg cm}^{-2} \, \text{s}^{-1}$, one obtains

$$-\lambda_n \frac{\partial T}{\partial z}\bigg|_{\infty} + \lambda_n \frac{\partial T}{\partial z}\bigg|_{150 \, \text{km}} = \int\limits_{150 \, \text{km}}^{\infty} \bar{Q}_{\text{heat}} dz \approx 0.3 \approx 45(750)^{3/4} \frac{\partial T}{\partial z}\bigg|_{150 \, \text{km}}$$

$$\therefore \frac{\partial T}{\partial z}\bigg|_{150 \, \text{km}} \approx 4.65 \, \text{K km}^{-1}, \tag{10.79}$$

where the fact that the heat inflow from the top is zero was used, and where it was assumed that the temperature at 150 km is 750 K. The thermal conductivity value corresponding to an even mixture of atomic oxygen and molecular nitrogen was also used. Some rather complex theoretical equations and observational data for the thermal conductivity of various neutral gas species of interest do exist.[20-22] However, it has been common practice by aeronomers to adopt the following simple expression for the neutral gas thermal conductivity:

$$\lambda_n = AT^s. \tag{10.80}$$

Table 10.1 gives the values of A and s for a variety of neutral gas constituents, obtained from theory or measurements.

Table 10.1 Constants for the expression of thermal conductivity given by Equation (10.80).

	N_2		O_2	O	CO_2	CH_4	H		H_2	
A^a	5.63	($T < 150$)	36	54	0.82	5.63	235	103		($T < 150$)
	36	($T > 150$)						223		($T > 150$)
s	1.12	($T < 150$)	0.75	0.75	1.28	1.12	0.75	0.92		($T < 150$)
	0.75	($T > 150$)						0.77		($T > 150$)

^{a}The values of A are in erg (cm s K)$^{-1}$.

The above highly simplified calculation can also be carried out going in the other direction (Problem 10.8), which leads to an estimate of the global mean upper thermospheric temperature if one assumes that the heating rate varies exponentially with altitude,

$$\bar{Q}_{\text{heat}} = Q_0 \exp\left\{-\frac{z - z_0}{H}\right\} \Rightarrow \int_z^\infty \bar{Q}_{\text{heat}} dz = Q_0 H \exp\left\{-\frac{z - z_0}{H}\right\}$$

$$\therefore \lambda_n \frac{\partial T}{\partial z}\bigg|_z = Q_0 H \exp\left\{-\frac{z - z_0}{H}\right\}, \tag{10.81}$$

where H is a characteristic length. Integrating, using the same parameters as in the previous example, and taking H to be 50 km, the calculated temperature at the top of the atmosphere comes out to be about 956 K. Even in this simple calculation, the argument is somewhat circular, because H must be related to the temperature.

Clearly, these highly simplified calculations only provide very crude, order-of-magnitude estimates of the mean global temperature. Temperatures calculated by a comprehensive one-dimensional, coupled, thermospheric and ionospheric model,[23] for solar cycle maximum and minimum conditions for the terrestrial thermosphere, are shown in Figure 10.15. As a comparison, Figure 10.16 shows a calculated temperature profile for Jupiter, along with some indirectly measured values.[24] Highly sophisticated three-dimensional numerical models called *thermosphere and ionosphere general circulation models* (TIGCMs) are now in existence, which solve the coupled continuity, momentum, and energy equations for the terrestrial thermosphere, mesosphere, and ionosphere.[25-27] The results of a representative set of calculations are shown in Figure 10.17, which presents neutral gas temperatures and wind velocities at a constant pressure surface, corresponding to approximately 286 km.[25] Such TIGCMs also exist for Venus, Mars, Jupiter, Saturn, and Titan.[28-31]

Over the last couple of decades, a great deal of observational data on terrestrial thermospheric temperatures and composition have been gathered by satellite-borne neutral mass spectrometers and ground-based incoherent scatter radars. These results have been used to obtain an empirical model of the thermosphere called the *MSIS*,

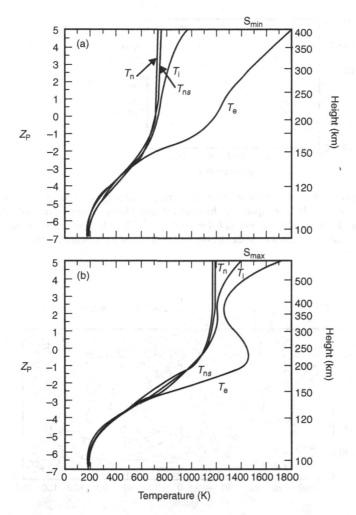

Figure 10.15 Calculated terrestrial mean neutral gas, T_n, electron, T_e, and ion, T_i, temperature profiles for solar cycle maximum and minimum conditions, obtained from a one-dimensional model. The curves marked T_{ns} correspond to the neutral temperatures from the empirical MSIS model.[23]

mass spectrometer incoherent scatter model,[32] which gives temperature and composition values as a function of altitude, time, geographic location, and geomagnetic conditions. It has become a widely used standard reference model. A new, expanded and improved version of this model, called NRLMSISE-00 is now also available.[33] A representative set of density and temperature values is given in Appendix K. In Figure 10.17 the calculated temperatures are compared with the MSIS values, showing reasonably good agreement. An observation-based reference model of densities and temperatures, sometimes called the *VIRA*, *Venus international reference atmosphere model*,[34, 35] has also been developed for Venus. Representative VIRA model values are also presented in Appendix K. Finally an empirical model of Titan's upper

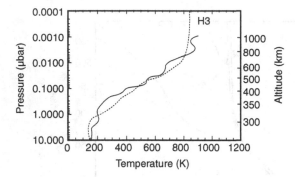

Figure 10.16 Calculated and observed neutral gas temperature profiles for Jupiter. The curve marked H3 corresponds to the values obtained from a one-dimensional model, which assumed a heating rate of 0.4 erg cm^{-2} s^{-1}. The other curve corresponds to the temperature extracted from the density profile obtained by the *Galileo* probe.[24]

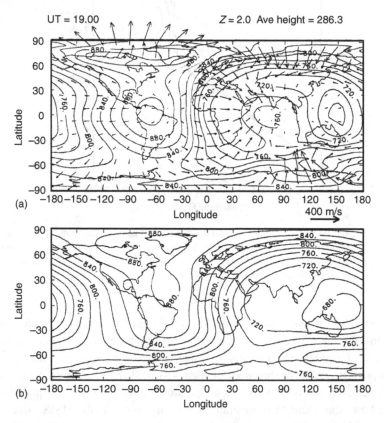

Figure 10.17 Terrrestrial neutral gas temperatures and velocities at a constant pressure height, corresponding to an altitude of about 286 km, calculated by a numerical model (top) and an empirical model (bottom).[25]

atmosphere has also been constructed.[36] It used a Legendre polynomial expansion to fit the densities measured by the *Cassini* ion neutral mass spectrometer.[37]

10.10   Exosphere

The uppermost regions of planetary atmospheres are called *exospheres*. In the thermosphere the collision frequency among the neutral gas constituents is sufficiently large that the particles have, to a good approximation, a Maxwellian velocity distribution. The collision frequency decreases with increasing altitude and the distribution eventually deviates significantly from a Maxwellian. This transition altitude is called the *critical level* or *exobase* and the region above it is the exosphere. In the exosphere, the mean-free-path is long enough to allow particles to follow ballistic trajectories, and those with sufficient energy escape from the planet's gravitational field. In the classical definition of the critical level, particles experience no further collisions in the region above this altitude. Therefore, this critical level or exobase is located at a radial distance, r_c, where the probability of a collision for an upward moving particle is unity;

$$\int_{r_c}^{\infty} n(r)\sigma \, \mathrm{d}r = 1 \cong \sigma n(r_c)H(r_c) = \frac{H(r_c)}{\lambda(r_c)}, \tag{10.82}$$

where σ is the collision cross section and $\lambda = 1/n\sigma$ is the vertical mean-free-path (Equation 4.3). In other words, the exobase is at an altitude where the mean-free-path is equal to the atmospheric scale height. Of course, this description is a highly simplified one. For one thing, the collision cross section is energy dependent, so the exobase altitude varies with energy even in this simple definition. The more important issue is that, in most classical calculations of exospheric densities and escape fluxes, the velocity distribution function at the critical level is assumed to be Maxwellian, which cannot be true because of the low collision frequencies and the escaping particles. Numerous Monte Carlo calculations have been used to study the exospheres and escape fluxes at Earth, Mars, and Venus.[38–40] In most of these calculations, sharp transition boundaries are generally not assumed and the transition between the thermosphere and exosphere comes about "naturally." However, the errors introduced by using a Maxwellian distribution at an appropriate critical level are typically not very large, and the analytic solutions thus obtained are instructive and given later in this section. There is another frequently used approach,[41] in which the exobase distribution function (or flux) is calculated using a multi-stream method (see Section 9.4) and the Liouville theorem used to determine the densities and velocities in the exosphere. This is discussed in more detail in the next section.

The particles found above the critical level can be divided into the following categories:

(*a*)   Particles that cross the critical level; and
(*b*)   Particles that never reach the critical level.

Some of the particles moving upward above the critical level have velocities sufficiently large to escape from the planet's gravitational pull along parabolic or hyperbolic trajectories. The *escape velocity* at the critical level, r_c, is

$$v_{esc} = \left(\frac{2GM}{r_c}\right)^{1/2},$$ (10.83)

where G is the gravitational constant and M is the mass of the planetary body. Recognizing the fact that some of the particles can escape means that a further subdivision is possible. Namely, particles in category (a) consist of

1.   Particles moving upward with a velocity less than the escape velocity;
2.   Particles moving downward with a velocity less than the escape velocity;
3.   Particles moving upward on ballistic escape trajectories; and
4.   Particles moving downward on ballistic trajectories from interplanetary space.

There are two subgroups making up category (b):

1.   Particles in satellite orbits whose periapses are above the critical level; and
2.   Particles on ballistic orbits from interplanetary space.

The classical *Jeans* or *thermal escape flux*[42] is obtained by assuming a discrete critical level and a Maxwellian velocity distribution at that level. The escape flux can then be obtained by integrating over the appropriate velocity limits

$$\Gamma_{esc} = n(r_c)\left(\frac{m}{2\pi kT}\right)^{3/2} \int_{v_{esc}}^{\infty} \int_0^{\pi/2} \int_0^{2\pi} v\cos\theta\, v^2 \exp\left(-\frac{mv^2}{2kT}\right)\sin\theta\, d\phi\, d\theta\, dv$$

$$= \frac{n(r_c)}{2}\sqrt{\frac{2kT}{\pi m}}\left(1 + \frac{GMm}{kTr_c}\right)\exp\left(-\frac{GMm}{kTr_c}\right)$$

$$= \frac{n(r_c)v_{mp}}{2\pi^{1/2}}\left(1 + \frac{v_{esc}^2}{v_{mp}^2}\right)\exp\left(-\frac{v_{esc}^2}{v_{mp}^2}\right),$$ (10.84)

where v_{mp} is the most probable speed, as given by Equation (H.24).

This Jeans thermal escape flux is highly temperature dependent and is important in the terrestrial case. However, at Venus and Mars, where the thermospheric temperatures are low, it is negligible, except for light gases, such as hydrogen. In the terrestrial case, the neutral gas temperature near the critical level is high enough to ensure that the thermal escape flux alone is, in general, sufficiently large so that the hydrogen escape rate is limited by the low altitude fluxes (Section 10.8). This hydrogen escape flux is of the order of $2-3 \times 10^8 \text{cm}^{-1}\text{s}^{-1}$.

There are a number of escape mechanisms, besides thermal escape, that are potentially very important.[43] The term "hydrodynamic escape" is associated with a global expansion of the atmosphere, resulting from a very large heat input.[44] One way of understanding this mechanism is as follows. In Jeans escape a small fraction of atoms in the high energy wing of the Maxwellian distribution have sufficient energy to escape, whereas in hydrodynamic escape basically the full Maxwellian distribution is lost. It is believed that during the early phase of the evolution of the terrestrial atmosphere about 4 billion years ago ($\sim$ 4 Gyr b.p.) when the solar EUV flux was much greater, hydrodynamic escape played an important role.

Photochemical escape is another important mechanism to consider.[45, 46] Dissociative recombination (Section 8.4) of molecular ions can create energetic neutral atoms, which may have enough energy to escape certain solar system bodies. For example, the dissociative recombination of molecular oxygen ions at Mars does make a significant contribution to the total oxygen escape from that planet. Charge exchange (Section 8.3) is also a potentially important escape mechanism. In this process a hot, energetic ion charge exchanges with a cold neutral. The resulting energetic neutral is not constrained by magnetic field lines, and may have enough energy to escape. For example, hot hydrogen neutrals can be created by the following two charge exchange reactions:

$$H_{hot}^+ + H \rightarrow H^+ + H_{hot}, \tag{10.85}$$

$$H_{hot}^+ + O \rightarrow O^+ + H_{hot}. \tag{10.86}$$

Thus, for example, if the ion energy is greater than 0.63 eV and the ion is moving away from the Earth, the resulting neutral hydrogen atom will be able to escape from the terrestrial atmosphere. Figure 10.18 shows an estimate of relative importance of the Jeans and charge exchange hydrogen escape fluxes, as a function of the exobase temperature, for the terrestrial case.[47]

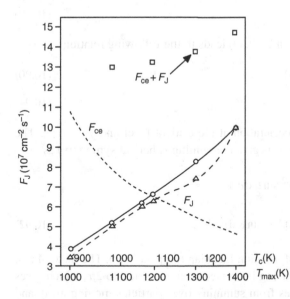

Figure 10.18 A plot of the estimated Jeans and charge exchange escape fluxes as a function of the Earth's exobase temperature. The two curves associated with the Jeans flux are based on different assumptions regarding the symmetry of the hydrogen distribution. The sum of the two fluxes, marked by the squares, is nearly constant over the indicated temperature range.[47]

Ion escape and ionospheric outflow are processes that also need to be considered. At the Earth, high-speed, polar wind type (Sections 5.8 and 12.12) of outflows can result in actual escape. However, current calculations indicate that this type of escape is not important in the terrestrial case. On the other hand, ionospheric outflows into the tail do make a significant contribution to the total escape at Mars.[48] Significant ion outflows can also result from the photo or electron impact ionization of neutrals, especially at nonmagnetic solar system bodies, such as Venus and Mars. In this case the newly created ions are picked up by the shocked solar wind and move and escape with it. Finally, the term sputtering has been introduced to describe the process by which ions impact the atmosphere and the collision results in the ejection of a neutral atmospheric particle which may have enough energy to escape.[49] Here again, this mechanism may make an important contribution to the escape of atomic oxygen from the atmosphere of Mars. The ion escape rate from Mars, during solar cycle minimum conditions, has recently been measured and found to be about 3.1×10^{23} s^{-1};[50] model calculations also obtain values consistent with this observation.[51]

A variety of exospheric density studies have used a simple form of Liouville's theorem to obtain estimates of exospheric densities.[41] According to this theorem one can write

$$f(r, v) = f(r_c, v_c), \tag{10.87}$$

where f is the velocity distribution function, and v and v_c are the particle speeds at an arbitrary radial position r and at the critical level, r_c, respectively. The relationships between v and v_c and θ and θ_c, the angles with respect to the local vertical, are given by the conservation of energy and angular momentum relations;

$$\frac{1}{2}mv^2 - \frac{GMm}{r} = \frac{1}{2}mv_c^2 - \frac{GMm}{r_c}, \tag{10.88}$$

$$rv \sin \theta = r_c v_c \sin \theta_c. \tag{10.89}$$

Rearranging Equations (10.88) and (10.89) leads to the following relations:

$$v_c^2 - v^2 = 2g_c r_c (1 - y), \tag{10.90}$$

$$v \sin \theta = yv_c \sin \theta_c, \tag{10.91}$$

where g_c is the gravitational acceleration at the critical level and $y = r_c/r$. The density $n(r)$ at a radial distance, r, is given (assuming spherical symmetry) as

$$n(r) = 4\pi \iint f(r, v)v^2 \sin \theta \, dv \, d\theta$$

$$= 4\pi \iint f(r_c, v_c)Jv^2 \sin \theta \, dv_c \, d\theta_c, \tag{10.92}$$

where J is the appropriate transformation Jacobian (Appendix C). The factor 4π in Equation (10.92) comes from having integrated over π from 0 to 2π, which gives 2π, and the factor of two comes from summing over particles moving away and toward the critical level. This latter factor of two is appropriate for "symmetric" distributions, but must be modified for other cases. Substituting for the Jacobian, which is obtained from the conservation relations (10.88) and (10.89), one obtains

$$n(r) = 4\pi y^2 \iint \frac{f(r_c, v_c)v_c^3 \cos\theta_c \sin\theta_c \, d\theta_c \, dv_c}{\left[v_c^2(1 - y^2 \sin^2\theta_c) - 2g_c r_c(1 - y)\right]^{1/2}}. \tag{10.93}$$

The integral over θ_c needs to be carried out from 0 to $\pi/2$. The integration limits over v_c can be obtained by noting that at the altitude level r, the particle must have an upward velocity of zero or greater, therefore

$$v^2 \cos^2\theta = v^2(1 - \sin^2\theta) \geq 0. \tag{10.94}$$

Use of Equations (10.90) and (10.91) leads to

$$v_c^2 - 2g_c r_c(1 - y) - v_c^2 y^2 \sin^2\theta_c \geq 0$$

$$\therefore v_c \geq v_{ce}u, \tag{10.95}$$

where the escape velocity v_{ce} at r_c is given by Equation (10.83) and can also be written as

$$v_{ce} = \sqrt{2g_c r_c} \tag{10.96}$$

and

$$u^2 = \frac{1 - y}{1 - y^2 \sin^2\theta_c}. \tag{10.97}$$

Given Equation (10.93), the density at a location, r, can be calculated if the velocity distribution function, $f(r_c, v_c)$, is known. Analytic expressions for $n(r)$ have been obtained assuming various distributions at the exobase, such as complete or truncated Maxwellian distributions. For example, the expression for $n(r)$, given a full Maxwellian distribution at the critical level, is[52]

$$n(r) = n(r_c) \exp\left[-E(1 - y)\right]\left[1 - (1 - y^2)^{1/2} \exp\left(-\frac{Ey^2}{1 + y}\right)\right], \tag{10.98}$$

where $E = mv_{ce}^2/2kT$. Note that at infinity ($y = 0$), Equation (10.98) gives zero density, which is different from that obtained assuming a hydrostatic approximation.

10.11   Hot atoms

The previous discussion of mechanisms leading to escape fluxes mentioned that a variety of different processes can lead to atoms having significant kinetic energies. Significant populations of such "hot" atoms have been predicted to be present in the upper atmospheres of solar system bodies. The only planet where such a hot atom population has been definitely detected is Venus.[53] However, there are some clear indications of their presence at Earth[54, 55] and Mars[40, 56] and they are expected to be present around some of the outer planet moons, namely Europa[57] and Titan.[58] These hot atoms may play a significant role in setting escape fluxes at planets with low thermospheric temperatures, as mentioned in the previous section, and may also play a role in the chemistry or energetics of the thermospheres and ionospheres.[54, 59] Finally, they can influence the solar wind interaction with weakly or nonmagnetized planets, such as Mars and Venus[60] and the interaction of magnetospheric plasmas with moons, such as Titan.[61]

In this section we will briefly examine the case of oxygen and carbon at Venus as representative examples. The three sources most commonly considered for hot oxygen atoms are the dissociative recombination of ions (Equation 8.40) and the following two charge exchange reactions:

$$O_{hot}^+ + O \rightarrow O^+ + O_{hot}, \tag{10.99}$$

$$O_{hot}^+ + H \rightarrow H^+ + O_{hot}. \tag{10.100}$$

Calculations have shown that, in effect, dissociative recombination is the major source of hot oxygen for Venus, Earth, and Mars.[40, 41] Hot carbon atoms have also been seen at Venus and the likely sources that have been examined are the dissociative recombination of CO^+, the photodissociation of CO and collisions with hot oxygen atoms.[62] The calculations indicated that the dominant processes are the first two of these.[63]

Exospheric densities and escape fluxes have been calculated using either Monte Carlo techniques[38–40] or a combination of the two-stream[56] (Section 9.4) and Monte Carlo[46] approaches to obtain the exobase fluxes, which are then used to calculate exospheric densities using Liouville's theorem. The exobase fluxes are usually calculated as a function of energy, which means that to use Liouville's theorem, Equation (10.93) has to be recast in terms of energy

$$n(r) = \frac{1}{2} \int_{\theta_c} \int_{E_c} F(r_c, E_c) \frac{\bar{v}_c y^2 \sin\theta_c \cos\theta_c}{\left[\bar{v}_c(1 - y^2 \sin^2\theta_c) - (1 - y)\right]^{1/2}} \, dE_c \, d\theta_c, \tag{10.101}$$

where F is the energy distribution function and $\bar{v}_c = v_c/v_{ce}$ is the normalized velocity at r_c. After some lengthy algebra it was shown that the number of particles at a radial distance, r, associated with an energy increment, ΔE_c, depending on the corresponding velocity range, is:[64]

$\bar{v}_c > 1$ (escaping particles),

$$\Delta n(r) = \frac{F(r_c, E_c)}{2\bar{v}_c} \left([\bar{v}_c^2 - (1 - y)]^{1/2} - [\bar{v}_c^2(1 - y^2) - (1 - y)]^{1/2}\right) \Delta E_c, \tag{10.102}$$

$$\frac{1}{[1+y]^{1/2}} < \bar{v}_c < 1,$$

$$\Delta n(r) = \frac{F(r_c, E_c)}{\bar{v}_c} \left(\left[\bar{v}_c^2 - (1-y)\right]^{1/2} - \left[\bar{v}_c^2(1-y^2) - (1-y)\right]^{1/2} \right) \Delta E_c,$$

$$(10.103)$$

$$[1-y]^{1/2} < \bar{v}_c < \frac{1}{[1+y]^{1/2}},$$

$$\Delta n(r) = \frac{F(r_c, E_c)}{\bar{v}_c} \left[\bar{v}_c^2 - (1-y)\right]^{1/2} \Delta E_c.$$

$$(10.104)$$

Equations (10.102–104) allow the total population of particles at a given altitude to be calculated, given the variation of F over the energy range of relevance. Examples of such calculations are shown in Figures 10.19 and 10.20. Figure 10.19

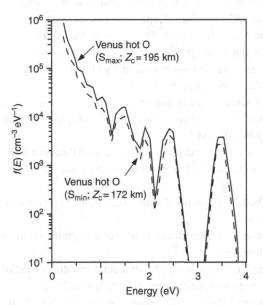

Figure 10.19 Calculated energy distribution of escaping oxygen ions at the exobase of Venus for solar cycle maximum and minimum conditions.[64]

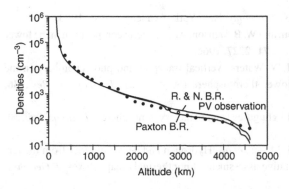

Figure 10.20 Calculated and observed hot oxygen densities at Venus. The solid dots represent observations and the solid lines correspond to model calculations based on two different assumptions concerning the branching ratio of O_2^+ recombination.[41]

shows the calculated energetic oxygen flux at the critical level for solar cycle minimum and maximum daytime conditions at Venus.[64] The calculated exospheric densities for solar cycle maximum daytime conditions are shown in Figure 10.20, along with the densities deduced from measurements of the OI 130.4 nm airglow emission.[41] There is good agreement between calculations and observations. It must be pointed out that these calculations were one-dimensional and that horizontal transport is not negligible, especially near the terminator, so more complex calculations (e.g., multi-dimensional Monte Carlo calculations) are necessary for improved accuracy.[40,65]

10.12   Specific references

1. Chapman, S. C., and T. G. Cowling, *The Mathematical Theory of Non-Uniform Gases*, New York: Cambridge University Press, 1970.
2. Hines, C. O., *The Upper Atmosphere in Motion: A Selection of Papers with Annotation*, Geophys. Monogr., **18**, Washington, DC: American Geophysical Union, 1974.
3. Chapman, S. C., and R. S. Lindzen, *Atmospheric Tides*, Dordrecht: D. Reidel, 1970.
4. Hines, C. O., Internal AGWs at ionospheric heights, *Can. J. Phys.*, **38**, 1441, 1960.
5. Siebert, M., Atmospheric tides, in *Advances in Geophysics*, ed. by H. E. Landsberg and J. Van Meigham, **105**, New York: Academic Press, 1961.
6. Hough, S. S., On the application of harmonic analysis to the dynamical theory of tides, Part I. On Laplace's "Oscillations of the first species," and on the dynamics of ocean currents, *Phil. Trans. Roy. Soc.*, **A189**, 201, 1897.
7. Hough, S. S., On the application of harmonic analysis to the dynamical theory of tides, Part II. On the general integration of Laplace's dynamical equations, *Phil. Trans. Roy. Soc.*, **A191**, 139, 1898.
8. Forbes, J. M., and H. B. Garrett, Theoretical studies of atmospheric tides, *Rev. Geophys. Space Phys.*, **17**, 1951, 1979.
9. Tohmatsu, T., *Compendium of Aeronomy*, Dordrecht, Holland: Terra Scientific Publishing Co., Kluwer Academic Publishers, 1990.
10. George, J. D., S. P. Zimmerman, and T. J. Keneshea, The altitudinal variation of major and minor neutral species in the upper atmosphere, *Space Res.*, **XII**, 695, 1972.
11. Chamberlain, J. W., and D. M. Hunten, *Theory of Planetary Atmospheres*, New York: Academic Press, 1987.
12. Banks, P. M., and G. Kockarts, *Aeronomy*, New York: Academic Press, 1973.
13. Colegrove, F. D., F. S. Johnson, and W. B. Hanson, Atmospheric composition in the lower thermosphere, *J. Geophys. Res.*, **71**, 2227, 1966.
14. Allen, M., Y. L. Yung, and J. W. Waters, Vertical transport and photochemistry in the terrestrial mesosphere and lower thermosphere (50–120 km), *J. Geophys. Res.*, **86**, 3617, 1981.
15. Hunten, D. M., The escape of light gases from planetary atmospheres, *J. Atmos. Sci.*, **30**, 1481, 1973.
16. Yung, Y. L., W. Jun-Shan, J. I. Moses, *et al.*, Hydrogen and deuterium loss from the terrestrial atmosphere: A quantitative assessment of nonthermal escape fluxes, *J. Geophys. Res.*, **94**, 14971, 1989.

17. Torr, M. R., P. G. Richards, and D. G. Torr, A new determination of the ultraviolet heating efficiency of the thermosphere, *J. Geophys. Res.*, **85**, 6819, 1980.

18. Fox, J. L., Heating efficiencies in the thermosphere of Venus reconsidered, *Planet. Space Sci.*, **36**, 37, 1988.

19. Rees, M. H., *Physics and Chemistry of the Upper Atmosphere*, Cambridge, UK: Cambridge University Press, 1989.

20. Hilsenrath, J. *et al.*, *Tables of Thermodynamic and Transport Properties*, Pergamon Press, 1960.

21. Reid, R. C., J. M. Prausnitz, and T. K. Sherwood, *The Properties of Gases and Liquids*, New York: McGraw-Hill Book Co., 1977.

22. Lide, D. R., Editor-in Chief, *CRC Handbook of Chemistry and Physics*, Boca Raton, FL: CRC Press, 1997.

23. Roble, R. G., E. C. Ridley, and R. E. Dickinson, On the global mean structure of the thermosphere, *J. Geophys. Res.*, **92**, 8745, 1987.

24. Waite, J. H., G. R. Gladstone, W. S. Lewis, *et al.*, Equatorial X-ray emissions: implications for Jupiter's high exospheric temperatures, *Science*, **276**, 104, 1997.

25. Roble, R. G., E. C. Ridley, A. D. Richmond and R. E. Dickinson, A coupled thermosphere and ionosphere general circulation model, *Geophys. Res. Lett.*, **15**, 1325, 1988.

26. Fuller-Rowell, T. and D. Rees, A three-dimensional, time dependent, global model of the thermosphere, *J. Atmos. Sci.*, **37**, 2545, 1980.

27. Ridley, A. J., Y. Deng, and G. Tóth, The global ionosphere-thermosphere model, *J. Atm. Sol.-Terr. Phys.*, **68**, 839, 2006.

28. Bougher, S. W., Simulations of the upper atmosphere of the terrestrial planets, in *Atmospheres in the Solar System: Comparative Aeronomy*, Geophysical Monograph 130, American Geophysical Union, Washington, 2002.

29. Bougher, S. W., J. H. Waite, T. Majeed, and G. R. Gladstone, Jupiter thermospheric general circulation model (JTGCM): Global structure and dynamics driven by auroral and Joule heating, *J. Geophys. Res.*, **110**, E04008, doi:10.1029/2003JE002230, 2005.

30. Müller-Wodarg, I. C. F., M. Mendillo, R. V. Yelle, and A. D. Aylward, A global circulation model of Saturn's thermosphere, *Icarus*, **180**, 147, 2006.

31. Müller-Wodarg, I. C. F. and R. V. Yelle, The effects of dynamics on the composition of Titan's upper atmosphere, *Geophys. Res. Lett.*, **29**, 2139, doi:10.1029/2002GL016100, 2002.

32. Hedin, A. E., MSIS-86 thermospheric model, *J. Geophys. Res.*, **92**, 4649, 1987.

33. Picone, J. M., A. E. Hedin, D. P. Drob, and A. C. Akin, NRLMSISE-00 empirical model of the atmosphere: Statistical comparisons and scientific issues, *J. Geophys. Res.*, **107**, A121468. doi:10.1029/2002JA009430, 2002.

34. Hedin, A. E., H. B. Niemann, W. T. Kasprzak, and A. Seil, Global empirical model of the Venus thermosphere, *J. Geophys. Res.*, **88**, 73, 1983.

35. Keating, G. M., J. L. Bertaux, S. W. Bougher, R. V. Yelle, J. Cui, and J. H. Waite, Models of Venus neutral upper atmosphere: structure and composition, *Adv. Space Res.*, **5**, 117, 1985.

36. Müller-Wodarg, I. C. F., *et al.*, Horizontal structures and dynamics in Titan's thermosphere, *J. Geophys. Res.*, **113**, E10005, doi:10.1029/2007JE003033, 2008.

37. Waite, J. H., W. S. Lewis, W. T. Kasprzak, *et al.*, The Cassini ion and neutral mass spectrometer (INMS) investigation, *Space Sci. Rev.*, **114**, 113, 2004.

38. Hodges, R. R., Monte Carlo simulation of the terrestrial hydrogen exosphere, *J. Geophys. Res.*, **99**, 23229, 1994.

39. Shematovich, V. I., D. V. Bisikalo, and J. C. Gerard, A kinetic model of the formation of the hot oxygen geocorona 1. Quiet geomagnetic conditions, *J. Geophys. Res.*, **99**, 23217, 1994.

40. Hodges, R. R., Distributions of hot oxygen for Venus and Mars, *J. Geophys. Res.*, **105**, 6971, 2000.

41. Nagy, A. F., J. Kim, and T. E. Cravens, Hot hydrogen and oxygen atoms in the upper atmospheres of Venus and Mars, *Ann. Geophysicae*, **8**, 251, 1990.

42. Jeans, J. H., *The Dynamical Theory of Gases, 4th Edition*, Cambridge, UK: Cambridge University Press, 1925.

43. Chassefiere, E. and F. Leblanc, Mars atmospheric escape and evolution; interaction with the solar wind, *Planet. Space Sci.*, **52**, 1039, 2004.

44. Hunten, D. M., The escape of light gases from planetary atmospheres, *J. Atmos. Sci.*, **30**, 148, 1973.

45. McElroy, M. B., Mars an evolving atmosphere, *Science*, **175**, 443, 1972.

46. Fox, J. L. and A. Hac, Spectrum of hot O at the exobases of terrestrial planets, *J. Geophys. Res.*, **102**, 24005, 1997.

47. Bertaux, J. L., Observed variations of the exospheric hydrogen density with the exospheric temperature, *J. Geophys. Res.*, **80**, 639, 1975.

48. Ma, Y., A. F. Nagy, I. V. Sokolov, and K. C. Hasnsen, 3D, multi-species, high spatial resolution MHD studies of the solar wind interaction with Mars, *J. Geophys. Res.*, **109**, A07211, doi:10.1029/2003JA010367, 2004.

49. Leblanc, F., J. G. Luhmann, R. E. Johnson, and E. Chassefiere, Some expected impacts of a solar energetic particle event at Mars, *J. Geophys. Res.*, **107**, A5, doi:10.1029/2001JA900178, 2002.

50. Barabash, S., A. Fedoror, R. Lundin, and J.-H. Sauvand, Martian atmospheric erosion rates, *Science*, **315**, 501, 2007.

51. Ma, Y.-J. and A. F. Nagy, Ion escape fluxes from Mars, *Geophys. Res. Lett.*, **34**, L08201, doi:10.1029/2006GL029208, 2007.

52. Shen, C. S., Analytic solution for density distribution in a planetary exosphere, *J. Atm. Sci.*, **20**, 69, 1963.

53. Nagy, A. F., T. E. Cravens, J.-H. Yee, and A. I. F. Stewart, Hot oxygen atoms in the upper atmosphere of Venus, *Geophys. Res. Lett.*, **8**, 629, 1981.

54. Oliver, W. L., Hot oxygen and the ion energy budget, *J. Geophys. Res.*, **102**, 2503, 1997.

55. Cotton, D. M., G. R. Gladstone, and S. Chakrabarti, Sounding rocket observation of a hot atomic geocorona, *J. Geophys. Res.*, **98**, 21651, 1993.

56. Kim, J., A. F. Nagy, J. L. Fox, and T. E. Cravens, Solar cycle variability of hot oxygen atoms at Mars, *J. Geophys. Res.*, **103**, 29339, 1998.

57. Nagy, A. F., J. Kim, T. E. Cravens, and A. J. Kliore, Hot oxygen corona at Europa, *Geophys. Res. Lett.*, **25**, 4153, 1998.

58. De La Haye, V., J. H. Waite, Jr., T. E. Cravens, *et al.*, Titan's corona: The contribution of exothermic chemistry, *Icarus*, **191**, 236, 2007.

59. Fox, J. L. and A. Dalgarno, Ionization luminosity and heating of the upper atmosphere of Mars, *J. Geophys. Res.*, **86**, 629, 1981.

60. Nagy, A. F., D. Winterhalter, K. Sauer, *et al.*, The plasma environment of Mars, *Space Sci. Rev.*, **111**, 33, 2004.

61. Ma, Y., A. F. Nagy, T. E. Cravens, *et al.*, Comparison between MHD model calculations and observations of Cassini flybys of Titan, *J. Geophys. Res.*, **11** A05207, doi:10.1029/2005JA011481, 2006.

62. Paxton, L. J., Pioneer Venus Orbiter ultraviolet spectrometer limb observations: Analysis and interpretation of the 166- and 156-nm data, *J. Geophys. Res.*, **90**, 5089, 1985.

63. Liemohn, M. W., J. L. Fox, A. I. Nagy, and X. Fang, Hot carbon densities in the exosphere of Venus, *J. Geophys. Res.*, **109**, A10307, doi:10.1029/2004JA010643, 2004.

64. Kim, J., *Model Studies of the Ionosphere of Venus: Ion Composition, Energetics and Dynamics*, Ph.D. Thesis, University of Michigan, Ann Arbor, 1991.

65. Valeille, A., M. R. Combi, V. Tenishev, S. W. Bougher, and A. F. Nagy, A study of suprathermal oxygen atoms in Mars upper atmosphere and exosphere over the range of limiting conditions, *Icarus*, in press, doi:10.1016/j.icarus.2008.08.018, 2008.

10.13   General references

Banks, P. M., and G. Kockarts, *Aeronomy*, New York: Academic Press, 1973.

Bauer, S. J., and H. Lammer, *Planetary Aeronomy*, Berlin: Springer Verlag, 2004.

Chamberlain, J. W., and D. M. Hunten, *Theory of Planetary Atmospheres*, New York: Academic Press, 1987.

Innis, J. L. and M. Conde, High-latitude thermosphereic vertical wind activity from Dynamics Explorer 2 wind and temperature spectrometer observations, Indications of a source region for polar cap gravity waves, *J. Geophys. Res.*, **107**, A8, doi:10.1029/200/JA009130, 2002.

Jarvis, M. J., Planetary wave trends in the lower thermosphere – evidence for 22-year solar modulation of the quasi 5-day wave, *J. Atmos. Solar-Terr. Phys.*, **68**, 1902, 2006.

Kato, S., *Dynamics of the Upper Atmosphere*, Boston: D. Reidel, 1980.

Lăstovička, J., Forcing of the ionosphere by waves from below, *J. Atmos. Solar-Terr. Phys.*, **68**, 479, 2006.

Müller-Wodarg, I. *et al.*, Neutral atmospheres, in *Comparative Aeronomy*, Space Sciences Series of ISSI, New York: Springer, 2008.

Nappo, C. J., *An Introduction to Atmospheric Gravity Waves*, International Geophysics Series, **85**, New York: Academic Press, 2002.

Ortland, D. A. and M. J. Alexander, Gravity wave influence on the global structure of the diurnal tide in the mesosphere and lower thermosphere, *J. Geophys. Res.*, **111**, A10S10, doi:10.1029/2005JA011467, 2006.

Rees, M. H., *Physics and Chemistry of the Upper Atmosphere*, Cambridge, UK: Cambridge University Press, 1989.

Rishbeth, H., F-region links with the lower atmosphere?, *J. Atmos. Solar-Terr. Phys.*, **68**, 469, 2006.

Shizgal, B. D., and G. G. Arkos, Nonthermal escape of the thermospheres of Venus, Earth and Mars, *Rev. Geophys.*, **34**, 483, 1996.

Tohmatsu, T., *Compendium of Aeronomy*, Terra Scientific Publishing Co., Dordrecht, Holland: Kluwer Academic Publishers, 1990.

10.14 Problems

Problem 10.1 Show that Equations (10.10a) and (10.10c) lead to Equation (10.11).

Problem 10.2 Calculate the Coriolis and centripetal accelerations (Equation 10.9) for Venus, Earth, and Mars at 45° latitude and 300 km altitude. Assume $u_{rot} = 200$ m s^{-1}.

Problem 10.3 Calculate the Brunt–Väisälä frequency (Equation 10.35) for Venus and Earth at 200 km at both noon and midnight using the reference atmospheres in Appendix K.

Problem 10.4 Obtain expressions for ρ_1, p_1, and $(\mathbf{K} \cdot \mathbf{u}_1)$ using Equations (10.31a–c) and show that the same dispersion relation is obtained in all three cases.

Problem 10.5 Show that the solution of Equations (10.30a–c) leads to the matrix given by Equation (10.36).

Problem 10.6 Show that the continuity (10.25a), momentum (10.41), and energy (10.25c) equations reduce to the tidal equations given by Equations (10.42a–e) when the four assumptions given in the paragraph that precedes these equations are adopted.

Problem 10.7 Assume that the diffusive time constant is defined by the following relation:

$$\tau_{i,\text{diff}} = \int\limits_{z_1}^{z_2} \frac{dz}{u_{i,\text{diff}}}$$

where the diffusion velocity for a minor species, i, flowing through a stationary atmosphere, $u_{i,\text{diff}}$ is given by

$$u_{i,\text{diff}} = -\frac{D}{H}\left[-X(1+\beta) + \beta + \frac{m_i}{m}\right],$$

where X is a parameter of the order of unity, β is the altitude slope of the scale height H, and $\langle m \rangle$ is the mean mass. Furthermore, assume that the height variation of H can be represented as

$$H = H_0 + \beta z \equiv H_0 \exp(\beta \zeta),$$

which defines a new height variable ζ. Using this new height variable one can write

$$\frac{n}{n_0} = \left(\frac{H}{H_0}\right)^{-[1+\beta]/\beta} = \exp\left[-(1+\beta)\zeta\right].$$

Finally, the altitude variation of D can be written as:

$$D = \frac{3}{8}\sqrt{\frac{g}{2\pi}}\left(1 + \frac{\overline{m}}{m_1}\right)^{1/2}\frac{H^{1/2}}{n\sigma^2} = D_0^*\frac{H^{1/2}}{n} = D_0\exp\left(1 + \frac{3}{2}\beta\right)\zeta.$$

Obtain an expression for the time constant as defined above.

Problem 10.8 Starting with Equation (10.55), show that if

$$\beta = \frac{dH_k}{dz} = \text{constant},$$

then

$$\frac{p_k(z_b)}{p_k(z_a)} = \left[\frac{H_k(z_b)}{H_k(z_a)}\right]^{-1/\beta}$$

and

$$\frac{n_k(z_b)}{n_k(z_a)} = \left[\frac{H_k(z_b)}{H_k(z_a)}\right]^{-(1+\beta)/\beta}.$$

To obtain this last relationship you need to assume that g does *not* vary with altitude.

Problem 10.9 In a hypothetical, isothermal atmosphere of 1500 K the relative ratio of helium ($m = 4$) to argon ($m = 40$) is 3.0×10^{-4} at ground level. It is known that there is complete mixing up to a certain altitude (the turbopause) and experimentally it was found that the helium to argon density ratio is 1.0×10^{-3} at 155 km. Assuming that g is 980 cm s^{-2} and a constant and that the helium to argon change from "mixed" to a "diffusive" atmosphere takes place abruptly at a given height, find the altitude of the turbopause.

Problem 10.10 Show that Equation (10.76) is the same as Equation (10.62) if the temperature gradient can be neglected.

Problem 10.11 On the planet Imaginus the photodissociation process, in the wavelength region 100 nm $< \lambda <$ 150 nm, results in two ground state X atoms, with the excess energy going into kinetic energy

$$X_2 + h\nu(100\,\text{nm} < \lambda < 150\,\text{nm}) \rightarrow X + X + \text{KE}.$$

For wavelengths less than 100 nm, one of the atoms is left in an excited state, with an excitation energy of 4 eV, and the rest of the surplus energy goes to kinetic energy

$$X_2 + h\nu(\lambda < 100\,\text{nm}) \rightarrow X^*(4\,\text{eV}) + X + KE.$$

Further, it can be assumed that the excited atoms, X^*, are rapidly and radiatively de-excited and that radiation as well as the energy from the recombination of X and photoelectron loss processes do not contribute to the heating of the ambient neutral gas. If collisions among the atmospheric neutral gas particles are frequent and all the above assumptions are applicable, what fraction of the total absorbed energy goes toward heating the ambient neutral gas in these two given wavelength regions? (The dissociation threshold for X_2 is at 150 nm. Also assume that the mean energy of the photons is at the midpoint of the wavelength intervals.)

Problem 10.12 Obtain an expression for the steady-state altitude variation of the neutral gas temperature on the planet Imaginus, assuming that thermal conduction is the only dominant energy transport mechanism and that the net heating rate is given by:

$$Q(z) = Q(z_0) \exp\left(-\frac{z - z_0}{H}\right)$$

where z_0 is a reference altitude and H is a constant. Assume no heat inflow from the top, the temperature at the reference altitude, z_0, is T_0, and the thermal conductivity, λ, is

$$\lambda = AT^{3/2},$$

where A is a constant.

Problem 10.13 The electron gas on the planet Imaginus receives a heat inflow of 10^{10} eV cm^{-2} s^{-1} at the top of the ionosphere and its temperature is 500 K at the surface of the planet. Assume that there are no energy sources and sinks for the electrons inside the ionosphere and that vertical thermal heat conduction is the only energy transport process.

1.   Obtain a general expression for the steady state electron temperature in the ionosphere given that the thermal conductivity, λ_e, is

$$\lambda_e = 7.7 \times 10^5 T_e^{5/2} \text{ eV cm}^{-1}\text{s}^{-1}\text{K}^{-1}.$$

2.   Evaluate the electron temperature at 400 km.

Problem 10.14 Calculate the energy per unit mass (in eV) that a gas particle must have, at an altitude of 200 km, to escape from Venus and Mars.

Chapter 11

The terrestrial ionosphere at middle and low latitudes

The plasma parameters in the Earth's ionosphere display a marked variation with altitude, latitude, longitude, universal time, season, solar cycle, and magnetic activity. This variation results not only from the coupling, time delays, and feedback mechanisms that operate in the ionosphere–thermosphere system, but also from the ionosphere's coupling to the other regions in the solar–terrestrial system, including the Sun, the interplanetary medium, the magnetosphere, and the mesosphere. The primary source of plasma and energy for the ionosphere is solar EUV, UV, and X-ray radiation; but magnetospheric electric fields and particle precipitation also have a significant effect on the ionosphere. The strength and form of the magnetospheric effect are primarily determined by the solar wind dynamic pressure and the orientation of the interplanetary magnetic field (IMF), i.e., by the state of the interplanetary medium. Also, tides and gravity waves that propagate up from the mesosphere directly affect the neutral densities in the lower thermosphere, and their variation then affects the plasma densities. The different external driving mechanisms, coupled with the radiative, chemical, dynamical, and electrodynamical processes that operate in the ionosphere, act to determine the global distributions of the plasma densities, temperatures, and drifts.

As noted in Section 2.3, the ionosphere is composed of different regions and, therefore, it is instructive to show the regions in which the different external processes operate. Figure 11.1 indicates the altitudes where the various external processes are most effective. Solar radiation leads to ion–electron production and heating via photoelectron energy degradation, with EUV wavelengths dominating in the lower thermosphere (E and F_1 regions) and UV and X-ray wavelengths dominating in the mesosphere (D region). These processes occur over the entire sunlit side of the Earth. On the night side, resonantly scattered solar radiation and starlight are important sources of ionization for the E region. At high latitudes, the main momentum and energy sources for the ionosphere are magnetospheric electric fields

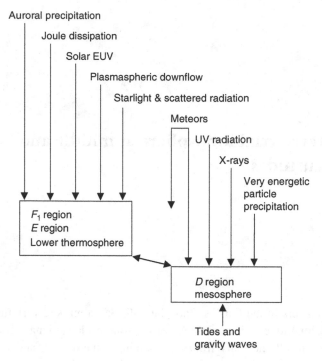

Figure 11.1 External processes that operate on the terrestrial ionosphere.[1]

and particle precipitation. The production of ionization due to auroral precipitation and the Joule heating that is associated with convection electric fields are maximized in the E and F_1 regions (Chapters 2 and 12). These magnetospheric processes affect not only the high-latitude ionosphere, but also the middle and low latitudes, particularly during storms and substorms. The magnetosphere also affects the lower ionosphere via very energetic particle precipitation from the radiation belts, which can produce ionization at all latitudes in the D region, and via a downward plasmaspheric flow, which helps maintain the nocturnal F region at mid-latitudes. With regard to the stratosphere, it has a significant effect on the lower ionosphere because upward-propagating tides and gravity waves from this region deposit most of their energy at E–F_1 region altitudes owing to wave breaking and dissipation. Finally, in a sporadic fashion, the ablation of impacting meteors produces neutral metal atoms, which are then ionized by charge transfer with molecular ions and by photoionization.[2]

This chapter focuses on the processes that affect the ionosphere at middle and low latitudes, where the plasma essentially co-rotates with the Earth, while those at high latitudes are discussed in Chapter 12. The topics covered in this chapter include the geomagnetic field, magnetic disturbances, the Sq and L current systems, the formation of ionospheric layers, nighttime maintenance processes, large-scale ionospheric features (light ion trough, sub-auroral red arcs, Appleton anomaly), plasma transport processes, dynamo electric fields, and the plasma thermal structure.

Also included in this chapter is a discussion of small-scale and irregular density features, such as sporadic E, intermediate layers, traveling ionospheric disturbances (TIDs), gravity waves, and equatorial plasma bubbles.

11.1   Dipole magnetic field

The effect that the various external processes have on the ionosphere is determined, to a large degree, by plasma transport processes, which are affected by the Earth's intrinsic magnetic field. At ionospheric altitudes, this *internal field* can be approximated by an Earth-centered dipole, with the dipole axis tilted with respect to the Earth's rotational axis by about 11.5° (Section 11.2). If $\mathbf{m}$ is the *dipole moment* at the Earth's center ($m \approx 7.9 \times 10^{15}$ T m^3) and if a spherical coordinate system (r, θ, ϕ) is adopted, with the polar axis parallel to the dipole axis (Figure 11.2), then the magnetic scalar potential (Φ_m) is given by

$$\Phi_m(r, \theta, \phi) = \frac{m \cos \theta}{r^2}. \tag{11.1}$$

The magnetic field ($\mathbf{B}$) is obtained by taking the gradient of the scalar potential, and it is given by

$$\mathbf{B} = -\nabla \Phi_m = \frac{2m \cos \theta}{r^3} \mathbf{e}_r + \frac{m \sin \theta}{r^3} \mathbf{e}_\theta, \tag{11.2}$$

where $\mathbf{e}_r$ and $\mathbf{e}_\theta$ are unit vectors in the radial and polar directions, respectively. The magnitude of this dipole magnetic field can be expressed as

$$B = \frac{m}{r^3} \left(1 + 3 \cos^2 \theta \right)^{1/2}. \tag{11.3}$$

There are several other dipole parameters that are useful for later applications. First, it is convenient to express m in terms of the magnetic field, B_E, on the Earth's surface ($r = R_E$) at the equator ($\theta = 90°$), for which Equation (11.3) yields $m = B_E R_E^3$. Next, it is useful to introduce a unit vector along $\mathbf{B}$, and this is simply

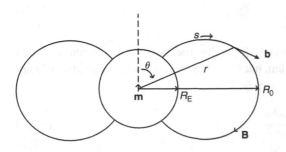

Figure 11.2 Geometry and parameters associated with a dipole magnetic field line.

given by

$$\mathbf{b} = \frac{\mathbf{B}}{B} = \frac{2\cos\theta\,\mathbf{e}_r + \sin\theta\,\mathbf{e}_\theta}{(1 + 3\cos^2\theta)^{1/2}}. \tag{11.4}$$

It is also convenient to introduce the *dip angle* or *inclination angle*, I, which is the angle that $\mathbf{B}$ makes with the horizontal direction. The important functions for I are

$$-\sin I = |\mathbf{b} \times \mathbf{e}_\theta| = \frac{2\cos\theta}{(1 + 3\cos^2\theta)^{1/2}}, \tag{11.5}$$

$$\cos I = \mathbf{b} \cdot \mathbf{e}_\theta = \frac{\sin\theta}{(1 + 3\cos^2\theta)^{1/2}}, \tag{11.6}$$

$$-\sin I \cos I = \frac{2\sin\theta\cos\theta}{1 + 3\cos^2\theta}. \tag{11.7}$$

In addition, the equation for a dipole field line is needed, and this can be obtained from the expression

$$\frac{r\,d\theta}{dr} = \frac{B_\theta}{B_r} = \frac{\tan\theta}{2}, \tag{11.8}$$

which relates differential arc lengths in the spherical coordinate system to the B-field components (Equation 11.2). The solution of Equation (11.8) is

$$r = R_0 \sin^2\theta, \tag{11.9}$$

where R_0 is the radial distance at which the dipole field line crosses the equator ($\theta = 90°$). The solution (11.9) can be easily verified by substituting it into Equation (11.8).

It is also useful to express derivatives along a magnetic field line in terms of r and θ derivatives. Letting s represent the distance along $\mathbf{B}$, then

$$\frac{\partial}{\partial s} = \mathbf{b} \cdot \nabla = -\sin I \frac{\partial}{\partial r} + \frac{\cos I}{r}\frac{\partial}{\partial\theta}, \tag{11.10}$$

where the equations for $\mathbf{b}$ (11.4), $\sin I$ (11.5), and $\cos I$ (11.6) have been used. Now, the cross sectional area, A, of a magnetic flux tube varies as $A \propto 1/B$, or, from Equation (11.3),

$$A = C\frac{r^3}{\left(1 + 3\cos^2\theta\right)^{1/2}}, \tag{11.11}$$

where C is a normalization constant. Typically, the constant is chosen such that $A = 1\,\text{cm}^2$ at an altitude of 1000 km, but for the present purposes the constant is not important. A quantity that is important is

$$\frac{1}{A}\frac{\partial A}{\partial s} = \frac{9\cos\theta + 15\cos^3\theta}{r(1 + 3\cos^2\theta)^{3/2}}, \tag{11.12}$$

which is derived by applying the $\partial/\partial s$ operator in Equation (11.10) to the expression for A (11.11) and then dividing by A. Note that near the poles of a dipole magnetic field ($\theta \approx 0°$), $(1/A)\partial A/\partial s \approx 3/r$. This result was used in the polar wind discussion (Section 5.8).

In the low-latitude ionosphere, co-rotation and dynamo electric fields exist that are directed perpendicular to $\mathbf{B}$. These fields cause the plasma to $\mathbf{E} \times \mathbf{B}$ drift across $\mathbf{B}$ (Section 11.11). The *co-rotation electric field* points toward the Earth, and the resulting $\mathbf{E} \times \mathbf{B}$ drift is eastward at a speed that matches the Earth's rotational speed. The main dynamo $\mathbf{E}$ field, on the other hand, is in the azimuthal (east–west) direction, and its associated electrodynamic drift is either toward or away from the Earth. Drifts away from the Earth result in an expanding plasma and a density decrease, while the reverse occurs for $\mathbf{E} \times \mathbf{B}$ drifts toward the Earth. This effect is important and must be taken into account in the continuity equation.

Given an azimuthal electric field, $\mathbf{E} = E_\phi \mathbf{e}_\phi$, and a dipole magnetic field (11.2), the electrodynamic drift, $\mathbf{u}_E = \mathbf{E} \times \mathbf{B}/B^2$, becomes

$$\mathbf{u}_E = \frac{r^3 E_\phi}{m(1 + 3\cos^2\theta)}\left(-\sin\theta\,\mathbf{e}_r + 2\cos\theta\,\mathbf{e}_\theta\right). \tag{11.13}$$

It is customary to express Equation (11.13) in terms of the plasma drift (u_{E_0}) at the magnetic equator, which is where the drift is typically measured. At the equator, $\theta = 90°$ and $r = R_0$ (Figure 11.2), and Equation (11.13) yields a radial drift of $u_{E_0} = -R_0^3 E_{\phi 0}/m$, where $E_{\phi 0}$ is the electric field at $r = R_0$. Using this result to derive an expression for the dipole moment m and then substituting this m into Equation (11.13) yields the following equation for $\mathbf{u}_E$:

$$\mathbf{u}_E = \frac{-u_{E_0}}{1 + 3\cos^2\theta}\frac{r^3}{R_0^3}\frac{E_\phi}{E_{\phi 0}}\left(-\sin\theta\,\mathbf{e}_r + 2\cos\theta\,\mathbf{e}_\theta\right). \tag{11.14}$$

Now, $r^3/R_0^3 = \sin^6\theta$ (Equation 11.9), and it is left as an exercise to show that $E_\phi/E_{\phi 0} = 1/\sin^3\theta$. Using these results, Equation (11.14) can be expressed in the form

$$\mathbf{u}_E = \frac{u_{E_0}\sin^3\theta}{1 + 3\cos^2\theta}\left(\sin\theta\,\mathbf{e}_r - 2\cos\theta\,\mathbf{e}_\theta\right). \tag{11.15}$$

This expression describes the $\mathbf{E} \times \mathbf{B}$ drift of plasma across dipole magnetic field lines. The drift is the largest at the equator, where $\theta = 90°$ and $\mathbf{u}_E = u_{E_0}\mathbf{e}_r$. The $\mathbf{E} \times \mathbf{B}$ drift decreases as the Earth is approached along a dipole field line.

The divergence of $\mathbf{u}_E$ can be obtained by taking the divergence of Equation (11.15), which is left as an exercise. However, it can also be obtained by taking the divergence of $\mathbf{E} \times \mathbf{B}/B^2$, which is the procedure used in Section 12.1. From Equation (12.3)

$$\nabla \cdot \left(\frac{\mathbf{E} \times \mathbf{B}}{B^2}\right) = (\mathbf{E} \times \mathbf{B}) \cdot \nabla\left(\frac{1}{B^2}\right) = -2\mathbf{u}_E \cdot \left(\frac{1}{B}\nabla B\right). \tag{11.16}$$

The term in the parentheses can be obtained from Equation (11.3), and the result is

$$\frac{1}{B} \nabla B = -\frac{3}{r} \left(\mathbf{e}_r + \frac{\sin\theta\cos\theta}{1 + 3\cos^2\theta} \mathbf{e}_\theta \right). \tag{11.17}$$

Substituting Equations (11.17) and (11.15) into Equation (11.16), and remembering that $r = R_0 \sin^2\theta$ (11.9), the divergence of the electromagnetic drift can be expressed as

$$\nabla \cdot \mathbf{u}_E = \frac{6u_{E_0}}{R_0} \frac{\sin^2\theta(1 + \cos^2\theta)}{(1 + 3\cos^2\theta)^2}. \tag{11.18}$$

Note that near the poles of a dipole ($\theta \approx 0°$), $\nabla \cdot \mathbf{u}_E \approx 0$. That is, the $\mathbf{E} \times \mathbf{B}$ drift is basically incompressible. This result is important for the high-latitude ionosphere (Section 12.1). Near the magnetic equator ($\theta = 90°$), $\nabla \cdot \mathbf{u}_E \approx 6u_{E_0}/R_0$, which corresponds to the maximum expansion or contraction rate.

It is convenient for some applications to introduce orthogonal dipolar coordinates (q_d, p_d), which are given by

$$q_d = \frac{R_E^2 \cos\theta}{r^2}, \tag{11.19}$$

$$p_d = \frac{r}{R_E \sin^2\theta}. \tag{11.20}$$

The coordinate q_d replaces the arc length s along $\mathbf{B}$ and p_d is the coordinate perpendicular to $\mathbf{B}$. The ultimate goal is to express $\partial/\partial s$, which appears in the transport equations, in terms of $\partial/\partial q_d$. This is accomplished by first noting that

$$\frac{\partial}{\partial q_d} = \frac{dr}{dq_d} \frac{\partial}{\partial r} + \frac{d\theta}{dq_d} \frac{\partial}{\partial \theta}. \tag{11.21}$$

When calculating dr/dq_d and $d\theta/dq_d$, it is important to remember that r and θ are related along a dipole magnetic field line (Equation 11.9). Therefore, q_d can be expressed in the form, $q_d = (R_E/R_0)^2 \cos\theta/\sin^4\theta$, and from this expression it can be shown that

$$\frac{d\theta}{dq_d} = -\frac{R_0^2}{R_E^2} \frac{\sin^5\theta}{1 + 3\cos^2\theta} = -\frac{r^2 \sin\theta}{R_E^2(1 + 3\cos^2\theta)}, \tag{11.22}$$

where Equation (11.9) was used again to obtain the second expression in Equation (11.22). Likewise, using Equation (11.9), q_d can also be expressed in terms of just r, $q_d = (R_E/r)^2(1 - r/R_0)^{1/2}$, and from this expression it follows that

$$\frac{dr}{dq_d} = -\frac{2r^3 \cos\theta}{R_E^2(1 + 3\cos^2\theta)}. \tag{11.23}$$

Now, combining Equations (11.21–23), $\partial/\partial q_d$ can be written in the following form:

$$\frac{\partial}{\partial q_d} = -\frac{r^3/R_E^2}{(1+3\cos^2\theta)^{1/2}}\left[\frac{2\cos\theta}{(1+3\cos^2\theta)^{1/2}}\frac{\partial}{\partial r} + \frac{\sin\theta}{(1+3\cos^2\theta)^{1/2}}\frac{1}{r}\frac{\partial}{\partial\theta}\right],$$

(11.24)

where the quantity in the square brackets is just $\partial/\partial s$ (Equation 11.10). Solving for $\partial/\partial s$, one obtains

$$\frac{\partial}{\partial s} = -\frac{R_E^2(1+3\cos^2\theta)^{1/2}}{r^3}\frac{\partial}{\partial q_d}.$$

(11.25)

The second derivative, $\partial^2/\partial s^2$, is also needed when the transport equations are solved in the so-called diffusion approximation (Section 5.7). This derivative can be obtained as follows:

$$\frac{\partial^2}{\partial s^2} = \frac{\partial}{\partial s}\left(\frac{\partial}{\partial s}\right) = \alpha^2\frac{\partial^2}{\partial q_d^2} + \alpha\frac{\partial\alpha}{\partial q_d}\frac{\partial}{\partial q_d},$$

(11.26)

where

$$\alpha = -\frac{R_E^2(1+3\cos^2\theta)^{1/2}}{r^3},$$

(11.27)

$$\frac{\partial\alpha}{\partial q_d} = -\frac{3\cos\theta(3+5\cos^2\theta)}{r(1+3\cos^2\theta)^{3/2}}.$$

(11.28)

11.2   Geomagnetic field

A magnetic dipole is a reasonable approximation for the geomagnetic field at low and middle latitudes. The simplest approximation is the *axial-centered dipole*, for which the Earth's magnetic and rotational axes coincide. The next approximation is a *tilted dipole*, with the dipole axis intersecting the Earth's surface at 78.5° N, 291° E, and 78.5° S, 111° E geographic. A better approximation is the *eccentric dipole*, for which the dipole axis is displaced from the Earth's center by a distance of about 500 km in the direction 21° N, 147° E. The eccentric dipole intersects the Earth's surface at 82° N, 270° E, and 75° S, 119° E.[3]

The most accurate representation of the geomagnetic field is the one obtained when the magnetic scalar potential is expanded in a spherical harmonic series of the form

$$\Phi_m(r',\theta',\phi') = R_E\sum_{n=1}^{\infty}\sum_{m=0}^{n}\left(\frac{R_E}{r'}\right)^{n+1}(g_n^m\cos m\phi' + h_n^m\sin m\phi')P_n^m(\cos\theta'),$$

(11.29)

where (r', θ', ϕ') are *geographic coordinates*, and where r' increases in the outward radial direction, θ' is co-latitude measured from the northern geographic pole, and ϕ' is east longitude.[4] In Equation (11.29), $P_n^m(\cos\theta')$ is the Schmidt form of the associated Legendre polynomial of degree n and order m, and g_n^m and h_n^m are expansion coefficients. The expansion coefficients are obtained by fitting the magnetic potential (11.29) to a global distribution of both ground-based and satellite magnetometer measurements. This fitting procedure is done at various times because the intrinsic magnetic field changes with time (the *secular variation*). The outcome of this effort is the *International Geomagnetic Reference Field* (IGRF).[5] The axial-centered dipole approximation is given by the ($n = 1, m = 0$) term:

$$\Phi_m(r', \theta', \phi') = R_E \left(\frac{R_E}{r'}\right)^2 g_1^0 \cos\theta' \tag{11.30}$$

and the tilted dipole approximation is given by the ($n = 1, m = 0, 1$) terms:

$$\Phi_m(r', \theta', \phi') = R_E \left(\frac{R_E}{r'}\right)^2 \left[g_1^0 \cos\theta' + (g_1^1 \cos\phi' + h_1^1 \sin\phi')\sin\theta'\right]. \tag{11.31}$$

Some of the commonly used angles and vector components of the geomagnetic field are shown in Figure 11.3. The relations between the different quantities are

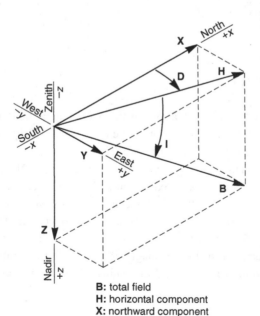

Figure 11.3 Vector components and angles associated with the geomagnetic field.[6]

B: total field
H: horizontal component
X: northward component
Y: eastward component
Z: vertical component
D: declination
I: inclination

given by

$$H = (X^2 + Y^2)^{1/2}, \tag{11.32}$$

$$B = (H^2 + Z^2)^{1/2}, \tag{11.33}$$

$$X = H \cos D, \tag{11.34}$$

$$Y = H \sin D, \tag{11.35}$$

$$D = \tan^{-1}(Y/X), \tag{11.36}$$

$$I = \tan^{-1}(Z/H), \tag{11.37}$$

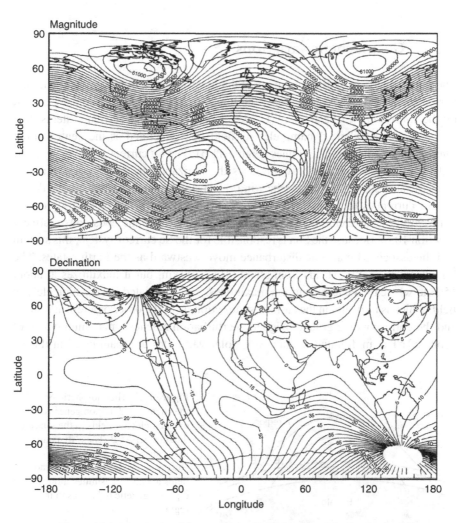

Figure 11.4 Magnitude of the geomagnetic field (top) and declination angle in degrees (bottom) at the Earth's surface.[6]

where B is the magnitude of the geomagnetic field, H is the magnitude of the horizontal component, and (X, Y, Z) are the Cartesian components of **B** in the northward, eastward, and downward directions, respectively. The angle D is the *declination*, which is the deflection of the geomagnetic field from the geographic pole. The angle I is the *inclination* or *dip angle* of the magnetic field from the horizontal (Equation 11.5). Note that the magnitude of the geomagnetic field is not uniform over the Earth's surface (Figure 11.4). In general, it is weaker in the equatorial region and stronger in the polar regions, but there are distinct regions where it reaches extreme values (e.g., the South Atlantic anomaly). Likewise, the declination and inclination angles are not uniform over the Earth's surface as shown, for example, in Figure 11.4, where the declination is plotted. As expected, the largest declination angles occur in the regions close to the magnetic poles.

11.3   Geomagnetic variations

The geomagnetic field displays an appreciable variation during both quiet and disturbed times.[6,7] During quiet times, the magnetic variations are primarily caused by the *solar-quiet* (Sq) and *lunar* (L) *current systems*. These current systems flow in the E region, where dynamo electric fields are generated as the neutrals drag ions across geomagnetic field lines (Figure 11.5). The Sq current is driven by solar EUV radiation, which not only produces the ionization in the E region but also heats the atmosphere and causes the wind. The primary wind component that drives the Sq current is the diurnal tide $(1, -2)$, which has a small phase progression with altitude, and therefore, the contribution of each altitude adds constructively (Section 10.6). Because the Sun is responsible for the Sq current system, this system and the associated magnetic disturbance move westward as the Earth rotates. The Sq current typically extends from about 90 to 200 km, but it maximizes at about 150 km where the Pedersen current maximizes. The associated polarization electric field, which is basically in the east–west direction, is of the order of a few mV m^{-1}, and the corresponding ground magnetic perturbation reaches a maximum value of about 20 nT at mid-latitudes. Naturally, solar variations are manifested in the Sq

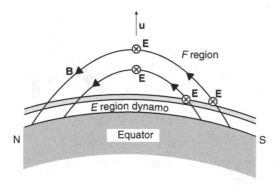

Figure 11.5 Thermospheric winds in the equatorial E region generate dynamo electric fields as the ions are dragged across **B**. These dynamo fields are responsible for the equatorial electrojet. The dynamo fields are also transmitted along the dipole magnetic field lines to the F region.

current system, and hence, the Sq currents display strong seasonal and solar cycle dependencies.

The Sq current increases substantially, by about a factor of four, in a narrow latitudinal band about the magnetic equator. This band of enhanced current, called the *equatorial electrojet*, is a consequence of the nearly horizontal field lines at the equator. An eastward electric field, in combination with the northward geomagnetic field, acts to drive a Hall current in the vertical direction. However, the E region conductivity is bounded in the vertical direction and this inhibits the Hall current. The net result is that a vertical polarization electric field is created, which induces a Hall current in the eastward direction. This latter Hall current augments the original Pedersen current, thereby enhancing the effective conductivity in the eastward electric field direction (*Cowling conductivity*). At latitudes just off the magnetic equator, the slight tilt of the geomagnetic field lines is sufficient to allow the polarization charges to partially drain, thus reducing the Cowling conductivity.[8]

The L current system and its associated magnetic disturbance are generated in the same manner as the corresponding Sq features, except that the driving winds are produced by gravitationally excited lunar tides. The prominent tide is the semi-diurnal (2, 2) mode (Section 10.6). The magnitude of the magnetic perturbation associated with the L current is about an order of magnitude smaller than that associated with the Sq current. The magnetic perturbation tends to follow the lunar day, which is 24 hours and 50 minutes on average.

In addition to the regular variations caused by the Sq and L current systems, the geomagnetic field can be disturbed by magnetospheric processes. The *disturbance field D'* is the magnetic field that results after the steady and quiet-variation fields have been subtracted from the total field. The characteristic times associated with most disturbances extend from minutes to days, and almost all of them can be traced to the effects that solar wind perturbations have on the magnetosphere. The largest disturbances are called *magnetic storms*. Typically, the disturbance field is separated into two components $D' = Dst + Ds$, where Dst is the *storm-time variation* and Ds is the *disturbance-daily variation*. The Dst component results from the magnetic disturbance generated by the ring current (Figure 2.10), while the Ds component is the magnetic field associated with the ionospheric currents generated by auroral particles precipitated from the ring current.

A magnetic storm generally has three phases; *initial, main*, and *recovery phases*. The initial phase results from a compression of the magnetosphere due to the arrival of a solar wind discontinuity (shock, CME) at the Earth. Frequently, storms begin abruptly and this is called a *sudden storm commencement* (SSC), but storms can also begin gradually without an SSC. Sometimes an impulsive change in the magnetic field occurs, but a storm does not develop, and this is called a *sudden impulse* (SI). The initial phase of a storm typically lasts 2–8 hours, during which Dst is increased, owing to the compression of the magnetosphere. During the main phase, Dst is decreased, often by more than $100\,nT$, relative to prestorm values. This decrease occurs because magnetic storms are generally associated with a southward interplanetary magnetic field, which allows for an efficient energy coupling of the solar wind and magnetosphere (Section 12.1). The net result is an intensification of the ring current, which is the westward current that encircles the Earth at equatorial latitudes (Figure 2.10). The enhanced westward current induces a horizontal magnetic field H that is southward (opposite to the Earth's dipole field), and this accounts for the negative *Dst* during the main phase of a storm. The recovery phase, which can last more than a day, is a time when *Dst* gradually increases to its prestorm value. This occurs because the source of the enhanced ring current subsides and the excess particles are lost via several different mechanisms.

Several indices have been used to describe magnetic activity in addition to *Dst*, which is calculated at low latitudes and describes the ring current. The AE, AL, and AU indices are calculated at auroral latitudes and primarily describe the auroral electrojet intensity. The K indices are calculated at all latitudes and are the most widely used of all the indices.[7] The three-hour K index provides a measure of the magnetic deviations from the regular daily variation during a three-hour period. The information about magnetic activity is provided via a semi-logarithmic numerical code that varies from 0 to 9, with the different numbers corresponding to different magnetic activity levels. The K indices from twelve observatories are combined to produce a *three-hour planetary index*, K_p, which provides information on the average level of magnetic activity on a worldwide basis. The K_p index is specified to one-third of a unit, with the steps varying as follows: $0_0, 0_+, 1_-, 1_0, 1_+, \ldots 8_-,$ $8_0, 8_+, 9_-, 9_0$. However, almost all of the magnetic observatories that are used to produce K_p are in the northern hemisphere at mid-latitudes, and therefore, K_p is not truly a worldwide index.

11.4　Ionospheric layers

The terrestrial ionosphere at all latitudes has a tendency to separate into layers, despite the fact that different processes dominate in different latitudinal domains (Figure 2.16). However, the layers (D, E, F_1, and F_2) are distinct only in the daytime ionosphere at mid-latitudes. The different layers are generally characterized by a density maximum at a certain altitude and a density decrease with altitude on both sides of the maximum. The E layer was the first layer to be detected, followed by the F and D layers (see Section 1.2). Typically, the E and F layers are described by critical frequencies (f_0E, f_0F_1, f_0F_2), peak heights (h_mE, h_mF_1, h_mF_2), and half-thicknesses (y_mE, y_mF_1, y_mF_2), as shown in Figure 11.6. The *critical frequency*, which is proportional to $n_e^{1/2}$ (Equation 2.6), is the maximum frequency that can be reflected from a layer. Electromagnetic waves with a higher frequency, transmitted from below the layer, will penetrate it and propagate to higher altitudes. Associated with each critical frequency is a *peak density* (N_mE, N_mF_1, N_mF_2) and a *peak height*, which is the altitude of the density maximum. Also, it is customary to define a *half-thickness* for each layer, which is obtained by fitting a parabola to the electron density profile in an altitude range centered at the density maximum. All of the layers occur during the daytime, but the F_1 layer decays at night and a distinct *E–F valley* can appear that separates the E and F_2 layers.

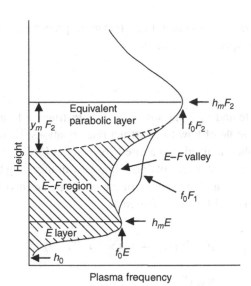

Figure 11.6 Schematic diagram of an electron density profile showing critical frequencies, peak heights, and half-thicknesses for the E, F_1, and F_2 layers. The curve labeled E–F valley is a nighttime profile and the one labeled f_0F_1 is a daytime profile.[9]

The dominant ions in the E region are molecular (NO^+, O_2^+, N_2^+) and the chemical time constants are short enough that plasma transport processes can be neglected. In this case, photochemistry prevails (Chapter 8). Although there is a large number of minor ion species in the E region, the dominant ions can be described, to a good approximation, by just a few photochemical processes. The main processes, which include photoionization, ion–molecule reactions, and electron–ion recombination, are given by

$$O + h\nu \rightarrow O^+ + e^-, \tag{11.38}$$

$$O_2 + h\nu \rightarrow O_2^+ + e^-, \tag{11.39}$$

$$N_2 + h\nu \rightarrow N_2^+ + e^-, \tag{11.40}$$

$$O^+ + N_2 \rightarrow NO^+ + N, \tag{11.41}$$

$$O^+ + O_2 \rightarrow O_2^+ + O, \tag{11.42}$$

$$N_2^+ + O_2 \rightarrow O_2^+ + N_2, \tag{11.43}$$

$$N_2^+ + O \rightarrow O^+ + N_2, \tag{11.44}$$

$$N_2^+ + O \rightarrow NO^+ + N, \tag{11.45}$$

$$NO^+ + e^- \rightarrow N + O, \tag{11.46}$$

$$O_2^+ + e^- \rightarrow O + O, \tag{11.47}$$

$$N_2^+ + e^- \rightarrow N + N, \tag{11.48}$$

where the rate constants (in cm^3 s^{-1}) are given in Tables 8.3 and 8.5 and the ionization frequencies can be found in Table 9.2.

When photochemistry is more important than transport processes, the ion continuity equation (5.22a) for species s reduces to

$$\frac{dn_s}{dt} = P_s - L_s, \tag{11.49}$$

where P_s is the production rate and L_s is the loss rate. It is instructive to use one of the ions, say N_2^+, in an example of how to construct rate equations. During the daytime, N_2^+ is produced via photoionization at a rate $P(N_2^+)$ and is lost in reactions with O_2, O, and electrons. The loss rates are obtained simply by multiplying the N_2^+ density by both the rate coefficient and the density of the other species involved in the reaction. For N_2^+, Equation (11.49) can be written as

$$\frac{dn(N_2^+)}{dt} = P(N_2^+) - \left[5 \times 10^{-11} n(O_2) + 1.4 \times 10^{-10} n(O) \right.$$
$$\left. + 2.2 \times 10^{-7} \left(\frac{300}{T_e} \right)^{0.39} n_e \right] n(N_2^+), \tag{11.50}$$

where the rate coefficients were obtained from Tables 8.3 and 8.5 and where $P(N_2^+)$ can be calculated from Equation (9.25). Similar equations hold for the other molecular ions (NO^+ and O_2^+). The important facts to note are that the ion equations are coupled and nonlinear because the electron density is the sum of the ion densities. Nevertheless, these photochemical rate equations can be readily solved with numerical techniques.

An analytical expression for the electron density in an ionosphere dominated by photochemical reactions can be obtained with the aid of a few simplifying assumptions. First, assume that one of the molecular ions is dominant, which means that the ion and electron densities are equal. Also, throughout most of the daytime, the time derivative term in the continuity equation (11.49) is negligible, and this equation then reduces to

$$P_e = k_d n_e^2, \tag{11.51}$$

where k_d is the ion–electron recombination rate. Now, assume that the Chapman production function (9.21) can be used to describe the ionization rate as a function of altitude, z, and solar zenith angle, χ. Using Equation (9.21), and the derivation that led to this equation, it follows that

$$P_c(z, \chi) = I_\infty \eta \sigma^{(a)} n(z) e^{-\tau}, \tag{11.52}$$

where

$$\tau = H\sigma^{(a)} n(z) \sec \chi, \tag{11.53}$$

$$n(z) = n_0 e^{-(z-z_0)/H}, \tag{11.54}$$

and where $n(z)$ is the neutral density (9.4), H is the neutral scale height, $\sigma^{(a)}$ is the absorption cross section, I_∞ is the flux of radiation incident on the top of the atmosphere, η is the ionization efficiency, τ is the optical depth, and z_0 is a reference altitude. If the reference altitude is chosen to be the level of unit optical depth ($\tau = 1$) for overhead sun ($\chi = 0°$), the production rate at z_0 becomes

$$P_{co} = I_\infty \eta \sigma^{(a)} n_0 e^{-1}. \tag{11.55}$$

Now, the substitution of Equations (11.53–55) into Equation (11.52) yields

$$P_c(z, \chi) = P_{co} \exp\left[1 - \frac{z - z_0}{H} - \exp\left(\frac{z_0 - z}{H}\right) \sec \chi \right], \tag{11.56}$$

where $H\sigma^{(a)} n_0 = 1$ from Equation (11.53) because $\tau = 1$ and $\chi = 0°$ at the reference altitude z_0. Finally, the substitution of the Chapman production function (11.56) into the continuity equation (11.51) leads to the expression for the *Chapman layer*;

$$n_e(z, \chi) = \left(\frac{P_{co}}{k_d}\right)^{1/2} \exp\left\{ \frac{1}{2}\left[1 - \frac{z - z_0}{H} - \exp\left(\frac{z_0 - z}{H}\right) \sec\chi \right] \right\}. \tag{11.57}$$

Note that near the peak of the layer ($z \approx z_0$), the exponentials can be expanded in a Taylor series, and the expression for n_e reduces to

$$n_e(z, 0°) \approx \left(\frac{P_{co}}{k_d}\right)^{1/2} \left[1 - \frac{(z - z_0)^2}{4H^2} \right] \tag{11.58}$$

for overhead sun ($\chi = 0°$). Hence, the electron density profile is parabolic near the peak of the layer, which is why half-thicknesses are defined with reference to a parabolic shape.

The F region is usually divided into three subregions. The lowest region, where photochemistry dominates, is called the F_1 *region*. The region where there is a transition from chemical to diffusion dominance is called the F_2 *region*, and the upper F region, where diffusion dominates, is called the *topside ionosphere*. In the F_1 region, the photochemistry simplifies because only one ion (O^+) dominates. The important reactions are photoionization of neutral atomic oxygen (11.38) and loss in reactions with N_2 and O_2 (Equations 11.41 and 11.42). However, transport processes become important in the F_2 and upper F regions, including ambipolar diffusion and wind-induced drifts along **B** (Equation 5.54) and electrodynamic drifts across **B** (Equation 5.99). In the mid-latitude ionosphere, the magnetic field is basically straight and uniform at F region altitudes, but it is inclined to the horizontal at an angle I (see Figure 11.7; Section 11.1). However, the mid-latitude ionosphere is horizontally stratified, which means that the density and temperature gradients (and gravity) are in the vertical direction. The inclined **B** field, therefore, reduces the effectiveness of diffusion because the charged particles are constrained to diffuse

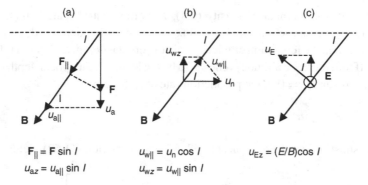

$$F_{\parallel} = F \sin I$$
$$u_{az} = u_{a\parallel} \sin I$$

$$u_{w\parallel} = u_n \cos I$$
$$u_{wz} = u_{w\parallel} \sin I$$

$$u_{Ez} = (E/B)\cos I$$

Figure 11.7 Geometry associated with induced vertical plasma drifts due to (a) field-aligned plasma diffusion, $\mathbf{u}_{a\parallel}$, driven by a vertical force $\mathbf{F}$, (b) an equatorward meridional neutral wind, $\mathbf{u}_n$, and (c) an eastward electric field, $\mathbf{E}$. The components u_{az}, u_{wz}, and u_{Ez} are the induced vertical drifts due to diffusion, the wind, and the electric field. The angle I is the inclination of the $\mathbf{B}$ field.

along $\mathbf{B}$ as a result of the small collision-to-cyclotron frequency ratios (Section 5.10). The inclined $\mathbf{B}$ field also affects the wind-induced and electrodynamic plasma drifts.

If $\mathbf{u}_{a\parallel}$ is the ambipolar diffusion part of the ion velocity along $\mathbf{B}$ (Equation 5.54), then for a vertical force $\mathbf{F}$ $(g, dT/dz, dn/dz)$, it is the component of $\mathbf{F}$ along $\mathbf{B}$ that drives diffusion, so that $\mathbf{u}_{a\parallel} \propto F \sin I$ (Figure 11.7). The vertical component of the ambipolar diffusion velocity, which enters the continuity equation, is $u_{az} = u_{a\parallel} \sin I$, so that $u_{az} \propto F \sin^2 I$. A vertical plasma drift is also induced by both a meridional neutral wind and a zonal electric field. For an equatorward neutral wind ($\mathbf{u}_n$), the induced plasma drift along $\mathbf{B}$ is $u_n \cos I$ and the vertical component of this plasma velocity is $u_n \sin I \cos I$. Finally, for an eastward electric field, the electrodynamic drift has a vertical component that is equal to $(E/B) \cos I$. When all three plasma drifts are taken into account, the ion diffusion equation (5.54) can be expressed in the form

$$u_{iz} = \frac{E}{B} \cos I + u_n \sin I \cos I - \sin^2 I D_a \left(\frac{1}{n_i} \frac{\partial n_i}{\partial z} + \frac{1}{T_p} \frac{\partial T_p}{\partial z} + \frac{1}{H_p} \right),$$

$$(11.59)$$

where $H_p = 2kT_p/(m_i g)$ is the plasma scale height (Equation 5.59) and where, for simplicity, the $\nabla \cdot \boldsymbol{\tau}_i$ term is neglected.

Equation (11.59) is the "classical" ambipolar diffusion equation for the F_2 region. It is applicable at both middle and high latitudes. However, for many applications, additional terms must be taken into account. For example, the $\nabla \cdot \boldsymbol{\tau}_i$ term is important at high latitudes in the regions where the convection electric fields are greater than about 40 mV m^{-1}, because it introduces a temperature anisotropy (Sections 5.2 and 5.13). Also, in deriving Equations (5.54) and (11.59), the heat flow collision terms in the momentum equation were neglected. These collision terms, which account for thermal diffusion and provide corrections to ordinary diffusion, are important in the upper F region and topside ionosphere at mid-latitudes (Section 5.14). Finally, the expressions for induced vertical plasma drifts due to electric fields and neutral winds become more complicated when the magnetic field declination is also taken into account.

Neutral winds and electric fields do not affect the basic shape of the F layer and, therefore, it is convenient temporarily to ignore their influence. In the daytime F_1 region at mid-latitudes, diffusion is not important, and during the daytime, the time variations are slow. For these conditions, the O^+ (or electron) density can be obtained simply by equating the production (11.38) and loss (11.41 and 11.42) terms in the O^+ continuity equation, which yields

$$n(O^+) = \frac{P_{ts}(O^+)}{1.2 \times 10^{-12}n(N_2) + 2.1 \times 10^{-11}n(O_2)}, \tag{11.60}$$

where the chemical rate constants were taken from Table 8.3 and P_{ts} is given by Equation (9.25). Equation (11.60) is the *chemical equilibrium* expression for O^+. When chemical equilibrium prevails, the O^+ density increases exponentially with altitude (Figure 11.8). This occurs because the O^+ photoionization rate, $P_{ts}(O^+)$, is directly proportional to the atomic oxygen density, which decreases exponentially with altitude, but at a slower rate than the decrease of the N_2 and O_2 densities. The net result is that the O^+ density increases exponentially with altitude. On the other hand, in the upper F region diffusion dominates and the O^+ density, in general, follows a *diffusive equilibrium* profile, which is obtained by setting the quantity in the parentheses in Equation (11.59) to zero (Section 5.5). Hence, the O^+ density decreases exponentially with altitude at a rate governed by both the plasma temperature gradient and scale height. The F region peak density occurs at the altitude

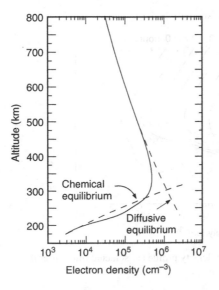

Figure 11.8 Representative O^+ density profile for the daytime F region at mid-latitudes. Also shown are the associated chemical and diffusive equilibrium profiles.

where the diffusion and chemical processes are of equal importance, i.e., where the chemical and diffusion time constants are equal (Section 8.1 and Equation 10.67).

As noted above, the charged particles are constrained to move along $\mathbf{B}$ at F region altitudes. As a consequence, a poleward neutral wind induces a downward plasma drift, while an equatorward wind induces an upward plasma drift. Likewise, a westward electric field induces a downward plasma drift and an eastward electric field induces an upward plasma drift. The effect of such induced drifts on the daytime F region is shown in Figure 11.9. For the upward plasma drift, the F layer moves to higher altitudes, where the O^+ loss rate is lower and, therefore, both N_mF_2 and h_mF_2 increase. The reverse occurs for a downward plasma drift.

Photoionization does not occur at night and, therefore, the ionosphere decays. This decay is shown in Figure 11.10 for the idealized situation where nighttime sources

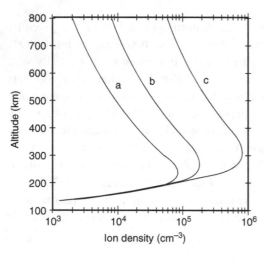

Figure 11.9 O^+ density profiles calculated for the daytime ionosphere at mid-latitudes. The profiles are for an induced downward plasma drift (curve a), no induced drift (curve b), and an induced upward plasma drift (curve c).[10]

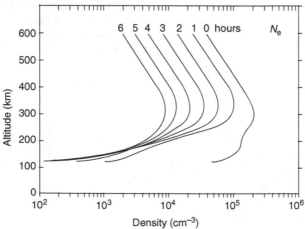

Figure 11.10 Calculated electron density profiles at selected times after the photoionization rates are set to zero.[11]

of ionization are ignored. The calculations start with a typical daytime ionosphere at mid-latitudes ($t = 0$). Subsequently, the photoionization rates are set to zero and the decays of the E and F regions are followed for several hours. The E region, which is populated by the molecular ions NO^+, O_2^+, and N_2^+, decays very rapidly because of the fast dissociative recombination rates. The O^+ density in the F region decays exponentially with time in a shape-preserving fashion. The time constant for the exponential decay is approximately equal to the inverse of the O^+ *loss frequency* (11.61) at the height of the F region peak. This fact can be deduced by considering the O^+ continuity equation. At night, photoionization is absent and near the peak of the layer the variation with altitude is small ($\partial/\partial z \rightarrow 0$). Consequently, the O^+ continuity equation reduces to

$$\partial n/\partial t = -k_\beta n, \tag{11.61}$$

where $k_\beta = [1.2 \times 10^{-12} n(N_2) + 2.1 \times 10^{-11} n(O_2)]$ is the O^+ loss frequency evaluated at the peak altitude, $h_m F_2$. The solution of this equation is $n \sim \exp(-k_\beta t)$. For the case shown in Figure 11.10, the initial O^+ density decays by a factor of ten in about four hours.

The ionospheric decay shown in Figure 11.10 occurs in the absence of ionization sources. However, this situation is not representative of the true nighttime conditions because ionization sources other than direct photoionization exist at night. Specifically, the nocturnal E region is maintained by production due to both starlight and resonantly scattered solar radiation (H Lyman α and β). The nocturnal F region is partially maintained by a downward flow of ionization from the overlying plasmasphere (discussed in Section 11.5). Also, the basic flow of the neutral atmosphere is around the globe from the subsolar point on the day side to the night side. Therefore, on the night side, the meridional neutral wind is generally toward the equator. This equatorward wind induces an upward plasma drift that raises the F layer and, hence, slows its decay. None of these nocturnal processes were included in the calculations shown in Figure 11.10.

The *D region*, which covers the altitude range from about 60 to 100 km, is discussed last because it is the most difficult region to observe and to model. Like the E region, the D region is controlled by chemical processes and the dominant species are molecular ions and neutrals. However, unlike the E region, the D region is composed of both positive and negative ions and water cluster ions; in addition, three-body chemical reactions are important. The cluster ions dominate the D region at altitudes below about 85 km and they are formed via hydration starting from the primary ions NO^+ and O_2^+. Also, in addition to the usual neutrals that are found in the E and lower F regions (N_2, O_2, O, N), several important minor neutral species [NO, CO_2, H_2O, O_3, OH, NO_2, HO_2, $O_2(^1\Delta g)$] must be taken into account. Nitric oxide, in particular, plays a crucial role in the D region ion chemistry because it can be ionized by Lyman α radiation. Unfortunately, the densities of the minor neutral species and many of the chemical reaction rates are not well known.

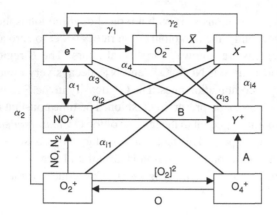

Figure 11.11 The Mitra–Rowe six-ion chemical scheme for the D and E regions.[13]

Cluster ions were first measured in 1963 with a rocket-borne mass spectrometer,[12] and subsequently, increasingly more complex models were developed to explain the measurements. Figure 11.11 shows the Mitra–Rowe six-ion chemical scheme for the D and E regions.[13] The various reactions and rates are given elsewhere[14] and are not repeated here.[14] Some of the important features of this simplified model are: (1) NO^+ and O_2^+ are the precursor ions; (2) clustering occurs through NO^+ above about 70 km and through O_2^+ below this altitude; (3) O_4^+ is included explicitly because the back reaction to O_2^+ inhibits clustering from the O_2^+ channel above about 85 km; (4) all cluster ions are lumped under a common ion called Y^+ (this is the main simplifying assumption); (5) all negative ions, except O_2^-, are lumped under X^-. The main advantage of the six-ion model is its computational efficiency, but another advantage is that numerous reactions with uncertain rate coefficients are lumped together.

The most sophisticated chemical model of the D and E regions that has been developed to date is the Sodankylä ion chemistry (SIC) model, which includes 24 positive ions and 11 negative ions.[14] The chemical scheme upon which the model is based is shown in Figure 11.12. Note that the water cluster ions are primarily of the form $H^+(H_2O)_n$, $NO^+(H_2O)_n$, and $O_2^+(H_2O)_n$, where n can be as large as eight. The SIC model takes account of both two- and three-body positive ion–neutral reactions, recombination of positive ions with electrons, photodissociation of positive ions, both two- and three-body negative ion–neutral reactions, electron photodetachment of negative ions, photodissociation of negative ions, electron attachment to neutrals, and ion–ion recombination. Overall, there are 174 chemical reactions in the SIC model. The specific reactions and their rates are given in the literature[14] and are not repeated here.[14]

A comparison of electron density profiles calculated from the Mitra–Rowe and SIC models with measurements clearly shows the current state of the art with regard to D and E region modeling. Such a comparison has been made using EISCAT incoherent scatter radar measurement of n_e over the altitude range of 80–120 km.[14] The comparison was limited to the daytime, summer ionosphere and quiet geomagnetic conditions. To start the model and data comparison, it was necessary to adopt

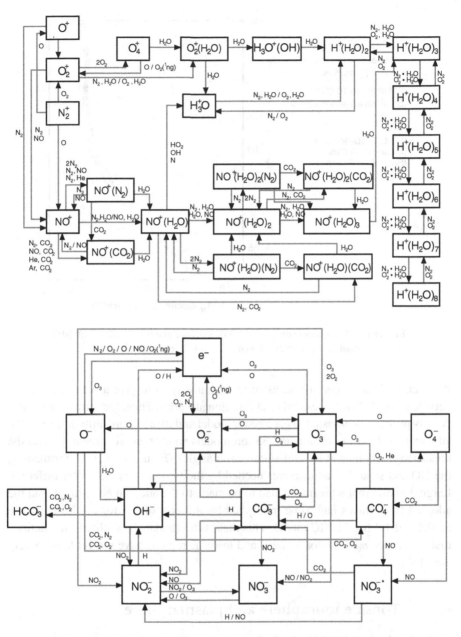

Figure 11.12 The 35-ion chemical scheme for the D and E regions developed at the Sodankylä Geophysical Observatory.[14]

a reference solar spectrum, a representative Lyman α flux, density profiles for the main neutral species (N_2, O_2, O, N, He, H), and altitude profiles for the minor neutral species [NO, CO_2, H_2O, O_3, OH, NO_2, HO_2, $O(^1\Delta_g)$], which were obtained from separate measurements. As expected, the initial comparison of both the six-ion and 35-ion models with the n_e measurements were not very successful. To get agreement,

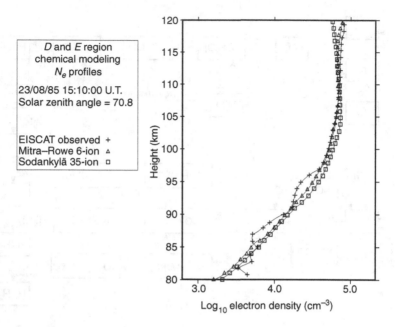

Figure 11.13 The comparison of the Mitra–Rowe and SIC models with EISCAT electron density measurements taken at 1510 UT on August 23, 1985.[14]

the solar EUV flux was first adjusted in such a way as to give a better agreement between modeled and measured n_e at E region altitudes. Then, the nitric oxide (NO) density profile was adjusted to get a better model and data fit at all altitudes. With the adjustments, both the six-ion and 35-ion models were in good agreement with the measured n_e profiles at several solar zenith angles (Figure 11.13). Unfortunately, the NO and solar flux adjustments needed by the two models were very different. Large, but different adjustments had to be made to the adopted NO profile, and the adopted EUV fluxes had to be multiplied by a factor of 2.5 for the Mitra–Rowe model and 1.3 for the SIC model. This means the fundamental chemical reactions that govern the n_e behavior in the D and lower E regions have not yet been clearly established.

11.5   Topside ionosphere and plasmasphere

The ionospheric layers were first studied in the 1925 to 1930 time period using ground-based radio sounding techniques (Section 14.5). However, high-frequency radio waves transmitted from the Earth are reflected only from altitudes up to the F_2 region peak. Above the F_2 peak, it was assumed that the electron density decreased exponentially with altitude until it merged with the solar wind. This view persisted up to 1953, when lightning-generated low-frequency radio waves (*whistlers*), which propagate along **B**, were used to deduce the presence of appreciable electron concentrations ($\sim 10^3$ cm^{-3}) up to altitudes as high as three or four Earth radii.[15] However,

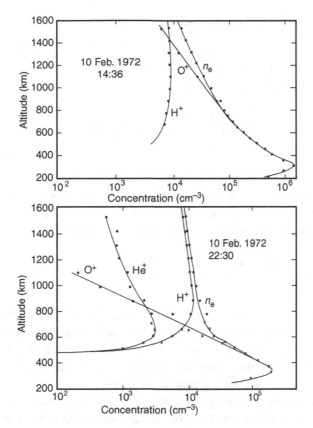

Figure 11.14 Altitude profiles of ion composition for the daytime (top panel) and nighttime (bottom panel) ionosphere. The profiles were measured with the incoherent scatter radar at Arecibo, Puerto Rico.[17]

it was not until 1960 that a plausible explanation was advanced to explain the high plasma densities. In 1960, it was suggested that the reversible charge exchange reaction $O^+ + H \Leftrightarrow H^+ + O$ could produce large quantities of H^+ ions that could then diffuse upward along geomagnetic field lines to high altitudes.[16]

It is now well known that the plasma environment that surrounds the Earth is composed of both a topside ionosphere and protonosphere. The *topside ionosphere* is defined to be that region above the F_2 peak ($h_m F_2$) where O^+ is the dominant ion; it extends from about 600 to 1500 km at mid-latitudes. The region above this, where H^+ becomes dominant, is referred to as the *protonosphere*. Figure 11.14 shows typical ion density profiles in the topside ionosphere and protonosphere, as measured by the incoherent scatter radar at Arecibo, Puerto Rico.

The H^+ ions at low altitudes are in chemical equilibrium with O^+ and their density is controlled by the charge exchange reaction (also see Section 8.3)

$$O^+ + H \underset{k_r}{\overset{k_f}{\longleftrightarrow}} H^+ + O, \tag{11.62}$$

where k_f and k_r are the forward and reverse reaction rates,[18]

$$k_f = 2.5 \times 10^{-11} \left[T_n + \frac{T(O^+)}{16} + 1.2 \times 10^{-8} \left(u(O^+) - u_n \right)^2 \right]^{1/2},$$

(11.63a)

$$k_r = 2.2 \times 10^{-11} \left[T(H^+) + \frac{T_n}{16} + 1.2 \times 10^{-8} \left(u(H^+) - u_n \right)^2 \right]^{1/2},$$

(11.63b)

and where the temperatures are in kelvins, the field-aligned velocities in cm s^{-1}, and the rate coefficients in cm^3 s^{-1}. At these altitudes, the H$^+$ density can be obtained simply by equating the H$^+$ production and loss terms, which yields

$$n(H^+) = 1.13 \frac{n(O^+) n(H)}{n(O)},$$

(11.64)

where the temperatures are assumed to be comparable and relatively high (Section 8.3), and the flow terms are negligible in the chemical equilibrium domain, so that $k_f / k_r \approx 1.13$. Therefore, when chemical equilibrium prevails, the H$^+$ density increases exponentially with altitude, because the O$^+$ and H densities decrease exponentially with altitude more slowly than the O density.

Eventually, chemical equilibrium gives way to diffusive equilibrium, and this occurs when H$^+$ is still a minor ion. However, under these circumstances, the H$^+$ density continues to increase with altitude, at a rate that is almost the same as that which occurs for chemical equilibrium (Section 5.7). When the H$^+$ density becomes greater than the O$^+$ density, it then decreases exponentially with altitude with a diffusive equilibrium scale height that is characteristic of a major ion (Equation 5.59).

When diffusive equilibrium controls the density structure of the topside ionosphere, thermal diffusion can be important (Section 5.14). Thermal diffusion arises as a result of the effect that heat flow has on the momentum balance when different gases collide. It is particularly strong in fully ionized gases when there are substantial ion or electron temperature gradients. In a plasma composed of O$^+$, H$^+$, and electrons, thermal diffusion acts to drive the light and heavy ions in opposite directions. The heavy ions are driven toward hotter regions, i.e., toward higher altitudes. This effect is illustrated in Figure 11.15, where theoretical ion and electron density profiles are shown for calculations with and without thermal diffusion. Three cases are illustrated, corresponding to plasma temperatures typical of the nighttime and daytime ionosphere at mid-latitudes and those found in sub-auroral red arcs (SAR arcs; Section 11.6). For all three cases, both T_e and T_i increase with altitude, with progressively larger temperature gradients in going from the nighttime to the daytime and then to the SAR arc case. Note that thermal diffusion acts to increase the O$^+$ density at high altitudes and decrease the H$^+$ density, and this can result in a substantial change in the O$^+$/H$^+$ transition altitude (by as much as 400 to 500 km).

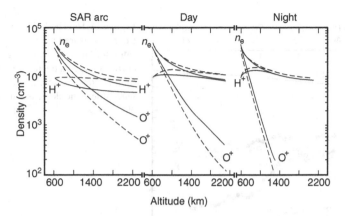

Figure 11.15 Electron and ion density profiles calculated with (solid) and without (dashed) allowance for thermal diffusion. The three cases correspond to plasma temperature profiles that are representative of SAR arc, daytime, and nighttime conditions.[19]

Diffusive equilibrium prevails when the upward H^+ flow speed is much smaller than the H^+ thermal speed. In general, diffusive equilibrium prevails at low latitudes and at the lower end of the mid-latitude domain. It can also prevail throughout the entire mid-latitude domain during magnetically quiet periods. In the more typical situation, the upward H^+ speed increases with latitude because the H^+ upward flow is determined by the pressure in the overlying *plasmasphere* (also called the protonosphere), which is a torus-shaped volume composed of closed, basically dipolar, magnetic field lines (Figure 2.10). As latitude increases, the volume of the plasmaspheric flux tubes increases, and consequently, the H^+ density and pressure at high altitudes tend to decrease.

Several situations are possible, depending on the solar cycle, seasonal, and geomagnetic activity conditions. First, when the ionosphere and plasmasphere are in equilibrium, diffusive equilibrium prevails and there is a gentle ebb and flow of ionization between the two regions. The flow is upward from the ionosphere during the day, when the O^+ density is relatively high, and downward at night, when the O^+ density decays. The downflowing H^+ ions charge exchange with O to produce O^+, and this process helps to maintain the nighttime F region. A different situation can occur near the solstices, where the flow can be interhemispheric. In this case, it is upward and out of the topside ionosphere throughout the day and night in the summer hemisphere. In the winter hemisphere, the flow is upward during the day and downward at night. Of course, the direction of the flow determines whether the flow is a source or sink for the ionosphere.

Another flow situation occurs after geomagnetic storms and substorms.[20] During these events, the plasma in the outer plasmasphere is convected away owing to enhanced magnetospheric electric fields (Section 12.1). The high-altitude depletions can be very substantial, and the consequent reductions in plasma pressure induce ionospheric upflows. The upflows typically occur throughout the day and night in

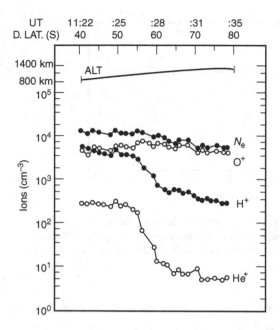

Figure 11.16 The light-ion trough measured by the *OGO-2* satellite. The light-ion trough corresponds to a decrease of the light ion densities with latitude, with little or no decrease in the electron density.[21]

both hemispheres, and they can last for many days after the storm or substorm. The flux tubes at low latitudes refill fairly quickly because their volumes are small. However, the flux tubes in the outer plasmasphere can take many days to refill, which is longer than the average time between geomagnetic storms. Therefore, the outer plasmaspheric flux tubes are always in a partially depleted state. The net result is that at a fixed altitude the densities of the light ions (H^+ and He^+) decrease with increasing latitude (Figure 11.16). This feature is known as the light-ion trough.

11.6   Plasma thermal structure

In Chapter 9, the general flow of energy in the Earth's upper atmosphere was discussed and the various heating and cooling rates for the electrons and ions were presented (Sections 9.6 and 9.7). As noted in Chapter 9, the photoelectrons provide the main source of energy for the thermal electrons at all latitudes, but precipitating auroral electrons are an important additional source of energy at high latitudes. The photoelectron energy is transferred to the ionospheric electrons by both direct and indirect processes. First, the low-energy photoelectrons (≤ 2 eV) transfer energy directly to the thermal electrons via Coulomb collisions in a region close to where they are created. This leads to a bulk heating of the thermal electron gas, with the bulk heating rate peaking in the 150 to 300 km altitude region depending on the geophysical conditions (Figure 9.12). However, the more energetic photoelectrons can also heat the ionospheric electrons by an indirect mechanism. These photoelectrons can escape the ionosphere, and they lose energy to the thermal electrons at high altitudes as they escape. This energy is then conducted down into the ionosphere along geomagnetic field lines.

The thermal electrons can lose energy to the various ion and neutral species via both elastic and inelastic collisional processes. At middle and low latitudes, the dominant electron cooling results from excitation of the fine structure levels of atomic oxygen at low altitudes and from both Coulomb collisions with ions and downward thermal conduction at high altitudes (Figure 9.17). However, when the electron temperature is high, as it is in the auroral oval and in sub-auroral red arcs, other inelastic collisional processes also become effective in cooling the thermal electrons.

The equation that governs the electron energy balance in a partially ionized plasma was derived earlier and is given by Equation (5.135c). In general, the viscous heating of the electron gas is small and, therefore, the $\tau_e : \nabla \mathbf{u}_e$ term in Equation (5.135c) can be neglected. On the other hand, additional terms must be added on the right-hand side of this equation to account for external heat sources and inelastic cooling processes. With these modifications, Equation (5.135c) becomes

$$
\frac{D_e}{Dt}\left(\frac{3}{2}p_e\right) + \frac{5}{2}p_e(\nabla \cdot \mathbf{u}_e) + \nabla \cdot \mathbf{q}_e = \sum Q_e - \sum L_e
$$

$$
- \sum_i \frac{\rho_e \nu_{ei}}{m_i} 3k(T_e - T_i)
$$

$$
- \sum_n \frac{\rho_e \nu_{en}}{m_n} 3k(T_e - T_n), \quad (11.65)
$$

where $\sum Q_e$ is the sum of the external heating rates (photoelectrons (9.51), auroral electrons, etc.), $\sum L_e$ is the sum of the inelastic cooling rates (Equations 9.52–70), and $D_e/Dt = \partial/\partial t + \mathbf{u}_e \cdot \nabla$ is the convective derivative. The energy equation (11.65) can be expressed in a more convenient form by using the source-free continuity equation, $\partial n_e/\partial t + \nabla \cdot (n_e \mathbf{u}_e) = 0$, and the result is

$$
\frac{3}{2}n_e k \frac{\partial T_e}{\partial t} = -n_e k T_e \nabla \cdot \mathbf{u}_e - \frac{3}{2}n_e k \mathbf{u}_e \cdot \nabla T_e - \nabla \cdot \mathbf{q}_e + \sum Q_e - \sum L_e
$$

$$
- \sum_i \frac{\rho_e \nu_{ei}}{m_i} 3k(T_e - T_i) - \sum_n \frac{\rho_e \nu_{en}}{m_n} 3k(T_e - T_n). \quad (11.66)
$$

The first term on the right-hand side of Equation (11.66) represents adiabatic expansion and the second term accounts for advection. These processes are negligible in the terrestrial ionosphere. Also, at middle and high latitudes, the electron heat flow is along $\mathbf{B}$, while the dominant temperature gradients are in the vertical, z, direction. This leads to the appearance of a $\sin^2 I$ term in the expression for $\nabla \cdot \mathbf{q}_e$ (the same as for diffusion; Equation 11.59). Therefore, for the terrestrial ionosphere at middle and high latitudes, with no field-aligned current ($\mathbf{J}_\parallel = 0$), the electron energy equation reduces to

$$\frac{3}{2}n_e k \frac{\partial T_e}{\partial t} = \sin^2 I \frac{\partial}{\partial z}\left(\lambda_e \frac{\partial T_e}{\partial z}\right) + \sum Q_e - \sum L_e,$$

$$- \sum_i \frac{\rho_e \nu_{ei}}{m_i} 3k(T_e - T_i) - \sum_n \frac{\rho_e \nu_{en}}{m_n} 3k(T_e - T_n), \qquad (11.67)$$

where the appropriate expression for q_e is given in Equation (5.141) and where the thermal conductivity, λ_e, is given by Equation (5.146).

Typically, the electron temperature in the ionosphere responds rapidly (a few seconds) to changing conditions and, therefore, the electron temperature is generally in a quasi-steady state ($\partial/\partial t \to 0$). Furthermore, at low altitudes, thermal conduction is not important because the neutrals are effective in inhibiting the flow of heat (Equation 5.146). Under these circumstances, the electron temperature is determined by a balance between local heating and cooling processes. The altitude below which a local thermal equilibrium prevails varies from 150 to 350 km, depending on local time, season, and solar cycle. On the other hand, the electron thermal balance at high altitudes is dominated by thermal conduction and the plasma is effectively fully ionized. In this case, the electron energy equation (11.67) reduces to

$$\frac{d}{dz}\left(7.7 \times 10^5 T_e^{5/2} \frac{dT_e}{dz}\right) = 0, \qquad (11.68)$$

where the fully ionized expression, $\lambda_e = 7.7 \times 10^5 T_e^{5/2}$ eV cm^{-1} s^{-1} K^{-1} (the *Spitzer conductivity*), was used (Equation 5.146). Equation (11.68) can be easily solved to obtain an analytical expression for the T_e profile at the altitudes where thermal conduction dominates, and the solution is

$$T_e = \left[T_{eb}^{7/2} - \frac{7}{2}\left(\frac{q_{et}}{7.7 \times 10^5}\right)(z - z_b)\right]^{2/7}, \qquad (11.69)$$

where T_{eb} (K) is the electron temperature at the bottom boundary of the thermal conduction regime, q_{et} (in eV cm^{-2} s^{-1}) is the electron heat flow through the top boundary, and z_b (in cm) is the altitude of the bottom boundary. Equation (11.69) shows that if there is a downward heat flow through the top boundary ($q_{et} < 0$), then T_e increases with altitude. If $q_{et} = 0$, then $T_e = T_{eb}$ at all altitudes, i.e., T_e is isothermal.

The basic physics described above is reflected in Figure 11.17, where calculated T_e profiles are plotted for both day and night local times.[22, 23] The profiles pertain to the ionosphere over Millstone Hill on March 23–24, 1970. For the daytime conditions, photoelectron heating and oxygen fine structure cooling dominate the electron thermal balance below 300 km, and the peak in T_e is associated with the peak in the photoelectron heating rate. Above 300 km, the T_e profile is dominated by thermal conduction, with cooling to the ions playing a minor role. The steep gradient in T_e is caused by the imposition of a large downward electron heat flow at the upper

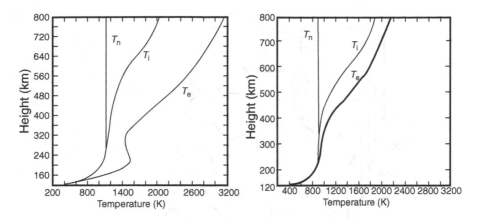

Figure 11.17 Calculated electron, ion, and neutral temperature profiles for the ionosphere over Millstone Hill on March 23–24, 1970. The left panel is for 1422 LT and the right panel for 0222 LT.[22]

boundary (800 km), which was required to bring the calculated T_e profiles into agreement with the measured profiles (not shown). At night, the photoelectron heat source is absent, and $T_e = T_n$ at low altitudes, owing to the strong collisional coupling of the electrons and neutrals. However, above about 250 km, thermal conduction becomes important and T_e increases with altitude in response to the imposed downward heat flux at 800 km. This latter downward heat flow results from either energy stored at high altitudes during the day or from the effects of wave–particle interactions at high altitudes.

The ion thermal balance is straightforward because temporal variations, advection, adiabatic expansion, and thermal conduction are not too important at midlatitudes. Typically, thermal conduction is only important for the ions above about 700 km. Therefore, the ion energy balance is simply governed by collisional coupling to both the hot thermal electrons and the cold neutrals. At low altitudes, coupling to the neutrals dominates and $T_i = T_n$ during both the day and night. As altitude increases, coupling to the electrons becomes progressively more important and T_i increases. However, ion–electron coupling is never complete and, therefore, T_i does not attain thermal equilibrium with T_e (Figure 11.17).

The ability of photoelectrons to escape the topside ionosphere leads to some interesting thermal phenomena. As noted above, the photoelectrons lose energy to the thermal electrons at high altitudes as they stream along **B**. The length of a dipole field line increases as latitude increases, and consequently, more photoelectron energy is transferred to the thermal electrons on longer, higher latitude, field lines than on shorter, lower latitude, field lines. The net effect is that the downward electron heat flow into the ionosphere and, hence, the ionospheric electron temperature increase with latitude at the altitudes where thermal conduction dominates. This phenomenon is illustrated in Figure 11.18, where T_e distributions at 1000 km are shown as a function of latitude and local time. The temperatures are from an empirical model

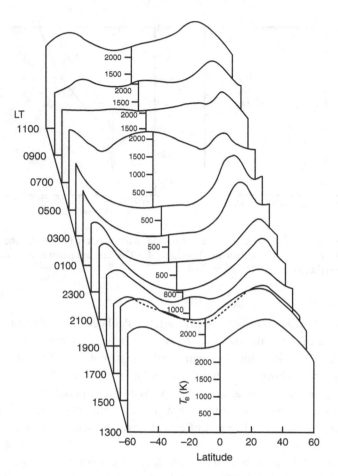

Figure 11.18 Electron temperatures obtained from an empirical model of the averaged latitudinal and local time behavior of T_e at 1000 km for winter solstice in 1964.[24]

that is based on Explorer 22 satellite measurements.[24] Note that there are T_e maxima in both hemispheres at about 50° latitude, which is the approximate upper limit of the mid-latitude domain.

Another interesting thermal phenomenon is known as the *predawn effect*, whereby T_e begins to increase rapidly before local sunrise. This feature is shown in Figure 11.19, where T_e measurements at 375 km are plotted as a function of local time for Millstone Hill. Note that T_e begins to increase at about 0230 LT, while local sunrise occurs at about 0530 LT. This early onset of the T_e increase occurs at a time that corresponds to sunrise in the magnetically conjugate ionosphere and the heating is caused by photoelectrons arriving from the conjugate hemisphere.

A thermal phenomenon that does not involve photoelectrons occurs during geomagnetic disturbances and is known as *sub-auroral red* (SAR) arcs or *stable auroral red* (SAR) arcs.[26, 27] These arcs correspond to a band of red emission that is narrow in latitude, but extended in longitude. The band of emission appears to extend

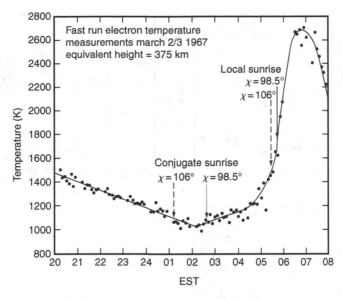

Figure 11.19 Electron temperature measurements obtained with the Millstone Hill incoherent scatter radar on 2–3 March 1967. Clearly evident is the predawn electron heating caused by photoelectrons arriving from the conjugate hemisphere.[25]

completely around the Earth, and it generally occurs in both hemispheres simultaneously. The stability of the emission led to the term stable auroral red arc, but SAR arcs occur equatorward of the auroral oval and, to avoid confusion, the term sub-auroral red arc was introduced. Sub-auroral red arcs are a manifestation of a thermal phenomenon and arise in the following manner. During geomagnetic disturbances, the cold, high density plasma in the plasmasphere comes into contact with the hot, tenuous plasma in the ring current. As a result of Coulomb collisions or wave–particle interactions (via ion cyclotron or hydromagnetic waves), energy is transferred from the ring current particles to the thermal electrons in the interaction region. This energy is then conducted down to the ionosphere along **B**, producing elevated electron temperatures (4000–10 000 K). At altitudes between 300 to 400 km, there is a sufficient number of hot electrons in the tail of the thermal electron velocity distribution to collisionally excite atomic oxygen to a higher electronic state. The excitation is from the $O(^3P)$ to $O(^1D)$ state, which requires an energy of 1.97 eV (Figure 8.1). The excited atoms subsequently emit 630 nm photons, which corresponds to the red line of atomic oxygen.

11.7 Diurnal variation at mid-latitudes

The ionosphere undergoes a marked diurnal variation as the Earth rotates into and out of sunlight. This diurnal variation is shown in Figure 11.20 for a typical mid-latitude location. The figure shows contours of n_e, T_e, and T_i as a function of altitude

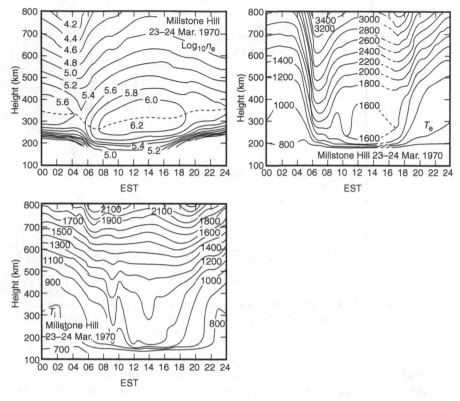

Figure 11.20 Contours of electron density (top left), electron temperature (top right), and ion temperature (bottom) measured by the Millstone Hill incoherent scatter radar on 23–24 March 1970.[28] n_e is in cm^{-3} and the temperatures are in K.

and Eastern Standard Time (EST). The measurements were made with the Millstone Hill incoherent scatter radar on March 23–24, 1970.[22, 23, 28] The physical processes that control the diurnal variation of the electron density change with both local time and altitude. At sunrise, the electron density begins to increase rapidly owing to photoionization (Section 9.3). After this initial sunrise increase, n_e displays a slow rise throughout the day, and then it decays at sunset as the photoionization source disappears. The ionization below the F region peak is under strong solar control, reaching its maximum value near noon, when the solar zenith angle is the smallest, and then decreasing symmetrically away from noon. This behavior results from the fact that photochemistry dominates at altitudes below the F region peak and the chemical time constants are short (Sections 8.2 to 8.4). The electron density above the F region peak is influenced by other processes, including diffusion, interhemispheric flow, and neutral winds. Therefore, the electron density contours at altitudes above the peak do not display a strong solar zenith angle dependence, and the maximum ionization occurs late in the afternoon close to the time when the neutral exospheric temperature peaks. At night, the plasma transport processes control the ionization decay. However, the height of the F region peak is primarily determined by the meridional neutral wind, which induces a plasma flow along the inclined geomagnetic field lines. It is upward at night and downward during the day (Section 11.4).

The electrons are heated by photoelectrons that are created in the photoionization process and by a downward flow of heat from the magnetosphere. However, the electron temperature is also influenced by both elastic and inelastic collisions of the thermal electrons with the ions and neutrals (Sections 9.6 and 9.7). In addition, for a given electron heating rate, the electron temperature is inversely related to the electron density. These facts help explain the diurnal variation of T_e shown in Figure 11.20. At sunrise, T_e increases rapidly, with a time constant of the order of seconds, because of photoelectron heating. The photoelectron heating rate does not vary appreciably during the early morning hours, but the electron density continues to increase. As the electron density increases, the electron temperature decreases (between 07 and 10 EST), because of the increasing heat capacity of the electron gas and the stronger coupling to the relatively cold ions. From about 10 to 16 EST, T_e is nearly constant, and then T_e decreases at sunset when the photoelectron heat source disappears. However, T_e stays elevated above T_n because of an energy flow from the plasmasphere, which produces the positive gradient in the nocturnal electron temperature above 200 km.

The diurnal variation of the ion temperature is more straightforward than those for the electron density and temperature. Below about 400 km, the ion temperature basically follows the neutral temperature, and its diurnal variation determines the T_i diurnal variation. Above this height, T_i increases with altitude, owing primarily to the increased thermal coupling to the hotter electrons, but there is also a small downward ion heat flow from the magnetosphere. Nevertheless, the diurnal variation of T_i is controlled by the diurnal variation of T_e above 400 km.

11.8 Seasonal variation at mid-latitudes

The ionosphere exhibits strong seasonal and solar cycle variations because the main source of ionization and energy for the ionosphere is photoionization. Therefore, whenever either the solar zenith angle or the solar radiation flux change, the ionosphere will change. The ionosphere's seasonal variation is related to a solar zenith angle change, while its solar cycle variation is related to a change in the solar EUV and X-ray radiation fluxes. However, the ionospheric variations are not always simple because the ionosphere is closely coupled to the thermosphere, which also undergoes seasonal and solar cycle changes. For example, Figure 11.21 shows the seasonal variation of the daytime ionosphere at mid-latitudes.[29] The important feature to note is that N_mF_2 in winter is greater than N_mF_2 in summer despite the fact that the solar zenith angle is smaller in summer. This phenomenon, which is called the *seasonal anomaly*, occurs because of the seasonal changes in the neutral atmosphere. Specifically, the summer-to-winter neutral circulation results in an increase in the O/N_2 ratio in the winter hemisphere and a decrease in the summer hemisphere.

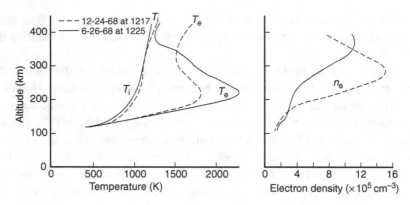

Figure 11.21 Summer (solid curves) and winter (dashed curves) profiles of T_e, T_i, and n_e measured by the Arecibo incoherent scatter radar during the daytime.[29]

The increased O and decreased N_2 densities in winter act to increase the O^+ densities, due to the relative increase in the production rate and decrease in the loss rate. This effect is more than enough to offset the tendency for decreased O^+ densities due to a larger solar zenith angle. The net result is that the O^+ densities in winter are larger than those in summer at F region altitudes. In turn, the higher electron densities in winter result in lower electron temperatures (Figure 11.21), owing to the inverse relationship between the electron density and temperature. Also, it should be noted that the neutral helium density displays a strong seasonal dependence and this leads to a seasonal dependence for He^+. However, this topic is discussed later in connection with the polar wind (Section 12.12).

11.9   Solar cycle variation at mid-latitudes

The solar cycle variation of the electron density and temperature is shown in Figure 11.22 for the daytime mid-latitude ionosphere at equinox. At solar maximum, the solar EUV fluxes and the atomic oxygen densities are greater than those at solar minimum, and these conditions lead to higher electron densities and lower electron temperatures. The higher electron densities at solar maximum are simply a result of an increased production, while the lower electron temperatures are a result of the inverse relationship between the electron density and temperature. With regard to the shape of the T_e profile, a pronounced peak occurs at about 250 km at solar maximum, while T_e increases monotonically with altitude at solar minimum. The T_e peak at solar maximum is again a consequence of the high electron densities, which lead to a dominance of electron–ion energy coupling over thermal conduction at altitudes between about 250 to 400 km. The stronger electron coupling to the cold ions causes the decrease in T_e over this altitude range. Above 400 km, thermal conduction dominates and T_e increases with altitude in response to a downward heat flow from the magnetosphere.

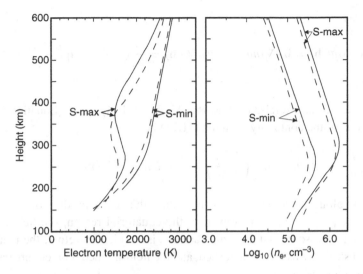

Figure 11.22 Electron temperature and density profiles for the daytime mid-latitude ionosphere at equinox for both solar minimum and maximum conditions. The solid curves are profiles measured by the Millstone Hill incoherent scatter radar, while the dashed curves are calculated.[30]

11.10   Plasma transport in a dipole magnetic field

The large-scale flow of plasma in the equatorial region of the terrestrial ionosphere can be described in terms of a flow along $\mathbf{B}(u_\parallel \mathbf{b})$ and an electrodynamic drift across $\mathbf{B}(\mathbf{u}_E)$, such that

$$\mathbf{u} = \mathbf{u}_E + u_\parallel \mathbf{b}. \tag{11.70}$$

In this case, the continuity equation (3.57) for an electrically neutral ($n_e = n_i = n$), current-free ($\mathbf{u}_e = \mathbf{u}_i = \mathbf{u}$), single-ion O^+-electron plasma can be expressed in the form

$$\frac{Dn}{Dt} + \nabla \cdot (nu_\parallel \mathbf{b}) = P - L'n - n(\nabla \cdot \mathbf{u}_E), \tag{11.71}$$

where L' is the O^+ loss frequency and

$$\frac{D}{Dt} = \frac{\partial}{\partial t} + \mathbf{u}_E \cdot \nabla \tag{11.72}$$

is the convective derivative that pertains to motion across $\mathbf{B}$ and where $\nabla \cdot \mathbf{u}_E$ is given by Equation (11.18). The second term on the left-hand side of Equation (11.71) can be written as

$$\nabla \cdot (nu_\parallel \mathbf{b}) = \mathbf{b} \cdot \nabla(nu_\parallel) + nu_\parallel(\nabla \cdot \mathbf{b}) = \frac{\partial}{\partial s}(nu_\parallel) + nu_\parallel\left(\frac{1}{A}\frac{\partial A}{\partial s}\right),$$

(11.73)

where it is left as an exercise to show that $\nabla \cdot \mathbf{b} = (1/A)\partial A/\partial s$ (Equation 11.12). With this result, the continuity equation (11.71) becomes

$$\frac{Dn}{Dt} + \frac{\partial}{\partial s}(nu_\parallel) + nu_\parallel\left(\frac{1}{A}\frac{\partial A}{\partial s}\right) = P - L'n - n(\nabla \cdot \mathbf{u_E}).$$

(11.74)

The flow along $\mathbf{B}$ can be obtained by taking the scalar product of $\mathbf{b}$ with the momentum equation (3.58). However, in the equatorial region, the field-aligned flow is usually subsonic and the temperature is isotropic. Therefore, the nonlinear inertial and stress terms can be neglected, and the field-aligned momentum equation for the ions reduces to

$$\frac{\partial p_i}{\partial s} + n_i m_i g \sin I - n_i e E_\parallel = n_i m_i \nu_i (\mathbf{u_n} - \mathbf{u_i})_\parallel,$$

(11.75)

where $\mathbf{u_n}$ is the neutral wind velocity, which is assumed to be the same for all neutral species, and where ν_i is the total ion–neutral collision frequency

$$\nu_i = \sum_n \nu_{in}.$$

(11.76)

The polarization electrostatic field is determined by the electron motion along $\mathbf{B}$ and is given by (Equation 5.61)

$$eE_\parallel = -\frac{1}{n_e}\frac{\partial p_e}{\partial s}.$$

(11.77)

Substituting Equation (11.77) into Equation (11.75) and noting that $p_e = n_e k T_e$, $p_i = n_i k T_i$, and $n_e = n_i = n$, the field-aligned momentum equation for the single-ion plasma can be written in the form of a classical diffusion equation

$$nu_\parallel = nu_{n\theta}\cos I - D_a\left[\frac{\partial n}{\partial s} + n\left(\frac{1}{T_p}\frac{\partial T_p}{\partial s} + \frac{m_i g \sin I}{2kT_p}\right)\right],$$

(11.78)

where $D_a = 2kT_p/(m_i\nu_i)$ is the ambipolar diffusion coefficient (Equation 5.55), $T_p = (T_e+T_i)/2$ is the plasma temperature (Equation 5.56), and $u_{n\theta}$ is the meridional component of the neutral wind. In deriving Equation (11.78), the contribution of the vertical neutral wind was neglected, because it is generally small.

The substitution of the expression for the field-aligned plasma flux (11.78) into the continuity equation (11.74) leads to a second-order, parabolic, partial differential equation in the coordinate s. The coordinate s is then typically replaced with the dipole coordinate q_d by using Equations (11.19), (11.25), and (11.26).[31–33]

11.11   Equatorial F region

The dynamo electric fields that are generated in the equatorial E region by thermospheric winds are transmitted along the dipole magnetic field lines to F region altitudes because of the high parallel conductivity (Figure 11.5). During the daytime, the dynamo electric fields are eastward, which causes an upward $E \times B$ plasma drift, while the reverse occurs at night. The plasma that is lifted during the daytime then diffuses down the magnetic field lines and away from the equator due to the action of gravity. This combination of electromagnetic drift and diffusion produces a fountain-like pattern of plasma motion (Figure 11.23), which is called the *equatorial fountain*. A result of the fountain motion is that ionization peaks are formed in the subtropics on both sides of the magnetic equator; this feature is termed the *equatorial anomaly* or *Appleton anomaly*. Figure 11.24 shows the Appleton anomaly, as calculated with a numerical model for December solstice conditions.[35] The figure shows the conditions corresponding to 2000 LT, which is when the upward $E \times B$ drift raises the F layer at the magnetic equator to 600 km. This leads to ionization peaks on both sides of the magnetic equator via the fountain effect. The asymmetry in the peaks is a result of a meridional neutral wind that blows from the southern (summer) hemisphere to the northern (winter) hemisphere. Such a wind acts to transport plasma up the field line in the southern hemisphere and down the field line in the northern hemisphere. Four hours later, the $E \times B$ drift is downward, the height of the F layer at the magnetic equator drops to 400 km, and the ionization peaks move closer to the equator (not shown). The asymmetry is also decreased because the northern peak, which is at a lower altitude, decays at a faster rate than the southern peak.[36, 37]

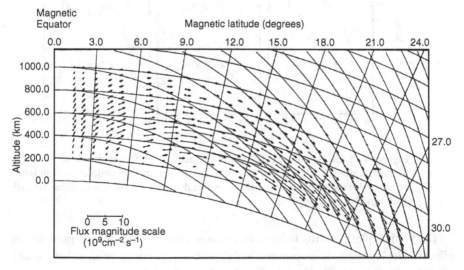

Figure 11.23 Plasma drift pattern at low latitudes due to the combined action of an upward $E \times B$ drift near the magnetic equator and a downward diffusion along B.[34]

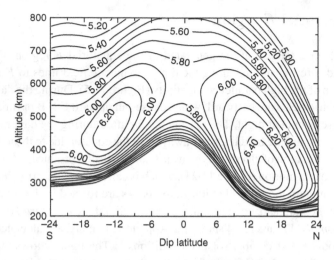

Figure 11.24 Calculated electron density contours ($\log_{10} n_e$) as a function of altitude and dip latitude at 2000 LT for December solstice conditions.[35]

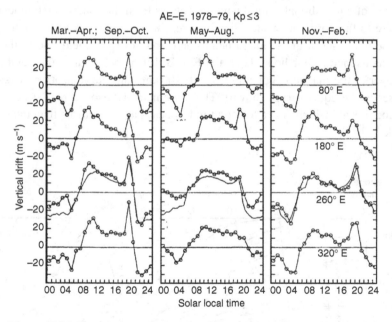

Figure 11.25 Empirical model of vertical plasma drifts in four longitude sectors and for three seasons. The results are for low magnetic activity and moderate to high solar activity. Also shown are the seasonal Jicamarca drift patterns for similar solar flux and geomagnetic conditions.[38]

The vertical plasma drifts induced by dynamo electric fields have a pronounced effect on the low-latitude ionosphere, and therefore, it is not surprising that a major effort has been devoted to obtaining empirical models of this drift component. Currently, the most comprehensive empirical model of vertical plasma drifts (zonal electric fields) is the one based on Atmosphere Explorer E satellite measurements.[38] The model includes diurnal, seasonal, solar cycle, and longitudinal dependencies. Figure 11.25 shows the vertical plasma drifts as a function of local time in four longitude sectors and for three seasonal periods. As noted earlier, the vertical drifts are upward during the day and downward at night, with typical magnitudes in the range of 10–30 m s^{-1}. A feature that is evident in most longitude sectors and seasons is the *pre-reversal enhancement* in the upward plasma drift near dusk ($\sim$18 LT). This feature is linked to equatorial spread F (Section 11.12).

The vertical drifts shown in Figure 11.25 correspond to the average drifts that occur in the low-latitude ionosphere. However, when magnetic activity changes rapidly, which occurs during storms and substorms, *disturbance electric fields* appear in the equatorial region.[39] These electric fields result from the prompt penetration of magnetospheric electric fields from high to low latitudes and from the dynamo action of storm-generated neutral winds. The direct penetration electric fields have a lifetime of about 1 hour. The disturbance dynamo (wind-driven) electric fields have a longer lifetime and amplitudes that are proportional to the energy input into the ionosphere–thermosphere system at high latitudes.

11.12   Equatorial spread F and bubbles

Plasma irregularities and inhomogeneities in the F region caused by plasma instabilities manifest as *spread F echoes* (Figure 11.26). The scale sizes of the density irregularities range from a few centimeters to a few hundred kilometers, and the irregularities can appear at all latitudes. However, spread F in the equatorial region can be particularly severe. At night, fully developed spread F is characterized by *plasma bubbles*, which are vertically elongated wedges of depleted plasma that drift upward from beneath the bottomside F layer to altitudes as high as 1500 km. The individual flux tubes in a vertical wedge are typically depleted along their entire north–south extents. The east–west extent of a disturbed region can be several thousand kilometers, with the horizontal distance between separate depleted regions being tens to hundreds of kilometers.[41, 42] When bubbles form, they drift upward with a speed that generally varies from 100 to 500 m s^{-1}. However, fast bubbles, with speeds in the range of from 500 m s^{-1} to 5 km s^{-1}, occur 40% of the time that bubbles are detected.[43] The plasma density in the bubbles can be up to two orders of magnitude lower than that in the surrounding medium. When spread F ends, the upward drift ceases and the bubbles become *fossil bubbles*. The fossil bubbles then drift toward the east with the background plasma, but the high-altitude bubbles tend to lag behind.

Figure 11.27 shows a schematic diagram of the evolution of equatorial spread F and bubbles that is consistent with simultaneous HF radar, rocket, and Jicamarca VHF radar measurements on March 14–15, 1983.[44] Near the dusk terminator, the equatorial F layer rises due to the action of dynamo electric fields and subsequently it descends. On the day the measurements were made, the layer was in the process

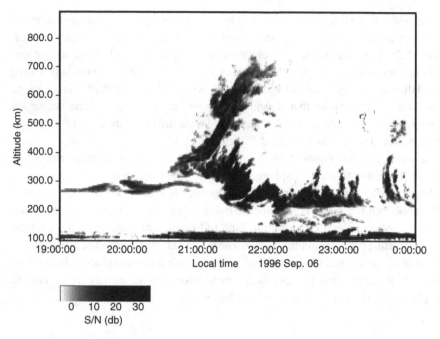

Figure 11.26 Spread F event seen by the JULIA coherent scatter radar on September 6, 1996. A range time intensity (RTI) plot of coherent backscatter signal-to-noise ratios is shown.[40] Note that density irregularities extend to 800 km at 2100 LT.

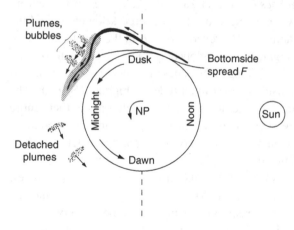

Figure 11.27 Schematic diagram showing the evolution of equatorial spread F and plasma bubbles that is consistent with multi-instrument measurements of these phenomena.[44]

of moving downward when spread F occurred. Plasma bubbles formed on the bottomside of the F layer and drifted to higher altitudes as the entire disturbed region convected toward midnight. Past midnight, the spread F disturbance ceased, but the bubbles (detached plumes) persisted. When a satellite traverses bubbles, the measured ambient plasma density can decrease by more than an order of magnitude (Figure 11.28). In this figure, the electron density variation along the polar orbiting DMSP F-10 satellite track is shown for three orbits on day 74 of 1991.[45] The satellite

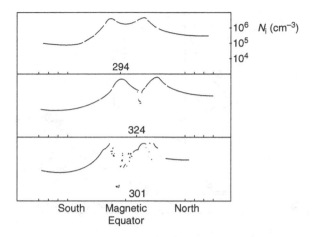

Figure 11.28 Plasma density variation along three orbits of the DMSP satellite F-10 for day 74 of 1991. The numbers in the plots correspond to the longitude at the magnetic equatorial crossing. The solar zenith angle at the magnetic equator varies from 113° to 116°, corresponding to a local solar time of about 1940.[45]

altitude varied from 745 to 855 km and data were taken every 2 s. Note that on two orbits there were large density depletions and the density in the depleted region was irregular. These depleted regions are equatorial bubbles.

The commonly accepted scenario for the formation of spread F and plasma bubbles is as follows. During the day, the thermospheric wind generates a dynamo electric field in the lower ionosphere that is eastward, and this field is mapped to F region altitudes along $\mathbf{B}$. The eastward electric field, in combination with the northward $\mathbf{B}$ field, produces an upward $\mathbf{E} \times \mathbf{B}$ drift of the F region plasma. As the ionosphere co-rotates with the Earth toward dusk, the zonal (eastward) component of the neutral wind increases, with the wind blowing predominantly across the terminator from the day side to the night side. The increased eastward wind component, in combination with the sharp day–night conductivity gradient across the terminator, leads to the pre-reversal enhancement in the eastward electric field (Figure 11.25). The F layer therefore rises as the ionosphere co-rotates into darkness. In the absence of sunlight, the lower ionosphere rapidly decays and a steep vertical density gradient develops on the bottomside of the raised F layer (Figure 11.29). This produces the classical configuration for the *Rayleigh–Taylor (R–T) instability*, in which a heavy fluid is situated above a light fluid.

A density perturbation can trigger the R–T instability on the bottomside of the F layer under certain conditions. Once triggered, density irregularities develop, and the field-aligned depletions then bubble up through the F layer. However, the F layer height and bottomside density gradient are not the only conditions necessary for the R–T instability and spread F. Upward propagating gravity waves, which induce vertical winds, can trigger the R–T instability both by providing an initial perturbation and by affecting the instability condition. However, a meridional neutral wind, which produces a north–south density asymmetry along $\mathbf{B}$ (Figure 11.24), can

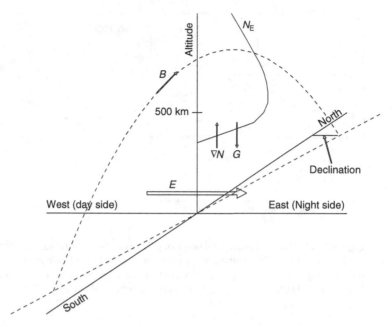

Figure 11.29 Schematic diagram showing the classical configuration for the Rayleigh–Taylor instability. The F layer rises in response to an enhanced eastward electric field and a steep density gradient develops on the bottomside of the F layer. The density gradient is opposed by gravity. (Courtesy of V. Eccles.)

stabilize the plasma. Also, the field-line integrated conductivity is important because spread F has been shown to display seasonal and longitudinal dependencies. Clearly, the longitudinal dependence is related to the B field declination (Figure 11.4) and the associated conductivity differences at the two ends of the B field line.

The basic physics underlying the R–T instability can be derived by considering the simple configuration depicted in Figure 11.30. In this figure, a plasma that is supported by a strong magnetic field sits on top of a vacuum. In the initial equilibrium state, the density–vacuum interface is smooth, a density gradient exists in the z-direction, $\nabla n_0 = (\partial n_0/\partial z)\mathbf{e}_z$, $\mathbf{G} = -g\mathbf{e}_z$, and $\mathbf{B}_0 = -B_0\mathbf{e}_x$, where ($\mathbf{e}_x$, $\mathbf{e}_y$, $\mathbf{e}_z$) are Cartesian unit vectors. For simplicity, the plasma is assumed to be cold ($T_e = T_i = 0$), there are no electric fields initially ($\mathbf{E}_0 = 0$), and the variation of gravity with altitude is ignored. As it turns out, the R–T instability for a plasma is an electrostatic mode that can be described by the hydrodynamic equations. In the equilibrium state, three fluid drifts are possible, including the electromagnetic, diamagnetic, and gravitational drifts (Section 5.10). However, only the gravitational drift is relevant because the plasma is cold ($\nabla p = 0$) and $\mathbf{E}_0 = 0$. For the configuration in Figure 11.30, the gravitational drift is given by (5.101)

$$\mathbf{u}_{G_s} = \pm\frac{g}{\omega_{c_s}}\mathbf{e}_y, \tag{11.79}$$

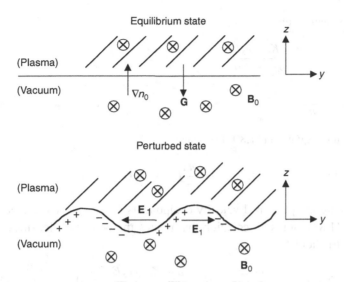

Figure 11.30 Schematic diagram for a simple Rayleigh–Taylor configuration. Relative to the Earth's equatorial ionosphere at dusk, x points to the south, y points to the east, and z points upward.

where $\omega_{c_s} = |e_s|B_0/m_s$ is the cyclotron frequency (2.7) for species s and the $\pm$ signs correspond to ions and electrons, respectively.

The stability of the plasma is determined by the procedure described in Section 6.2. The equilibrium state is perturbed (Equations 6.31a–d), the continuity and momentum equations are linearized with respect to the perturbed quantities (Equations 6.33 and 6.35), and plane wave solutions (6.36) are assumed. The resulting transport equations are given by Equations (6.37) and (6.38), except for the appearance of an additional term in the continuity equation that contains $\partial n_0/\partial z$. Also, for the problem at hand, $\mathbf{u}_{s0} = \mathbf{u}_{Gs}$ (a constant), and the perturbation propagates along the plasma–vacuum interface, so that $\mathbf{K} = K\mathbf{e}_y$. Therefore, the relevant continuity and momentum equations become

$$(\omega - Ku_{s0})n_{s1} - n_{s0}\mathbf{K} \cdot \mathbf{u}_{s1} + i\mathbf{u}_{s1} \cdot \nabla n_{s0} = 0, \tag{11.80}$$

$$i(\omega - Ku_{s0})\mathbf{u}_{s1} + \frac{e_s}{m_s}(\mathbf{E}_1 + \mathbf{u}_{s1} \times \mathbf{B}_0) = 0, \tag{11.81}$$

where subscript 0 refers to the equilibrium state and subscript 1 to the perturbed state. Note that charge neutrality prevails in the equilibrium state ($n_{i0} = n_{e0} \equiv n_0$).

The perturbed velocities, $\mathbf{u}_{s1}$, can have both y and z components, but it is assumed that the perturbed electric field, $\mathbf{E}_1$, is only in the y-direction (Figure 11.30). Also, as it turns out, the R–T instability satisfies the condition $\omega_{c_s}^2 \gg (\omega - Ku_{s0})^2$. Using this information, the solution of the momentum equation (11.81) for the individual velocity components yields

$$(u_{s1})_y = -i\frac{\omega - Ku_{s0}}{\omega_{c_s}^2}\frac{e_sE_{1y}}{m_s}, \qquad (11.82a)$$

$$(u_{s1})_z = \frac{E_{1y}}{B_0}, \qquad (11.82b)$$

and the continuity equation (11.80) becomes

$$(\omega - Ku_{s0})n_{s1} - Kn_0(u_{s1})_y + i\frac{\partial n_0}{\partial z}(u_{s1})_z = 0. \qquad (11.83)$$

Now, substituting the perturbed velocity components (11.82a,b) into the continuity equation (11.83) yields an equation that relates the perturbed densities to the perturbed electric field;

$$(\omega - Ku_{s0})n_{s1} + i\frac{\partial n_0}{\partial z}\frac{E_{1y}}{B_0} + iKn_0\frac{\omega - Ku_{s0}}{\omega_{c_s}^2}\frac{e_sE_{1y}}{m_s} = 0. \qquad (11.84)$$

When applied to the electrons, Equation (11.84) simplifies because of the small electron mass ($\mathbf{u}_{e0} \to 0$, $\omega_{c_e} \to \infty$), and it reduces to

$$\frac{E_{1y}}{B_0} = i\frac{\omega n_{e1}}{\left(\frac{\partial n_0}{\partial z}\right)}. \qquad (11.85)$$

Also, charge neutrality prevails not only in the equilibrium state but in the perturbed state as well ($n_{e1} = n_{i1}$) because the frequency of the perturbation is low. Now, using this fact and Equation (11.85), Equation (11.84) for the ions can be expressed in the form

$$\omega^2 - \omega Ku_{i0} + \frac{\omega_{c_i}u_{i0}}{n_0}\frac{\partial n_0}{\partial z} = 0. \qquad (11.86)$$

This quadratic equation can be easily solved, and the solution is

$$\omega = \frac{1}{2}Ku_{i0} \pm \left(\frac{1}{4}K^2u_{i0}^2 - \frac{g}{n_0}\frac{\partial n_0}{\partial z}\right)^{1/2}, \qquad (11.87)$$

where $\omega_{c_i}u_{i0} = g$ (Equation 11.79). Therefore, when the equilibrium density gradient is sufficiently large, the second term in the square root dominates and the plasma is unstable. In this case, the situation that develops is shown in the bottom panel of Figure 11.30. The perturbation at the plasma–vacuum interface leads to a polarization electric field that causes density depletions to $\mathbf{E} \times \mathbf{B}$ drift into the plasma and density enhancements to $\mathbf{E} \times \mathbf{B}$ drift into the vacuum. The situation is unstable and the perturbations grow.

The above mathematical analysis only provides linear growth rates, and a full nonlinear treatment is needed to describe the complete evolution of the plasma. In this regard it should be noted that the time-dependent evolution of equatorial spread *F* has been simulated via two-dimensional numerical solutions of the nonlinear hydrodynamic equations. The initial simulations showed that the R–T instability does indeed lead to bottomside spread *F*, which then evolves into plasma bubbles.[46] Further simulations showed the dependence of spread *F* on the *F* region peak altitude, the bottomside density gradient, zonal and vertical winds, electric fields, gravity waves, and the *E* region conductivities in the conjugate hemispheres.

11.13   Sporadic *E* and intermediate layers

Sporadic *E* layers are ionization enhancements in the *E* region at altitudes between 90 and 120 km (also see Section 13.5).[47] The layers tend to occur sporadically and can be seen at all latitudes. The layer densities can be up to an order of magnitude greater than background densities and the primary ions in the layers are metallic (e.g., Fe^+, Mg^+). Neutral metal atoms are created during meteor ablation, and their subsequent ionization via photoionization and charge exchange yields the long-lived metallic ions.[2] A characteristic feature of sporadic *E* layers is that they are very narrow (0.6–2 km wide). At times, multiple layers can occur simultaneously, separated by 6–10 km in altitude, and after formation the layers tend to descend at a slow speed (0.6–4 m s^{-1}). Sometimes the sporadic *E* layers are flat and uniform in the horizontal direction, while at other times they are like clouds (2–100 km in size) that move horizontally at speeds of 20–130 m s^{-1}.[47]

An example of a sporadic *E* layer is shown in Figure 11.31. The figure shows electron density profiles as a function of altitude at different times, as measured by the

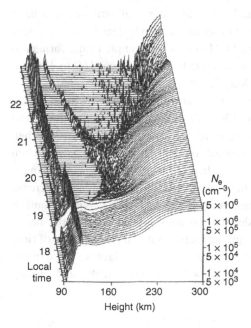

Figure 11.31 Electron density profiles versus height at selected times that show both a sporadic *E* layer and a descending intermediate layer. The profiles were measured with the Arecibo incoherent scatter radar on May 7, 1983.[48]

Arecibo incoherent scatter radar on May 7, 1983.[48] During the early evening, from 1710 to 1910 Atlantic Standard Time (AST), a sporadic E layer was present at 116 km with a peak electron density of about 5×10^5 cm^{-3}. After sunset (1810 AST), the densities below the F region decayed rapidly and a deep valley formed. However, the sporadic E layer persisted, but it descended to 114 km and its peak density decreased to about 1×10^4 cm^{-3}. After 1910 AST, the layer continued to descend and it remained weak until 2148 AST, at which time it reached 105 km. Subsequently, the layer density started to increase.

Sporadic E layers at mid-latitudes are primarily a result of wind shears, but they can also be created by diurnal and semi-diurnal tides as well as by gravity waves.[47] The layers are formed when the vertical ion drift changes direction with altitude, and the layers occur at the altitudes where the ion drift converges. In the E region, the zonal neutral wind is primarily responsible for inducing vertical ion drifts, which result from a $\mathbf{u}_n \times \mathbf{B}$ dynamo action ($\mathbf{u}_n$ is the zonal wind and $\mathbf{B}$ is the geomagnetic field). Hence, a reversal of the zonal neutral wind with altitude will result in ion convergence and divergence regions. The ions accumulate in the convergence regions, but since the molecular ions (NO^+, O_2^+, N_2^+) rapidly recombine, it is the long-lived metallic ions that survive and dominate the sporadic E layers. At equatorial latitudes, gradient instabilities also play an important role in creating sporadic E layers, while at high latitudes they can be created by convection electric fields.

In contrast to sporadic E layers, intermediate layers are broad (10–20 km wide), are composed of molecular ions (NO^+, O_2^+), and occur in the altitude range of 120–180 km.[49] They frequently appear at night in the valley between the E and F regions, but they can also occur during the day. They tend to form on the bottomside of the F region and then slowly descend throughout the night toward the E region. As with sporadic E layers, intermediate layers can occur at all latitudes, can have a large horizontal extent, and can have an order of magnitude density enhancement relative to background densities. Figure 11.31 shows an example of the formation and subsequent downward descent of an intermediate layer, from 160 to 120 km, which appeared at about 2030 AST on May 7, 1983.[48]

Intermediate layers are primarily a result of wind shears connected with the semi-diurnal tide.[47] In the $E–F$ region valley (130–180 km), the meridional neutral wind is mainly responsible for inducing the upward and downward ion drifts. When the wind blows toward the poles a downward ion drift is induced, whereas when it blows toward the equator, an upward ion drift is induced. If the wind changes direction with altitude (a wind shear), the plasma will either diverge and decrease its density or converge and increase its density (layer formation). When a null in the wind shear moves down in altitude, the ion convergence region, and hence intermediate layer, also descend. Although the meridional wind component of tidal motion is the primary mechanism for creating intermediate layers, the dynamics of these layers can be affected by zonal winds, electric fields, and gravity waves.

11.14 F_3 layer and He^+ layer

The existence of additional layers above the F_2 peak at low latitudes was established more than fifty years ago,[50] but the additional layers were simply attributed to traveling ionospheric disturbances. In the 1990s, however, ionosonde measurements and modeling have helped identify some of the important characteristics of the additional layers, now called F_3 layers.[51–54] The basic physics underlying the formation of an F_3 layer is related to photochemical processes, electric fields, and neutral winds. During the day, the F_2 layer forms near the equator by photochemistry and diffusion, but as the F_2 layer drifts upward in response to the upward $\mathbf{E} \times \mathbf{B}$ drift and neutral wind the F_3 layer is formed at a higher altitude (above $\sim$500 km). Subsequently, a normal F_2 layer forms again at a lower altitude by the standard photochemical and transport processes. At times, the F_3 layer peak density can be greater than the F_2 peak density and at those times it can be measured with ionosondes. The F_3 layer has been observed on both the summer and winter sides of the magnetic equator, and it becomes weaker and less evident with increasing solar activity. The F_3 layer has also been observed to form in association with a magnetic storm and the consequent sudden change in the southward Interplanetary Magnetic Field.[55]

Measurements at the Arecibo incoherent scatter radar facility have shown that an He^+ layer can form near the O^+ to H^+ transition altitude at mid-latitudes during the night.[56, 57] The layer starts to develop just before midnight and its build-up is associated with the nighttime collapse of the topside ionosphere.[56] The He^+ layer occurs during both solar maximum and minimum. At solar maximum, the He^+ layer is clearly evident, with He^+ relative abundances that can be more than 50% at the altitude of the He^+ peak. The measurements also indicate that at solar maximum there are regions of He^+ dominance in the topside ionosphere between 750 to 1200 km.[57] However, at solar minimum, He^+ is never the dominant ion at the peak of the He^+ layer, where typical He^+ relative abundances are between 10 and 20%.

11.15   Tides and gravity waves

Tides and gravity waves play an important role in the dynamics and energetics of the thermosphere, particularly in the altitude range from 100 to 250 km.[58, 59] These waves are generated *in situ* by solar UV and EUV heating as well as by temporally varying auroral processes (precipitation, currents, convection). Tides and gravity waves are also generated in the lower atmosphere and then propagate up to thermospheric heights. For example, the heating associated with the absorption of solar radiation by H_2O in the troposphere and by O_3 in the stratosphere generates upward propagating tides that penetrate the lower thermosphere. Although the upward propagating tides and gravity waves have a significant effect on the lower thermosphere, they are difficult to include in numerical models in a realistic manner, owing to the lack of global measurements of the forcing function.

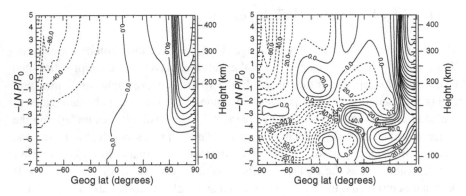

Figure 11.32 Variation of the meridional neutral wind versus altitude and latitude for 70° W at 1800 UT on a quiet day. The variation is shown both without (left panel) and with (right panel) tidal effects. Solid contours are for winds blowing toward the south and dashed contours correspond to northward winds.[63] The contour interval is 10 m s^{-1}.

As noted in Section 10.6, migrating tides are synchronized with the Sun (or Moon) and appear to travel westward on the rotating Earth. However, *nonmigrating* tides also exist in the mesosphere and lower thermosphere. These tides can be excited in the troposphere by latent heat release associated with tropical convection and by nonlinear interactions of global-scale waves.[60–62] The nonmigrating tides can propagate toward the east or west, and can also have standing modes. The interaction of a wavenumber 3 nonmigrating tide with the zonal wind in the lower thermosphere (diurnal tide) can act to modify the equatorial ionization anomaly, as will be shown.

The mathematical description of tides was discussed in Section 10.6, where it was shown that both diurnal and semi-diurnal tidal components exist. However, the semi-diurnal tide is the more important component in the lower thermosphere. An example of the effect that semi-diurnal tides can have on the thermosphere is shown in Figure 11.32. The results in this figure are from the NCAR Thermospheric General Circulation Model (TGCM), which simulated the magnetically quiet period of September 18–19, 1984. The figure shows the variation of the meridional neutral wind versus latitude for the 70° W longitude at 1800 universal time (UT). The left panel shows the wind without semi-diurnal tides, while the right panel shows the wind with tidal effects. Obviously, semi-diurnal tides can be very important in the lower thermosphere. The wind structure below 300 km is complex, with reversals of the wind direction clearly evident. The semi-diurnal tides also have a similar effect on the neutral temperature and densities. The tidal-induced perturbations in the neutral parameters then affect the ionosphere at D and E region altitudes because the time constant for chemical reactions is short (Section 8.2).

As noted already, the interaction of upward propagating nonmigrating tides and the diurnal tide can act to modify the equatorial ionization anomaly. As the tides propagate upwards, their amplitudes grow and at E region altitudes they dominate the locally driven tides, but the upward propagating tides are damped before they

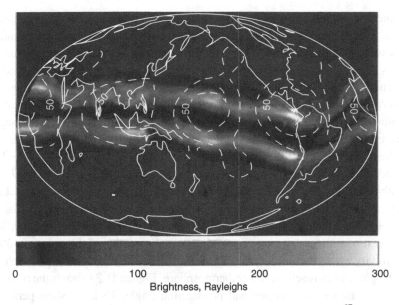

0   100   200   300
Brightness, Rayleighs

Figure 11.33 Tidal-induced peaks in the equatorial ionization anomaly.[67]

reach the *F* region. However, these tides can induce a longitudinal variation in the *E* region dynamo electric field as the tidal winds drag ions across geomagnetic field lines. The modulated electric field can then affect the *F* region ionization via the associated electromagnetic plasma drift.[64–69]

The connection between weather in the tropics, upward propagating tides, dynamo electric fields in the *E* region, and the equatorial ionization anomaly was made by examining night side far ultraviolet (FUV) observations from the *IMAGE* satellite.[67] One month (20 March–20 April, 2002) of FUV data was averaged as a function of local time (LT) and magnetic latitude and longitude. The averaged data displayed four distinct tidal-induced peaks in the equatorial ionization anomaly (Figure 11.33). The variation in n_e from peak to trough that is associated with the brightness in Figure 11.33 is about 20%. However, this study was limited to equinox and solar maximum conditions.

A more comprehensive study of this phenomenon was conducted with a 13-year data set of vertical total electron content (TEC) measurements from the *TOPEX* and *Jason* satellites.[70] The dataset was used to establish the local time, seasonal, solar cycle, and geomagnetic activity dependence of the wavenumber-4 longitudinal pattern under these conditions. The study showed that the wavenumber-4 pattern is created during the daytime hours at equinox and June solstice, but is either absent or washed out by other processes during December solstice. During equinox, the wavenumber-4 pattern is created around noon with well-defined enhancements in the low-latitude TEC. These enhancements, which are symmetric about the geomagnetic equator during this season, last for many hours. The longitudinal patterns are found to be nearly identical between the vernal (March/April) and autumnal

(September/October) equinoxes and are largely independent of the solar cycle conditions. The wavenumber-4 pattern is also observed during active magnetic conditions, indicating that the processes that create this pattern are also present during active times. During the June solstice, the wavenumber-4 pattern is also observed in the afternoon hours, but in contrast to the equinox cases, it exhibits a strong hemispheric asymmetry and is not observed during the night. The low-latitude TEC exhibits clear longitudinal variations during the December solstice, with large daytime enhancements over the East-Asian and Pacific regions and a third enhancement in the afternoon over the Atlantic ocean, but a clear wavenumber-4 pattern is not observed during this season. Although the equatorial and low-latitude TEC values exhibit clear longitudinal patterns during all seasons, a significant amount of scatter remains in the TEC data that is not accounted for by changes in the solar cycle, season, local time, or by the longitudinal variability. This remaining scatter in TEC is largest near the poleward edges of the anomalies and is of the order of 40%.[70]

Gravity waves that are generated in the Earth's lower atmosphere (mesosphere and stratosphere) can reach the lower thermosphere. Figure 11.34 shows the relationship between gravity wave period and propagation angle. Usually, short-period (high frequency) waves propagate at a steeper angle than the long-period (low frequency) waves. The gravity waves that reach the lower thermosphere typically have periods of less than ten minutes.[71]

In general, the situation is more complicated because of *wind filtering*. In the derivation of the gravity wave dispersion relation (10.38), it was assumed that the atmosphere was stationary ($\mathbf{u}_0 = 0$). If there is a background wind, then the gravity

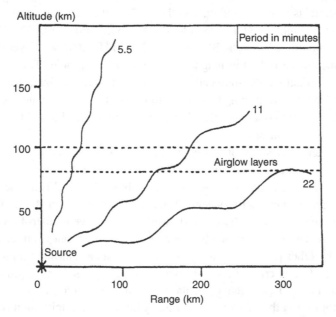

Figure 11.34 Relationship between gravity wave period and propagation angle.[71]

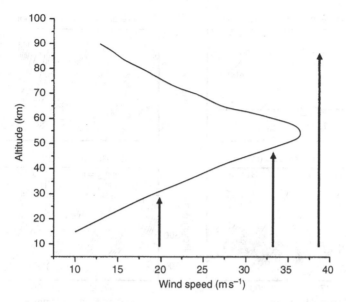

Figure 11.35 An example of gravity wave filtering. The arrows indicate upward propagating gravity waves with horizontal phase speeds of 20, 33, and 38 m s^{-1}. The curve shows a profile of a horizontal wind versus altitude. Gravity waves with horizontal phase speeds less than 36 m s^{-1} are absorbed into the background atmosphere.[72]

wave frequency is Doppler shifted ($\omega \to \omega - \mathbf{K} \bullet \mathbf{u}_0$) and a wind can act to filter the gravity waves. Specifically, if a gravity wave propagates in the same direction as the wind, the gravity wave frequency can be Doppler shifted to zero and the wave energy can be absorbed into the background atmosphere, as illustrated in Figure 11.35. In this figure, a horizontal wind versus altitude is shown. For this example, gravity waves that primarily propagate horizontally in the direction of the wind with phase speeds less than the maximum wind speed (36 m s^{-1}) will be filtered out of the wave spectrum that is generated at low altitudes. However, gravity waves that propagate in a direction opposite to the background wind are not filtered, and this can lead to an anisotropy in the gravity wave spectrum.

The mathematical description of atmospheric gravity waves (AGWs) is given in Section 10.5, where it is noted that AGWs are responsible for *traveling iono-spheric disturbances* (TIDs). Large-scale TIDs have periods of the order of one hour, wavelengths of about 1000 km, and horizontal speeds greater than 250 m s^{-1}. Figure 11.36 shows an imposed large-scale AGW and its effect on the ionosphere, as calculated with a numerical ionospheric model.[73] The calculations are for the location of the EISCAT incoherent scatter radar (Table 14.1), a fall equinox day (September 6, 1988), and moderate solar activity ($F_{10.7} = 152$). The simulation covers the period 1600 to 1900 UT, where the local time at the EISCAT site is obtained by adding 1.25 hours to the universal time. The top panel shows the AGW perturbation imposed on the thermosphere. The AGW has a 1-hour period, a

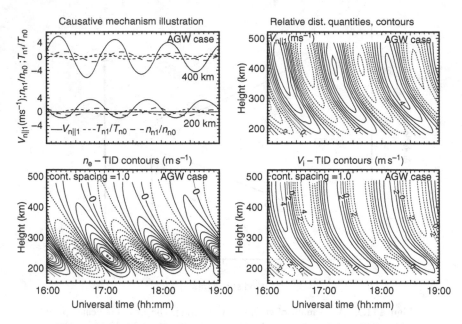

Figure 11.36 Imposed AGW perturbations (top panel) and the calculated ionospheric response (bottom panel). The calculations are for the location of the EISCAT incoherent scatter radar, a fall equinox day, and moderate solar activity.[73]

1000 km horizontal wavelength, a southward-downward phase propagation, and a wave-associated neutral wind perturbation of 5 m s^{-1} at the bottom of the F region (160 km). The AGW perturbation is shown via the change in the field-aligned neutral velocity ($v_{n\|1}$, m s^{-1}), and the perturbed-to-background density (n_{n1}/n_{n0}, %) and temperature (T_{n1}/T_{n0}, %) ratios. The top-left plot shows the temporal evolution of the perturbations at 200 and 400 km and the top-right plot shows contours of $v_{n\|1}$ versus altitude and time. The bottom plots show the ionospheric response to the imposed AGW via perturbations in the electron density and field-aligned ion velocity. Note that the perturbations in v_n are greater than those in T_n and n_n, and that the phase of the T_n and n_n perturbations are advanced or delayed by about 60° relative to the phase of the v_n perturbation. Also note that the contour plots of $v_{n\|1}$, n_e, and v_i show the inclined wave phase fronts that are characteristic of AGW perturbations. The largest ionospheric perturbations occur at about 250 km for the adopted AGW.

11.16   Ionospheric storms

As noted in Section 11.3, geomagnetic storms can result from a compression of the magnetosphere due to the arrival of a discontinuity in the solar wind. During the growth phase, the magnetospheric electric fields, currents, and particle precipitation increase, while the reverse occurs during the recovery phase. The net result is that a large amount of energy is deposited into the ionosphere–thermosphere system at high latitudes during a storm. In response to this energy input, the auroral E region electron densities increase, and there is an overall enhancement in the electron and ion temperatures at high latitudes. In addition, neutral composition changes occur, wind speeds increase, and equatorward propagating gravity waves are excited. At mid-latitudes, the equatorward propagating waves drive the F region plasma toward higher altitudes, which can result in ionization enhancements. Behind the wave disturbance are enhanced meridional neutral winds, and these diverging winds cause upwellings and decreases in the O/N_2 density ratio. The latter, in turn, leads to decreased electron densities in the F region. For big storms, the enhanced neutral winds and composition changes can penetrate all the way to the equatorial region. In general, when the electron density increases as a result of storm dynamics, it is called a *positive ionospheric storm*, whereas a decrease in electron density is called a *negative ionospheric storm*.

Strong electric fields can penetrate to the low-latitude ionosphere during the early stages of a magnetic storm, creating a region of strong plasma drift known as a *sub-auroral polarization stream (SAPS)*. The plasma in the stream drifts in a northwest direction across the USA and in the stream there are *storm-enhanced densities (SEDs)*, as shown in Figure 11.37.[74, 75] The SEDs appear in the evening–afternoon

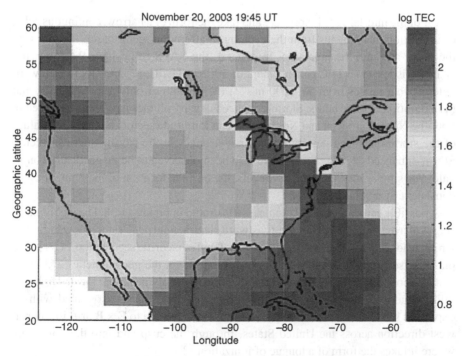

Figure 11.37 Vertical total electron content (TEC) values obtained from Global Positioning System (GPS) satellites and ground receivers during the 20 November 2003 storm. Note the ridge of enhanced density that extends across the eastern United States and into Canada. Log10 (TEC) is shown in TECU (1 TECU = 1×10^{16} el/m^2). The dark region shows the high-density ridge.[75]

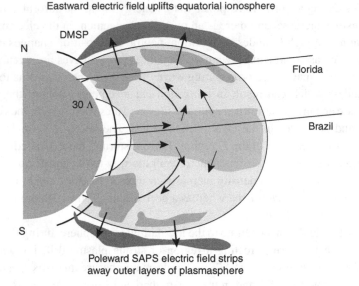

Eastward electric field uplifts equatorial ionosphere

Poleward SAPS electric field strips
away outer layers of plasmasphere

Figure 11.38 A schematic diagram that shows a mechanism that can create SEDs
(Courtesy of J. C. Foster).

sector of the mid-latitude ionosphere and take the form of a narrow, continuous ridge
of ionization that extends from the evening to afternoon sector at mid-latitudes, to
the cusp, and into the polar cap. In general, the SAPS-SEDs have the following
characteristics: (1) a TEC enhancement of 50–100 TECU; (2) an F region sunward
velocity that ranges from 50–1000 m s^{-1}; (3) a width of only a few degrees; and
(4) a duration of three hours. Storm-enhanced density plumes have been observed in
both the North European and North American longitude sectors but the TEC values
of the storm-enhanced density plumes are stronger in the North American sector.
Storm-enhanced density plumes have also been observed to be magnetically conjugate. For similar magnetic activity levels, the latitude location of the plume base is
consistent over solar cycle.[74–77]

A possible cause of the SED plumes is illustrated in Figure 11.38. During the early
stages of a magnetic storm, a strong eastward electric field penetrates to low latitudes,
which leads to a relative large upward ($\mathbf{E} \times \mathbf{B}$) plasma drift (see Section 11.11). The
plasma rises to high altitudes before it diffuses down the geomagnetic field lines,
and as a consequence, the Equatorial Ionization Anomaly peaks are located at lower
mid-latitudes. Subsequently, a northward electric field that is associated with an
expanded storm-time convection pattern causes the plasma to $\mathbf{E} \times \mathbf{B}$ drift in a northwest direction across the United States, through the cusp and into the polar cap,
where it takes the form of a tongue of ionization.[75]

The response of the ionosphere–thermosphere system to different geomagnetic
storms can be significantly different, and even for a given storm, the system's
response can be very different in different latitudinal and longitudinal regions.[78, 79]
Nevertheless, it is instructive to show the ionospheric response to the large magnetic

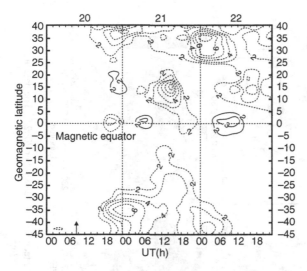

Figure 11.39 Contours of f_0F_2 deviations vs. latitude and time for the Asian and Pacific sector during October 20–22, 1989. The diurnal variation of f_0F_2 on October 19 was used as a quiet time reference and it was subtracted from the f_0F_2 variations on October 20–22 to yield the f_0F_2 deviations. The solid curves correspond to positive deviations and the dashed curves to negative deviations. The interval between adjacent contours is 1 MHz.[80]

storm that was triggered by a solar flare which appeared at 1229 UT on October 19, 1989. Associated with this flare was an enhanced solar wind speed of about 2000 km s^{-1}, and the IMF turned southward two times (between 1250–1340 UT and 1650–1900 UT). As a consequence, a sudden storm commencement occurred at 0917 UT on October 20, and after an initial phase the storm displayed two periods of enhanced activity during the following 48 hours. Auroral glows were seen down to about 29° N geomagnetic latitude in the United States during the height of the storm. In response to the storm, there were long-lasting electron density depletions at high latitudes, as measured by a worldwide network of ionosondes (Figure 11.39). In the equatorial region, both negative and positive storm effects occurred at different times. In addition, large-scale TIDs were observed on two nights, with equatorward propagation velocities in the range of 330 to 680 m s^{-1}.

Medium-scale and large-scale TIDs have been observed in association with both storms and substorms.[81–84] Figure 11.40 shows a TID passing over North America on 1–2 October 2002.[81] This period was marked by a magnetic storm ($K_p \sim 7$; $Dst \sim -175$ nT) and a series of periodic substorms, which acted to excite a series of large-scale TIDs that propagated toward the equator. Figure 11.40 corresponds to snapshots of the perturbation displayed in vertical total electron content (TEC). The perturbation is of the order of 2 TEC units (TECU) and is associated with the neutral wind of a large-scale traveling atmospheric disturbance (TAD). The TEC measurements were made with the North American network of 500 Global Positioning System (GPS) receivers.

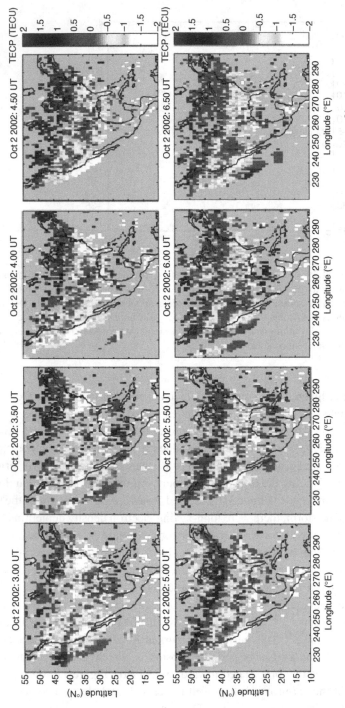

Figure 11.40 Images of TEC perturbations over North America obtained from a GPS mapping technique on 2 October 2002.[81] The data are shown from 0300 to 0630 UT at selected times (shown at the top of each panel). 1 TECU-10[16]el/m[2].

11.17   Specific references

1. Schunk, R. W., and J. J. Sojka, The lower ionosphere at high latitudes, in the upper mesosphere and lower thermosphere, *Geophys. Monograph*, **87**, 37, 1995.

2. Kapp, E., On the abundance of metal ions in the lower ionosphere, *J. Geophys. Res.*, **102**, 9667, 1997.

3. Fraser-Smith, A. C., Centered and eccentric geomagnetic dipoles and their poles, 1600–1985, *Rev. Geophys.*, **25**, 1, 1987.

4. Chapman, S., and J. Bartels, *Geomagnetism*, **2**, London: Oxford University Press, 1940.

5. Barraclough, D., *Phys. Earth Planet. Int.*, **48**, 279, 1987.

6. Knecht, D. J., and B. M. Shuman, The geomagnetic field, in *Handbook of Geophysics and the Space Environment* ed. by A. S. Jursa, Boston: Air Force Geophysics Laboratory, 4.1, 1985.

7. Menville, M., The *K*-derived planetary indices: description and availability, *Rev. Geophys.*, **29**, 415, 1991.

8. Rishbeth, H., and O. K. Garriott, *Introduction to Ionospheric Physics*, New York: Academic Press, 1969.

9. Bradley, P. A., and J. R. Dudeney, A simple model of the vertical distribution of the electron concentration in the ionosphere, *J. Atmos. Terr. Phys.*, **35**, 2131, 1973.

10. Schunk, R. W., and W. J. Raitt, Atomic nitrogen and oxygen ions in the daytime high-latitude *F* region, *J. Geophys. Res.*, **85**, 1255, 1980.

11. Schunk, R. W., P. M. Banks, and W. J. Raitt, Effects of electric fields and other processes upon the nighttime high latitude *F* layer, *J. Geophys. Res.*, **81**, 3271, 1976.

12. Narcisi, R. S., and A. D. Bailey, Mass spectrometer measurements of positive ions at altitudes from 64 to 112 kilometers, *J. Geophys. Res.*, **70**, 3687, 1965.

13. Mitra, A. P., and J. N. Rowe, Ionospheric effects of solar flares – VI. Changes in *D* region ion chemistry during solar flares, *J. Atmos. Terr. Phys.*, **34**, 795, 1972.

14. Burns, C. J., J. K. Hargreaves, E. Turunen, H. Matreinen, and H. Ranta, Chemical modelling of the quiet summer *D* and *E* regions using EISCAT electron density profiles, *J. Atmos. Terr. Phys.*, **53**, 115, 1991.

15. Storey, L. R. O., An investigation of whistling atmospherics, *Philos. Trans. R. Soc. London, Ser. A*, **246**, 113, 1953.

16. Johnson, F. S., The ion distribution above the *F*2-maximum, *J. Geophys. Res.*, **65**, 577, 1960.

17. Hagen, J. B., and P. Hsu, The structure of the protonosphere above Arecibo, *J. Geophys. Res.*, **79**, 4269, 1974.

18. Barakat, A. R., R. W. Schunk, T. E. Moore, and J. H. Waite, Jr., Ion escape fluxes from the terrestrial high-latitude ionosphere, *J. Geophys. Res.*, **92**, 12255, 1987.

19. Schunk, R. W., and J. C. G. Walker, Thermal diffusion in the topside ionosphere for mixtures which include multiply-charged ions, *Planet. Space Sci.*, **17**, 853, 1969.

20. Dent, Z. C., I. R. Mann, J. Goldstein, F. W. Menk, and L. G. Ozeke, Plasmaspheric depletion, refilling, and plasmapause dynamics: a coordinated ground-based and IMAGE satellite study, *J. Geophys. Res.*, 111, A03205, doi:10.1029/2005JA011046, 2006.

21. Taylor, H. A., and W. J. Walsh, The light ion trough, the main trough, and the plasmapause, *J. Geophys. Res.*, **77**, 6716, 1972.

22. Roble, R. G., The calculated and observed diurnal variation of the ionosphere over Millstone Hill on March 23–24, 1970, *Planet. Space Sci.*, **23**, 1017, 1975.

23. Lei, J., R. G. Roble, W. Wang, B. A. Emery, and S.-R. Zhang, Electron temperature climatology at Millstone Hill and Arecibo, *J. Geophys. Res.*, **112**, A02302, doi:10.1029/2006JA012041, 2007.

24. Brace, L. H., B. M. Reddy, and H. G. Mayr, Global behavior of the ionosphere at 1000 km altitude, *J. Geophys. Res.*, **72**, 265, 1967.

25. Evans, J. V., Mid-latitude electron and ion temperatures at sunspot minimum, *Planet. Space Sci.*, **15**, 1557, 1967.

26. Kozyra, J. U., A. F. Nagy, and D. W. Slater, High-altitude energy source(s) for stable auroral red arcs, *Rev. Geophys.*, **35**, 155, 1997.

27. Gurgiolo, C., B. R. Sandel, J. D. Perez, *et al.*, Overlap of the plasmasphere and ring current: relation to subauroral ionospheric heating, *J. Geophys. Res.*, **110**, A12217, doi:10.1029/2004JA010986, 2005.

28. Roble, R. G., A. I. Stewart, M. R. Torr, D. W. Rusch, and R. H. Wand, The calculated and observed ionospheric properties during Atmosphere Explorer-C satellite crossing over Millstone Hill, *J. Atmos. Terr. Phys.*, **40**, 21, 1978.

29. Swartz, W. E., and J. S. Nisbet, Incompatibility of solar EUV fluxes and incoherent scatter measurements at Arecibo, *J. Geophys. Res.*, **78**, 5640, 1973.

30. Roble, R. G., Solar EUV flux variation during a solar cycle as derived from ionospheric modeling considerations, *J. Geophys. Res.*, **81**, 265, 1976.

31. Sterling, D. L., W. B. Hanson, R. J. Moffett, and R. G. Baxter, Influence of electromagnetic drifts and neutral air winds on some features of the $F2$ region, *Radio Sci.*, **4**, 1005, 1969.

32. Anderson, D. N., A theoretical study of the ionospheric F region equatorial anomaly: I, Theory, *Planet. Space Sci.*, **21**, 409, 1973.

33. Bailey, G. J., and N. Balan, A low-latitude ionosphere-plasmasphere model, in *STEP Handbook of Ionosphere Models* ed. by R. W. Schunk, **173**, Logan: Utah State University Press, 1996.

34. Hanson, W. B., and R. J. Moffett, Ionization transport effects in the equatorial F region, *J. Geophys. Res.*, **71**, 5559, 1966.

35. Anderson, D. N., and R. G. Roble, Neutral wind effects on the equatorial F-region ionosphere, *J. Atmos. Terr. Phys.*, **43**, 835, 1981.

36. Abdu, M. A., K. N. Iyer, R. T. de Madeiros, I. S. Batista, and J. H. A. Sobral, Thermospheric meridional wind control of equatorial spread F and evening prereversal electric field, *Geophys. Res. Lett.*, **33**, L07106, doi:10.1029/2005GL024835, 2006.

37. Maus, S., P. Alken, and H. Lühr, Electric fields and zonal winds in the equatorial ionosphere inferred from CHAMP satellite magnetic measurements, *Geophys. Res. Lett.*, **34**, L23102, doi:10.1029/2007GL030859, 2007.

38. Fejer, B. G., E. R. de Paula, R. A. Heelis, and W. B. Hanson, Global equatorial ionospheric vertical plasma drifts measured by the AE-E satellite, *J. Geophys. Res.*, **100**, 5769, 1995.

39. Fejer, B. G., and L. Scherliess, Time dependent response of equatorial ionospheric electric fields to magnetospheric disturbances, *Geophys. Res. Lett.*, **22**, 851, 1995.

40. Hysell, D. L., and J. D. Burcham, JULIA radar studies of equatorial spread F, *J. Geophys. Res.*, **103**, 29155, 1998.

41. Hei, M. A., R. A. Heelis, and J. P. McClure, Seasonal and longitudinal variation of large-scale topside equatorial plasma depletions, *J. Geophys. Res.*, **110**, A12315, doi:10.1029/2005JA011153, 2005.

42. Ma, G., and T. Maruyama, A super bubble detected by dense GPS network at east Asian longitudes, *Geophys. Res. Lett.*, **33**, L21103, doi:10.1029/2006GL027512, 2006.

43. Hanson, W. B., W. R. Coley, R. A. Heelis, and A. L. Urguhart, Fast equatorial bubbles, *J. Geophys. Res.*, **102**, 2039, 1997.

44. Argo, P. E., and M. C. Kelley, Digital ionosonde observations during equatorial spread *F*, *J. Geophys. Res.*, **91**, 5539, 1986.

45. Hanson, W.B, and A. L. Urquhart, High altitude bottomside bubbles, *Geophys. Res. Lett.*, **21**, 2051, 1994.

46. Scannapieco, A. J., and S. L. Ossakow, Nonlinear equatorial spread *F*, *Geophys. Res. Lett.*, **3**, 451, 1976.

47. Whitehead, J. D., Recent work on mid-latitude and equatorial sporadic *E*, *J. Atmos. Terr. Phys.*, **51**, 401, 1989.

48. Riggin, D., W. E. Swartz, J. Providakes, and D. T. Farley, Radar studies of long-wavelength waves associated with mid-latitude sporadic *E* layers, *J. Geophys. Res.*, **91**, 8011, 1986.

49. Mathews, J. D., Y. T. Morton, and Q. Zhou, Observations of ion layer motions during the AIDA campaign, *J. Atmos. Terr. Phys.*, **55**, 447, 1993.

50. Sen, H. Y., Stratification of the F2-layer of the ionosphere over Singapore, *J. Geophys. Res.*, **54**, 363, 1949.

51. Balan, N., G. J. Bailey, M. A. Abdu, *et al.*, Equatorial plasma fountain and its effects over three locations: evidence for an additional layer, the F_3 layer, *J. Geophys. Res.*, **102**, 2047, 1997.

52. Balan, N., J. S. Batista, M. A. Abdu, J. MacDougall, and G. J. Bailey, Physical mechanism and statistics of occurrence of an additional layer in the equatorial ionosphere, *J. Geophys. Res.*, **103**, 29 169, 1998.

53. Uemoto, J., T. Ono, T. Maruynma, *et al.*, Magnetic conjugate observation of the F3 layer using the SEALION ionosonde network, *Geophys. Res. Lett.*, **34**, L02110, 2007.

54. Thampi, S. V., N. Balan, S. Ravindran, *et al.*, An additional layer in the low-latitude ionosphere in Indian longitudes: total electron content observations and modeling, *J. Geophys. Res.*, **112**, A06301, 2007.

55. Paznukhove, V. V., B. W. Reinisch, P. Song, *et al.*, Formation of an F3 layer in the equatorial ionosphere: a result from strong IMF changes, *J. Atmos. Solar-Terr. Phys.*, **69**, 1292, 2007.

56. Gonzalez, S. A., and M. P. Sulzer, Detection of He^+ layering in the topside ionosphere over Arecibo during equinox solar minimum conditions, *Geophys. Res. Lett.*, **23**, 2509, 1996.

57. Wilford, C. R., R. J. Moffett, J. M. Rees, G. J. Bailey, and S. A. Gonzalez, Comparison of the He^+ layer observed over Arecibo during solar maximum and solar minimum with CTIP model results, *J. Geophys. Res.*, **108**, A12 1452, doi:10.1029/2003JA009940, 2003.

58. Fesen, C. G., R. G. Roble, and M.-L. Duboin, Simulations of seasonal and geomagnetic activity effects at Saint Santin, *J. Geophys. Res.*, **100**, 21377, 1995.

59. Forbes, J.M., F.A. Marcos, and F. Kamalabadi, Wave structures in lower thermosphere density from the Satellite Electrostatic Triaxial Accelerometer measurements, *J. Geophys. Res.*, **100**, 14693, 1995.

60. Hagan, M.E., and J.M. Forbes, Migrating and nonmigrating diurnal tides in the middle and upper atmosphere excited by tropospheric latent heat release, *J. Geophys. Res.*, **107** (D24), 4754, doi:10.1029/2001JD001236, 2002.

61. Hagan, M.E., and J.M. Forbes, Migrating and nonmigrating semidiurnal tides in the upper atmosphere excited by tropospheric latent heat release, *J. Geophys. Res.*, **108** (A2), 1062, doi:10.1029/2002JA009466, 2003.

62. Forbes, J.M., M.E. Hagan, S. Miyahara, Y. Miyoshi, and X. Zhang, Diurnal nonmigrating tides in the tropical lower thermosphere, *Earth Planets Space*, **55**, 419, 2003.

63. Crowley, G., B.A. Emery, R.G. Roble, H.C. Carlson, and D.J. Knipp, Thermosphere dynamics during September 18–19, 1984. 1. Model simulations, *J. Geophys. Res.*, **94**, 16925, 1989.

64. Millward, G.H., I.E.F. Müller-Wodarg, A.D. Aylward, *et al.*, An investigation into the influence of tidal forcing on F region equatorial vertical ion drift using a global ionosphere–thermosphere model with coupled electrodynamics, *J. Geophys. Res.*, **106**, 24, 733, 2001.

65. Jadhav, G., M. Rajaram, and R. Rajaram, A detailed study of equatorial electrojet phenomenon using Ørsted satellite observations, *J. Geophys. Res.*, **107**(A8), 1175, doi:10.1029/2001JA000183, 2002.

66. Sagawa, E., T.J. Immel, H.U. Frey, and S.B. Mende, Longitudinal structure of the equatorial anomaly in the nighttime ionosphere observed by IMAGE/FUV, *J. Geophys. Res.*, **110**, A11302, doi:10.1029/2004JA010848, 2005.

67. Immel, T.J., E. Sagawa, S.L. England, *et al.*, Control of equatorial ionospheric morphology by atmospheric tides, *Geophys. Res. Lett.*, **33**, L15108, doi:10.1029/2006GL026161, 2006.

68. England, S.L., T.J. Immel, E. Sagawa, *et al.*, Effect of atmospheric tides on the morphology of the quiet time, postsunset equatorial ionospheric anomaly, *J. Geophys. Res.*, **111**, A10519, doi:10.1029/2006JA011795, 2006.

69. England, S.L., S. Maus, T.J. Immel, and S.B. Mende, Longitudinal variation of the *E*-region electric fields caused by atmospheric tides, *Geophys. Res. Lett.*, **33**, L21105, doi:10.1029/2006GL027465, 2006.

70. Scherliess, L., D.C. Thompson and R.W. Schunk, Longitudinal variability of low-latitude total electron content: tidal influences, *J. Geophys. Res.*, **113**, A01311, doi:10.1029/2007JA012480, 2008.

71. Taylor, M.J., Observation and analysis of wave-like structures in the lower thermosphere nightglow emissions, Ph.D. dissertation, University of Southampton, UK, 1985.

72. Nielsen, K., Climatology and case studies of mesospheric gravity waves observed at polar latitudes, Ph.D. dissertation, Utah State University, Logan, USA, 2007.

73. Kirchengast, G., Characteristics of high-latitude TIDs from different causative mechanisms deduced by theoretical modeling, *J. Geophys. Res.*, **102**, 4597, 1997.

74. Foster, J.C., and H.B. Vo, Average characteristics and activity dependence of the subauroral polarization stream, *J. Geophys. Res.*, **107**, 1475, doi:10.1029/2002JA009409, 2002.

75. Foster, J. C., A. J. Coster, P. J. Erickson, *et al.*,Multiradar observations of the polar tongue of ionization, *J. Geophys. Res.*, **110**, A09S31, doi:10.1029/2004JA010928, 2005.

76. Oksavik, K., R. A. Greenwald, J. M. Ruohoniemi *et al.*, First observations of the temporal/spatial variation of the sub-auroral polarization stream from the SuperDARN Wallops HF radar, *Geophys. Res. Lett.*, **33**, L12104, doi:10.1029/2006GL026256, 2006.

77. Coster, A. J., M. J. Colerico, J. C. Foster, W. Rideout, and F. Rich, Longitude sector comparisons of storm enhanced density, *Geophys. Res. Lett.*, **34**, L18105, doc:10.1029/2007GL030682, 2007.

78. Rishbeth, H., *F*-region storms and thermospheric dynamics, *J. Geomagn. Geoelectr.*, **43**, 513, 1991.

79. Prölss, G. W., Ionospheric *F*-region storms, in *Handbook of Atmospheric Electrodynamics*, vol. 2 ed. by H. Volland, **195**, Boca Raton, FL: CRC Press, 1995.

80. Ma, S., L. Xu, and K. C. Yeh, A study of ionospheric electron density deviations during two great storms, *J. Atmos. Terr. Phys.*, **57**, 1037, 1995.

81. Nicolls, M. J., M. C. Kelley, A. J. Coster, S. A. González, and J. J. Makela, Imaging the structure of a large-scale TID using ISR and TEC data, *Geophys. Res. Lett.*, **31**, L09812, doi:10.1029/2004GL019797, 2004.

82. Tsugawa, T., Y. Otsuka, A. J. Coster, and A. Saito, Medium-scale traveling ionospheric disturbances detected with dense and wide TEC maps over North America, *Geophys. Res. Lett.*, **34**, L22101, doi:10.1029/2007GL031663, 2007.

83. Ding, F., W. Wan, B. Ning, and M. Wang, Large-scale traveling ionospheric disturbances observed by GPS total electron content during the magnetic storm of 29-30 October 2003, *J. Geophys. Res.*, **112**, A06309, doi:10.1029/2006JA012013, 2007.

84. Shiokawa, K., G. Lu, Y. Otsuka, *et al.*, Ground observation and AMIE-TIEGCM modeling of a storm-time traveling ionospheric disturbance, *J. Geophys. Res.*, **112**, A05308, doi:10.1029/2006JA011772, 2007.

11.18   General references

Banks, P. M., and G. Kockarts, *Aeronomy*, New York: Academic Press, 1973.

Carovillano, R. L., and J. M. Forbes, *Solar-Terrestrial Physics*, Dordrecht, Netherlands: D. Reidel, 1983.

Demars, H. G., and R. W. Schunk, Temperature anisotropies in the terrestrial ionosphere and plasmasphere, *Rev. Geophys.*, **25**, 1659, 1987.

Goldstein, J., Plasmasphere response: tutorial and review of recent imaging results, *Space Sci. Rev.*, **124**, 203, 2006.

Heelis, R. A., Electrodynamics in the low and middle latitude ionosphere: a tutorial, *J. Atmos. Solar-Terr. Phys.*, **66**, 825, 2004.

Kelley, M. C., *The Earth's Ionosphere*, San Diego, CA: Academic Press, 1989.

Kelley, M. C., and C. A. Miller, Mid-latitude thermospheric plasma physics and electrodynamics: a review, *J. Atmos. Solar-Terr. Phys.*, **59**, 1643, 1997.

Kohl, H., R. Rüster, and K. Schlegel, *Modern Ionospheric Science*, Lindau, Germany: Max-Planck-Institute für Aeronomie, 1996.

Kozyra, J. U., A. F. Nagy, and D. W. Slater, High-altitude energy source(s) for stable auroral red arcs, *Rev. Geophys.*, **35**, 155, 1997.

Lemiare, J. F., and K. I. Gringauz, *The Earth's Plasmasphere*, Cambridge, UK: Cambridge University Press, 1998.

Mendillo, M., Storms in the ionosphere: patterns and processes for total electron content, *Rev. Geophys*, 2006.

Prölss, G. W., Ionospheric *F*-region storms, in *Handbook of Atmospheric Electrodynamics*, vol. 2 ed. by H. Volland, **195**, Boca Raton, FL: CRC Press, 1995.

Raghavarao, R., and R. Suhasini, Equatorial temperature and wind anomaly (ETWA): a review, *J. Atmos. Solar-Terr. Phys.*, **64**, 1371, 2002.

Rees, M. H., *Physics and Chemistry of the Upper Atmosphere*, Cambridge, UK: Cambridge University Press, 1989.

Rishbeth, H., and O. K. Garriott, *Introduction to Ionospheric Physics*, New York: Academic Press, 1969.

Schunk, R. W., and A. F. Nagy, Electron temperatures in the *F*-region of the ionosphere: theory and observations, *Rev. Geophys. Space Phys.*, **16**, 355, 1978.

Schunk, R. W., and A. F. Nagy, Ionospheres of the terrestrial planets, *Rev. Geophys. Space Phys.*, **18**, 813, 1980.

Stening, R. J., Modeling the low latitude *F*-region, *J. Atmos. Terr. Phys.*, **54**, 1387, 1992.

Titheridge, J. E., Winds in the ionosphere – a review, *J. Atmos. Terr. Phys.*, **57**, 1681, 1995.

Vadas, S. L., Horizontal and vertical propagation and dissipation of gravity waves in the thermosphere from lower atmospheric and thermospheric sources, *J. Geophys. Res.*, **112**, A06305, doi:10.1029/2006JA011845, 2007.

11.19 Problems

Problem 11.1 Calculate the variation of the magnitude of **B** and $\sin I$ over the altitude range from 200 to 1000 km at a dipole magnetic latitude of 45°.

Problem 11.2 Show that $\nabla \cdot \mathbf{b} = (1/A)\partial A/\partial s$ by taking the divergence of **b** (Equation 11.4) in spherical coordinates.

Problem 11.3 Show that $E_\phi/E_{\phi 0} = 1/\sin^3 \theta$, where E_ϕ is the magnitude of an azimuthal electric field that is perpendicular to a north–south dipole magnetic field. $E_{\phi 0}$ is the electric field at the equatorial crossing of the dipole **B** field. See Equations (11.13) to (11.15).

Problem 11.4 Calculate $\nabla \cdot \mathbf{u}_E$ by taking the divergence of Equation (11.15) and show that the result is Equation (11.18).

Problem 11.5 Calculate values of q_d (Equation 11.19) and p_d (Equation 11.20) for $\theta = 60°$, 90°, and 120° and for a dipole **B** field line that has an equatorial crossing altitude of 3000 km.

Problem 11.6 Show that Equation (11.28) is correct.

Problem 11.7 When O^+ is in chemical equilibrium (Equation 11.60), show that the O^+ density increases exponentially with altitude. Calculate the scale height associated with the O^+ density increase. Neglect $n(O_2)$ and assume that P_{ts} is proportional to $n(O)$.

Problem 11.8 Using Equation (11.64), show that the H^+ density increases exponentially with altitude when the O^+, H, and O densities are in diffusive equilibrium. Calculate the scale height associated with the H^+ density increase.

Problem 11.9 Show that when H^+ is a minor ion and in diffusive equilibrium with the major ion O^+, the H^+ density increases exponentially with altitude with a scale height that is approximately equal to the chemical equilibrium scale height (Problem 11.8).

Problem 11.10 For $T_e = T_i = 1000$ K and $z = 1000$ km, compare the H^+, He^+, and O^+ diffusive equilibrium scale heights assuming that each ion is, separately, the major ion.

Problem 11.11 Assuming that $q_e = -\lambda_e dT_e/dz$ and $\lambda_e = 7.7 \times 10^5 T_e^{5/2}$ eV cm^{-1} s^{-1} K^{-1}, calculate q_e and λ_e for $T_e = 3000$ K and $dT_e/dz = 1$ and 3 K km^{-1}.

Problem 11.12 Derive an expression for the Cowling conductivity.

Problem 11.13 Derive an expression for the Rayleigh–Taylor dispersion relation (equivalent to Equation (11.87)) for the case when an initial constant electric field E_0 is perpendicular to the vacuum–plasma interface (E_0 points in the z-direction in Figure 11.30).

Problem 11.14 Obtain a numerical solution for the F region ionization in the terrestrial ionosphere at mid-latitudes using $\Delta z = 5$ km and $\Delta t = 300$ seconds (see Appendix O). Initially, assume that the electron (or O^+) density is 10^5 cm^{-3} at all altitudes and then run the model until a steady state is achieved. Let $z_B = 200$ km and $z_T = 800$ km. Set $T_e = T_i = T_n = 1000$ K at all altitudes.

Problem 11.15 Repeat Problem 11.14 with different values of Δt, including 0.001, 0.1, 1, 1000, and 10,000 seconds. Plot the temporal evolution at selected times. The steady state should be the same, but the temporal evolution could be different.

Problem 11.16 Repeat Problem 11.14 with different values of Δz, including 0.001, 0.1, 1, 10, and 1000 km. Plot the temporal evolution at selected times.

Problem 11.17 Repeat Problem 11.14 with different upper boundary altitudes; including 600, 1000, and 1500 km.

Problem 11.18 Repeat Problem 11.14 with the diffusion coefficient multiplied by a factor of two.

Problem 11.19 Repeat Problem 11.14 with $T_e = T_i = 3000$ K.

Chapter 12

The terrestrial ionosphere at high latitudes

The magnetosphere–ionosphere–atmosphere system at high latitudes is strongly coupled via electric fields, particle precipitation, field-aligned currents, heat flows, and frictional interactions, as shown schematically in Figure 12.1. Electric fields of magnetospheric origin induce a large-scale motion of the high-latitude ionosphere, which affects the electron density morphology. As the plasma drifts through the neutrals, the ion temperature is raised owing to ion–neutral frictional heating. The elevated ion temperature then alters the ion chemical reaction rates, topside plasma scale heights, and ion composition. Also, particle precipitation in the auroral oval acts to produce enhanced ionization rates and elevated electron temperatures, which affect the ion and electron densities and temperatures. These ionospheric changes,

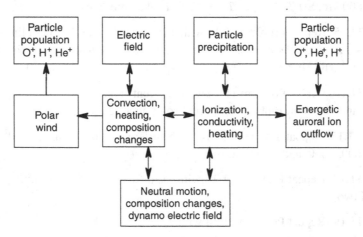

Figure 12.1 Coupling processes in the magnetosphere–ionosphere–atmosphere system.

in turn, have a significant effect on the thermospheric structure, circulation, and composition. At F region altitudes, the neutral atmosphere tends to follow, but lags behind, the convecting ionospheric plasma. The resulting ion–neutral frictional heating induces vertical winds and O/N_2 composition changes. These atmospheric changes then affect the ionospheric densities and temperatures.

The ionosphere–thermosphere system also has a significant effect on the magnetosphere. Precipitating auroral electrons produce conductivity enhancements, which can modify the convection electric field, large-scale current systems, and the electrodynamics of the magnetosphere–ionosphere system as a whole. Also, once the thermosphere is set into motion by convection electric fields, the large inertia of the neutral atmosphere will act to produce dynamo electric fields whenever the magnetosphere tries to change its electrodynamic state. Additional feedback mechanisms exist on polar cap and auroral field lines via a direct flow of plasma from the ionosphere to the magnetosphere. In the polar cap, there is a continual outflow of thermal plasma from the ionosphere (the polar wind) and it represents a significant source of mass, momentum, and energy for the magnetosphere. On auroral field lines, energized ionospheric plasma is injected into the magnetosphere via ion beams, conics, rings, and toroidal distributions.

This chapter elucidates the effect that the various magnetospheric processes have on the ionosphere–thermosphere system. The topics covered include the effects of convection electric fields, particle precipitation, field-aligned currents, geomagnetic storms, and substorms. This chapter also includes discussions concerning large-scale plasma structuring mechanisms, the polar wind, and energetic ion outflow.

12.1 Convection electric fields

Electrodynamical coupling is perhaps the most important process linking the magnetosphere, ionosphere, and thermosphere at high latitudes. This coupling arises as a result of the interaction of the magnetized solar wind with the Earth's geomagnetic field. When the supersonic solar wind first encounters the geomagnetic field, a free-standing *bow shock* is formed that deflects the solar wind around the Earth in a region called the *magnetosheath* (Figure 2.10). The subsequent interaction of the magnetosheath flow with the geomagnetic field leads to the formation of the *magnetopause*, which is a relatively thin boundary layer that acts to separate the solar wind's magnetic field from the geomagnetic field. The separation is accomplished via a magnetopause current system. However, the shielding is not perfect, and a portion of the solar wind's magnetic field (also known as the *interplanetary magnetic field* – IMF) penetrates the magnetopause and connects with the geomagnetic field.

This connection is shown in Figure 12.2 for the case of a southward IMF.[1] Note that the IMF has vector components (B_x, B_y, B_z). The B_x component is in the ecliptic plane directed along the Sun–Earth line (positive toward the Sun), B_y is in the ecliptic plane perpendicular to the Sun–Earth line (positive toward dusk), and B_z (the north–south component) is perpendicular to the ecliptic plane and positive to the north

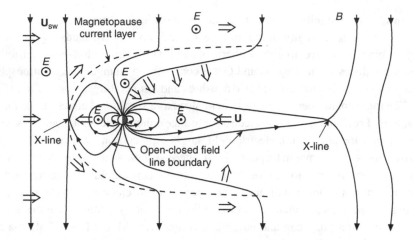

Figure 12.2 Schematic diagram showing the directions of the electric and magnetic fields, and the plasma flows, in the vicinity of the Earth. The Sun is to the left, north is at the top, and south is at the bottom. The electric field points from dawn to dusk (out of the plane of the figure).[1]

(Figures 2.9 and 2.10). The connection of the IMF and the geomagnetic field occurs in a circular region known as the *polar cap*, and the connected field lines are referred to as *open field lines*. At latitudes equatorward of the polar cap, the geomagnetic field lines are closed. The auroral oval is an intermediate region that lies between the open field line region (polar cap) and the low-latitude region that contains *dipolar field lines*. The field lines in the auroral oval are closed, but they are stretched deep in the magnetospheric tail (Figure 12.2).

The solar wind is a highly conducting, collisionless, magnetized plasma that, to lowest order, can be described by the ideal MHD equations (7.45a–g). Therefore, the electric field in the solar wind is governed by the relation $\mathbf{E} = -\mathbf{u}_{sw} \times \mathbf{B}$ (Equation 7.45d), where $\mathbf{u}_{sw}$ is the solar wind velocity. When the radial solar wind, with a southward IMF component, interacts with the Earth's magnetic field (Figure 12.2), an electric field is imposed that points in the dawn-to-dusk direction across the polar cap. This imposed electric field, which is directed perpendicular to $\mathbf{B}$, maps down to ionospheric altitudes along the highly conducting geomagnetic field lines. In the ionosphere, this electric field causes the plasma in the polar cap to $\mathbf{E} \times \mathbf{B}$ drift in an antisunward direction. Further from the Earth, the plasma on the *open* polar cap field lines exhibits an $\mathbf{E} \times \mathbf{B}$ drift that is toward the equatorial plane (Figure 12.2). In the distant magnetospheric tail, the field lines reconnect, and the flow on these closed field lines is toward and around the Earth.

The existence of an electric field across the polar cap implies that the boundary between open and closed magnetic field lines is charged. The charge is positive on the dawn side and negative on the dusk side, as shown in Figure 12.3. This figure displays the same configuration as that in Figure 12.2, except that the view is from the magnetotail looking toward the Sun. The solar wind is out of the plane of the figure

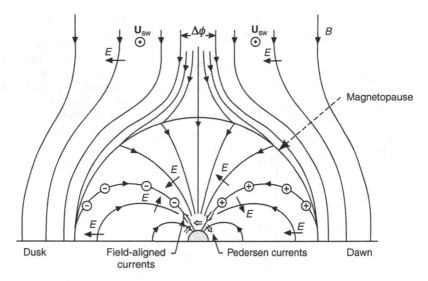

Figure 12.3 Schematic diagram showing the electric and magnetic fields in the vicinity of the Earth. The view is from the magnetotail looking toward the Sun. The solar wind is toward the observer and north is at the top.[1]

and north is at the top. The charges on the polar cap boundary act to induce electric fields on nearby closed field lines that are opposite in direction to the electric field in the polar cap. These oppositely directed electric fields are situated in the regions just equatorward of the dawn and dusk sides of the polar cap (Figure 12.3). As with the polar cap electric field, the electric fields on the closed field lines map down to ionospheric altitudes and cause the plasma to $\mathbf{E} \times \mathbf{B}$ drift in a *sunward* direction. On the field lines that separate the oppositely directed electric fields, field-aligned (or *Birkeland*) currents flow between the ionosphere and magnetosphere. The current flow is along $\mathbf{B}$ and toward the ionosphere on the dawn side, across the ionosphere at low altitudes, and then along $\mathbf{B}$ and away from the ionosphere on the dusk side.

The net effect of the electric field configuration shown in Figure 12.3 is as follows. Closed dipolar magnetic field lines connect to the IMF at the day side magnetopause. When this connection occurs, the ionospheric foot of the field line is at the day side boundary of the polar cap. After connection, the open field line and attached plasma convect in an antisunward direction across the polar cap. When the ionospheric foot of the open field line is at the night side polar cap boundary, the magnetospheric end is in the equatorial plane of the distant magnetotail (Figure 12.2). The open field line then reconnects, and subsequently, the newly closed and stretched field line convects around the polar cap and toward the day side magnetopause. The direction of the $\mathbf{E} \times \mathbf{B}$ drift in the ionosphere that is associated with the magnetospheric electric field is shown in Figure 12.4. This figure displays electrostatic potential contours in a magnetic latitude–local time coordinate system, with the magnetic pole at the center. Note that the electrostatic potential contours coincide with the streamlines of the flow when there is only an $\mathbf{E} \times \mathbf{B}$ drift. The flow pattern exhibits a two-cell

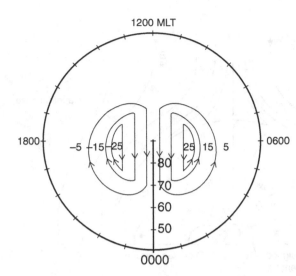

1200 MLT

1800

0600

0000

Figure 12.4 Contours of the magnetospheric electrostatic potential in a magnetic latitude-MLT reference frame. The contours display a symmetric two-cell pattern of the Volland-type.[2] The total potential drop is 64 kV. Courtesy of M.D. Bowline.

character, with antisunward flow over the polar cap and return (sunward) flow at latitudes equatorward of the polar cap.

Magnetospheric electric fields are not the only source of ionospheric drifts and, therefore, it is important to determine the relative contributions of the various sources. A general expression for the cross-field transport of plasma was derived in Chapter 5 (Equation 5.103). At altitudes above about 150 km, the ratio of the collision-to-cyclotron frequencies is very small for all of the charged particles, and the expression for the cross-field transport of plasma reduces to (Equation 5.98):

$$\mathbf{u}'_{j\perp} = \frac{\mathbf{E}' \times \mathbf{B}}{B^2} - \frac{1}{n_j e_j} \frac{\nabla p_j \times \mathbf{B}}{B^2} + \frac{m_j}{e_j} \frac{\mathbf{G} \times \mathbf{B}}{B^2}, \tag{12.1}$$

where $\mathbf{u}'_{j\perp} = \mathbf{u}_{j\perp} - \mathbf{u}_{n\perp}$ and $\mathbf{E}' = \mathbf{E} + \mathbf{u}_n \times \mathbf{B}$. At ionospheric altitudes (300 km), the magnetospheric electric field typically varies from about 10 to 200 mV m^{-1}, which corresponds to $\mathbf{E} \times \mathbf{B}$ drifts that vary from about 200 m s^{-1} to 4 km s^{-1}. Also, at these altitudes, typical values for the O$^+$ density and temperature are 10^5 cm^{-3} and 1000 K, respectively. These values can be used to compare the three drifts in Equation (12.1). For O$^+$, the gravity term $(m_j G/e_j)$ is equivalent to an electric field of the order of 10^{-3} mV m^{-1} and is, therefore, negligible. The pressure gradient term $(\nabla p_j/n_j e_j)$ is equivalent to a 10 mV m^{-1} electric field when the pressure scale length is of the order of 10 meters. In other words, the diamagnetic drift will only be important for scale lengths less than 10 meters, and hence, is negligible. Finally, a neutral wind of 200 m s^{-1} is equivalent to a 10 mV m^{-1} electric field at F region altitudes.

The above analysis indicates that the electrodynamic drift dominates the plasma motion at altitudes above approximately 150 km. However, the net electrodynamic drift is driven by both magnetospheric and co-rotational electric fields. Specifically, the ionosphere at low and middle latitudes is observed to co-rotate with the Earth, and this motion is driven by a *co-rotational electric field*. At high latitudes, the plasma also has a tendency to co-rotate, and this must be taken into account when calculating the plasma convection paths. The co-rotational electric field causes the plasma to drift around the Earth once every 24 hours and, as a consequence, the plasma remains above the same geographic location at all times. In a geographic inertial frame, with the geographic pole at the center, the drift trajectories are concentric circles about the geographic pole. The magnetospheric potential pattern, on the other hand, maps to the ionosphere along magnetic field lines and, therefore, the location of the magnetic pole is relevant. Unfortunately, the geographic and magnetic poles do not coincide. The offset is 11.5° in the northern hemisphere and 14.5° in the southern hemisphere (Section 11.2). For magnetospheric convection (Figure 12.4), the appropriate coordinate system to use is a quasi-inertial magnetic reference frame, with the magnetic pole at the center and the noon–midnight direction taken as one of the axes. In this magnetic frame, the magnetospheric convection pattern stays aligned with the noon–midnight axis as the magnetic pole rotates about the geographic pole. As it turns out, co-rotation in the geographic inertial frame is equivalent to co-rotation in this quasi-inertial magnetic frame.[3] Therefore, the contours of the co-rotational electric potential are concentric circles about the magnetic pole in the quasi-inertial magnetic frame.

When the co-rotational and magnetospheric electric potentials are combined, the plasma drift trajectories take the form shown in Figure 12.5. Eight representative trajectories are shown, along with the corresponding circulation times. The plasma following the outer trajectories 1 and 2 essentially co-rotates with the Earth. For these trajectories, the plasma drift is eastward and a complete traversal takes about one day. For trajectories just poleward of trajectory 2, the eastward co-rotational drift is opposed by the westward (sunward) magnetospheric drift. Consequently, the plasma slows down and a stagnation region appears. Plasmas following trajectories that enter this region have circulation times that are longer than a day (trajectory 4). For the trajectories that are confined to the polar cap (3, 5–8), the circulation times are less than a day because the trajectories are short and the $\mathbf{E} \times \mathbf{B}$ drift speeds are high (Figure 12.6). Another important aspect of magnetospheric convection concerns the vertical drift. The magnetospheric electric field is perpendicular to $\mathbf{B}$, but the magnetic field is not vertical. Consequently, there is a vertical $\mathbf{E} \times \mathbf{B}$ component that is upward on the day side of the polar cap and downward on the night side (Figure 12.6).

These convection features are relevant to a magnetospheric convection pattern that is constant for about 1.5 days. Clearly, if the magnetospheric convection pattern varies with time, the trajectories that the plasma elements follow will be more complex. Also, even for a constant magnetospheric convection pattern, the trajectories will appear to be more complex in a geographic inertial frame because of the motion of the magnetic pole about the geographic pole. This is illustrated in Figure 12.7, where two representative plasma trajectories are shown that cover a 24-hour period. Also shown in this figure are the positions of the terminator at winter solstice (W), equinox (E), and summer solstice (S). The trajectory in panel A corresponds to trajectory 3 in Figure 12.5. The plasma following this trajectory has a circulation

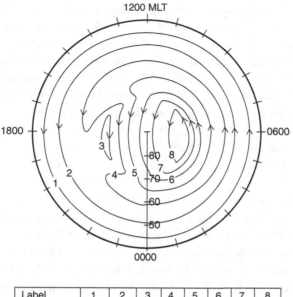

Label	1	2	3	4	5	6	7	8
Circulation period (day)	1.00	1.01	0.10	1.34	0.50	0.31	0.18	0.11

Figure 12.5 Plasma drift trajectories in the polar region viewed in a magnetic quasi-inertial frame. These trajectories are for a symmetric two-cell convection pattern with co-rotation added. The potential drop across the polar cap is 64 kV, and the circulation periods are tabulated at the bottom.[3]

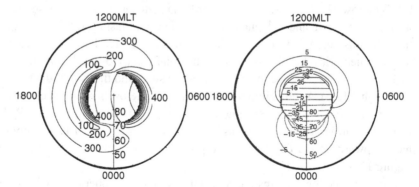

Figure 12.6 Contours of the horizontal drift speeds (left) and vertical drifts (right) for the plasma convection pattern shown in Figure 12.5. The speeds are in m s^{-1}. For the vertical drifts, solid contours are for upward drifts and dashed contours are for downward drifts. Courtesy of M.D. Bowline.

period of about two hours and, hence, it executes many cycles per day. Depending on the location of the terminator, the plasma may drift entirely in sunlight, entirely in darkness, or move in and out of sunlight many times during the course of a day. The trajectory in panel B corresponds to trajectory 4 in Figure 12.5, and its circulation

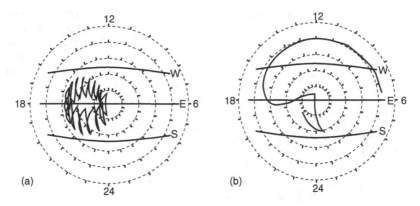

Figure 12.7 Plasma drift trajectories in the geographic inertial frame. The trajectories are for a 24-hour period. The trajectories in panels A and B correspond to trajectories 3 and 4 in Figure 12.5. The curves labeled W, E, and S show the locations of the terminator at winter solstice, equinox, and summer solstice, respectively.[3]

period is longer than 24 hours. For equinox conditions, the plasma following this trajectory crosses the terminator three times in a 24-hour period.

An important feature of the plasma motion induced by the co-rotational and magnetospheric electric fields is that the flow is incompressible.[4] This can be shown by taking the divergence of the electrodynamic drift (12.1)

$$\nabla \cdot \mathbf{u}_E = \nabla \cdot \left(\mathbf{E}' \times \mathbf{B}/B^2 \right). \tag{12.2}$$

The divergence of the cross product can be expanded by using one of the vector identities given in Appendix B, and the result is

$$\nabla \cdot \mathbf{u}_E = \frac{1}{B^2} \left[\mathbf{B} \cdot (\nabla \times \mathbf{E}') - \mathbf{E}' \cdot (\nabla \times \mathbf{B}) \right] + (\mathbf{E}' \times \mathbf{B}) \cdot \nabla \left(\frac{1}{B^2} \right). \tag{12.3}$$

For an electrostatic field, $\nabla \times \mathbf{E}' = 0$. Also, $\nabla \times \mathbf{B} \propto \mathbf{J}$ and $\mathbf{J}$ is either zero or parallel to $\mathbf{B}$ at F region altitudes. Therefore, the term $\mathbf{E}' \cdot (\nabla \times \mathbf{B}) = 0$, because $\mathbf{E}' \perp \mathbf{B}$. The last term in Equation (12.3) represents compression (rarefaction) as the plasma drifts into a region of greater (smaller) $\mathbf{B}$, and it can be shown that this term is small at high latitudes.[4] Therefore, $\nabla \cdot \mathbf{u}_E \approx 0$, and the flow is essentially incompressible. This means that when the plasma approaches a convection throat its speed increases and a density build-up does not occur.

12.2 Convection models

The simple two-cell convection pattern shown in Figure 12.4 does indeed exist at certain times. Figure 12.8 shows drift velocities measured along two Dynamics Explorer 2 (*DE 2*) orbits as the satellite passed through the high-latitude region of the northern hemisphere. It is evident that the measured flow directions are basically

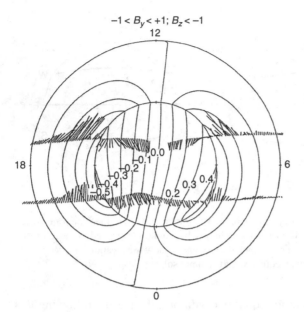

Figure 12.8 Drift velocities measured along two *DE 2* northern hemisphere passes and the streamlines associated with a symmetric two-cell convection pattern.[5]

consistent with a symmetric two-cell convection pattern. In general, however, the magnetospheric convection pattern is more complex than that shown in Figure 12.8. In fact, it is now well known that the magnetospheric electric field is strongly correlated with magnetic activity, K_p (see Section 11.3), and that it depends on the solar wind dynamic pressure and the direction of the IMF (B_x, B_y, B_z). During the last 30 years, a major effort has been devoted to obtaining empirical or statistical patterns of plasma convection for a wide range of conditions. Typically, these *empirical models* are constructed from data collected over many months or years from numerous ground-based sites or satellite orbits. The data are synthesized, binned, and then fitted with simple analytical expressions. As a consequence, the empirical convection models represent average magnetospheric conditions, not instantaneous patterns. Also, the convection boundaries that exist in these models are smooth, whereas the instantaneous convection boundaries can be fairly sharp.

When the IMF is southward ($B_z < 0$), plasma convection at high latitudes exhibits a two-cell pattern with antisunward flow over the polar cap and return flow equatorward of the polar cap. The potential drop across the polar cap, which determines the convection speed, varies with the solar wind dynamic pressure. However, the potential drop can be distributed uniformly or asymmetrically between the two cells depending on the IMF B_y component. For $B_y \approx 0$, the convection cells are symmetric (Figure 12.8). For other values of B_y, the two-cell convection pattern is asymmetric, with enhanced convection in the dawn cell for $B_y > 0$ and enhanced convection in the dusk cell for $B_y < 0$ in the northern hemisphere (Figure 12.9). Also, the entry of the flow into the polar cap is in the prenoon sector for $B_y > 0$ and in the postnoon sector for $B_y < 0$. Finally, it should be noted that for a given sign of B_y, the asymmetry is reversed in the southern hemisphere.

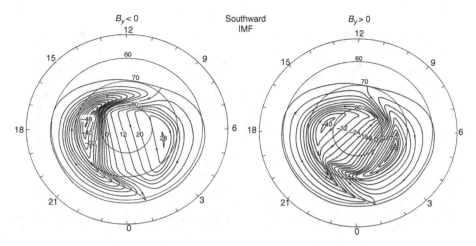

Figure 12.9 Plasma convection patterns in the northern polar region for southward IMF and for both negative (left dial) and positive (right dial) IMF B_y components. Co-rotation is not included.[6]

When the IMF is northward ($B_z > 0$), the plasma convection patterns are more complex than those found for southward IMF. In particular, measurements have shown that when the IMF is northward, the convection in the polar cap can be sunward.[7] The sunward convection was first interpreted to be a signature of a four-cell convection pattern, but such patterns were clearly seen only on the sunlit side of the polar region.[8] Subsequently, it was suggested that three-cell convection patterns can occur for northward IMF, depending on the direction of the B_y component.[9] Figure 12.10 shows the proposed convection patterns in the southern polar region for $B_z > 0$ and three B_y cases. For $B_y = 0$, a four-cell convection pattern occurs. When B_y becomes either positive or negative, one of the convection cells in the polar cap expands and the other shrinks. The net result is that for large B_y values, the convection pattern appears to have just three cells. On the other hand, the sunward convection in the polar cap has been interpreted in terms of a severely distorted two-cell convection pattern, as shown in Figure 12.11. Although the form that the convection pattern takes for northward IMF is controversial, the consensus of the scientific community appears to be leaning toward multi-cell convection patterns, rather than distorted two-cell patterns. However, this issue is still not completely settled.

Another new empirical model of magnetospheric electric fields (or plasma convection) has been constructed from a large database of satellite measurements.[10, 11] This model yields electric field patterns for all IMF (B_y, B_z) combinations and for several ranges of the magnitude of the IMF. Typical patterns are shown in Figure 12.12. Note that for northward IMF, the new empirical model yields multi-cell convection patterns.

The empirical convection models discussed above are useful for many applications, but some caveats should be noted. First, as noted above, the empirical models

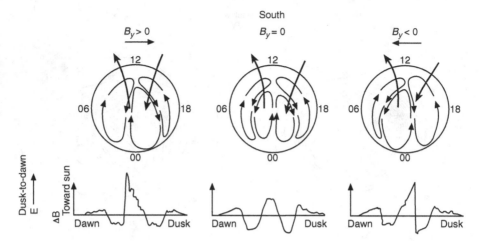

Figure 12.10 Schematic diagram of the plasma convection patterns and NBZ (northward B_z) Birkeland current directions in the southern polar region for a northward IMF. The patterns are shown for $B_y > 0$ (left dial), $B_y = 0$ (middle dial), and $B_y < 0$ (right dial). The traces of $\Delta\mathbf{B}$ and electric field observed by a dawn–dusk orbiting satellite are shown at the bottom. Co-rotation is not included.[9]

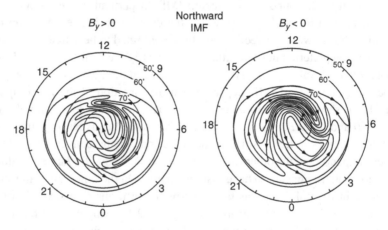

Figure 12.11 Distorted two-cell convection patterns for a strongly northward IMF and for $B_y > 0$ (left dial) and $B_y < 0$ (right dial) in the northern hemisphere. Co-rotation is not included.[6]

provide average patterns, not instantaneous pictures, and sharp convection boundaries tend to get smoothed in the model construction. Furthermore, when the IMF changes direction, the convection pattern is in a transitory state, and that state is probably not captured by empirical models. Finally, at times, the convection pattern appears to be turbulent, as shown in Figure 12.13. The ion drift velocities shown in this figure were measured by the *DE 2* satellite during a crossing of the northern polar region when the IMF was northward. The traversal of the polar region took

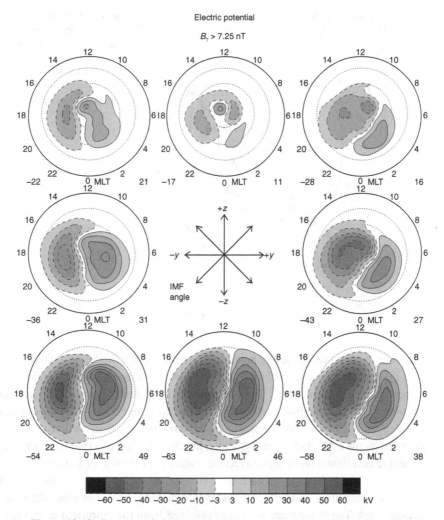

Figure 12.12 Contours of electric potential from an empirical convection model. The results are for the case when the magnitude of the IMF is greater than 7.25 nT. Co-rotation is not included.[10, 11]

only 12 minutes and, therefore, the highly structured drift velocities probably represent spatial structure in the convection pattern and not time variations. A careful examination of the figure indicates that there are nine reversals of the flow direction. Although this case corresponds to an extreme case of electric field structure, it does indicate what is missing from the empirical convection models.

Recently, a study was conducted to determine how well empirical convection models represent instantaneous convection.[13] The Weimer empirical convection model was selected for this determination because it was the most comprehensive model of its time.[11] The cross-track velocities measured by the *Defense Meteorological Satellite Program (DMSP) F13* satellite were compared with the corresponding

Ion drift meter DE-2
University of Texas at Dallas
October 17, 1981
1634-1646 UT

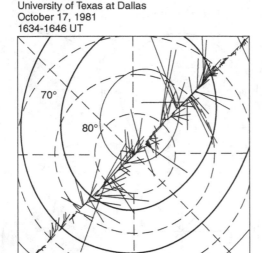

1 km s⁻¹

Figure 12.13 Plasma convection velocities in the high-latitude *F* region in a magnetic latitude-MLT reference frame. The data were obtained with the ion drift meter on the *Dynamics Explorer 2* satellite.[12]

velocities obtained from the Weimer empirical convection model. The comparisons were made for nearly a year of satellite crossings of the northern polar region (4430 successive crossings in 1998). The comparisons indicated that the empirical model was able to capture the gross structure of the plasma convection pattern, but it could not adequately capture the mesoscale spatial structure and convection magnitudes observed by the *DMSP* satellite. Typically, the empirical convection model was able to capture real (instantaneous) convection features in only 6% of the satellite crossings. Figure 12.14 shows a representative comparison of the modeled and measured cross-track velocities. A multi-cell convection pattern is evident from the satellite measurement. The cross-track velocities obtained from the empirical model are off by more than a factor of two and at certain places the flow directions are wrong.[13]

12.3   Effects of convection

The effect that convection electric fields have on the ionosphere depends on altitude, as shown in Figure 12.15. At ionospheric altitudes, the electron–neutral collision frequency is much smaller than the electron cyclotron frequency (Chapter 4), and hence, the combined effect of the perpendicular electric field, **E**, and the geomagnetic field, **B**, is to induce an electron drift in the $\mathbf{E} \times \mathbf{B}$ direction. For the ions, on the other hand, the ion–neutral collision frequencies are greater than the corresponding cyclotron frequencies at low altitudes (*E* region), with the result that the ions drift in the direction of the perpendicular electric field. As altitude increases, the ion drift velocity rotates toward the $\mathbf{E} \times \mathbf{B}$ direction because the ion–neutral

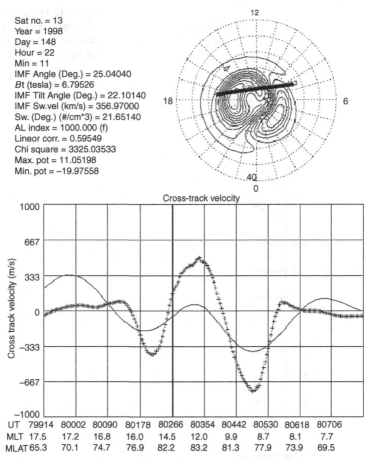

Sat no. = 13
Year = 1998
Day = 148
Hour = 22
Min = 11
IMF Angle (Deg.) = 25.04040
Bt (tesla) = 6.79526
IMF Tilt Angle (Deg.) = 22.10140
IMF Sw.vel (km/s) = 356.97000
Sw. (Deg.) (#/cm*3) = 21.65140
AL index = 1000.000 (f)
Lineor corr. = 0.59549
Chi square = 3325.03533
Max. pot = 11.05198
Min. pot = −19.97558

Cross-track velocity: Solid line predicted from Weimer, plus sign measured by DMSP SAT no. = 13

Figure 12.14 Comparison of the cross-track velocities measured by the *DMSP F13* satellite (plus sign) and the corresponding velocities obtained from the Weimer empirical model (solid curve).[13] The associated Weimer convection pattern is shown in the upper dial.

collision frequencies decrease with altitude. At F region altitudes (≥ 150 km), both the ions and electrons drift in the $\mathbf{E} \times \mathbf{B}$ direction, and therefore, it is below this altitude where the horizontal ionospheric currents flow (Section 12.5). At altitudes above about 800 km, the plasma begins to flow out of the topside ionosphere with a speed that increases with altitude, and this phenomenon is known as the polar wind (Section 12.12).

The convecting ionosphere can be a significant source of momentum and energy for the thermosphere via ion–neutral collisions. The resulting interactions act to modify the thermospheric circulation, temperature, and composition, and this, in turn, affects the ionosphere. The extent of the coupling, however, depends on the plasma density. For plasma densities of 10^3 to 10^6 cm^{-3}, the characteristic time constant for

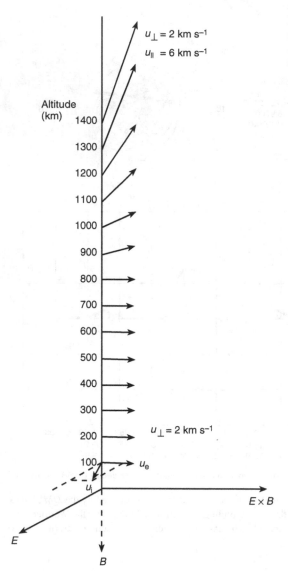

Figure 12.15 Ion and electron drift velocities as a function of altitude in the high-latitude ionosphere.[14]

accelerating the thermosphere ranges from 200 hours (several days) to 10 minutes. Therefore, when the plasma density is high or when the ionospheric driving source persists for a long time, a significant thermospheric response can be expected.

Satellite measurements have been extremely useful for elucidating the extent of the ion–neutral coupling at high latitudes.[15, 16] Figure 12.16 (left dial) shows neutral wind vectors along the track of the *DE 2* satellite for three orbits that crossed the southern (summer) polar region.[15] The orbits are evenly distributed in universal time and thus cross the southern auroral oval in different regions. In the polar cap, the wind direction is from day to night, but the magnitude of the wind is typically much greater than expected if solar heating was the only process driving the flow (about

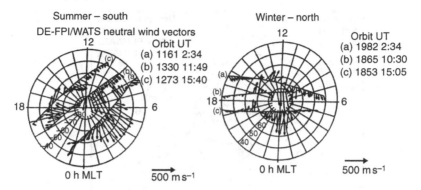

Figure 12.16 Neutral wind vectors measured along the track of the *DE 2* satellite for several passes over the summer, southern polar region (left dial) and the winter, northern polar region (right dial).[15]

200 m s^{-1} for solar heating alone). Also, at lower latitudes, either the magnitude of the antisunward flow is reduced or there is a reversal to sunward flow. These features strongly suggest that the convecting high-latitude ionosphere has a significant effect on the thermospheric circulation. The evidence for convection-driven winds is also clear in the northern (winter) hemisphere (Figure 12.16, right dial), but the momentum forcing does not appear to be as strong in the winter hemisphere as it is in the summer hemisphere. This trend is consistent with the seasonal variation one would expect if ionospheric convection controls the thermospheric circulation. In the northern winter hemisphere, the bulk of the polar cap is in darkness, and consequently, the electron densities are lower than those found in the summer hemisphere. The lower electron densities, in turn, yield a weaker momentum source.

The neutral wind vectors shown in Figure 12.16 are consistent with a two-cell plasma convection pattern, which occurs when the IMF is southward. However, when the IMF is northward, multi-cell plasma convection patterns can exist, and if the conditions are right, the multi-cell signature should also be reflected in the thermospheric circulation pattern. Figure 12.17 shows neutral winds and ion drifts measured along a *DE 2* track in the northern hemisphere at a time when the IMF was northward.[16] Although the ion drift velocities are highly structured, a clear multi-cell convection pattern can be seen with some sunward flow in the polar cap. To a certain extent, the neutral circulation pattern mimics the ion convection pattern. The neutral flow is sunward in the morning side of the central polar cap, but the wind speed is much smaller than the ion convection speed. Also, the neutral reversal regions are co-located with the ion reversal regions. These results provide further evidence for the strong coupling of the ions and neutrals in the polar regions.

The ions are frictionally heated, via ion–neutral collisions, as they convect through the slower moving neutral gas, and this acts to raise the ion temperature. At high latitudes, ion–electron energy coupling is not as important as it is at middle and low latitudes because the electron densities are generally smaller at high latitudes. Consequently, to a good approximation, the ion temperature at *F* region altitudes

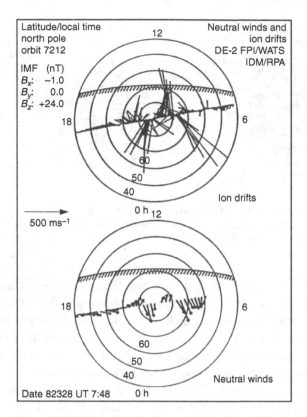

Figure 12.17 Neutral winds (bottom dial) and ion drifts (top dial) along a *DE 2* track in the northern hemisphere for a northward IMF. The curved line represents the solar terminator (90° solar zenith angle). Note the scale difference of a factor of two for the ion and neutral velocities.[16]

can be obtained simply by considering ion–neutral collisional coupling, which yields (Equation 5.36)

$$T_i = T_n + \frac{m_n}{3k}(\mathbf{u}_i - \mathbf{u}_n)^2, \tag{12.4}$$

where only one neutral species is considered and the subscripts n and i refer to neutrals and ions, respectively. The ion–neutral relative velocity along **B** is generally small, and therefore, the velocity term in Equation (12.4) can be calculated by assuming that $\mathbf{E} \times \mathbf{B}$ motion dominates. With this assumption, T_i can be expressed directly in terms of the electric field. Above 150 km, where $\nu_{in}/\omega_{ci} \ll 1$, the expression for T_i reduces to

$$T_i = T_n + \frac{m_n}{3k}\left(\frac{E'}{B}\right)^2, \tag{12.5}$$

where $\mathbf{E}' = \mathbf{E} + \mathbf{u}_n \times \mathbf{B}$ is the effective electric field (Equation 5.37). This relation indicates that for large electric fields, $T_i \propto (E')^2$.

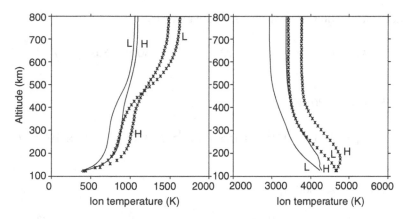

Figure 12.18 Daytime O^+ temperature profiles for 0 (left panel) and $100\,mV\,m^{-1}$ (right panel) meridional electric fields. The profiles were calculated for solar minimum conditions, for summer (x curves) and winter (solid curves), and for high (H) and low (L) geomagnetic activity. Note that the two panels have different temperature scales.[17]

The effect of frictional heating on the ion temperature profile is shown in Figure 12.18. The profiles in this figure were calculated for daytime, steady-state conditions at solar minimum for both summer and winter solstices and for atmospheric conditions characteristic of both high and low geomagnetic activity.[17] The calculations included thermal conduction and ion–electron coupling in addition to ion–neutral coupling. The left panel shows the results for $E' = 0$ and the right panel is for a meridional (north–south) electric field of $100\,mV\,m^{-1}$. Without the electric field, T_i is equal to T_n at altitudes below 400 km for all of the geophysical cases. Above 400 km, T_i increases with altitude because of ion collisions with the hotter electrons (T_e is not shown). Thermal conduction is important only above about 600 km, and it acts to produce isothermal profiles. With the $100\,mV\,m^{-1}$ electric field, T_i is significantly enhanced at all altitudes for all of the geophysical cases. In each case, the highest ion temperature ($\sim$4000 K) occurs at low altitudes, where the ion–neutral frictional heating is the greatest. The ion temperature decreases with altitude in the F region because of the decrease of the neutral density and, hence, frictional heating rate. Associated with the negative temperature gradient is an upward heat flow, which acts to raise T_i at altitudes above about 600 km.

The elevated ion temperatures shown in Figure 12.18 act to alter the ion composition in the lower ionosphere through temperature-dependent chemical reaction rates. For example, the most important chemical reactions for O^+ are

$$O^+ + N_2 \rightarrow NO^+ + N, \qquad k_1, \qquad\qquad (12.6)$$

$$O^+ + O_2 \rightarrow O_2^+ + O, \qquad k_2, \qquad\qquad (12.7)$$

$$O^+ + NO \rightarrow NO^+ + O, \qquad k_3, \qquad\qquad (12.8)$$

where the reaction rates (in cm^3 s^{-1}) are given by:

$$k_1 = 1.533 \times 10^{-12} - 5.92 \times 10^{-13}(T/300) + 8.60 \times 10^{-14}(T/300)^2$$

$$\text{(12.9a)}$$

for $350 \leq T \leq 1700\,\text{K}$;

$$k_1 = 2.73 \times 10^{-12} - 1.155 \times 10^{-12}(T/300) + 1.483 \times 10^{-13}(T/300)^2$$

$$\text{(12.9b)}$$

for $1700 < T < 6000\,\text{K}$;

$$k_2 = 2.82 \times 10^{-11} - 7.74 \times 10^{-12}(T/300) + 1.073 \times 10^{-12}(T/300)^2$$
$$- 5.17 \times 10^{-14}(T/300)^3 + 9.65 \times 10^{-16}(T/300)^4 \qquad \text{(12.10)}$$

for $350 \leq T \leq 6000\,\text{K}$;

$$k_3 = 8.36 \times 10^{-13} - 2.02 \times 10^{-13}(T/300) + 6.95 \times 10^{-14}(T/300)^2$$

$$\text{(12.11a)}$$

for $320 < T < 1500\,\text{K}$; and

$$k_3 = 5.33 \times 10^{-13} - 1.64 \times 10^{-14}(T/300) + 4.72 \times 10^{-14}(T/300)^2$$
$$- 7.05 \times 10^{-16}(T/300)^3 \qquad \text{(12.11b)}$$

for $1500 < T < 6000\,\text{K}$. In Equations (12.9–11), T is the effective temperature, which can be expressed in the form[18, 19]

$$T = T(\text{O}^+) + \frac{m(\text{O}^+)}{m(\text{O}^+) + m_r} \frac{m_r - m_b}{3k} u_\perp^2(\text{O}^+), \qquad \text{(12.12)}$$

where

$$m_b = \frac{\displaystyle\sum_n \frac{m_n \nu(\text{O}^+, n)}{m(\text{O}^+) + m_n}}{\displaystyle\sum_n \frac{\nu(\text{O}^+, n)}{m(\text{O}^+) + m_n}}, \qquad \text{(12.13)}$$

and where m_r is the reactant mass (N$_2$, O$_2$, or NO) and $T(\text{O}^+)$ is the O$^+$ temperature. The effective temperature is different for the three reactions in Equations (12.9–11) because of the presence of m_r.

The expression for the effective temperature takes a particularly simple form at altitudes above about 200 km, where $\nu_{in}/\omega_{ci} \ll 1$ and where atomic oxygen is the main neutral species impeding the flow of O^+ $[m_b \rightarrow m(O)]$. Setting $m_r = m(N_2)$, $B = 0.5$ gauss, and assuming that the O^+ perpendicular drift is due to $\mathbf{E} \times \mathbf{B}$ motion, the expression for T becomes

$$T = T_n + 0.33E'^2, \tag{12.14}$$

where E' is in mV m^{-1}. For large electric field strengths, $T \propto E'^2$ and $k_1 \propto E'^4$. Therefore, a factor of two increase in the electric field results in a factor of 16 increase in the $O^+ + N_2 \rightarrow NO^+ + N$ reaction rate.

The above analysis indicates that in the regions where the convection electric field is large, the associated frictional heating should lead to a rapid conversion of O^+ into NO^+. This effect is shown in Figure 12.19, where ion and electron density profiles are given for convection electric fields of 0 and 100 mV m^{-1}. The profiles were calculated for daytime steady-state conditions.[19] With no electric field, the molecular ions dominate in the E region and O^+ is the dominant ion in the F region.

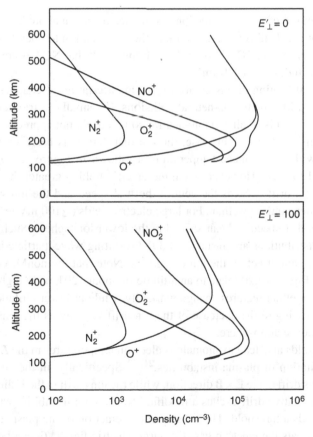

Figure 12.19 Ion and electron density profiles calculated for the daytime high-latitude ionosphere and for meridional electric fields of 0 (top panel) at 100 mV m^{-1} (bottom panel).[19]

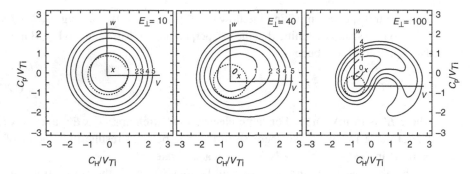

Figure 12.20 Ion velocity distributions in the E region for meridional electric fields of 10 (left panel), 40 (middle panel), and $100\,\mathrm{mV\,m}^{-1}$ (right panel). The contours are shown in the plane perpendicular to **B**. They indicate the points where the ion distribution has decreased by a factor of e^{α}, where α is the number attached to each curve. The x marks the location of the ion drift velocity, and the dashed circle shows the region of velocity space occupied by most of the ions after collisions.[20]

The transition from molecular to atomic ion dominance occurs at about 225 km. On the other hand, for a $100\,\mathrm{mV\,m}^{-1}$ electric field, the elevated ion temperature leads to an increased conversion of O^{+} into NO^{+}, with the result that NO^{+} becomes the dominant ion at altitudes up to 330 km.

The ion frictional heating discussed above is a manifestation of changes in the ion velocity distribution due to ion–neutral collisions. For small electric fields, the ion–neutral relative drift is small and ion–neutral collisions do not appreciably alter the ion velocity distribution. In this case, the ion distribution is basically a drifting Maxwellian with an enhanced temperature, as shown in Figure 12.20 for an altitude of about 120 km.[20] However, when the electric field is greater than about $40\ \mathrm{mV\,m}^{-1}$, the ion drift exceeds the neutral thermal speed and the ion velocity distribution becomes nonMaxwellian. For large electric fields ($\geq 100\ \mathrm{mV\,m}^{-1}$), the ion distribution tends to become bean-shaped in the lower ionosphere. Such highly nonMaxwellian distributions are unstable, and the resulting wave–particle interactions have a significant effect on the ion energetics. Note that the nonMaxwellian features shown in Figure 12.20 relate to an altitude of about 120 km. At higher altitudes, the nonMaxwellian features change markedly, while at lower altitudes they rapidly disappear owing to the decrease of the ion drift velocity as the ions try to penetrate a more dense atmosphere.

Large electric fields also lead to anomalous electron temperatures in the E region owing to the excitation of plasma instabilities.[21, 22] Specifically, in the auroral E region the electrons drift in the $\mathbf{E} \times \mathbf{B}$ direction, while the ions drift in the $\mathbf{E}$ direction. This ion–electron relative drift excites a modified two-stream instability when the electric field exceeds a threshold. The subsequent interaction of the plasma waves and the electrons heats the electron gas. For large electric fields, T_e can be much greater than T_n in the lower ionosphere. This is illustrated in Figure 12.21, where EISCAT radar measurements of the electric field, electron and ion temperatures,

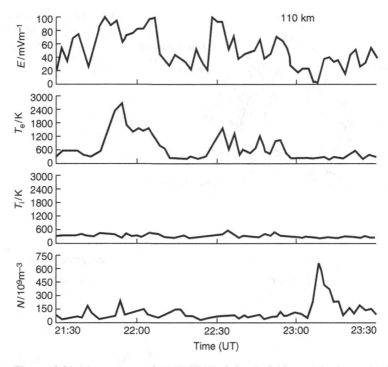

Figure 12.21 Measurements from EISCAT of electric field strength, electron and ion temperatures, and electron density as a function of time. The measurements were made at an altitude of 110 km on September 13, 1990 between 2130 and 2330 UT.[22]

and electron density are shown versus time at an altitude of 110 km.[22] Note that T_i, and probably T_n, remain below 600 K throughout the observing period, but that T_e is significantly enhanced at certain times. The peaks in the electron temperature coincide with electric field enhancements. However, not all of the electric field enhancements produce T_e increases, but this is probably because of the need to satisfy certain threshold conditions for the plasma instability.

12.4   Particle precipitation

Particle precipitation is another important mechanism that links the magnetosphere, ionosphere, and thermosphere at high latitudes. Energetic electron precipitation in the auroral oval is not only the source of optical emissions, but also a source of ionization due to electron impact with the neutral atmosphere, a source of bulk heating for both the ionosphere and atmosphere, and a source of heat that flows down from the lower magnetosphere into the ionosphere. For a southward IMF, the electron precipitation occurs in distinct regions in the auroral oval, as shown schematically in Figure 12.22. In addition to *diffuse auroral precipitation*, there are discrete arcs in the nocturnal oval, low-energy *polar rain precipitation* in the polar

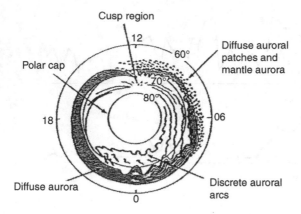

Figure 12.22 Schematic diagram showing the different particle precipitation regions in the auroral oval for southward IMF.[23]

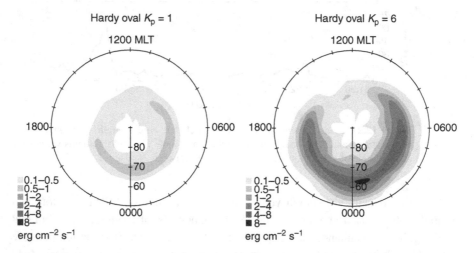

Figure 12.23 The auroral electron energy flux in the northern polar region for both quiet ($K_p = 1$) and active ($K_p = 6$) geomagnetic conditions from an empirical model.[24]

cap, *soft precipitation* in the cusp, and *diffuse auroral patches* in the morning oval. For northward IMF, there are *sun-aligned arcs* in the polar cap (Section 12.9).

The energy flux and characteristic energy of the auroral electron precipitation have been extensively measured via particle detectors on polar orbiting satellites and several empirical models are currently available to describe these parameters.[24, 25] Figure 12.23 shows representative auroral electron energy fluxes in the northern hemisphere for both quiet ($K_p = 1$) and active ($K_p = 6$) magnetic conditions (see Section 11.3).[24] For quiet magnetic conditions, the largest energy fluxes occur in the midnight–dawn sector of the auroral oval and the maximum energy flux is about $1 \ \mathrm{erg \ cm^{-2} \ s^{-1}}$. For active magnetic conditions, the precipitation is more intense, with the maximum energy flux reaching $8 \ \mathrm{ergs \ cm^{-2} s^{-1}}$. Also, for active magnetic

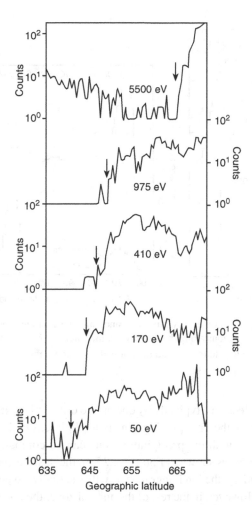

Figure 12.24 Counts in selected channels of the SSJ/3 detector showing the variation in auroral zone boundary location with electron energy.[26]

conditions, the auroral oval has a greater latitudinal width, extending from about 50° to 80° in latitude.

The contours of the characteristic energy of electron precipitation have a morphological form similar to the energy flux contours shown in Figure 12.23. However, electrons with different characteristic energies generally have sharp spatial boundaries. Figure 12.24 shows the auroral boundaries in the dusk sector as a function of electron energy. The data were obtained with the particle detector (SSJ/3) on the *DMSP/F2* satellite as it crossed the northern polar region. At each energy, the equatorward boundary of the auroral region can be identified by a factor of 10 increase in the electron number flux (counts) over a range of 0.1° to 1° in latitude. For the energy range shown (50 to 5500 eV), the different boundaries extend over a 2.5° latitude range.

Ion precipitation also occurs in the auroral zone and, on average, the ion precipitation pattern varies systematically with magnetic latitude, magnetic local time, and magnetic activity (K_p).[27] The integral number flux of the precipitating ions is always

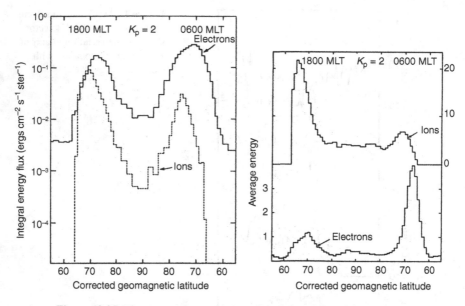

Figure 12.25 The average integral energy flux (left panel) and the average energy (right panel) for electrons and ions along the dawn–dusk meridian for $K_p = 2$. For the right panel, the electron scale is to the left and the ion scale is to the right.[27]

much smaller than that for the electrons, typically by one or two orders of magnitude. Therefore, the current carried by the precipitating ions is negligible. On the other hand, the energy flux associated with the precipitating ions can be comparable to that for the precipitating electrons, as shown in Figure 12.25. In the dusk sector, near the electron equatorward boundary, the ion integral energy flux actually exceeds the electron integral energy flux. However, in the rest of the auroral zone, the ion energy flux is comparable to, but smaller than, the electron energy flux. For example, the ratio of the ion to electron integral energy fluxes is 0.28 at midnight, 0.14 at dawn, and 0.43 at noon for $K_p = 2$. Along the dawn–dusk meridian, the pattern for the ion energy flux is displaced equatorward of that for the electrons on the dusk side and poleward on the dawn side. Also, along this meridian, the highest ion energies occur at dusk, while the highest electron energies occur at dawn. Note that the average energy of the precipitating ions is substantially greater than that for the precipitating electrons.

In general, the particle precipitation in the auroral zone is structured and highly time dependent. This should act to produce structure in the ionization created by the precipitation as well as important temporal variations. The rapid build-up of ionization structure in response to ongoing auroral precipitation is shown in Figure 12.26. The measurements were made with the Chatanika incoherent scatter radar when it was in the auroral oval. Two altitude–latitude scans are shown, separated by about 10 minutes. Note the rapid enhancement in the electron density, particularly in the lower ionosphere.

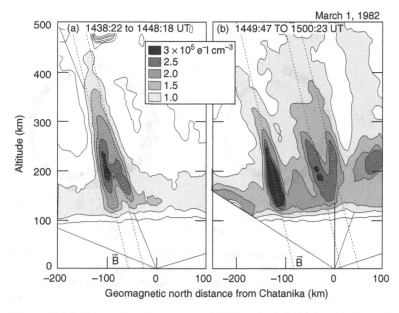

Figure 12.26 Electron densities measured with the Chatanika radar when it was in the auroral oval. The two altitude–latitude scans are separated by about 10 minutes.[28]

12.5 Current systems

The precipitating auroral electrons are responsible for the upward field-aligned (Birkeland) current. Associated with these precipitating magnetospheric electrons are upflowing ionospheric electrons, which provide for a return current. These upward and downward field-aligned currents have been extensively measured with satellite-borne magnetometers and their average properties have been incorporated into empirical models.[29] Figure 12.27 shows statistical patterns of Birkeland currents for southward IMF and for both quiet (left dial) and active (right dial) magnetic conditions. The field-aligned currents are concentrated in two principal areas that encircle the geomagnetic pole. The poleward (*Region 1*) currents exhibit current flow into the ionosphere in the morning sector and away from the ionosphere in the evening sector, while the equatorward (*Region 2*) currents contain current flows in the opposite directions at a given local time. The basic field-aligned current flow pattern is the same during geomagnetically quiet and active periods. The magnitudes of the currents in the poleward and equatorward regions are not well known, but it appears that the net current is inward on the morning side and outward on the evening side in the northern hemisphere.

The Region 1 and Region 2 field-aligned currents display both annual and semi-annual variations with regard to their location and intensity.[30] For example, on the day side the field-aligned current moves poleward in the summer hemisphere and equatorward in the winter hemisphere, while the night side field-aligned current displays an opposite seasonal dependence. The average day side field-aligned current

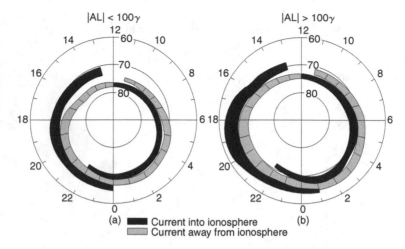

Figure 12.27 The distribution and flow directions of large-scale field-aligned currents determined from (a) data obtained from 439 passes of the *Triad* satellite during weakly disturbed geomagnetic conditions and (b) data obtained from 366 *Triad* passes during active periods.[29] The poleward currents are the Region 1 currents and the equatorward currents are the Region 2 currents.

intensity is larger in the summer hemisphere than in the winter hemisphere, and this is related to the seasonal variation of the ionospheric conductivity. The primary cause of the annual variation of the field-aligned current is related to the dipole tilt of the geomagnetic field, while the semiannual variation appears to be related to the fact that geomagnetic activity tends to be greater around the equinoxes.[30]

In addition to the Region 1 and 2 current systems, there is another current system associated with the cusp region (Figure 12.28). The cusp field-aligned currents are located poleward of the Region 1 and 2 currents in the 0930 to 1430 magnetic local-time (MLT) sector and are statistically distributed between 78° and 80° invariant latitudes during weak magnetic activity. These currents generally flow away from the ionosphere in the prenoon sector (0930–1200 MLT) and into the ionosphere in the postnoon sector (1200–1430 MLT).

When the IMF is northward, an additional field-aligned current system exists in the polar cap, which is called the *NBZ current system*. The NBZ currents are concentrated on the sunlit side of the polar cap and the intensity of the currents increases as the magnitude of B_z increases. The statistical distribution of the NBZ currents is shown in Figure 12.29 for strongly northward B_z ($\geq 5\,\mathrm{nT}$) conditions.[32] The NBZ currents are poleward of the Region 1 currents and are in opposite direction to the Region 1 currents at a given local time. The NBZ currents are nearly as intense as the Region 1 and 2 currents. When the NBZ currents are present, the Region 1 and 2 currents continue to exist, although their intensity is diminished. In the southern hemisphere, the NBZ currents flow into the ionosphere on the dusk side of the polar cap and away from the ionosphere on the dawn side. These currents also display a distinct B_y dependence.

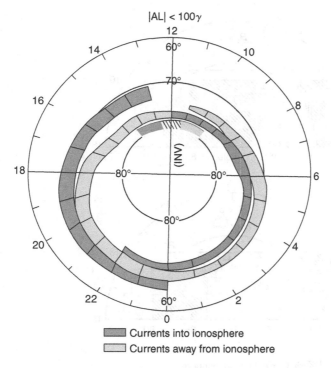

Figure 12.28 The distribution and flow directions of the large-scale field-aligned currents for southward IMF and including the cusp currents.[31]

The field-aligned currents that flow into and out of the ionosphere are connected via horizontal currents that flow in the lower ionosphere, as shown schematically in Figure 12.30. These large-scale currents, the auroral conductivity enhancements due to precipitating electrons, and the convection electric fields are not independent, but instead are related via Ohm's law and the current continuity equation (Sections 5.11 and 7.2). Numerous model studies, based on these equations, have been conducted over the years. These studies have shown that for southward IMF the Region 1 and 2 current systems (Figure 12.27), in combination with conductivity distributions obtained from empirical precipitation models (Figure 12.23), are consistent with the basic two-cell pattern of plasma convection (Figure 12.4).

12.6 Large-scale ionospheric features

The magnetospheric electric fields, particle precipitation, and field-aligned currents act in concert to produce several large-scale ionospheric features. These include polar holes, ionization troughs, tongues of ionization, plasma patches, auroral ionization enhancements, and electron and ion temperature hot spots. However, whether a feature occurs and also the detailed characteristics of a feature depend on the phase of the solar cycle, season, time of day, type of convection pattern, and the strength of

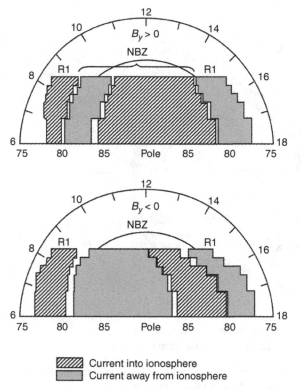

Current into ionosphere
Current away from ionosphere

Figure 12.29 Spatial distribution and flow directions of the large-scale NBZ Birkeland currents in the southern polar region for strongly northward IMF and for $B_y > 0$ (top panel) and $B_y < 0$ (bottom panel). The statistical distribution was determined from an analysis of 146 *MAGSAT* satellite orbits.[32]

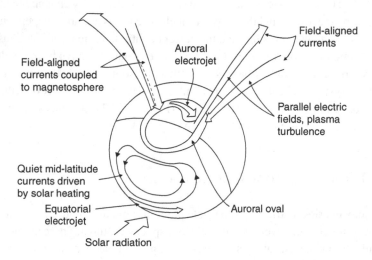

Figure 12.30 Schematic diagram showing the current systems in the terrestrial ionosphere. (From *Space Plasma Physics: The Study of Solar-System Plasmas*, National Academy of Sciences, 1978.)

convection. Because of the myriad of possibilities, only the basic physics governing the formation of certain large-scale ionospheric features is described here.

Most of the large-scale features that have been identified occur for southward IMF. In this case, a two-cell convection pattern exists, with antisunward flow over the polar cap and sunward flow at lower latitudes. The effect of the antisunward flow is to transport the high-density day side plasma into the polar cap. However, the effect of this process depends on the speed of the antisunward flow and the location of the solar terminator. For low antisunward speeds ($\sim 200\,\mathrm{m\,s^{-1}}$), the plasma travels 720 km ($7.2°$ of latitude) in one hour, while for moderate antisunward speeds ($\sim 1\,\mathrm{km\,s^{-1}}$) it travels 3600 km ($36°$ of latitude). In summer, when the bulk of the polar cap is sunlit, the difference in convection speeds is not significant because the plasma density tends to be uniform. In winter, on the other hand, the difference in antisunward convection speeds is important.

Figure 12.31 shows the ionospheric feature that occurs for slow convection in winter. After the plasma convects across the solar terminator, it decays, owing to the absence of sunlight coupled with ordinary ionospheric recombination. The e-folding decay time for N_mF_2 is about half an hour. When the convection speed is low, the plasma density can decay to very low values ($N_mF_2 \sim 10^3\,\mathrm{cm^{-3}}$) just before the plasma enters the nocturnal oval. In the oval, the density is enhanced because of impact ionization due to precipitating electrons. The net result is a *polar hole*, which is situated just poleward of the nocturnal oval. On the other hand, when the antisunward convection speed is high, the high-density day side plasma can be transported great distances before it decays appreciably. The net result is a *tongue of ionization* that extends across the polar cap from the day side to the night side (Figure 12.32). Measurements have clearly established the existence of both the polar hole and the tongue of ionization, as well as their characteristics, for different seasonal and solar cycle conditions.

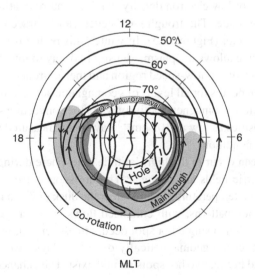

Figure 12.31 Schematic illustration of the Earth's polar region showing the plasma convection trajectories at 300 km (solid lines) in a magnetic local time (MLT), invariant latitude reference frame. Also shown are the locations of the high-latitude ionization hole, the main plasma trough, and the quiet-time auroral oval.[33]

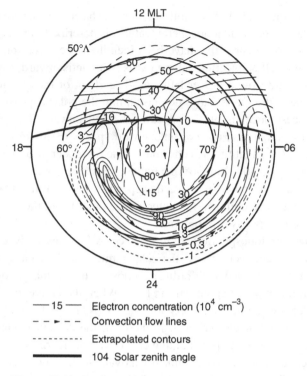

Figure 12.32 Contours of N_mF_2 in the northern polar region, as calculated from a numerical model. Also shown are the streamlines of the convecting plasma. The spatial coordinates are magnetic local time (MLT) and invariant latitude.[34]

Another interesting feature that is evident in winter is the *main or mid-latitude electron density trough*. This trough, which is situated just equatorward of the nocturnal auroral oval, is a region of low electron density that has a narrow latitudinal extent, but is extended in longitude. The trough's existence can be traced to the low-speed region in the dusk sector (Figure 12.5). In winter, this region is in darkness, and the long residence time allows the plasma density to decay to low values. Eventually, the plasma drifts out of this low-speed region and then co-rotates around the night side. The main trough occurs for all levels of geomagnetic activity, but it is especially pronounced during low geomagnetic activity when the convection speeds are slow. Trough electron densities as low as 10^3 cm^{-3} have been measured at 300 km during quiet geomagnetic conditions.[35]

Ion temperature hot spots can occur in the high-latitude ionosphere during periods when the convection electric fields are strong.[36] The hot spots correspond to localized regions of elevated ion temperatures located near the dusk or dawn meridians. For asymmetric convection patterns, with enhanced flows in either the dusk or dawn sectors of the polar region, a single hot spot occurs in association with the strong convection cell. However, on geomagnetically disturbed days, two strong convection cells can occur, and hence, two hot spots should exist. The enhanced ion

Millstone Hill radar

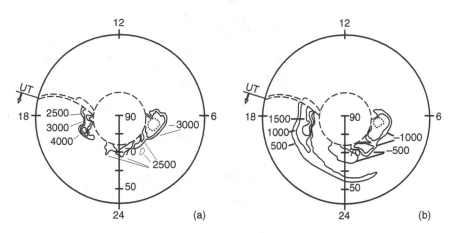

Figure 12.33 Contours of ion temperature (panel a) and line-of-sight ion drift velocity (panel b) observed at an altitude of 500 km from Millstone Hill over a 24-hour period on October 10–11, 1980. The temperatures are in K and the velocities are in m s^{-1}, with $\pm$ being toward and away from the radar. The panels are polar diagrams with local time indicated on the outer circle and dip latitude on the vertical scale.[36]

temperatures are a consequence of the increased ion–neutral frictional heating that is associated with the elevated convection speeds. Figure 12.33 provides experimental evidence for the existence of ion temperature hot spots. The figure shows contours of the ion temperature and line-of-sight plasma convection velocity at 500 km altitude, as observed via the Millstone Hill incoherent scatter radar over a 24-hour period on October 10–11, 1980. The line marked UT corresponds to the local time the observations began. During this day, K_p remained above five from the time the measurements began until about 1000 LT. Therefore, it is highly probable that the basic convection pattern was a two-cell pattern and that it persisted during this time period. The measurements show two distinct regions where the north–south, line-of-sight convection velocity exceeds 1 km s^{-1}. When the full vector is constructed with the aid of a two-cell convection model, these line-of-sight convection velocities are consistent with two strong convection cells, one at 0600 LT and the other at 1800 LT. Horizontal speeds in excess of 2 km s^{-1} are obtained in both of the convection cells. Associated with the large convection speeds are high ion temperatures, with T_i reaching 4000 K in a small region near the dusk terminator. The enhanced ion temperatures are confined to the general region where the line-of-sight velocities are large, which yields two distinct hot spots in the high-latitude ionosphere.

Electron temperature hot spots are also prevalent in the high-latitude ionosphere. The main source of T_e hot spots is electron precipitation. The precipitating electrons transfer energy to the thermal electrons via Coulomb collisions and they create energetic secondary electrons, thereby raising the temperature of the thermal electrons. Low-energy (soft) precipitation is most effective in raising T_e, because of the velocity dependence of the Coulomb cross section (Equation 4.51). Consequently,

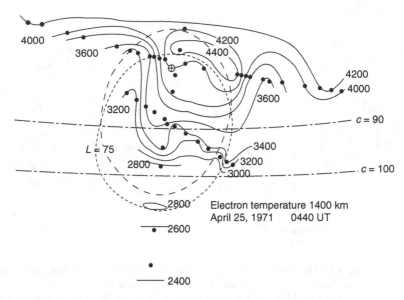

Figure 12.34 Contours of electron temperature deduced from data obtained from the *Alouette 1, Isis 1*, and *Isis 2* satellites. Note the elevated electron temperatures (4200–4400 K) in the cusp region.[37]

the cusp is expected to appear as an electron temperature hot spot, and that is indeed the case (Figure 12.34). However, localized T_e enhancements also occur in association with patches of precipitation, sun-aligned polar cap arcs, and auroral arcs. In addition, T_e hot spots can occur when the low-density plasma in the main trough convects into sunlight (Figure 12.31), which can occur in the morning or evening sectors. The photoelectron heating rate is spatially uniform, but the heat capacity of the low-density plasma is lower than that of the surrounding plasma. Therefore, T_e is elevated in the low-density region relative to the surrounding plasma and, hence, a T_e hot spot appears.[38]

12.7   Propagating plasma patches

Plasma patches are regions of enhanced plasma density and 630 nm emission that occur at polar latitudes. They have been observed for more than 15 years via optical, digisonde, and *in situ* satellite measurements.[39–42] Patches typically appear when the IMF turns southward. They have been observed in summer and winter at both solar maximum and minimum. They seem to be created either in the day side cusp or just equatorward of the cusp. Once formed, they convect in an antisunward direction across the polar cap at the prevailing convection speed, which typically varies from $300 \, \mathrm{m \, s^{-1}}$ to $1 \, \mathrm{km \, s^{-1}}$. Patch densities are a factor of 3–10 greater than background densities and their horizontal dimensions vary from 200 to 1000 km. As they convect across the polar cap, the associated electron temperatures are low, which indicates

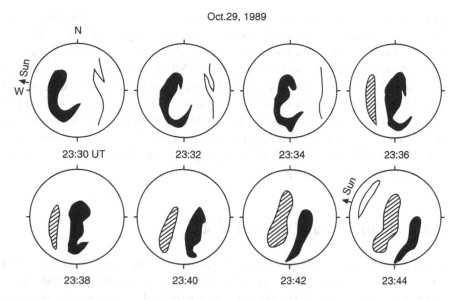

Figure 12.35 Propagating plasma patches observed at Qaanaaq on October 29, 1989. The dials represent a digitization of all-sky images (630 nm) taken at two-minute intervals. The solid and shaded areas show two plasma patches moving in an antisunward direction.[39]

an absence of particle precipitation. However, intermediate-scale irregularities (1–10 km) and scintillations are usually associated with propagating plasma patches.[43] Figure 12.35 shows an example of plasma patches observed at Qaanaaq, Greenland, on October 29, 1989.[39] The figure corresponds to a digitization of a sequence of all-sky photographs (630 nm) taken at two-minute intervals. The direction of the sun is indicted by an arrow on the first and last photographs. At 23:30 UT, a patch that is extended in the dawn–dusk direction is observed and it subsequently moves in an antisunward direction. Six minutes later, another patch appears in the all-sky camera's field-of-view and it also moves in an antisunward direction. The velocity of the patches is about 730 m s^{-1}.

Several mechanisms have been proposed to explain the appearance of plasma patches.[44] One mechanism suggested is that the patches are created in the cusp by pulsating soft electron precipitation, and then the patches convect into the polar cap. Another mechanism suggested is that the patches are created as a result of the sudden expansion and then contraction of the convection pattern. When the convection pattern expands, high-density plasma from the sunlit ionosphere is transported through the cusp and into the polar cap. When the convection pattern contracts, high-density plasma no longer flows into the polar cap, and the high-density plasma already there becomes isolated, forming a plasma patch. Although both of these mechanisms can, in principle, account for the formation of propagating plasma patches, the most likely cause of them is time-dependent changes in the B_y component of the IMF.[40, 45] With this mechanism, the tongue of ionization that normally extends through the cusp and into the polar cap (Figure 12.32) is broken into patches as the convection throat moves in response to B_y changes. However, it was also suggested that the tongue of ionization is broken by the sudden appearance of a *fast plasma jet*.[40] The appearance of the plasma jet coincides with a change in the IMF B_y component. The plasma jet is latitudinally narrow (300 km), extended in the east–west direction (2000 km), and contains eastward velocities in excess of $2 \, \text{km s}^{-1}$. The plasma jet is located just poleward of the cusp and perpendicular to the tongue of ionization. The jet causes a rapid depletion of the ionization because of the increased $O^+ + N_2$ reaction rate that is associated with ion–neutral frictional heating (Section 12.3). This process breaks the tongue of ionization into patches.

12.8   Boundary and auroral blobs

Boundary and auroral blobs are regions of enhanced plasma density that are located either inside or on the equatorward edge of the auroral oval. Figure 12.36 shows examples of such features. The figure shows contours of the electron density measured on November 11, 1981, by the Chatanika incoherent scatter radar.[28, 46] The contours are plotted as a function of altitude and geomagnetic north distance from

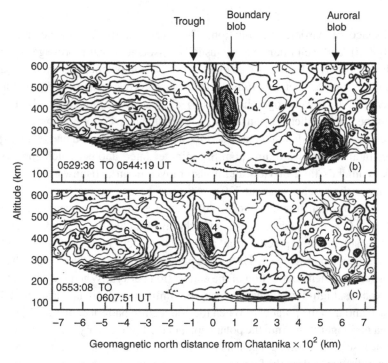

Figure 12.36 Contours of electron density measured on November 11, 1981 by the Chatanika incoherent scatter radar. The contours are plotted as a function of altitude and geomagnetic north distance from the radar (in 100 km units).[46]

the radar (in 100 km units). Two 15-minute radar scans are shown that are close to each other in time. The auroral blob is seen in the first scan and is located about 500 km north of the radar. The structure extends from 180–300 km in altitude and is about 200 km wide. The structure is no longer evident in the second scan. The boundary blob appears in both radar scans and is situated just equatorward of the auroral E layer and poleward of the mid-latitude trough. The auroral E layer is evident in the second scan as enhanced densities north of the radar at about 130 km altitude, while the mid-latitude trough is located south of the radar and is the narrow latitudinal region of low plasma densities. At still lower latitudes, a classical F region is clearly evident. Although not shown in Figure 12.36, boundary blobs can persist for many hours and can extend over large longitudinal distances.

Auroral blobs are thought to be produced by nonuniform particle precipitation in the auroral oval. Indeed, the measurements in Figure 12.27 reveal that a substantial ionization enhancement can occur in both the E and F regions within 10 minutes after precipitation commences. After the precipitation ceases, the E region ionization rapidly decays via recombination, leaving an auroral blob in the F region. Boundary blobs, on the other hand, are not created locally. They are polar cap patches that have convected through the night side auroral oval and around toward dusk.[47] Figure 12.37 shows the calculated evolution of plasma in a circular region in the polar cap when the ionospheric dynamics is governed by a two-cell convection pattern. Starting with plasma in a circular region in the dusk convection cell, the subsequent plasma

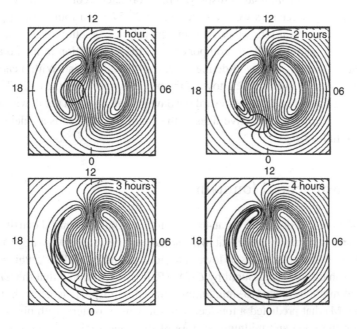

Figure 12.37 Distortion of a circular patch of ionization as it convects from the polar cap through and around the nocturnal oval toward dusk. A two-cell convection pattern was used.[47]

convection distorts the circle as the elements inside the circle move along different convection trajectories. After three hours, the circle transforms into a structure that is narrow in latitude but extended in longitude. The structure is located on the equatorward edge of the dusk side auroral oval, which is where boundary blobs are located.

12.9   Sun-aligned arcs

Sun-aligned polar cap arcs are discrete 630 nm emission structures in the polar cap.[48] The arcs appear when the IMF is near zero or northward and are a result of electron precipitation, with the characteristic energy varying from 300 eV to 5 keV and the energy flux varying from 0.1 to a few ergs cm^{-2} s^{-1}. They are relatively narrow (≤ 300 km), but are extended along the noon–midnight direction (1000–3000 km). Under conditions of large (>10 nT) northward IMF, a single arc can form that extends all the way from the day side to the night side auroral oval, with the associated optical emission forming the Greek letter theta when viewed from space.[11] Typically, however, the arcs do not completely extend across the polar cap, and frequently, multiple arcs are observed. Once formed, the arcs tend to drift toward either the dawn or dusk side of the polar cap at speeds of a few hundred meters per second. Figure 12.38 shows the temporal evolution of multiple polar cap arcs observed at Qaanaaq, Greenland, on February 19, 1989. The arcs are reconstructions of 630 nm images displayed in a corrected geomagnetic coordinate system.[49] Initially, three arcs were visible, but at 22:57 UT a fourth arc appeared, which then drifted toward the other arcs. In general, the direction of motion of the arcs depends on both the IMF B_y component and the arc location in the polar cap. For a given value of B_y, two well-defined regions (or cells) exist. Within each cell, the arcs move in the same direction toward the boundary between the cells. The arcs located in the dusk side cell move toward the dawn, while those in the dawn side cell move toward the dusk. The relative sizes of the dawn and dusk cells are determined by the magnitude of B_y.

12.10   Cusp neutral fountain

The *CHAMP* satellite observations at 400 km indicate that the neutral density in the day side cusp can be nearly twice that in the adjacent regions.[50] It was suggested that Joule heating of the neutral gas at lower altitudes caused upwelling, which led to the density enhancement measured by *CHAMP*. It was also noted that the *CHAMP* observations might be related to simulations with a thermosphere general circulation model (TGCM) that predicted a four-cell neutral density pattern with high-density structures in the noon and midnight sectors of the polar region.[51] These modeled neutral density cells, which were 1000–2000 km in diameter, were caused by a combination of Joule heating and ion drag. However, the modeled cells were only

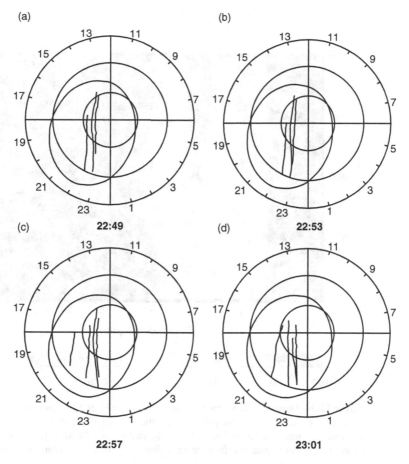

Figure 12.38 Multiple polar cap arcs observed at Qaanaaq, Greenland, on February 19, 1989. The arcs are displayed in a corrected geomagnetic coordinate system.[49] The oval marks the polar cap boundary.

evident at low altitudes (140–300 km) and the neutral density variation at 400 km was relatively smooth. In a follow-up study with a more comprehensive *CHAMP* data set,[52] numerous neutral density peaks and troughs were observed in the polar region. The width of the structures, either maxima or minima, varied from a few hundred km to 2000 km and the amplitudes of the structures approached 50% of the ambient density. The maxima clustered around the cusp, while the minima tended to cluster around the pole. In further comparisons of the observed density peaks and troughs with the predictions of global ionosphere–thermosphere models, it was concluded that these models do not produce neutral density structures at the *CHAMP* satellite altitude, and therefore, are not consistent with the *CHAMP* measurements.

A possible problem with the previous global ionosphere–thermosphere simulations could simply be that the spatial resolution used in the simulations was not

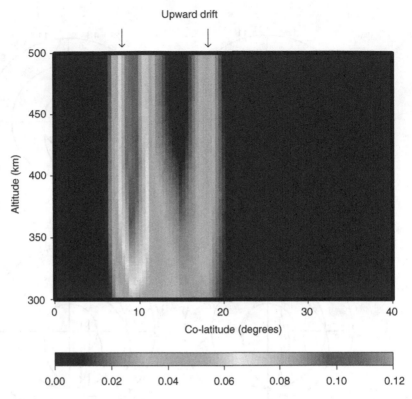

Figure 12.39 Upward component of neutral drift velocity as a function of altitude and geographic co-latitude from the pole, through the cusp, to 40 degrees on the day side. The region of imposed heating lies between the two arrows shown at the top of the figure. The upward neutral drifts have been divided by the mean radius of the Earth.[53]

adequate and that the Joule heating and ion drag force were not strong enough in the vicinity of the cusp. To test this hypothesis, a high-resolution (50 km × 50 km horizontal resolution in the polar region), global, thermosphere–ionosphere model was used to simulate the thermosphere's response to ion heating in the day side cusp.[53] With a series of relatively simple simulations, it was found that increased ion–neutral frictional heating and ion drag in the cusp results in the formation of a *cusp neutral fountain* with upwelling of the neutral gas in the heated region and a divergence and gradual subsidence of the gas outside of the heated region at higher altitudes. For frictional heating in the cusp that results in a doubling of neutral density at 400 km relative to background densities, the fountain contains a 200 m s^{-1} vertical wind and a 1500 K neutral temperature (Figure 12.39). However, it is possible that ion–neutral frictional heating is not the only mechanism responsible for elevated neutral densities in the cusp. Other possible mechanisms include heating via particle precipitation, field-aligned currents, solar EUV radiation, and the dissipation of atmospheric gravity waves.[52]

12.11   Neutral density structures

The cusp neutral fountain is a prominent thermospheric structure. However, the polar region also contains other thermospheric structures that are associated with mesoscale (100–1000 km) plasma structures, which can appear in the form of propagating plasma patches, auroral and boundary blobs, and ionization channels associated with polar cap arcs, discrete auroral arcs, and storm-enhanced densities (SEDs). Mesoscale plasma structures can also be associated with density depletions, including sub-auroral ion drift (SAIDs) events and equatorial plasma bubbles. These plasma structures have been observed to have a pronounced effect on the thermosphere.[54] The observations have shown the thermosphere to be highly structured both during geomagnetic storms and near discrete auroral features, with spatial scales varying from 50 to 500 km. The thermosphere was also observed to exhibit fairly rapid temporal variations, with time scales as short as 10–30 minutes. The plasma structures not only affect the local thermosphere, but the cumulative effect of multiple plasma structures can alter the global mean circulation and temperature of the thermosphere.[55]

There have been several modeling studies of the effect that ionospheric structures have on the thermosphere. Time-dependent, high-resolution simulations have been conducted to study the thermosphere's response to discrete auroral arcs, sub-auroral ion drift events, single and multiple propagating plasma patches, circular and cigar-shaped plasma patches, single and multiple sun-aligned polar cap arcs, theta aurora, and equatorial plasma bubbles.[56–62] In all cases, the neutral disturbances induced by the plasma structures were characterized by neutral density, temperature, wind, and composition enhancements and depletions. Figure 12.40 shows the thermosphere's response to a series of propagating plasma patches, which act as a collisional "snowplow", creating a hole in the thermosphere in and behind the plasma patches and a neutral density enhancement at the front of the patches. The neutral disturbance that is induced by the propagating plasma patches moves along with the patches, and at

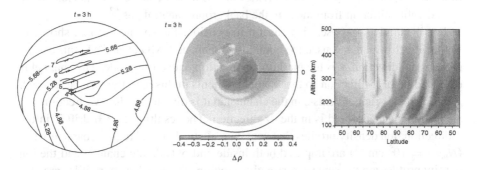

Figure 12.40 Effect of multiple propagating plasma patches on the thermosphere. Shown are n_e at 300 km (left), the neutral density perturbation at 300 km (middle), and the neutral density perturbation versus altitude and latitude across the polar cap from noon to midnight (right).[60]

300 km the disturbance is characterized by a neutral density perturbation (30–35%), an increased wind speed (100–150 m s^{-1}), a temperature enhancement (100–400 K), neutral gas upwelling, and O/N_2 composition changes. In general, the neutral gas perturbations due to plasma structures can be comparable to the day–night change in the neutral gas at 300 km (see Appendix K).

12.12   Neutral response to convection channels

As noted in Sections 12.1 and 12.2, the plasma convection pattern at high latitudes can exhibit a relatively simple cellular structure (Figures 12.8 to 12.12). Frequently, however, the convection pattern can contain narrow channels (200–1000 km wide) of high-speed flow (Figures 12.3 and 12.17) and the plasma density in the channels can be either higher or lower than the plasma density outside the channels, depending on where the channels form. These convection channels can form in the polar cap, in the auroral oval, and equatorward of the oval at all local times. An example of a well-known convection channel is the one associated with a sub-auroral ion drift (SAID) event. These SAID events correspond to narrow regions of rapid westward ion drifts that are located in the evening sector just equatorward of the auroral oval. The SAID events have a narrow latitudinal width, but are extended in longitude (Section 12.15).

The enhanced electric field in a narrow convection channel will have a significant effect on the plasma in the channel. In addition to the rapid plasma flow, there will be elevated ion temperatures due to ion–neutral collisions, $O^+ \rightarrow NO^+$ conversions and electron density depletions due to chemical reactions, and plasma upflows due to ion heating (Section 12.3). There is also a significant thermospheric response to a narrow channel of rapidly convecting plasma and the disturbance occurs both inside and outside the channel. Some of the important characteristics of the neutral flow in and near a convection channel have been simulated with two-dimensional, high-resolution models of the thermosphere–ionosphere system, and these simulations indicate that the flow characteristics in and near a convection channel can be considerably different from those in the background neutral gas.[63–67]

For illustrative purposes, it is useful to consider the simple channel shown in Figure 12.41, which was used to simulate an idealized SAID event.[66] In this case, x is the zonal direction, y is the meridional direction, and z is altitude. The channel is extended in the x-direction so that spatial gradients in this direction can be ignored. A constant electric field exists in the channel that is directed in the $-y$-direction and a constant magnetic-field is in the $-z$-direction. The resulting $\mathbf{E} \times \mathbf{B}$ drift is in the x-direction. Ion density profiles that contain realistic E and F region characteristics ($N_mF_2 = 10^5$ cm^{-3}) are imposed both inside and outside the channel and the ion density profiles are the same across a given region. The ion density profiles are also kept constant with time. Consider the case where the channel is 150 km wide, the electric field in the channel is $E = 100$ mV m^{-1} (2 km s^{-1} $\mathbf{E} \times \mathbf{B}$ drift), $E = 0$ outside the channel, the electron density in the channel is 10 times larger than that outside the

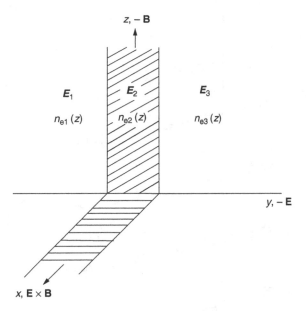

Figure 12.41 Schematic diagram of an idealized plasma convection channel.[66]

channel, and the neutral atmosphere is initially stationary. After the electric field is turned on and a steady state is reached (20–60 minutes), the horizontal neutral velocities in and near the convection channel are as shown in Figures 12.42 and 12.43. For the case considered, the ion collision-to-cyclotron frequency ratio is much less than one at altitudes above 150 km, so that the ions and electrons $\mathbf{E} \times \mathbf{B}$ drift together in the x-direction (Equation 5.110, Figure 12.15). Therefore, the neutral flow pattern shown in Figure 12.42 is a result of the competition between the following processes; the drifting ions tend to drag the neutrals in the $\mathbf{E} \times \mathbf{B}$ (or x) direction; the Coriolis force acts to deflect the flow to the negative y-direction; and vertical and horizontal viscosity act to decrease vertical and horizontal velocity gradients, respectively, and, hence, act to transmit velocity information upward and out of the plasma convection channel. It is apparent that the Coriolis force dominates in the channel below about 250 km.

Above 250 km, the ion drag force, which decreases with altitude, is balanced by vertical and horizontal viscosity. Vertical viscosity acts to smooth vertical gradients and increase the flow speed at high altitudes, while horizontal viscosity acts to transmit flow information outside the channel and reduce the flow speed. At altitudes above 300 km, horizontal viscosity dominates, and therefore, both the meridional and zonal neutral velocities decrease with altitude. This is in sharp contrast to what happens in the background thermosphere, where the velocity profiles tend to go constant with altitude owing to the importance of vertical viscosity (Figure 12.43). Also, outside the channel the horizontal and vertical viscosity forces tend to balance each other, and as a consequence, the Coriolis force has an appreciable effect on the momentum balance to altitudes up to about 400 km. Furthermore, horizontal

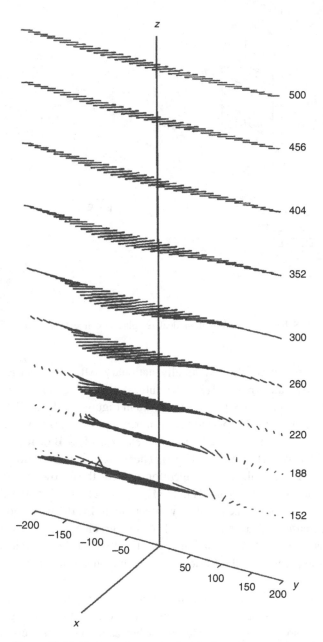

Figure 12.42 Horizontal neutral velocities in and near a convection channel. The channel width is 150 km, $E = 100 \, \text{mV m}^{-1}$ in the channel, and the electron density is 10 times larger than that outside the channel.[66]

viscosity acts to induce a thermospheric motion over a region that extends up to 1000 km on both sides of the electric field region.[65, 66] Because of the importance of horizontal viscosity, viscous heating ($-\tau_n : \nabla u_n$; viscous term brought to the right-hand-side of Equation 10.20c) can be comparable to or more important

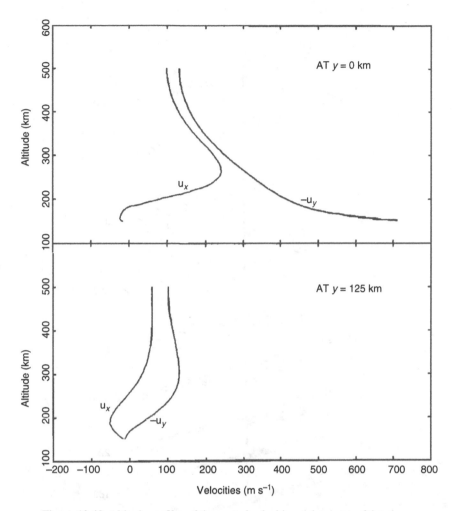

Figure 12.43 Altitude profiles of the neutral velocities at the center of the plasma convection channel (top) and at a selected location outside of the channel (bottom); u_x is in the $\mathbf{E} \times \mathbf{B}$ direction and; $-u_y$ is in the direction of the Coriolis force.[66]

than Joule heating (ion–neutral frictional heating) in and near plasma convection channels.[65–67]

The simulation shown in Figures 12.42 and 12.43 corresponds to a case when there is no wind in the background thermosphere. However, if there is a large-scale pressure gradient that drives a meridional wind across the convection channel in the $+y$-direction (toward the equator) then the thermospheric flow pattern changes from that in Figure 12.42 to the one shown in Figure 12.44. In this case the large-scale pressure gradient tends to balance the Coriolis force, which results in larger neutral velocities in the $\mathbf{E} \times \mathbf{B}$ (zonal) direction.

With regard to the thermal structure in and near a plasma convection channel, thermal conduction in the vertical and horizontal directions plays a role that is

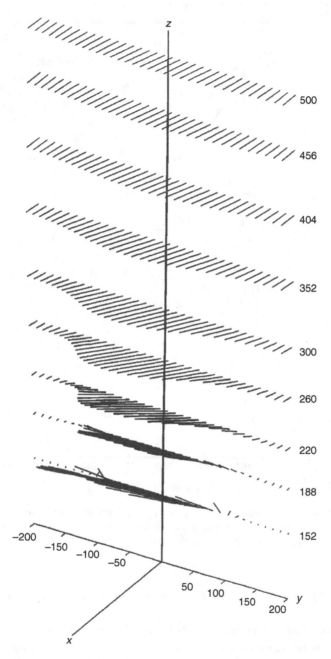

Figure 12.44 Horizontal neutral velocities in and near a convection channel. The conditions are the same as for Figure 12.42, except that a meridional wind blows across the channel in the +y-direction.[66]

similar to viscosity in that heat is conducted upward and out of the convection channel. The tendency is for T_n to decrease with altitude and to increase outside of the channel due to thermal conduction. However, advection of heat out of the channel by cross-channel winds can also occur.[67] In a more general simulation, the various processes discussed here will occur simultaneously and the net outcome will depend on several parameters, including the channel width, the location of the channel, the electric field and electron density profiles inside and outside of the channel, the temporal evolution of the electron density profiles, the lifetime of the event, and the background thermospheric conditions. Nevertheless, the basic characteristics of the neutral flow in and near convection channels will be similar to that discussed in this section.

12.13   Supersonic neutral winds

Dynamics Explorer 2 (DE 2) satellite data indicate that supersonic neutral winds frequently occur in the high-latitude thermosphere.[68] The supersonic neutral winds occur most often in the dawn sector of the polar cap between 75° and 85° magnetic latitude and at altitudes from 300–400 km, but they have been observed as high as 600 km. The Mach numbers of the supersonic neutral winds fall in the range $1 < M < 2$. The average horizontal extent of the supersonic flow regions is 140 km in a roughly dawn–dusk direction. The Mach number and neutral temperature were found to be anti-correlated.[68]

A time-dependent, high-resolution model of the global thermosphere–ionosphere system was used to study the geophysical conditions that give rise to supersonic neutral winds.[69] Simulations were conducted for a wide range of solar cycle, seasonal, and geomagnetic activity conditions. Since high-speed neutral winds are primarily driven by antisunward plasma convection (see Figure 12.16), only southward IMF conditions ($B_z < 0$) were considered. However, three B_y cases (0, positive, negative) and three values for the cross-polar-cap potential (100, 150, 200 kV) were considered. In a typical simulation, a diurnally reproducible thermosphere was first calculated for quiet conditions, and then at a selected time the convection pattern was changed to mimic a southward turning of the IMF and enhanced convection. The simulations indicated that the main factor controlling the neutral wind speed is the magnitude of the cross-polar-cap potential. For a sufficiently large cross-polar-cap potential ($\gtrsim 150$ kV), supersonic flow of the neutral gas can occur. Typically, the neutral gas is accelerated to supersonic speeds in less than three hours. In general, the region of supersonic flow occurs about 70–90 degrees magnetic latitude in the dawn–midnight quadrant. The spatial extent of the supersonic flow region and the maximum Mach numbers that occur there are seen to increase with altitude, but the increases are very small above 400 km. For cross-polar-cap potentials in the range of 150–200 kV, the dawn–dusk extent of the supersonic flow region is about 150 km, and at an altitude of 400 km the Mach numbers vary from $1 < M < 2$, where $M = u_n/(kT_n/m_n)^{1/2}$. The simulations also indicate that the Mach number and neutral temperature are anti-correlated. This occurs because the ion drag force is proportional to $(u_i - u_n)$, while the ion–neutral frictional heating is proportional to $(u_i - u_n)^2$. As the neutral gas is accelerated, the heating decreases more rapidly than the drag force. Note that all of the features predicted by the global thermosphere–ionosphere model[69] are in good agreement with the measured supersonic flow events.[68]

Figure 12.45 shows simulation results for moderate solar activity ($F10.7 = 150$) and autumn equinox (day 268) conditions in the northern polar region.[69] The simulation was conducted with the plasma convection defined by a symmetric two-cell convection pattern similar to that shown in Figure 12.4.[2] Initially, a

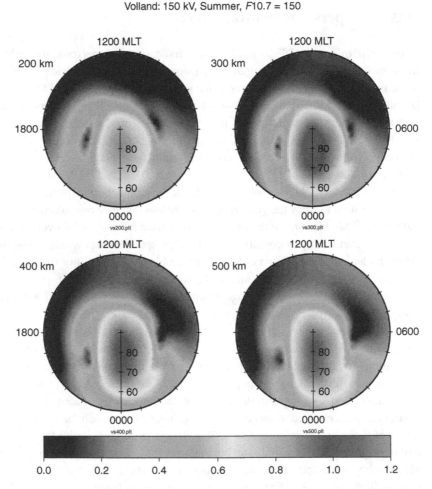

Figure 12.45 Contour plots of the Mach number in the northern hemisphere at high latitudes and at 200, 300, 400, and 500 km. The case is for equinox, moderate solar activity, and a Volland convection pattern with a cross-polar-cap potential of 150 kV.[69]

diurnally reproducible thermosphere–ionosphere system was calculated with a 40 kV cross-polar-cap-potential, and then at 0 UT the potential was increased to 150 kV. Figure 12.45 is a snapshot of the simulation results three hours after the increase in the potential. Figure 12.46 shows ion drift velocities (left panel) and neutral velocities (middle panel), both at 400 km and three hours. Note that the variation of the neutral velocities across the polar cap is similar to the measurements displayed in Figure 12.16. The right panel of Figure 12.46 shows the Mach number versus distance along the dusk–dawn path through the region containing large Mach numbers. Mach numbers curves are shown at 200, 300, and 400 km. Mach numbers greater than one are clearly evident in the midnight–dawn sector of the polar cap at 400 km. Typically, Mach numbers greater than one occur at and above 300 km for cross-polar-cap potentials greater than about 150 kV.[69]

12.14   Geomagnetic storms

Geomagnetic storms occur when there is a large sudden change in the solar wind dynamic pressure at the magnetopause, which occurs when it is impacted by a coronal mass ejection (Figures 2.2 and 2.5) or solar flare material (Figure 2.4). The storms can be particularly strong when the increased solar wind pressure is associated with a large southward IMF component. A *sudden storm commencement* (SSC) is followed sequentially by *initial*, *main*, and *recovery phases*. During the growth phase, the plasma convection and particle precipitation patterns expand, the electric fields become stronger, and the precipitation intensifies. These changes are accompanied by substantial increases in the Joule and particle heating rates and the electrojet currents. The energy input to the upper atmosphere maximizes during the main phase, while during the recovery phase the geomagnetic activity and energy input decrease. Large storms can significantly modify the density, composition, and circulation of the ionosphere–thermosphere system on a global scale, and the modifications can persist for several days after the geomagnetic activity ceases. If the electron density increases as a result of storm dynamics, it is called a *positive ionospheric storm*, while a decrease in electron density is called a *negative ionospheric storm*. During a sudden storm commencement, gravity waves can be excited at high latitudes and their subsequent propagation toward lower latitudes leads to a *traveling ionospheric disturbance (TID)*. Unfortunately, the response of the ionosphere–thermosphere system to different geomagnetic storms can be significantly different, and even for a given storm the system's response can be very different in different latitudinal and longitudinal regions.

The sequence of events that occurs during a geomagnetic storm is as follows.[70] In response to the large energy input at high latitudes, auroral E region densities increase, day side high-density plasma convects into the polar cap at F region altitudes, the main trough moves equatorward, the neutral and charged particle temperatures increase, the thermospheric wind speed increases, the O/N_2 ratio

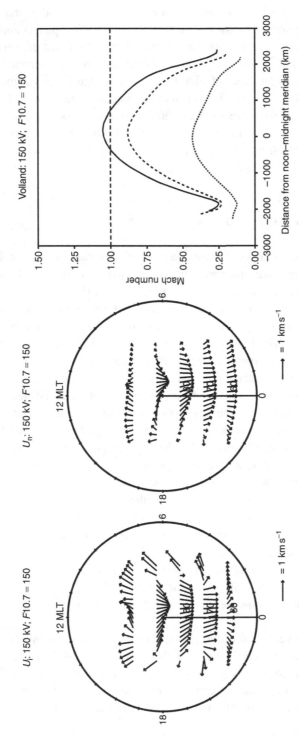

Figure 12.46 Ion drift velocity (left), neutral velocity (middle), and Mach number profiles (right) for the case shown in Figure 12.45. The velocity vectors are at an altitude of 400 km. The Mach number profiles are along a horizontal line from dusk (left) to dawn (right) that passes through the peak of the Mach number distribution. The profiles are for altitudes of 200 (dotted), 300 (dashed), and 400 (solid) km.[69]

decreases, and equatorward propagating gravity waves are excited. At mid-latitudes, the equatorward propagating waves drive the F region ionization toward higher altitudes, which results in an ionization enhancement (positive storm effect). Behind the wave disturbance are enhanced meridional winds. These diverging winds cause upwelling and neutral composition (O/N_2) changes, which then lead to decreased electron densities (negative storm effect). For major magnetic storms, the composition changes and winds can penetrate all the way to the magnetic equator, but that is rare. However, in the mid–low latitude region, the enhanced winds can generate dynamo electric fields that can affect the equatorial ionosphere (Sections 11.11 and 11.16).

A significant effort has been devoted to modeling magnetic storms, including weak, moderate, large, and super storms.[71–78] The results of these studies indicate that electric fields, neutral winds, and O/N_2 composition changes all contribute to the positive phase of a storm, while the negative phase of a storm is due to N_2 upwelling (decrease in the O/N_2 ratio). In one study, the thermosphere–ionosphere–mesosphere–electrodynamics general circulation model (TIME-GCM) was used to simulate the response of the global thermosphere–ionosphere system to the onset of the November 20, 2003 magnetic storm.[72, 73] The model simulated the dynamic changes in the neutral winds, temperatures, and composition as well as in the ionospheric densities, temperatures, and drifts. The inputs to the model were the solar flux, tides, and high-latitude electric fields and particle precipitation, which were obtained with the aid of satellite measurements. The simulation ran on November 19–22, 2003.

Unfortunately, there were very few data available for validating the calculated neutral parameters. However, the GUVI instrument on the NASA-*TIMED* satellite provided remote sensing observations of the column-integrated O/N_2 ratio during the storm.[72, 73] The GUVI measurements were made in a narrow swath below the satellite, and therefore, to obtain a global picture it was necessary to accumulate data from about 15 orbits, which takes a day. On November 19–21, the GUVI measurements were all made close to local noon. Figure 12.47 shows the column-integrated O/N_2 ratios obtained from the GUVI instrument (top panel) and the corresponding ratios calculated from the TIME-GCM model (bottom panel). The results are shown as a function of latitude and longitude, and time runs from right to left. The black ellipse on November 20 shows the European and North African regions near 1200 UT. In this ellipse the O/N_2 values on the storm day are generally larger than those on the previous quiet day (November 19). However, as the storm proceeds (November 20), there is a reduction in the O/N_2 values relative to the prestorm values (November 19). The location of these reduced O/N_2 values moves equatorward as the storm develops. There are also elevated O/N_2 values in the region equatorward of the depleted O/N_2 values. The TIME-GCM simulation indicates that the model can capture the general stormtime behavior, although the magnitudes of the O/N_2 ratios are not accurate.

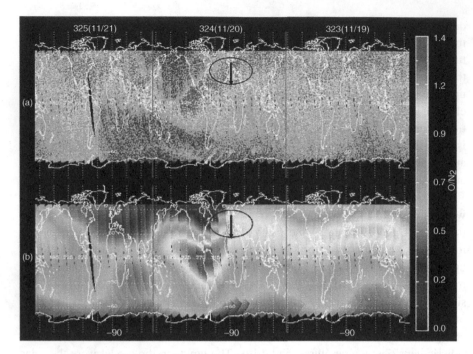

Figure 12.47 Column integrated O/N$_2$ ratio measured by the GUVI instrument (a) and predicted by the TIME-GCM (b). The vertical lines indicate day boundaries. Note that time runs from right to left.[73]

12.15   Substorms

Substorms correspond to the explosive release of energy in the auroral region near midnight MLT.[79] After substorm onset, there are growth, expansion, and recovery phases, with the expansion phase typically lasting about 30 minutes and the entire substorm two to three hours. When the substorm is viewed via the associated optical emission, it first appears as a region of bright emission that is located on the poleward edge of the auroral oval near midnight MLT. This *bulge* is part of a *westward traveling surge* that occurs during the expansion phase of a substorm. Associated with substorms are localized regions of enhanced electric fields, particle precipitation, and both field-aligned and electrojet currents. Discrete auroral arcs also usually appear near the poleward and westward fronts of the bulge. Eventually, the disturbances associated with substorms encompass the entire high-latitude region.[80]

Numerous models have been invoked to explain substorms, but to date, the measurements have not been able to determine which of the models is correct.[81–84] However, a possible sequence of events is shown in Figure 12.48 for the simple case of an isolated substorm.[85] The sequence of events begins with a southward turning of the IMF, which leads to an increased rate of magnetic merging of the Earth's field and the IMF at the day side magnetopause. The newly opened field lines are then convected across the polar cap and into the magnetotail on the night side. After

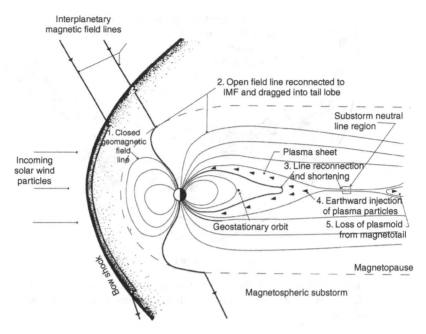

Figure 12.48 Possible sequence of events involved in an isolated substorm. The Earth's magnetosphere is viewed in the noon–midnight, north–south plane.[85]

about 30–60 minutes, the increased magnetic stress in the tail leads to a thinning of the plasma sheet and then to magnetic reconnection. When the oppositely directed magnetic fields above and below the equatorial plane reconnect in an x-line configuration, there is a sudden conversion of magnetic energy into particle acceleration in the plasma sheet (the *expansion phase*). The flow is toward the Earth on the near-Earth side of the x-line and away from the Earth on the other side. The subsequent injection of hot plasma into the Earth's upper atmosphere is responsible for the substorm's effects on the ionosphere.

Recently, the electrodynamic parameters in the nighttime ionosphere were examined during 35 substorms in an effort to determine the characteristic features of a substorm.[86] Data from the *DE 1* and *2* satellites were studied with respect to a generic aurora for 35 isolated substorm events; in this way it was possible to identify specific electrodynamic features during substorm expansion. Figure 12.49 shows the synthesis of the results, including the field-aligned currents, electric fields, and auroras. Note that at the poleward boundary of the bulge, upward and downward field-aligned currents are present in association with a narrow eastward or antisunward plasma flow.

During the recovery phase of a substorm, a sub-auroral ion drift (SAID) event can occur. These SAID events correspond to relatively narrow regions of rapid westward ion drifts located in the evening sector just equatorward of the auroral oval.[87] In SAID events, the ion drifts can reach $4\,\mathrm{km\,s^{-1}}$. The latitudinal width of the region varies from $0.1°$ to $2°$, and the lifetime of the event ranges from less than

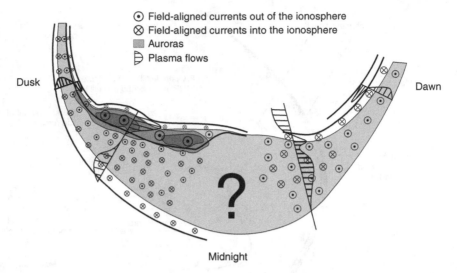

Figure 12.49 Schematic diagram showing the distributions, field-aligned currents, plasma flows, and auroras associated with a generic aurora during a bulge-type substorm.[86]

30 minutes to 3 hours. The SAID events are extended in longitude, but usually are confined to the 1800–2400 LT sector. These events are commonly thought to occur because of a separation between ion and electron drift paths in the plasma sheet that develops during the recovery phase of substorms.

12.16   Polar wind

The suggestion that light ionospheric ions (H^+ and He^+) might be able to escape the Earth's gravitational field can be traced to studies in the 1960s. At that time, it was recognized that the Earth's geomagnetic field is stretched into a long comet-like tail on the night side that extends past the Moon's orbit (Figures 2.10, 2.15). The magnetic field lines that form the tail originate in the polar region, and because the pressure in the ionosphere is much greater than the pressure in the distant tail, it was suggested that a continual escape of thermal plasma should occur along these open field lines.[88, 89] The early suggestions of light ion outflow were based on the well-known theory of thermal evaporation, which had been successfully applied to the escape of neutral gases from planetary atmospheres (Section 10.10). As a result of thermal evaporation, the light ions should escape the topside ionosphere with velocities close to their thermal speeds, and then they should flow along magnetic field lines to the magnetospheric tail. However, it was subsequently argued that the ion outflow should be supersonic, and it was termed the polar wind in analogy to the solar wind.[90, 91] Measurements later confirmed the supersonic nature of the outflow by both direct and indirect means.[92]

After 30 years of intensive study, it is now well-known that the *classical polar wind* is an ambipolar outflow of thermal plasma from the high-latitude ionosphere. As the light ion plasma flows up and out of the topside ionosphere along diverging geomagnetic field lines, it undergoes four major transitions, including a transition from chemical to diffusion dominance, a transition from subsonic to supersonic flow, a transition from collision-dominated to collisionless regimes, and a transition from a heavy (O^+) to a light (H^+) ion. At times, however, O^+ can remain the dominant ion to very high altitudes in the polar cap. Another important aspect of the flow concerns its horizontal motion. Because of magnetospheric electric fields (Figure 12.5), the high-latitude ionosphere and polar wind are in a continual state of horizontal motion, convecting into and out of the cusp, polar cap, nocturnal auroral oval, nighttime trough, and sunlit hemisphere. This horizontal motion is significant because the time it takes the polar wind to flow up and out of the topside ionosphere is comparable to the transit time across the polar cap and, hence, the *local* conditions are constantly changing.

Owing to the complicated nature of the flow, numerous mathematical approaches have been used over the years to model the classical polar wind. These include hydrodynamic, hydromagnetic, generalized transport, kinetic, semikinetic, and macroscopic particle-in-cell models. Also, numerous studies have been conducted of the *nonclassical* polar wind, which may contain, for example, ion beams or hot electrons. Polar wind studies have been conducted to examine its supersonic nature, its anisotropic thermal structure, its evolution through the collision-dominated to collisionless transition region, its stability in the presence of nonthermal plasma components, and its seasonal and solar cycle dependencies. Studies have also been carried out to understand the extent to which various processes can affect the polar wind, including charge exchange between O^+ and H, photoelectrons, elevated thermal electron and ion temperatures, ion heating transverse to **B**, hot electrons and ions of magnetospheric origin, centrifugal acceleration, wave–particle interactions in the polar cap, and field-aligned auroral currents.

The purpose of this list is simply to indicate that a myriad of processes could be operating in the polar wind and that an extensive amount of work has been done to date. Further details concerning these processes can be found in the comprehensive review articles listed in the General references section. Here, the goal is to elucidate the basic physics of the ion outflow and not to discuss all of the relevant processes in detail. However, at the end of this section, a summary of all the possible polar wind processes puts them in perspective.

The early polar wind studies were based on a hydrodynamic formulation that contained relatively simple continuity, momentum, and energy equations. In these studies, one-dimensional, steady state solutions were obtained that included ion production and loss processes. However, at that time, it was necessary to use a mixture of both linear and nonlinear collision terms, because general collision terms were not available for all of the moment equations. As a consequence, nonlinear collision terms were used for the exchange of momentum (4.124b) and energy (4.124c) between the interacting species, but linear collision terms were used for the stress tensor (4.129f) and heat flow vector (4.129g). Despite these simplifications, these hydrodynamic equations were able to describe the basic polar wind characteristics in the altitude range from 200 to 3000 km.

The hydrodynamic equations adopted in these early studies were obtained from Equations (3.57) to (3.59) and are given by

$$\frac{d}{dr}(n_s u_s) = P_s - L'_s n_s,$$ (12.15)

$$n_s m_s u_s \frac{du_s}{dr} + \frac{dp_s}{dr} + \frac{d\tau_{s\|}}{dr} - n_s e_s E_\| + n_s m_s g_\| = n_s m_s \sum_t \nu_{st}(u_t - u_s)\Phi_{st},$$ (12.16)

$$\frac{3}{2}kn_s u_s \frac{dT_s}{dr} + \frac{3}{2}kT_s \frac{d}{dr}(n_s u_s) + n_s kT_s \frac{du_s}{dr} - \frac{d}{dr}\left(\lambda_s \frac{dT_s}{dr}\right)$$
$$= \sum_t \frac{n_s m_s \nu_{st}}{m_s + m_t}\left[3k(T_t - T_s)\Psi_{st} + m_t(\mathbf{u}_s - \mathbf{u}_t)^2 \Phi_{st}\right],$$ (12.17)

where the $\|$ signs denote quantities parallel to $\mathbf{B}$, r is the radial distance along $\mathbf{B}$, Φ_{st} and Ψ_{st} are the velocity-dependent correction factors (Equations 4.125, 4.126, and 4.127), P_s is the production rate, L'_s is the loss frequency, and subscript s can be used for either O^+, H^+, or electrons. The summations in Equations (12.16) and (12.17) are over all charged and neutral species (O^+, H^+, e^-, N_2, O_2, O, He, H). The collision terms on the right-hand sides of both equations are the nonlinear terms. The frictional term in the energy equation (12.17), which is proportional to $(\mathbf{u}_s - \mathbf{u}_t)^2$, accounts not only for heating due to a relative H^+–O^+ flow along $\mathbf{B}$, but also for ion heating as the plasma drifts horizontally through the slower moving neutral atmosphere due to magnetospheric electric fields. The parallel component of the stress tensor, $\tau_{s\|}$, and the heat flow vector used in the early studies were obtained from the collision-dominated expressions for the Navier–Stokes stress tensor (5.130) and the thermal conductivity (5.131), respectively, which are obtained from linear collision terms. Note that for a three-component plasma composed of ions, electrons, and neutrals, the momentum equation (12.16) is equivalent to the Mach number equation (5.80 or 5.87) derived previously, except for the $d\tau_{s\|}/dr$ and dA/dr terms. The stress tensor term was included in the early polar wind studies because it removes the singularity at $M = \pm 1$ (Equation 5.87). The dA/dr term is discussed later.

Typical results obtained from the hydrodynamic equations (12.15) to (12.17) are shown in Figures 12.50 and 12.51 for the case when horizontal transport due to magnetospheric electric fields is not considered. The different sets of density, drift velocity, and temperature profiles correspond to different H^+ escape velocities at the upper boundary of 3000 km, when O^+ is assumed to be gravitationally bound. The H^+ ions are produced via the accidentally resonant charge exchange reaction $O^+ + H \longleftrightarrow H^+ + O$ (Equation 8.3), and then they diffuse upward to higher altitudes. The upward H^+ speed increases at altitudes above 600 km as the assumed H^+ escape velocity at 3000 km is increased (Figure 12.50). Curve (a) corresponds

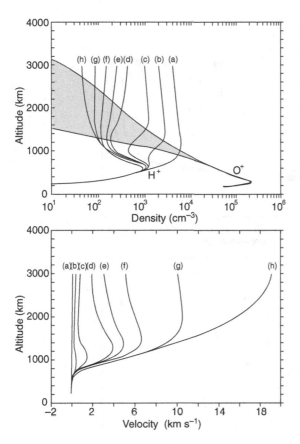

Figure 12.50 Theoretical H^+ density and field-aligned drift velocity profiles for the Earth's daytime high-latitude ionosphere. The different curves correspond to different H^+ escape velocities at 3000 km: (a) 0.06, (b) 0.34, (c) 0.75, (d) 2.0, (e) 3.0, (f) 5.0, (g) 10.0, and (h) 20.0 km s^{-1}. The shaded region shows the range of O^+ densities.[93]

to a near diffusive equilibrium situation, with H^+ becoming the dominant ion at 900 km (Section 5.7). The O^+ density in this case follows the lower curve of the shaded region. As the H^+ escape velocity is increased, the H^+ density is progressively reduced, with a peak in the H^+ density profile occurring in the 600–700 km region. Curves (b–e) correspond to subsonic outflow, while curves (g–h) correspond to supersonic outflow. Curve (f) is for a transonic flow, with the Mach number increasing to 1.17 at 1400 km, and then decreasing to 0.89 at 3000 km. For curve (h), which is for an H^+ escape velocity of 20 km s^{-1} at 3000 km, the H^+ escape flux is 8.5×10^7 cm^{-2} s^{-1}.

The temperatures of both H^+ and O^+ are affected by the H^+ outflow. The behavior of the O^+ temperature is straightforward in that it decreases at high altitudes as the H^+ escape velocity increases. This behavior results because the H^+ density decreases as the H^+ escape velocity increases, and O^+ then becomes more tightly coupled to the relatively cold neutrals. For H^+, on the other hand, the variation of the temperature with escape velocity is more complicated. As the escape velocity is increased, the H^+ temperature at high altitudes first decreases, then increases, and then decreases again. This behavior is related to the relative contributions made to the H^+ thermal balance by convection, advection, thermal conduction, frictional heating, and collisional

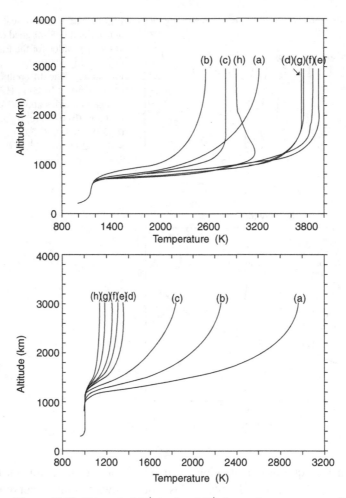

Figure 12.51 Theoretical H^+ (top) and O^+ (bottom) temperature profiles for the Earth's daytime high-latitude ionosphere. The profiles correspond to the density and drift velocity profiles shown in Figure 12.50.[93]

cooling.[93] However, the general trend of increasing H^+ temperatures at high altitudes as the H^+ escape velocity increases in the subsonic regime (curves b–e) is due primarily to enhanced frictional heating as H^+ moves through a gravitationally bound O^+ population with an increasing speed. The decrease in the H^+ temperature with increasing H^+ escape velocity in the supersonic regime (curves f–h) is due both to a decrease in frictional heating as the plasma becomes collisionless and to a change in the shape of the velocity profile, which acts to increase the importance of convective cooling.

An interesting feature of the H^+ outflow is its flux-limiting character. As the H^+ escape velocity increases, the H^+ escape flux increases to a saturation limit. This behavior is shown in Figure 12.52 in terms of the H^+ density at 3000 km. For sufficiently high H^+ densities, the H^+ flux is downward (negative), but as the H^+

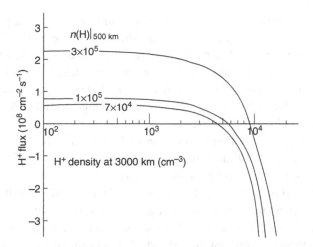

Figure 12.52 The H^+ escape flux for different H^+ boundary densities and neutral hydrogen densities.[94]

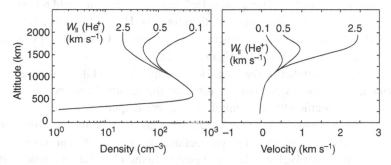

Figure 12.53 He^+ density (left panel) and drift velocity (right panel) profiles versus altitude for He^+ upper boundary velocities of 0.1, 0.5, and 2.5 km s^{-1}.[95]

density at 3000 km is lowered, an outward H^+ flux is established, and it quickly saturates in magnitude. Physically, this occurs because the production of H^+ is limited by the charge exchange reaction $O^+ + H \Leftrightarrow H^+ + O$, and in the steady state, the escape rate depends on the production rate. This implies that the H^+ escape flux is directly proportional to both the O^+ and H densities and inversely proportional to the O density. The limiting H^+ escape fluxes for several atomic hydrogen densities are shown in Figure 12.52, which demonstrates the direct relationship between the H^+ escape flux and the atomic hydrogen density.

The hydrodynamic equations (12.15–17) were also solved to obtain steady state polar wind solutions for He^+, assuming that He^+ was a minor ion at all altitudes between 200 and 2000 km.[95] Figure 12.53 shows the calculated He^+ density and drift velocity profiles for three upper boundary He^+ escape velocities (0.1, 0.5, and 2.5 km s^{-1}). For these calculations, the O^+ density at the F_2 peak was 2.1×10^5 cm^{-3}, the H^+ escape velocity at 3000 km was 10 km s^{-1}, and the convection electric field was neglected. The He^+ ions are created by photoionization of neutral helium, He, and lost in chemical reactions with molecular nitrogen, N_2. In general, the characteristics of the He^+ density profiles are similar to those of H^+. For all three profiles, there is a region below 600 km where chemistry dominates, whereas at high altitudes diffusion dominates. At intermediate altitudes, the competition between chemistry and diffusion yields a peak in the He^+ density profile at about 600 km. Also, as was found for H^+, the He^+ density at high altitudes decreases when the He^+ escape velocity is increased.

The He^+ drift velocity profiles are also similar to those of H^+. Specifically, the He^+ flow is downward at low altitudes, becomes positive in the vicinity of the He^+ density peak, and then increases to a peak value before falling back to the imposed upper boundary value. However, a notable difference in the He^+ and H^+ drift velocity profiles is that the rapid increase in the outflow velocity with altitude occurs at a higher altitude for He^+ (1300 km) than for H^+ (800 km). This difference is primarily due to the smaller diffusion coefficient for He^+, which allows chemistry to dominate to higher altitudes for He^+ than for H^+. The smaller He^+ diffusion coefficient, and the greater mass, account for the generally lower He^+ escape velocities compared to those obtained for H^+. These lower He^+ velocities, in turn, yield relatively small He^+–O^+ frictional heating rates and, therefore, the He^+ temperature is not significantly elevated above the O^+ temperature (not shown). This latter result is in sharp contrast to that obtained for H^+ (Figure 12.51).

Measurements indicate that the He^+ escape flux exhibits a large seasonal variation,[92] and this is primarily a result of the seasonal changes in the neutral atmosphere. Specifically, the neutral atmosphere displays a *winter helium bulge*, wherein the He densities in winter are 20 times greater than those in summer. This, in turn, yields much greater He^+ production rates in winter than in summer, and hence, much greater He^+ densities and escape fluxes. Calculations indicate that the limiting He^+ escape fluxes in winter are greater than those in summer, with the winter/summer ratio varying between 20 and 30. Also, the He^+ escape fluxes at solar maximum are typically 1.5–2 times greater than those at solar minimum. The net result is that the limiting He^+ escape flux varies by two orders of magnitude over the extremes of geophysical conditions. It varies from about 10^5 cm^{-2} s^{-1} for solar minimum, summer, and high magnetic activity conditions to about 1–2×10^7 cm^{-2} s^{-1} for solar maximum, winter, and low magnetic activity conditions.

The H^+ and He^+ outflow cases discussed above correspond to situations where O^+ is gravitationally bound. However, it is now well known that O^+ is an important magnetospheric constituent and that O^+ energization occurs over a range of altitudes in the ionosphere. When O^+ is energized at some altitude, the O^+ ions escape and the O^+ density then decreases at that altitude. The energization process triggers an O^+ upflow from lower altitudes, and the consequent reduction in the O^+ density then affects the H^+ escape flux because the two ions are coupled via the $O^+ + H \Leftrightarrow H^+ + O$ reaction (Equations 8.3 and 11.62). Figure 12.54 provides a summary of four possible outflow situations. Panel (a) shows the case of no ion outflow. In this case, both ion species are in diffusive equilibrium at altitudes above the F region peak, and consequently, the lighter ion H^+ becomes the dominant ion

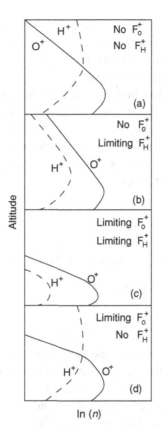

Figure 12.54 Schematic representation of ion density profiles for four possible outflow situations: (a) no ion outflow; (b) H^+ limiting outflow and no O^+ outflow; (c) both H^+ and O^+ limiting outflows; and (d) limiting O^+ outflow and no H^+ outflow.[96] (where F_O^+ and F_H^+ = Flux/ion outflow for O and H respectively)

in the topside ionosphere. For panel (b), H^+ is in a flux-limiting situation and O^+ is gravitationally bound (no O^+ escape). In this case, H^+ remains a minor ion at high altitudes because of the outflow. Also, because O^+ is the dominant ion species impeding the H^+ escape, the H^+ density scale height approaches the major ion (O^+) scale height. Panel (c) corresponds to the case of limiting escape fluxes for both H^+ and O^+. When O^+ is in a saturated outflow state, O^+ is depleted at high altitudes, and its density scale height approaches that of neutral atomic oxygen because it is the dominant species impeding the O^+ outflow. Likewise, the H^+ scale height approaches the O^+ scale height because O^+ is the dominant species impeding its outflow. Finally, panel (d) shows the case where O^+ approaches its limiting escape flux, but the H^+ escape flux is negligibly small. In this case, O^+ has a density scale height equal to the neutral atomic oxygen scale height, while H^+ is in a state of diffusive equilibrium.

The polar wind solutions presented above are valid at the altitudes where the H^+, He^+, and O^+ gases are collision-dominated. As a rough guide, the ion gases are effectively collision-dominated when $u_i/(H_i\nu_i) \ll 1$, where u_i is the ion field-aligned drift velocity, H_i is the ion density scale height, and ν_i is the ion collision frequency. For H^+, this condition generally begins to break down at about 1500 km and is clearly violated at 2000 km. However, He^+ and O^+ can remain collision dominated to altitudes as high as 3000 km. When the plasma is not collision dominated, the ion pressure (or temperature) distributions can become anisotropic and the ion heat flow vectors are not simply proportional to the ion temperature gradients. Also, in the collisionless regime above about 2000–3000 km, the divergence of the geomagnetic field becomes progressively more important as altitude increases.

The effect of an anisotropic ion pressure distribution on the momentum balance enters via the stress term in Equation (3.58), and the variation of the geomagnetic field with altitude comes into play when the divergence operator is used (Appendix B). With allowance for these effects, the steady state continuity (3.57) and momentum (3.58) equations become

$$\frac{1}{A}\frac{d}{dr}(An_s u_s) = P_s - L'_s n_s, \tag{12.18}$$

$$n_s m_s u_s \frac{du_s}{dr} + \frac{d}{dr}\left(n_s k T_{s\parallel}\right) - n_s e_s E_{\parallel} + n_s m_s g_{\parallel} + n_s k\left(T_{s\parallel} - T_{s\perp}\right)\frac{1}{A}\frac{dA}{dr}$$

$$= n_s m_s \sum_t v_{st}(u_t - u_s)\Phi_{st}, \tag{12.19}$$

where $T_{s\parallel}$ and $T_{s\perp}$ are the ion temperatures parallel and perpendicular to **B**, respectively. The quantity $(1/A)\, dA/dr$ accounts for the divergence of the magnetic field with distance. For a spherically symmetric flow (solar wind), $A \sim r^2$, while near the poles of a dipole magnetic field, $A \sim r^3$ (Section 11.1).

The collisionless characteristics of the polar wind can be described by kinetic, hydromagnetic, and generalized transport models.[97] For supersonic flow, these models produce density and drift velocity profiles that are similar to those obtained from the hydrodynamic equations. However, the ion temperature distributions are different, with the collisionless models yielding large temperature anisotropies at high altitudes. Typical results are shown in Figure 12.55, where the H^+ and O^+ temperatures parallel and perpendicular to the geomagnetic field are plotted as a function of altitude for collisionless, supersonic H^+ outflow. The ion temperature distributions were calculated with both kinetic and hydromagnetic (collisionless transport) models and the results are similar. The parallel ion temperatures are essentially constant with altitude at high altitudes, while the perpendicular ion temperatures decrease monotonically with altitude. The decrease of the perpendicular temperatures occurs because of the conservation of the ion adiabatic invariant, $mv_\perp^2/2B = $ constant (Chapter 5). Because of this decrease, the parallel-to-perpendicular temperature anisotropy grows with altitude, reaching nearly a factor of 50 for H^+ at a distance of 10 Earth radii.

The temperature anisotropies shown in Figure 12.55 were calculated for the collisionless regime above 4500 km, while the isotropic temperatures shown in Figure 12.51 were calculated for the collision-dominated regime below about 1500 km (the region where the temperatures are valid). At intermediate altitudes, the polar wind passes through a relatively narrow transition region where the ion velocity distributions evolve from Maxwellians in the low-altitude collision-dominated

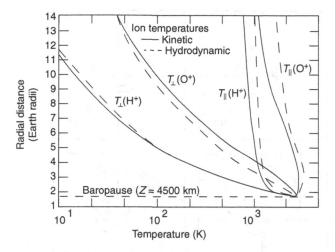

Figure 12.55 O^+ and H^+ temperatures parallel and perpendicular to the geomagnetic field obtained from kinetic (solid curves) and hydromagnetic (dashed curves) models of the collisionless, supersonic polar wind.[98]

regime to highly nonMaxwellian distributions in the high-altitude collisionless regime. Figure 12.56 shows this evolution for H^+, where contours of the ion velocity distribution are plotted at six altitudes from 230 to 1850 km.[99] The distributions were calculated with a Monte Carlo technique for the case of a steady state flow of H^+ through a stationary background plasma composed of O^+ and electrons. At low altitudes (230 km), the H^+ velocity distribution is a drifting Maxwellian (panel a). As the H^+ gas drifts upward, the high-speed ions in the tail of the distribution are accelerated by the H^+ pressure gradient and the polarization electric field created by the major ions (O^+) and electrons (Sections 5.6–5.8). The low-speed H^+ ions in the core of the distribution are more strongly coupled to the nondrifting O^+ ions than the high-speed H^+ ions in the tail because of the velocity dependence of the Coulomb cross section (Equation 4.50). The net result is that an extended tail forms on the H^+ velocity distribution in the upward direction (panel b). As the H^+ gas continues its upward drift, only the high-speed H^+ ions reach high altitudes because the ions in the core of the distribution remain coupled to O^+. This leads to the formation of a minimum in the distribution that separates the high- and low-speed components, and the distribution becomes double-humped (panels c and d). The high-speed component grows with altitude, while the low-speed component decreases (panels e and f). At 1850 km, the low-speed component disappears and the H^+ velocity distribution becomes kidney shaped. Note that at this altitude the perpendicular H^+ temperature is greater than the parallel H^+ temperature, which is evident from the width of the distribution in these two directions. As the H^+ ions drift to still higher altitudes, the velocity distribution changes shape again, becoming basically bi-Maxwellian, with $T_{i\parallel} > T_{i\perp}$ (Figure 12.55). As noted above, this change results from the conservation of the first adiabatic invariant in a diverging magnetic field.

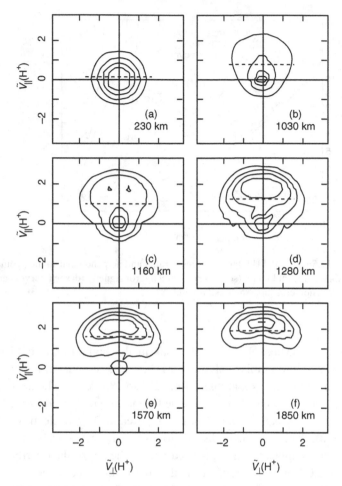

Figure 12.56 Contours of the H^+ velocity distribution at six altitudes. The altitudes extend from low altitudes (230 km), through the transition region (1030, 1160, 1280, 1570 km), to the top of the transition region (1850 km). The contours are plotted with respect to the normalized velocity components $\tilde{V}_\parallel$ and $\tilde{V}_\perp$, where the normalization is $[2kT(O^+)/m(O^+)]^{1/2}$. The contour levels are $0.9f_{max}$, $0.8f_{max}$, $0.7f_{max}$, etc., where f_{max} is the maximum value of the distribution function. The dotted line shows the H^+ drift velocity.[99]

Up to this point, the focus of the polar wind discussion has been on steady-state solutions obtained from one-dimensional models applied to fixed geographical locations. These solutions were useful for elucidating the basic polar wind characteristics. However, in reality, the polar wind is rarely, if ever, in a steady state, and the ionosphere–polar wind system continually convects across the polar region due to magnetospheric electric fields. Indeed, three-dimensional time-dependent simulations of the global ionosphere and polar wind have shown that, during changing geomagnetic activity, the temporal variations and horizontal plasma convection have a significant effect on the polar wind structure and dynamics.[100] The three-dimensional model used in these studies covered the altitude range from 90 to 9000 km for latitudes greater than 50° magnetic in the northern hemisphere. At low altitudes (90 to 800 km), three-dimensional density (NO^+, O_2^+, N_2^+, N^+, O^+), drift velocity, and temperature (T_e, $T_{i\parallel}$, $T_{i\perp}$) distributions were obtained from a numerical solution of the appropriate continuity, momentum, and energy equations. At high altitudes (800–9000 km), the time-dependent, nonlinear, hydrodynamic equations for H^+ and O^+ (Equations 12.18 and 12.19; with time derivatives) were solved self-consistently with the ionospheric equations.

The global ionosphere–polar wind model was used to study the system's response to an idealized geomagnetic storm for different seasonal and solar cycle conditions. The modeled geomagnetic storm, which commenced at 0400 UT when K_p was one, contained a one-hour growth phase, a one-hour main phase, and a four-hour decay phase. During increasing magnetic activity, the plasma convection and auroral precipitation patterns expanded, convection speeds increased, and particle precipitation became more intense. The reverse occurred during decreasing magnetic activity. The global simulations produced the following interesting results:

1.   Plasma pressure disturbances in the ionosphere, due to variations in either T_e, T_i, or electron density, are mimicked in the polar wind, but there are time delays because of the propagation time required for a disturbance to move from low to high altitudes.

2.   Plasma convection through the auroral oval and regions of high electric fields produces transient O^+ upflows and downflows. Typically, the H^+ upward flow is enhanced when the plasma convects into these regions and is reduced when the plasma convects out of them.

3.   The density structure in the polar wind can be considerably more complicated than in the underlying ionosphere because of horizontal convection and changing vertical propagation speeds due to spatially varying ionospheric temperatures. For example, transient H^+ downflows can occur at intermediate altitudes (3000–6000 km) even though the H^+ flow is upward from the ionosphere and upward at high altitudes (9000 km).

4. O^+ upflows typically occur in the auroral oval at all local times and downflows occur in the polar cap. However, during increasing magnetic activity, O^+ upflows can occur in the polar cap. The O^+ upflows are generally the strongest in the cusp at the location of the day side convection throat, where both T_e and T_i are elevated.

5.   During increasing magnetic activity, O^+ can be the dominant ion to altitudes as high as 9000 km throughout the bulk of the polar region.

6.   For winter, solar-minimum conditions, an H^+ *blowout* can occur throughout the bulk of the polar region shortly after a storm commences, and then the H^+ density slowly recovers when the storm subsides. However, the O^+ density variation is opposite to this. There is an increase in the O^+ density above 1000 km during the storm's main phase, and then the O^+ density decreases during the recovery phase.

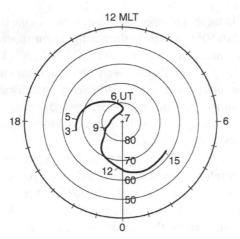

Figure 12.57 Convection trajectory for a representative flux tube of plasma during changing geomagnetic activity. The tick marks along the trajectory indicate the times in UT hours.[100]

7.   For summer, solar-maximum conditions, the O^+ and H^+ temporal morphologies are in phase, but the ion density variations at high altitudes are opposite to those at low altitudes. During the peak of the storm, the H^+ and O^+ densities increase at high altitudes and decrease at low altitudes.

Some of these results can be seen by following an individual flux tube of plasma as it convects across the polar region. Figure 12.57 shows a representative convection trajectory in a magnetic latitude–MLT reference frame. At the start of the simulation (0300 UT), the plasma flux tube following this trajectory was located at about 1900 MLT and 67° magnetic latitude, and the geomagnetic activity level was low ($K_p = 1$). Subsequently, the plasma flux tube moved sunward and entered the day side "storm" oval, passed through the convection throat, moved antisunward across the polar cap, entered the "quiet" nocturnal oval, and then exited the evening oval near the end of the trajectory.

Figure 12.58 shows the temporal variations of the plasma and neutral parameters at 500 km that are associated with the flux tube that followed the trajectory in Figure 12.57. This altitude was selected for presentation because the H^+ outflow typically begins above this level and, hence, the variations at 500 km show the drivers of the polar wind. The increase in T_e to almost 4800 K between 0400 and 0600 UT was a result of heating due to electron precipitation in the storm auroral oval. The increase in T_i during this time period was primarily a result of ion–neutral frictional heating in the throat region, where the electric fields were large. These increases in T_e and T_i resulted in a substantial O^+ upflow, with T_e being the main driver in this case because T_e was substantially larger than T_i. Associated with the O^+ upflow was an O^+ density enhancement, but its peak lagged behind the peak in the upward O^+ velocity. The H^+ ions were in chemical equilibrium at 500 km and, hence, the H^+ drift velocity was negligibly small at all times. When the flux tube exited the day side oval, there was a rapid decrease in both T_e and T_i and, as a consequence, the O^+ flow turned downward as the topside ionosphere collapsed. The O^+ flow remained downward at 500 km for most of the time that the flux tube

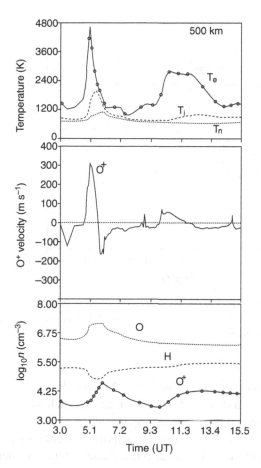

Figure 12.58 Temporal variations of the temperatures (top), O^+ field-aligned drift velocity (middle), and densities (bottom) at 500 km for the plasma flux tube that followed the convection trajectory shown in Figure 12.57.[100]

was in the polar cap, and associated with this downflow was a slow decay of the O^+ density (between 0600 and 0900 UT). The storm-enhanced O^+ densities in the polar cap, which persisted after the storm subsided, produced enhanced H^+ densities and escape fluxes at higher altitudes, and this, in turn, led to a time delay in the build-up of the maximum global H^+ escape rate. The maximum H^+ escape rate from the entire polar region (1.7×10^{25} ion s^{-1}) occurred at 0700 UT, which was one hour after the storm's main phase. Finally, between about 0930 and 1400 UT, the convecting flux tube of plasma was in the quiet nighttime auroral oval. Here, T_e and T_i were elevated, the O^+ flow was initially upward, and there was a slow build-up of the O^+ density at 500 km. However, the increases were smaller than those in the daytime storm oval because of the smaller convection speeds and smaller electron precipitation fluxes.

Figure 12.59 shows the temporal variations of the ion drift velocities and densities at 2500 km that are associated with the flux tube trajectory in Figure 12.57. When the flux tube first entered the day side storm oval, both the H^+ and O^+ flows were upward at this altitude, with a drift velocity of about 16 km s^{-1} for H^+ and 3.5 km s^{-1} for O^+ (both ions were supersonic). After this initial surge, the upward H^+ velocity first

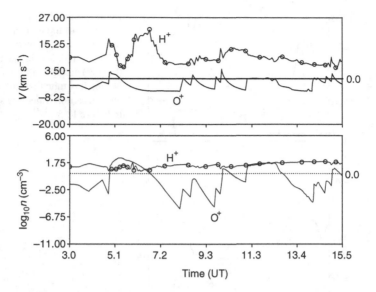

Figure 12.59 Temporal variations of the ion field-aligned drift velocities (top) and densities (bottom) at 2500 km for the plasma flux tube that followed the convection trajectory shown in Figure 12.47.[100]

decreased to 3.5 km s^{-1} and then increased to more than 20 km s^{-1}. In contrast, the O$^+$ flow at 2500 km was downward shortly after 0500 UT, as it was at lower altitudes (Figure 12.58). Despite this reversal in flow direction, O$^+$ was the dominant ion at 2500 km during most of the time that the magnetic activity was enhanced (from about 0430 to 0630 UT). The O$^+$ density was also comparable to the H$^+$ density in the quiet nocturnal oval, where T_e was elevated. As the plasma flux tube drifted along the trajectory, the H$^+$ flow remained upward at 2500 km and the H$^+$ density displayed a relatively slow variation, with $n(H^+) \sim 60$ cm^{-3}. On the other hand, the O$^+$ flow was both upward and downward, and the O$^+$ density varied by more than six orders of magnitude.

The ionosphere–polar wind simulation discussed here represented the classical polar wind, which is driven by thermal processes in the lower ionosphere. However, the polar wind may be affected by other processes not included in the classical picture of the polar wind, as shown schematically in Figure 12.60. Specifically, in sunlit regions, escaping photoelectrons may provide an additional ion acceleration at high altitudes ($\geq$7000 km) as they drag the thermal ions with them. This process would act to increase the O$^+$ and H$^+$ drift velocities in the polar regions where the ion flows are upward.[101] Cusp ion beams and conics that have convected into the polar cap can destabilize the polar wind when they pass through it at high altitudes. The resulting wave–particle interactions act to heat both O$^+$ and H$^+$ in a direction perpendicular to **B**, which then affects the escape velocities and fluxes.[102] The interaction of hot magnetospheric electrons (polar rain, showers, and squall) with the cold, upflowing, polar wind electrons can result in a double-layer potential drop over the polar

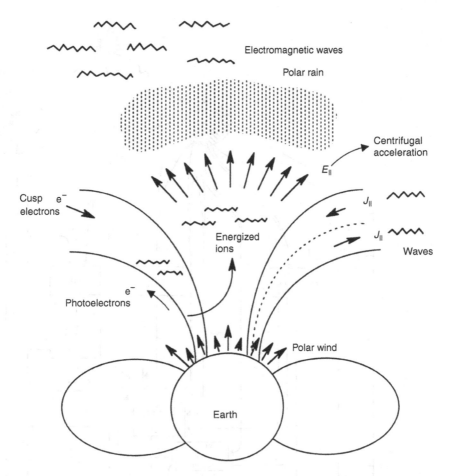

Figure 12.60 Schematic diagram showing nonclassical processes that may affect the polar wind.[100]

cap (at about 4000 km), which can energize the O^+ and H^+ ions. The H^+ energy gain varies from a few eV to about two keV, depending on the hot electron density and temperature.[103] At altitudes above 6000 km in the polar cap, electromagnetic turbulence can significantly affect the ion outflow via perpendicular heating through wave–particle interactions.[104] Also, above about 3000 km in the polar cap, centrifugal acceleration will increase ion upward velocities, which may affect ion densities at high altitudes.[105] Finally, anomalous resistivity on auroral field lines can affect the polar wind as the plasma convects through the nocturnal auroral oval.[106]

12.17   Energetic ion outflow

In addition to the coupling of the ionosphere and magnetosphere via convection electric fields, field-aligned currents, and energetic particle precipitation, which have

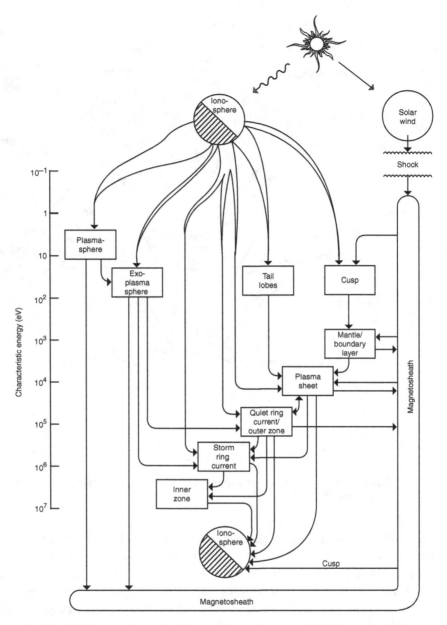

Figure 12.61 Schematic diagram showing the plasma paths between the ionosphere and magnetosphere.[112]

been discussed earlier, the two regions are strongly linked via upflowing ionospheric ions.[107–11] Figure 12.61 is a schematic diagram that shows how the upflowing ionospheric ions populate the different regions of the magnetosphere and vice versa.[112] The ionospheric ions feed the magnetosphere at all latitudes. Upflowing ions from the day side cusp populate the mantle or boundary layer and plasma sheet regions.

Ions escaping from the polar cap can populate both the plasma sheet and magneto-tail lobes. Ionospheric ions energized in the nocturnal auroral oval can populate the plasma sheet and ring current regions. At lower latitudes, upflowing thermal ions can populate both the inner and outer plasmaspheric regions. Except for H^+ outflow, nearly all upflowing ion events require an energization source in addition to ordinary ionospheric heating (solar heating, exothermal chemical reactions, ion–neutral frictional heating, etc.). Therefore, energetic ion outflow is discussed separately in this section.

The first measurements of outflowing energetic heavy ions from the polar ionosphere were made in the mid-1970s.[113] Instruments on the *S3-3* satellite at about 5000 km detected field-aligned upflowing O^+ beams with keV energies. Subsequently, upflowing O^+ conics with keV energies were detected. Since these pioneering discoveries, significant progress has been made in elucidating the characteristics of energetic ion outflow events. It is now well known that energetic ions are produced in the cusp, polar cap, and nocturnal auroral oval. It has also been clearly established that the production of energetic ions varies systematically with the solar cycle, season, magnetic activity, and local time. In addition, it has been shown that the altitude of the energization spans the range of 500 to 8000 km, that both parallel and transverse (to **B**) energization can occur, and that the energy of the escaping ionospheric ions varies from 10 eV to tens of keV.[114, 115]

In the day side cusp, ionospheric ions (O^+, H^+, He^+, N^+, and O^{++}) are heated in a direction transverse to the geomagnetic field (due to wave–particle interactions) to energies of 10–50 eV. The heated ions are then driven upward by the gradient-**B** force; they also convect in an antisunward direction across the polar cap due to magnetospheric electric fields. The lower energy heavy ions ultimately fall back to the Earth, while the more energetic ions convect to the plasma sheet. The net result is the so-called *cleft ion fountain*. Evidence for this fountain is shown in Figure 12.62. In this figure, segments of *DE 1* orbits are shown in which O^+ ions were observed in the polar cap contiguous with upwelling ion events. The data were binned with regard to the magnetic activity index K_p. For low K_p, the upwelling O^+ ions occurred only on the day side, while for high K_p the O^+ ions extended across the polar cap.

The polar cap contains both relatively cool (0.1–10 eV) polar wind ions and warm (10–50 eV) ions that have convected into this region from the day side cusp. However, more energetic ($\sim$1 keV) upflowing ions have also been detected at high altitudes above the polar cap, and these probably have a source region in the polar cap, based on the time it would take the ions to convect from the cusp to the satellite for typical electric field strengths. In particular, when the IMF is northward, sun-aligned arcs occur in the polar cap and upflowing energetic ions have been observed in association with these arcs.[116]

In the nocturnal auroral region, both parallel and perpendicular acceleration of the ionospheric ions can occur through a range of mechanisms and over a range of altitudes. Consequently, virtually all ionospheric ion species participate in energetic ion outflow events, and the escaping ions have velocity distributions in the form of

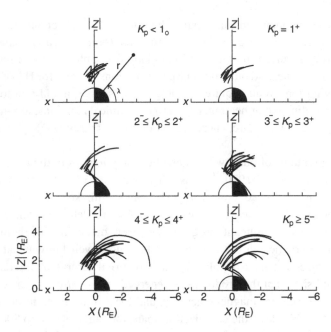

Figure 12.62 *DE 1* observations of the locations of O^+ ions in the polar cap contiguous with upwelling ion events. The data are plotted for six K_p ranges.[114]

beams, conics, rings, and toroidal distributions.[115, 117] This is an area of research that is ongoing and very active, with the main emphasis on identifying the acceleration mechanisms, altitudes of acceleration, and reasons why certain mechanisms dominate at certain times.

Several statistical studies relate to the characteristics of the acceleration mechanisms leading to energetic ion outflow. In particular, one study of the long-term variation in the energetic (0.01–17 keV) H^+ and O^+ outflow rates used *DE 1* ion composition data acquired in the auroral and polar regions between September 1981 and May 1986.[118] This period began near the maximum of solar cycle 21 and ended near the minimum; $F_{10.7}$ varied over the range $70\text{–}250 \times 10^{-22}$ W m^{-2} Hz^{-1}. Figure 12.63 shows the H^+ and O^+ outflow rates as a function of $F_{10.7}$ for three K_p ranges. In each K_p range, the variation with $F_{10.7}$ is the same. For O^+, there is a factor of five increase in the outflow rate in going from near solar minimum to near solar maximum. For H^+, on the other hand, there is about a factor of two decrease in the outflow rate over the same $F_{10.7}$ range, which may or may not be statistically significant. The H^+ and O^+ outflow rates as a function of K_p are shown in Figure 12.64 for three $F_{10.7}$ ranges. For all three $F_{10.7}$ ranges, the outflow variations with K_p are similar. For O^+, there is a factor of 20 increase in the outflow rate as K_p varies from 0 to 6, while for H^+ the increase is a factor of four over this K_p range. These results imply that there is an order of magnitude increase in the O^+/H^+ composition ratio in energetic ion outflow events in going from solar minimum to solar maximum and in going from quiet to active magnetic conditions.

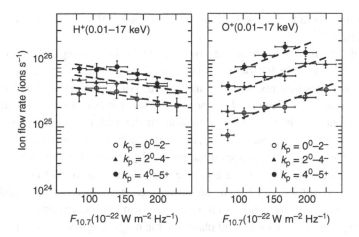

Figure 12.63 Energetic H^+ and O^+ outflow rates from the auroral and polar cap regions as a function of the solar radio flux $F_{10.7}$. The data are shown for three K_p ranges.[118]

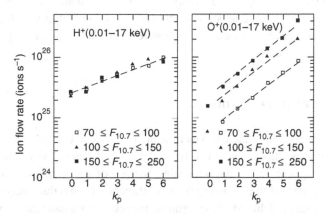

Figure 12.64 Energetic H^+ and O^+ outflow rates from the auroral and polar cap regions as a function of the magnetic index K_p. The data are shown for three $F_{10.7}$ ranges.[118]

This statistical study did not include ions with energies less than about 10 eV. However, a study was conducted of the relative contributions to the total ion outflow of ion fluxes below and above 10 eV.[119] The study was limited to the cleft ion fountain and solar maximum conditions. It was found that for O^+ the "less than 10 eV flux" was four times greater than the "greater than 10 eV flux," while the reverse was found for H^+. However, one caution should be noted because O^+ ions with energies less than 10 eV may not escape the ionosphere if they do not gain additional energy at high altitudes. Nevertheless, measured ion upflow rates from several statistical studies and from different ionospheric regions were used, in combination with ion residence-time estimates for the different magnetospheric regions, to calculate typical ion densities for the magnetospheric region. These calculated densities suggest that plasma outflow from the ionosphere alone can account for all of the plasma in the magnetosphere.[119]

Although significant progress has been made in identifying the ionospheric regions where energetic ion outflow events are likely to occur, much work still remains before these events are fully understood. Numerous parallel and perpendicular acceleration mechanisms have been proposed, but because the experimental data are incomplete, it has not been possible to determine, conclusively, which mechanisms dominate at specific times. Also, although it is known that the neutral atmosphere has a strong influence on the ion outflow rates, it is not clear why the acceleration mechanisms are more effective at solar maximum than at solar minimum. Additional satellite measurements of ion escape fluxes are needed and more theoretical work needs to be done to improve the estimates concerning the percentage of ionospheric versus solar wind plasma in the magnetosphere. Specifically, the transport and loss processes for ionospheric plasma in the magnetosphere are not well known, and hence, the residence time estimates are in question. Also, there may be a cold (less than 10 eV) ionospheric population in the magnetosphere that has gone undetected because of spacecraft charging problems. Finally, it is not known whether the heavy ionospheric ions affect the dynamics and stability of the magnetosphere. Global numerical models have now reached the point where they can begin to include such features.

12.18   Neutral polar wind

A significant discovery was made when an instrument on the *NASA IMAGE* satellite measured large escape fluxes ($\sim 1-4 \times 10^9 \, \mathrm{cm^{-2} \, s^{-1}}$) of neutral atoms from the high-latitude ionosphere.[120, 121] An interesting feature of the measurements was that the neutrals appeared to be coming from all directions. This discovery was in conflict with previous modeling studies that predicted relatively small neutral particle escape rates at high latitudes.[122, 123] However, in the previous studies, simplified steady state ionosphere models were adopted, and the contribution of the polar wind was either ignored or taken into account only in an approximate way. On the other hand, global ionosphere–polar wind simulations indicated that the magnitudes of the upward H^+ and O^+ fluxes increased markedly during geomagnetic storms, and that they were spatially nonuniform and highly time dependent.[124–8] This implied that during geomagnetic storms substantial fluxes of H and O can be created via charge exchange in the polar wind.

The H^+ and O^+ ions execute three characteristic motions. They spiral about the geomagnetic field, flow up and out of the top side ionosphere with a velocity that eventually becomes supersonic, and they drift horizontally across the polar region, moving into and out of sunlight, the cusp, polar cap, and nocturnal auroral oval. During this motion, the ions can undergo charge exchange reactions with the background neutral atmosphere, which is composed of both thermal neutrals and a hot neutral geocorona (Figure 12.65). For example, an upflowing O^+ ion can

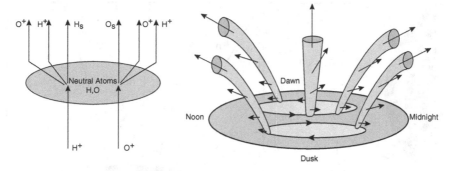

Figure 12.65 A schematic diagram that shows the production of neutral stream particles as the upflowing polar wind traverses a background neutral atmosphere (left panel). The right panel is a schematic diagram of the three-dimensional nature of the neutral polar wind and the production of low-energy neutrals flowing in all directions.[129]

undergo a charge exchange reaction with either H or O, and this would yield an up-flowing O atom. This reaction would also produce a nonflowing H^+ or O^+ ion, and subsequently, this ion would be accelerated upwards by the polarization electric field in the polar wind. The initial velocity of a neutral particle created in the polar wind is equal to the velocity of the H^+ or O^+ parent ion just before the charge exchange process. Consequently, at high altitudes, neutral streams of H and O are created that predominantly flow in the vertical direction (the *neutral polar wind*), while at low altitudes the neutrals tend to *move in all directions* owing to ion gyration and plasma convection.[129-131]

Substantial particle fluxes are created in the ion and neutral polar winds during geomagnetic storms at altitudes above about 500 km, with the neutral fluxes larger than the ion fluxes. As the H^+ and O^+ ions drift upward in response to storm heating, they are accelerated to velocities as high as 10–20 km s^{-1} for H^+ and 3–5 km s^{-1} for O^+. Charge exchange of the upflowing ions with the background neutral atmosphere (thermal and hot geocoronal neutrals) acts to produce energetic streaming H and O neutrals. Both H^+ and H have sufficient energy to escape, but most of the O^+ and O atoms eventually reverse direction and head toward the Earth (Figure 12.66). A similar situation occurs in the plasmasphere. The plasma in this region co-rotates with the Earth, but it can flow along **B** from one hemisphere to the other (Section 11.5). During solstice conditions, the flow is primarily from the summer to the winter hemisphere. However, during geomagnetic storms and substorms, the outer regions of the plasmasphere are peeled away, and then the ionospheric flow is upwards from both hemispheres day and night as the plasmasphere refills, which takes about 10 days. Typically, storms and substorms occur frequently, and therefore, the outer plasmasphere is in a continual state of refilling. As the H^+ and O^+ ions flow upwards along **B**, they can exchange charge with the background neutral atmosphere, including thermal and hot geocoronal neutrals, and this acts to create energetic neutral streams in a manner analogous to what occurs in the neutral polar wind. Initially,

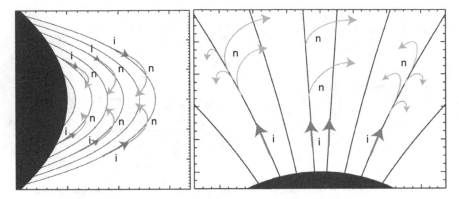

Figure 12.66 Not to scale. Upflowing ions in the plasmasphere (left) and polar wind (right). Charge exchange and the resulting downward flux of neutrals provides momentum and energy to the thermosphere.[132]

the neutral O and H atoms move upward because they acquire the velocity of their parent ions, and they form *neutral plasmaspheric streams*. However, the O atoms do not have sufficient energy to escape Earth's gravity and they subsequently follow parabolic trajectories back to Earth, as they do at high latitudes (Figure 12.66). The global distribution of downward streaming O atoms, which pass through the exobase at 500–600 km, corresponds to a continuous *neutral rain on the thermosphere*, and this could be an important source of momentum and energy for the thermosphere.[132]

12.19   Specific references

1. Lyons, L. R., Formation of auroral arcs via magnetosphere–ionosphere coupling, *Rev. Geophys.*, **30**, 93, 1992.
2. Volland, H., A model of the magnetospheric electric convection field, *J. Geophys. Res.*, **83**, 2695, 1978.
3. Sojka, J. J., W. J. Raitt, and R. W. Schunk, Effect of displaced geomagnetic and geographic poles on high-latitude plasma convection and ionospheric depletions, *J. Geophys. Res.*, **84**, 5943, 1979.
4. Rishbeth, H., and W. B. Hanson, A comment on plasma "pile-up" in the F-region, *J. Atmos. Terr. Phys.*, **36**, 703, 1974.
5. Lu, G., P. H. Reiff, J. L. Karty, M. R. Hairston, and R. A. Heelis, Distribution of convection potential around the polar cap boundary as a function of the interplanetary magnetic field, *J. Geophys. Res.*, **94**, 13447, 1989.
6. Heppner, J. P., and N. C. Maynard, Empirical high-latitude electric field models, *J. Geophys. Res.*, **92**, 4467, 1987.
7. Maezawa, K., Magnetospheric convection induced by the positive and negative z components of the interplanetary magnetic field: quantitative analysis using polar cap magnetic records, *J. Geophys. Res.*, **81**, 2289, 1976.

8. Burke, W. J., M. C. Kelley, R. C. Sagalyn, M. Smiddy, and S. T. Lai, Polar cap electric field structures with a northward interplanetary magnetic field, *Geophys. Res. Lett.*, **6**, 21, 1979.

9. Potemra, T. A., L. J. Zanetti, P. F. Bythrow, A. T. Y. Lui, and T. Iijima, By-dependent convection patterns during northward interplanetary magnetic field, *J. Geophys. Res.*, **89**, 9753, 1984.

10. Weimer, D. R., Models of high-latitude electric potentials derived with a least error fit of spherical harmonic coefficients, *J. Geophys. Res.*, **100**, 19595, 1995.

11. Weimer, D. R., An improved model of ionospheric electric potential including substorm perturbation and application to the Geospace Environment Modeling November 24, 1996 event, *J. Geophys. Res.*, **106**, 407, 2001.

12. Frank, L. A., The theta aurora, *J. Geophys. Res.*, **91**, 3177, 1986.

13. Bekerat, H. A., R. W. Schunk, and L. Scherliess, Evaluation of statistical convection patterns for real-time ionospheric specifications and forecasts, *J. Geophys. Res.*, **108**, A12, 1413, doi:10.1029/2003JA009945, 2003.

14. Schunk, R. W., The terrestrial ionosphere, in *Solar-Terrestrial Physics*, ed. by R. L. Carovillano and J. M. Forbes, **609**, Dordrecht, Netherlands: D. Reidel, 1983.

15. Hays, P. B., T. L. Killeen, N. W. Spencer, *et al.*, Observations of the dynamics of the polar thermosphere, *J. Geophys. Res.*, **89**, 5597, 1984.

16. Killeen, T. L., P. B. Hays, R. A. Heelis, W. B. Hanson, and N. W. Spencer, Neutral motions in the polar thermosphere for northward interplanetary magnetic field, *Geophys. Res. Lett.*, **12**, 159, 1985.

17. Schunk, R. W. and J. J. Sojka, Ion temperature variations in the daytime high-latitude F region, *J. Geophys. Res.*, **87**, 5169, 1982.

18. McFarland, M., D. L. Albritton, F. C. Fehsenfeld, E. E. Ferguson, and A. L. Schmettekopf, Flow-drift technique for ion mobility and ion-molecule reaction rate constant measurements, II, Positive ion reactions of N^+, O^+, and N_2^+ with O_2 and O^+ with N_2 from thermal to $\sim$2 eV, *J. Chem. Phys.*, **59**, 6620, 1973.

19. Schunk, R. W., W. J. Raitt, and P. M. Banks, Effect of electric fields on the daytime high-latitude E and F regions, *J. Geophys. Res.*, **80**, 3121, 1975.

20. St-Maurice, J.-P., and R. W. Schunk, Behavior of ion velocity distributions for a simple collision model, *Planet. Space Sci.*, **22**, 1, 1974.

21. St-Maurice, J.-P., K. Schlegel, and P. M. Banks, Anomalous heating of the polar E region by unstable plasma waves. 2, Theory, *J. Geophys. Res.*, **86**, 1453, 1981.

22. Williams, P. J. S., B. Jones, and G. O. L. Jones, The measured relationship between electric field strength and electron temperature in the auroral E region, *J. Atmos. Terr. Phys.*, **54**, 741, 1992.

23. Akasofu, S.-I., Recent progress in studies of DMSP auroral photographs, *Space Sci. Rev.*, **19**, 169, 1976.

24. Hardy, D. A., M. S. Gussenhoven, and E. Holeman, A statistical model of auroral electron precipitation, *J. Geophys. Res.*, **90**, 4229, 1985.

25. Evans, D. S., Global statistical patterns of auroral phenomena, in *Proceedings of Quantitative Models of Magnetospheric-Ionospheric Coupling Processes*, **325**, Kyoto, Japan: Kyoto University, 1987.

26. Gussenhoven, M. S., D. A. Hardy, and W. J. Burke, DMSP/F2 electron observations of equatorward auroral boundaries and their relationship to magnetospheric electric fields, *J. Geophys. Res.*, **86**, 768, 1981.

27. Hardy, D. A., M. S. Gussenhoven, and D. Brautigam, A statistical model of auroral ion precipitation, *J. Geophys. Res.*, **94**, 370, 1989.

28. Tsunoda, R. T., High-latitude F region irregularities: A review and synthesis, *Rev. Geophys.*, **26**, 719, 1988.

29. Iijima, T. and T. A. Potemra, Large-scale characteristics of field-aligned currents associated with substorms, *J. Geophys. Res.*, **83**, 599, 1978.

30. Ohtani, S., G. Ueno, T. Higuchi, and H. Kawano, Annual and semiannual variations in the location and intensity of large-scale field-aligned currents, *J. Geophys. Res.*, **110**, A01216, doi:10.1029/2004JA010634, 2005.

31. Iijima, T. and T. A. Potemra, Field-aligned currents in the dayside cusp observed by Triad, *J. Geophys. Res.*, **81**, 5971, 1976.

32. Iijima, T., T. A. Potemra, L. J. Zanetti, and P. F. Bythrow, Large-scale Birkeland currents in the dayside polar region during strongly northward IMF: a new Birkeland current system, *J. Geophys. Res.*, **89**, 7441, 1984.

33. Brinton, H. C., J. M. Grebowsky, and L. H. Brace, The high-latitude winter F region at 300 km: thermal plasma observations from AE-C, *J. Geophys. Res.*, **83**, 4767, 1978.

34. Knudsen, W. C., P. M. Banks, J. D. Winningham, and D. M. Klumpor, Numerical model of the convecting F2 ionosphere at high latitudes, *J. Geophys. Res.*, **82**, 4784, 1977.

35. Rodger, A. S., R. J. Moffett, and S. Quegan, The role of ion drift in the formation of ionization troughs in the mid- and high-latitude ionosphere – a review, *J. Atmos. Terr. Phys.*, **54**, 1, 1992.

36. Schunk, R. W. and J. J. Sojka, Ionospheric hot spot at high latitudes, *Geophys. Res. Lett.*, **9**, 1045, 1982.

37. Whitteker, J. H., L. H. Brace, E. J. Maier, *et al.*, A snapshot of the polar ionosphere, *Planet. Space Sci.*, **24**, 25, 1976.

38. Schunk, R. W., J. J. Sojka, and M. D. Bowline, Theoretical study of the electron temperature in the high latitude ionosphere for solar maximum and winter conditions, *J. Geophys. Res.*, **91**, 12041, 1986.

39. Fukui, K., J. Buchau, and C. E. Valladares, Convection of polar cap patches at Qaanaaq, Greenland during the winter of 1989–1990, *Radio Sci.*, **29**, 231, 1994.

40. Valladares, C. E., H. C. Carlson, and F. Fukui, Experimental evidence for the formation and entry of patches into the polar cap, *Radio Sci.*, **29**, 167, 1994.

41. Gondarenko, N. A. and P. N. Guzdar, Plasma patch structuring by the nonlinear evolution of the gradient drift instability in the high-latitude ionosphere, *J. Geophys. Res.*, **109**, A09301, doi:10.1029/2004JA01054, 2004.

42. Carlson, H. C., Role of neutral atmospheric dynamics in cusp density and ionospheric patch formation, *Geophys. Res. Lett.*, **24**, L13101, doi:10.1029/2007GL029316, 2007.

43. Basu, S., S. Basu, J. J. Sojka, R. W. Schunk, and E. MacKenzie, Macroscopic modeling and mesoscale observations of plasma density structures in the polar cap, *Geophys. Res. Lett.*, **22**, 881, 1995.

44. Schunk, R. W. and J. J. Sojka, Ionosphere-thermosphere space weather issues, *J. Atmos. Terr. Phys.*, **58**, 1527, 1996.

45. Sojka, J. J., M. D. Bowline, R. W. Schunk, *et al.*, Modeling polar cap F-region patches using time varying convection, *Geophys. Res. Lett.*, **20**, 1783, 1993.

46. Rino, C. L., R. C. Livingston, R. T. Tsunoda, *et al*., Recent studies of the structure and morphology of auroral zone F region irregularities, *Radio Sci.*, **18**, 1167, 1983.

47. Robinson, R. M., R. T. Tsunoda, J. F. Vickrey, and L. Guerin, Sources of F region ionization enhancements in the nighttime auroral zone, *J. Geophys. Res.*, **90**, 7533, 1985.

48. Zhu, L., J. J. Sojka, and R. W. Schunk, Active ionospheric role in small-scale aurora structuring, *J. Atmos. Solar-Terr. Phys.*, **67**, 687, 2005.

49. Valladares, C. E., H. C. Carlson, and F. Fukui, Interplanetary magnetic field dependency of stable sun-aligned polar cap arcs, *J. Geophys. Res.*, **99**, 6247, 1994.

50. Luhr, H., M. Rother, W. Köhler, P. Ritter, and L. Grunwaldt, Thermospheric up-welling in the cusp region: evidence from CHAMP observations, *Geophys. Res. Lett.*, **31**, L06805, 2004.

51. Schoendarf, J., G. Crowley, and R. G. Roble, Neutral density cells in the high latitude thermosphere – 1. Solar maximum cell morphology and data analysis, *J. Atmos. Solar-Terr. Phys.*, **58**, 1751, 1996.

52. Schlegel, K., H. Lühr, G. Crowley, and C. Hackert, Thermospheric density structures over the polar regions observed with CHAMP, *Ann. Geophys.*, **23**, 1659, 2005.

53. Demars, H. G. and R. W. Schunk, Thermospheric response to ion heating in the dayside cusp, *J. Atmos. Solar-Terr. Phys.*, **69**, 649, 2007.

54. Conde, M. and R. W. Smith, Spatial structure in the thermospheric horizontal wind above Poker Flat, Alaska, during solar minimum, *J. Geophys. Res.*, **103** (A5), 9449, doi:10.1029/97JA03331, 1998.

55. Smith, R. W., The global-scale effect of small-scale thermospheric disturbances, *J. Atmos. Solar-Terr. Phys.*, **62**, 1623, 2000.

56. Straus, J. M., R. L. Walterscheid, and K. E. Taylor, The response of the high-latitude thermosphere to geomagnetic substorms, *Adv. Space Res.*, **5**, 289, 1985.

57. Chang, C. A., and J.-P. St-Maurice, Two-dimensional high latitude thermospheric modeling: a comparison between moderate and extremely disturbed conditions, *Can. J. Phys.*, **69**, 1007, 1991.

58. Keskinen, M. J., and P. Satyanarayana, Nonlinear unstable auroral-arc driven thermospheric winds in an ionosphere-magnetosphere coupled model, *Geophys. Res. Lett.*, **20**, 2687, 1993.

59. Ma, T.-Z. and R. W. Schunk, Effect of polar cap patches on the polar thermosphere, *J. Geophys. Res.*, **100**, 19, 701, 1995.

60. Ma, T.-Z. and R. W. Schunk, The effects of multiple propagating plasma patches on the polar thermosphere, *J. Atmos. Solar-Terr. Phys.*, **63**, 355, 2001.

61. Schunk, R. W. and H. G. Demars, Effect of equatorial plasma bubbles on the thermosphere, *J. Geophys. Res.*, **108**, No. A6, 1245, 2003.

62. Schunk, R. W. and H. G. Demars, Thermospheric weather due to mesoscale ionospheric structures, *2005 Ionospheric Effects Symposium*, 799, ed. by J. M. Goodman, JMG Space Associates, 2005.

63. Cole, K. D., Electrodynamic heating and movement of the thermosphere, *Planet. Space Sci.*, **19**, 59, 1971.

64. Roble, R. G. and R. E. Dickinson, Time-dependent behavior of a stable auroral red arc excited by an electric field, *Planet. Space Sci.*, **20**, 591, 1972.

65. Wu, S. T. and K. D. Cole, Transient thermospheric heating and movement caused by an auroral electric field, *Planet. Space Sci.*, **24**, 727, 1976.

66. St-Maurice, J.-P. and R. W. Schunk, Ion-neutral momentum coupling near discrete high-latitude ionospheric features, *J. Geophys. Res.*, **86**, 11299, 1981.

67. Fuller-Rowell, T. J., A two-dimensional, high-resolution, nested-grid model of the thermosphere. 1. Neutral response to an electric field 'spike', *J. Geophys. Res.*, **89**, 2971, 1984.

68. Balthazor, R. L. and G. J. Bailey, G. J., Transonic neutral wind in the thermosphere observed by the DE 2 satellite, *J. Geophys. Res.*, **111**, A01301, doi:10.1029/2004JA010622, 2006.

69. Demars, H. G. and R. W. Schunk, Modeling supersonic flow in the high-latitude thermosphere, *J. Geophys. Res.*, submitted, 2008.

70. Prölss, G. W., Ionospheric F region storms, in *Handbook of Atmospheric Electrodynamics*, vol. **2**, ed. by H. Volland, **195**, Boca Raton, FL: CRC Press, 1995.

71. Huang, C.-S., J. C. Foster, L. P. Goncharenko, *et al.*, A strong positive phase of ionospheric storms observed by the Millstone Hill incoherent scatter radar and global GPS network, *J. Geophys. Res.*, **110**, A06303, doi:10.1029/2004JA010865, 2005.

72. Meier, R. R., G. Crowleg, D. J. Strickland, *et al.*, First look at the November 20, 2003 super storm with TIME/GUVI, *J. Geophys. Res.*, **110**, A09S41, doi:10.1029/2004JA010990, 2005.

73. Crowley, G., C. L. Hackert, R. R. Meier, *et al.*, Global thermosphere-ionosphere response to onset of 20 November 2003 magnetic storm, *J. Geophys. Res.*, **111**, A10S18, doi:10.1029/2005JA011S18, 2006.

74. Burns, A. G., W. Wang, T. L. Killeen, S. C. Solomon, and M. Wiltberger, Vertical variations in the N_2 mass mixing ratio during a thermospheric storm that have been simulated using a coupled magnetosphere-ionosphere-thermosphere model, *J. Geophys. Res.*, **111**, A11309, doi:10.1029/2006JA011746, 2006.

75. Bruinsma, S., J. M. Forbes, R. S. Nerem, and X. Zhang, Thermosphere density response to the 20–21 November 2003 solar and geomagnetic storm from CHAMP and GRACE accelerometer data, *J. Geophys. Res.*, **111**, A06303, doi:10.1029/2005JA011284, 2006.

76. Goncharenko, L., J. Salah, G. Crowley, *et al.*, Large variations in the thermosphere and ionosphere during minor geomagnetic disturbances in April 2002 and their association with IMF B_y, *J. Geophys. Res.*, **111**, A03303, doi:10.1029/2004JA010683, 2006.

77. Mishin, E. V., F. A. Marcos, W. J. Burke, *et al.*, Prompt thermospheric response to the 6 November 2001 magnetic storm, *J. Geophys. Res.*, **112**, A05313, doi:10.1029/2006JA011783, 2007.

78. Burke, W. J., C. Y. Huang, F. A. Marcos, and J. O. Wise, Interplanetary control of thermospheric densities during large magnetic storms, *J. Atmos. Solar-Terr. Phys.*, **69**, 279, 2007.

79. Akasofu, S.-I., The development of the auroral substorm, *Planet. Space Sci.*, **12**, 273, 1964.

80. Sánchez, E. R., Toward an observational synthesis of substorm models: precipitation regions and high latitude convection reversal observed in the nightside auroral oval by DMSP satellite and HF radars, *J. Geophys. Res.*, **101**, 19801, 1996.

81. Zhu, L., R. Schunk, J. Sojka, and M. David, Model study of ionospheric dynamics during a substorm, *J. Geophys. Res.*, **105**, 15 807, 2000.

82. Fox, N. J., S. W. H. Cowley, J. A. Davies, *et al.*, Ionospheric ion and electron heating at the poleward boundary of a poleward expanding substorm-disturbed region, *J. Geophys. Res.*, **106**, 12 845, 2001.

83. Akasofu, S.-I., Several "controversial" issues on substorms, *Space Sci. Rev.*, **113**, 1, 2004.

84. Liou, K., C.-I. Meng, and C.-C. Wu, On the interplanetary magnetic field B_y control of substorm bulge expansion, *J. Geophys. Res.*, **111**, A09312, doi:10.1029/2005JA011556, 2006.

85. Baker, D. N. and R. D. Belian, Impulsive ion acceleration in Earth's outer magnetosphere, in *Ion Acceleration in the Magnetosphere and Ionosphere*, ed. by T. Chang, *Geophys. Monograph*, **38**, 375, Washington, DC: American Geophysical Union, 1986.

86. Fujii, R., R. A. Hoffman, P. C. Anderson, *et al.*, Electrodynamic parameters in the nighttime sector during auroral substorms, *J. Geophys. Res.*, **99**, 6093, 1994.

87. Anderson, P. C., W. B. Hanson, R. A. Heelis, *et al.*, A proposed production model of rapid subauroral ion drifts and their relationship to substorm evolution, *J. Geophys. Res.*, **98**, 6069, 1993.

88. Bauer, S. J., The structure of the topside ionosphere, in *Electron Density Profiles in Ionosphere and Exosphere*, ed. by J. Frihagen, **52**, New York: North-Holland, 1966.

89. Dessler, A. J. and F. C. Michel, Plasma in the geomagnetic tail, *J. Geophys. Res.*, **71**, 1421, 1966.

90. Axford, W. I., The polar wind and the terrestrial helium budget, *J. Geophys. Res.*, **73**, 6855, 1968.

91. Banks, P. M. and T. E. Holzer, Features of plasma transport in the upper atmosphere, *J. Geophys. Res.*, **74**, 6304, 1969.

92. Hoffman, J. H. and W. H. Dodson, Light ion concentrations and fluxes in the polar regions during magnetically quiet times, *J. Geophys. Res.*, **85**, 626, 1980.

93. Raitt, W. J., R. W. Schunk, and P. M. Banks, A comparison of the temperature and density structure in the high and low speed thermal proton flows, *Planet. Space Sci.*, **23**, 1103, 1975.

94. Banks, P. M., Behavior of thermal plasma in the magnetosphere and topside ionosphere, in *Critical Problems of Magnetospheric Physics*, ed. by E. R. Dyer, **157**, Washington, DC: National Academy of Sciences, 1972.

95. Raitt, W. J., R. W. Schunk, and P. M. Banks, Helium ion outflow from the terrestrial ionosphere, *Planet. Space Sci.*, **26**, 255, 1978.

96. Barakat, A. R., R. W. Schunk, T. E. Moore, and J. H. Waite Jr., Ion escape fluxes from the terrestrial high-latitude ionosphere, *J. Geophys. Res.*, **92**, 12255, 1987.

97. Lemaire, J. and M. Scherer, Kinetic models of the solar and polar winds, *Rev. Geophys. Space Phys.*, **11**, 427, 1973.

98. Holzer, T. E., J. A. Fedder, and P. M. Banks, A comparison of kinetic and hydrodynamic models of an expanding ion-exosphere, *J. Geophys. Res.*, **76**, 2453, 1971.

99. Barakat, A. R., I. A. Barghouthi, and R. W. Schunk, Double-hump H^+ velocity distribution in the polar wind, *Geophys. Res. Lett.*, **22**, 1857, 1995.

100. Schunk, R. W. and J. J. Sojka, The global ionosphere-polar wind system during changing magnetic activity, *J. Geophys. Res.*, **102**, 11625, 1997.

101. Khazanov, G. V., M. W. Liemohn, and T. E. Moore, Photoelectron effects on the self-consistent potential in the collisionless polar wind, *J. Geophys. Res.*, **102**, 7509, 1997.

102. Barakat, A. R. and R. W. Schunk, Stability of H^+ beams in the polar wind, *J. Geophys. Res.*, **94**, 1487, 1989.

103. Barakat, A. R. and R. W. Schunk, Effect of hot electrons on the polar wind, *J. Geophys. Res.*, **89**, 9771, 1984.

104. Lundin, R., G. Gustafsson, A. I. Eriksson, and G. Marklund, On the importance of high-altitude low-frequency electric fluctuations for the escape of ionospheric ions, *J. Geophys. Res.*, **95**, 5905, 1990.

105. Cladis, J. B., Parallel acceleration and transport of ions from polar ionosphere to plasma sheet, *Geophys. Res. Lett.*, **13**, 893, 1986.

106. Ganguli, S. B., P. N. Guzdar, V. V. Gavrishchaka, W. A. Krueger, and P. E. Blanchard, Cross-field transport due to low-frequency oscillations in the auroral region: a three-dimensional simulation, *J. Geophys. Res.*, **104**, 4297, 1999.

107. Huddleston, M. M., C. R. Chappell, D. C. Delcourt, *et al.*, An examination of the process and magnitude of ionospheric plasma supply to the magnetosphere, *J. Geophys. Res.*, **110**, A12202, doi:10.1029/2004JA010401, 2005.

108. Winglee, R. M., W. Lewis, and G. Lu, Mapping of the heavy ion outflow as seen by IMAGE and multifluid global modeling for the 17 April 2002 storm, *J. Geophys. Res.*, **110**, A12524, doi:10.1029/2004JA010909, 2005.

109. Chen, S.-H. and T. E. Moore, Magnetospheric convection and thermal ions in the dayside outer magnetosphere, *J. Geophys. Res.*, **111**, A03215, doi:10.1029/2005JA011084, 2006.

110. Yizengaw, E., M. B. Moldwin, P. L. Dyson, B. J. Fraser, and S. Morley, First tomographic image of ionospheric outflows, *Geophys. Res. Lett.*, **33**, L20102, doi:10.1029/2006GL027698, 2006.

111. Fok, M.-C., T. E. Moore, P. S. Brandt, *et al.*, Impulsive enhancements of oxygen ions during substorms, *J. Geophys. Res.*, **111**, A10222, doi:10.1029/2006JA011839, 2006.

112. Young, D. T., Experimental aspects of ion acceleration in the Earth's magnetosphere, in *Ion Acceleration in the Magnetosphere and Ionosphere, Geophys. Monogr.*, **38**, 17, Washington, DC: American Geophysical Union, 1986.

113. Sharp, R. D., R. G. Johnson, E. G. Shelley, and K. K. Harris, Energetic O^+ ions in the magnetosphere, *J. Geophys. Res.*, **79**, 1844, 1974.

114. Lockwood, M., J. H. Waite, T. E. Moore, C. R. Chappell, and M. O. Chandler, The cleft ion fountain, *J. Geophys. Res.*, **90**, 9736, 1985.

115. Collin, H. L., W. K. Peterson, and E. G. Shelley, Solar cycle variation of some mass dependent characteristics of upflowing beams of terrestrial ions, *J. Geophys. Res.*, **92**, 4757, 1987.

116. Shelley, E. G., W. K. Peterson, A. G. Ghielmetti, and J. Geiss, The polar ionosphere as a source of energetic magnetospheric plasma, *Geophys. Res. Lett.*, **9**, 941, 1982.

117. Klumpar, D. M., W. K. Peterson, and E. G. Shelley, Direct evidence for two-stage (bimodal) acceleration of ionospheric ions, *J. Geophys. Res.*, **89**, 10779, 1984.

118. Yau, A. W., W. K. Peterson, and E. G. Shelley, Quantitative parameterization of energetic ionospheric ion outflow, in *Modeling Magnetospheric Plasma, Geophys. Monogr.*, **44**, 229, Washington, DC: American Geophysical Union, 1988.

119. Chappell, C. R., T. E. Moore, and J. H. Waite, The ionosphere as a fully adequate source of plasma for the Earth's magnetosphere, *J. Geophys. Res.*, **92**, 5896, 1987.

120. Wilson G. R., T. E. Moore, and M. R. Collier, Low-energy neutral atoms observed near the Earth, *J. Geophys. Res.*, **108** (A4), 1142, doi:10.1029/2002JA009643, 2003.

121. Wilson G. R. and T. E. Moore, Origins and variation of terrestrial energetic neutral atoms outflow, *J. Geophys. Res.*, **110**, A02207, doi:10.1029/2003JA010356, 2005.

122. Tinsley, B. A., R. R. Hodges, Jr., and R. P. Rohrbaugh, Monte Carlo models for the terrestrial exosphere over a solar cycle, *J. Geophys. Res.*, **91**, 13631, 1986.

123. Hodges, R. R. Jr., Monte Carlo simulation of the terrestrial hydrogen exosphere, *J. Geophys. Res.*, **99** (A12), 23229-23248, 10.1029/94JA02183, 1994.

124. Schunk, R. W. and J. J. Sojka, A three-dimensional time-dependent model of the polar wind, *J. Geophys. Res.*, **94**, 8973, 1989.

125. Demars, H. G. and R. W. Schunk, Seasonal and solar cycle variations of the polar wind, *J. Geophys. Res.*, **106**, 8157, 2001.

126. Demars, H. G. and R. W. Schunk, Three-dimensional velocity structure of the polar wind, *J. Geophys. Res.*, **107** (A9), 1250, doi: 10.1029/2001JA000252, 2002.

127. Demars, H. G. and R. W. Schunk, Seasonal and solar cycle variations of propagating polar wind jets, *J. Atmos. Solar-Terr. Phys.*, 68, 1791, 2006.

128. Barakat, A. R. and R. W. Schunk, A three-dimensional model of the generalized polar wind, *J. Geophys. Res.*, **111**, A12314, doi:10.1029/2006JA011662, 2006.

129. Gardner, L. C. and R. W. Schunk, Neutral polar wind, *J. Geophys. Res.*, **109** A05301, doi: 10.1029/2003JA010291, 2004.

130. Gardner, L. C. and R. W. Schunk, Global neutral polar wind model, *J. Geophys. Res.*, **110**, A10302, doi:10.1029/2005JA011029, 2005.

131. Gardner, L. C. and R. W. Schunk, Ion and neutral polar winds for northward interplanetary magnetic field conditions, *J. Atmos. Sol.-Terr. Phys.*, **68**, 1279, 2006.

132. Gardner, L. C. and R. W. Schunk, Ion and neutral streams in the ionosphere and plasmasphere, *2008 Ionospheric Effects Symposium*, 2008.

12.20   General references

Banerjee, S. and V. V. Gavrishchaka, Multimoment convecting flux tube model of the polar wind system with return current and microprocesses, *J. Atmos. Solar-Terr. Phys.*, **69**, 2071, 2007.

Banks, P. M. and G. Kockarts, *Aeronomy*, New York: Academic Press, 1983.

Brekke, A., *Physics of the Upper Polar Atmosphere*, New York: Wiley, 1997.

Burns, A. G., S. C. Solomon, W. Wang, and T. L. Killeen, The ionospheric and thermospheric response to CMEs: Challenges and successes, *J. Atmos. Solar-Terr. Phys.*, **69**, 77, 2007.

Carovillano, R. L. and J. M. Forbes, *Solar-Terrestrial Physics*, Dordrecht, Netherlands: D. Reidel, 1983.

Chang. T. and J. R. Jasperse, *Physics of Space Plasmas (1998)*, vol. **15**, Cambridge, MA: MIT, 1998.

Cravens, T. E., *Physics of Solar System Plasmas*, Cambridge, UK: Cambridge University Press, 1997.

Ganguli, S. B., The polar wind, *Rev. Geophys.*, **34**, 311, 1996.

Gavrishchaka, V. V., S. Banerjee, and P. N. Guzdar, Large-scale oscillations and transport processes generated by multiscale inhomogeneities in the ionospheric field-aligned flows: a 3-D simulation with a dipole magnetic field, *J. Atmos. Solar-Terr. Phys.*, **69**, 2058, 2007.

Hargreaves, J. K., *The Solar-Terrestrial Environment*, Cambridge, UK: Cambridge University Press, 1992.

Horwitz, J. L. *et al.*, The polar cap environment of outflowing O^+, *J. Geophys. Res.*, **97**, 8361, 1992.

Kelley, M. C., *The Earth's Ionosphere*, San Diego, CA: Academic Press, 1989.

Lemaire, J. F., W. K. Peterson, T. Chang, *et al.*, History of kinetic polar wind models and early observations, *J. Atmos. Solar-Terr. Phys.*, **69**, 1901, 2007.

Rees, M. H., *Physics and Chemistry of the Upper Atmosphere*, Cambridge, UK: Cambridge University Press, 1989.

Rishbeth, H. and O. K. Garriott, *Introduction to Ionospheric Physics*, New York: Academic Press, 1969.

Sandholt, P. E., H. C. Carlson, and A. Egeland, *Dayside and Polar Cap Aurora*, Netherlands: Kluwer Academic Publishers, Dordrecht, 2002.

Schunk, R. W., Polar wind tutorial, *Space Plasma Physics*, **8**, 81, 1988.

Schunk, R. W., Theoretical developments on the causes of ionospheric outflow, *J. Atmos. Solar-Terr. Phys.*, **62**, 399, 2002.

Schunk, R. W., Time-dependent simulations of the global polar wind, *J. Atmos. Solar-Terr. Phys.*, **69**, 2028, 2007.

Schunk, R. W. and A. F. Nagy, Electron temperatures in the F-region of the ionosphere: theory and observations, *Rev. Geophys. Space Phys.*, **16**, 355, 1978.

Schunk, R. W. and A. F. Nagy, Ionospheres of the terrestrial planets, *Rev. Geophys. Space Phys.*, **18**, 813, 1980.

Sojka, J. J., Global scale, physical models of the F region ionosphere, *Rev. Geophys.*, **27**, 371, 1989.

St-Maurice, J.-P. and R. W. Schunk, Ion velocity distributions in the high latitude ionosphere, *Rev. Geophys. Space Phys.*, **17**, 99, 1979.

Tam, S. W. Y., T. Chang, and V. Pierrard, Kinetic modeling of the polar wind, *J. Atmos. Solar-Terr. Phys.*, **69**, 1984, 2007.

Tsunoda, R. T., High-latitude F region irregularities: a review and synthesis, *Rev. Geophys.*, **26**, 719, 1988.

Yau, A. W., T. Abe, and W. K. Peterson, The polar wind: recent observations, *J. Atmos. Solar-Terr. Phys.*, **69**, 1936, 2007.

12.21   Problems

Problem 12.1 A plasma element is initially located at an altitude of 300 km, at 70° latitude and 12 MLT in a magnetic-latitude–MLT coordinate system. The plasma element then convects in an antisunward direction across the polar region along the noon–midnight meridian. Calculate the time it takes the plasma element to $\mathbf{E} \times \mathbf{B}$ drift from 70° on the day side to 70° on the night side for electric field strengths of 50, 100, and 200 mV m^{-1}. Assume the altitude does not change.

Problem 12.2 The magnetospheric electric field decreases equatorward of the auroral oval and eventually the co-rotational electric field dominates. Calculate the magnitude of the co-rotational electric field at both 60° and 70° magnetic latitude at 300 km altitude.

Problem 12.3 Consider a dipole magnetic field, as given by Equations (11.1–11.4). A plasma element is located at 300 km altitude, 70° latitude, and 12 MLT. An electric field points from dawn-to-dusk ($\perp \mathbf{B}$). Calculate the vertical component of the $\mathbf{E} \times \mathbf{B}$ drift for electric field strengths of 50, 100, and 200 mV m^{-1}.

Problem 12.4 The Earth's magnetic pole is located about 11.5° from the geographic pole in the northern hemisphere. Consider a quasi-inertial magnetic reference frame, with the magnetic pole at the center. As the magnetic pole rotates about the geographic pole, the noon–midnight meridian of the magnetic coordinate system stays aligned with the Sun. Show that co-rotation in the geographic coordinate system leads to co-rotation in the quasi-inertial magnetic reference frame.

Problem 12.5 Ion–neutral frictional heating is the process that controls the ion energy balance at F region altitudes when the convection electric field is large. Calculate T_i from Equation (12.5) for $T_n = 800$ K, $\theta = 70°$ and for effective electric fields of 100, 200, and 300 mV m^{-1}.

Problem 12.6 Calculate the rate coefficient for the $O^+ + N_2$ reaction (Equations 12.9a,b) for effective temperatures, T, of 350, 500, 1000, 2000, 3000, and 6000 K. Sketch k_1 versus T.

Problem 12.7 Calculate the effective temperature, T, using Equation (12.12) for $u_\perp(O^+) = 0.5$, 1, and 4 km s^{-1}. Assume $T(O^+) = 1000$ K, $m_r = m(N_2)$, and $m_b = m(O)$.

Problem 12.8 For Problem 12.6, calculate the $O^+ + N_2$ loss rate at 300 km for solar-minimum, winter (Table K.3) and solar-maximum, summer (Table K.5) conditions at noon.

Problem 12.9 Assume that O^+ begins to convect through an initially stationary atmosphere with a speed of 1 km s^{-1}. Calculate the ion drag force on the atmosphere at 300 km, noon, and both solar-minimum, winter (Table K.3) and solar-maximum, summer (Table K.5) conditions.

Problem 12.10 The upward H^+ flux at 1000 km is 10^8 cm^{-2} s^{-1} near the magnetic pole. Calculate the change in flux with altitude for a steady state flow in which production and loss processes are negligible. Calculate the H^+ flux at 2000, 4000, and 8000 km.

Chapter 13

Planetary ionospheres

This chapter summarizes our current understanding of the various ionospheres in the solar system. The order of presentation of the planetary ionospheres follows their position with respect to the Sun, that is, it starts with Mercury and ends with Pluto. The amount of information currently available varies widely, from a reasonably good description for Venus to just a basic guess for Pluto. In the last section of this chapter, the ionospheres of the various moons and that of Comet Halley are described. Here again the existing data are limited, with the exception of Titan, which is currently undergoing extensive exploration by the *Cassini Orbiter*.

13.1   Mercury

Mercury does not have a conventional gravitationally bound atmosphere, as indicated in Section 2.4. The plasma population caused by photo and impact ionization of the neutral constituents, which is present in the neutral exosphere, is an ion exosphere, not a true ionosphere. No quantitative calculations of the plasma densities have been carried out to date. The global Na^+ production rate was estimated to be a few times 10^{23} ions s^{-1}, but no other studies have been published and there are no observations concerning the thermal plasma densities.[1] The *Messenger* spacecraft is currently on its way and will be placed in orbit around Mercury in 2011. Our understanding of Mercury's environment will increase significantly with data from a successful *Messenger* mission.

13.2   Venus

Of all the nonterrestrial thermospheres and ionospheres in the solar system, those of Venus have been the most studied and best understood, mainly because of the

Pioneer Venus Orbiter (PVO) spacecraft, which made measurements over the 14-year period from 1978 to 1992. Comprehensive published reviews of the aeronomy of Venus are listed in the General References.

The major source of daytime ionization at Venus is solar EUV radiation. The photoionization rate peaks at around 140 km above the surface of the planet. At this altitude the major neutral atmospheric constituent is CO_2, along with about 10–20% of atomic oxygen (Figure 2.19). The predominance of CO_2 led to early predictions that the main ion in the Venus ionosphere is CO_2^+; however it was realized, even before direct measurements could confirm it, that chemical reactions quickly transform CO_2^+ to O_2^+.[2] The main chemical reactions affecting the major ion species in the altitude region where chemistry dominates (≤ 180 km) are

$$CO_2 + h\nu \rightarrow CO_2^+ + e^-, \tag{13.1}$$

$$CO_2^+ + O \rightarrow O_2^+ + CO, \tag{13.2}$$

$$\rightarrow O^+ + CO_2, \tag{13.3}$$

$$O^+ + CO_2 \rightarrow O_2^+ + CO, \tag{13.4}$$

$$O_2^+ + e^- \rightarrow O + O. \tag{13.5}$$

The last reaction, dissociative recombination of O_2^+, is the major terminal loss process for ions. A block diagram of the main ion chemistry of Venus is shown in Figure 13.1. Figure 13.2 shows modeled and measured ion densities for the day side ionosphere, indicating that (1) the peak total ion (and electron) density is near 140 km, (2) the major ion is O_2^+, (3) CO_2^+ is truly a minor ion, and (4) O^+ becomes the major ion and peaks near 200 km.[4] The figure also shows that there are many other

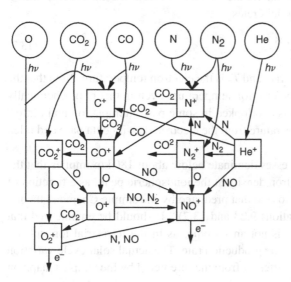

Figure 13.1 Ion chemistry scheme appropriate for Venus and Mars.[3]

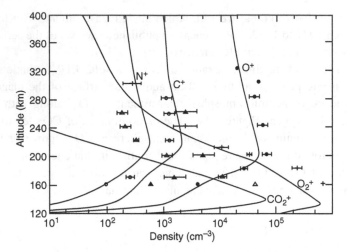

Figure 13.2 Measured and calculated daytime ion densities at Venus.[4]

ion species present, which again are the result of a large variety of photochemical processes, some of which involve metastable species.[5]

As mentioned before, the dominant ion loss mechanism is dissociative recombination of O_2^+, and this combined with the fact that O_2^+ is the major ion below about 180 km, allows one to approximate the total ion or electron loss rate as

$$L_i = k_d n_i^2, \tag{13.6}$$

where k_d is the dissociative recombination rate (Table 8.5), and n_i is the total ion density. Note that the dissociative recombination rate of O_2^+ is electron temperature dependent. In the altitude region below about 180 km, where chemical processes dominate, the following expression for the total ion or electron densities is obtained, by equating the production and loss rates

$$n_i = n_e = 277\sqrt{P_e}T_e^{0.35}, \tag{13.7}$$

where P_e is the total ionization rate and T_e is the electron temperature. Note that the electron density depends on the electron temperature even in this photochemically controlled region. This fact was overlooked in some past attempts to obtain information on the neutral gas temperature from the electron density data base and it led to incorrect conclusions.

The fact that chemical processes dominate below about 180 km implies that the variations in the day side electron density at the ionospheric peak, as a function of solar zenith angle, should be close to that predicted by the simple Chapman theory, which yields $(\cos \chi)^{1/2}$ (Equations 9.23 and 13.7).[6] It should be emphasized that the ionospheric peak at Venus is not an F_2 type, as in the terrestrial ionosphere; instead it results from a peak in the production rate. The actual solar cycle variation in the electron density peak is different from that predicted by the simple Chapman theory in terms of the $F_{10.7}$ solar flux. This is not surprising because $F_{10.7}$ is only a crude proxy for the true ionization flux and the neutral atmospheric density and temperature also change with solar activity. Furthermore, the recombination rate is a function of the electron temperature, which is also solar cycle dependent. A fit to 115 electron density profiles found the following relation for the day side peak electron density as a function of the $F_{10.7}$ flux and solar zenith angle[7]

$$n_{e,max}(F10.7, \chi) = (5.92 \pm 0.03) \times 10^5 (F_{EUV}/150)^{0.376}(\cos \chi)^{0.511},$$

(13.8)

where $F10.7$ is the value of the 10.7-cm flux, corrected to the orbital position of Venus. Finally, it should be mentioned that the altitude of the peak does not rise with solar zenith angle, as predicted by the Chapman theory (Equation 9.22), but remains near 140 km. This invariance of the peak altitude is the result of the drop of the neutral atmosphere as a function of zenith angle.[6]

Above 200 km, the chemical lifetime becomes long enough to allow transport processes (owing to diffusion or bulk plasma drifts) to dominate. Venus has no significant intrinsic magnetic field, although at times of high solar wind dynamic pressure, a significant ($\sim$100 nT) induced horizontal magnetic field is present in the ionosphere; examples of both situations are shown in Figure 13.3.[8] Note that narrow flux ropes are generally present even in the nonmagnetized situation. Given these conditions, the plasma can move freely in both vertical and horizontal directions, except when the induced field is significant. The vertical distribution of the ion or electron density near the subsolar region is believed to be controlled mainly by vertical diffusion, while horizontal plasma flows become dominant at larger zenith angles. Ion velocity measurements have indicated that the horizontal plasma

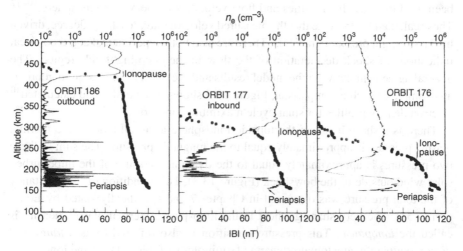

Figure 13.3 Measured altitude variations of magnetic field strength and electron densities corresponding to unmagnetized and magnetized ionospheric conditions at Venus.[8]

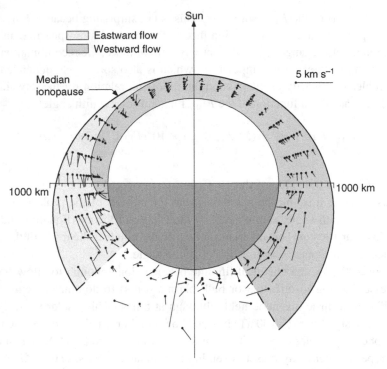

Figure 13.4 Measured ion drift velocities at Venus. Note that the altitude scale is exaggerated by a factor of four relative to the planetary radius.[9]

velocities increase with altitude and solar zenith angles, reaching a few km s^{-1} at the terminator, and becoming supersonic on the night side (Figure 13.4).[9] A variety of one- and multi-dimensional hydrodynamic and magnetohydrodynamic models have been used to study the densities and flow velocities in the Venus ionosphere.[10–12] These calculations indicate that the measured velocities are, to a large degree, driven by day-to-night pressure gradients. There are also both experimental and theoretical indications of shock deceleration of the flow in the deep night side region. The general agreement between the model results and the observations is quite good; as an example of such a comparison, Figure 13.5 shows measured[13] and calculated[10] electron density profiles for solar cycle maximum conditions.

There is a sharp break in the topside ionosphere at an altitude where thermal plasma pressure is approximately equal to the magnetic pressure. The sum of these two pressures is approximately equal to the dynamic pressure of the unperturbed solar wind outside of the bowshock (Figure 13.3). This condition of the constancy of the total pressure was discussed in Chapter 7 and specifically stated by Equation (7.51). The very sharp gradient in the ionospheric thermal plasma density is called the *ionopause*. This pressure transition is also referred to as a *tangential discontinuity* in magnetohydrodynamic terminology (Table 7.1). At the ionopause, there is a transition from an ionospheric plasma pressure to a magnetic pressure dominated region in an altitude increment of only a few tens of kilometers when the

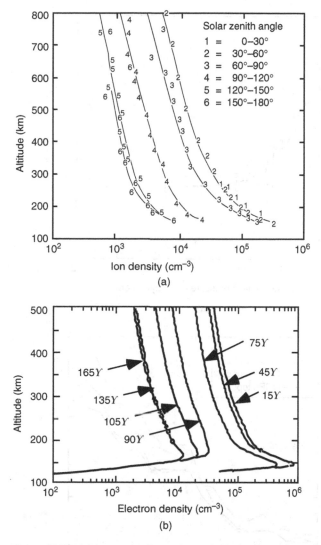

Figure 13.5 (a) Measured solar cycle maximum ion (electron) densities as a function of zenith angle.[13] (b) Calculated solar cycle maximum electron densities as a function of zenith angle.[10]

ionosphere is not magnetized. However, the transition region is much broader in the case of a magnetized ionosphere, as seen in Figure 13.3. Given that the ionopause is at an altitude where the ionospheric thermal pressure balances the solar wind dynamic pressure, its location must change as the solar wind and ionospheric conditions change. For example, as the solar wind pressure increases, the ionopause height decreases, but it actually levels off at around 300 km when the pressure exceeds about 4×10^{-8} dyne cm^{-2}. Also, the mean ionopause height rises from about 350 km at the subsolar location to about 900 km at a solar zenith angle of 90°, as indicated in Figure 13.6.[14]

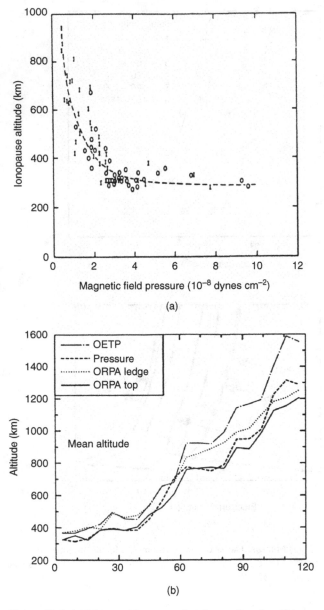

Figure 13.6 (a) Measured ionopause height as a function of the magnetic pressure above the ionopause.[14] (b) Measured ionopause height as a function of zenith angle.[14]

The effective night on Venus lasts about 58 Earth days (the solar rotation period is 117 Earth days), during which time the ionosphere could be expected to disappear because no new photoions and electrons are created to replace the ones lost by recombination. Therefore, it was very surprising, at first, when *Mariner 5* found a significant night side ionosphere at Venus.[15] Subsequently, extensive measurements have confirmed the presence of a significant, but highly variable, night side ionosphere. Figure 13.5 shows the average measured night side electron densities during solar cycle maximum conditions. Plasma flow from the day side, along with impact ionization caused by precipitating electrons, is responsible for the observed nighttime densities. The relative importance for a given ion species depends on the solar wind pressure and solar conditions (e.g., during solar cycle maximum conditions day-to-night transport is the main source of plasma for the night side ionosphere.[16]) It must be emphasized that the electron density profiles shown in Figure 13.5 are mean values. The night side electron densities are extremely variable both with time and location. Order of magnitude changes have been seen by the instruments on *PVO* along a single path through the ionosphere and subsequent passes. Terms such as *disappearing ionospheres*, *ionospheric holes*, *tail rays*, *troughs*, *plasma clouds*, etc., have been introduced to classify the apparently different situations encountered. For example, Figure 13.7 shows two ionospheric holes observed during orbit 530 of *PVO*.[17] Strong radial magnetic fields found to be present in these holes allow easy escape of the ionospheric thermal plasma into the tail, presumably causing the sharp drops in density.

The observed solar cycle maximum ion and electron temperatures, for different solar zenith angle increments, are shown in Figure 13.8.[13] These plasma temperatures are significantly higher than the neutral gas temperature (Figure 2.20) and cannot be explained in terms of EUV heating and classical thermal conduction, as is the case for the mid-latitude terrestrial ionosphere (Section 11.6). The two

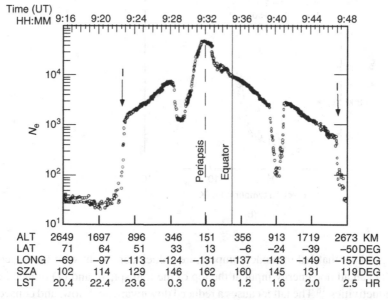

Figure 13.7 Electron densities measured by the electron temperature probe carried aboard the *Pioneer Venus Orbiter* during Orbit 530. The satellite altitude (ALT), latitude (LAT), longitude (LONG), solar zenith angle (SZA), and local solar time (LST) are given along the abscissa. The location of the ionopauses encountered during entry and exit of the ionosphere are indicated by *I*.[17]

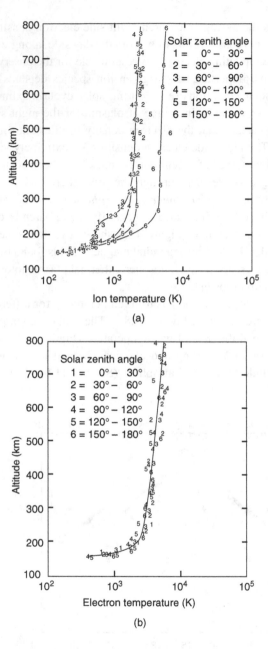

Figure 13.8 (a) Measured solar cycle maximum ion temperatures as a function of zenith angle. (b) Measured solar cycle maximum electron temperatures as a function of zenith angle.[13]

suggestions that led to model temperature values consistent with the observations are (1) an ad hoc energy input at the top of the ionosphere and (2) reduced thermal conductivities.[18] The latter causes a reduced downward heat flow, and consequently, a decreased energy loss to the neutrals at the lower altitudes. There are reasons to believe that both mechanisms are present, but there is insufficient information available to establish which is dominant, when, and why. Measurements by the *PVO* plasma wave instrument indicate significant wave activity at and above the

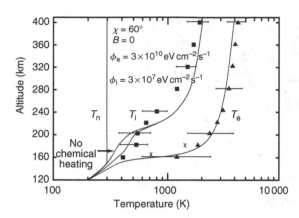

Figure 13.9 Measured (solid squares and triangles) and calculated (solid lines) electron and ion temperatures for zero magnetic field and 60° solar zenith angle. The assumed heat inputs at the upper boundary are indicated.[19]

ionopause, and different estimates of the heat input into the ionosphere, from these waves, all lead to values of the order of 10^{10} eV cm^{-2} s^{-1}, which is about the magnitude necessary to explain the observed plasma temperatures.[19] Some studies also suggest that the shocked solar wind plasma from the tail region can move along draped field lines toward the ionosphere, when it is magnetized, and provides the necessary energy to explain the observed temperatures.[20] Other studies claim that a reduction in the thermal conductivity from its classical value (Equation 5.146) can be justified because of the presence of fluctuating magnetic fields in the ionosphere. The associated conductivity values result in temperatures consistent with the measured ones.[19] A one-dimensional model calculation used classical electron and ion conductivities and topside electron and ion heat flows of 3×10^{10} and 3×10^{7} eV cm^{-2} s^{-1}, respectively, to produce temperatures reasonably close to the observed values (Figure 13.9). On the other hand, a one-dimensional model calculation that assumed no topside heat inflow, but incorporated reduced thermal conductivities resulting from magnetic field fluctuations, also led to calculated values consistent with the measured ones (Figure 13.10). The parameter λ in Figure 13.10 is the correlation length of the assumed fluctuations. Note that while the electrons are strongly affected by these fluctuations, the effect is smaller on the ions; this is the result of the significant difference in the respective gyroradii.

A small bump observed in the day side ion temperatures just below 200 km can be accounted for by considering either chemical or Joule heating processes. Also, the mechanisms controlling the temperatures on the night side are even less understood. It is certainly reasonable to assume that energy is transported from the day side to the night side by heat flow and advection and that heat input from above or from the tail is also present.[21] However, the specific roles of these different potential energy sources have not been elucidated. It was also observed that the H$^+$ temperatures are lower than the O$^+$ ones on the night side. This appears to be caused by the differences in thermal conductivities resulting from ion–neutral collisional effects,[22] similar to the electron thermal conductivity situation (Equation 5.146).

At this time, there is no clear understanding of the mechanisms controlling the energetics of the ionosphere of Venus,[18] and further progress is unlikely until more

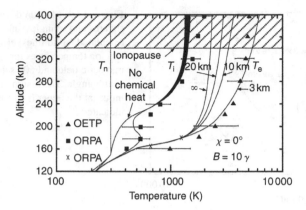

Figure 13.10 Measured (solid squares and triangles) and calculated (solid lines) electron and ion temperatures for zero heat input, 0° solar zenith angle and a constant magnetic field of 10 nT. The influence of the assumed fluctuating magnetic field is indicated by the various effective mean-free-paths.[19]

direct information becomes available from either Venus or, possibly, Mars, because of the assumed similarities between the two planets. It is clear that conventional EUV heating and classical thermal conductivity lead to temperature values well below the observed ones. To remedy this situation, the two main suggestions invoke either additional heat sources or reduced thermal conductivities. Unfortunately, there is insufficient direct information to establish their validity and distinguish between these hypotheses. It is most likely that both processes play a role, but whether one or the other dominates is unclear. Also, it is not known whether other processes, not yet considered, are important.

13.3   Mars

The ionosphere of Mars is believed to be similar to that of Venus, but the amount of information available concerning its structure and behavior is much more limited. Except for two vertical profiles of ion densities and temperatures and electron temperatures, obtained from the retarding potential analyzers (Section 14.3) carried by the two *Viking* landers,[23, 24] all the information concerning the ionospheric properties of Mars comes from radio occultation or radar observations (see Sections 14.6 and 14.7) by the various US, ESA, and USSR Mars missions.

Photochemical processes control the behavior of the main ionospheric layer of Mars, just as is the case for Venus; the block diagram of the main ion chemistry shown in Figure 13.1 applies to both Venus and Mars. A typical value for the day side peak plasma density is about of the order of 10^5 cm^{-3} and the height of the maximum is about 135 km. One of the ion density profiles obtained by the *Viking* lander *retarding potential analyzer* (RPA) instrument is shown in both Figures 2.23 and 13.11. These measurements clearly established that the principal ion in the Martian ionosphere

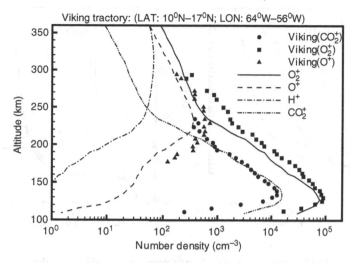

Figure 13.11 Ion density profiles measured by the *Viking 1* RPA[23] and calculated in a self-consistent manner by a three-dimensional MHD model.[25]

does not correspond to the main ionizable neutral constituent, CO_2, but is O_2^+, in a manner totally analogous to the Venus ionosphere. The *Viking* RPA results also established the presence of CO_2^+ and O^+, with O^+ becoming comparable in concentration to that of O_2^+ at altitudes at and above about 300 km (Figures 2.23 and 13.11).

Analysis of the appropriate time constants and more sophisticated models indicate that transport processes become more important than photochemistry somewhere between 170 to 200 km in the day side ionosphere of Mars. Comprehensive models have been developed to describe the behavior of the ionosphere in both the photochemical and transport controlled regions. One such model[26] solved the coupled continuity, momentum, and energy equations to study the chemistry and energetics of the Martian ionosphere. This model was successful in matching the observed ion densities, as indicated in Figure 2.23. However, remember that no direct measurements of the neutral atomic oxygen density have been made, therefore, the oxygen profile used in this model was obtained by forcing a best fit to the observed ion densities. Furthermore, the calculations were carried out using mixed upper-boundary flow conditions. The model results are relatively independent of these upper boundary values below about 250 km, but at higher altitudes the densities depend strongly on transport, for which no data are available at this time. A more recent model solved the coupled continuity and momentum equations and used measured electron and ion temperatures.[27] A good fit to the data was also achieved in these calculations as long as significant upward fluxes were assumed to be present at the upper boundary. A three-dimensional, multi-species, magnetohydrodynamic (MHD) model developed to study the solar wind interaction with Mars, has been successful in reproducing the observed density profiles, without any ad hoc assumptions regarding ionospheric fluxes, as shown in Figure 13.11.[25]

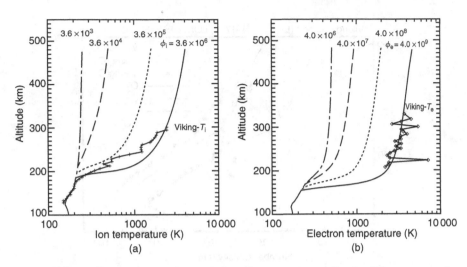

Figure 13.12 Calculated ion and electron temperature profiles for different assumed topside heat inflows. The ion and electron temperatures measured by the *Viking 1* RPA[22] are also shown by the curves marked with crosses and diamonds, respectively.[28]

The daytime ion and electron temperatures measured by the RPAs[24] depart from the neutral gas temperature ($T_n \sim 200$ K; Section 2.4) at altitudes above the ionospheric peak ($\sim$135 km) (Figure 13.12). Just as for the Venus ionosphere, EUV heating alone predicted ion and electron temperatures considerably lower than the measured ones.[26, 28] The one-dimensional models constructed to study the energetics of the ionosphere of Mars came to similar conclusions as those reached for Venus. Namely, that to arrive at electron and ion temperatures consistent with the RPA measured day side values either a topside heat source or reduced thermal conductivities must be invoked.[26, 28] Figure 13.12 shows calculated ion and electron temperatures, which were obtained assuming different topside ion heat inflows and which led to temperature values close to the measured ones.[28]

A very large number of ionospheric electron density profiles have been obtained by radio occultation methods (Section 14.6), mainly from using the *Viking Orbiter*, *Mars Global Surveyor* (MGS) and *Mars Express* (MEX) spacecrafts. The density profiles thus obtained are basically limited to altitudes below about 300 km, but nevertheless they have provided a great deal of information concerning ionospheric variables[29, 30] and have been used to provide indirect information on the thermosphere.[31] The MEX radio occultation observations also established the presence of a sporadic third electron density peak in the altitude range between 65 and 110 km, below the main peak which is around 135 km and a secondary peak around 110 km.[32]

Mars Express spacecraft carried a radar system designed to study both the subsurface and the ionosphere of Mars. The topside ionospheric electron density profiles thus obtained extend to about 400 km and provide information beyond what was available earlier.[33] These measurements, as well as earlier radio occultation data,[34]

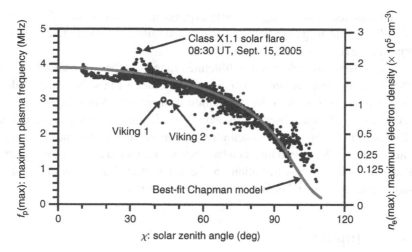

Figure 13.13 Solar zenith angle variation of the peak electron densities measured by the *Mars Express* radar system[33] (also indicated is the theoretical Chapman variation of $[\cos \chi]^{0.5}$).

showed that the variation of the electron density peak with solar zenith angle is close to that predicted by the ideal Chapman theory ($[\cos \chi]^{0.5}$; see Equations 9.23 and 13.7); the *MEX* radar results are shown in Figure 13.13.

There is only limited information available on the densities in the night side ionosphere of Mars. The radio occultation measurements obtained by *Mars 4* and *5* and *Viking 1* and *2* showed that the observed peak densities are highly variable; at times none were detected. The highly variable structure, along with ionospheric holes, observed by the *Mars Express* radar,[33] are very reminiscent of the Venus situation. The mean peak density is about 5×10^3 cm^{-3}, with a peak altitude around 160 km. The rotation period of Mars is relatively short, close to that of Earth, therefore, the observed small densities do not seem to be especially difficult to account for. There are some indirect indications that electron impact ionization may be an important nighttime ionization source, as well as day-to-night transport processes, again in a manner similar to the Venus situation. Auroral emission have been seen by the spectrometer on *Mars Express*,[35] which was explained to be the result of accelerated electrons,[36] consistent with the suggestion that impact ionization is present.[37]

The question of whether an intrinsic magnetic field is present at Mars was debated for many years. Observations prior to 1997 established that the mean intrinsic field, if present, must be very weak, leading to a field of less than about 40 nT in the ionosphere. The *Mars Global Surveyor* (MGS) was the first spacecraft that carried a magnetometer (see Section 14.5), which made measurements deep in the ionosphere, thus capable of resolving this question unambiguously. These measurements clearly established the presence of localized patches of relatively strong crustal magnetic fields, but no intrinsic field of significance ($< 2 \times 10^{11}$ T m^3).[38] These crustal fields are located mainly, but not exclusively, in the southern hemisphere. Good models of these measured fields, using harmonic expansion, are available.[39, 40] The presence of these crustal magnetic fields means that significant spatial structures must be present, the result of these changing fields.

The presence of a clear and well-defined ionopause has been established by the *PVO* measurements at Venus. This is not the case at Mars. No clear evidence of an ionopause was seen in the occultation data prior to *Mars Express*. The radio occultation and radar observations with *Mars Express*, provide some indications of an ionopause, but so far no definitive or consistent detection has been reported. Finally, the *MGS* electron flux data has shown a transition layer in the suprathermal electron fluxes and this transition has been interpreted as the boundary between shocked solar wind and planetary ionospheric plasma.[41]

13.4   Jupiter

The presently available direct information regarding the ionosphere of Jupiter is based on the *Pioneer 10* and *11*, *Voyager 1* and *2*, and *Galileo* radio occultation measurements. Some indirect information, mainly optical remote sensing observations, also provide insight into certain ionospheric processes. Given that Jupiter's upper atmosphere consists mainly of molecular hydrogen, as indicated in Section 2.5, the major primary ion, which is formed by either photoionization or particle impact, is H_2^+. In the equatorial and low-latitude regions, electron–ion pair production is due mainly to solar EUV radiation, while at higher latitudes impact ionization by precipitating particles is believed to become very important. The actual equilibrium concentration of the major primary ion, H_2^+, is very small because it undergoes rapid charge transfer reactions. The rest of the discussion in this section is based, for the sake of brevity, on photoionization and photodissociation only, because particle ionization leads to similar products and processes. Solar radiation, with appropriate wavelengths, leads to

$$H_2 + h\nu \rightarrow H + H, \tag{13.9}$$

$$\rightarrow H_2^+ + e^-, \tag{13.10}$$

$$\rightarrow H^+ + H + e^-. \tag{13.11}$$

The resulting neutral atomic hydrogen can also be ionized:

$$H + h\nu \rightarrow H^+ + e^-. \tag{13.12}$$

At high altitudes, where hydrogen atoms are the dominant neutral gas species, H^+ can only recombine directly via radiative recombination, which is a very slow process (Table 8.4). It was suggested that H^+ could charge exchange with H_2 excited to a vibrational state of $v > 4$.[42] The vibrational distribution of H_2 is not known, but calculations indicate that the vibrational temperature is elevated at Jupiter;[43, 44] however, it is not very clear how important this effect is.

H_2^+ is very rapidly transformed into H_3^+, especially at the lower altitudes where H_2 is dominant:

$$H_2^+ + H_2 \rightarrow H_3^+ + H. \tag{13.13}$$

H_3^+ is likely to undergo dissociative recombination

$$H_3^+ + e^- \rightarrow H_2 + H. \tag{13.14}$$

Significant uncertainties have been associated with the dissociative recombination rate of H_3^+. However, recent measurements have shown that the rate is rapid (Table 8.5), even if the ion is in its lowest vibrational state.[45]

The primary ions in the middle ionosphere can be rapidly lost by reactions with upflowing methane. However, the importance of this process depends on the rate at which methane is transported up from lower altitudes, which in turn depends on the eddy diffusion coefficient, which is not well known. Direct photoionization of hydrocarbon molecules at lower altitudes can lead to a relatively thin hydrocarbon ion layer around 300 km.[46]

The early hydrogen-based models predicted an ionosphere composed predominantly of H^+ because of its long lifetime ($\sim 10^6$ s). In these models H^+ is removed by downward diffusion to the vicinity of the homopause (~ 1100 km), where it undergoes charge exchange with heavier gases, mostly hydrocarbons such as methane. The hydrocarbon ions, in turn, are lost rapidly via dissociative recombination. The *Voyager* and *Galileo* electron density profiles indicated the presence of an ionosphere with peak densities between 10^4 and 10^5 cm^{-3}, as indicated in Figures 2.25 and 13.14. These electron density profiles seem to fall into two general classes. One group has the peak electron density located at an altitude around 2000 km and the other group has the electron density peak near 1000 km. The two groups also exhibit different topside scale heights, with the high-altitude peaks associated with the larger

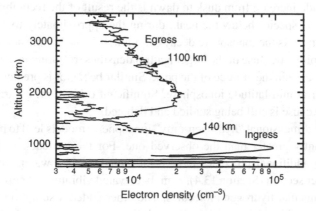

Figure 13.14 Measured electron density profiles of Jupiter's ionosphere near the terminator.[47]

scale heights. There appears to be no clear latitudinal or temporal association with these separate groups of profiles. The different peaks may be the result of a combination of different major ionizing sources (EUV versus X-ray or particle impact) and different ion chemistries. A number of different models of the ionosphere, both one- and multi-dimensional ones,[48-50] have been developed. The limited observational data base, combined with significant variations in the observed density profiles (as shown in Figures 2.25 and 13.14), along with the large uncertainties associated with such important parameters as the relevant reaction rates, drift velocities, degree of vibrational excitation and the magnitude and nature of the precipitating particles, means that there are too many free parameters to allow a definitive model of the ionosphere to be developed at this time.

It has been a fairly well-accepted assumption that the major source of Jupiter's magnetospheric plasma is Io.[51] However, the question has been raised some time ago whether the ionosphere, as in the case of the Earth, could also be a significant contributor. Calculations[52] and recent observations[53] have shown that the ionosphere may also make a significant contribution.

13.5   Saturn, Uranus, Neptune, and Pluto

A handful of electron density profiles of Saturn were obtained by the *Pioneer 11* and *Voyager 1* and *2* spacecraft using radio occultation techniques more than two decades ago. The number of such density profiles available has increased significantly with the insertion of *Cassini* in orbit around Saturn.[54] So far the *Cassini* data has been restricted to near equatorial latitudes, but it has still been helpful in advancing our understanding of Saturn's ionosphere. Figure 13.15 shows the averaged equatorial electron density profiles for dawn and dusk conditions (note that orbital conditions only allow near dusk and dawn occultations). The peak densities for the dusk and dawn results are 5.4×10^3 and 1.7×10^3 cm^{-3}, respectively, while the corresponding altitudes are 1880 and 2360 km, respectively. It has been suggested that the density decrease and altitude increase from dusk to dawn is the result of the recombination of the molecular ion species below the peak, during the approximately five hours of Saturn's nighttime. As the ionosphere decays at night, it is the bottomside which decreases more rapidly, because of the larger neutral densities, resulting in a drop in density and an increase in the altitude of the peak. Similar behavior is predicted and seen in the terrestrial mid-latitude ionosphere. Significant orbit to orbit variations were seen, but the cause is still being studied and debated.

The simple one-dimensional "hydrogen-only" ionospheric models lead to plasma densities significantly greater than the observed one. For these models to fit the observations, two additional processes have been included. One was mentioned earlier in the Jupiter section (Section 13.4), namely elevated vibrational temperature for the neutral molecular hydrogen molecules. The other added assumption is that water from the rings is transported into Saturn's upper atmosphere, which then modifies the chemistry in the ionosphere.[55] The presence of H_2O results in the

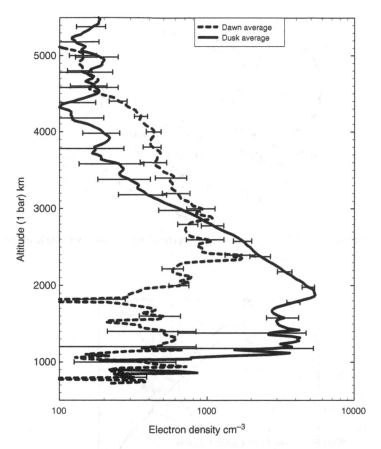

Figure 13.15 Average dawn and dusk, near equatorial electron density profiles obtained from radio occultation measurements at Saturn by *Cassini*.[54]

following catalytic process:

$$H^+ + H_2O \rightarrow H_2O^+ + H, \tag{13.15}$$

$$H_2O^+ + H_2O \rightarrow H_3O^+ + OH, \tag{13.16}$$

$$H_3O^+ + e^- \rightarrow H_2O + H. \tag{13.17}$$

A block diagram of the chemistry scheme, involving water is shown in Figure 13.16. The results of such a one-dimensional model calculation[56] along with the observed density profile is shown in Figure 13.17. There are no data indicating what the actual ion composition is in Saturn's ionosphere. There are differences in the various model predictions,[57, 58] but in general they predict H^+ to dominate at the higher altitudes and the presence of significant population of H_3^+ and H_3O^+ ions at the lower altitudes.

The low-frequency cut-off of the Saturn electrostatic discharges (SEDs) observed by *Voyager*, which originate in the equatorial atmosphere from lightning, was used to

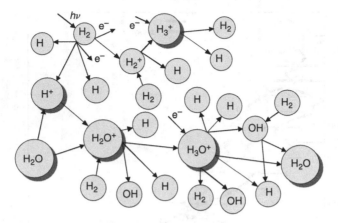

Figure 13.16 A block diagram of the ion chemistry scheme involving hydrogen and water.

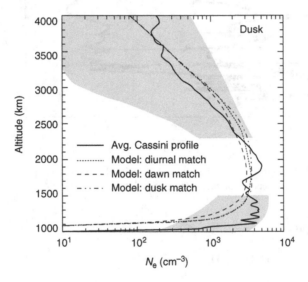

Figure 13.17 Measured and calculated electron densities at Saturn.[56]

provide indirect information on the diurnal variation in the electron density peaks.[59] This work indicated that the diurnal variations exceed two orders of magnitude, which no model came even close to being able to explain. The dawn-to-dusk variations seen by *Cassini* radio occultation are only about a factor of three; clearly the midday to midnight variations are greater, but still much less than what was implied by the SEDs. A recent paper[60] suggested that the interpretation of the SEDs in terms of peak electron densities may be incorrect, because of ring shadowing effects. However, some new *Cassini* results[61] contradict this suggestion; thus this issue is still open.

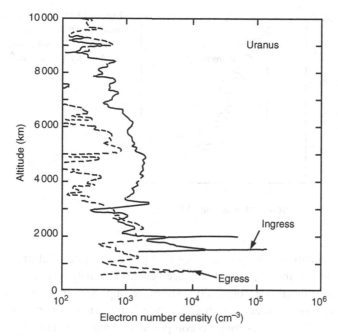

Figure 13.18 Electron density profile obtained from radio occultation measurements at Uranus by *Voyager 2*.[65]

Besides the conventional ionosphere of Saturn, *Cassini* ion–neutral mass spectrometer also detected an ionosphere, in the broadest sense, around the A-ring of Saturn.[62] It saw signatures of molecular and atomic oxygen ions and of protons. The likely explanation for these ions is the photo or impact ionization of neutral molecular oxygen associated with a tenuous ring atmosphere or cloud, which in turn is probably the result of energetic ion sputtering from the cloud.

Finally, it should be mentioned that the question whether Saturn's ionosphere is a significant source of plasma for the magnetosphere has also been raised, as it was for Jupiter. Here again, model calculations indicate that it may make a significant contribution,[63] although Enceladus and the rings are likely to be the major sources.[64]

The only direct information concerning the ionospheres of Uranus and Neptune come from the *Voyager 2* radio occultation measurements. The ionospheric densities measured at the two planets are shown in Figures 13.18 and 13.19 respectively.[65, 66] The observed day side UV emissions[67] from Jupiter, Saturn, and Uranus indicate that a column integrated energy flux of about 0.1–0.3 erg cm^{-2} s^{-1}, due to soft (< 15 eV) electrons, may be present; this has been referred to as *electroglow*. However, alternative explanations of the observed emissions have also been put forward.[68] Simple one-dimensional model calculations of the ionospheres of Uranus and Neptune have been published; some of them included ionization caused by the electroglow electrons. All the calculated peak electron densities found by these models exceeded the measured values; this result has been interpreted as an indication of a significant influx of water molecules, similar to the situation of Saturn.

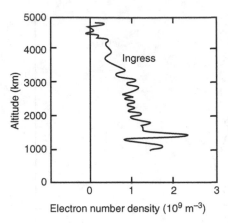

Figure 13.19 Electron density profile obtained from radio occultation measurements at Neptune by *Voyager 2*.[66]

The radio occultation data from all the giant planets (Jupiter, Saturn, Uranus, and Neptune) indicate the presence of enhanced electron density layers in the lower ionosphere. These layers are potentially extremely important in establishing the integrated ionospheric conductivities, because they lie in the appropriate ion–neutral collision regime. These layers may be composed of long-lived metallic ions of meteoric or satellite origin, somewhat analogous to the terrestrial sporadic E layers, formed by wind shears (Section 11.13). Gravity waves may also play a role in creating these narrow, multiple layers[47] (Section 10.5).

No measurements of Pluto's ionosphere have been obtained to date. Model calculations suggest that the peak ionospheric density is less than $10^3\,cm^{-3}$ and that the major ions are likely to be $HCNH^+$ and CH_5^+.[69] However, as suggested for Titan (Section 13.6), more complex hydrocarbon molecules may be synthesized, because of their higher proton affinity.

13.6   Satellites and comets

The first direct indication of an ionosphere around Io was the radio occultation observations by *Pioneer 10* in 1973.[70] A number of further electron density profiles have been obtained using *Galileo* radio occultation measurements and representative results are shown in Figure 13.20.[71] The densities vary dramatically between the leading and trailing hemispheres, showing the influence of the rapidly flowing torus plasma on Io's atmosphere and ionosphere. Plasma near Io's equatorial plane was observed to be moving away from Io at high velocities, which increased from $30\,km\,s^{-1}$ at three Io radii up to the co-rotation speed of $57\,km\,s^{-1}$ at seven Io radii.

The vapor pressure of SO_2 exhibits such a strong dependence on temperature that an atmosphere in equilibrium with surface frost could result in many orders of magnitude difference between day and night atmospheric densities. This picture would be considerably modified if significant amounts of a noncondensing gas, such as O_2, were present. Another outstanding issue is the source of atmospheric

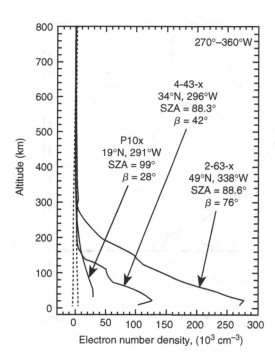

Figure 13.20 Electron density profiles obtained from radio occultation measurements at Io by *Galileo*. Dashed lines show typical ± sigma uncertainties due to thermal noise. The latitude, longitude, solar zenith angle, and β, the angle measured from the center of the upstream hemisphere, corresponding to each profile are shown.[71]

Na species. Potential sodium species, such as NaO_2 and Na_2O, which have lower saturation vapor pressures than SO_2, do not sublime easily. It is also very important to recognize the fact that the interaction of the *Galilean satellites* with the magnetosphere of Jupiter is certain to influence the nature and variability of the respective ionospheres. Given all these uncertainties, ionospheric modeling is very uncertain because of the lack of constraints on many of the crucial parameters. Figure 13.21 shows the results of a specific model,[72] which leads to a reasonable agreement with the *Pioneer 10* day side profile. Finally, it should be emphasized that Io's atmosphere and ionosphere are very likely to be highly variable, both spatially and temporally, given the nature of the volcanic sources. Ionospheres have also been detected by the *Galileo* radio occultation measurements at Europa, Ganymede, and Callisto. The peak electron densities are seen near the surface and have values of about 1, 0.4, and 0.1×10^4 cm^{-3}, respectively.[73, 74]

Titan, the largest satellite of Saturn, is surrounded by a substantial atmosphere and, therefore, one expects a correspondingly significant ionosphere. Until recently, when the *Cassini* spacecraft began its orbital mission around Saturn, the only information concerning Titan's ionosphere came from the *Voyager 1* radio occultation observation. After a careful reanalysis of these data, an electron density peak of about 2.7×10^3 cm^{-3} at an altitude near 1190 km, for a solar zenith angle of 90° was found.[75] The *Cassini* spacecraft has now observed this ionosphere on numerous occasions, using a Langmuir probe,[76] an ion–neutral mass spectrometer,[77] and the radio occultation technique.[78] These measurements reconfirmed the presence of a day side ionosphere with a peak density of a few times 10^3 cm^{-3} at an altitude around

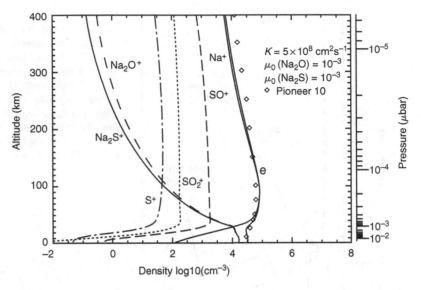

Figure 13.21 Calculated ion densities for Io. The day side electron density profile obtained from radio occultation measurements by *Pioneer 10* is also shown for comparison.[72]

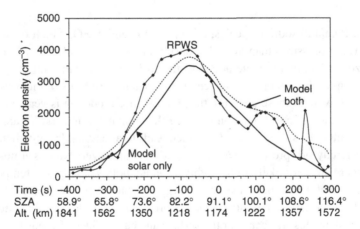

Figure 13.22 Comparison of measured and calculated electron densities along the T_a flyby of Titan by *Cassini*. The calculations assumed chemical equilibrium conditions. One of the calculations considered only the solar sources of plasma, while the other also included impact ionization by magnetospheric electrons.[81]

1200 km. The radio occultation observations also found the presence of a shoulder in the electron density profiles around 1000 km, caused by solar X-ray ionization,[78] and an intermittent, but significant second density peak in the 500–600 km region, the likely result of energetic particle precipitation[79] or ablation of meteorites.[80] Figure 13.22 shows the results of some of these measurements. As mentioned in Section 2.6 and indicated in Figure 13.23 given the orbit of Titan around Saturn and the fact that its orbital speed is less than the co-rotation velocity it is rammed

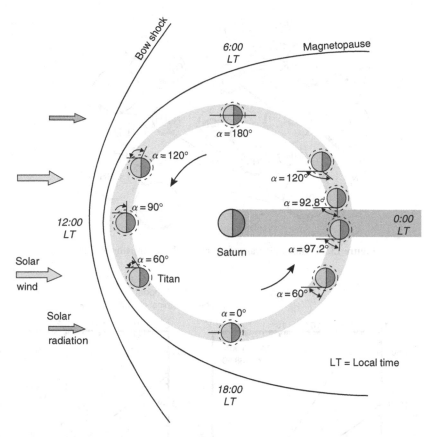

Figure 13.23 Diagram indicating the changing solar illumination and ram direction at Titan as it orbits around Saturn (courtesy of H. Backes).

by the magnetospheric plasma. This means that the ram side is at times sunlit, in the dark or somewhere in between, leading to the dominance of different ionization sources as the orbit changes. Under nominal conditions the calculations indicate that photoionization is the main source of ionization for the day side ionosphere,[81] followed by photoelectron impact, and finally impact by magnetospheric electrons. Of course, magnetospheric electrons must dominate in the night side ionosphere. These comments apply to the case when Titan is inside the magnetosphere of Saturn, but in the case of high solar wind pressure Titan may find itself in the magnetosheath, or rarely even in the unshocked solar wind and the importance of impact ionization is likely to increase significantly.

A variety of one- and multi-dimensional models have been developed which calculate the electron density profiles, and they all lead to reasonably good agreement with the observed values.[81, 82] An example of calculated and observed densities, corresponding to a Titan flyby is shown in Figure 13.22. The major initial ionospheric ion, given Titan's neutral atmosphere, is N_2^+ (see Figure 2.28), but these ions quickly undergo a number of ion–neutral reactions, leading to $HCNH^+$ and other more complex hydrocarbon ions, as indicated in Figure 13.24.

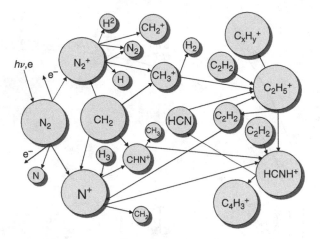

Figure 13.24 Block diagram of the major chemical paths of Titan.

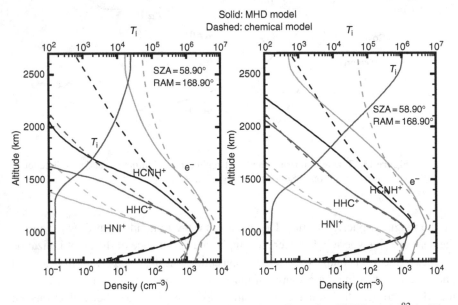

Figure 13.25 Ion densities calculated by a three-dimensional MHD model[82] (solid lines) and a comprehensive one-dimensional chemical equilibrium model[81] (dashed lines).

A variety of studies looked at the issue of the transition from chemical to transport control in the ionosphere of Titan. Simple time constant considerations, as well as detailed multi-dimensional calculations, have indicated that the transition from chemical to diffusive control takes place in the altitude region around 1400–1500 km. The comparison between chemical equilibrium and a three-dimensional MHD model calculation is shown in Figure 13.25, and it clearly indicates where the chemical equilibrium assumption breaks down. In general, the magnetospheric plasma velocity ($\sim$ 120 km s^{-1}) is just below the magnetosonic velocity and thus no bow shock forms and the magnetospheric plasma slows gradually as it moves through Titan's exosphere, by mass loading. Titan has no intrinsic magnetic field, so as the magnetospheric plasma moves in, carrying Saturn magnetic field, a pile-up results and eventually the field drapes around Titan. Of course, there are times when the plasma flow is fast enough for a weak shock to form. The situation at Titan has many similarities to the solar wind interaction with Venus and Mars.

The plasma temperatures in the ionosphere of Titan have been a topic of interest for some time before *Cassini*'s arrival. Model studies have indicated that the expected plasma temperatures will be significantly higher on the ram side than on the anti-ram side.[83] This is because the draped magnetic field is nearly horizontal and thus inhibits vertical heat flow. The *Langmuir* probe carried aboard the *Cassini* spacecraft[76] has provided the first direct measurements of electron temperatures at Titan. The measured altitude variation of the electron temperature from an orbit near the terminator and about 90° with respect to the ram is shown in Figure 13.26. A model,[84] which used the measured suprathermal electron fluxes to calculate heating rates and then substituted these values into an electron energy equation, was successful in obtaining values consistent with the measured temperatures in the 1200–1300 km altitude region.

As indicated in Section 2.6 there is a significant neutral gas plume originating in the southern hemisphere of Enceladus, and thus the ionization of this gas, mainly

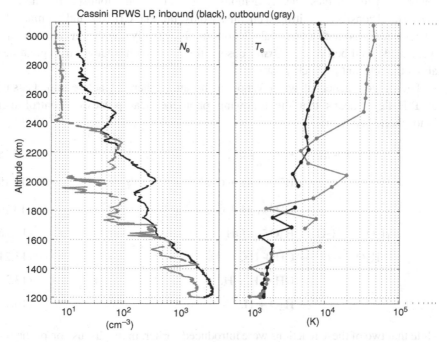

Figure 13.26 Electron density and temperatures altitude profiles measured by the Langmuir probe on Cassini.[76] (Courtesy of J.-E. Wahlund)

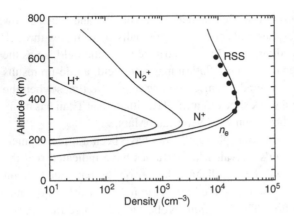

Figure 13.27 Calculated and measured ion and electron density profiles for Triton.[85] The model calculations assumed an ionization source of 3×10^8 ions cm^{-2} s^{-1}. The solid dots are the electron density values obtained during the ingress of *Voyager 2*.[66]

H$_2$O, will result in a plasma population. However, at this time no measurements of this plasma have yet been reported.

A well-established ionosphere has been observed at Triton, the major satellite of Neptune, by the *Voyager 2* radio occultation measurements[66] (Figure 13.27). These *Voyager* measurements prompted the development of a number of ionospheric models, which assumed, consistent with the airglow observations,[86] that the main sources of ionospheric plasma are photoionization by solar EUV radiation and magnetospheric electron impact ionization. A one-dimensional model calculation,[85] which solved the coupled continuity and momentum equations for the more important neutral and ion species, clearly demonstrated that Triton's ionosphere cannot be understood considering nitrogen chemistry only, but that CH$_4$, H, and H$_2$ must also be considered. Note that these calculations used relatively simple ion chemistry and were made well before the *Cassini* observations, which has taught us a great deal about the relevant photochemistry.

The predominance of water vapor in the atmosphere of active comets, such as P/Halley, means that the following photochemical processes control their ionospheric behavior:

$$H_2O + h\nu \rightarrow H_2O^+ + e^-, \tag{13.18}$$

$$\rightarrow H^+ + OH + e^-, \tag{13.19}$$

$$\rightarrow OH^+ + H + e^-, \tag{13.20}$$

$$H_2O^+ + H_2O \rightarrow H_3O^+ + OH, \tag{13.16}$$

$$H_3O^+ + e^- \rightarrow OH + H_2, \tag{13.21}$$

$$\rightarrow OH + H + H, \tag{13.22}$$

$$\rightarrow H_2O + H. \tag{13.17}$$

Note that two of these reactions were introduced earlier, in the discussion of the role of inflowing water vapor into Saturn's ionosphere. The chemistry scheme indicated in Figure 13.16 can also be used to understand the ion chemistry of water-dominated

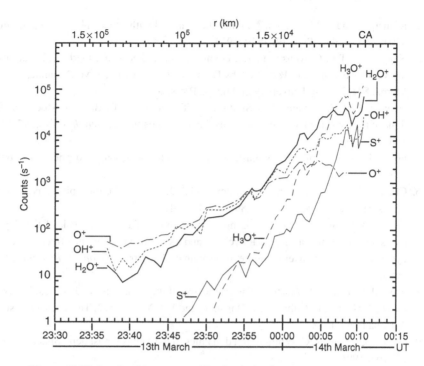

Figure 13.28 Ion densities measured by the *Giotto* ion mass spectrometer at Comet P/Halley. The distance, r, corresponds to the location of the spacecraft from the nucleus.[89]

cometary ionospheres. The very rapid rate at which H_2O^+ transforms into H_3O^+ means that in comets with water-dominated atmospheres, such as Halley, H_3O^+ is the dominant ionospheric constituent. Model calculations have also shown that the electron density varies roughly as $1/r$, where r is the radial distance from the nucleus, under both photochemical and transport controlled conditions, as long as the transport velocity is constant.[87]

The *Giotto* spacecraft carried two spectrometers that were capable of measuring the ion composition in Halley's ionosphere. The neutral spectrometer, operating in its ion mode, found that the H_3O^+ to H_2O^+ ratio increases with decreasing distance from the nucleus and it exceeds unity at distances less than about $20\,000\,\text{km}$.[88] The variations of the different ion densities measured by the ion mass spectrometer, carried aboard the *Giotto* spacecraft as it flew by comet P/Halley,[89] are shown in Figure 13.28. Model calculations of the ion composition and structure[87] are in qualitative agreement with these measurements.

13.7   Specific references

1. Ip, W.-H., The sodium exosphere and magnetosphere of Mercury, *Geophys. Res. Lett.*, **13**, 423, 1986.

2. Kumar, S., and D. M. Hunten, An ionospheric model with an exospheric temperature of 350 K, *J. Geophys. Res.*, **79**, 2529, 1974.

3. Nagy, A. F., T. E. Cravens, and T. I. Gombosi, Basic theory and model calculations of the Venus ionosphere, in *Venus*, ed. by D. M. Hunten, L. Colin, T. M. Donahue, and V. I. Moroz, **841**, Tucson: University of Arizona Press, 1983.

4. Nagy, A. F., T. E. Cravens, S. G. Smith, H. A. Taylor, and H. C. Brinton, Model calculations of the dayside ionosphere of Venus: ionic composition, *J. Geophys. Res.*, **85**, 7795, 1980.

5. Fox, J. L., The chemistry of metastable species in the Venusian ionosphere, *Icarus*, **51**, 248, 1982.

6. Cravens, T. E., J. U. Kozyra, A. F. Nagy, and A. J. Kliore, The ionospheric peak on the Venus dayside, *J. Geophys. Res.*, **86**, 11323, 1981.

7. Kliore, A. J., and L. F. Mullen, Long term behavior of the main peak of the dayside ionosphere of Venus during solar cycle 21 and its implications on the effect of the solar cycle upon the electron temperature in the main peak region, *J. Geophys. Res.*, **94**, 13339, 1989.

8. Russell, C. T., and O. Vaisberg, The interaction of the solar wind with Venus, *Venus*, ed. by D. M. Hunten, L. Colin, T. M. Donahue, and V. I. Moroz, **873**, Tucson: University of Arizona Press, 1983.

9. Miller, K. L., and R. C. Whitten, Ion dynamics in the Venus ionosphere, *Space Sci. Rev.*, **55**, 165, 1991.

10. Shinagawa, H., A two-dimensional model of the Venus ionosphere, 1. Unmagnetized ionosphere, *J. Geophys. Res.*, **101**, 26911, 1996.

11. Tanaka, T., and K. Murawski, Three-dimensional MHD simulation of solar wind interaction with the ionosphere of Venus: results of two-component reacting plasma simulation, *J. Geophys. Res.*, **102**, 19, 805, 1997.

12. Bauske, R., A. F. Nagy, T. I. Gombosi, *et al.*, 3D multiscale mass loaded MHD simulations of the solar wind interaction with Venus, *J. Geophys. Res.*, **103**, 23625, 1998.

13. Miller, K. L., C. W. Knudsen, K. Spenner, R. C. Whitten, and V. Novak, Solar zenith angle dependence of ionospheric ion and electron temperatures and densities on Venus, *J. Geophys. Res.*, **85**, 7759, 1980.

14. Brace, L. H., and A. J. Kliore, The structure of the Venus ionosphere, *Space Sci. Rev.*, **55**, 81, 1991.

15. Kliore, A. J., G. S. Levy, D. L. Cain, G. Fjeldbo, and S. I. Rasool, Atmosphere and ionosphere of Venus from the Mariner 5 S-band radio occultation measurement, *Science*, **158**, 1683, 1967.

16. Dobe, Z., A. F. Nagy, and J. L. Fox, A theoretical study concerning the solar cycle dependence of the nightside ionosphere of Venus, *J. Geophys. Res.*, **100**, 14507, 1995.

17. Brace, L. H., R. F. Theis, H. G. Mayr, S. A. Curtis, and J. G. Luhmann, Holes in the nightside ionosphere of Venus, *J. Geophys. Res.*, **87**, 199, 1982.

18. Nagy, A. F., and T. E. Cravens, Ionosphere: energetics, *Venus II*, ed. by S. W. Bougher, D. M. Hunten, and R. J. Phillips, **189**, Tucson: University of Arizona Press, 1997.

19. Cravens, T. E., T. I. Gombosi, J. Kozyra, *et al.*, Model calculations of the dayside ionosphere of Venus: Energetics, *J. Geophys. Res.*, **85**, 7778, 1980.

20. Gan, L., T. E. Cravens, and M. Horanyi, Electrons in the ionopause boundary layer of Venus, *J. Geophys. Res.*, **95**, 19023, 1990.

21. Nagy, A. F., A. Korosmezey, J. Kim, and T. I. Gombosi, A two-dimensional shock cap-turing, hydrodynamic model of the Venus ionosphere, *Geophys. Res. Lett.*, **18**, 801, 1991.

22. Knudsen, W. C., A. F. Nagy, and K. Spenner, Lack of thermal equilibrium between H^+ and O^+ temperatures in the Venus nightside ionosphere, *J. Geophys. Res.*, **102**, 2185, 1997.

23. Hanson, W. B., S. Sanatani, and D. R. Zuccaro, The Martian ionosphere as observed by the Viking retarding potential analyzers, *J. Geophys. Res.*, **82**, 4351, 1977.

24. Hanson, W. B., and G. P. Mantas, Viking electron temperature measurements: evidence for a magnetic field in the Martian ionosphere, *J. Geophys. Res.*, **93**, 7538, 1988.

25. Ma, Y., A. F. Nagy, I. V. Sokolov, and K. C. Hansen, 3D, multi-species, high spatial resolution MHD studies of the solar wind interaction with Mars, *J. Geophys. Res.*, **109**, A07211, doi:10.1029/2003JA010367, 2004.

26. Chen, R. H., T. E. Cravens, and A. F. Nagy, The Martian ionosphere in light of the Viking observations, *J. Geophys. Res.*, **83**, 3871, 1978.

27. Fox, J. L., The production and escape of nitrogen atoms on Mars, *J. Geophys. Res.*, **98**, 3297, 1993.

28. Choi, Y. W., J. Kim, K. W. Min, A. F. Nagy, and K. I. Oyama, Effect of the magnetic field on the energetics of Mars's ionosphere, *Geophys. Res. Lett.*, **25**, 2753, 1998.

29. Withers, P., M. Mendillo, H. Rishbeth, D. P. Hinson, and J. Arkani-Hamed, Iono-spheric characteristics above Martian crustal anomalies, *Geophys. Res. Lett.*, **32**, L16204, doi:10.1029/2005GL023483, 2005.

30. Mendillo, M., P. Withers, D. Hinson, H. Rishbeth, and B. Reinisch, Effects of solar flares on the ionosphere of Mars, *Science*, **311**, 1135, 2006.

31. Breus, T. K., A. M. Krymskii, D. H. Crider, *et al.*, Effect of the solar radiation in the topside atmosphere/ionosphere of Mars: Mars Global Surveyor observations, *J. Geophys. Res.*, **109**, A09310, doi:10.1029/2004JA010431, 2004.

32. Pätzold, M., S. Tellmann, B. Häusher, *et al.*, A sporadic third layer in the ionosphere of Mars, *Science*, **310**, 837, 2005.

33. Gurnett, D. A., D. L. Kirchner, R. L. Huff, *et al.*, Radar soundings of the ionosphere of Mars, *Science*, **310**, 1929, 2005.

34. Zou, H., J.-S. Wang, and E. Nielsen, Reevaluating the relationship between the Mar-tian ionospheric peak density and the solar radiation, *J. Geophys. Res.*, **111**, A07305, doi:10.1029/2005JA011580, 2006.

35. Bertaux, J.-L., F. Leblanc, O. Witasse, *et al.*, Discovery of an aurora on Mars, *Nature*, **435**, 790, 2005.

36. Lundin, R., D. Winningham, S. Barabash, *et al.*, Plasma acceleration above Martian magnetic anomalies, *Science*, **311**, 980, 2006.

37. Fox, J. L., J. F. Brannon, and H. S. Porter, Upper limits to the nightside ionosphere of Mars, *Geophys. Res. Lett.*, **20**, 1339, 1993.

38. Acuna, M. H., J. E. P. Connerney, P. Wasilewski, *et al.*, Magnetic field and plasma obser-vations at Mars: Initial results of the Mars Global Surveyor mission, *Science*, **279**, 1676, 1998.

39. Cain, J. C., B. B. Ferguson, and D. Mozzoni, An $n = 90$ internal potential func-tion of the Martian crustal magnetic field, *J. Geophys. Res.*, **108**, E2, 5008, doi:10.1029/2000JE001487, 2003.

40. Arkani-Hamed, J., A coherent model of the crustal magnetic field of Mars, *J. Geophys. Res.*, **109**, E09005, doi:10.1029/2004JE002265, 2004.

41. Mitchell, D. L., R. P. Lin, H. Rème, *et al.*, Oxygen Auger electrons observed in Mars's ionosphere, *Geophys. Res. Lett.*, **27**, 1871, 2000.

42. McElroy, M. B., The ionospheres of the major planets, *Space Sci. Rev.*, **14**, 460, 1973.

43. Cravens, T. E., Vibrationally excited molecular hydrogen in the upper atmosphere of Jupiter, *J. Geophys. Res.*, **92**, 11083, 1987.

44. Hallett, J. T., D. E. Shemansky, and X. Lin, A rotational-level hydrogen physical chemistry model for general astrophysical application, *Astrophys. J.*, **624**, 448, 2005.

45. Sundstrom, G., J. R. Mowat, H. Danarad, *et al.*, Destruction rate of H_3^+ by low energy electrons measured in a storage-ring experiment, *Science*, **263**, 785, 1994.

46. Kim, Y. H., and J. L. Fox, The chemistry of hydrocarbon ions in the Jovian ionosphere, *Icarus*, **112**, 310, 1994.

47. Hinson, D. P., F. M. Flasar, A. J. Kliore, *et al.*, Jupiter's ionosphere: Results from the first Galileo radio occultation experiment, *Geophys. Res. Lett.*, **24**, 2107, 1997.

48. Majeed, T., and J. C. McConnell, The upper ionospheres of Jupiter and Saturn, *Planet. Space Sci.*, **39**, 1715, 1991.

49. Achilleos, N., S. Miller, J. Tennyson, *et al.*, JIM: a time-dependent, three dimensional model of Jupiter's thermosphere and ionosphere, *J. Geophys. Res.*, **103**, 20, 089, 1998.

50. Bougher, S. W., J. H. Waite, T. Majeed, and G. R. Gladstone, Jupiter's thermospheric general circulation model (JTGCM): Global structure and dynamics driven by auroral and Joule heating, *J. Geophys. Res.*, **110**, E04008, doi:10.1029/2003JE002230, 2005.

51. Bagenal, F., The magnetosphere of Jupiter: coupling the equator to the poles, *J. Atm. Sol. Terr. Phys.*, **69**, 387, 2007.

52. Nagy, A. F., A. R. Barakat and R. W. Schunk, Is Jupiter's ionosphere a significant plasma source for its magnetosphere? *J. Geophys. Res.*, **91**, 351, 1986.

53. McComas, D. J., F. Allegrini, F. Bagenal, *et al.*, Diverse plasma populations and structures in Jupiter's magnetotail, *Science*, **318**, 217, 2007.

54. Nagy, A. F., A. J. Kliore, E. Marouf, *et al.*, First results from the ionospheric radio occultations of Saturn by the Cassini spacecraft, *J. Geophys. Res.*, **111**, A06310, doi:10.1029/2005JA011519, 2006.

55. Connerney, J. E. P., and J. H. Waite, New model of Saturn's ionosphere with an influx of water, *Nature*, **312**, 136, 1984.

56. Moore, L., A. F. Nagy, A. J. Kliore, *et al.*, Cassini radio occultation of Saturn's iono-sphere: model comparisons using a constant water flux, *Geophys. Res. Lett.*, **33**, L22202, doi:10.1029/2006GL027375, 2006.

57. Moses, J. J. and S. F. Bass, The effects of external material on the chemistry and structure of Saturn's ionosphere, *J. Geophys. Res.*, **105**, 7013, 2000.

58. Moore, L. E., M. Mendillo, I. C. F. Müller-Wodarg, and D. L. Murr, Modeling of global variations and ring shadowing in Saturn's ionosphere, *Icarus*, **172**, 503, 2004.

59. Kaiser, M. L., M. D. Desch, and J. E. P. Connerney, Saturn's ionosphere: inferred electron densities, *J. Geophys. Res.*, **89**, 2371, 1984.

60. Mendillo, M., L. Moore, J. Clarke, *et al.*, Effects of ring shadowing on the detection of electrostatic discharges at Saturn, *Geophys. Res. Lett.*, **32**, L05107, doi:10.1029/2004GL021934, 2005.

61. Fischer, G., W. S. Kurth, U. A. Dyudina, *et al.*, Analysis of a giant lightning storm on Saturn, *Icarus*, **190**, 528, 2007.

62. Waite, J. H., T. E. Cravens, W.-H. Ip, *et al.*, Oxygen ions observed near Saturn's A ring, *Science*, **307**, 1260, 2005.

63. Glocer, A., T. I. Gombosi, G. Toth, *et al.*, The polar wind outflow model: Saturn results, *J. Geophys. Res.*, **112**, A01304, doi:10.1029/2006JA011755, 2007.

64. Waite, J. H., M. R. Combi, W.-H. Ip, *et al.*, Cassini ion and neutral mass spectrometer: Enceladus plume composition and structure, *Science*, **311**, 1419, 2006.

65. Lindal, G. F., J. R. Lyons, D. N. Sweetnam, V. R. Eshleman, and D. P. Hinson, The atmosphere of Uranus: results of radio occultation measurements with Voyager 2, *J. Geophys. Res.*, **92**, 14987, 1987.

66. Tyler, G. L., D. N. Sweetnam, J. D. Anderson, *et al.*, Voyager radio science observations of Neptune and Triton, *Science*, **246**, 1466, 1989.

67. Broadfoot, A. L., F. Herbert, J. B. Holberg, *et al.*, Ultraviolet spectrometer observations of Uranus, *Science*, **233**, 74, 1986.

68. Yelle, R. V., B. R. Sandel, A. L. Broadfoot, and J. C. McConnell, The dependence of electroglow on the solar flux, *J. Geophys. Res.*, **92**, 15110, 1987.

69. Lara, L. M., W.-H. Ip, and R. Rodrigo, Photochemical models of Plutos atmosphere, *Icarus*, **130**, 16, 1997.

70. Kliore, A. J., G. Fjeldbo, B. L. Seidel, *et al.*, The atmosphere of Io from Pioneer 10 radio occultation measurements, *Icarus*, **24**, 407, 1975.

71. Hinson, D. P., A. J. Kliore, F. M. Flasar, *et al.*, Galileo radio occultation measurements of Io's ionosphere and plasma wave, *J. Geophys. Res.*, **103**, 29343, 1998.

72. Summers, M. E., and D. F. Strobel, Photochemistry and vertical transport in Io's atmosphere and ionosphere, *Icarus*, **120**, 290, 1996.

73. Kliore, A. J., D. P. Hinson, F. M. Flasar, A. F. Nagy, and T. E. Cravens, The ionosphere of Europa from Galileo radio occultations, *Science*, **277**, 355, 1997.

74. Kliore, A. J., A. Anabtawi, R. G. Herrera, *et al.*, The ionosphere of Callisto from Galileo radio occultation observations, *J. Geophys. Res.*, **107**, doi: 11.1407/2002JA009365, 2002.

75. Bird, M. K., R. Dutta-Roy, S. W. Asmar, and T. A. Rebold, Detection of Titan's ionosphere from Voyager 1 radio occultation observations, *Icarus*, **130**, 426, 1997.

76. Wahlund, J. E., R. Boström, G. Gustafsson, *et al.*, Cassini measurements of cold plasma in the ionosphere of Titan, *Science*, **308**, 986, 2005.

77. Cravens, T. E., I. P. Robertson, J. H. Waite, *et al.*, Composition of Titan's ionosphere, *Geophys. Res. Lett.*, **33**, L07105, doi: 10.1029/2005GL025575, 2006.

78. Kliore, A. J., A. F. Nagy, E. A. Marouf, *et al.*, First Results from the Cassini radio occultation of the Titan ionosphere, *J. Geophys. Res.*, **113**, doi:10.1029/2007JA012965, 2008.

79. Cravens, T. E., J. P. Robertson, S. A. Ledvina, *et al.*, Energetic ion precipitation at Titan, *Geophys. Res. Lett.*, **35**, L03103, doi:10.1029/2007GL032451, 2008.

80. Molina-Cuberos, G. J., H. Lammer, W. Stumptner, *et al.*, Ionospheric layer induced by meteoric ionization in Titan's atmosphere, *Planet. Space Sci.*, **49**, 143, 2001.

81. Cravens, T. E., I. P. Robertson, J. Clark, *et al.*, Titan's ionosphere: model comparisons with Cassini Ta data, *Geophys. Res. Lett.*, **32**, L12108, doi:10.1029/2005GL023249, 2005.

82. Ma, Y.-J., A. F. Nagy, T. E. Cravens, *et al.*, Comparisons between model calculations and observations of Cassini flybys of Titan, *J. Geophys. Res.*, 111, 5, A05207, doi: 10.1029/2005JA011481, 2006.

83. Roboz, A., and A. F. Nagy, The energetics of Titan's ionosphere, *J. Geophys. Res.*, **99**, 2087, 1994.

84. Galand, M., R. V. Yelle, A. J. Coates, H. Backes, and J.-E. Wahlund, Electron temperatures of Titan's sunlit ionosphere, *Geophys. Res. Lett.*, **33**, L21101 doi: 10.1029/2006 GL027488, 2006.

85. Majeed, T., J. C. McConnell, D. F. Strobel, and M. E. Summers, The ionosphere of Triton, *Geophys. Res. Lett.*, **17**, 1721, 1990.

86. Broadfoot, A. L., S. K. Atreya, J. L. Bertaux, *et al.*, Ultraviolet spectrometer observations of Neptune and Triton, *Science*, **246**, 1459, 1989.

87. Körösmezey, A., T. E. Cravens, T. I. Gombosi, *et al.*, A new model of cometary ionospheres, *J. Geophys. Res.*, **92**, 7331, 1987.

88. Krankowsky, D., P. Lammerzahl, I. Herrwerth, *et al.*, *In situ* gas and ion measurements at Comet Halley, *Nature*, **321**, 326, 1986.

89. Balsiger, H., K. Altwegg, F. Buhler, *et al.*, Ion composition and dynamics at Comet Halley, *Nature*, **321**, 330, 1986.

13.8 General references

Atreya, S. K., *Atmospheres and Ionospheres of the Outer Planets and Their Satellites*, Berlin: Springer-Verlag, 1986.

Bauer, S. J., and H. Lammer, *Planetary Aeronomy*, Berlin: Springer-Verlag, 2004.

Brace, L. H., *et al.*, The ionosphere of Venus: observations and their interpretation, in *Venus*, ed. by D. M. Hunten, L. Colin, T. M. Donahue, and V. I. Moroz, **779**, Tucson: University of Arizona Press, 1983.

Brace, L. H., and A. J. Kliore, The structure of the Venus ionosphere, *Space Sci. Rev.*, **55**, 81, 1991.

Chamberlain, J. W., and D. M. Hunten, *Theory of Planetary Atmospheres*, New York: Academic Press, 1987.

Cravens, T. E., *et al.*, Composition and structure of the ionosphere and thermosphere, in *Titan After Cassini-Huygens*, Berlin, Springer-Verlag, 2009.

Fox, J. L., and A. J. Kliore, Ionosphere: solar cycle variations, in *Venus II*, ed. by S. W. Bougher, D. M. Hunten, and R. J. Phillips, **161**, Tucson: University of Arizona Press, 1997.

Fox, J. L., Advances in aeronomy of Venus and Mars, *Adv. Space Res.,*, **33**, 132, 2004.

Kar, J., Recent advances in planetary ionospheres, *Space Sci. Rev.*, **77**, 193, 1996.

Mahajan, K. K., and J. Kar, Planetary ionospheres, *Space Sci. Rev.*, **47**, 193, 1988.

Nagy, A. F., T. E. Cravens, and T. I. Gombosi, Basic theory and model calculations of the Venus ionosphere, in *Venus*, ed. by D. M. Hunten, L. Colin, T. M. Donahue and V. I. Moroz, **841**, Tucson: University of Arizona Press, 1983.

Nagy, A. F., and T. E. Cravens, Ionosphere: energetics, in *Venus II*, ed. by S. W. Bougher, D. M. Hunten, and R. J. Phillips, Tucson: University of Arizona Press, 1997.

Nagy, A. F., and T. E. Cravens, Titan's ionosphere: a review, *Planet. Space Sci.*, **46**, 1149, 1998.

Nagy, A. F., K. Winterhalter, T. E. Sauer, *et al.*, The plasma environment of Mars, *Space Sci. Rev.*, **111**, 33, 2004.

Nagy, A. F., *et al.*, Saturn: upper atmosphere and ionosphere, in *Saturn After Cassini-Huygens*, Springer-Verlag, 2009.

Schunk, R. W., and A. F. Nagy, Ionospheres of the terrestrial planets, *Rev. Geophys. Space Phys.*, **18**, 813, 1980.

Witasse, O., and A. F. Nagy, Outstanding aeronomy problems at Venus, *Planet. Space Sci.*, **54**, 1381, 2006.

Witasse, O., T. Cravens, M. Mendillo, *et al.*, Solar system ionospheres, *Space Sci. Rev.*, Solar system **139**, 235, 2008.

13.9   Problems

Problem 13.1 Assume that the ionospheric peak at Venus is formed at unit optical depth, where the CO_2 and O densities are 1×10^{11} and 1×10^{10} cm^{-3}, respectively. Write down the chemical equilibrium continuity equations for CO_2^+, O^+, and O_2^+. Solve for the solar maximum steady state value of O_2^+, assuming that its value is approximately equal to that of the total electron density, or in other words that it is the major ion. Check if this assumption is consistent with your answer. You may assume that the electron temperature is 300 K.

Problem 13.2 At around 210 km altitude on the day side, solar-maximum, ionosphere of Venus, the controlling chemical loss of O^+ is the reaction indicated by Equation (13.4). Estimate the relevant chemical and diffusive time constants to confirm that the transition from chemical to diffusive control takes place in this general altitude region. Are the time constants of the same order of magnitude?

Problem 13.3 Using the information provided in this chapter and Chapter 2, calculate the maximum daytime ionospheric thermal plasma pressure of Venus, for solar cycle maximum (*PVO*) conditions. Compare this result with the total, unperturbed solar wind pressure at Venus. Repeat these calculations for low solar cycle (*Viking*) conditions at Mars (assume that the peak ionospheric pressure is at 160 km at both planets). Are these ionospheres capable of holding off the solar wind?

Problem 13.4 Assume that the major ion in the ionosphere of Venus is O^+ above 200 km and that it is in diffusive equilibrium with a density of 10^5 cm^{-3} at 200 km. Assume altitude-independent electron and ion temperatures of 3000 K and a constant g of 800 cm s^{-2}. If the solar wind density and velocity are 10 cm^{-3} and 400 km s^{-1}, respectively, at what altitude will the subsolar ionopause be located? Repeat the calculation for an increased solar wind velocity of 500 km s^{-1}.

Problem 13.5 Assume that the electron energy transport is only controlled by thermal conduction in the upper ionosphere of Venus. If the electron temperature is

1500 K at 200 km and a topside heat inflow of 10^{10} eV cm^{-2} s^{-1} is imposed at 600 km, what is the electron temperature at 400 km? Assume that the thermal conductivity is given by Equation (5.146) with the denominator equal to unity. Repeat the calculations for an increased topside heat inflow of 5×10^{10} eV cm^{-2} s^{-1}. Also, repeat the calculations for the original heat flow value (10^{10} eV cm^{-2} s^{-1}), but a thermal conductivity reduced by a factor of 10.

Problem 13.6 Assume that in the day side mid-latitude ionosphere of Jupiter photoionization dominates and the optical depth is zero at and above 1500 km. If 50% of the photoionization of H_2 leads to H^+, calculate the chemical equilibrium density of H^+ at 2000 km, if the H_2 density and electron temperature at that altitude are 10^8 cm^{-3} and 1500 K, respectively, and if radiative recombination is the only possible loss process. Assume solar maximum conditions and that H^+ is the major ion. Calculate the optical depth of H_2 for 30.378 nm at 2000 km given a neutral temperature of 1000 K, in order to test the zero optical depth assumption. Also, repeat the calculations for the H^+ density assuming that 10% of H_2 is in a vibrational state of $v > 4$ and so charge exchange with H_2 has the rate given in Table 8.3 and the resulting H_2^+ is in its ground state and is lost by dissociative recombination. H^+ is now not the major ion: to show this, calculate both the H^+ and H_2^+ densities. Choose all necessary parameters from the information presented in the book, and compare the two potential loss rates of H^+, to see if one is negligible.

Problem 13.7 Show that the steady state ratio of $[H^+]/[H_3^+]$ in the chemically controlled region of Jupiter can be written as:

$$\frac{[H^+]}{[H_3^+]} = \frac{J_{11}\alpha n_e}{J_{10}k_v[H_2]_v}$$

where J_{10} and J_{11} correspond to ionization frequencies associated with Equations (13.10) and (13.11), respectively, $[H_2]_v$ is the number density of molecular hydrogen in a vibrationally excited state, $v > 4$, α is the dissociative recombination rate corresponding to Equation (13.14) and k_v is the reaction rate for

$$H^+ + H_2(v > 4) \rightarrow H_2^+ + H.$$

What, if any, assumption did you have to make to arrive at your answer?

Problem 13.8 In the photochemically controlled region of a certain comet, the only neutral gas constituent is H_2O, with a density of 10^7 cm^{-3}. You can assume that the only resulting ions are H_2O^+ and H_3O^+. The comet is located at 1 AU during solar-minimum conditions, the optical depth is zero, and the electron temperature is 300 K. What is the resulting equilibrium electron density? (Neglect dissociative recombination of H_2O^+.)

Chapter 14

Ionospheric measurement techniques

This chapter describes the various measurement techniques that are directly applicable to the determination of ionospheric parameters. This discussion is restricted to the most commonly used methods, which measure the thermal plasma densities, temperatures, and velocities, as well as magnetic fields (currents). In general, these techniques can be grouped as remote or direct (*in situ*) ones. Topics related to direct measurement techniques are described in the first five sections and the rest of the chapter deals with remote sensing. The remote, radio sensing methods rely on the fact that the ionospheric plasma is a dispersive media (Section 6.8) while the relevant radar measurements use the reflective properties of the plasma. The direct *in situ* measurement techniques discussed here are restricted to those that are applicable to altitudes where the mean-free-path is greater than the characteristic dimension of the instrument.

14.1   Spacecraft potential

In situ measurements of ionospheric densities and temperatures are based on the laboratory technique developed and discussed by Irving Langmuir and co-workers over eighty years ago.[1] These so-called *Langmuir probes*, or *retarding potential analyzers* (RPAs), have been used for many years in laboratory plasmas before they were adopted for space applications.[2] On a rocket or a satellite, the voltage applied to an instrument has to be driven against the potential of the vehicle, and therefore, it is appropriate to begin with a discussion of the factors that affect the value of this potential. The equilibrium potential is the one that a floating (conducting) body immersed in a plasma acquires in order to cause the net collected current to be zero. Assuming comparable ion and electron temperatures, the ions, due to their much larger mass, have a thermal velocity (Equations 3.14. H.21, and H.22) that is considerably less than that of the electrons. The ion and electron densities are in general the same; therefore, the body accumulates more negative than positive charges initially. Eventually, the body attains a negative potential that is just large enough to repel enough of the electrons and attract a sufficient number of ions so that equal numbers of ions and electrons reach it. This negative potential is called the *equilibrium* or *floating potential*. It follows that the electron density in the immediate vicinity of the probe is lower than in the undisturbed plasma, resulting in a net positive charge in this region. The magnitude of this total net positive charge is equal to the negative charge on the body. This region of net positive charge, referred to as the *positive ion sheath*, shields the floating body potential from the rest of the ambient plasma. The thickness of the ion sheath is related to the fundamental length parameter of a plasma, the Debye length (Equation 2.4). Typical Debye lengths in the terrestrial ionosphere are about 1 cm.

The equilibrium potential, V_S, of a stationary body immersed in a plasma, if only ambient thermal particle effects are considered, is given by

$$V_S = -\frac{kT_e}{e} \log\left(\frac{I_{oe}}{I_{oi}}\right), \tag{14.1}$$

where k is the Boltzmann constant, e is the magnitude of the electronic charge, I_{oe} is the random electron current, and I_{oi} is the random ion current. The *random current* is the rate at which charged particles cross an area, A, in a plasma with a nondrifting Maxwellian velocity distribution. This random electron and ion current is (Equation H.26)

$$I_{os} = en_s A \sqrt{\frac{kT_s}{2\pi m_s}} = en_s A \frac{\langle c_s \rangle_M}{4}, \tag{14.2}$$

where n_s, T_s, and m_s are the density, temperature, and mass of the given charge carriers and $\langle c_s \rangle_M = (8kT_s/\pi m_s)^{1/2}$ is the mean thermal velocity (Equation H.21). Typical random electron current densities in the terrestrial ionosphere ($T_e \approx 1500$ K; $n_e \approx 10^5$ cm^{-3}) are of the order of 1×10^{-3} A m^{-2}.

The actual equilibrium potential that a moving body (rocket or satellite) acquires in the ionosphere depends on a number of factors. Among them are the ratio of the thermal velocity of the ionospheric particles to the satellite velocity, *photoemission* resulting from the interaction of solar radiation with the surface, and secondary electron emission resulting from the impact of energetic particles. Therefore, the satellite potential depends on the effective areas for these various processes (e.g., the effective area for the photoemission current is only a fraction of the area for the thermal electron current). The satellite potential, more generally, is given by

$$V_S = -\frac{kT_e}{e} \log\left(\frac{\int j_e dS_e}{\int j_i dS_i + \int j_p dS_p + \int j_s dS_s}\right), \tag{14.3}$$

where j_e is the electron, j_i the positive ion, j_p the photoemission current densities, and j_S the current density due to secondary electrons from energetic particle bombardment.

The geomagnetic field induces a potential gradient in a moving spacecraft ($\mathbf{U} \times \mathbf{B}$ effect) and this can be especially important when the spacecraft has long booms. In such a case, it can no longer be assumed that a satellite moving in an ionosphere is an equipotential surface and the vehicle potential may vary along the spacecraft. A typical value for the induced $\mathbf{U} \times \mathbf{B}$ potential difference is about $0.2\,\mathrm{Vm^{-1}}$ in the terrestrial ionosphere. The $\mathbf{U} \times \mathbf{B}$ effect is directly proportional to body dimension; therefore, long booms (>20 m), as used on many satellites, can lead to substantial potential differences.

All present observations of spacecraft potential fall into one of the following three categories:

1. Small negative or positive values, $|V_S| \lesssim 2V$.
2. Significant negative values, $V_S \approx -10V$, resulting from the presence, on the spacecraft, of exposed areas with large positive potentials that collect large electron currents that drive the spacecraft negative.
3. Occasionally large (~ 1 keV) negative potentials, on solar-eclipsed satellites at very high altitudes (outside the plasmasphere), where the thermal particle density is small and the energetic particle population is significant.

14.2 Langmuir probes

The total electron current density collected by a *Langmuir probe* is given by

$$j_e = \frac{I_e}{A} = e \iiint v_n f(v)\, d^3 v, \tag{14.4}$$

where I_e is the total electron current, A is the probe area, v_n is the particle velocity component normal to the probe surface, and $f(v)$ is the velocity distribution function. Note that flux is defined as

$$n\langle c \rangle \equiv \iiint v_n f(v)\, d^3 v, \tag{14.5}$$

so that the current density j_e, or I_e/A, represents a measure of the charged particle flux.

For a Maxwellian distribution, the electron current collected by the probe (Figure 14.1a) in the *electron-retarding region* ($V_p < V_o$) is given by[1]

$$I_e = e n_e A \left(\frac{kT_e}{2\pi m_e} \right)^{1/2} \exp\left(-\frac{e|V_p - V_o|}{kT_e} \right), \tag{14.6}$$

where n_e is the electron density, T_e is the electron temperature, V_p is the potential applied to the probe relative to the vehicle potential, V_S, and V_o is the plasma potential relative to V_S. This relationship between electron current and retarding

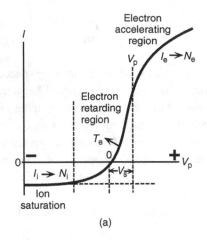

Figure 14.1 (a) Sketch of current collected by a cylindrical Langmuir probe, showing the electron retarded and accelerated current regions. (Courtesy of L. H. Brace.) (b) Current versus retarding potential data points from one of the cylindrical Langmuir probes carried by the *Pioneer Venus Orbiter*. The solid line is a fit to the retarding region of the curve, which is used to determine the electron temperature. The electron temperature value corresponding to the fit is also indicated.[3]

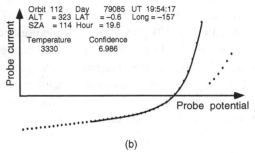

potential is valid regardless of probe shape and it also holds for moving probes as long as the probe velocity is small compared to the electron thermal velocity.[4] The probe current, however, is reduced when a magnetic field is present.[5]

Taking the logarithm of Equation (14.6) gives

$$\log I_e = -\frac{e}{kT_e}|V_p - V_o| + \log I_{oe}. \tag{14.7}$$

Taking the derivative of Equation (14.7) with respect to the probe potential leads to

$$\frac{d(\log I_e)}{dV_p} = -\frac{e}{kT_e}. \tag{14.8}$$

Thus, a linear dependence of $\log I_e$ with respect to V_p indicates a Maxwellian distribution and the electron temperature can be determined from the logarithmic slope of the I_e versus V_p characteristic. A nonlinear $\log I_e$ versus V_p dependence indicates a nonMaxwellian energy distribution or multiple Maxwellian populations. The electron temperature can also be obtained from the retarding portion of the I–V characteristic, by the so-called *Druyvestyn-type analysis*[6]

$$T_e = -\frac{e}{k}\frac{dI_e}{dV_p}\bigg/\frac{d^2I_e}{dV_p^2}. \tag{14.9}$$

The plasma potential corresponds to the value of V_p at which $d^2 I_e / dV_p^2 = 0$. When the energy distribution function is not Maxwellian one needs to establish the actual energy distribution function. Druyvestyn[6] demonstrated that the energy distribution function, $f(E)$, of the ambient charged particles is related to the current collected by a stationary probe with retarding potentials through the following relation:

$$f(E)dE = \frac{2}{Ae}\left(\frac{2mV}{e}\right)^{1/2} \frac{d^2 I}{dV_r^2}\, dE, \tag{14.10}$$

where A is the probe area, m/e is the mass-to-charge ratio of the charged particle, $V_r = V_p - V_o$ is the retarding potential of the probe relative to the plasma, I is the collected current, and E is in eV. This approach of reconstructing the energy distribution function has been used on a few occasions.[7]

The total current collected by a probe operating in the electron retarding region, I_{et}, is given by

$$I_{et} = I_{oe}\exp\left(\frac{-e|V_p - V_o|}{kT}\right) - I_i - I_{sp}, \tag{14.11}$$

where I_i is the positive ion current to the probe, and I_{sp} is the sum of all other "spurious" currents, such as that due to a photoelectron current to the probe. As long as I_i and I_{sp} are small, the total probe current in the retarding region can be interpreted as the electron current and the electron temperature can then be determined in a straightforward fashion. The ion current I_i will generally be more than an order of magnitude lower than I_{oe}. However, at high altitudes, where the ambient density is low, I_{sp} may become a significant contributor to the probe current and must be taken into account. Gridded Langmuir probes, commonly called retarding potential analyzers (RPAs), have the advantage of eliminating the unwanted currents I_i and I_{sp} directly and are, therefore, preferable in the low-density regions of any ionosphere. Retarding potential analyzers are discussed in the next section.

The electron density is usually derived from the value of the current at either the plasma potential from the random current (14.2) or the current collected in the electron accelerating region (Figure 14.1a). The value of the random current is usually obtained by establishing the transition point in the I–V curve, separating the retarding and accelerating regions. The electron density measurement is thus a weak function of T_e. In the case of a cylindrical probe, the electron density can also be determined without an a-priori knowledge of T_e, by measuring the current in the accelerating region ($V_p > V_o$) of the I_e–V_p characteristic. When the diameter of the collector is small compared with the Debye length, which is indicative of the sheath dimension, the probe operates under the so-called orbital motion limited condition. In this case the accelerated electron current collected by such a cylindrical probe is[1]

$$I_e = \frac{2Aen_e}{\pi^{1/2}}\left(\frac{kT_e}{2\pi m_e}\right)^{1/2}\left(1 + \frac{e|V_p - V_o|}{kT_e}\right)^{1/2}. \tag{14.12}$$

When $(e[V_p - V_o])/kT_e \gg 1$, the current collected by the probe varies as the square root of the applied voltage and is independent of T_e. Therefore, long cylindrical probes have the added practical advantage that they can be operated at a fixed positive potential to make continuous measurements of n_e without interruptions in order to obtain the electron temperature values.

Such cylindrical probes have been used widely. A measured volt–ampere characteristic from a cylindrical probe flown on the *Pioneer Venus Orbiter* and the theoretical fit with the deduced electron temperature are shown in Figure 14.1b.[3] In most spacecraft applications the available data rate is limited. Therefore, instead of simply telemetering the full measured volt–ampere characteristics, a variety of data compression schemes have been employed. Some of the simpler systems transmitted the full curves only intermittently and in between curve transmissions they telemetered some indicators of the desired quantities. A clever and relatively simple scheme employed with the *AE*, *DE*, and *Pioneer Venus* Langmuir probes relied on automatic gain and voltage sweep adjustments as indicators of the ion densities and electron temperatures, respectively.[3] A very different approach was used for the Langmuir probes built for the *Akebono* and *Nozomi* satellites. In the *Akebono* instrument a small 3 kHz ac signal was applied to the probe along with the usual dc sweep voltage. The second harmonic component of the current to the probe, which is proportional to the second derivative of the current with respect to the sweep voltage and thus to the energy distribution function (Equation 14.10), was monitored.[8] In the *Nozomi* instrument a sinusoidal signal was applied to the probe at floating potential. The resulting shift in the floating potential was monitored and, since this shift is proportional to the electron temperature, values of T_e could be obtained.[9]

14.3   Retarding potential analyzers

Langmuir probe theory applies to positive ion measurements as well as to the electron measurements discussed in the preceding section. The basic difference between the two is the fact that for ions the motion of the spacecraft through the plasma generally cannot be neglected. The random current density for ions is also reduced compared with that for electrons by the mass ratio $(m_e/m_i)^{1/2}$, so that the ion current will be smaller than the electron current. Furthermore, ion measurements may also be masked by photoemission currents, resulting from the interaction between solar radiation and the probe. However, this effect can be eliminated or at least reduced with the use of grids. The term *retarding potential analyzers* (RPAs) refers to charged particle collectors with a screening aperture and grids that allow for instrumental rejection of particles of either polarity (electron and ion modes of operation), as well as for suppression of photo and secondary electron emission effects. Two different geometrical configurations of gridded analyzers have been used, namely planar collectors with circular openings and spherical collectors. The spherical traps on some rocket flights and satellite missions were operated in the electron mode to give electron density and temperature information.[10,11] Planar RPAs have also been used in both ion and electron modes in some recent applications.[12] However, in most cases these RPAs are used to measure ion density and temperature.[13,14] Also, planar RPAs are widely used because of their ease of mounting on a spacecraft.

The equation for the positive ion current to a moving planar collector operating in a retarding potential mode, in a multi-species ionosphere, is[14,15]

$$I_i = \alpha e A U \cos\theta \sum_j n_j \left(\frac{1}{2} + \frac{1}{2}\mathrm{erf}(y) + \frac{\exp\{-y^2\}}{2\pi^{1/2}\lambda_j} \right), \tag{14.13}$$

where α is the total transparency of the grids, A is the collecting area, U is the spacecraft velocity, θ is the angle between the velocity vector and the normal to the planar collecting surface, n_j is the density of the jth ion species, $y = \lambda_j - (eV/kT_j)^{1/2}$, $V = V_p - V_0$, $\lambda_j = (U\cos\theta)/(\sqrt{2kT_j/m_j})$, and T_j and m_j are the ion temperature and mass, respectively. The current in the ion accelerating potential mode is independent of the applied potential and is also given by Equation (14.13) by simply setting $V = 0$. Figure 14.2 shows the results from a retarding potential scan of the RPA carried by the *DMSP F10* satellite (R. A. Heelis, private communication). The figure also shows the fit to the data and the deduced ionospheric parameters.

As indicated in Figure 14.2, and implied by Equation (14.13), an RPA provides information on the energy of the ions in the spacecraft frame of reference, and thus, in a multiconstituent medium, it can yield ion composition and ion temperature data.

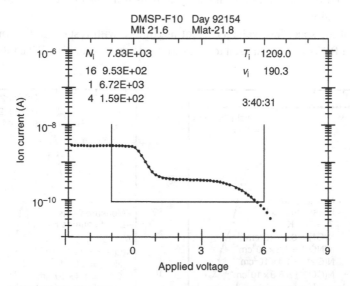

Figure 14.2 Ion current versus retarding potential characteristic measured by the retarding potential analyzer carried by the *DMSP F10* satellite. The curve-fitting procedure leads to the total ion density and relative composition, shown in the left column; the deduced ion temperature, the ion drift velocity along the sensor look direction, and the universal time the measurement was made are shown in the right column. (Courtesy of R. A. Heelis.)

The energy of a stationary ion of mass m_j, in the frame of reference of an orbiting satellite is

$$E_j = \frac{1}{2}m_j U^2, \tag{14.14}$$

where U is the satellite velocity with respect to the plasma. An ion with a unit positive charge will be prevented from reaching the collector and contributing to the current when the potential is more positive than E_j/e. Because E_j is a function of m_j, sweeping the collector potential to larger positive values will result in decreases of the ion current collected at the various voltages, corresponding to the different ion masses. Therefore, this type of instrument can be used as a low-resolution ion mass spectrometer. This description is, of course, a simplified one because it ignores the effect of the ion thermal velocity. The inclusion of thermal velocity effects causes the drop in the current to be less abrupt; the degree of sharpness of this drop in the current depends on the ion temperature, as indicated in Figure 14.2.

More sophisticated data handling approaches, beyond simply telemetering the I–V curves, have also been introduced and used by a variety of RPA experimenters. For example, ion composition, temperature, and instrument potential can also be obtained by taking the derivative of the I–V curve at the satellite and telemetering this information back to Earth. The RPA carried by the *Pioneer Venus Orbiter* could be operated in the full I–V mode or it could be commanded to transmit $\Delta I/\Delta V$ information, which yields density and temperature information.[12] Figure 14.3 shows data points and theoretical fits obtained by that instrument in these two different modes of operation.

In general the spacecraft velocity is not negligible with respect to the ion drift velocity, and RPAs have been used to measure these drift velocities. When the ions

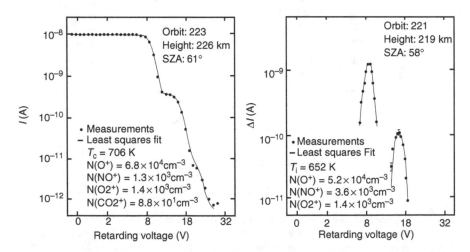

Figure 14.3 Ion current and ΔI values versus retarding potential as measured and telemetered by the RPA that was carried by the *Pioneer Venus Orbiter*. Least square fits and the resulting parameters are also shown.[12]

have a net drift velocity, the peak current will occur not when the sensor is looking parallel to the spacecraft velocity, but when it looks parallel to the total velocity vector. Thus, if the spacecraft velocity is well established, the ion drift velocity component along the normal to the collector axis can be derived from the shifted position of the maximum current. Such a procedure has been used by numerous scientists using RPA data.

A clever modification of simple planar RPAs has been developed to measure plasma drifts and is now widely used.[16] These so-called *drift meters* have special four-segment equal area collectors. If the mean velocity of the ions entering the sensor is perpendicular to the collecting surfaces, then the currents to all four segments are the same. When the entry velocity is no longer perpendicular, the currents to the various segments will be different. Therefore, if the spacecraft orientation with respect to its velocity vector is accurately known, the ion drift velocity in the plane of the four planar segments can be derived from the measured current ratios. The ion drift velocity in the direction normal to the collector surface is determined from the conventional RPA operation, as characterized by Equation (14.13) and indicated in Figure 14.2, and thus the total velocity vector can be obtained.

14.4   Thermal ion mass spectrometers

The first spectrometers used successfully in the space program were radio frequency (RF/Bennett) instruments.[17] Since the mid 1950s, this type of instrument has been used widely on rockets and satellites both in the United States and the USSR.[18,19] The general principle of operation of this instrument is illustrated with the aid of Figure 14.4, which is a cross-sectional view of the spectrometer used on both the *Atmosphere Explorer* satellites and the *Pioneer Venus Orbiter*.[20] Ambient ions enter the instrument orifice through the guard-ring grid and are accelerated down the axis of the spectrometer by a slowly varying negative dc sweep potential V_A. Corresponding to each ion mass there is a value of V_A which accelerates the ion to the instrument's resonant velocity. These resonant ions traverse the analyzer stages in phase with the applied RF potentials and gain enough energy to overcome the retarding dc potential V_R. The relationship between resonant ion mass, sweep potential, and frequency, assuming that the instrument is at rest with respect to the plasma, is given as

$$m_i = 0.266|V_A|/(S^2f^2),\eqno(14.15)$$

where m_i is the ion mass in amu, V_A is the sweep potential in volts, S is the analyzer intergrid spacing in cm, and f is the frequency applied to the analyzer in MHz.

Since the early 1960s, a variety of instruments have been used in which ions with different e/m ratios are separated using deflection caused by a magnetic field. Figure 14.5 shows a drawing of the magnetic ion mass spectrometer built for the *Atmospheric Explorer (AE) C, D*, and *E* satellites,[21] which clearly indicates the simple principles involved in such a device. Ions accelerated through a potential

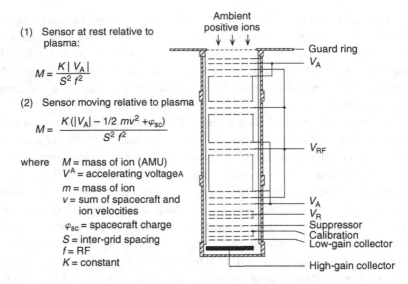

Figure 14.4 A schematic diagram of the Bennett ion mass spectrometer. The mass analysis equations are also indicated.[20]

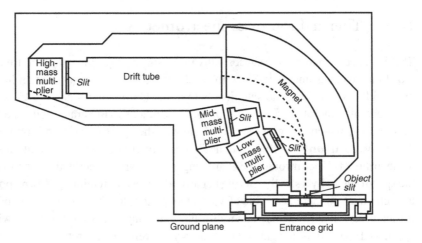

Figure 14.5 A schematic drawing of the magnetic deflection ion mass spectrometer carried by the *Atmosphere Explorer C, D,* and *E* satellites. The collector slits were placed so as to enable the simultaneous collection of ions with mass ratios of 1:4:16.[21]

difference V_a and injected into a magnetic field of strength B will move with a radius of curvature, that is given by

$$r_0 = \left(\frac{2m_i V_a}{eB^2} \right)^{1/2}, \qquad (14.16)$$

where m_i is the ion mass and e is the electronic charge. In the specific spectrometer shown in Figure 14.5, the ions are accelerated through the entrance aperture into the

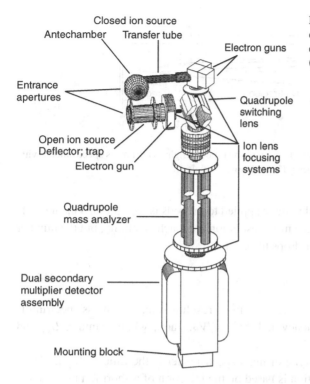

Closed ion source
Antechamber   Transfer tube
Electron guns

Entrance
apertures

Quadrupole
switching
lens

Open ion source
Deflector; trap
Electron gun

Ion lens
focusing
systems

Quadrupole
mass analyzer

Dual secondary
multiplier detector
assembly

Mounting block

Figure 14.6 A schematic diagram of an ion–neutral quadrupole spectrometer. (Courtesy of H. B. Niemann.)

analyzer system by a negative sweep voltage; three parallel detection systems are employed to measure high, mid, and low mass numbers simultaneously. Mass spectrometers that use a combination of electrostatic and magnetic deflection sections have also been used.

The quadrupole mass spectrometer was developed in the 1950s for isotope separation,[22] but since then it has been widely used for space applications, generally for neutral gas spectrometry[23] and recently as a combined ion–neutral mass spectrometer.[24] Figure 14.6 shows a schematic diagram of a quadrupole spectrometer, which basically consists of four rod-shaped electrodes with hyperbolic cross sections and spaced a distance of R_0 from the central (long) axis. Opposite pairs of electrodes are electrically connected and a combination of RF and dc potentials applied to them. The ions are injected along the axis of the poles. For a particular combination of voltages, ions within a small range of e/m_i ratios have stable trajectories that oscillate closely around the axis, while the other ions follow unstable trajectories that strike the rods, and thus, are unable to reach the collector.

The condition for collection of ions with a mass m_i is given by

$$m_i = \frac{1.385 \times 10^7 V_{RF}}{R_0^2 f^2}, \qquad (14.17)$$

where m_i is the ion mass in AMU, R_0 is half the diametrical spacing between rods in meters, f is the frequency of the applied RF voltage in Hz, and V_{RF} is the peak

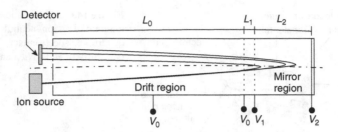

Figure 14.7 A schematic diagram of the time-of-flight (TOF) mass spectrometer with a two-stage gridded mirror. The drift space is field free.[25]

RF voltage in volts. The total voltage applied to the rods is the sum of V_{RF} and a dc potential U_{dc}. Theoretically the mass resolution approaches infinity, but the number of ions reaching the collector drops to zero for

$$U_{dc} = 0.168V_{RF}. \tag{14.18}$$

In practice the analyzer is operated at a finite resolution and the mass spectrum is obtained by fixing the frequency and the U_{dc}/V_{RF} ratio, while scanning U_{dc} and V_{RF}.

A relatively more recent type of mass spectrometer is the time-of-flight (TOF) one.[25] Its principle of operation is based on the injection of a short ion pulse into a combination of electric fields and field-free drift regions, as indicated in Figure 14.7. Ions leave the so-called source region, with an energy spread of ΔE_s, and are accelerated to an energy $eV_o \gg \Delta E_s$, before entering the actual spectrometer. During their field-free drift they disperse in time according to their m_i/e ratio. After being turned around in the mirror region they are focused at the detector. The measurement of their dispersed arrival time at the detector yields the m_i/e information. The resolution depends on the ion path length and is limited by the initial spread in ion energies. The ions are accelerated through a potential V_0, gaining energy before entering the TOF spectrometer; the measured time of flight is related to m_i/e by:

$$m_i/e = 2V_0T^2/L^2. \tag{14.19}$$

Taking the logarithmic derivative of Equation (4.19) gives:

$$m_i/\Delta m_i = T/\Delta T. \tag{14.20}$$

Equation (14.20) shows that to get good mass resolution a TOF spectrometer must have a very good time resolution.

A significant improvement in TOF spectrometers was made with the introduction of an electrostatic mirror, usually called a "reflectron" (see Figure 14.7). The more energetic and faster electrons penetrate deeper into the mirror field and spend more time being turned around than the lower energy ones. This corrects to a first order the flight time problems associated with the energy spread. As indicated in Figure 14.7, the drift region of length, L_0, is held at the accelerating potential, V_0, which is followed by a retarding lens of length L_1, and the mirror depth is L_2 with a potential of V_2 at the back. Given this configuration the mass resolution, which takes into account the energy spread of ΔE_s, so that the total energy of the entering ion, $E = eV_0, +\Delta E_s$, can be written as:

$$m_i/\Delta m_i = 2\{L_0 + 2L_2(E/q - V_1)/V_2\}/\{(L_0)(\Delta E_s/eV_0)^{1/2}\}. \quad (14.21)$$

A problem common to all ion mass spectrometers on spacecraft (in fact, common to all direct-measurement devices of charged particles) is the conversion of measured collector currents to actual ion densities. For ion mass spectrometers mounted on rockets and satellites, processes inside the sensor, as well as those controlling the effective collection area associated with the aperture of the sensor, have to be considered. The conversion from ion currents to ion densities is generally performed by resorting to a laboratory calibration of the instrument or by normalizing to ion densities obtained from a total ion collector (RPA) using the appropriate formulas discussed in Section 14.3.

14.5   Magnetometers

The motion of ionospheric charged particles is constrained by the magnetic field, therefore a detailed knowledge of the spatial and temporal variations of this field is of primary importance. Magnetometers have been widely used for such measurements aboard terrestrial[26] and planetary space missions.[27] These measurements can also provide information on low frequency waves. The two most commonly used basic types of magnetometers are (1) the fluxgate magnetometers and (2) the helium magnetometers.

The principle of operation of a fluxgate magnetometer is based on the nonlinear magnetic saturation characteristics of its core material. Consider a saturable transformer whose characteristics can be represented by straight line segments, as shown in Figure 14.8. The external magnetic field of intensity ΔH biases the core, as indicated, and a triangular-shaped driving signal of frequency $1/T$ is applied to the primary of the transformer, which causes the core to saturate when H reaches H_C. In general practice the driving waveform is sinusoidal; the triangular waveform is used here only to simplify the discussion. The output signal induced into the secondary is, according to Faraday's Law, proportional to the time rate of the change of flux, which is alternatively driven between plus and minus saturation. The output signal, therefore, consists of nonuniformly spaced positive and negative pulses, where the separation is related to the external ambient magnetic field.

Fourier analysis of the output waveform gives the results shown in Figure 14.8. Even harmonics of the primary driving frequency appear in the output only in the presence of an external magnetic field. The ratio, r, of the second to the first

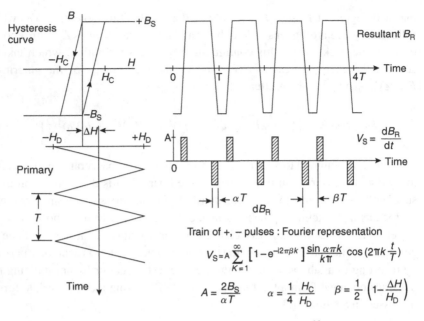

Figure 14.8 Principle of operation of a fluxgate magnetometer.[28]

harmonic is:

$$r = \left[\frac{1 - \exp\left(-i\pi\{1 - \Delta H/H_{\mathrm{D}}\}^2\right)}{1 - \exp\left(-i\pi\{1 - \Delta H/H_{\mathrm{D}}\}\right)} \right] \frac{\sin 2\pi\alpha}{2\sin\pi\alpha}, \tag{14.22}$$

where the various symbols are defined in Figure 14.8. If $\Delta H \ll H_{\mathrm{D}}$, and $H_{\mathrm{D}} \gg H_{\mathrm{C}}$, the ratio r is approximately given as:

$$r \approx i\pi \frac{\Delta H}{H_{\mathrm{D}}}. \tag{14.23}$$

Thus the second harmonic is $\pm 90°$ out of phase with the primary signal and its amplitude is a linear function of the ambient component parallel to the core axis. To measure the three vector components, three such systems have to be used. Beyond the difficulty associated with measuring the small second harmonic component, the spurious magnetic fields associated with the spacecraft have to be minimized, both by ensuring magnetic "cleanliness" and by placing the magnetometers on a boom away from the spacecraft interferences. It has also been a common practice to place two magnetometers separated by some distance on the boom, to be able to use the $1/r^3$ dependence of the spacecraft-generated field to account for it properly.

A helium magnetometer can be designed to operate in either a scalar or vector mode. The operation of the scalar magnetometer is based on the Zeeman effect, i.e., field dependent light absorption, and optical pumping to sense the field. Helium in an absorption cell is excited by a radio frequency (RF) discharge to maintain a population of metastable, long-lived atoms. The cell is made to resonate in a manner

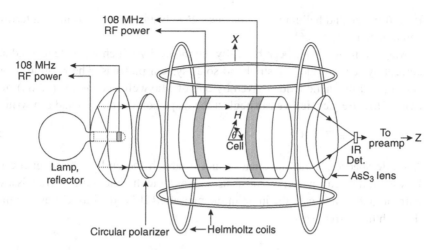

Figure 14.9 Schematic diagram of a helium magnetometer.

similar to a class of magnetometers called the alkali vapor magnetometers, e.g., rubidium. This magnetometer works best in very large fields such as those on the surface of the Earth, but is subject to corrections that are functions of the direction of the magnetic field that are relatively larger for smaller fields. Also the magnetometer has a directional dead zone in which it does not work, a cone with half angle of about 45°, centered on the optical axis of the detector.

In vector mode the helium magnetometer does not resonate the helium cell at the Larmor frequency, but uses the cell as a null detector. As shown in Figure 14.9, circularly polarized light at a wavelength of 1083 nm is passed through the cell. The ambient magnetic field affects the pumping efficiency of the metastable helium population. The optical pumping efficiency is proportional to $\cos^2 \vartheta$ where ϑ is the angle between the optical axis and the direction of the magnetic field. The Helmholtz coils around the cell are driven to create a rotating magnetic field alternately in two planes. The light received by the detector depends on the vector sum of the rotating sweep field and the ambient field. The detector signal is then amplified and phase detected using the sweep field as a reference. This signal is then applied to feedback coils in each axis. The feedback keeps the helium cell in zero field. The current required to null the field is then proportional to the ambient field in each axis, just like the fluxgate magnetometer. This magnetometer, as implemented on *Ulysses* and *Cassini* spacecrafts, has good zero-level stability. However, it cannot achieve the bandwidth of a fluxgate magnetometer. Typically a vector helium magnetometer has a sample rate of about 1 Hz whereas fluxgates are often run about 100 Hz.

The vector helium magnetometer, unlike the scalar helium magnetometer, works well in low fields. Thus, it is sometimes switched from vector to scalar mode where the spacecraft enters magnetic fields above 250 nT. It should be noted that, because of the directional variation of the sensitivity of the scalar helium magnetometer, in most applications it needs to be operated in parallel with a vector magnetometer.

Both fluxgate and helium magnetometers have been widely used in both terrestrial and planetary studies.[26,27]

Magnetometers have been basically the means by which we have learned about current systems in the terrestrial and solar system plasmas. The relation between current and magnetic field is given by one of Maxwell's equations (3.76d); in the case of low frequencies we can neglect the displacement current and can write

$$\mu_0 \mathbf{J} = \nabla \times \mathbf{B}. \tag{14.24}$$

A single spacecraft measurement cannot establish the curl of the magnetic field. However, it is reasonable, for most cases, to assume that the actual current is a sheet current, and in that case the jump in the measured $\mathbf{B}$ is equal to $\mu_0 \mathbf{J}$ as one moves through the current sheet.

14.6   Radio reflection

The first indication of the presence of the terrestrial ionosphere was by "remote sensing," as described in Chapter 1. All the early radio techniques were based on the fact that the refractive index, μ, of a weakly ionized plasma is proportional to the free electron number density. The so-called Appleton–Hartree equations give the general value of μ in the presence of collisions and a magnetic field. In the highly simplified case that neglects collisions and magnetic field effects, the *refractive index* is simply

$$\mu^2 = 1 - \frac{\omega_p^2}{\omega^2}, \tag{14.25}$$

where ω_p is the plasma frequency (see Equations 2.6 or 6.43) and ω is the frequency of the propagating wave.

An *ionosonde* or *ionospheric sounder*, the oldest ionospheric remote sensing device and one that is still widely used, transmits a radio pulse vertically and measures the time it takes for the signal to return. The reflection takes place, to a first order, where $\omega_p = \omega$. Thus, the time delay is used to determine the altitude of reflection, and the frequency is an indicator of the electron density at that location. In actuality, the interpretation of an *ionogram*, the delay time versus frequency characteristics, is more complicated. One complication is that the radio wave travels at the group velocity and not at the constant velocity of light, and this group velocity is itself a function of the refractive index.

If one neglects magnetic field effects and collisions, the so-called round-trip delay time for a horizontally stratified ionosphere, as a function of frequency, is given by:

$$\Delta t(f) = \frac{2}{c} \int_0^{z(f_p)} \frac{dz}{\sqrt{1 - \frac{\omega_p^2(z)}{\omega^2}}}, \tag{14.26}$$

where integration is carried out from the transmitter location to the altitude of the reflection location, $z(f_p)$.

The height calculated assuming that the waves travel with the velocity of light is called the *virtual height*. A further complication pertains to the effect of the geomagnetic field, which leads to multiple values of the refractive index. This, in turn, results in different propagation paths and velocities, giving rise to the so-called ordinary and extraordinary waves (Section 6.9).

Despite all these inversion difficulties associated with arriving at an electron density profile from the return signal, ionosondes have been the workhorse for monitoring the terrestrial electron densities below the altitude of the peak density. The highest frequency that can be reflected, at vertical incidence, is called the *critical frequency*. Thus, ground-based transmitters are limited to making measurements only up to an altitude corresponding to the maximum electron density (F_2 peak). A transmitter on a satellite that orbits at high enough altitudes can make measurements down to the altitude of the peak density.[30] Such transmitters have flown in the past carried by both terrestrial and planetary orbiters (e.g., *Alouette*, *ISIS*, *EXOS*, *IMAGE*, and *Mars Express* satellites) and will undoubtedly be used again. Although the terminology has not been widely used, an ionosonde is a monostatic radar system in which the transmitter and receiver are co-located. Thus, such a device can have multiple uses; as an example the radar system on *Mars Express* was designed to study both the ionosphere and the surface and subsurface of Mars.[31]

Modern ionosondes are sophisticated, digital instruments that automatically scale the ionograms and provide the ionospheric parameters in real time.[32] A representative digital ionogram is shown in Figure 14.10. The symbols E_o, F_{1o}, F_{2o} are the critical frequencies for ordinary wave reflections from the E, F_1, and F_2 regions,

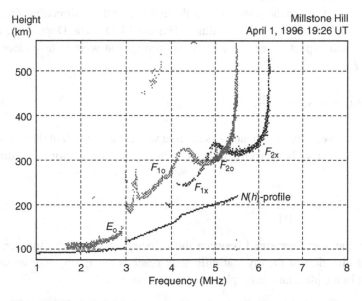

Figure 14.10 A digital ionogram, with an incorporated automatic scaling procedure, taken at Millstone Hill, MA., on April 1, 1996. The deduced electron density altitude profile is plotted, in terms of the corresponding plasma frequency. The various symbols are defined in the text. The autoscaled electron density profile is also shown.[32]

respectively, and F_{1x} and F_{2x} are for extraordinary reflections from the F_1 and F_2 regions, respectively. The autoscaled electron density profile is also indicated in Figure 14.10.

14.7   Radio occultation

The simplest and the most common ionospheric remote sensing technique that has been used outside the Earth is the *radio occultation technique*.[33–36] This method is based on the fact that radio waves transmitted from a satellite, as it flies behind a solar system body (e.g., planet or moon), pass through an atmosphere and ionosphere and undergo refractive bending, which introduces a *Doppler shift* in addition to its free space value. This difference, commonly called *Doppler residual*, Δf_d, is proportional to the refractive index of the media through which the wave travels.

To convert the time-varying Doppler shift into a quantity suitable for inversion, the trajectory or *ephemeris* of the spacecraft needs to be used. For each time t_j for which the value of the Doppler residual, Δf_d, is available, the spacecraft ephemeris provides a position vector relative to the receiving station on the Earth, as well as position and velocity vectors relative to the center of the planet. The component of the planet-centered spacecraft velocity in the direction of the Earth, $\mathbf{v_e}$, one light propagation interval after the time t_j, is given by

$$\mathbf{v_e} = (\mathbf{v} \cdot \hat{\mathbf{e}})\hat{\mathbf{e}}, \tag{14.27}$$

where $\mathbf{v}$ is the planet-centered velocity in the plane containing the spacecraft and the centers of the Earth and the planet and $\hat{\mathbf{e}}$ is the unit vector in the direction of the Earth one light propagation time after the time t_j (Figure 14.11). The Doppler frequency that one would expect to see at the Earth, if propagation were to take place in the direction $\hat{\mathbf{e}}$, is

$$\Delta f_e = \frac{f}{c}|\mathbf{v_e}| = \frac{|\mathbf{v_e}|}{\lambda}, \tag{14.28}$$

where c is the velocity of light and λ is the free space wavelength of the transmitted signal. The angle Ψ_e between the velocity vector, $\mathbf{v}$, and the vector in the direction of the Earth is

$$\Psi_e = \cos^{-1}\frac{|\mathbf{v_e}|}{|\mathbf{v}|}. \tag{14.29}$$

However, the angle measured between the velocity vector of the spacecraft and the actual direction of the ray that ultimately reaches the Earth, after undergoing refraction in the planetary atmosphere and ionosphere, is[36]

$$\Psi = \cos^{-1}\left[\frac{c}{f|\mathbf{v}|}(\Delta f_e + \Delta f_d)\right]. \tag{14.30}$$

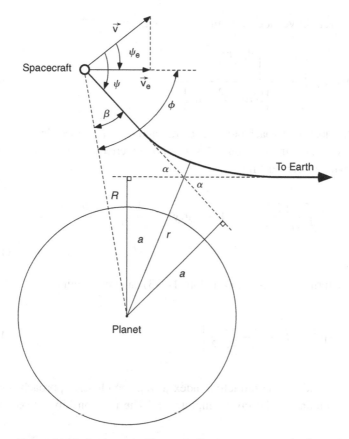

Figure 14.11 A schematic diagram indicating a representative ionosphere occultation geometry. The solid line shows the ray path followed by the radio signal propagating from the spacecraft to Earth, which lies in the plane containing the transmitting and receiving antennas and the local center of curvature of the planet.

The *refractive bending angle*, α, is then given by

$$\alpha = \psi - \psi_{\mathrm{e}}. \tag{14.31}$$

Finally, as can be seen from Figure 14.11, the ray asymptote distance, a, is

$$a = R \sin \beta, \tag{14.32}$$

where R is the distance from the center of the planet to the spacecraft, $\beta = \phi - \alpha$, and ϕ is the angle subtended by the Earth and the center of the planet at the spacecraft. Thus, one now has, in effect, the refractive bending angle as a function of the ray asymptote distance. However, the desired result is the refractive index, $\mu(r)$, which can be obtained from $\alpha(a)$ by an inversion procedure. It can be shown that the bending angle, α, corresponding to a ray passing through a spherically symmetrical medium with a refractive index variation, $\mu(r)$, is given by[36-9]

$$\alpha(a) = 2a \int_{r_0}^{\infty} \frac{\left[\dfrac{d\mu(r)}{dr}\right] dr}{\mu(r) \left\{ [r\mu(r)]^2 - a^2 \right\}^{1/2}}, \tag{14.33}$$

where r_0 is the closest approach point of the ray and $a = \mu_0 r_0$ (see Equation 14.36 below). The expression in Equation (14.33) can be inverted using the Abel integral transform,[36] which states that if

$$f(w) = k \int_0^w \frac{g'(y)\, dy}{(w - y)^{1/2}} \quad \text{then} \quad g(y) = \frac{1}{k\pi} \int_0^y \frac{f(w)\, dw}{(y - w)^{1/2}} + g_0. \tag{14.34}$$

Using this transform relationship, Equation (14.33) can be written as

$$\mu_j = \exp\left[\frac{1}{\pi} \int_{a_j}^{\infty} \frac{\alpha(a)\, da}{(a^2 - a_j^2)^{1/2}}\right], \tag{14.35}$$

where μ_j corresponds to the refractive index at r_{0j}, the closest approach point of the ray, corresponding to the ray asymptote, a_j. The relationship between these parameters is

$$r_{0j} = \frac{a_j}{\mu_j(r_{0j})}, \tag{14.36}$$

where $\mu_j(r_{0j})$ is the refractive index at the closest approach for this ray. The *refractivity*, N, is defined as

$$N = (\mu - 1) \times 10^6. \tag{14.37}$$

The relationship between the electron density, n_e, in units of cm^{-3} and refractivity is

$$n_{ej} = \frac{-f^2 N_j}{4.03 \times 10^3}, \tag{14.38}$$

where N_j is given in units of μ_j and f is in Hz.

A significant improvement in the sensitivity of radio occultation measurements can be achieved by using two harmonically related frequencies. In this way, any nondispersive (not frequency dependent) effects are eliminated when the signals from the two frequencies are differenced. This allows the elimination of the uncertainties in the motion of the spacecraft and the propagation through the neutral atmosphere of the Earth, for example. The two frequencies used by NASA and

JPL for radio occultation measurements are the S-band ($\sim$2.4 GHz) and X-band ($\sim$8.8 GHz), which are related by

$$f_X = \frac{11}{3} f_S \tag{14.39}$$

and so the differential Doppler residual is

$$\Delta f_{SX} = \Delta f_S - \left(\frac{3}{11}\right) \Delta f_X. \tag{14.40}$$

This differential Doppler is converted to a bending angle and inverted in the conventional manner described above to obtain the differential refractivity, N_{SXj}. The relationship between the electron density and N_{SXj} is

$$n_{ej} = \frac{-f_S^2 N_{SXj}}{4.03 \times 10^{13}} \left[1 - \left(\frac{3}{11}\right)^2\right]. \tag{14.41}$$

An example of the measured Doppler residuals, Δf_S, from *Mariner 6* S-band results at Mars[37] is shown in Figure 14.12, along with the refractivity profile obtained by the type of inversion outlined above. The negative and positive refractivities correspond to the ionosphere and neutral atmosphere, respectively.

The inversion of the radio occultation measurements to a refractive index or, in effect, to the electron density altitude profile has been made in all past cases by assuming spherical symmetry in the regions probed by the signal. This can introduce significant errors, especially near the terminator region and in cases where patchy layers are present. Unfortunately, the radio occultation measurements at the outer planets are obtained from near the terminator, and sharp layers appear to be

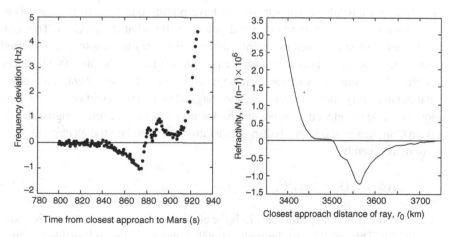

Figure 14.12 The measured S-band frequency residuals, Δf_d, from the *Mariner 6* entry data, and the corresponding refractivity profile, obtained by the appropriate inversion.[37]

present. This makes the interpretations difficult, but nevertheless still very useful. Furthermore, the signal has to pass through the interplanetary medium and the terrestrial ionosphere before it is received on the ground. It is reasonable to assume that during the short period of occultation these do not change and, therefore, do not introduce a significant error.

In the past, radio occultation has been used exclusively for planetary exploration. However, recently, the availability of a fleet of *global positioning system* (GPS) satellites has allowed column density measurements of the terrestrial ionosphere over a very wide geographic range. These data are then used, with *tomographic inversion* methods, to provide near real-time ionospheric distributions.[40] This information has become an important component of space weather activities.

14.8   Incoherent (Thomson) radar backscatter

About a century ago, J. J. Thomson established that single electrons are capable of scattering electromagnetic waves.[41] The radar cross section corresponding to such a single electron scattering event is

$$\sigma_e = 4\pi (r_e \sin \psi)^2, \tag{14.42}$$

where r_e is the classical electron radius and ψ is the angle between the direction of the incident electric field and the direction of the observer. If the only density fluctuations in the ionospheric plasma come from random thermal motion, the resulting cross section, σ, for energy backscatter by a unit volume in the ionosphere is simply[42]

$$\sigma = n_e \sigma_e, \tag{14.43}$$

where n_e is the electron density. The scattered radar return signal from the electrons in a finite volume of the ionosphere will have phases that vary in time and bear no relation to each other, and the signal powers will add at the receiver. The term *incoherent scatter* was used to describe this process. This terminology is still used today, even though we now know that the presence of ions in the plasma introduces a degree of coherence. There was an attempt to introduce the name *Thomson scatter*, but it did not gain general acceptance. It was argued in the 1950s that powerful radars should be able to detect this incoherent backscatter and that the return signal should have a Gaussian shape with a half-power width determined by the electron thermal motion, as given by

$$\Delta f_e = \frac{1}{\lambda} (8kT_e/m_e)^{1/2}, \tag{14.44}$$

where k is the Boltzmann constant, T_e is the electron temperature, and λ is the radar wavelength. This means that the return signals were expected to be widely spread in frequency. It was on this basis that the construction of the *Arecibo Observatory* was proposed.

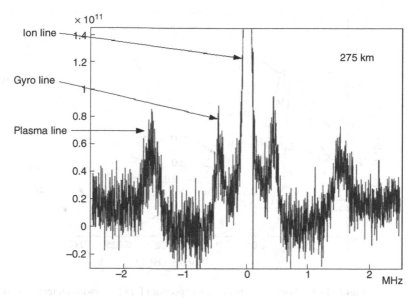

Figure 14.13 Ion, gyro and plasma lines observed at Arecibo (courtesy of A. N. Bhatt).

The first successful detection of backscattered radar signals was achieved by Bowles in 1958, using a high power transmitter with a large dipole array in Illinois.[43] However, contrary to the initial expectations, the observed bandwidth of the return signal was much narrower than predicted by Equation (14.44), and it was found to be related to the ion motion in the plasma. This led to a number of comprehensive theoretical papers, which all demonstrated that when the radar wavelength is much longer than the Debye length, λ_D (Equation 2.4), the scattering arises from density fluctuations resulting from longitudinal oscillations in the plasma. The main wave components are ion-acoustic waves and electron-induced waves at the plasma and electron gyrofrequency (see Figure 14.13). The power spectrum of the scattered signal is given by an extremely complex relation, which is not given here, but can be found in a variety of references.[44–47] The parameter α_D, charactarizing the ratio of the Debye length, λ_D, to the radar wavelength, λ, is defined as

$$\alpha_D = 4\pi \lambda_D / \lambda. \tag{14.45}$$

When α_D is large, the wavelength is small compared with the Debye length and the scattering is from individual electrons. In this case, the nature of the return signal is as originally anticipated. As this parameter decreases and becomes much less than unity, the amount of power in the electronic component of the return spectrum decreases and appears as a single line, Doppler shifted by approximately the plasma frequency of the scattering volume. Now the major portion of the returned signal is concentrated in the ionic component, which is caused by the organized motion (oscillations) of the plasma. The width of this ion component in the return spectrum

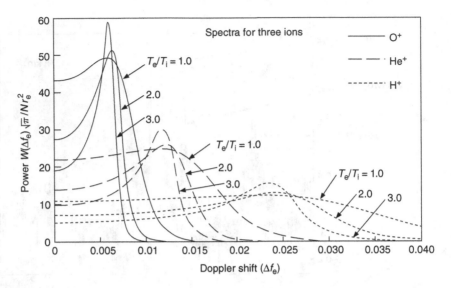

Figure 14.14 Normalized backscatter spectrum for electron-to-ion temperature ratios of 1.0, 2.0, and 3.0, and for O^+, He^+, and H^+. The frequency scale is normalized to Δf_e, therefore the spectra show the true relative widths.[48]

is of the order of the Doppler shift, Δf_i, corresponding to the mean speed of the ions

$$\Delta f_i = \frac{1}{\lambda}(8kT_i/m_i)^{1/2}. \tag{14.46}$$

The fact that the return signal is concentrated in such a narrow spectral region makes this method a practical and powerful technique for exploring the terrestrial ionosphere, without the need of extremely high power radar facilities. This *ion line* has been used to extensively study the ionosphere, and the remaining discussion focuses on this component of the return signal.

Figure 14.14 shows the spectra of the three main ionospheric species for different ratios of the electron-to-ion temperature, T_e/T_i, for $\alpha_D \rightarrow 0$.[47,48] If the mean ion velocity of the plasma in the scattering volume is not zero, the returned spectrum is Doppler shifted with respect to the transmitter frequency. Thus, the return, or echo, signal carries information about the electron density and temperature, the ion mass and temperature, and the mean ion velocity along the line-of-sight of the radar, inside the scattering volume. Least square fits of the returned spectra to theoretical ones have been used at a number of radar facilities. Figure 14.15 shows such an observation obtained at Millstone Hill, Massachusetts.

In the other commonly used method, pairs of short pulses are transmitted, separated by a short interval, τ, which allows the *autocorrelation function* between the echoes from the altitude of interest to be computed as τ is varied over the appropriate range $[0 \leq \tau \leq (\Delta f_i)^{-1}]$. This method has been used extensively at the Arecibo Observatory in Puerto Rico. Figure 14.16 is an example of data obtained

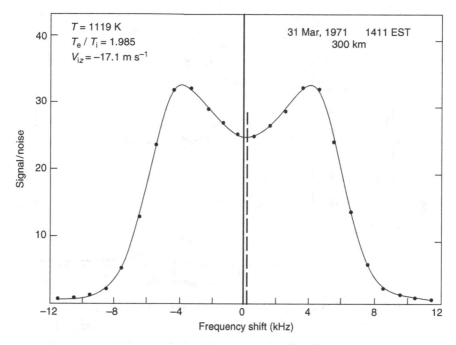

Figure 14.15 Radar backscatter spectrum, from an altitude of 300 km, measured at Millstone Hill, MA., on March 31, 1971. The solid line shows the theoretical fit to the data points. The dashed vertical line indicates the mean shift of the returned spectrum due to the mean motion of the ionospheric plasma. The temperature values and the line-of-sight velocity deduced are also indicated. (Courtesy of J. M. Holt.)

at Arecibo, displayed and fitted in both the frequency (spectra) and time domain (autocorrelation).

At altitudes below about 120 km, collisions of the electrons and ions with the neutral background gas become important.[49] The parameter of significance is the ratio of the radar wavelength to the mean-free-paths of the electrons and ions, λ_e and λ_i, respectively. The ratio, usually denoted by the symbol Ψ, is written as

$$\Psi_{e,i} = \lambda/(4\pi\lambda_{e,i}). \tag{14.47}$$

The mean-free-path of the electrons is about an order of magnitude larger than the ion mean-free-path at the same altitude. Therefore, the ion–neutral collisions are of greater significance in this case. Figure 14.17 shows how the return spectrum changes as this Ψ parameter increases. As the ion–neutral mean-free-path becomes comparable to or smaller than $\lambda/4\pi$, the double-humped spectrum disappears and it is no longer possible to establish the T_e/T_i ratio from the spectrum. The total scattered power remains the same, independent of the collision frequency.

In general, both the theory and data analyses of ionospheric incoherent radar measurements assume that the plasma has a Maxwellian distribution. This is a good assumption in most cases. However, in situations when this is not true, the use

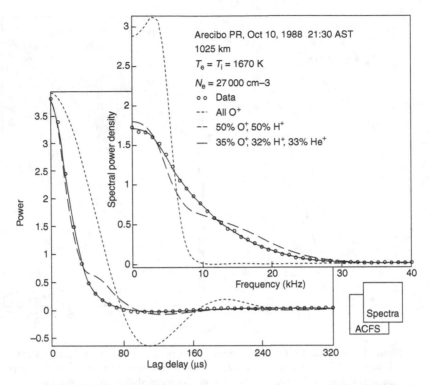

Figure 14.16 Radar backscatter data, corresponding to an altitude of 1025 km, obtained at Arecibo, Puerto Rico, on October 10, 1988. The data points are shown by open circles and the different possible fits are also indicated. The data have been analyzed both in the frequency and time domains. (Courtesy of M. P. Sulzer.)

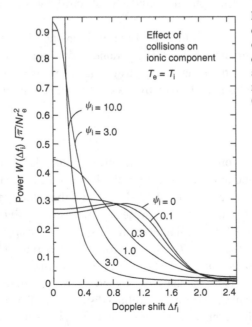

Figure 14.17 The effect of collisions on the ionic component in the case of $\alpha_D \to 0$ and $T_e = T_i$. The plots are for different values of the ratio Ψ_i; Ψ_e has been taken to be $0.1\ \Psi_i$.[49]

Table 14.1 Operating radar facilities.

Facility	Geographic latitude	Geographic longitude	Dip latitude	L-value	Peak transmitter power (MW)	Transmitting frequency (MHz)
Jicamarca, Peru, 1963–	11.9°S	76.9°W	1.1°N	1.1	5–6	50
Arecibo, Puerto Rico, 1963–	18.3°N	66.7°W	30.0°N	1.4	2	430
Millstone Hill, MA, USA 1960–	42.6°N	71.5°W	53.2°N	2.8	2.5	440
Kharkov, Ukraine	48.5°N	36.0°E	49.5°N	2.05	2.5	150
Irkutsk, Russia	52.2°N	104.5°E	71.0°N	11.9	2.5	150
Sondrestrom, Greenland, 1983–	67.0°N	51.0°W	71.0°N	>15.0	3.5	1290
EISCAT, Tromso, Norway 1981–	69.6°N	19.2°E	66.9°N	6.2	2.2 3.0	928.5 224
EISCAT, Svalbard, Norway 1996–	78.2°N	16.0°E	74.2°N	14.7	1	500
AMISR, Poker Flat, Alaska	65.1°N	147.5°W	66.74°N	5.71	1.5–2	450
AMISR, Resolute Bay, Canada	74.7°N	95.0°W	86.8°N	73	1.5–2	450
ALTAIR, Roi-Namur, Marshall Islands	9.39°N	167.47°E	–	–	4	422

of a Maxwellian distribution leads to incorrect results. For example, at high latitudes, in the presence of significant electric fields, the velocity distribution of the plasma becomes nonMaxwellian. Theoretical calculations have established how the return spectrum is modified in these cases and appropriate care needs to be taken to analyze such data.[50] It should also be mentioned that the discussion in this section is applicable to a magnetized plasma, except when the probing direction is nearly perpendicular to the magnetic field.

At present there are a number of operating radar facilities that are fully or partially dedicated to ionospheric research. All but the UHF EISCAT facility are pulsed, monostatic radars, with the transmitter and receiver co-located. The UHF EISCAT system is a tristatic one; the transmitter is located near Tromso, Norway, and two steerable receiving antennas are located near Kiruna, Sweden, and Södönkylä, Finland. Some details about the location, transmitting power, and frequency of all currently operating facilities are shown in Table 14.1.

The latest backscatter facility to be constructed is the *Advanced Modular Incoherent Scatter Radar (AMISR)* system. It is a phased array incoherent scatter radar with unique features that allow efficient and cost-effective dismantling, shipping, and reassembly. Three identical antenna faces have been constructed, each with a sensitivity somewhat better than the incoherent scatter radar currently operating in Sondre Stromfjord, Greenland. An AMISR antenna face is approximately 35 meters square and contains more than 3000 individual radiating elements with a transmitting power of 500 W each. The first of the three faces began operation in 2007 at Poker Flat, Alaska. The second and third faces will be deployed at Resolute Bay, Nunavut, in Arctic Canada.

14.9   Specific references

1. Mott-Smith, H. M., and I. Langmuir, The theory of collectors in gaseous discharges, *Phys. Rev.*, **28**, 727, 1926.
2. Dow, W. G. and A. F. Reifman, Dynamic probe measurements in the ionosphere, *Phys. Rev.*, **76**, 987, 1949.
3. Krehbiel, J. P., L. H. Brace, R. F. Theish *et al.*, Pioneer Venus Orbiter electron temperature probe, *IEEE Trans. Geosci. Rem. Sen.*, **GE-18**, 49, 1980.
4. Kanal, M., Theory of current collection of moving cylindrical probes, *J. Appl. Phys.*, **35**, 1697, 1964.
5. Dote, T., H. Amemiya, and T. Ichimiya, Effect of the geomagnetic field on an ionospheric sounding probe, *J. Geophys. Res.*, **70**, 2258, 1965.
6. Druyvestyn, M. J., Der Niedervoltbogen, *Z. Phys.*, **64**, 781, 1930.
7. Hays, P. B., and A. F. Nagy, Thermal electron energy distribution measurements in the ionosphere, *Planet. Space Sci.*, **21**, 1301, 1973.
8. Abe, T., T. Okuzawa, K.-I. Oyama, H. Amemiya, and S. Watanake, Measurements of temperature and velocity distribution of thermal electrons by the Akebono (EXOS-D) satellite, *J. Geomagn. Geoelectr.*, **42**, 537, 1990.
9. Oyama, K., Electron temperature measurements carried out by Japanese scientific satellites, *Adv. Space Res.*, **11(10)**, 149, 1991.
10. Nagy, A. F., L. H. Erace, G. R. Carignan, and M. Kanal, Direct measurements bearing on the extent of the thermal nonequilibrium in the ionosphere, *J. Geophys. Res.*, **68**, 6401, 1963.

11. Sagalyn, R. C., and R. H. Wand, Daytime rocket and Thomson scatter studies of the lower ionosphere, *J. Geophys. Res.*, **76**, 3783, 1971.

12. Knudsen, W. C., J. Bakke, K. Spenner, and V. Novak, Pioneer Venus Orbiter retarding potential analyzer plasma experiment, *Trans. IEEE Trans. Geosci. Rem. Sens.*, **GE-18**, 54, 1980.

13. Hanson, W. B., and D. D. McKibbin, An ion trap measurement of the ion concentration profile above the F2 peak, *J. Geophys Res.*, **66**, 1667, 1961.

14. Knudsen, W. C., Evaluation and demonstration of the use of retarding potential analyzers for measuring several ionospheric quantities, *J. Geophys. Res.*, **71**, 4669, 1966.

15. Whipple, E. C., The ion trap results in exploration of the upper atmosphere with the help of the third Soviet Sputnik, *Proc. IRE*, **47**, 2023, 1959.

16. Hanson, W. B., D. R. Zuccaro, C. R. Lippincott, and S. Sanatani, The retarding-potential analyzer on Atmosphere Explorer, *Radio Sci.*, **8**, 333, 1973.

17. Bennett, W. H., Radio frequency mass spectrometer, *J. Appl. Phys.*, **21(2)**, 143, 1950.

18. Johnson, C. Y., and E. B. Meadows, First investigation of ambient positive ion composition to 219 km by rocket-borne spectrometer, *J. Geophys. Res.*, **60**, 193, 1955.

19. Istomin, V. G., Investigation of the ionosphere composition of the Earth's atmosphere on geophysical rockets, *Planet. Space Sci.*, **9**, 179, 1962.

20. Taylor, H. A., H. C. Brinton, T. C. G. Wagner, B. H. Blackwell, and G. R. Cordier, Bennett ion mass spectrometers on the Pioneer Venus Bus and Orbiter, *IEEE Trans. Geosci. Rem. Sen.* **GE-18**, 44, 1980.

21. Hoffman, J. H., W. B. Hanson, C. R. Lippincott, and E. E. Ferguson, The magnetic ion-mass spectrometer on Atmosphere Explorer, *Radio Sci.*, **8**, 315, 1973.

22. Paul, W., H. P. Reinhard, and U. Von Zahn, Das elektrische Massenfilter als Massenspektrometer und Iosotopentrenner, *Z. Phys.*, **152**, 143, 1958.

23. Niemann, H. B., J. R. Booth, J. E. Cooley, *et al.*, Pioneer Venus Orbiter Neutral Gas Mass Spectrometer Experiment, *IEEE Trans. Geosci. Rem. Sens.*, **GE-18**, 60, 1980.

24. Kasprzak, W. T., H. Niemann, D. Harpold *et al.*, Cassini orbiter ion and neutral mass spectrometer instrument, *Proc. Soc. Photo-Optical Instr. Eng.*, **2803**, 129, 1996.

25. Young, D. T., Mass spectrometry for planetary science, in *Atmospheres in the Solar System*, ed. by M. Mendillo, A. Nagy, and J. H. Waite, Geophysical Monograph **130**, American Geophysical Union, 2002.

26. Elphic, R. C., J. D. Means, R. C. Snare, *et al.*, Magnetic field instruments for the FAST auroral snapshot explorer, *Space Sci. Rev.*, **98**, 151, 2001.

27. Dougherty, M. K., S. Kellock, D. J. Southwood, *et al.*, The Cassini magnetic field investigation, *Space Sci. Rev.*, **114**, 331, 2004.

28. Bauer, S. J. and A. F. Nagy, Ionospheric direct measurement techniques, *Proc. IEEE*, **63**, 230, 1975.

29. Hunsucker, R. D. *Radio Techniques for Probing the Terrestrial Ionosphere*, Berlin: Springer-Verlag, 1991.

30. Jackson, J. E., E. R. Schmerling, and J. H. Whitteker, Mini-review on topside sounding, *IEEE Trans. Antennas Prop.*, **AP-28**, 284, 1980.

31. Picardi, G., *et al.*, MARSIS: Mars advanced radar for subsurface and ionosphere sounding, in *Mars Express: A European Mission to the Red Planet*, Eur. Space Agency Spec. Publ., **SP-1240**, 51, 2004.

32. Reinisch, B. W., Modern ionosondes, in *Modern Ionospheric Science*, ed. by H. Kohl, R. Ruster, and K. Schlegel, **440**, Lindau, Germany: Max-Planck-Institüt für Aeronomie, 1996.

33. Hinson, D. P., F. M. Flasar, A. J. Kliore, *et al.*, Jupiter's ionosphere: results from the first Galileo radio occultation experiment, *Geophys. Res. Lett.*, **24**, 2107, 1997.

34. Pätzold, M., S. Tellmann, B. Häusler, *et al.*, A sporadic third layer in the ionosphere of Mars, *Science*, **310**, 837, 2005.

35. Nagy, A. F., A. J. Kliore, E. Marouf, *et al.*, First results from the ionospheric radio occultations of Saturn by the Cassini Spacecraft, *J. Geophys. Res.*, **111**, No. A6, A06310, 10.1029/2005JA011519, 2006.

36. Fjeldbo, G., A. J. Kliore, and V. R. Eshleman, The neutral atmosphere of Venus as studied with the Mariner V radio occultation experients, *Astron. J.*, **76**, 123, 1971.

37. Kliore, A. J., Current methods of radio occultation data inversion, *Proc. Workshop on the Mathematics of Profile Inversion*, ed. by L. Colin, NASA TM-X-62, 1972.

38. Phinney, R. A., and D. L. Anderson, On the radio occultation method for studying planetary atmospheres, *J. Geophys. Res.*, **73**, 1819, 1968.

39. Eshleman, V. R., The radio occultation method for the study of planetary atmospheres, *Planet. Space Sci.*, **21**, 1521, 1973.

40. Mannucci, A. J., B. D. Wilson, D. N. Yuan, *et al.*, A global mapping technique for GPS-derived ionospheric total electron content measurements, *Radio Sci.*, **33**, 565, 1998.

41. Thomson, J. J., *Conduction of Electricity Through Gases*, Cambridge, UK: Cambridge University Press, 1906.

42. Fejer, J., Scattering of radio waves by an ionized gas in thermal equilibrium, *Can. J. Phys.*, **38**, 1114, 1960.

43. Bowles, K. L., Observations of vertical incidence scatter from the ionosphere at 41 Mc/sec, *Phys. Rev. Lett.*, **1**, 454, 1958.

44. Dougherty, J. P., and D. T. Farley, A theory of incoherent scattering of radio waves by a plasma, *Proc. Roy. Soc. (London)*, **A259**, 79, 1960.

45. Renau, J., Scattering of electromagnetic waves from a non-degenerate ionized gas, *J. Geophys. Res.*, **65**, 3631, 1960.

46. Fejer, J. A., Scattering of radiowaves by an ionized gas in thermal equilibrium in the presence of a uniform magnetic field, *Can. J. Phys.*, **39**, 716, 1961.

47. Evans, J. V., Theory and practice of ionosphere study by Thomson scatter radar, *Proc. IEEE*, **57**, 496, 1969.

48. Moorcroft, D. R., On the determination of temperature and ionic composition by electron backscattering from the ionosphere and magnetosphere, *J. Geophys. Res.*, **69**, 955, 1964.

49. Dougherty, J. P. and D. T. Farley, A theory of incoherent scattering of radio waves by a plasma. III: scattering in a partly ionized gas, *J. Geophys. Res.*, **66**, 5473, 1963.

50. Raman, Venkat R. S., J. P. St-Maurice, and R. S. B. Ong, Incoherent scattering of radar waves in the auroral ionosphere, *J. Geophys. Res.*, **86**, 4751, 1981.

14.10   General references

Bauer, S. J., *Physics of Planetary Ionospheres*, Berlin: Springer-Verlag, 1973.

Bauer, S. J., and A. F. Nagy, Ionospheric direct measurement techniques, *Proc. IEEE*, **63**, 230, 1975.

Brace, L. H., Langmuir probe measurements in the ionosphere, in *Measurement Techniques in Space Plasmas: Particles*, ed. by J. Borovsky, R. Pfaff, and D. Young, AGU Monograph, **102**, Washington, DC, 1998.

Evans, J. V., Theory and practice of ionosphere study by Thomson scatter radar, *Proc. IEEE*, **57**, 496, 1969.

Hunsucker, R. D., *Radio Techniques for Probing the Terrestrial Ionosphere*, Berlin: Springer-Verlag, 1991.

Kelley, M. C., *The Earth's Ionosphere*, New York: Academic Press, 1989.

Pfaff, R. F., *In-situ* measurement techniques for ionospheric research, in *Modern Ionospheric Science*, ed. by H. Kohl, R. Ruster, and K. Schlegel, 459, Lindau, Germany: Max-Planck-Institüt für Aeronomie, 1996.

Pfaff, R. F., J. E. Borovsky, and D. T. Young (eds.), *Measurement Techniques in Space Plasmas*, Geophysical Monograph, **102**, AGU, 1998.

Reinisch, B. W., Modern ionosondes, in *Modern Ionospheric Science*, ed. by H. Kohl, R. Ruster, and K. Schlegel, 440, Lindau, Germany: Max-Planck-Institüt für Aeronomie, 1996.

Willmore, A. P., Electron and ion temperatures in the ionosphere, *Space Sci. Rev.*, **11**, 607, 1970.

Appendix A

Physical constants and conversions

A.1 Physical constants

e	1.602×10^{-19} C	Fundamental charge
k	1.381×10^{-23} J K^{-1}	Boltzmann constant
c	2.998×10^{8} m s^{-1}	Speed of light
h	6.626×10^{-34} J·s	Planck constant
ε_0	8.854×10^{-12} C^2 N^{-1} m^{-2}	Permittivity of vacuum
μ_0	$4\pi \times 10^{-7}$ N A^{-2}	Permeability of vacuum
m_e	9.109×10^{-31} kg	Electron mass
m_i	1.673×10^{-27} kg	Proton mass
M_i	1.6605×10^{-27} kg	Mass of unit atomic weight
m_p/m_e	1836	Mass ratio
G	6.674×10^{-11} m^3 s^{-2} kg^{-1}	Gravitational constant
N_0	6.022×10^{23} mol^{-1}	Avogadro number
S_0	1340 W m^{-2}	Solar constant
a_0	5.29×10^{-11} m	Radius of first Bohr orbit
r_0	2.82×10^{-15} m	Classical electron radius

A.2 Conversions

1 AU	=	1.496×10^{11} m
1 meter	=	100 cm
1 angstrom	=	10^{-10} m
1 joule	=	10^7 erg
1 joule	=	0.2389 calorie
1 watt	=	10^7 erg s^{-1}

1 newton	=	10^5 dyne
1 kilogram	=	10^3 grams
1 coulomb	=	3×10^9 statcoulomb
1 volt	=	$(1/300)$ statvolt
1 volt m^{-1}	=	$(1/3) \times 10^{-4}$ statvolt cm^{-1}
1 tesla	=	10^4 gauss
1 tesla	=	10^9 gamma
1 gamma	=	10^{-5} gauss
1 weber	=	10^8 maxwell
1 farad	=	9×10^{11} esu
1 amp	=	3×10^9 statamp
1 ohm	=	$(1/9) \times 10^{-11}$ s cm^{-2}
1 henry	=	$(1/9) \times 10^{-11}$ s^2 cm^{-1}
1 eV	=	1.602×10^{-19} joule
1 eV	=	1.602×10^{-12} erg
1 eV	=	11 610 K
1 eV	=	3.827×10^{20} calorie
1 eV photon	=	1239.8 nm
1 eV photon	=	12398 Å
1 eV/particle	=	2.305×10^4 calorie mol^{-1}
1 eV/particle	=	9.649×10^4 joule mol^{-1}
1 pascal	=	1 newton m^{-2} = 10 dyne cm^{-2}
1 bar	=	10^5 newton m^{-2}
1 atm	=	1.013×10^5 newton m^{-2}
1 barn	=	10^{-28} m^2

Appendix B

Vector relations and operators

B.1   Vector relations

$$\mathbf{A} \cdot (\mathbf{B} \times \mathbf{C}) = \mathbf{B} \cdot (\mathbf{C} \times \mathbf{A}) = \mathbf{C} \cdot (\mathbf{A} \times \mathbf{B})$$

$$\mathbf{A} \times (\mathbf{B} \times \mathbf{C}) = (\mathbf{A} \cdot \mathbf{C})\mathbf{B} - (\mathbf{A} \cdot \mathbf{B})\mathbf{C}$$

$$(\mathbf{A} \times \mathbf{B}) \cdot (\mathbf{C} \times \mathbf{D}) = (\mathbf{A} \cdot \mathbf{C})(\mathbf{B} \cdot \mathbf{D}) - (\mathbf{A} \cdot \mathbf{D})(\mathbf{B} \cdot \mathbf{C})$$

$$(\mathbf{A} \times \mathbf{B}) \times (\mathbf{C} \times \mathbf{D}) = (\mathbf{A} \times \mathbf{B} \cdot \mathbf{D})\mathbf{C} - (\mathbf{A} \times \mathbf{B} \cdot \mathbf{C})\mathbf{D}$$

$$\nabla(\phi\psi) = \phi\nabla\psi + \psi\nabla\phi$$

$$\nabla(\mathbf{A} \cdot \mathbf{B}) = \mathbf{A} \times (\nabla \times \mathbf{B}) + \mathbf{B} \times (\nabla \times \mathbf{A})$$
$$+ (\mathbf{A} \cdot \nabla)\mathbf{B} + (\mathbf{B} \cdot \nabla)\mathbf{A}$$

$$\nabla \cdot (\phi\mathbf{A}) = \phi\nabla \cdot \mathbf{A} + \mathbf{A} \cdot \nabla\phi$$

$$\nabla \cdot (\mathbf{A} \times \mathbf{B}) = \mathbf{B} \cdot \nabla \times \mathbf{A} - \mathbf{A} \cdot \nabla \times \mathbf{B}$$

$$\nabla \cdot \nabla\phi = \nabla^2\phi$$

$$\nabla \cdot (\nabla \times \mathbf{A}) = 0$$

$$\nabla \times (\phi\mathbf{A}) = \phi\nabla \times \mathbf{A} + \nabla\phi \times \mathbf{A}$$

$$\nabla \times (\mathbf{A} \times \mathbf{B}) = \mathbf{A}(\nabla \cdot \mathbf{B}) - \mathbf{B}(\nabla \cdot \mathbf{A}) + (\mathbf{B} \cdot \nabla)\mathbf{A} - (\mathbf{A} \cdot \nabla)\mathbf{B}$$

$$(\nabla \times \mathbf{A}) \times \mathbf{B} = (\mathbf{B} \cdot \nabla)\mathbf{A} - (\nabla\mathbf{A}) \cdot \mathbf{B}$$

$$\nabla \times (\nabla \times \mathbf{A}) = \nabla(\nabla \cdot \mathbf{A}) - \nabla^2\mathbf{A}$$

$$\nabla \times \nabla\phi = 0$$

B.2   Vector operators

Cartesian (x, y, z)

$$\mathrm{d}\mathbf{r} = \mathrm{d}x\,\mathbf{e}_x + \mathrm{d}y\,\mathbf{e}_y + \mathrm{d}z\,\mathbf{e}_z$$

$$\mathrm{d}V = \mathrm{d}x\,\mathrm{d}y\,\mathrm{d}z$$

$$\nabla\psi = \mathbf{e}_x\frac{\partial\psi}{\partial x} + \mathbf{e}_y\frac{\partial\psi}{\partial y} + \mathbf{e}_z\frac{\partial\psi}{\partial z}$$

$$\nabla\cdot\mathbf{A} = \frac{\partial A_x}{\partial x} + \frac{\partial A_y}{\partial y} + \frac{\partial A_z}{\partial z}$$

$$\nabla\times\mathbf{A} = \mathbf{e}_x\left(\frac{\partial A_z}{\partial y} - \frac{\partial A_y}{\partial z}\right) + \mathbf{e}_y\left(\frac{\partial A_x}{\partial z} - \frac{\partial A_z}{\partial x}\right) + \mathbf{e}_z\left(\frac{\partial A_y}{\partial x} - \frac{\partial A_x}{\partial y}\right)$$

$$\nabla^2\psi = \frac{\partial^2\psi}{\partial x^2} + \frac{\partial^2\psi}{\partial y^2} + \frac{\partial^2\psi}{\partial z^2}$$

Cylindrical $(\rho, \theta, \mathbf{z})$

$$\mathrm{d}\mathbf{r} = \mathrm{d}\rho\,\mathbf{e}_\rho + \rho\mathrm{d}\theta\,\mathbf{e}_\theta + \mathrm{d}z\,\mathbf{e}_z$$

$$\mathrm{d}V = \rho\mathrm{d}\rho\,\mathrm{d}\theta\,\mathrm{d}z$$

$$\nabla\psi = \mathbf{e}_\rho\frac{\partial\psi}{\partial\rho} + \mathbf{e}_\theta\frac{1}{\rho}\frac{\partial\psi}{\partial\theta} + \mathbf{e}_z\frac{\partial\psi}{\mathrm{d}z}$$

$$\nabla\cdot\mathbf{A} = \frac{1}{\rho}\frac{\partial}{\partial\rho}(\rho A_\rho) + \frac{1}{\rho}\frac{\partial A_\theta}{\partial\theta} + \frac{\partial A_z}{\partial z}$$

$$\nabla\times\mathbf{A} = \mathbf{e}_\rho\left(\frac{1}{\rho}\frac{\partial A_z}{\partial\theta} - \frac{\partial A_\theta}{\partial z}\right) + \mathbf{e}_\theta\left(\frac{\partial A_\rho}{\partial z} - \frac{\partial A_z}{\partial\rho}\right)$$

$$+ \mathbf{e}_z\frac{1}{\rho}\left[\frac{\partial}{\partial\rho}(\rho A_\theta) - \frac{\partial A_\rho}{\partial\theta}\right]$$

$$\nabla^2\psi = \frac{1}{\rho}\frac{\partial}{\partial\rho}\left(\rho\frac{\partial\psi}{\partial\rho}\right) + \frac{1}{\rho^2}\frac{\partial^2\psi}{\partial\theta^2} + \frac{\partial^2\psi}{\partial z^2}$$

Spherical (r, θ, ϕ)

$$\mathrm{d}\mathbf{r} = \mathrm{d}r\,\mathbf{e}_r + r\mathrm{d}\theta\,\mathbf{e}_\theta + r\sin\theta\mathrm{d}\phi\,\mathbf{e}_\phi$$

$$\mathrm{d}V = r^2\sin\theta\mathrm{d}r\,\mathrm{d}\theta\,\mathrm{d}\phi$$

$$\nabla \psi = \mathbf{e}_r \frac{\partial \psi}{\partial r} + \mathbf{e}_\theta \frac{1}{r} \frac{\partial \psi}{\partial \theta} + \mathbf{e}_\phi \frac{1}{r \sin \theta} \frac{\partial \psi}{\partial \phi}$$

$$\nabla \cdot \mathbf{A} = \frac{1}{r^2} \frac{\partial}{\partial r} (r^2 A_r) + \frac{1}{r \sin \theta} \frac{\partial}{\partial \theta} (\sin \theta A_\theta) + \frac{1}{r \sin \theta} \frac{\partial A_\phi}{\partial \phi}$$

$$\nabla \times \mathbf{A} = \mathbf{e}_r \frac{1}{r \sin \theta} \left[\frac{\partial}{\partial \theta} (\sin \theta A_\phi) - \frac{\partial A_\theta}{\partial \phi} \right] + \mathbf{e}_\theta \left[\frac{1}{r \sin \theta} \frac{\partial A_r}{\partial \phi} - \frac{1}{r} \frac{\partial}{\partial r} (r A_\phi) \right]$$

$$+ \mathbf{e}_\phi \frac{1}{r} \left[\frac{\partial}{\partial r} (r A_\theta) - \frac{\partial A_r}{\partial \theta} \right]$$

$$\nabla^2 \psi = \frac{1}{r^2} \frac{\partial}{\partial r} \left(r^2 \frac{\partial \psi}{\partial r} \right) + \frac{1}{r^2 \sin \theta} \frac{\partial}{\partial \theta} \left(\sin \theta \frac{\partial \psi}{\partial \theta} \right) + \frac{1}{r^2 \sin^2 \theta} \frac{\partial^2 \psi}{\partial \phi^2}$$

Dipole (q, p, φ)

The dipole coordinate system typically used in aeronomy is the right-handed orthogonal system (q, p, φ), which is defined in terms of spherical coordinates (r, θ, φ) as (Equations 11.19 and 11.20)

$$q = \frac{R_E^2 \cos \theta}{r^2} \qquad p = \frac{r}{R_E \sin^2 \theta} \qquad \varphi = \varphi$$

where the subscript d has been dropped on q and p.[1-3] For this system $\mathbf{B} = -B\mathbf{e}_q$ and[4]

$$d\mathbf{r} = \mathbf{e}_q \frac{r^3}{R_E^2 \left(1 + 3 \cos^2 \theta \right)^{1/2}} dq + \mathbf{e}_p \frac{R_E \sin^3 \theta}{\left(1 + 3 \cos^2 \theta \right)^{1/2}} dp$$

$$+ \mathbf{e}_\varphi r \sin \theta \, d\varphi$$

$$dV = \frac{r^4 \sin^4 \theta}{R_E \left(1 + 3 \cos^2 \theta \right)} dq \, dp \, d\varphi$$

$$\nabla \psi = \mathbf{e}_q \frac{R_E^2 \left(1 + 3 \cos^2 \theta \right)^{1/2}}{r^3} \frac{\partial \psi}{\partial q} + \mathbf{e}_p \frac{\left(1 + 3 \cos^2 \theta \right)^{1/2}}{R_E \sin^3 \theta} \frac{\partial \psi}{\partial p}$$

$$+ \mathbf{e}_\varphi \frac{1}{r \sin \theta} \frac{\partial \psi}{\partial \varphi}$$

$$\nabla \cdot \mathbf{A} = \frac{R_E^2 \left(1 + 3 \cos^2 \theta \right)}{r^4 \sin^4 \theta} \frac{\partial}{\partial q} \left[\frac{r \sin^4 \theta}{\left(1 + 3 \cos^2 \theta \right)^{1/2}} A_q \right]$$

$$+ \frac{\left(1 + 3 \cos^2 \theta \right)}{R_E r^4 \sin^4 \theta} \frac{\partial}{\partial p} \left[\frac{r^4 \sin \theta}{\left(1 + 3 \cos^2 \theta \right)^{1/2}} A_p \right] + \frac{1}{r \sin \theta} \frac{\partial A_\varphi}{\partial \varphi}$$

$$\nabla \times \mathbf{A} = \mathbf{e}_q \left[\frac{\left(1 + 3\cos^2\theta\right)^{1/2}}{R_E r \sin^3\theta} \frac{\partial}{\partial p} \left(A_\varphi r\right) - \frac{1}{r \sin\theta} \frac{\partial A_p}{\partial \varphi} \right]$$

$$+ \mathbf{e}_p \left[\frac{1}{r \sin\theta} \frac{\partial A_q}{\partial \varphi} - \frac{R_E^2 (1 + 3\cos^2\theta)^{1/2}}{r^4 \sin\theta} \frac{\partial}{\partial q} \left(A_\varphi \sin^3\theta\right) \right]$$

$$+ \mathbf{e}_\varphi \left[\frac{R_E^2 \left(1 + 3\cos^2\theta\right)}{r^3 \sin^3\theta} \frac{\partial}{\partial q} \left(\frac{A_p \sin^3\theta}{\left(1 + 3\cos^2\theta\right)^{1/2}} \right) \right.$$

$$\left. - \frac{(1 + 3\cos^2\theta)^{1/2}}{R_E r^3 \sin^3\theta} \frac{\partial}{\partial p} \left(A_q r^3\right) \right]$$

$$\nabla^2 \psi = \frac{R_E^4 \left(1 + 3\cos^2\theta\right)}{r^4 \sin^4\theta} \frac{\partial}{\partial q} \left[\frac{\sin^4\theta}{r^2} \frac{\partial\psi}{\partial q} \right]$$

$$+ \frac{\left(1 + 3\cos^2\theta\right)}{R_E^2 r^4 \sin^4\theta} \frac{\partial}{\partial p} \left[\frac{r^4}{\sin^2\theta} \frac{\partial\psi}{\partial p} \right] + \frac{1}{r^2 \sin^2\theta} \frac{\partial^2\psi}{\partial\varphi^2}$$

B.3   Specific references

1. Orens, J. H., T. R. Young, E. S. Oran, and T. P. Coffey, Vector operations in a dipole coordinate system, NRL Memo. Rep. 3984, Washington DC: Naval Research Laboratory, 1979.

2. Swisdok, M., Notes on the dipole coordinate system, *arXiv: physics/0606044* **1** 5 June 2006.

3. Kageyama, A., T. Sugiyama, K. Watanabe, and T. Sato, A note on the dipole coordinates, *Computers and Geoscience*, **32**, 265, 2006.

4. Wohlwend, C. S., Modeling the electrodynamics of the low-latitude ionosphere, Ph.D. dissertation, Logan, Utah: Utah State University, 2008.

Appendix C

Integrals and transformations

C.1   Integral relations

Divergence theorem

$$
\int_V dV \nabla \cdot \mathbf{A} = \oint_S da\, \hat{\mathbf{n}} \cdot \mathbf{A}
$$

$$
\int_V dV \nabla \phi = \oint_S da\, \hat{\mathbf{n}} \phi
$$

$$
\int_V dV \nabla \times \mathbf{A} = \oint_S da\, \hat{\mathbf{n}} \times \mathbf{A}
$$

S is a closed surface surrounding a volume V and $\hat{\mathbf{n}}$ is an outwardly directed unit normal on the surface. Note that these integral relations are valid both in configuration space and in velocity space.

Stokes theorem

$$
\int_S da\, (\nabla \times \mathbf{A}) \cdot \hat{\mathbf{n}} = \oint_C \mathbf{A} \cdot d\boldsymbol{\ell}
$$

$$
\int_S da\, \hat{\mathbf{n}} \times \nabla \phi = \oint_C \phi d\boldsymbol{\ell}
$$

S is an open surface that is surrounded by a closed curve C and $\hat{\mathbf{n}}$ is a unit vector that is perpendicular to the open surface. The direction of integration around the closed curve and the direction of $\hat{\mathbf{n}}$ are related by the right-hand rule.

C.2 Important integrals

$$\int_0^\infty dx\, e^{-\alpha^2 x^2} = \frac{1}{2\alpha}\sqrt{\pi}$$

$$\int_0^\infty dx\, x^{2n+1} e^{-\alpha^2 x^2} = \frac{n!}{2\alpha^{2n+2}}$$

$$\int_0^\infty dx\, x^{2n} e^{-\alpha^2 x^2} = \frac{(2n-1)(2n-3)\cdots 1}{2^{n+1}\alpha^{2n+1}}\sqrt{\pi}$$

where n is a positive integer or zero and $0! = 1$.

$$\int_0^x dt\, e^{-\alpha^2 t^2} = \frac{\sqrt{\pi}}{2\alpha}\mathrm{erf}(\alpha x)$$

$$\int_0^x dt\, t e^{-\alpha^2 t^2} = \frac{1}{2\alpha^2}\left(1 - e^{-\alpha^2 x^2}\right)$$

$$\int_0^x dt\, t^2 e^{-\alpha^2 t^2} = \frac{1}{2\alpha^2}\left[\frac{\sqrt{x}}{2\alpha}\mathrm{erf}(\alpha x) - x e^{-\alpha^2 x^2}\right]$$

$$\int_0^x dt\, t^{n+2} e^{-\alpha^2 t^2} = -\frac{1}{2\alpha}\frac{d}{d\alpha}\int_0^x dt\, t^n e^{-\alpha^2 t^2}$$

$$\int_0^x dt\, t e^{t^2}\mathrm{erf}(t) = \frac{1}{2}e^{x^2}\mathrm{erf}(x) - \frac{x}{\sqrt{\pi}}$$

$$\int_0^x dt\, t^3 e^{t^2}\mathrm{erf}(t) = \frac{1}{2}(x^2-1)e^{x^2}\mathrm{erf}(x) + \frac{x}{\sqrt{\pi}} - \frac{x^3}{3\sqrt{\pi}}$$

$$\int_0^x dt\, t^5 e^{t^2}\mathrm{erf}(t) = \frac{1}{2}(x^4 - 2x^2 + 2)e^{x^2}\mathrm{erf}(x) - \frac{2x}{\sqrt{\pi}} + \frac{2x^3}{3\sqrt{\pi}} - \frac{x^5}{5\sqrt{\pi}}$$

C.3   Integral transformations

In the course of evaluating multiple integrals, it is frequently necessary to change the integration variables from one coordinate system to another. Consider a multiple integral in terms of the variables (v_1, v_2, v_3),

$$\iiint dv_1 \, dv_2 \, dv_3 f(v_1, v_2, v_3).$$

The transformation of this multiple integral to one in terms of the variables (c_1, c_2, c_3) is accomplished with the aid of a Jacobian determinant, J,

$$\iiint dc_1 \, dc_2 \, dc_3 |J| f(c_1, c_2, c_3),$$

where

$$J = \frac{\partial(\mathbf{v})}{\partial(\mathbf{c})} = \frac{\partial(v_1, v_2, v_3)}{\partial(c_1, c_2, c_3)} = \begin{vmatrix} \dfrac{\partial v_1}{\partial c_1} & \dfrac{\partial v_2}{\partial c_1} & \dfrac{\partial v_3}{\partial c_1} \\[2mm] \dfrac{\partial v_1}{\partial c_2} & \dfrac{\partial v_2}{\partial c_2} & \dfrac{\partial v_3}{\partial c_2} \\[2mm] \dfrac{\partial v_1}{\partial c_3} & \dfrac{\partial v_2}{\partial c_3} & \dfrac{\partial v_3}{\partial c_3} \end{vmatrix},$$

and where the new integral extends over the range of values of the variables (c_1, c_2, c_3) that correspond to the range of values of the original variables (v_1, v_2, v_3). Note that what actually appears in the transformed integral is the magnitude of the Jacobian. Also, the Jacobian transformation is valid for both spatial and velocity integrals.

As an example, consider the transformation of the density integral from a Cartesian coordinate system in velocity space to a spherical coordinate system

$$n = \int_{-\infty}^{\infty} \int_{-\infty}^{\infty} \int_{-\infty}^{\infty} dv_x \, dv_y \, dv_z \, f(v_x, v_y, v_z).$$

The transformation is from (v_x, v_y, v_z) to (v, θ, ϕ), where

$$v_x = v \sin \theta \cos \phi,$$
$$v_y = v \sin \theta \sin \phi,$$
$$v_z = v \cos \theta,$$

$$J = \frac{\partial(v_x, v_y, v_z)}{\partial(v, \theta, \phi)} = \begin{vmatrix} \dfrac{\partial v_x}{\partial v} & \dfrac{\partial v_y}{\partial v} & \dfrac{\partial v_z}{\partial v} \\[2mm] \dfrac{\partial v_x}{\partial \theta} & \dfrac{\partial v_y}{\partial \theta} & \dfrac{\partial v_z}{\partial \theta} \\[2mm] \dfrac{\partial v_x}{\partial \phi} & \dfrac{\partial v_y}{\partial \phi} & \dfrac{\partial v_z}{\partial \phi} \end{vmatrix},$$

$$J = \begin{vmatrix} \sin\theta \cos\phi & \sin\theta \sin\phi & \cos\theta \\ v\cos\theta \cos\phi & v\cos\theta \sin\phi & -v\sin\theta \\ -v\sin\theta \sin\phi & v\sin\theta \cos\phi & 0 \end{vmatrix},$$

$$|J| = v^2 \sin\theta.$$

Because the original integral is over all velocities, in the new variables the limits of integration are $0 \le v < \infty$, $0 \le \theta \le \pi$, and $0 \le \phi \le 2\pi$. The transformed integral becomes

$$n = \int_0^\infty v^2 dv \int_0^\pi \sin\theta \, d\theta \int_0^{2\pi} d\phi f(v, \theta, \phi).$$

Appendix D

Functions and series expansions

D.1   Important functions

Error function

$$\operatorname{erf}(x) = \frac{2}{\sqrt{\pi}} \int\limits_0^x \mathrm{d}t \, \mathrm{e}^{-t^2}$$

$$= \frac{2}{\sqrt{\pi}} \left(x - \frac{x^3}{3} + \frac{x^5}{10} - \cdots \right) \quad \text{for } x \le 1$$

$$= 1 - \frac{2}{\sqrt{\pi}} \mathrm{e}^{-x^2} \left(\frac{1}{2x} - \frac{1}{4x^3} + \cdots \right) \quad \text{for } x \to \infty$$

$$\operatorname{erf}(-x) = -\operatorname{erf}(x)$$

Gamma function

$$\Gamma(x) = \int\limits_0^\infty \mathrm{d}t \, t^{x-1} \mathrm{e}^{-t}$$

$$= \sqrt{2\pi} \, x^{(x-1/2)} \mathrm{e}^{-x} \quad \text{for } x \to \infty$$

$$\Gamma(x+1) = x\Gamma(x)$$

$$\Gamma(1/2) = \sqrt{\pi}$$

$$\Gamma(n) = (n-1)! \quad \text{for } n = \text{integer}$$

D.2   Series expansions for small arguments

$$\sin x = x - \frac{x^3}{3!} + \frac{x^5}{5!} - \cdots$$

$$\cos x = 1 - \frac{x^2}{2!} + \frac{x^4}{4!} - \cdots$$

$$e^x = 1 + x + \frac{x^2}{2!} + \frac{x^3}{3!} + \cdots$$

$$\ln(1 + x) = x - \frac{x^2}{2} + \frac{x^3}{3} - \frac{x^4}{4} + \cdots$$

$$(1 + x)^m = 1 + mx + m(m - 1)\frac{x^2}{2!} + m(m - 1)(m - 2)\frac{x^3}{3!} + \cdots$$

$$I_m(x) = \frac{1}{\Gamma(m + 1)}\left(\frac{x}{2}\right)^m; \quad \text{modified Bessel function } (m \geq 0; x \ll 1)$$

$$f(x + \Delta x) = f(x) + \frac{df}{dx}\Delta x + \frac{1}{2}\frac{d^2 f}{dx^2}(\Delta x)^2 + \cdots$$

$$f(\mathbf{r} + \Delta\mathbf{r}) = f(\mathbf{r}) + \Delta\mathbf{r} \cdot \nabla f + \frac{1}{2}\Delta\mathbf{r}\Delta\mathbf{r} : \nabla\nabla f + \cdots$$

Appendix E

Systems of units

Throughout the book all equations and formulas are expressed in the MKSA system of units. However, Gaussian-cgs units are still frequently used by many scientists. In this latter system, all electrical quantities are in electrostatic units (esu) except for **B**, which is in electromagnetic units (emu). Most formulas that are in MKSA units can be converted to Gaussian-cgs units by replacing **B** with $(\mathbf{B}/c)$ and ε_0 by $1/4\pi$, where $c = (\varepsilon_0\mu_0)^{-1/2}$ is the speed of light.

For easy reference, the formulas in Table E.1 are given in both MKSA and Gaussian-cgs units. The last four equations are known as the Maxwell equations and, as given here, pertain to a vacuum.

Table E.1 Widely used formulas.

Quantity	Symbol	MKSA	Gaussian-cgs
Plasma frequency	$\omega_{p\alpha}$	$\left(\dfrac{n_\alpha e^2}{\varepsilon_0 m_\alpha}\right)^{1/2}$	$\left(\dfrac{4\pi n_\alpha e^2}{m_\alpha}\right)^{1/2}$
Cyclotron frequency	$\omega_{c\alpha}$	$\dfrac{eB}{m_\alpha}$	$\dfrac{eB}{m_\alpha c}$
Debye length	λ_D	$\left(\dfrac{\varepsilon_0 k T_e}{n_e e^2}\right)^{1/2}$	$\left(\dfrac{k T_e}{4\pi n_e e^2}\right)^{1/2}$
Larmor radius	r_α	$\dfrac{(2kT_\alpha/m_\alpha)^{1/2}}{\omega_{c\alpha}}$	$\dfrac{(2kT_\alpha/m_\alpha)^{1/2}}{\omega_{c\alpha}}$
$\mathbf{E} \times \mathbf{B}$ drift	u_E	$\dfrac{E}{B}$	$\dfrac{cE}{B}$
Lorentz force	$\mathbf{F}$	$q_\alpha[\mathbf{E} + \mathbf{v}_\alpha \times \mathbf{B}]$	$q_\alpha\left[\mathbf{E} + \dfrac{1}{c}\mathbf{v}_\alpha \times \mathbf{B}\right]$

Table E.1 *(Continued)*

Quantity	Symbol	MKSA	Gaussian-cgs
Pressure ratio	β	$\dfrac{n_i k(T_e + T_i)}{(B^2/2\mu_0)}$	$\dfrac{n_i k(T_e + T_i)}{(B^2/8\pi)}$
Alfvén speed	V_A	$\dfrac{B}{(\mu_0 n_i m_i)^{1/2}}$	$\dfrac{B}{(4\pi n_i m_i)^{1/2}}$
Acoustic speed	V_S	$\left[\dfrac{k(T_e + 3T_i)}{m_i}\right]^{1/2}$	$\left[\dfrac{k(T_e + 3T_i)}{m_i}\right]^{1/2}$
Gauss' law	—	$\nabla \cdot \mathbf{E} = \rho_c/\varepsilon_0$	$\nabla \cdot \mathbf{E} = 4\pi\rho_c$
Faraday's law	—	$\nabla \times \mathbf{E} = -\dfrac{\partial \mathbf{B}}{\partial t}$	$\nabla \times \mathbf{E} = -\dfrac{1}{c}\dfrac{\partial \mathbf{B}}{\partial t}$
No monopoles	—	$\nabla \cdot \mathbf{B} = 0$	$\nabla \cdot \mathbf{B} = 0$
Ampère's law	—	$\nabla \times \mathbf{B} = \mu_0 \mathbf{J} + \varepsilon_0\mu_0\dfrac{\partial \mathbf{E}}{\partial t}$	$\nabla \times \mathbf{B} = \dfrac{4\pi}{c}\mathbf{J} + \dfrac{1}{c}\dfrac{\partial \mathbf{E}}{\partial t}$

Appendix F

Maxwell transfer equations

In Chapter 3, the general transport equations were derived by taking velocity moments of the Boltzmann equation (3.24) with respect to the velocity $\mathbf{v}_s$. Although this is a straightforward procedure and easy to follow, most of the important moments are in terms of the random velocity $\mathbf{c}_s$. Therefore, an alternative way to derive the transport equations is first to express the Boltzmann equation in terms of $\mathbf{c}_s$ and then take a general velocity moment $\xi_s(\mathbf{c}_s)$. The resulting equation is known as the Maxwell transfer equation in terms of $\mathbf{c}_s$. It can also be obtained in terms of $\mathbf{v}_s$ (see end of this appendix).

To transform the Boltzmann equation, it is necessary to change from the independent variables $(\mathbf{r}, \mathbf{v}_s, t)$ to $(\mathbf{r}, \mathbf{c}_s, t)$, where

$$\mathbf{c}_s = \mathbf{v}_s - \mathbf{u}_s(\mathbf{r}, t). \tag{F.1}$$

The derivative terms in the Boltzmann equation become

$$\frac{\partial f_s(\mathbf{r}, \mathbf{v}_s, t)}{\partial t} = \frac{\partial f_s(\mathbf{r}, \mathbf{c}_s, t)}{\partial t} + \nabla_c f_s(\mathbf{r}, \mathbf{c}_s, t) \cdot \frac{\partial \mathbf{c}_s}{\partial t}$$

$$= \frac{\partial f_s(\mathbf{r}, \mathbf{c}_s, t)}{\partial t} - \nabla_c f_s(\mathbf{r}, \mathbf{c}_s, t) \cdot \frac{\partial \mathbf{u}_s}{\partial t}, \tag{F.2}$$

$$\nabla f_s(\mathbf{r}, \mathbf{v}_s, t) = \nabla f_s(\mathbf{r}, \mathbf{c}_s, t) + \nabla_c f_s(\mathbf{r}, \mathbf{c}_s, t) \cdot (\nabla \mathbf{c}_s)$$

$$= \nabla f_s(\mathbf{r}, \mathbf{c}_s, t) - \nabla_c f_s(\mathbf{r}, \mathbf{c}_s, t) \cdot (\nabla \mathbf{u}_s), \tag{F.3}$$

$$\nabla_v f_s(\mathbf{r}, \mathbf{v}_s, t) = \nabla_c f_s(\mathbf{r}, \mathbf{c}_s, t), \tag{F.4}$$

where the chain rule was used in taking derivatives. Substituting Equations (F.2), (F.3), and (F.4) into the Boltzmann equation (3.24) yields

$$\left(\frac{\partial f_s}{\partial t} - \frac{\partial \mathbf{u}_s}{\partial t} \cdot \nabla_c f_s\right) + (\mathbf{c}_s + \mathbf{u}_s) \cdot [\nabla f_s - (\nabla \mathbf{u}_s) \cdot \nabla_c f_s] + \mathbf{a}_s \cdot \nabla_c f_s = \frac{\delta f_s}{\delta t}.$$

$$(\text{F.5})$$

After rearranging the terms, the equation becomes

$$\frac{\partial f_s}{\partial t} + (\mathbf{c}_s + \mathbf{u}_s) \cdot \nabla f_s - \frac{D_s \mathbf{u}_s}{Dt} \cdot \nabla_c f_s - \mathbf{c}_s \cdot (\nabla \mathbf{u}_s) \cdot \nabla_c f_s + \mathbf{a}_s \cdot \nabla_c f_s = \frac{\delta f_s}{\delta t},$$

$$(\text{F.6})$$

where

$$\frac{D_s}{Dt} = \frac{\partial}{\partial t} + \mathbf{u}_s \cdot \nabla, \tag{F.7}$$

$$\mathbf{a}_s = \mathbf{G} + \frac{e_s}{m_s}[\mathbf{E} + (\mathbf{c}_s + \mathbf{u}_s) \times \mathbf{B}]. \tag{F.8}$$

Equation (F.6) is the Boltzmann equation expressed in the new independent variables $(\mathbf{r}, \mathbf{c}_s, t)$.

The next step in the derivation of the Maxwell transfer equation is to multiply Equation (F.6) by $\xi_s(\mathbf{c}_s)$, where ξ_s is an arbitrary function of velocity, and then integrate over all velocities. Considering each term separately, the first term becomes

$$\int d^3 c_s \xi_s \frac{\partial f_s}{\partial t} = \frac{\partial}{\partial t} \int d^3 c_s \xi_s f_s = \frac{\partial}{\partial t}[n_s \langle \xi_s \rangle], \tag{F.9}$$

where now $\partial/\partial t$ does not operate on $\mathbf{c}_s$ because $\mathbf{r}$, $\mathbf{c}_s$, and t are independent variables.

The second term can be manipulated as follows:

$$\int d^3 c_s \xi_s (\mathbf{c}_s + \mathbf{u}_s) \cdot \nabla f_s = \int d^3 c_s \xi_s \mathbf{c}_s \cdot \nabla f_s + \int d^3 c_s \xi_s \mathbf{u}_s \cdot \nabla f_s$$

$$= \int d^3 c_s (\mathbf{c}_s \cdot \nabla)(f_s \xi_s) + \int d^3 c_s (\mathbf{u}_s \cdot \nabla)(f_s \xi_s)$$

$$= \nabla \cdot \int d^3 c_s \mathbf{c}_s f_s \xi_s + \mathbf{u}_s \cdot \nabla \int d^3 c_s f_s \xi_s$$

$$= \nabla \cdot [n_s \langle \mathbf{c}_s \xi_s \rangle] + \mathbf{u}_s \cdot \nabla[n_s \langle \xi_s \rangle], \tag{F.10}$$

where ∇ does not operate on $\mathbf{c}_s$. Also, it should be noted the ξ_s can be any function of velocity, including a scalar, vector, or tensor of arbitrary order. Therefore, in the manipulations it must be remembered that the "dot" products are between ∇ and $\mathbf{c}_s$ in the first term and between ∇ and $\mathbf{u}_s$ in the second term. The velocity moment ξ_s is not involved in this vector operation.

The third term involves some extra steps

$$-\int d^3c_s \xi_s \frac{D_s \mathbf{u}_s}{Dt} \cdot \nabla_c f_s = -\int d^3 c_s \nabla_c \cdot \left(\frac{D_s \mathbf{u}_s}{Dt} \xi_s f_s \right)$$

$$+ \int d^3 c_s f_s \nabla_c \cdot \left(\frac{D_s \mathbf{u}_s}{Dt} \xi_s \right),$$

where the following mathematical identity was used:

$$\nabla_c \cdot \left(\frac{D_s \mathbf{u}_s}{Dt} f_s \xi_s \right) = \xi_s \frac{D_s \mathbf{u}_s}{Dt} \cdot \nabla_c f_s + f_s \nabla_c \cdot \left(\frac{D_s \mathbf{u}_s}{Dt} \xi_s \right).$$

When ξ_s is a scalar, this is the well-known expression for the divergence of a vector times a scalar. The expression is still valid if tensors are involved, but again, it is important to keep track of what vector is involved in the "dot" product. Because of the divergence theorem, which also holds if tensors are involved, the volume integral of a pure divergence can be converted into a surface integral that surrounds the volume;

$$\int d^3 c_s \nabla_c \cdot \left(\frac{D_s \mathbf{u}_s}{Dt} \xi_s f_s \right) = \int_{S_c} dA_c \, \hat{\mathbf{n}}_c \cdot \frac{D_s \mathbf{u}_s}{Dt} f_s \xi_s = 0,$$

where the surface is at infinity in velocity space and where it is assumed that $f_s \to 0$ as $c_s \to \infty$ at a rate fast enough to insure that $f_s \xi_s \to 0$ for any ξ_s. Therefore,

$$-\int d^3 c_s \xi_s \frac{D_s \mathbf{u}_s}{Dt} \cdot \nabla_c f_s = \int d^3 c_s f_s \nabla_c \cdot \left(\frac{D_s \mathbf{u}_s}{Dt} \xi_s \right) = \int d^3 c_s f_s \frac{D_s \mathbf{u}_s}{Dt} \cdot \nabla_c \xi_s$$

$$= \frac{D_s \mathbf{u}_s}{Dt} \cdot \int d^3 c_s f_s \nabla_c \xi_s = n_s \frac{D_s \mathbf{u}_s}{Dt} \cdot \langle \nabla_c \xi_s \rangle. \quad \text{(F.11)}$$

The fourth term must be manipulated carefully, to keep track of "dot" products. This can be easily done by temporarily introducing index notation, where a repeated Greek letter implies a summation over the coordinate indices,

$$-\int d^3 c_s \xi_s \mathbf{c}_s \cdot \nabla \mathbf{u}_s \cdot \nabla_c f_s = -\int d^3 c_s \xi_s c_{s\alpha} \frac{\partial u_{s\beta}}{\partial x_\alpha} \frac{\partial f_s}{\partial c_{s\beta}}$$

$$= -\frac{\partial u_{s\beta}}{\partial x_\alpha} \int d^3 c_s \xi_s c_{s\alpha} \frac{\partial f_s}{\partial c_{s\beta}}$$

$$= -\frac{\partial u_{s\beta}}{\partial x_\alpha} \left[\int d^3 c_s \frac{\partial}{\partial c_{s\beta}} (\xi_s c_{s\alpha} f_s) \right.$$

$$\left. - \int d^3 c_s f_s \frac{\partial}{\partial c_{s\beta}} (c_{s\alpha} \xi_s) \right]$$

$$= \frac{\partial u_{s\beta}}{\partial x_\alpha} \left\langle \frac{\partial}{\partial c_{s\beta}} (c_{s\alpha} \xi_s) \right\rangle n_s$$

$$= (\nabla \mathbf{u}_s) : \langle \nabla_c (\mathbf{c}_s \xi_s) \rangle n_s, \tag{F.12}$$

where the same vector identity was used as for the third term and where one of the volume integrals was converted to a surface integral at infinity and then set to zero.

The last term on the left-hand side of Equation (F.6) becomes

$$\int d^3 c_s \xi_s \mathbf{a}_s \cdot \nabla_c f_s = \int d^3 c_s \xi_s \nabla_c \cdot (\mathbf{a}_s f_s)$$

$$= \int d^3 c_s \nabla_c \cdot (\mathbf{a}_s f_s \xi_s) - \int d^3 c_s f_s \mathbf{a}_s \cdot \nabla_c \xi_s$$

$$= -n_s \langle \mathbf{a}_s \cdot \nabla_c \xi_s \rangle, \tag{F.13}$$

where $\nabla_c \cdot \mathbf{a}_s = 0$ and where the manipulations are similar to those done for the third and fourth terms.

Collecting the terms given by Equations (F.9) to (F.13) yields the *Maxwell transfer equation*

$$\frac{\partial}{\partial t} [n_s \langle \xi_s \rangle] + \nabla \cdot [n_s \langle \mathbf{c}_s \xi_s \rangle] + \mathbf{u}_s \cdot \nabla [n_s \langle \xi_s \rangle]$$

$$+ n_s \frac{D_s \mathbf{u}_s}{Dt} \cdot \langle \nabla_c \xi_s \rangle - n_s \langle \mathbf{a}_s \cdot \nabla_c \xi_s \rangle$$

$$+ n_s (\nabla \mathbf{u}_s) : \langle \nabla_c (\mathbf{c}_s \xi_s) \rangle = \int d^3 c_s \xi_s \frac{\delta f_s}{\delta t}. \tag{F.14}$$

The general transport equations can be obtained from Equation (F.14) by selecting the appropriate velocity moments. For example, setting ξ_s equal to one, $m_s \mathbf{c}_s$, $(1/2) m_s c_s^2$, $m_s \mathbf{c}_s \mathbf{c}_s$, and $(1/2) m_s c_s^2 \mathbf{c}_s$ yields the continuity, momentum, energy, pressure tensor, and heat flow equations, respectively, for species s.

As an example, the simple case of $\xi_s = 1$ is considered

$$\langle \xi_s \rangle = 1,$$

$$\langle \mathbf{c}_s \xi_s \rangle = \langle \mathbf{c}_s \rangle = 0,$$

$$\nabla_c \xi_s = 0,$$

$$\nabla_c (\mathbf{c}_s \xi_s) = \nabla_c \mathbf{c}_s = \mathbf{I},$$

where $\mathbf{I}$ is the unit tensor ($\delta_{\alpha\beta}$ in index notation). Substituting these quantities into the Maxwell transfer equation (F.14) yields:

$$\frac{\partial n_s}{\partial t} + \mathbf{u}_s \cdot \nabla n_s + n_s \nabla \cdot \mathbf{u}_s = \frac{\delta n_s}{\delta t},$$

or

$$\frac{\partial n_s}{\partial t} + \nabla \cdot (n_s \mathbf{u}_s) = \frac{\delta n_s}{\delta t}, \tag{F.15}$$

where

$$\nabla \mathbf{u}_s : \mathbf{I} = \frac{\partial u_\alpha}{\partial x_\beta} \delta_{\alpha\beta} = \frac{\partial u_\alpha}{\partial x_\alpha} = \nabla \cdot \mathbf{u}_s.$$

The Maxwell transfer equation is sometimes derived in terms of a function $\xi_s(\mathbf{v}_s)$ instead of $\xi_s(\mathbf{c}_s)$. In this case, it is not necessary to express the Boltzmann equation in terms of $\mathbf{c}_s$, as was done above. Instead, the transfer equation is derived in a manner similar to that used in Chapter 3 to derive the continuity, momentum, and energy equations. Specifically, the Boltzmann equation (3.24) is multiplied by $\xi_s(\mathbf{v}_s)$ and then the resulting equation is integrated over all velocities. After algebraic manipulations similar to those described above, the Maxwell transfer equation for the general velocity moment $\xi_s(\mathbf{v}_s)$ can be expressed in the form

$$\frac{\partial}{\partial t}[n_s\langle\xi_s\rangle] + \nabla \cdot [n_s\langle\mathbf{v}_s\xi_s\rangle] - n_s\langle\mathbf{a}_s \cdot \nabla_v\xi_s\rangle = \int d^3 v_s \xi_s \frac{\delta f_s}{\delta t}. \tag{F.16}$$

Equation (F.16) is equivalent to Equation (F.14). However, when using Equation (F.16), remember that the higher velocity moments (T_s, $\boldsymbol{\tau}_s$, $\mathbf{P}_s$, $\mathbf{q}_s$, etc.) are defined relative to the random velocity $\mathbf{c}_s$, whereas in Equation (F.16) $\xi_s = \xi_s(\mathbf{v}_s)$. Therefore, additional algebra is required when Equation (F.16) is used instead of Equation (F.14) to obtain the transport equations for the higher velocity moments.

Appendix G

Collision models

G.1 Boltzmann collision integral

The collision term $\delta f_s/\delta t$ describes the rate of change of the velocity distribution as a result of collisions. The effect of a collision is instantaneously to change the velocity of a particle, and hence, collisions cause the sudden appearance and disappearance of particles in velocity space. Consider a spatial volume element d^3r about a position $\mathbf{r}$ and a velocity volume element d^3v_s about a velocity $\mathbf{v}_s$ (Figure 3.1). If the rate of change of f_s due to particles scattered into d^3v_s is denoted by $\delta f_s^+/\delta t$ and the rate of change f_s due to particles scattered out of d^3v_s is denoted by $\delta f_s^-/\delta t$, then

$$\frac{\delta f_s}{\delta t} = \frac{\delta f_s^+}{\delta t} - \frac{\delta f_s^-}{\delta t}. \tag{G.1}$$

The velocity–space production and loss terms in Equation (G.1) were calculated by Boltzmann assuming binary elastic collisions between particles possessing symmetric force fields.[1] In addition, Boltzmann based his derivation on the assumption of *molecular chaos*, which means that there is no correlation between the positions and velocities of the different particles before collisions. Considering first the loss term, $\delta f_s^-/\delta t$, the number of s particles in the spatial volume d^3r and the velocity element d^3v_s is

$$f_s d^3v_s d^3r. \tag{G.2}$$

Some of these s particles will be scattered out of the velocity element d^3v_s by t particles that are in the same spatial element d^3r and in a velocity element d^3v_t. With binary elastic collisions, an individual s particle is exposed to a flux of t particles with a relative velocity $\mathbf{g}_{st}$, with impact parameters between b and $b + db$, and with collision planes between ε and $\varepsilon + d\varepsilon$ (Figure G.1). In a time dt, the flux of

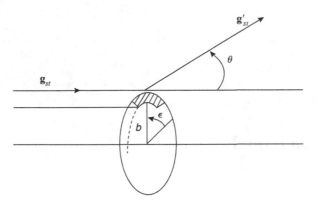

Figure G.1 Coordinate system used to calculate $\delta f_s/\delta t$. The relative velocities before and after the collision are $\mathbf{g}_{st}$ and $\mathbf{g}'_{st}$, respectively, b is the impact parameter, ε is the azimuthal angle that defines the plane of the collision, and θ is the scattering angle. The cross-hatched area is $b\,db\,d\varepsilon$.

t particles occupies a cylindrical spatial volume of $(g_{st}dt)b\,db\,d\varepsilon$, and the number of t particles in this volume is

$$(f_t d^3 v_t)(g_{st} b\,db\,d\varepsilon\,dt). \tag{G.3}$$

This number also corresponds to the number of collisions a single s particle has with the t particles in the cylindrical volume element and in a time dt. Therefore, the number of collisions that scatter s particles out of the velocity element $d^3 v_s$ due to interactions with t particles in element $d^3 v_t$ in a time dt is obtained by multiplying (G.3) by the number of s particles in the phase space element $d^3 v_s d^3 r$ (Equation G.2)

$$dN_c^- = (f_s d^3 v_s\,d^3 r)(f_t d^3 v_t\,g_{st} b\,db\,d\varepsilon\,dt). \tag{G.4}$$

Therefore, the total number of s particles scattered out of velocity element $d^3 v_s$ in a time dt, N_c^-, is obtained by integrating Equation (G.4) over all t particle velocities, all impact parameters, and all collision plane orientations

$$N_c^- = f_s d^3 v_s\,d^3 r\,dt \iiint d^3 v_t\,d\varepsilon\,b\,db\,g_{st} f_t. \tag{G.5}$$

When this total number of scattered s particles is divided by $d^3 v_s d^3 r dt$, the result is

$$\frac{\delta f_s^-}{\delta t} = \iiint d^3 v_t\,d\varepsilon\,b\,db\,g_{st} f_t f_s, \tag{G.6}$$

where f_s can be put under the integrals because it is not involved in the integrations.

To calculate the term $\delta f_s^+/\delta t$, which accounts for the appearance of s particles in $d^3 v_s$ due to collisions with t particles, it is necessary to consider *inverse collisions*.

An inverse collision is one in which an s particle with an initial velocity in $d^3v'_s$ about $\mathbf{v}'_s$ is scattered into d^3v_s about $\mathbf{v}_s$ due to a collision with a t particle having an initial velocity in $d^3v'_t$ about $\mathbf{v}'_t$. Also, in an inverse symmetric collision, the t particle is in the same impact parameter range (between b and $b + db$) and the same collision plane range (between ε and $\varepsilon + d\varepsilon$) as in the previous case where the s particle was scattered out of d^3v_s. Note, it can be shown that inverse collisions always exist for binary elastic collisions when the interparticle force field is symmetric.[2]

The calculation of $\delta f_s^+/\delta t$ proceeds in a manner similar to that described above for $\delta f_s^-/\delta t$. An individual s particle in volume elements d^3r and $d^3v'_s$ will, in a time dt, collide with all the t particles in a cylindrical spatial volume of size $(g'_{st}\,dt)(b\,db\,d\varepsilon)$, and this number is

$$(f'_t d^3v'_t)(g'_{st} b\,db\,d\varepsilon\,dt), \tag{G.7}$$

where $f'_t = f_t(\mathbf{r}, \mathbf{v}'_t, t)$. The number of s particles in $d^3r\,d^3v'_s$ that are scattered into $d^3r\,d^3v_s$ in a time dt is obtained by multiplying Equation (G.7) by $f'_s d^3v'_s\,d^3r$, which yields

$$dN_c^+ = (f'_s d^3v'_s\,d^3r)(f'_t d^3v'_t\,g'_{st} b\,db\,d\varepsilon\,dt). \tag{G.8}$$

Therefore, the total number of s particles scattered into d^3v_s in a time dt is obtained by integrating over all t particle velocities, all impact parameters, and all collision plane orientations, which yields

$$N_c^+ = f'_s d^3v'_s d^3r dt \iiint d^3v'_t\,d\varepsilon\,b\,db\,g'_{st} f'_t. \tag{G.9}$$

In an elastic collision, $g'_{st} = g_{st}$ (Equation 4.17). The volume elements $d^3v'_s d^3v'_t$ can be related to $d^3v_s d^3v_t$ with the aid of a Jacobian (Appendix C)

$$d^3v'_s\,d^3v'_t = |J|d^3v_s\,d^3v_t, \tag{G.10}$$

where

$$J = \frac{\partial(\mathbf{v}'_s, \mathbf{v}'_t)}{\partial(\mathbf{v}_s, \mathbf{v}_t)} = 1. \tag{G.11}$$

The evaluation of the Jacobian of the transformation is accomplished with the aid of Equations (4.6) to (4.17). Using Equations (G.10) and (G.11), Equation (G.9) can be expressed in the form

$$N_c^+ = f'_s d^3v_s\,d^3r\,dt \iiint d^3v_t\,d\varepsilon\,b\,db\,g_{st} f'_t. \tag{G.12}$$

Dividing Equation (G.12) by $d^3v_s d^3r\, dt$ yields

$$\frac{\delta f_s^+}{\delta t} = \iiint d^3v_t\, d\varepsilon\, b\, db\, g_{st} f_t' f_s', \tag{G.13}$$

where $f_s' = f_s(\mathbf{r}, \mathbf{v}_s', t)$ can be put under the integrals because it is not involved in the integrations.

The Boltzmann collision integral can now be obtained by substituting Equations (G.6) and (G.13) into Equation (G.1), which yields

$$\frac{\delta f_s}{\delta t} = \iiint d^3v_t\, d\varepsilon\, b\, db\, g_{st}(f_s' f_t' - f_s f_t). \tag{G.14}$$

An alternative form of Equation (G.14) can be obtained by relating $b\, db\, d\varepsilon$ to the differential scattering cross section (Equation 4.44)

$$b\, db\, d\varepsilon = \sigma_{st}(g_{st}, \theta) \sin\theta\, d\theta\, d\varepsilon$$
$$= \sigma_{st}(g_{st}, \theta)\, d\Omega, \tag{G.15}$$

which yields the following form for the Boltzmann collision integral:

$$\frac{\delta f_s}{\delta t} = \iint d^3v_t\, d\Omega\, \sigma_{st}(g_{st}, \theta) g_{st}(f_s' f_t' - f_s f_t). \tag{G.16}$$

An important quantity is the rate of change of the mean value of a transport property, $\xi_s(\mathbf{v}_s)$, as a result of collisions between the s and t particles. For example, the transport property of interest could be the momentum, $\xi_s = m_s \mathbf{v}_s$, or the energy, $\xi_s = m_s v_s^2/2$. The rate of change of $\xi_s(\mathbf{v}_s)$ can be obtained by multiplying Equation (G.16) by $\xi_s d^3v_s$ and then integrating over all s particle velocities. However, a more convenient form can be obtained by going back to Equation (G.5). This equation gives the number of s particles scattered out of element d^3v_s in a time dt due to collisions with the t particles. As a result of a collision, the transport property, $\xi_s(\mathbf{v}_s)$, is altered by an amount, $\xi_s(\mathbf{v}_s') - \xi_s(\mathbf{v}_s) \equiv \xi_s' - \xi_s$. Multiplying this change by the number of collisions for s particles in d^3v_s, in a time dt, with all t particles, Equation (G.5) yields

$$(\xi_s' - \xi_s) f_s d^3r\, d^3v_s\, dt \iiint d^3v_t\, d\varepsilon\, b\, db\, g_{st} f_t. \tag{G.17}$$

The integration of Equation (G.17) over all s particle velocities gives the total change of ξ_s in a time dt and spatial element d^3r due to collisions. However, this resulting integral is equal to $(n_s d^3r) dt\, \bar{\xi}_s$, so that

$$n_s d^3r\, dt\, \bar{\xi}_s = d^3r\, dt \iiiint d^3v_s\, d^3v_t\, d\varepsilon\, b\, db\, g_{st} f_s f_t (\xi_s' - \xi_s). \tag{G.18}$$

where $\bar{\xi}_s$ is the average change of ξ_s in a time dt and spatial element d^3r due to collisions. Dividing by $n_s d^3r\, dt$ and using Equation (G.15) for $b\, db\, d\varepsilon$ yields the following form:

$$\bar{\xi}_s = \frac{1}{n_s} \iiint d^3v_s\, d^3v_t\, d\Omega\, \sigma_{st}(g_{st}, \theta) g_{st} f_s f_t (\xi_s' - \xi_s). \tag{G.19}$$

Finally, note that Equation (G.19) can also be written in terms of the random velocities $\mathbf{c}_s$ and $\mathbf{c}_t$ because the Jacobian of the velocity transformation from $(\mathbf{v}_s, \mathbf{v}_t)$ to $(\mathbf{c}_s, \mathbf{c}_t)$ is $|J| = 1$, so that $d^3v_s\, d^3v_t = d^3c_s\, d^3c_t$. When this change is made, Equation (G.19) becomes the same as Equation (4.60).

G.2 Fokker–Planck collision term

The Boltzmann collision integral can be applied to charged particle interactions, but the complexity of this expression resulted in a search for simpler collision models. The motivation for simplifying the Boltzmann collision integral in the case of Coulomb interactions is that these are long-range interactions and, therefore, the change in velocity of a particle due to a collision, $\Delta\mathbf{v}_s$, is small for most collisions. In this case, the distribution functions evaluated after the collision, f_s' and f_t', can be expressed in terms of those evaluated before the collision, f_s and f_t, by means of a Taylor series expansion, with $\Delta\mathbf{v}_s$ as the small parameter. The Fokker–Planck collision operator is obtained if only those terms proportional to $\Delta\mathbf{v}_s$ and $\Delta\mathbf{v}_s\Delta\mathbf{v}_s$ are retained[3]

$$\frac{\delta f_s}{\delta t} = -\nabla_v \cdot [f_s \langle \mathbf{v}_s \rangle] + \frac{1}{2} \nabla_v \nabla_v : [f_s \langle \Delta\mathbf{v}_s \Delta\mathbf{v}_s \rangle], \tag{G.20}$$

where

$$\langle \Delta\mathbf{v}_s \rangle = \iint d^3v_t\, d\Omega\, g_{st}\sigma_{st}(g_{st}, \theta) f_t \Delta\mathbf{v}_s, \tag{G.21}$$

$$\langle \Delta\mathbf{v}_s \Delta\mathbf{v}_s \rangle = \iint d^3v_t\, d\Omega g_{st}\sigma_{st}(g_{st}, \theta) f_t \Delta\mathbf{v}_s \Delta\mathbf{v}_s, \tag{G.22}$$

and where the double-dot product is defined as $\sum_{\alpha,\beta} \partial^2/\partial v_\alpha \partial v_\beta (f_s \langle \Delta v_s \Delta v_s \rangle_{\alpha\beta})$. The quantities α and β are the coordinate indices.

The Fokker–Planck collision term is used to describe small angle collisions. That is, a given particle collides consecutively with many particles, and the effect of such collisions is that the velocity vectors of the colliding particles only change by a small amount. These multiple small-angle collisions can be thought of as causing a continuous flow of phase points in velocity space. The quantity $\langle \Delta\mathbf{v}_s \rangle$, which is called *dynamical friction*, slows down the s particles as a result of collisions with the t particles. The quantity $\langle \Delta\mathbf{v}_s \Delta\mathbf{v}_s \rangle$ provides for a *diffusion in velocity space*. In practical applications, the Fokker–Planck collision term, which appears to be just as difficult to evaluate as the Boltzmann collision integral, can be reduced to relatively simple forms in many specific cases.[3,4] This can be a real advantage. On the other hand, in a partially ionized plasma, several different types of collision need to be included, all of which can be described by the Boltzmann collision integral. Therefore, the advantage gained by using Fokker–Planck collision terms for Coulomb collisions is often offset by the mathematical inconvenience of using different collision terms.

G.3   Charge exchange collision integral

Resonant charge exchange is an important process for a collision between an ion and its parent neutral. As noted in Chapter 4, this process is pseudo-elastic because both energy and momentum are approximately conserved in a collision. As a consequence, the Boltzmann collision integral can be used to describe such collisions. Starting from the Boltzmann collision integral and assuming that the resonant charge exchange cross section, Q_E, given in Equation (4.148) is constant, the following simplified collision term for resonant charge exchange has been derived[5]

$$\frac{\delta f_i}{\delta t} = Q_E f_n(\mathbf{v}_i) \int d^3 v_n\, f_i(\mathbf{v}_n) |\mathbf{v}_i - \mathbf{v}_n|$$

$$- Q_E f_i(\mathbf{v}_i) \int d^3 v_n\, f_n(\mathbf{v}_n) |\mathbf{v}_i - \mathbf{v}_n|, \tag{G.23}$$

where subscripts i and n distinguish ions and neutrals. A collision term similar to Equation (G.23) has also been used to describe accidentally resonant charge exchange collisions, such as those in H^+ and O interactions.

The advantage of the collision term (G.23) is that it is easier to use than the full Boltzmann collision integral. However, this collision term implies a constant charge exchange cross section, while the actual cross section (4.148) is energy dependent. Therefore, in general, it is better to use the Boltzmann collision integral to include energy dependence.

G.4   Krook collision models

Numerous, relatively simple, collision models have been used over the years in an effort to include the effects of collisions while avoiding mathematical complications. These *relaxation collision models* take the simple form[6]

$$\frac{\delta f_s}{\delta t} = -\nu_0 (f_s - f_0), \tag{G.24}$$

where f_0 is a local equilibrium distribution function and ν_0 is the relaxation collision frequency. The effect of the collision term (G.24) is to drive the distribution function, f, to the equilibrium distribution, f_0, at a rate governed by the collision time, $\tau = v_0^{-1}$. Consider a simple situation in which there are no spatial gradients or forces in a plasma, but initially the distribution is not in equilibrium. In this case, the Boltzmann equation reduces to

$$\frac{\partial f_s}{\partial t} = -v_0(f_s - f_0), \tag{G.25}$$

or

$$\frac{\partial f_s}{\partial t} + v_0 f_s = v_0 f_0. \tag{G.26}$$

Assuming that v_0 is constant, the solution of Equation (G.26) is

$$f_s(\mathbf{v}, t) = f_0 + [f_s(\mathbf{v}, 0) - f_0]e^{-v_0 t}. \tag{G.27}$$

This solution indicates that the velocity distribution relaxes from the initial distribution to the equilibrium distribution in an exponential manner, with a time constant of v_0^{-1}.

Relaxation collision models have been used to describe collisions between identical particles as well as between ions and neutrals.[7] In the latter case, collisions in a weakly ionized gas were described by the relaxation model given as

$$\frac{\delta f_i}{\delta t} = -v_{in}(f_i - f_{i0}), \tag{G.28}$$

where

$$f_{i0} = n_i \left(\frac{m_i}{2\pi k T_n}\right)^{3/2} \exp\left(-\frac{m_i v_i^2}{2k T_n}\right) \tag{G.29}$$

is a Maxwellian distribution with a neutral temperature, T_n, and v_{in} is a velocity-independent ion–neutral collision frequency. An advantage of the relaxation model (G.28) is that an exact solution to the Boltzmann equation can be obtained for a homogeneous plasma subjected to perpendicular electric and magnetic fields. Also, with a judicious choice for v_{in}, the momentum and energy collision terms obtained with the relaxation model can be made to agree with those obtained from the more rigorous Boltzmann collision integral.

The main advantage of the relaxation collision models is their simplicity, but they can have serious deficiencies. First, the different macroscopic velocity moments (density, drift velocity, temperature, stress, heat flow, etc.) have different relaxation times and this feature is not properly described by a simple relaxation collision model. Also, some transport properties are sensitive to the nature of the collision process. For example, thermal diffusion does not occur for nonresonant ion–neutral interactions, but is very strong in fully ionized gases. Therefore, the use of a simple relaxation model could inadvertently eliminate an important process. Finally, even when the relaxation model can be configured to yield the correct momentum and energy collision terms, the collision terms for the higher velocity moments are not properly described. Typically, a relaxation collision model tends to overestimate the higher velocity moments, which also means that it overestimates the deviations from a Maxwellian velocity distribution.[7]

G.5   Specific references

1. Boltzmann, L., *Vorlesungen über Gastheorie*, Vol. **1**, 1896.
2. Chapman, S. and T. G. Cowling, *The Mathematical Theory of Non-Uniform Gases*, New York: Cambridge University Press, 1970.
3. Tanenbaum, B. S., *Plasma Physics*, New York: McGraw Hill, 1967.
4. Burgers, J. M., *Flow Equations for Composite Gases*, New York: Academic, 1969.
5. Banks, P. M. and G. J. Lewak, Ion velocity distributions in a partially ionized plasma, *Phys. Fluids*, **11**, 804, 1968.
6. Bhatnagar, P. L., E. P. Gross, and M. Krook, A model for collision processes in gases, I. Small amplitude processes in charged and neutral one-component systems, *Phys. Rev.*, **94**, 511, 1954.
7. St-Maurice, J.-P. and R. W. Schunk, Behavior of ion velocity distributions for a simple collision model, *Planet. Space Sci.*, **22**, 1, 1974.

Appendix H

Maxwell velocity distribution

Collisions drive a gas toward an *equilibrium state*. In such a state, the velocity distribution of the particles in the gas is independent of both position and time and, therefore, collisions no longer affect the velocity distribution. Considering a simple gas of identical particles, collisions drive the gas to a state in which the density, n_0, drift velocity, $\mathbf{u}_0$, and temperature, T_0, are constants. However, this state will be reached only if no other forces act on the gas. Under these circumstances, all of the terms on the left-hand side of the Boltzmann equation (3.7) are zero and this equation reduces to (G.16)

$$\frac{\delta f}{\delta t} = \iint d^3 v_2 \, d\Omega \, g_{12} \sigma_{12}(g_{12}, \theta) \left[f(\mathbf{v}_1') f(\mathbf{v}_2') - f(\mathbf{v}_1) f(\mathbf{v}_2) \right] = 0, \quad \text{(H.1)}$$

where subscripts 1 and 2 distinguish the identical particles in this gas.

The equilibrium velocity distribution is obtained from the solution of Equation (H.1). However, the solution of Equation (H.1) does not necessarily mean that the integrand is zero, because the integrand can be positive or negative and these contributions to the integral may simply cancel when the integration is performed. On the other hand, Boltzmann showed, using the *H theorem*, that the integral in Equation (H.1) vanishes if, and only if

$$f(\mathbf{v}_1') f(\mathbf{v}_2') = f(\mathbf{v}_1) f(\mathbf{v}_2) \tag{H.2}$$

for all values of $\mathbf{v}_1$ and $\mathbf{v}_2$. Therefore, Equation (H.2) is both a *necessary and sufficient* condition for equilibrium.[1] Taking the logarithm of Equation (H.2) yields the alternate form

$$\ln f(\mathbf{v}_1') + \ln f(\mathbf{v}_2') = \ln f(\mathbf{v}_1) + \ln f(\mathbf{v}_2). \tag{H.3}$$

575

In a binary elastic collision, the mass, momentum, and energy are conserved (Equations 4.14 and 4.15), and these quantities are known as *collisional invariants*. Given the initial relative position and velocity of the particles before the collision, the conservation of mass, momentum, and energy are all that are needed to determine completely these quantities after the collision. Therefore, all other collisional invariants can be expressed as a linear sum of the invariants m, $m\mathbf{v}$, and $mv^2/2$. An inspection of Equation (H.3) indicates that $\ln f(\mathbf{v})$ is a collisional invariant, and hence, it can be expressed in the form

$$\ln f = \alpha_0 + \boldsymbol{\alpha}_1 \cdot \mathbf{v} + \alpha_2 v^2, \tag{H.4}$$

or

$$f = e^{\alpha_0} e^{(\boldsymbol{\alpha}_1 \cdot \mathbf{v} + \alpha_2 v^2)}, \tag{H.5}$$

where α_0, $\boldsymbol{\alpha}_1$, and α_2 are constants. Equation (H.5) can also be expressed in the form

$$f = a_0 e^{-a_2 (\mathbf{v} - \mathbf{a}_1)^2}, \tag{H.6}$$

where a_0, $\mathbf{a}_1$, and a_2 are new constants that are introduced after the square of the velocity in Equation (H.5) is computed. These unknown constants can be expressed in terms of the gas parameters n_0, $\mathbf{u}_0$, and T_0 by using the definitions of these quantities (Equations 3.10, 3.11, and 3.15);

$$n_0 = \int \mathrm{d}^3 v f = a_0 \int \mathrm{d}^3 v\, e^{-a_2 (\mathbf{v} - \mathbf{a}_1)^2} = a_0 \left(\frac{2\pi k T_0}{m} \right)^{3/2}, \tag{H.7}$$

$$\mathbf{u}_0 = \frac{1}{n_0} \int \mathrm{d}^3 v f \mathbf{v} = \mathbf{a}_1, \tag{H.8}$$

$$T_0 = \frac{m}{3k n_0} \int \mathrm{d}^3 v f (\mathbf{v} - \mathbf{u}_0)^2 = \frac{m}{2k a_2}, \tag{H.9}$$

where the integrations can be performed using either a spherical or Cartesian coordinate system in velocity space (as previously shown). The substitution of the constants a_0, $\mathbf{a}_1$, and a_2 into Equation (H.6) yields the equilibrium velocity distribution

$$f = n_0 \left(\frac{m}{2\pi k T_0} \right)^{3/2} \exp \left[-\frac{m(\mathbf{v} - \mathbf{u}_0)^2}{2k T_0} \right], \tag{H.10}$$

which is known as the *drifting Maxwell–Boltzmann distribution function* or a *drifting Maxwellian velocity distribution*.

In the ionospheres, forces are always at work and an equilibrium velocity distribution is rarely, if ever, obtained. Also, the ionospheres contain multiple ion and neutral species. Under these circumstances, collisions between both like and unlike particles must be considered. The effect of collisions between identical particles is to drive the species distribution function toward a local drifting Maxwellian:

$$f_s^{\mathrm{M}}(\mathbf{r}, \mathbf{v}_s, t) = n_s(\mathbf{r}, t) \left[\frac{m_s}{2\pi k T_s(\mathbf{r}, t)} \right]^{3/2}$$
$$\cdot \exp\{-m_s[\mathbf{v}_s - \mathbf{u}_s(\mathbf{r}, t)]^2 / 2k T_s(\mathbf{r}, t)\}, \tag{H.11}$$

where f_s^{M} takes this form at all positions in space and at all times because the macroscopic transport properties $(n_s, \mathbf{u}_s, T_s)$ vary with $\mathbf{r}$ and t. The effect of collisions between unlike particles, in the presence of forces, is to drive the distribution away from a local drifting Maxwellian. This is why the actual velocity distribution function is usually expanded in an orthogonal series about a local drifting Maxwellian (Chapter 3).

It is instructive to consider some of the properties of a local drifting Maxwellian because of its importance to transport theory. As noted in Chapter 3, this distribution is consistent with the general definitions of density (3.10), drift velocity (3.11), and temperature (3.15). Other macroscopic transport properties of interest are the heat flow vector (3.16) and the pressure tensor (3.17). Considering first the heat flow vector, for a Maxwellian this becomes

$$\mathbf{q}_s = \frac{m_s}{2} \int \mathrm{d}^3 v_s f_s^{\mathrm{M}} (\mathbf{v}_s - \mathbf{u}_s)^2 (\mathbf{v}_s - \mathbf{u}_s) = \frac{m_s}{2} \int \mathrm{d}^3 c_s f_s^{\mathrm{M}} c_s^2 \mathbf{c}_s$$
$$= \frac{n_s m_s}{2} \left(\frac{m_s}{2\pi k T_s} \right)^{3/2} \int \mathrm{d}^3 c_s \, e^{-m_s c_s^2 / (2k T_s)} c_s^2 \mathbf{c}_s, \tag{H.12}$$

where the random velocity, $\mathbf{c}_s = \mathbf{v}_s - \mathbf{u}_s$, is introduced and where $\mathrm{d}^3 c_s = \mathrm{d}^3 v_s$ (only the origin of velocity space is different). For a spherical coordinate system in velocity space, with polar angle θ and azimuthal angle ϕ, $\mathrm{d}^3 c_s = c_s^2 \sin\theta \, \mathrm{d}\theta \, \mathrm{d}\phi \, \mathrm{d}c_s$ and

$$\mathbf{c}_s = c_s(\sin\theta \cos\phi \, \mathbf{e}_1 + \sin\theta \sin\phi \, \mathbf{e}_2 + \cos\theta \, \mathbf{e}_3), \tag{H.13}$$

where $(\mathbf{e}_1, \mathbf{e}_2, \mathbf{e}_3)$ are Cartesian unit vectors. Substituting Equation (H.13) into Equation (H.12) and integrating over θ and ϕ yields

$$\mathbf{q}_s = 0. \tag{H.14}$$

For a Maxwellian, the pressure tensor (3.17) becomes

$$\mathbf{P}_s = m_s \int \mathrm{d}^3 v_s f_s^{\mathrm{M}} (\mathbf{v}_s - \mathbf{u}_s)(\mathbf{v}_s - \mathbf{u}_s) = m_s \int \mathrm{d}^3 c_s f_s^{\mathrm{M}} \mathbf{c}_s \mathbf{c}_s$$
$$= n_s m_s \left(\frac{m_s}{2\pi k T_s} \right)^{3/2} \int\limits_0^\infty \mathrm{d}c_s c_s^2 \int\limits_0^\pi \sin\theta \, \mathrm{d}\theta \int\limits_0^{2\pi} \mathrm{d}\phi \, \mathbf{c}_s \mathbf{c}_s e^{-m_s c_s^2 / (2k T_s)}, \tag{H.15}$$

where $\mathbf{c}_s\mathbf{c}_s$ is a second-order tensor obtained by multiplying Equation (H.13) by itself (the form is similar to Equation 4.76). All of the off-diagonal elements are zero after integration over solid angle, and Equation (H.15) reduces to

$$\mathbf{P}_s = n_s m_s \left(\frac{m_s}{2\pi k T_s}\right)^{3/2} \int\limits_0^\infty dc_s \int\limits_0^\pi \sin\theta \, d\theta \int\limits_0^{2\pi} d\phi \, c_s^4 e^{-m_s c_s^2/(2kT_s)}$$

$$\cdot \, (\sin^2\theta \cos^2\phi \mathbf{e}_1\mathbf{e}_1 + \sin^2\theta \sin^2\phi \, \mathbf{e}_2\mathbf{e}_2 + \cos^2\theta \, \mathbf{e}_3\mathbf{e}_3). \qquad (H.16)$$

The integration of the quantity in parentheses over the solid angle yields $(4\pi/3)\mathbf{I}$, where $\mathbf{I} = \mathbf{e}_1\mathbf{e}_1 + \mathbf{e}_2\mathbf{e}_2 + \mathbf{e}_3\mathbf{e}_3$ is the unit dyadic ($\delta_{\alpha\beta}$ in index notation). Therefore, Equation (H.16) becomes

$$\mathbf{P}_s = \frac{4\pi}{3} \mathbf{I} n_s m_s \left(\frac{m_s}{2\pi k T_s}\right)^{3/2} \int\limits_0^\infty dc_s c_s^4 e^{-m_s c_s^2/(2kT_s)}. \qquad (H.17)$$

The remaining integral in Equation (H.17) can be obtained from the formulas in Appendix C and is

$$\frac{3\sqrt{\pi}}{8}\left(\frac{2kT_s}{m_s}\right)^{5/2}.$$

Therefore, Equation (H.17) reduces to

$$\mathbf{P}_s = (n_s k T_s)\mathbf{I}. \qquad (H.18)$$

For a Maxwellian, the pressure tensor is diagonal and all three elements are equal to the scalar pressure, $p_s = n_s k T_s$.

In addition to the transport properties discussed above, there are several Maxwellian-averaged speeds that are frequently used. These average speeds are obtained in the usual manner by multiplying the drifting Maxwellian velocity distribution (H.11) by the desired velocity parameter, $\xi_s(\mathbf{c}_s)$, and then integrating over all velocities

$$\langle \xi_s(\mathbf{c}_s)\rangle_M = \frac{1}{n_s} \int d^3 c_s \, \xi_s f_s^M(\mathbf{r}, \mathbf{c}_s, t), \qquad (H.19)$$

where in most cases the integrals are most easily performed in a spherical coordinate system. The *root-mean-square speed* is obtained by setting $\xi_s = c_s^2$, and then Equation (H.19) yields

$$\langle c_s^2\rangle_M^{1/2} = \left(\frac{3kT_s}{m_s}\right)^{1/2}. \qquad (H.20)$$

For the *average speed*, $\xi_s = |\mathbf{c}_s|$, and for the *average speed in one direction*, $\xi_s = |\mathbf{c}_{sx}|$. For these speeds, the integrations defined in Equation (H.19) yield, respectively,

$$\langle|\mathbf{c}_s|\rangle_M = \left(\frac{8kT_s}{\pi m_s}\right)^{1/2}, \tag{H.21}$$

$$\langle|\mathbf{c}_{sx}|\rangle_M = \left(\frac{2kT_s}{\pi m_s}\right)^{1/2}. \tag{H.22}$$

Another useful distribution is the *Maxwell speed distribution*, $F^M(c_s)$, which depends only on the speed c_s. This distribution is obtained from the drifting Maxwellian distribution (H.11) by integrating over solid angle in a spherical coordinate system, such that

$$F_s^M(c_s)\,dc_s = 4\pi n_s\left(\frac{m_s}{2\pi kT_s}\right)^{3/2} e^{-m_s c_s^2/(2kT_s)}c_s^2 dc_s, \tag{H.23}$$

where the 4π results from the solid angle integration because F_s^M does not depend on the angles. The speed distribution does not peak at zero, but instead, peaks at the speed where $dF_s^M/dc_s = 0$. This occurs at the *most probable speed*, $(c_s)_{mps}$, which is given by

$$(c_s)_{mps} = \left(\frac{2kT_s}{m_s}\right)^{1/2}. \tag{H.24}$$

Finally, it is instructive to calculate the random or thermal flux that crosses an imaginary plane from one side to the other. Consider a nondrifting ($\mathbf{u}_s = 0$) Maxwellian plasma and assume that $x = 0$ defines the imaginary plane. The thermal flux of particles that crosses this plane from the negative to the positive side is given by

$$\Gamma_{sx} = n_s\left(\frac{m_s}{2\pi kT_s}\right)^{3/2}\int\limits_0^\infty dv_x v_x e^{-m_s v_x^2/(2kT_s)}\int\limits_{-\infty}^\infty dv_y e^{-m_s v_y^2/(2kT_s)}$$

$$\times \int\limits_{-\infty}^\infty dv_z e^{-m_s v_z^2/(2kT_s)}, \tag{H.25}$$

where only those particles with a positive v_x cross the plane in the desired direction. The integrals in Equation (H.25) can be readily evaluated using the formulas in

Appendix C and the result is

$$\Gamma_{sx} = n_s \left(\frac{kT_s}{2\pi m_s} \right)^{1/2} = \frac{n_s \langle c_s \rangle_{\mathrm{M}}}{4}.$$ (H.26)

where $c_s = |\mathbf{c}_s|$.

H.1   Specific reference

1. Chapman, S. and T. G. Cowling, *The Mathematical Theory of Non-Uniform Gases*, New York: Cambridge University Press, 1970.

Appendix I

Semilinear expressions for transport coefficients

I.1 Diffusion coefficients and thermal conductivities

The heat flow and ambipolar diffusion equations that contain the higher-order transport effects, such as thermal diffusion and diffusion thermal heat flow, are presented in Section 5.14. The transport coefficients that appear in these equations have been calculated using both the linear (4.129a–g) and semilinear (4.132a,b) collision terms.[1,2] Here, the more general semilinear transport coefficients are presented, which are valid for arbitrarily large temperature differences between the interacting species. These coefficients reduce to the linear coefficients in the limit of small temperature differences, i.e., when $(T_s - T_t)/T_{st} \ll 1$.

The general expressions for the ion and neutral heat flows are summarized as follows:

$$\mathbf{q}_s = -K'_{ts}\nabla T_s - K_{st}\nabla T_t + R_{st}(\mathbf{u}_s - \mathbf{u}_t), \tag{I.1}$$

$$\mathbf{q}_t = -K_{ts}\nabla T_s - K'_{st}\nabla T_t - R_{ts}(\mathbf{u}_s - \mathbf{u}_t), \tag{I.2}$$

where subscripts s and t refer to either ion or neutral species. The thermal conductivities and diffusion thermal coefficients in Equations (I.1) and (I.2) are given by

$$K'_{st} = -F_{st}J_t/H_{st}, \tag{I.3}$$

$$K_{st} = C_{st}J_t/H_{st}, \tag{I.4}$$

$$R_{st} = (C_{st}A_{ts} + F_{ts}A_{st})/H_{st}, \tag{I.5}$$

581

where

$$A_{st} = \frac{p_s v_{st} \mu_{st}}{m_t} \left\{ \frac{5}{2} \left[\frac{T_t}{T_{st}} + \frac{m_t}{m_s} \frac{T_{st}}{T_s} (1 - z_{st}) \right] + y_{st} \left(1 - \frac{T_t}{T_s} \right) - \frac{5}{2} \left(\frac{m_t}{\mu_{st}} \right) \right\},$$

(I.6)

$$F_{st} = - \left\{ \frac{2 z_{ss}''}{5} v_{ss} + v_{st} \left[3 \left(\frac{\mu_{st}}{m_t} \right)^2 \left(\frac{T_t}{T_{st}} \right)^2 \right. \right.$$
$$+ B_{st}^{(3)} \left(z_{st}' - \frac{5}{2} z_{st} \right) - B_{st}^{(1)} + \frac{\mu_{st}}{(m_s + m_t)}$$
$$\left. \left. \cdot \frac{T_t}{T_{st}} \left(\frac{4}{5} z_{st}'' - \frac{5}{2} \frac{T_s}{T_{st}} z_{st} \right) + \frac{5}{2} \frac{T_s}{T_{st}} \frac{\mu_{st}}{m_s} z_{st} \right] \right\},$$

(I.7)

$$C_{st} = v_{st} \frac{\rho_s}{\rho_t} \left[3 \left(\frac{\mu_{st}}{m_s} \right)^2 \left(\frac{T_s}{T_{st}} \right)^2 + B_{st}^{(3)} \left(z_{st}' - \frac{5}{2} z_{st} \right) + B_{st}^{(2)} - \frac{\mu_{st}}{(m_s + m_t)} \right.$$
$$\left. \cdot \frac{T_s}{T_{st}} \left(\frac{4}{5} \frac{m_t}{m_s} z_{st}'' + \frac{5}{2} \frac{T_t}{T_{st}} z_{st} \right) + \frac{5}{2} \frac{T_s}{T_{st}} \frac{\mu_{st}}{m_s} z_{st} \right],$$

(I.8)

$$J_s = \frac{5}{2} \frac{k p_s}{m_s},$$

(I.9)

$$H_{st} = F_{st} F_{ts} - C_{st} C_{ts}.$$

(I.10)

Note that a simple change of subscripts in Equations (I.3) to (I.10) yields the other transport coefficients that are needed. Also, in Equations (I.6) to (I.10), the parameters y_{st}, $B_{st}^{(1)}$, $B_{st}^{(2)}$, and $B_{st}^{(3)}$ are given by Equations (4.133a–d), $\mu_{st} = m_s m_t / (m_s + m_t)$ is the reduced mass (4.98) and $T_{st} = (m_s T_t + m_t T_s) / (m_s + m_t)$ is the reduced temperature (4.99). The quantities z_{st}, z_{st}', and z_{st}'' are pure numbers that are different for different collisional processes; values are given in Chapter 4 for the processes relevant to the ionospheres.

I.2   Fully ionized plasma

The transport coefficients Δ_{ij}, α_{ij}, α_{ij}^*, γ_i, and γ_j given in Equations (5.158) to (5.161) are expressed in terms of the thermal conductivities and diffusion thermal coefficients as follows:

$$\Delta_{ij} = \frac{z_{ij} \mu_{ij}}{\rho_i k T_{ij}} \left(R_{ij} + \frac{\rho_i}{\rho_j} R_{ji} \right),$$

(I.11)

$$\alpha_{ij} = \frac{z_{ij} v_{ij} \mu_{ij}}{k^2 T_{ij}} \frac{n_i + n_j}{n_i n_j} \left(K_{ji}' - \frac{\rho_i}{\rho_j} K_{ji} \right),$$

(I.12)

$$\alpha_{ij}^* = \frac{z_{ij} v_{ij} \mu_{ij}}{k^2 T_{ij}} \frac{n_i + n_j}{n_i n_j} \left(\frac{\rho_i}{\rho_j} K'_{ij} - K_{ij} \right), \tag{I.13}$$

$$\gamma_i = \frac{15\sqrt{2}}{8} \frac{n_j Z_i Z_j (Z_i - Z_j)}{\frac{13\sqrt{2}}{8}(n_i Z_i^2 + n_j Z_j^2) + n_i Z_i + n_j Z_j}, \tag{I.14}$$

$$\gamma_j = \frac{15\sqrt{2}}{8} \frac{n_i Z_i Z_j (Z_i - Z_j)}{\frac{13\sqrt{2}}{8}(n_i Z_i^2 + n_j Z_j^2) + n_i Z_i + n_j Z_j}. \tag{I.15}$$

I.3 Partially ionized plasma

For a three-component plasma composed of electrons, ions, and neutrals, and when $m_i = m_n$, the transport coefficients Δ_{in}, ω, and ω^* are given by

$$\Delta_{in} = \frac{z_{in}}{2 n_i k T_{in}} \left(R_{in} + \frac{n_i}{n_n} R_{ni} \right), \tag{I.16}$$

$$\omega = \frac{z_{in}}{4} \frac{v_{in} m_i}{n_i k^2 T_{in}} \left(K_{in} - \frac{n_i}{n_n} K'_{in} \right), \tag{I.17}$$

$$\omega^* = \frac{z_{in}}{4} \frac{v_{in} m_i}{n_i k^2 T_{in}} \left(K'_{ni} - \frac{n_i}{n_n} K_{ni} \right). \tag{I.18}$$

I.4 Specific references

1. St-Maurice, J.-P. and R. W. Schunk, Diffusion and heat flow equations for the mid-latitude topside ionosphere, *Planet. Space Sci.*, **25**, 907, 1977.
2. Conrad, J. R. and R. W. Schunk, Diffusion and heat flow equations with allowance for large temperature differences between interacting species, *J. Geophys. Res.*, **84**, 811, 1979.

Appendix J

Solar fluxes and relevant cross sections

Table J.1 Parameters for the EUVAC solar flux model.[1]

Interval	Å	F74113[a]	A_i[b]
1	50–100	1.200	1.0017(−02)
2	100–150	0.450	7.1250(−03)
3	150–200	4.800	1.3375(−02)
4	200–250	3.100	1.9450(−02)
5	256.32	0.460	2.7750(−03)
6	284.15	0.210	1.3768(−01)
7	250–300	1.679	2.6467(−02)
8	303.31	0.800	2.5000(−02)
9	303.78	6.900	3.3333(−03)
10	300–350	0.965	2.2450(−02)
11	368.07	0.650	6.5917(−03)
12	350–400	0.314	3.6542(−02)
13	400–450	0.383	7.4083(−03)
14	465.22	0.290	7.4917(−03)
15	450–500	0.285	2.0225(−02)
16	500–550	0.452	8.7583(−03)
17	554.37	0.720	3.2667(−03)
18	584.33	1.270	5.1583(−03)
19	550–600	0.357	3.6583(−03)
20	609.76	0.530	1.6175(−02)
21	629.73	1.590	3.3250(−03)
22	600–650	0.342	1.1800(−02)
23	650–700	0.230	4.2667(−03)
24	703.36	0.360	3.0417(−03)
25	700–750	0.141	4.7500(−03)
26	765.15	0.170	3.8500(−03)
27	770.41	0.260	1.2808(−02)

Table J.1 *(Continued)*

Interval	Å	F74113[a]	$A_i{}^b$
28	789.36	0.702	3.2750(−03)
29	750–800	0.758	4.7667(−03)
30	800–850	1.625	4.8167(−03)
31	850–900	3.537	5.6750(−03)
32	900–950	3.000	4.9833(−03)
33	977.02	4.400	3.9417(−03)
34	950–1000	1.475	4.4167(−03)
35	1025.72	3.500	5.1833(−03)
36	1031.91	2.100	5.2833(−03)
37	1000–1050	2.467	4.3750(−03)

[a] Multiply the F74133 reference flux values by 10^9 to yield photons $cm^{-2} s^{-1}$.

[b] Read 1.0017(−2) as 1.0017×10^{-2}.

Table J.2a Photoabsorption and photoionization cross sections[a,b] for N_2 and O_2.[1]

Interval	λ, Å	$N_{2\,abs}$	$N^+_{2\,total}$	N^+_2	N^+	$O_{2\,abs}$	$O^+_{2\,total}$	O^+_2	O^+
1	50–100	0.720	0.720	0.443	0.277	1.316	1.316	1.316	0.000
2	100–150	2.261	2.261	1.479	0.782	3.806	3.806	2.346	1.460
3	150–200	4.958	4.958	3.153	1.805	7.509	7.509	4.139	3.368
4	200–250	8.392	8.392	5.226	3.166	10.900	10.900	6.619	4.281
5	256.30	10.210	10.210	6.781	3.420	13.370	13.370	8.460	4.910
6	284.15	10.900	10.900	8.100	2.800	15.790	15.790	9.890	5.900
7	250–300	10.493	10.493	7.347	3.145	14.387	14.387	9.056	5.332
8	303.31	11.670	11.670	9.180	2.490	16.800	16.800	10.860	5.940
9	303.78	11.700	11.700	9.210	2.490	16.810	16.810	10.880	5.930
10	300–350	13.857	13.857	11.600	2.257	17.438	17.438	12.229	5.212
11	368.07	16.910	16.910	15.350	1.560	18.320	18.320	13.760	4.560
12	350–400	16.395	16.395	14.669	1.726	18.118	18.118	13.418	4.703
13	400–450	21.675	21.675	20.692	0.982	20.310	20.310	15.490	4.818
14	465.22	23.160	23.160	22.100	1.060	21.910	21.910	16.970	4.940
15	450–500	23.471	23.471	22.772	0.699	23.101	23.101	17.754	5.347
16	500–550	24.501	24.501	24.468	0.033	24.606	24.606	19.469	5.139
17	554.37	24.130	24.130	24.130	0.000	26.040	26.040	21.600	4.440
18	584.33	22.400	22.400	22.400	0.000	22.720	22.720	18.840	3.880
19	550–600	22.787	22.787	22.787	0.000	26.610	26.610	22.789	3.824
20	609.76	22.790	22.790	22.790	0.000	28.070	26.390	24.540	1.850
21	629.73	23.370	23.370	23.370	0.000	32.060	31.100	30.070	1.030
22	600–650	23.339	23.339	23.339	0.000	26.017	24.937	23.974	0.962
23	650–700	31.755	29.235	29.235	0.000	21.919	21.306	21.116	0.190
24	703.36	26.540	25.480	25.480	0.000	27.440	23.750	23.750	0.000
25	700–750	24.662	15.060	15.060	0.000	28.535	23.805	23.805	0.000
26	765.15	120.490	65.800	65.800	0.000	20.800	11.720	11.720	0.000
27	770.41	14.180	8.500	8.500	0.000	18.910	8.470	8.470	0.000
28	789.36	16.487	8.860	8.860	0.000	26.668	10.191	10.191	0.000
29	750–800	33.578	14.274	14.274	0.000	22.145	10.597	10.597	0.000

Table J.2a *(Continued)*

Interval	λ, Å	$N_{2\,abs}$	$N_{2\,total}^+$	N_2^+	N^+	$O_{2\,abs}$	$O_{2\,total}^+$	O_2^+	O^+
30	800–850	16.992	0.000	0.000	0.000	16.631	6.413	6.413	0.000
31	850–900	20.249	0.000	0.000	0.000	8.562	5.494	5.494	0.000
32	900–950	9.680	0.000	0.000	0.000	12.817	9.374	9.374	0.000
33	977.02	2.240	0.000	0.000	0.000	18.730	15.540	15.540	0.000
34	950–1000	50.988	0.000	0.000	0.000	21.108	13.940	13.940	0.000
35	1025.72	0.000	0.000	0.000	0.000	1.630	1.050	1.050	0.000
36	1031.91	0.000	0.000	0.000	0.000	1.050	0.000	0.000	0.000
37	1000–1050	0.000	0.000	0.000	0.000	1.346	0.259	0.259	0.000

[a] The cross section values are presented in units of Mb; to transform them to cm^2 multiply by 10^{-18}.
[b] $N_{2\,total}^+$ and $O_{2\,total}^+$ denote the sum of all ionization cross sections; individual ionization cross sections are also listed separately (e.g., $N_2 + h\nu \rightarrow N^+ + N$ denoted as N^+).

Table J.2b Photoabsorption and photoionization cross sections[a] for O and N.[1]

Interval	Å	^4S	^2D	^2P	^4P	$^2P*$	O^{++}	O_{abs}	N^+	N^{++}	N_{abs}
1	50–100	0.190	0.206	0.134	0.062	0.049	0.088	0.730	0.286	0.045	0.331
2	100–150	0.486	0.529	0.345	0.163	0.130	0.186	1.839	0.878	0.118	0.996
3	150–200	0.952	1.171	0.768	0.348	0.278	0.215	3.732	2.300	0.190	2.490
4	200–250	1.311	1.762	1.144	0.508	0.366	0.110	5.202	3.778	0.167	3.946
5	256.30	1.539	2.138	1.363	0.598	0.412	0.000	6.050	4.787	0.085	4.874
6	284.15	1.770	2.620	1.630	0.710	0.350	0.000	7.080	5.725	0.000	5.725
7	250–300	1.628	2.325	1.488	0.637	0.383	0.000	6.461	5.192	0.051	5.244
8	303.31	1.920	2.842	1.920	0.691	0.307	0.000	7.680	6.399	0.000	6.399
9	303.78	1.925	2.849	1.925	0.693	0.308	0.000	7.700	6.413	0.000	6.413
10	300–350	2.259	3.446	2.173	0.815	0.000	0.000	8.693	7.298	0.000	7.298
11	368.07	2.559	3.936	2.558	0.787	0.000	0.000	9.840	8.302	0.000	8.302
12	350–400	2.523	3.883	2.422	0.859	0.000	0.000	9.687	8.150	0.000	8.150
13	400–450	3.073	4.896	2.986	0.541	0.000	0.000	11.496	9.556	0.000	9.556
14	465.22	3.340	5.370	3.220	0.000	0.000	0.000	11.930	10.578	0.000	10.578
15	450–500	3.394	5.459	3.274	0.000	0.000	0.000	12.127	11.016	0.000	11.016
16	500–550	3.421	5.427	3.211	0.000	0.000	0.000	12.059	11.503	0.000	11.503
17	554.37	3.650	5.670	3.270	0.000	0.000	0.000	12.590	11.772	0.000	11.772
18	584.33	3.920	6.020	3.150	0.000	0.000	0.000	13.090	11.778	0.000	11.778
19	550–600	3.620	5.910	3.494	0.000	0.000	0.000	13.024	11.758	0.000	11.758
20	609.760	3.610	6.170	3.620	0.000	0.000	0.000	13.400	11.798	0.000	11.798
21	629.73	3.880	6.290	3.230	0.000	0.000	0.000	13.400	11.212	0.000	11.212
22	600–650	4.250	6.159	2.956	0.000	0.000	0.000	13.365	11.951	0.000	11.951
23	650–700	5.128	11.453	0.664	0.000	0.000	0.000	17.245	12.423	0.000	12.423
24	703.36	4.890	6.570	0.000	0.000	0.000	0.000	11.460	13.265	0.000	13.265
25	700–750	6.739	3.997	0.000	0.000	0.000	0.000	10.736	12.098	0.000	12.098
26	765.15	4.000	0.000	0.000	0.000	0.000	0.000	4.000	11.323	0.000	11.323
27	770.41	3.890	0.000	0.000	0.000	0.000	0.000	3.890	11.244	0.000	11.244
28	789.36	3.749	0.000	0.000	0.000	0.000	0.000	3.749	10.961	0.000	10.961
29	750–800	5.091	0.000	0.000	0.000	0.000	0.000	5.091	11.171	0.000	11.171
30	800–850	3.498	0.000	0.000	0.000	0.000	0.000	3.498	10.294	0.000	10.294
31	850–900	4.554	0.000	0.000	0.000	0.000	0.000	4.554	0.211	0.000	0.211
32	900–950	1.315	0.000	0.000	0.000	0.000	0.000	1.315	0.000	0.000	0.000

[a] The cross section values are presented in units of Mb; to transform them to cm^2, multiply by 10^{-18}.

Table J.2c Photoabsorption and photoionization cross sections[a,b] for CO_2 and CO. (J.A. Fennelly; private communication)

Interval	Å	CO_{2abs}	CO_{2total}^+	CO_2^+	CO^+	O^+	C^+	CO_2^{++}	CO_{abs}	CO_{total}^+	CO^+	O^+	C^+	C^{++}
1	50–100	1.550	1.550	0.447	0.163	0.626	0.306	0.009	0.866	0.866	0.291	0.247	0.282	0.046
2	100–150	4.616	4.616	2.083	0.510	1.320	0.658	0.045	2.391	2.391	1.074	0.600	0.672	0.045
3	150–200	9.089	9.089	4.960	1.052	1.929	1.033	0.115	4.671	4.671	2.459	1.029	1.156	0.027
4	200–250	14.361	14.320	8.515	1.618	2.622	1.433	0.132	7.011	7.011	4.082	1.411	1.514	0.004
5	256.30	16.505	16.114	11.113	1.467	2.260	1.168	0.106	8.614	8.614	5.449	1.572	1.593	0.000
6	284.15	19.016	18.602	13.004	1.640	2.572	1.287	0.098	10.541	10.541	7.713	1.687	1.141	0.000
7	250–300	17.518	17.140	11.906	1.539	2.382	1.219	0.095	9.424	9.424	6.361	1.561	1.502	0.000
8	303.31	21.492	21.387	14.390	1.959	3.271	1.706	0.061	11.867	11.867	9.209	1.582	1.076	0.000
9	303.78	21.594	21.435	14.414	1.968	3.280	1.715	0.058	11.900	11.900	9.246	1.581	1.073	0.000
10	300–350	23.574	23.629	15.954	2.442	3.426	1.794	0.013	13.441	13.441	11.532	0.946	0.963	0.000
11	368.07	25.269	25.557	18.271	3.040	3.128	1.104	0.015	15.259	15.259	13.980	0.509	0.771	0.000
12	350–400	24.871	25.518	17.982	2.995	3.224	1.310	0.006	14.956	14.956	13.609	0.533	0.814	0.000
13	400–450	28.271	27.172	21.082	3.369	2.597	0.124	0.000	17.956	17.956	16.876	0.118	0.962	0.000
14	465.22	29.526	28.755	24.378	2.247	2.130	0.000	0.000	20.173	20.173	19.085	0.058	1.029	0.000
15	450–500	30.254	30.578	27.163	1.504	1.911	0.000	0.000	20.574	20.574	19.669	0.009	0.895	0.000
16	500–550	31.491	32.595	30.138	0.820	1.636	0.000	0.000	21.085	21.085	20.454	0.000	0.631	0.000
17	554.37	33.202	33.211	31.451	0.409	1.351	0.000	0.000	21.624	21.624	21.565	0.000	0.060	0.000
18	584.33	34.200	33.858	32.382	0.305	1.170	0.000	0.000	22.000	22.000	22.000	0.000	0.000	0.000
19	550–600	34.913	34.959	33.482	0.306	1.171	0.000	0.000	21.910	21.895	21.895	0.000	0.000	0.000
20	609.76	35.303	35.303	34.318	0.135	0.850	0.000	0.000	22.100	21.918	21.918	0.000	0.000	0.000

(Continued)

Table J.2c *(Continued)*

Interval	Å	CO_{2abs}	CO_{2total}^+	CO_2^+	CO^+	O^+	C^+	CO_2^{++}	CO_{abs}	CO_{total}^+	CO^+	O^+	C^{++}
21	629.73	34.300	34.300	33.795	0.037	0.468	0.000	0.000	22.025	22.025	0.000	0.000	0.000
22	600–650	34.447	34.573	34.003	0.043	0.527	0.000	0.000	21.915	21.845	0.000	0.000	0.000
23	650–700	33.699	32.295	32.287	0.000	0.000	0.000	0.000	21.036	20.097	0.000	0.000	0.000
24	703.36	23.518	20.856	20.856	0.000	0.000	0.000	0.000	23.853	22.115	0.000	0.000	0.000
25	700–750	32.832	27.490	27.490	0.000	0.000	0.000	0.000	25.501	21.084	0.000	0.000	0.000
26	765.15	93.839	86.317	86.317	0.000	0.000	0.000	0.000	26.276	13.033	0.000	0.000	0.000
27	770.41	61.939	51.765	51.765	0.000	0.000	0.000	0.000	15.262	9.884	0.000	0.000	0.000
28	789.36	26.493	21.676	21.676	0.000	0.000	0.000	0.000	33.132	17.350	0.000	0.000	0.000
29	750–800	39.831	34.094	34.094	0.000	0.000	0.000	0.000	20.535	11.375	0.000	0.000	0.000
30	800–850	13.980	10.930	10.930	0.000	0.000	0.000	0.000	22.608	17.559	0.000	0.000	0.000
31	850–900	44.673	7.135	7.135	0.000	0.000	0.000	0.000	36.976	11.701	0.000	0.000	0.000
32	900–950	52.081	0.000	0.000	0.000	0.000	0.000	0.000	50.318	0.000	0.000	0.000	0.000
33	977.02	42.869	0.000	0.000	0.000	0.000	0.000	0.000	28.500	0.000	0.000	0.000	
34	950–1000	50.311	0.000	0.000	0.000	0.000	0.000	0.000	52.827	0.000	0.000	0.000	
35	1025.72	15.100	0.000	0.000	0.000	0.000	0.000	0.000	1.388	0.000	0.000	0.000	0.000
36	1031.91	14.200	0.000	0.000	0.000	0.000	0.000	0.000	1.388	0.000	0.000	0.000	0.000
37	1000–1050	18.241	0.000	0.000	0.000	0.000	0.000	0.000	8.568	0.000	0.000	0.000	0.000

[a] The cross section values are presented in Mb; to transform them to cm^2 multiply by 10^{-18}.

[b] CO_{2total}^+ and CO_{total}^+ denote the sum of all ionization cross sections; individual ionization cross sections are also listed separately (e.g., $CO_2 + h\nu \rightarrow CO^+ + O$ denoted as CO^+).

Table J.2d Photoabsorption and photoionization cross sections[a,b] for H_2O and He (J.A. Fennelly; private communication).

Interval	Å	H_2O_{abs}	$H_2O^+_{tot}$	H_2O^+	OH^+	H^+	O^+	He_{abs}	He^+
1	50–100	0.699	0.699	0.385	0.093	0.171	0.050	0.1441	0.1441
2	100–150	1.971	1.971	1.153	0.306	0.404	0.107	0.4785	0.4785
3	150–200	4.069	4.069	2.366	0.733	0.781	0.189	1.1571	1.1571
4	200–250	6.121	6.121	3.595	1.197	1.105	0.223	1.6008	1.6008
5	256.30	7.520	7.520	4.563	1.560	1.166	0.230	2.1212	2.1212
6	284.15	8.934	8.934	5.552	1.889	1.262	0.230	2.5947	2.5947
7	250–300	8.113	8.113	4.974	1.704	1.206	0.230	2.3205	2.3205
8	303.31	9.907	9.907	6.182	2.148	1.347	0.231	2.9529	2.9529
9	303.78	9.930	9.930	6.198	2.154	1.347	0.230	2.9618	2.9618
10	300–350	11.350	11.350	7.237	2.559	1.347	0.207	3.5437	3.5437
11	368.07	13.004	13.004	8.441	3.065	1.327	0.171	4.2675	4.2675
12	350–400	12.734	12.734	8.218	2.984	1.352	0.180	4.1424	4.1424
13	400–450	16.032	16.032	10.561	4.042	1.354	0.075	5.4466	5.4466
14	465.22	18.083	18.083	11.908	4.688	1.458	0.028	6.5631	6.5631
15	450–500	18.897	18.897	12.356	4.981	1.529	0.031	7.2084	7.2084
16	500–550	20.047	20.047	12.990	5.374	1.660	0.024	0.9581	0.9581
17	554.37	21.159	21.159	13.559	5.789	1.811	0.000	0.0000	0.0000
18	584.33	21.908	21.908	13.968	6.096	1.844	0.000	0.0000	0.0000
19	550–600	21.857	21.857	13.972	6.090	1.795	0.000	0.0000	0.0000
20	609.76	22.446	22.446	14.392	6.383	1.672	0.000	0.0000	0.0000
21	629.73	22.487	22.026	14.464	6.279	1.282	0.000	0.0000	0.0000
22	600–650	22.502	22.297	14.558	6.368	1.371	0.000	0.0000	0.0000
23	650–700	22.852	20.735	17.443	3.118	0.174	0.000	0.0000	0.0000
24	703.36	22.498	19.655	18.283	1.364	0.008	0.000	0.0000	0.0000
25	700–750	22.118	17.945	17.557	0.386	0.002	0.000	0.0000	0.0000
26	700–750	19.384	13.080	13.080	0.000	0.000	0.000	0.0000	0.0000
27	765.15	20.992	13.512	13.512	0.000	0.000	0.000	0.0000	0.0000
28	770.41	16.975	10.636	10.636	0.000	0.000	0.000	0.0000	0.0000
29	789.36	18.151	11.625	11.625	0.000	0.000	0.000	0.0000	0.0000
30	750–800	16.623	9.654	9.654	0.000	0.000	0.000	0.0000	0.0000
31	800–850	19.837	9.567	9.567	0.000	0.000	0.000	0.0000	0.0000
32	850–900	20.512	8.736	8.736	0.000	0.000	0.000	0.0000	0.0000
33	900–950	15.072	6.188	6.188	0.000	0.000	0.000	0.0000	0.0000
34	950–1000	15.176	4.234	4.234	0.000	0.000	0.000	0.0000	0.0000
35	1025.72	18.069	0.000	0.000	0.000	0.000	0.000	0.0000	0.0000
36	1031.91	15.271	0.000	0.000	0.000	0.000	0.000	0.0000	0.0000
37	1000–1050	8.001	0.000	0.000	0.000	0.000	0.000	0.0000	0.0000

[a] The cross section values are presented in Mb; to transform them to cm^2 multiply by 10^{-18}.

[b] $H_2O^+_{total}$ denotes the sum of all ionization cross sections; individual ionization cross sections are also listed separately (e.g., $H_2O + h\nu \rightarrow OH^+ + H$ denoted as OH^+).

Table J.2e Photoabsorption and photoionization cross sections[a,b] for CH_4 (J.A. Fennelly; private communication).

Interval	Å	$CH_{4\,abs}$	$CH_{4\,total}^+$	CH_4^+	CH_3^+	CH_2^+	CH^+	C^+	H_2^+	H^+
1	50–100	0.204	0.204	0.051	0.052	0.033	0.014	0.003	0.005	0.047
2	100–150	0.593	0.593	0.147	0.152	0.095	0.039	0.008	0.015	0.137
3	150–200	1.496	1.496	0.387	0.409	0.201	0.095	0.023	0.038	0.344
4	200–250	2.794	2.794	0.839	0.884	0.416	0.165	0.031	0.046	0.414
5	256.30	3.857	3.857	1.192	1.290	0.576	0.214	0.035	0.058	0.492
6	284.15	5.053	5.053	1.681	1.824	0.665	0.232	0.042	0.049	0.559
7	250–300	4.360	4.360	1.398	1.514	0.614	0.223	0.038	0.055	0.519
8	303.31	6.033	6.033	2.095	2.287	0.701	0.282	0.057	0.052	0.559
9	303.78	6.059	6.059	2.103	2.302	0.701	0.282	0.058	0.052	0.561
10	300–350	7.829	7.829	2.957	3.108	0.781	0.310	0.066	0.055	0.552
11	368.07	10.165	10.165	3.972	4.305	0.867	0.361	0.085	0.053	0.521
12	350–400	9.776	9.776	3.820	4.101	0.852	0.344	0.079	0.054	0.527
13	400–450	14.701	14.701	6.255	6.573	1.074	0.359	0.059	0.018	0.362
14	465.22	18.770	18.770	8.442	8.776	1.097	0.211	0.007	0.000	0.238
15	450–500	21.449	21.449	9.837	10.212	1.014	0.162	0.001	0.000	0.225
16	500–550	24.644	24.644	11.432	11.974	0.926	0.131	0.000	0.000	0.181
17	554.37	27.924	27.924	13.398	13.853	0.652	0.021	0.000	0.000	0.000
18	584.33	31.052	31.052	14.801	15.501	0.750	0.000	0.000	0.000	0.000
19	550–600	30.697	30.697	14.640	15.374	0.683	0.000	0.000	0.000	0.000
20	609.76	33.178	33.178	15.734	16.719	0.726	0.000	0.000	0.000	0.000
21	629.73	35.276	35.276	17.102	17.494	0.680	0.000	0.000	0.000	0.000
22	600–650	34.990	34.990	16.883	17.422	0.685	0.000	0.000	0.000	0.000
23	650–700	39.280	39.280	19.261	19.266	0.754	0.000	0.000	0.000	0.000
24	703.36	41.069	41.069	20.222	20.092	0.755	0.000	0.000	0.000	0.000
25	700–750	42.927	42.927	21.314	20.850	0.764	0.000	0.000	0.000	0.000
26	765.15	45.458	44.800	22.599	21.436	0.765	0.000	0.000	0.000	0.000
27	770.41	45.716	44.796	22.763	21.316	0.717	0.000	0.000	0.000	0.000
28	789.36	46.472	44.607	23.198	20.899	0.510	0.000	0.000	0.000	0.000
29	750–800	45.921	44.693	22.886	21.145	0.662	0.000	0.000	0.000	0.000
30	800–850	48.327	40.284	25.607	14.651	0.025	0.000	0.000	0.000	0.000
31	850–900	48.968	25.527	24.233	1.294	0.000	0.000	0.000	0.000	0.000
32	900–950	48.001	13.863	13.863	0.000	0.000	0.000	0.000	0.000	0.000
33	977.02	41.154	0.136	0.136	0.000	0.000	0.000	0.000	0.000	0.000
34	950–1000	38.192	0.475	0.475	0.000	0.000	0.000	0.000	0.000	0.000
35	1025.72	32.700	0.000	0.000	0.000	0.000	0.000	0.000	0.000	0.000
36	1031.91	30.121	0.000	0.000	0.000	0.000	0.000	0.000	0.000	0.000
37	1000–1050	29.108	0.000	0.000	0.000	0.000	0.000	0.000	0.000	0.000

[a] The cross section values are presented in Mb; to transform them to cm^2, multiply by 10^{-18}.

[b] $CH_{4\,total}^+$ denotes the sum of all ionization cross sections; individual ionization cross sections are also listed separately (e.g., $CH_4 + h\nu \rightarrow CH_3^+ + H$ denoted as CH_3^+).

Table J.2f Photoabsorption and photoionization cross sections[a,b] for SO_2 (J.A. Fennelly; private communication).

Interval	Å	$SO_{2\,abs}$	SO_{2tot}^+	SO_2^+	SO^+	S^+, O_2^+	O^+	SO^{++}
1	50–100	4.33	4.33	0.97	1.11	1.03	1.18	0.042
2	100–150	6.32	6.32	1.62	1.68	1.46	1.51	0.056
3	150–200	10.54	10.54	3.01	3.00	2.31	2.13	0.090
4	200–250	13.91	13.91	4.35	4.27	2.83	2.36	0.096
5	256.30	15.38	15.38	5.23	4.85	2.94	2.27	0.087
6	284.15	16.62	16.62	6.09	5.37	3.04	2.06	0.060
7	250–300	15.88	15.88	5.60	5.05	2.97	2.18	0.076
8	303.31	17.33	17.33	6.68	5.57	3.13	1.90	0.049
9	303.78	17.38	17.38	6.70	5.59	3.14	1.90	0.049
10	300–350	19.91	19.91	8.20	6.26	3.83	1.61	0.016
11	368.07	23.41	23.41	10.34	7.25	4.71	1.11	0.000
12	350–400	22.77	22.77	9.95	7.10	4.55	1.18	0.000
13	400–450	29.78	29.78	13.92	10.54	4.92	0.40	0.000
14	465.22	35.20	35.20	16.88	14.39	3.64	0.29	0.000
15	450–500	38.15	38.15	18.54	16.77	2.56	0.28	0.000
16	500–550	42.27	42.27	20.23	20.23	1.50	0.31	0.000
17	554.37	47.75	47.75	22.64	24.16	0.63	0.33	0.000
18	584.33	53.16	52.58	24.88	27.08	0.37	0.25	0.000
19	550–600	52.75	52.09	24.50	26.89	0.47	0.24	0.000
20	609.76	57.61	55.96	26.31	29.20	0.37	0.08	0.000
21	629.73	61.64	58.72	28.10	30.21	0.38	0.03	0.000
22	600–650	60.65	58.07	27.79	29.85	0.39	0.04	0.000
23	650–700	62.78	60.76	32.11	28.23	0.42	0.00	0.000
24	703.36	62.06	59.15	32.36	26.37	0.42	0.00	0.000
25	700–750	55.36	50.21	29.45	20.40	0.37	0.00	0.000
26	765.15	51.91	42.52	40.36	2.16	0.00	0.00	0.000
27	770.41	50.22	42.56	41.22	1.34	0.00	0.00	0.000
28	789.36	47.40	44.27	44.27	0.00	0.00	0.00	0.000
29	750–800	48.99	42.59	41.24	1.30	0.05	0.00	0.000
30	800–850	43.36	41.88	41.88	0.00	0.00	0.00	0.000
31	850–900	41.21	38.41	38.41	0.00	0.00	0.00	0.000
32	900–950	49.32	34.27	34.27	0.00	0.00	0.00	0.000
33	977.02	42.76	18.73	18.73	0.00	0.00	0.00	0.000
34	950–1000	41.81	17.58	17.58	0.00	0.00	0.00	0.000
35	1025.72	43.03	0.00	0.00	0.00	0.00	0.00	0.000
36	1031.91	44.80	0.00	0.00	0.00	0.00	0.00	0.000
37	1000–1050	45.47	0.26	0.26	0.00	0.00	0.00	0.000

[a] The cross section values are presented in Mb; to transform them to cm^2, multiply by 10^{-18}.

[b] $SO_{2\,total}^+$ denotes the sum of all ionization cross sections; individual ionization cross sections are also listed separately (e.g., $SO_2 + h\nu \rightarrow SO^+ + O$ denoted as SO^+).

Table J.2g Photoabsorption and photoionization cross sections[a] for H_2 and $H.^2$

Interval	Å	$H_{2\,abs}$	H^+	H_2^+	H_{abs}
1	50–100	0.0108	0.0011	0.0097	0.0024
2	100–150	0.0798	0.0040	0.0758	0.0169
3	150–200	0.2085	0.0075	0.2009	0.0483
4	200–250	0.4333	0.0305	0.4028	0.1007
5	256.30	0.6037	0.0527	0.5509	0.1405
6	284.15	0.8388	0.0773	0.7454	0.1913
7	250–300	0.7296	0.0661	0.6538	0.1676
8	303.31	1.0180	0.1005	0.8999	0.2324
9	303.78	1.0220	0.1011	0.9041	0.2334
10	300–350	1.4170	0.1200	1.2960	0.3077
11	368.07	1.9420	0.1577	1.7840	0.4152
12	350–400	1.9010	0.1594	1.7420	0.3984
13	400–450	3.0250	0.1255	2.8900	0.6163
14	465.22	3.8700	0.0925	3.7780	0.8387
15	450–500	4.5020	0.0944	4.0470	0.9739
16	500–550	5.3560	0.1020	5.2540	1.1990
17	554.37	6.1680	0.1184	6.0500	1.4190
18	584.33	7.0210	0.1208	6.9000	1.6620
19	550–600	6.8640	0.1237	6.7410	1.6200
20	609.76	7.8110	0.1429	7.6680	1.8880
21	629.73	8.4640	0.1573	8.2990	2.0790
22	600–650	8.4450	0.1524	8.2880	2.0760
23	650–700	9.9000	0.0287	9.7020	2.6410
24	703.36	10.7310	0.0000	10.7310	2.8970
25	700–750	11.3720	0.0000	9.7610	3.1730
26	765.15	10.7550	0.0000	8.6240	3.7300
27	770.41	8.6400	0.0000	7.0710	3.8070
28	789.36	7.3390	0.0000	5.0720	4.0930
29	750–800	8.7480	0.0000	6.6290	3.8680
30	800–850	8.2530	0.0000	0.0889	4.7840
31	850–900	0.4763	0.0000	0.0000	5.6700
32	900–950	0.1853	0.0000	0.0000	3.4690
33	977.02	0.0000	0.0000	0.0000	0.0000
34	950–1000	0.0456	0.0000	0.0000	0.0000
35	1025.72	0.0000	0.0000	0.0000	0.0000
36	1031.91	0.0000	0.0000	0.0000	0.0000
37	1000–1050	0.0000	0.0000	0.0000	0.0000

[a] The cross section values are presented in Mb; to transform them to cm^2, multiply by 10^{-18}.

J.1   Specific references

1. Richards, P. G., J. A. Fennelly, and D. G. Torr, EUVAC: A solar EUV flux model for aeronomic calculations, *J. Geophys. Res.*, **99**, 8981, 1994.
2. Maurellis, A. N., *Non-auroral Models of the Jovian Ionosphere*, Ph.D. thesis, University of Kansas, 1998.

Appendix K

Atmospheric models

K.1 Introduction

Empirical models of the Venus and terrestrial upper atmospheres have been developed. Tables K.1 and K.2 provide representative values of the Venus neutral temperature and densities for noon and midnight conditions, respectively. The values are from the *Venus International Reference Atmosphere* (VIRA) model.[1] Representative neutral temperatures and densities for the Earth's thermosphere are given in Tables K.3 to K.6. The tables provide typical values at noon and midnight for both solar maximum and minimum conditions, and for quiet geomagnetic activity. The neutral parameters are from the *Mass Spectrometer* and *Incoherent Scatter* (MSIS) empirical model.[2,3]

The latest version of the MSIS empirical model covers both the lower and upper atmosphere and includes diurnal, semi-diurnal, and terdiurnal migrating tidal modes. A reference atmosphere for Mars that is based on measurements has not been developed. However, an engineering-level Mars atmosphere model that is based on models is available. The *Mars Global Reference Atmospheric* Model (Mars-GRAM)[4] is based on the NASA *Ames Mars General Circulation* Model below 80 km[5] and the University of Michigan *Mars Thermospheric General Circulation* Model above 80 km.[6]

Table K.1 VIRA model of composition, temperature, and density (noon, $16°N$, $F_{10.7} = 150$).

Altitude (km)	T_n (K)	CO_2 (cm^{-3})	O (cm^{-3})	CO (cm^{-3})	He (cm^{-3})	N (cm^{-3})	N_2 (cm^{-3})	H (cm^{-3})
150	246.5	9.81(9)[a]	4.00(9)	2.34(9)	5.01(6)	4.65(7)	1.32(9)	8.88(4)
155	257.4	3.87(9)	2.78(9)	1.27(9)	4.51(6)	3.36(7)	7.20(8)	8.43(4)
160	265.0	1.60(9)	1.98(9)	7.18(8)	4.10(6)	2.49(7)	4.06(8)	8.09(4)
165	270.5	6.83(8)	1.43(9)	4.15(8)	3.75(6)	1.87(7)	2.34(8)	7.82(4)
170	274.4	2.98(8)	1.05(9)	2.43(8)	3.46(6)	1.42(7)	1.37(8)	7.59(4)
175	277.1	1.32(8)	7.76(8)	1.44(8)	3.19(6)	1.09(7)	8.15(7)	7.40(4)
180	279.0	5.90(7)	5.76(8)	8.63(7)	2.96(6)	8.40(6)	4.08(7)	7.23(4)
185	280.4	2.66(7)	4.30(8)	5.19(7)	2.74(6)	6.50(6)	2.93(7)	7.07(4)
190	281.4	1.21(7)	3.22(8)	3.14(7)	2.55(6)	5.04(6)	1.77(7)	6.93(4)
195	282.1	5.51(6)	2.42(8)	1.90(7)	2.37(6)	3.92(6)	1.07(7)	6.79(4)
200	282.5	2.52(6)	1.82(8)	1.15(7)	2.21(6)	3.05(6)	6.53(6)	6.67(4)
205	282.9	1.16(6)	1.37(8)	7.03(6)	2.05(6)	2.38(6)	3.97(6)	6.54(4)
210	283.1	5.32(5)	1.03(8)	4.29(6)	1.91(6)	1.86(6)	2.42(6)	6.43(4)
215	283.3	2.45(5)	7.77(7)	2.62(6)	1.78(6)	1.45(6)	1.48(6)	6.31(4)
220	283.4	1.13(5)	5.87(7)	1.60(6)	1.66(6)	1.13(6)	9.05(5)	6.20(4)
225	283.5	5.24(4)	4.43(7)	9.80(5)	1.55(6)	8.88(5)	5.54(5)	6.09(4)
230	283.6	2.43(4)	3.35(7)	6.01(5)	1.44(6)	6.95(5)	3.40(5)	5.98(4)
235	283.6	1.13(4)	2.53(7)	3.68(5)	1.35(6)	5.44(5)	2.08(5)	5.88(4)
240	283.6	5.23(3)	1.92(7)	2.26(5)	1.26(6)	4.26(5)	1.28(5)	5.78(4)
245	283.7	2.43(3)	1.45(7)	1.39(5)	1.17(6)	3.34(5)	7.85(4)	5.68(4)
250	283.7	1.13(3)	1.10(7)	8.55(4)	1.09(6)	2.62(5)	4.83(4)	5.58(4)

[a] $9.81(9) = 9.81 \times 10^9$.

Table K.2 VIRA model of composition, temperature, and density (midnight, 16°N, $F_{10.7} = 150$).

Altitude (km)	T_n (K)	CO_2 (cm^{-3})	O (cm^{-3})	CO (cm^{-3})	He (cm^{-3})	N (cm^{-3})	N_2 (cm^{-3})	H (cm^{-3})
150	127.4	7.10(7)^a	8.51(8)	7.24(7)	1.89(7)	5.80(6)	5.91(7)	1.64(7)
155	127.4	1.23(7)	4.50(8)	2.37(7)	1.81(7)	3.32(6)	1.94(7)	1.58(7)
160	127.4	2.13(6)	2.38(8)	7.78(6)	1.37(7)	1.90(6)	6.35(6)	1.51(7)
165	127.4	3.71(5)	1.26(8)	2.56(6)	1.17(7)	1.09(6)	2.09(6)	1.45(7)
170	127.4	6.48(4)	6.68(7)	8.42(5)	9.99(6)	6.26(5)	6.87(5)	1.40(7)
175	127.4	1.13(4)	3.54(7)	2.78(5)	8.53(6)	3.59(5)	2.27(5)	1.34(7)
180	127.4	1.99(3)	1.88(7)	9.18(4)	7.28(6)	2.07(5)	7.49(4)	1.29(7)
185	127.4	3.50(2)	1.00(7)	3.04(4)	6.21(6)	1.19(5)	2.48(4)	1.24(7)
190	127.4	6.18(1)	5.33(6)	1.01(4)	5.31(6)	6.84(4)	8.22(3)	1.19(7)
195	127.4	1.09(1)	2.84(6)	3.35(3)	4.53(6)	3.94(4)	2.73(3)	1.15(7)
200	127.4	1.94(0)	1.51(6)	1.11(3)	3.88(6)	2.28(4)	9.09(2)	1.10(7)
205	127.4	3.46(−1)	8.08(5)	3.71(2)	3.31(6)	1.31(4)	3.03(2)	1.06(7)
210	127.4	6.17(−2)	4.32(5)	1.24(2)	2.83(6)	7.59(3)	1.01(2)	1.02(7)
215	127.4	1.10(−2)	2.31(5)	4.15(1)	2.42(6)	4.39(3)	3.39(1)	9.81(6)
220	127.4	1.98(−3)	1.24(5)	1.39(1)	2.07(6)	2.54(3)	1.13(1)	9.43(6)
225	127.4	3.56(−4)	6.62(4)	4.67(0)	1.77(6)	1.47(3)	3.81(0)	9.07(6)
230	127.4	6.43(−5)	3.55(4)	1.57(0)	1.52(6)	8.54(2)	1.28(0)	8.73(6)
235	127.4	1.16(−5)	1.91(4)	5.29(−1)	1.30(6)	4.96(2)	4.32(−1)	8.40(6)
240	127.4	2.11(−6)	1.03(4)	1.79(−1)	1.11(6)	2.88(2)	1.46(−1)	8.08(6)
245	127.4	3.84(−7)	5.52(3)	6.04(−2)	9.52(5)	1.67(2)	4.92(−2)	7.77(6)
250	127.4	7.00(−8)	2.97(3)	2.04(−2)	8.16(5)	9.75(1)	1.67(−2)	7.47(6)

^{a} 7.10(7) = 7.10 × 10^7.

Table K.3 MSIS model of terrestrial neutral parameters (noon, 45°N, 0°E, $F_{10.7} = 70$, winter).

Altitude (km)	T_n (K)	N_2 (cm^{-3})	O_2 (cm^{-3})	O (cm^{-3})	He (cm^{-3})	H (cm^{-3})
100	192.0	9.22(12)^a	2.21(12)	4.26(11)	1.14(8)	2.41(7)
120	352.2	3.27(11)	4.83(10)	8.97(10)	3.55(7)	4.79(6)
140	525.2	5.22(10)	5.56(9)	2.81(10)	4.33(7)	1.49(6)
160	617.3	1.48(10)	1.33(9)	1.27(10)	3.53(7)	6.89(5)
180	668.4	5.17(9)	4.05(8)	6.72(9)	2.93(7)	4.41(5)
200	696.9	1.99(9)	1.37(8)	3.82(9)	2.51(7)	3.48(5)
220	712.8	8.11(8)	4.92(7)	2.26(9)	2.18(7)	3.07(5)
240	721.8	3.40(8)	1.83(7)	1.36(9)	1.92(7)	2.85(5)
260	726.8	1.45(8)	6.91(6)	8.35(8)	1.69(7)	2.71(5)
280	729.7	6.28(7)	2.65(6)	5.16(8)	1.50(7)	2.61(5)
300	731.3	2.74(7)	1.03(6)	3.21(8)	1.33(7)	2.52(5)
400	733.3	4.76(5)	1.00(4)	3.16(7)	7.44(6)	2.18(5)
500	733.4	9.37(3)	1.13(2)	3.35(6)	4.24(6)	1.89(5)
600	733.4	2.07(2)	1.44(0)	3.79(5)	2.46(6)	1.65(5)
700	733.4	5.08(0)	2.08(−2)	4.56(4)	1.45(6)	1.45(5)
800	733.4	1.38(−1)	3.39(−4)	5.81(3)	8.66(5)	1.27(5)
900	733.4	4.16(−3)	6.19(−6)	7.85(2)	5.25(5)	1.12(5)
1000	733.4	1.38(−4)	1.26(−7)	1.12(2)	3.23(5)	9.93(4)

^{a} $9.22(12) = 9.22 \times 10^{12}$.

Table K.4 MSIS model of terrestrial neutral parameters (midnight, 45°N, 0°E, $F_{10.7} = 70$, winter).

Altitude (km)	T_n (K)	N_2 (cm^{-3})	O_2 (cm^{-3})	O (cm^{-3})	He (cm^{-3})	H (cm^{-3})
100	189.3	9.70(12)^a	2.33(12)	4.15(11)	1.20(8)	2.96(7)
120	342.3	3.28(11)	4.83(10)	8.34(10)	3.61(7)	6.57(6)
140	510.1	5.04(10)	5.31(9)	2.52(10)	3.82(7)	2.29(6)
160	594.5	1.39(10)	1.23(9)	1.11(10)	3.08(7)	1.15(6)
180	638.7	4.69(9)	3.59(8)	5.67(9)	2.55(7)	7.73(5)
200	661.9	1.74(9)	1.16(8)	3.12(9)	2.18(7)	6.26(5)
220	674.1	6.76(8)	3.97(7)	1.79(9)	1.89(7)	5.58(5)
240	680.6	2.70(8)	1.39(7)	1.04(9)	1.65(7)	5.21(5)
260	684.0	1.10(8)	4.97(6)	6.19(8)	1.45(7)	4.97(5)
280	685.9	4.51(7)	1.80(6)	3.70(8)	1.27(7)	4.78(5)
300	686.8	1.87(7)	6.57(5)	2.22(8)	1.12(7)	4.61(5)
400	687.9	2.49(5)	4.74(3)	1.88(7)	6.04(6)	3.94(5)
500	688.0	3.79(3)	3.96(1)	1.72(6)	3.32(6)	3.40(5)
600	688.0	6.49(1)	3.79(−1)	1.68(5)	1.86(6)	2.94(5)
700	688.0	1.25(0)	4.15(−3)	1.76(4)	1.06(6)	2.55(5)
800	688.0	2.68(−2)	5.15(−5)	1.96(3)	6.11(5)	2.22(5)
900	688.0	6.40(−4)	7.21(−7)	2.32(2)	3.58(5)	1.95(5)
1000	688.0	1.69(−5)	1.13(−8)	2.90(1)	2.13(5)	1.71(5)

^{a} $9.70(12) = 9.70 \times 10^{12}$.

Table K.5 MSIS model of terrestrial neutral parameters (noon, 45°N, 0°E, $F_{10.7} = 220$, summer).

Altitude (km)	T_n (K)	N_2 (cm^{-3})	O_2 (cm^{-3})	O (cm^{-3})	He (cm^{-3})	H (cm^{-3})
100	220.8	5.83(12)^a	1.25(12)	2.55(11)	6.83(7)	8.70(6)
120	404.3	3.34(11)	3.46(10)	7.29(10)	1.81(7)	1.23(6)
140	754.9	5.89(10)	4.03(9)	2.16(10)	6.46(6)	2.30(5)
160	990.1	2.18(10)	1.31(9)	1.08(10)	4.72(6)	7.57(4)
180	1146.5	1.05(10)	5.81(8)	6.60(9)	3.96(6)	4.00(4)
200	1250.9	5.74(9)	2.95(8)	4.46(9)	3.49(6)	2.88(4)
220	1320.6	3.36(9)	1.61(8)	3.19(9)	3.14(6)	2.44(4)
240	1367.4	2.05(9)	9.24(7)	2.36(9)	2.88(6)	2.23(4)
260	1398.8	1.29(9)	5.45(7)	1.79(9)	2.67(6)	2.12(4)
280	1419.8	8.26(8)	3.28(7)	1.37(9)	2.49(6)	2.05(4)
300	1434.2	5.36(8)	2.00(7)	1.07(9)	2.33(6)	2.00(4)
400	1459.9	6.81(7)	1.90(6)	3.25(8)	1.72(6)	1.83(4)
500	1463.8	9.48(6)	2.00(5)	1.05(8)	1.30(6)	1.70(4)
600	1464.4	1.40(6)	2.25(4)	3.52(7)	9.86(5)	1.59(4)
700	1464.5	2.19(5)	2.69(3)	1.22(7)	7.56(5)	1.49(4)
800	1464.5	3.60(4)	3.43(2)	4.35(6)	5.85(5)	1.40(4)
900	1464.5	6.24(3)	4.61(1)	1.60(6)	4.55(5)	1.31(4)
1000	1464.5	1.13(3)	6.56(0)	6.02(5)	3.57(5)	1.23(4)

^{a} $5.83(12) = 5.83 \times 10^{12}$.

Table K.6 MSIS model of terrestrial neutral parameters (midnight, 45°N, 0°E, $F_{10.7} = 220$, summer).

Altitude (km)	T_n (K)	N_2 (cm^{-3})	O_2 (cm^{-3})	O (cm^{-3})	He (cm^{-3})	H (cm^{-3})
100	217.2	6.15(12)^a	1.31(12)	2.39(11)	7.20(7)	9.15(6)
120	394.0	3.35(11)	3.46(10)	6.45(10)	1.79(7)	1.30(6)
140	722.6	5.82(10)	3.94(9)	1.86(10)	4.95(6)	2.44(5)
160	915.1	2.14(10)	1.26(9)	9.16(9)	3.60(6)	8.16(4)
180	1026.8	1.00(10)	5.40(8)	5.46(9)	3.05(6)	4.39(4)
200	1092.0	5.24(9)	2.60(8)	3.58(9)	2.70(6)	3.19(4)
220	1130.0	2.90(9)	1.33(8)	2.47(9)	2.44(6)	2.73(4)
240	1152.3	1.66(9)	7.06(7)	1.75(9)	2.23(6)	2.51(4)
260	1165.5	9.69(8)	3.82(7)	1.27(9)	2.05(6)	2.39(4)
280	1173.2	5.72(8)	2.10(7)	9.29(8)	1.90(6)	2.32(4)
300	1177.8	3.41(8)	1.16(7)	6.88(8)	1.76(6)	2.26(4)
400	1184.0	2.76(7)	6.55(5)	1.61(8)	1.23(6)	2.06(4)
500	1184.5	2.42(6)	4.07(4)	4.00(7)	8.66(5)	1.88(4)
600	1184.6	2.28(5)	2.73(3)	1.04(7)	6.18(5)	1.73(4)
700	1184.6	2.30(4)	1.99(2)	2.80(6)	4.45(5)	1.60(4)
800	1184.6	2.47(3)	1.55(1)	7.82(5)	3.24(5)	1.47(4)
900	1184.6	2.83(2)	1.30(0)	2.26(5)	2.37(5)	1.36(4)
1000	1184.6	3.42(1)	1.17(−1)	6.78(4)	1.76(5)	1.26(4)

^{a} $6.15(12) = 6.15 \times 10^{12}$.

K.2   Specific references

1. Keating, G.M., J.L. Bertaux, S.W. Bougher, *et al.*, Models of Venus neutral upper atmosphere: structure and composition, *Adv. Space Res.*, **5**, 117, 1985.
2. Hedin, A.E., MSIS-86 thermospheric model, *J. Geophys. Res.*, **92**, 4649, 1987.
3. Picone, J.M., A.E. Hedin, D.P. Drob, and A.C. Aikin, NRLMSISE-00 empirical model of the atmosphere: statistical comparisons and scientific issues, *J. Geophys. Res.*, **107**, 1468, 2002.
4. Justus, C.G., A.L. Duvall, and D.L. Johnson, Global MGS TES data and Mars-GRAM validation, *Adv. Space Res.*, **35**, 4, 2005.
5. Barnes, J.R., J.B. Pollock, R.M. Haberk, *et al.*, Mars atmospheric dynamics as simulated by the NASA Ames General Circulaton Model. II – Transient baroclinic eddies, *J. Geophys. Res.*, **98**, 3125, 1993.
6. Bougher, S.W., S. Engel, R.G. Roble, and B. Foster, Comparative terrestrial planet thermospheres. II – Solar cycle variation of global structure and winds at equinox, *J. Geophys. Res.*, **104** (E7), 16591, 1999.

Appendix L

Scalars, vectors, dyadics, and tensors

Plasma physics is a subject where advanced mathematical techniques are frequently required to gain an understanding of the physical phenomena under consideration. This is particularly true in studies involving kinetic theory and plasma transport effects, where scalars, vectors, and multi-order tensors are needed (Chapters 3 and 4). Therefore, it is useful to review briefly some of the required mathematics.

A *scalar* is a single number that is useful for describing, say, the temperature of a gas. However, to describe the velocity of the gas, both a magnitude and direction are required (e.g., a vector). A *vector* is defined relative to some orthogonal coordinate system and three numbers, corresponding to the components of the vector, are required to define the vector. In a Cartesian coordinate system, the vector $\mathbf{a}$ is given as

$$\mathbf{a} = a_1 \mathbf{e}_1 + a_2 \mathbf{e}_2 + a_3 \mathbf{e}_3, \tag{L.1}$$

where $\mathbf{e}_1$, $\mathbf{e}_2$, and $\mathbf{e}_3$ are unit vectors along the x, y, and z axes, respectively. In index notation, the vector $\mathbf{a}$ is simply represented by a_α where α varies from 1 to 3. Suppose that another vector $\mathbf{b}$ exists, where

$$\mathbf{b} = b_1 \mathbf{e}_1 + b_2 \mathbf{e}_2 + b_3 \mathbf{e}_3. \tag{L.2}$$

It is then possible to take both the *scalar product* and *cross product* of the vectors $\mathbf{a}$ and $\mathbf{b}$, which are given by

$$\mathbf{a} \cdot \mathbf{b} = \mathbf{b} \cdot \mathbf{a} = a_1 b_1 + a_2 b_2 + a_3 b_3 = a_\alpha b_\alpha, \tag{L.3}$$

$$\mathbf{a} \times \mathbf{b} = (a_2 b_3 - a_3 b_2) \mathbf{e}_1 + (a_3 b_1 - a_1 b_3) \mathbf{e}_2 + (a_1 b_2 - a_2 b_1) \mathbf{e}_3$$

$$= \varepsilon_{\alpha\beta\gamma} a_\alpha b_\beta \mathbf{e}_\gamma \tag{L.4}$$

where the last expression in these equations is the result in index notation. In Equations (L.3), the *repeated indices* imply a summation, while in Equation (L.4) $\varepsilon_{\alpha\beta\gamma} = +1$ if α, β, γ are all unequal and in the order $123123\ldots$, -1 if the order is $213213\ldots$, and zero otherwise.

The vectors **a** and **b** can also be used to construct a *dyadic*, which is denoted by **ab**. A dyadic is composed of nine numbers, with each corresponding to a different orthogonal direction. The dyadic is therefore the extension of the vector concept to two dimensions and is equivalent to a second-order tensor. It can be represented in the three equivalent forms

$$\mathbf{ab} = a_\alpha b_\beta = \begin{pmatrix} a_1 b_1 & a_1 b_2 & a_1 b_3 \\ a_2 b_1 & a_2 b_2 & a_2 b_3 \\ a_3 b_1 & a_3 b_2 & a_3 b_3 \end{pmatrix}, \tag{L.5}$$

where $a_\alpha b_\beta$ corresponds to the index representation of the dyadic and the quantity on the right is its matrix representation. In a Cartesian coordinate system, **ab** is expressed as

$$\begin{aligned}
\mathbf{ab} &= (a_1\mathbf{e}_1 + a_2\mathbf{e}_2 + a_3\mathbf{e}_3)(b_1\mathbf{e}_1 + b_2\mathbf{e}_2 + b_3\mathbf{e}_3) \\
&= a_1 b_1 \mathbf{e}_1\mathbf{e}_1 + a_1 b_2 \mathbf{e}_1\mathbf{e}_2 + a_1 b_3 \mathbf{e}_1\mathbf{e}_3 \\
&\quad + a_2 b_1 \mathbf{e}_2\mathbf{e}_1 + a_2 b_2 \mathbf{e}_2\mathbf{e}_2 + a_2 b_3 \mathbf{e}_2\mathbf{e}_3 \\
&\quad + a_3 b_1 \mathbf{e}_3\mathbf{e}_1 + a_3 b_2 \mathbf{e}_3\mathbf{e}_2 + a_3 b_3 \mathbf{e}_3\mathbf{e}_3,
\end{aligned} \tag{L.6}$$

where quantities such as $\mathbf{e}_1\mathbf{e}_2$ are *unit dyadics*. Unit dyadics are an extension of the concept of unit vector to two dimensions. In comparing the matrix in Equation (L.5) with Equation (L.6), it is apparent that the unit dyadics are introduced to provide orthogonal directions to the nine elements in the matrix. When dealing with dyadics, the location of a particular vector in the dyadic is important. In general, $\mathbf{ab} \neq \mathbf{ba}$, and when a dyadic is operated on via a dot or cross product, it is the closest vector in the dyadic that is affected by the operation. For example,

$$\mathbf{ab} \cdot \mathbf{c} = \mathbf{a}(\mathbf{b} \cdot \mathbf{c}), \tag{L.7}$$

$$\mathbf{c} \cdot \mathbf{ab} = (\mathbf{c} \cdot \mathbf{a})\mathbf{b}, \tag{L.8}$$

$$\mathbf{ab} \times \mathbf{c} = \mathbf{a}(\mathbf{b} \times \mathbf{c}), \tag{L.9}$$

$$\mathbf{c} \times \mathbf{ab} = (\mathbf{c} \times \mathbf{a})\mathbf{b}. \tag{L.10}$$

Second-order tensors, which are composed of nine independent elements, can also be expressed in dyadic form. In analogy to Equation (L.5), a second-order tensor **W** can be expressed in the three equivalent forms

$$\mathbf{W} = W_{\alpha\beta} = \begin{pmatrix} W_{11} & W_{12} & W_{13} \\ W_{21} & W_{22} & W_{23} \\ W_{31} & W_{32} & W_{33} \end{pmatrix}, \tag{L.11}$$

and in Cartesian coordinates $\mathbf{W}$ becomes

$$
\begin{aligned}
\mathbf{W} = {} & W_{11}\mathbf{e}_1\mathbf{e}_1 + W_{12}\mathbf{e}_1\mathbf{e}_2 + W_{13}\mathbf{e}_1\mathbf{e}_3 \\
& + W_{21}\mathbf{e}_2\mathbf{e}_1 + W_{22}\mathbf{e}_2\mathbf{e}_2 + W_{23}\mathbf{e}_2\mathbf{e}_3 \\
& + W_{31}\mathbf{e}_3\mathbf{e}_1 + W_{32}\mathbf{e}_3\mathbf{e}_2 + W_{33}\mathbf{e}_3\mathbf{e}_3.
\end{aligned}
\tag{L.12}
$$

If the tensor is not symmetric, then

$$
\mathbf{a} \cdot \mathbf{W} \neq \mathbf{W} \cdot \mathbf{a}
\tag{L.13}
$$

$$
\mathbf{a} \times \mathbf{W} \neq \mathbf{W} \times \mathbf{a}
\tag{L.14}
$$

and in Equations (L.13) and (L.14) the dot and cross products operate on the closest vectors in Equation (L.12). That is, if the dot or cross product is on the left, then the left vectors in the unit dyadics are affected. For example,

$$
\begin{aligned}
\mathbf{a} \cdot \mathbf{W} = a_\alpha W_{\alpha\beta} = {} & a_1(W_{11}\mathbf{e}_1 + W_{12}\mathbf{e}_2 + W_{13}\mathbf{e}_3) \\
& + a_2(W_{21}\mathbf{e}_1 + W_{22}\mathbf{e}_2 + W_{23}\mathbf{e}_3) \\
& + a_3(W_{31}\mathbf{e}_1 + W_{32}\mathbf{e}_2 + W_{33}\mathbf{e}_3) \\
= {} & (a_1 W_{11} + a_2 W_{21} + a_3 W_{31})\mathbf{e}_1 \\
& + (a_1 W_{12} + a_2 W_{22} + a_3 W_{32})\mathbf{e}_2 \\
& + (a_1 W_{13} + a_2 W_{23} + a_3 W_{33})\mathbf{e}_3.
\end{aligned}
\tag{L.15}
$$

The second-order tensor $\mathbf{W}$ can also operate on another second-order tensor $\mathbf{Y}$. The dot product of the two tensors is given by

$$
\mathbf{W} \cdot \mathbf{Y} = W_{\alpha\beta} Y_{\beta\gamma}
\tag{L.16}
$$

so that the dot product of two second-order tensors is another second-order tensor. It is possible to have a *double dot* (or *scalar*) product of $\mathbf{W}$ and $\mathbf{Y}$, and the result is

$$
\mathbf{W} : \mathbf{Y} = W_{\alpha\beta} Y_{\beta\alpha} = \mathbf{Y} : \mathbf{W},
\tag{L.17}
$$

where the convention is that the two inner indices and the two outer indices are repetitive.

The *transpose* of a tensor $\mathbf{W}$ with components $W_{\alpha\beta}$ is denoted by $\mathbf{W}^{\mathrm{T}}$, and it is obtained simply by interchanging rows and columns so that its components are $W_{\beta\alpha}$. In general, $\mathbf{W} \neq \mathbf{W}^{\mathrm{T}}$, but when they are equal the tensor is *symmetric*.

The *unit* or *identity dyadic*, $\mathbf{I}$, is equivalent to the Kronecker delta in index notation and can be expressed in the three forms

$$
\mathbf{I} = \delta_{\alpha\beta} = \begin{pmatrix} 1 & 0 & 0 \\ 0 & 1 & 0 \\ 0 & 0 & 1 \end{pmatrix}.
\tag{L.18}
$$

In Cartesian coordinates, $\mathbf{I}$ becomes

$$\mathbf{I} = \mathbf{e}_1\mathbf{e}_1 + \mathbf{e}_2\mathbf{e}_2 + \mathbf{e}_3\mathbf{e}_3. \tag{L.19}$$

When $\mathbf{I}$ is dotted with either a vector or tensor, the quantity is recovered, so that

$$\mathbf{I} \cdot \mathbf{a} = \delta_{\alpha\beta}a_\beta = a_\alpha = \mathbf{a}, \tag{L.20}$$

$$\mathbf{I} \cdot \mathbf{W} = \delta_{\alpha\beta}W_{\beta\gamma} = W_{\alpha\gamma} = \mathbf{W}, \tag{L.21}$$

and

$$\mathbf{I} \cdot \mathbf{a} = \mathbf{a} \cdot \mathbf{I} = \mathbf{a}, \tag{L.22}$$

$$\mathbf{I} \cdot \mathbf{W} = \mathbf{W} \cdot \mathbf{I} = \mathbf{W}. \tag{L.23}$$

In addition, it is possible to take a double dot product of $\mathbf{I}$ with a second-order tensor $\mathbf{W}$. This yields the *trace of the tensor*, which is the sum of the diagonal elements, given by

$$\mathbf{I} : \mathbf{W} = \delta_{\alpha\beta}W_{\beta\alpha} = W_{\alpha\alpha}. \tag{L.24}$$

Also,

$$\mathbf{I} : \mathbf{I} = \delta_{\alpha\beta}\delta_{\beta\alpha} = \delta_{\alpha\alpha} = 3. \tag{L.25}$$

Two additional operations that are useful with regard to the material in the book are the divergence of a tensor, $\nabla \cdot \mathbf{W}$, and the construction of the dyadic $\nabla\mathbf{u}$. From Equation (L.12), the divergence of $\mathbf{W}$ is given by

$$\nabla \cdot \mathbf{W} = \left(\mathbf{e}_1\frac{\partial}{\partial x_1} + \mathbf{e}_2\frac{\partial}{\partial x_2} + \mathbf{e}_3\frac{\partial}{\partial x_3}\right) \cdot \mathbf{W}$$

$$= \left(\frac{\partial W_{11}}{\partial x_1}\mathbf{e}_1 + \frac{\partial W_{12}}{\partial x_1}\mathbf{e}_2 + \frac{\partial W_{13}}{\partial x_1}\mathbf{e}_3\right)$$

$$+ \left(\frac{\partial W_{21}}{\partial x_2}\mathbf{e}_1 + \frac{\partial W_{22}}{\partial x_2}\mathbf{e}_2 + \frac{\partial W_{23}}{\partial x_2}\mathbf{e}_3\right)$$

$$+ \left(\frac{\partial W_{31}}{\partial x_3}\mathbf{e}_1 + \frac{\partial W_{32}}{\partial x_3}\mathbf{e}_2 + \frac{\partial W_{33}}{\partial x_3}\mathbf{e}_3\right)$$

$$= \left(\frac{\partial W_{11}}{\partial x_1} + \frac{\partial W_{21}}{\partial x_2} + \frac{\partial W_{31}}{\partial x_3} \right) \mathbf{e}_1$$

$$+ \left(\frac{\partial W_{12}}{\partial x_1} + \frac{\partial W_{22}}{\partial x_2} + \frac{\partial W_{32}}{\partial x_3} \right) \mathbf{e}_2$$

$$+ \left(\frac{\partial W_{13}}{\partial x_1} + \frac{\partial W_{23}}{\partial x_2} + \frac{\partial W_{33}}{\partial x_3} \right) \mathbf{e}_3. \tag{L.26}$$

The dyadic $\nabla \mathbf{u}$ is constructed as follows

$$\nabla \mathbf{u} = \left(\mathbf{e}_1 \frac{\partial}{\partial x_1} + \mathbf{e}_2 \frac{\partial}{\partial x_2} + \mathbf{e}_3 \frac{\partial}{\partial x_3} \right) \left(u_1 \mathbf{e}_1 + u_2 \mathbf{e}_2 + u_3 \mathbf{e}_3 \right)$$

$$= \frac{\partial u_1}{\partial x_1} \mathbf{e}_1 \mathbf{e}_1 + \frac{\partial u_2}{\partial x_1} \mathbf{e}_1 \mathbf{e}_2 + \frac{\partial u_3}{\partial x_1} \mathbf{e}_1 \mathbf{e}_3 + \frac{\partial u_1}{\partial x_2} \mathbf{e}_2 \mathbf{e}_1 + \frac{\partial u_2}{\partial x_2} \mathbf{e}_2 \mathbf{e}_2 + \frac{\partial u_3}{\partial x_2} \mathbf{e}_2 \mathbf{e}_3$$

$$+ \frac{\partial u_1}{\partial x_3} \mathbf{e}_3 \mathbf{e}_1 + \frac{\partial u_2}{\partial x_3} \mathbf{e}_3 \mathbf{e}_2 + \frac{\partial u_3}{\partial x_3} \mathbf{e}_3 \mathbf{e}_3. \tag{L.27}$$

Therefore, $\nabla \mathbf{u}$ is a second-order tensor with nine independent elements. The transpose of $\nabla \mathbf{u}$, which is denoted by $(\nabla \mathbf{u})^{\mathrm{T}}$, is obtained by interchanging the elements in the rows with the elements in the columns, which yields

$$(\nabla \mathbf{u})^{\mathrm{T}} = \frac{\partial u_1}{\partial x_1} \mathbf{e}_1 \mathbf{e}_1 + \frac{\partial u_1}{\partial x_2} \mathbf{e}_1 \mathbf{e}_2 + \frac{\partial u_1}{\partial x_3} \mathbf{e}_1 \mathbf{e}_3$$

$$+ \frac{\partial u_2}{\partial x_1} \mathbf{e}_2 \mathbf{e}_1 + \frac{\partial u_2}{\partial x_2} \mathbf{e}_2 \mathbf{e}_2 + \frac{\partial u_2}{\partial x_3} \mathbf{e}_2 \mathbf{e}_3$$

$$+ \frac{\partial u_3}{\partial x_1} \mathbf{e}_3 \mathbf{e}_1 + \frac{\partial u_3}{\partial x_2} \mathbf{e}_3 \mathbf{e}_2 + \frac{\partial u_3}{\partial x_3} \mathbf{e}_3 \mathbf{e}_3. \tag{L.28}$$

Finally, the nonlinear inertial term $(\mathbf{u} \cdot \nabla)\mathbf{u}$ can be obtained either by first taking the dot product $(\mathbf{u} \cdot \nabla)$ and then operating on the vector $\mathbf{u}$ or by taking the dot product of $\mathbf{u}$ with the tensor $(\nabla \mathbf{u})$ given in Equation (L.27). In both cases the result is

$$(\mathbf{u} \cdot \nabla)\mathbf{u} = \mathbf{u} \cdot (\nabla \mathbf{u}) = \left(u_1 \frac{\partial u_1}{\partial x_1} + u_2 \frac{\partial u_1}{\partial x_2} + u_3 \frac{\partial u_1}{\partial x_3} \right) \mathbf{e}_1$$

$$+ \left(u_1 \frac{\partial u_2}{\partial x_1} + u_2 \frac{\partial u_2}{\partial x_2} + u_3 \frac{\partial u_2}{\partial x_3} \right) \mathbf{e}_2$$

$$+ \left(u_1 \frac{\partial u_3}{\partial x_1} + u_2 \frac{\partial u_3}{\partial x_2} + u_3 \frac{\partial u_3}{\partial x_3} \right) \mathbf{e}_3. \tag{L.29}$$

Appendix M

Radio wave spectrum

Table M.1

Extra low frequency (ELF)	30–330 Hz
Voice frequency (VF)	300–3000 Hz
Very low frequency (VLF)	3–30 kHz
Low frequency (LF)	30–300 kHz
Medium frequency (MF)	300–3000 kHz
High frequency (HF)	3–30 MHz
Very high frequency (VHF)	30–300 MHz
Ultra high frequency (UHF)	300–3000 MHz
Super high frequency (SHF)	3–30 GHz
Extremely high frequency (EHF)	30–300 GHz
L1 frequency	1.57542 GHz
L2 frequency	1.227 GHz

Note: $1\,Hz = 1$ cycle s^{-1}; kilo (k) $= 10^3$; mega (M) $= 10^6$; giga (G) $= 10^9$

Appendix N

Simple derivation of continuity equation

A rigorous derivation of the transport equations that is based on Boltzmann's equation is given in Chapter 3. However, physical insight can be gained by considering a derivation of the continuity equation via a more straightforward method. Consider a volume element in a three-dimensional fluid flow, as shown in Figure N.1. The density of the fluid is n and the drift velocity components are u_x, u_y, and u_z, all of which are a function of x, y, z, and t. The flux of particles moving in the x-direction is $nu_x(\mathrm{m}^{-2}\mathrm{s}^{-1})$ and the number of particles per second that enters the volume element across a plane at location x is $(nu_x)_x(\Delta y\,\Delta z)$. The number per second that leave the volume element at location $x + \Delta x$ is $(nu_x)_{x+\Delta x}(\Delta y\,\Delta z)$. Therefore the net increase in the number per second is:

$$\frac{\text{number}}{s} = \left\{(nu_x)_x - (nu_x)_{x+\Delta x}\right\}\Delta y\,\Delta z\ . \tag{N.1}$$

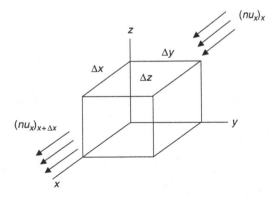

Figure N.1 Volume element showing a flux of particles entering the volume at x and leaving at $x + \Delta x$.

If Δx is small, the second term in Equation (N.1) can be expanded in a Taylor series, which yields:

$$\frac{\text{number}}{s} = \left[(nu_x)_x - \left((nu_x)_x + \frac{\partial}{\partial x}(nu_x)_x \Delta x \ldots \right) \Delta y \, \Delta z \right],$$ (N.2)

$$\frac{\text{number}}{s} = -\frac{\partial}{\partial x}(nu_x) \Delta x \, \Delta y \, \Delta z + \ldots,$$ (N.3)

where the subscript x on the curved brackets has been dropped and where the additional terms in the Taylor series are of order $(\Delta x)^2$ or higher.

The flow across the surfaces along the y and z axes yield similar results, so the net increase in the number of particles per second that enter the volume is:

$$\frac{\text{number}}{s} = \left[-\frac{\partial}{\partial x}(nu_x) - \frac{\partial}{\partial y}(nu_y) - \frac{\partial}{\partial z}(nu_z) \right] \Delta x \, \Delta y \, \Delta z + \ldots$$

$$= -\nabla \cdot (n\mathbf{u}) \Delta x \, \Delta y \, \Delta z + \ldots$$ (N.4)

However, the net increase in the number of particles in the volume is related to the density change with time:

$$\frac{\text{number}}{s} = \frac{\partial n}{\partial t} \Delta x \, \Delta y \, \Delta z \, .$$ (N.5)

Therefore,

$$\frac{\partial n}{\partial t} \Delta x \, \Delta y \, \Delta z = -\nabla \cdot (n\mathbf{u}) \Delta x \, \Delta y \, \Delta z + \ldots$$ (N.6)

Dividing both sides of Equation (N.6) by the volume element and then letting Δx, Δy, and Δz go to zero yields the continuity equation:

$$\frac{\partial n}{\partial t} + \nabla \cdot (n\mathbf{u}) = 0 \, .$$ (N.7)

Each species in the plasma will be governed by Equation (N.7) so it is appropriate to add a subject, j, to n and $\mathbf{u}$ to identify the different species. Also, Equation (N.7) was derived by only considering the flow into and out of a volume element. In reality, particles can be either created or destroyed by photoionization or chemical reactions, and hence, production, P_j, and loss, L_j, terms need to be added to the right-hand side of Equation (N.7), and the continuity equation then becomes:

$$\frac{\partial n_j}{\partial t} + \nabla \cdot (n_j \mathbf{u}_j) = P_j - L_j \, .$$ (N.8)

Appendix O

Numerical solution for F region ionization

The main ionization peak in the terrestrial ionosphere is associated with the F_2 layer and its basic characteristics are described in Section 11.4. However, it is instructive to obtain a numerical solution of the continuity and momentum equations that govern this region. Restricting the analyses to mid-latitudes, the plasma in this region is primarily produced by photoionization of neutral atomic oxygen and lost in chemical reactions with N_2 and O_2. If the ionosphere is assumed to be horizontally stratified (a reasonable assumption at mid-latitudes), the electron (or O^+) continuity equation becomes

$$\frac{\partial n_e}{\partial t} + \frac{\partial}{\partial z}(n_e u_{ez}) = P_e - k_\beta n_e, \tag{O.1}$$

where $n_e = n_i$, P_e is the electron production rate, k_β is the electron loss frequency (equation 11.61), z is the altitude, and subscript i corresponds to O^+. In the F region, the momentum equation reduces to the diffusion equation (11.59), and if, for simplicity, it is assumed that there is no neutral wind ($u_n = 0$), no imposed electric field ($E = 0$), and the magnetic field is vertical ($I = 90°$), Equation (11.59) becomes

$$n_e u_{ez} = -D_a \left[\frac{\partial n_e}{\partial z} + n_e \left(\frac{1}{T_p} \frac{\partial T_p}{\partial z} + \frac{1}{H_p} \right) \right], \tag{O.2}$$

where $H_p = 2kT_p/(m_i g)$ is the plasma scale height (Equation 5.59), $T_p = (T_e + T_i)/2$ is the plasma temperature (equation 5.56) and $D_a = 2kT_p/(m_i \nu_{in})$ is the ambipolar diffusion coefficient (Equation 5.55).

For illustrative purposes, it is useful to adopt the following simple expressions for the production rate, loss frequency, and ambipolar diffusion coefficient:

$$P_e = 4 \times 10^{-7} n(O), \tag{O.3}$$

$$k_\beta = 1.2 \times 10^{-12} n(N_2) + 2.1 \times 10^{-11} n(O_2), \tag{O.4}$$

$$D_a = \frac{3 \times 10^{17} T_p}{T_R^{1/2} n(O)} \tag{O.5}$$

where $T_R = (T_i + T_n)/2$, and where the densities are in cm^{-3}, T_p and T_R in K, P_e in $cm^{-3} s^{-1}$, k_β in s^{-1}, and D_a in $cm^2 s^{-1}$. Equation (O.3) takes account of the fact that at F region altitudes the atmosphere is optically thin, and hence, the production rate is simply proportional to the neutral oxygen density (the numerical coefficient is a typical value). Equation (O.4) is discussed in connection with Equation (11.60), and the O^+–O collision frequency used in Equation (O.5) is given in Table 4.5. At F region altitudes, the neutral densities are in hydrostatic equilibrium and can be calculated separately from Equation (9.12), where z_0 is some reference altitude and $H_n = kT_n/(m_n g)$ is the neutral scale height. Reference values for the N_2, O_2, and O densities are given in tables K.3 to K.6.

The substitution of Equation (O.2) into Equation (O.1) yields a second-order, parabolic, partial differential equation for the electron density,

$$\frac{\partial n_e}{\partial t} = A_1(z,t) \frac{\partial^2 n_e}{\partial z^2} + A_2(z,t) \frac{\partial n_e}{\partial z} + A_3(z,t) n_e + A_4(z,t), \tag{O.6}$$

where

$$A_1 = D_a, \tag{O.7a}$$

$$A_2 = \frac{\partial D_a}{\partial z} + D_a \left(\frac{1}{T_p} \frac{\partial T_p}{\partial z} + \frac{1}{H_p} \right), \tag{O.7b}$$

$$A_3 = D_a \left[\frac{1}{T_p} \frac{\partial^2 T_p}{\partial z^2} - \frac{1}{T_p^2} \left(\frac{\partial T_p}{\partial z} \right)^2 \right.$$
$$\left. + \frac{\partial}{\partial z} \left(\frac{1}{H_p} \right) \right] + \frac{\partial D_a}{\partial z} \left[\frac{1}{T_p} \frac{\partial T_p}{\partial z} + \frac{1}{H_p} \right] - k_\beta, \tag{O.7c}$$

$$A_4 = P_e. \tag{O.7d}$$

The solution of the second-order partial differential equation (O.6) requires an initial condition (e.g., a starting n_e altitude profile) and two boundary conditions. The lower boundary (z_B) is usually taken at an altitude where chemistry dominates, and hence, the electron density at this altitude can be obtained by equating local production and loss rates, which yields

$$n_e = P_e/k_\beta. \tag{O.8}$$

At the top boundary (z_T), diffusion dominates, and therefore, a flux boundary condition is appropriate. A simple boundary condition is that the particle flux through the top boundary is a constant (F_a), which yields

$$n_e u_{ez} = F_a .$$

(O.9)

Different values for F_a can be selected and different initial conditions can also be selected. These issues will be discussed later.

Equation (O.6) for the electron density is a linear, parabolic, partial differential equation and the completely implicit numerical technique provides a very stable and simple numerical solution.[1] The first step is to break the altitude range into a discrete set of grid points. Let z_B and z_T be the bottom and top grid points, respectively, and Δz the spatial step between grid points. Let the grid index j go from 1 at ($z_B + \Delta z$) to J at z_T. If t is the current time and Δt is the time interval, then $t + \Delta t$ is the future time. With the completely implicit numerical technique, the electron density terms on the right-hand side of Equation (O.6) are evaluated at the future time. Using space-centered spatial derivatives and a forward time derivative, the electron density terms in Equation (O.6) become

$$\frac{\partial^2 n}{\partial z^2} = \frac{n_{j+1}^{t+\Delta t} - 2n_j^{t+\Delta t} + n_{j-1}^{t+\Delta t}}{(\Delta z)^2},$$

(O.10a)

$$\frac{\partial n}{\partial z} = \frac{n_{j+1}^{t+\Delta t} - n_{j-1}^{t+\Delta t}}{2(\Delta z)},$$

(O.10b)

$$n = n_j^{t+\Delta t},$$

(O.10c)

$$\frac{\partial n}{\partial t} = \frac{n_j^{t+\Delta t} - n_j^t}{\Delta t},$$

(O.10d)

where the subscript e on n_e has been dropped for convenience. The quantities A_1 to A_4 are arrays that should be evaluated at the current time (t) and at each grid point (j). Note that in calculating numerical derivatives of quantities that are known and appear in A_2 and A_3, space-centered derivatives can only be calculated at the interior grid points ($2 \leq j \leq J - 1$). At $j = 1$, a forward derivative is needed [($\partial T/\partial z) = (T_2 - T_1)/\Delta z$], while at $j = J$ a backward derivative is used [($\partial T/\partial z)_J = (T_J - T_{J-1})/\Delta z$].

Substituting Equations (O.10a–d) into Equation (O.6) yields an equation of the form

$$-\alpha_j n_{j-1}^{t+\Delta t} + \beta_j n_j^{t+\Delta t} - \gamma_j n_{j+1}^{t+\Delta t} = S_j^t,$$

(O.11)

where

$$\alpha_j = 1 - \frac{\Delta z}{2} \frac{A_2(j)}{A_1(j)},$$

(O.12a)

$$\beta_j = 2 + \frac{(\Delta z)^2}{(\Delta t)A_1(j)} - \frac{(\Delta z)^2 A_3(j)}{A_1(j)}, \tag{O.12b}$$

$$\gamma_j = 1 + \frac{\Delta z}{2} \frac{A_2(j)}{A_1(j)}, \tag{O.12c}$$

$$S_j^t = \frac{(\Delta z)^2 A_4(j)}{A_1(j)} + \frac{(\Delta z)^2}{(\Delta t)A_1(j)} n_j^t, \tag{O.12d}$$

where j goes from 1 to J. Equation (O.11) constitutes a tri-diagonal matrix, except for the $j = 1$ and $j = J$ rows which are modified by the boundary conditions.

At the bottom boundary ($j = 0$), the density is specified via Equation (O.8), so that

$$n_0^{t+\Delta t} = n_0^t = (P_e/k_\beta)_0^t . \tag{O.13}$$

Equation (O.11) at the $j = 1$ grid point then becomes

$$\beta_1 n_1^{t+\Delta t} - \gamma_1 n_2^{t+\Delta t} = S_1^t + \alpha_1 n_0^t . \tag{O.14}$$

At the top boundary ($j = J$), the flux is specified via Equation (O.9), so that

$$(nu)_j^{t+\Delta t} = F_a, \tag{O.15}$$

where (nu) is given by Equation (O.2) and can be written as

$$(nu)_j^{t+\Delta t} = -A_1(j) \left[\frac{n_{j+1}^{t+\Delta t} - n_{j-1}^{t+\Delta t}}{2(\Delta z)} + n_j^{t+\Delta t} Q(j) \right], \tag{O.16}$$

$$Q(j) = \left[\frac{1}{T_p} \frac{\partial T_p}{\partial z} + \frac{1}{H_p} \right]_j . \tag{O.17}$$

At $j = J$, the flux becomes F_a and Equation (O.16) can be solved for the density value that is beyond the altitude range,

$$n_{j+1}^{t+\Delta t} = n_{j-1}^{t+\Delta t} - 2(\Delta z)Q(J)n_j^{t+\Delta t} - \frac{2(\Delta z)F_a}{A_1(J)} . \tag{O.18}$$

Setting $j = J$ in Equation (O.11) and eliminating $n_{J+1}^{t+\Delta t}$ with Equation (O.18) yields

$$-[\alpha_J + \gamma_J]n_{J-1}^{t+\Delta t} + [\beta_j + 2(\Delta z)\gamma_j Q(J)]n_J^{t+\Delta t} = S_J^t - \frac{2(\Delta z)\gamma_J F_a}{A_1(J)}.$$

$$(O.19)$$

For $j = 1$, Equation (O.14) applies, for $2 \le j \le J - 1$ Equation (O.11) applies, and for $j = J$, Equation (O.19) applies. Therefore, the matrix that needs to be inverted to obtain the future densities contains three elements in each row except for the first ($j = 1$) and last ($j = J$) rows. This matrix can be solved as follows. First fill the α_j, β_j, γ_j, and S_j^t arrays from $j = 1$ to $j = J$ using Equations (O.12a–d), where n_j^t is the electron density profile obtained at time t and is known from the previous matrix inversion. Next, modify α_j, β_j, γ_j, and S_j^t for $j = 1$ and $j = J$ as follows;

$$S_1^t \to S_1^t + \alpha_1 n_0^t, \tag{O.20a}$$

$$\alpha_1 \to 0, \tag{O.20b}$$

which comes from Equation (O.14) and

$$\alpha_J \to \alpha_J + \gamma_J = 2, \tag{O.21a}$$

$$\beta_J \to \beta_J + 2(\Delta z)\gamma_J Q(J), \tag{O.21b}$$

$$S_J^t \to S_J^t - \frac{2(\Delta z)\gamma_J F_a}{A_1(J)}, \tag{O.21c}$$

$$\gamma_J \to 0 \tag{O.21d}$$

which follows from Equation (O.19). Note that α_J and γ_J have to be used first before they are set to zero.

With the tri-diagonal matrix modified as described above, arrays r_j and e_j are filled from the bottom to the top, where r_j and e_j are given by

$$r_1 = \gamma_1/\beta_1, \tag{O.22a}$$

$$e_1 = S_1^t/\beta_1, \tag{O.22b}$$

$$r_j = \frac{\gamma_j}{\beta_j - \alpha_j r_{j-1}}, \qquad 2 \le j \le J, \tag{O.22c}$$

$$e_j = \frac{S_j^t + \alpha_j e_{j-1}}{\beta_j - \alpha_j r_{j-1}}, \qquad 2 \le j \le J. \tag{O.22d}$$

The electron densities are then calculated from the top to the bottom,

$$n_J^{t+\Delta t} = e_J, \tag{O.23a}$$

$$n_j^{t+\Delta t} = r_j n_{j+1}^{t+\Delta t} + e_j, \qquad j = J - 1, \dots, 1. \tag{O.23b}$$

This procedure is repeated in a time loop until the desired time-dependent solution is obtained. Note that before the next time loop is executed, $n_j^{t+\Delta t} \to n_j^t$ and a new density profile is again obtained.

O.1   Specific reference

1. Oran, E. S. and J. P. Boris, *Numerical Simulation of Reactive Flow*, Cambridge, UK: Cambridge University Press, 2001.

Appendix P

Monte Carlo methods

Monte Carlo (MC) techniques were developed in the 1940s and are based on a random sampling to determine the possible outcome of an event.[1-4] The name Monte Carlo comes from the famous casino in Monaco, where your success depends on "chance" events at the gambling tables. There are various Monte Carlo techniques and they are now widely used in almost all areas of science and mathematics, including statistics, quantum mechanics, kinetic theory, astrophysics, atmospheric physics, space science, nuclear physics, and thermodynamics. They are particularly useful for evaluating complex integrals and for solving partial differential equations.

Integrations provide a useful tool to show how the Monte Carlo method works. Consider a function $g(x, y, z)$ that is to be integrated over a volume V. With the Monte Carlo method, a random number generator is used to select N points (x_i, y_i, z_i) inside the volume V, where i goes from 1 to N. The arithmetic mean of the function g over the N selected points is given by:[1,2]

$$\langle g \rangle_a = \frac{1}{N} \sum_{i=1}^{N} g\,(x_i, y_i, z_i),$$ (P.1)

and the approximation to the volume integral is:

$$\int d^3 x g\,(x, y, z) \approx V \langle g \rangle_a$$ (P.2)

An estimate for the uncertainty in the integral is given by the one standard deviation error,

$$\pm V \left[\frac{\langle g^2 \rangle_a - \langle g \rangle_a^2}{N} \right]^{\frac{1}{2}},$$ (P.3)

where

$$\left\langle g^2 \right\rangle_a = \frac{1}{N} \sum_{i=1}^{N} g^2\left(x_i, y_i, z_i\right). \tag{P.4}$$

The volume integral (P.2) is simply the arithmetic mean of the function g in the volume multiplied by the volume. The uncertainty decreases as more random points are used to calculate the arithmetic mean. However, the uncertainty decreases as $N^{-\frac{1}{2}}$, so the convergence to the answer is slow. Nevertheless, with high-speed computers, the Monte Carlo method is a powerful tool for solving many problems.

The MC technique has been used in simulating space plasma for several decades.[5-8] The simulations increased in sophistication with time; from homogeneous to nonhomogeneous, steady state to time varying, and from single to multi-species. Here, a simple model will be presented, to elucidate the technique.

Following Barakat,[5-7] we consider the steady-state flow of a homogeneous, weakly ionized plasma in response to a homogeneous electric field. The ion–neutral collision is assumed to obey a simple relaxation model, for which the total collision cross section σ_T is inversely proportional to the relative velocity g between colliding particles (Section 4.5). Also, according to this model, the colliding particles exchange their velocities (or, equivalently, exchange identities) during the collision, which occurs during a charge exchange collision. This is equivalent to assuming that the ions and neutrals have the same mass (ions in their parent gas) and that the differential scattering cross section varies as $[\sigma\left(g, \theta\right) \to \delta\left(\theta - \pi\right)/g]$ where θ is the scattering angle in the center-of-mass frame of reference and $\delta\left(x\right)$ is the delta function. For this simple model, a closed-form expression exists,[5] which can be used in validating the numerical MC model. The neutrals are assumed to have a nondrifting Maxwellian with temperature T_n.

$$f_n\left(v\right) = n_n \left(\frac{m_n}{2\pi k T_n}\right)^{\frac{3}{2}} \exp\left[-\frac{v_n}{\left(v_{th}\right)_n}\right]^2, \tag{P.5}$$

where n_n, m_n, and $\left(v_{th}\right)_n$ are the density, mass, and thermal speed of the neutral gas, respectively. The ion–ion collisions are neglected in comparison with the ion–neutral collisions. The standard procedure of the MC simulation is to follow the motion of one ion for a large number of collisions, and to monitor its velocity. According to the ergodic theorem, the time average can be equated to the corresponding ensemble averages. In particular, the following steps are taken:

1. An initial ion velocity is picked (say $v_i = 0$). To "remove" the bias due to that choice, the first few hundred collisions are ignored.
2. The time interval between consecutive collisions (t_c) is found. Since the mean-free-time ($\tau = 1/\nu_{in}$) is constant here, the probability density

function of t_c is given by

$$p\left(t_c\right) = \frac{1}{\tau} e^{-t_c/\tau}, \tag{P.6}$$

and hence,[4]

$$t_c = -\tau \ell n\left(\gamma\right), \tag{P.7}$$

where γ is the value of a uniformly distributed (pseudo) random variable, such that

$$p\left(\gamma\right) = \begin{cases} 1 & \text{for } 0 < \gamma < 1 \\ 0 & \text{elsewhere} \end{cases}. \tag{P.8}$$

3.   The ion is followed between collisions as it moves in phase space (velocity space here) in a straight line with

$$\mathbf{v}_{i\perp} = \left(\mathbf{v}_{i\perp}\right)_0, \quad \mathbf{v}_{i\|} = \left(\mathbf{v}_{i\|}\right)_0 + \left(\frac{q_i\mathbf{E}}{m_i}\right) t, \tag{P.9}$$

where the subscript 0 indicates values at the start of the interval t_c.

4.   At the end of t_c, the ion collides with a neutral particle, which is picked randomly in accordance with Equation (P.5), specifically

$$v_{n\perp}^2 = -2\left(v_{th}\right)_n^2 \ell n\left(\gamma\right), \tag{P.10}$$

$$v_{n\|} = \left(v_{th}\right)_n \left[-2\ell n\left(\gamma_1\right)\right]^{\frac{1}{2}} \cos\left(2\pi\gamma_2\right), \tag{P.11}$$

where γ, γ_1, and γ_2 are statistically independent values with probability density functions similar to Equation (P.8).[4]

5.   After a collision, the ion assumes the neutral's velocity:

$$\mathbf{v}_i' = \mathbf{v}_n, \tag{P.12}$$

where the prime indicates the value after the collision.

6. $\mathbf{v}_i'$ is then used as the initial velocity and steps 2 through 5 are repeated for a very large number (say 10^6) of collisions.

A rectangular grid is used to record the probe ion velocities. The grid has $v_{i\|}$ and $v_{i\perp}$ as coordinates, while the azimuthal dependence is taken care of by virtue of cylindrical symmetry. The ion's velocity distribution function is proportional to:

$$f_i\left(\mathbf{v}_b\right) \rightarrow \sum_i \left(\Delta t_i\right)_b / V_b, \tag{P.13}$$

where $\mathbf{v_b}$ is the velocity of the bin's center, $(\Delta t_i)_b$ is the time the ion spends within the bin, V_b is the bin's size, and $\sum\limits_i$ indicates summation over intervals between collisions, after throwing out the first few hundred, as mentioned earlier.

The velocity moments (or density, temperature, etc.) can be found from the values of $f_i(\mathbf{v_b})$ as follows:

$$\left\langle v_{i\|}^{\alpha}\, v_{i\perp}^{\beta}\right\rangle = \frac{\sum\limits_b v_{b\|}^{\alpha} v_{b\perp}^{\beta} f_i\,(\mathbf{v_b})}{\sum\limits_b f_i\,(\mathbf{v_b})}, \tag{P.14}$$

where $\sum\limits_b$ indicates a summation over all bins. More accurate values of the moments can be achieved by using the individual trajectory segments of the test ion directly.[6]

P.1 Specific references

1. Cheney, W. and D. Kincaid, *Numerical Mathematics and Computing*, Monterey, California: Brooks/Cole Publishing Company, 1980.
2. Koonin, S. E., *Computational Physics*, Menlo Park, California: Benjamin/Cummings Publishing Company, 1986.
3. Bird, G. A., *Molecular Gas Dynamics and the Direct Simulation of Gas Flows*, Oxford, UK: Oxford University Press, 1994.
4. Sobol, I. M., *A Primer for the Monte Carlo Method*, Boca Raton, Florida: CRC Press, 1994.
5. Barakat, A. R. and R. W. Schunk, Comparison of transport equations based on Maxwellian and bi-Maxwellian distributions for anisotropic plasmas, *J. Phys. D: Appl. Phys.*, **15**, 1195, 1982.
6. Barakat, A. R. and R. W. Schunk, Comparison of Maxwellian and bi-Maxwellian expansions with Monte Carlo simulations for anisotropic plasmas, *J. Phys. D: Appl. Phys.*, **15**, 2189, 1982.
7. Barakat, A. R. and D. Hubert, Comparison of Monte Carlo simulations and polynomial expansions of auroral non-Maxwellian distributions. 2 The 1D representation, *Ann. Geophys.*, **8**, 697, 1990.
8. Combi, M. R., Time-dependent gas kinetics in tenuous planetary atmospheres, *Icarus*, **123**, 207, 1996.

Index

5-moment approximation, 92, 97, 119, 153, 164
10.7 cm radio flux, 259, 468, 485
13-moment approximation, 62, 64, 91, 98, 143, 274
20-moment approximation, 65
630 nm radiation, 249, 364

absorption cross section, 255, 256, 349, 584
accidental resonance, 232
activation energy, 237, 247
AE index, 346
airglow, **248**, 328
Akebono spacecraft, 522
AL index, 346
Alfvén wave speed, 20, 186, 223
Alfvén waves, **185, 221**
Alouette spacecraft, 533
ambipolar diffusion, 126, 151
ambipolar diffusion coefficient, 127, 152
ambipolar diffusion equations, 127, 133, 152, 350, 370, 608
ambipolar electric field (see polarization electric field)
ambipolar expansion, 135, 136
Ampère's Law, 68, 161, 214, 561
anisotropic ion temperatures, 64, 66, 124
anisotropic pressure tensor, 124, 220, 459
anomalous electron temperatures, 418
anomalous resistivity, 196
Appleton anomaly, 30, 371
arc length of magnetic field, 338
Archimedes' spiral, 17, 217
Arecibo incoherent scatter radar, 538, 543

Arrhenius equation, 238
associative detachment, 246
astronomical unit (AU), 19
Atmosphere Explorer satellites (AE), 259, 373, 522, 525
atmospheric gravity waves (AGW), **295**, 381
atmospheric models:
    empirical terrestrial (MSIS), 318, **594**
    empirical Titan, 319
    empirical Venus (VIRA), 319, **594**
atmospheric sputtering, 324
atomic oxygen red line, 233, 249, 364
attachment reaction, 245
AU index, 346
auroral blobs, **432**
auroral electrons (see particle precipitation)
auroral oval, 25, 400, 420, 421
average drift velocity, 53
average speed, 53, 579
axial-centered dipole, 341
azimuthal electric field, 339

β (of a plasma), 20, 215
B field divergence, 133, 338, 458
ballerina skirt model, 17
beam spreading, 270
Bennett ion mass spectrometer, 525
Bessel function, 559
bi-Maxwellian velocity distribution, 66, 219, 459
bimolecular reaction, 232, 236, 242
Birkeland current, 401, 424
Boltzmann collision integral, 52, 85, **567**

Boltzmann equation, 52, 55, 264, 562

Boltzmann H theorem, 575

Boltzmann relation, 129

boundary blobs, **432**

bow shock, 22, 399

branching ratio, 244, 262

Brunt–Väisälä frequency, 298

buoyancy frequency, 298

Burgers linear collision terms, 99

Burgers semilinear collision terms, 98

Callisto, 40, 503

Cassini spacecraft, 20, 37, 40, 498, 500, 501, 531

catalytic process, 498

center of mass, 74

central force, 76

centripetal acceleration, 125, 291, 451, 465

CH_4 38–42, 497, 506, 508, 590

Chapman, 54, 63, 484, 495

Chapman–Cowling collision integrals, 96, 103

Chapman function, 258

Chapman layer, 349

Chapman production function, 260, 348

characteristic energy of precipitating
            particles, 419

characteristic time, 120, 234, 345, 411

charge density, 68

charge exchange, 105, 232, 240, 326

charge exchange collision integral, **572**

charge exchange reaction rates, 240, 242

charge neutrality, 127, 161, 174, 177

charge separation, 127, 129, 130

Charon, 38

Chatanika incoherent scatter radar, 422, 432, 543

chemical equilibrium, 351, 358

chemical kinetics, **231**

chemical reactions, 231, 237, 275, 354

chemical time constant, 233

Chew–Goldberger–Low (CGL) approximation,
            63, 219

chromosphere, 12

circular polarization, 202

classical MHD equations, 206

cleft ion fountain, 468, 469

closure conditions, 58, **60**

CH_5^+, 502

CO, 34, 483, 587, 595

CO^+, 483

CO_2, 34, 483, 493, 587, 595

CO_2^+, 35, 37, 483, 493

coefficient of viscosity, 117, 292

cold plasma, 170, 197

collision cross section, **80**

collision-dominated flow, 63, 143, 191, 451, 457

collision frequency:
    electron–ion, 104
    electron–neutral, 108, 109
    ion–ion, 104
    ion–neutral, 105–107
    momentum, 89, 96, 102
    relaxation, 573

collision time, 73, 573

collisional de-excitation, 248, 314

collisional detachment, 246

collisional invariants, 576

collisionless flow, 17, 63, 135, 225, 458

collisionless shock, 22, 225, 228

column density, 257

coma, 42

Comet Hale–Bopp, 42

Comet P/Halley, 42, 508, 509

composition of atmospheres:
    comets, 42, 508
    Earth, 28, 597–598
    Enceladus, 40,507
    Io, 39, 502
    Jupiter, 38, 496
    Mars, 36, 492
    Mercury, 31
    Neptune, 38, 501
    Pluto, 38, 502
    Saturn, 38, 498
    Titan, 40, 503
    Uranus, 38, 501
    Venus, 334 483

configuration space, 51

conjugate hemispheres, 364, 379

conjugate ionosphere, 30, 254, 364

conservative form of transport equations, 56,
            192, 226

contact discontinuity, 228

contact surface, 42

continuity equation, 56, 62, 119, 124, 133, 136, 165,
            191, 208, 292, 301, 452, 458

convection:
    anisotropic temperatures, 63, 124, 458
    electromagnetic drift, 29, 138, 217, 339, 371,
            399, 410
    frictional heating, 121, 124, 414
    heat source for the thermosphere, 447
    increased chemical reaction rate, 416
    momentum source for the thermosphere, 413,
            414, 445

convection channels, 438

convection electric field, 29, 122, 138, 139, 399

convection models, 405

convection patterns:
    four-cell, 408
    three-cell, 408
    two-cell, 29, 402, 406, 407, 408, 409
    multi-cell, 410, 414
convective derivative, 57, 209, 361, 369
convective zone, 11
cooling rates, 276, 361
core, 11
Coriolis acceleration (force), 28, 125, 291
corona, 12
coronal holes, 13
coronal loops, 12
coronal mass ejection (CME), 15, 445
coronal streamer, 12
co-rotation speed, 502
co-rotational electric field, 339, 402, 404
Coulomb collisions (see also collision frequency),
        76, 83, 98, 360
Coulomb logarithm, 84
Cowling conductivity, 345
critical frequency, 346, 533
critical level, 321
cross **B** field plasma transport: 137–139
    diamagnetic drift, 138, 402
    electromagnetic drift, 138, 339
    gravitational drift, 138, 376
cross section (see absorption and ionization cross
        sections)
cross sectional area, 338, 458
current continuity, 213, 426
current density, 68, 142, 210, 519
current sheet, 17, 25
current systems: 423
    electrojet, 345, 445, 448
    lunar, 344
    solar-quiet, 344
currents:
    Birkeland, 401, 423
    cusp, 425
    NBZ, 424
    Region 1, 423
    Region 2, 423
cusp neutral fountain, 434
cut-off frequency, 160, 179, 182, 185
cyclotron frequency, 45, 121, 174, 202, 377, 560

D region, 30, 245, **353**
Debye length, 45, 84, 129, 171, 560
Debye shielding, 83
Debye sphere, 45, 84
declination, 343, 376
derivative of vectors in a rotating frame, 290

descending layers, 379
detached plumes, 374
detachment, 245, 354
diamagnetic cavity, 43
diamagnetic current, 215
diamagnetic drift, 138, 402
differential scattering cross section, 81, 570
diffuse auroral patches, 420
diffuse auroral precipitation, 420
diffusion coefficients:
    ambipolar diffusion, 127, 152
    classical, 115, 121, 305
    major ion, 127
    minor ion, 130
    perpendicular to **B**, 139
    thermal, 151, 582
diffusion equations:
    ambipolar, 127, 131, 152, 350
    classical, 115, 370
    magnetic, 216
    major ion, 127
    minor ion, 130
diffusion in velocity space, 571
diffusion thermal coefficient, 150, 581
diffusion thermal heat flow, 150
diffusion velocity, 207
diffusive equilibrium:
    classical, 306, 351
    minor ion, 132
dip angle, 338, 344
dipolar coordinates, 340
dipole magnetic field, 337
dipole moment, 23, 337
disappearing ionosphere, 489
discrete auroral arcs, 420
dispersion relations, 162, 170, 172–186, 201, 223
displacement current, 185
dissociative excitation, 247
dissociative recombination, 232, 243–245, 314, 353,
        483, 484, 497
distribution function, 51, 53, 59, 65, 273, 521
disturbance dynamo, 373
disturbance electric fields, 373
diurnal tide, 300, 344
divergence of electrodynamic drift, 405
divergence theorem, 57, 227, 554
DMSP satellites, 374, 409, 421, 523
Doppler residual, 534
Doppler shift, 534
double adiabatic energy equations, 219
double dot product, 58, 61, 571, 603
double layers, 196
drift energy, 199

drift meter, 525
drift motion, 265
drift velocity, 53, 57, 207, 410, 455, 524
drifting Maxwellian, 59, 92, 418, 459, 576
Druyvestyn-type analysis, 521
Ds index, 346
Dst index of geomagnetic activity, 346
dynamic pressure, 22, 215, 406, 486
dynamical friction, 571
Dynamics Explorer Satellites (DE), 405, 408, 379,
    412, 449, 522

$E \times B$ drift (see electrodynamic drift)
E region, 30, 347, 418, 432, 445
Earth, 22, 300, 306, 335, 398
eccentric dipole, 341
ecliptic plane, 19, 399
eddy diffusion, 307, 308, 497
E–F region valley, 347, 380
effective electric field, 138, 414
effective gravity, 305
effective temperature, 416
Einstein coefficient, 234, 250
EISCAT incoherent scatter radar, 354, 385, 418, 543
elastic collision, 73, 74
elastic electron–neutral collision, 90, 98, 108, 109
electric field, 52, 120, 128, 129, 137, 138, 145, 150,
    160, 211, 339, 370, 399
electrical conductivities:
    Cowling, 345
    Hall, 141
    parallel, 142
    Pedersen, 141
electrodynamic drift, 138, 339, 350, 369, 402, 405
electroglow, 501
electrojet, 344, 345
electromagnetic drift (see electrodynamic drift)
electromagnetic waves, 159–164, 177–179,
    179–185, 202
electron accelerating region, 521
electron current, 141, 201, 519
electron cyclotron frequency, 45, 141, 174, 410, 560
electron Debye length, 45, 129
electron density, 21, 31, 35, 37, 39, 44, 53, 128, 129,
    137, 150, 169, 212, 347–354, 357–369, 372,
    375, 379, 390, 415, 417, 423, 428, 431–432,
    463, 484–485, 489, 493, 495, 497, 499–508,
    560, 608
electron impact, 247, 249, 419, 489, 494, 501, 503
electron plasma frequency, 45, 170, 174, 179, 560
electron retarding region, 519
electron temperature, 44, 359, 360, 363, 366, 419,
    430, 489–492, 494, 507

electron thermal speed, 169, 202
electron–ion collision, 104
electron–neutral collision, 108, 109
electron–neutral cooling rate, 98, 276, 284
electrostatic double layers, 196
electrostatic potential, 129, 402
electrostatic waves, 159, 160, 168–176, 201
elementary reaction, 232
elliptic polarization, 180, 202
empirical atmosphere models:
    terrestrial atmosphere, 318, 319
    Titan atmosphere, 319
    Venus atmosphere, 319
empirical ionosphere models:
    Venus ionosphere (see VIRA)
ENA (energetic neutral atom), 271
endothermic reaction, 238
energetic ion outflow, **465**
energy deposition, 254, 270–272, 315
energy equations:
    electron, 361
    ion, 121–125, 416, 452
    neutral gas, 292, 294, 295, 315
energy flux, 420, 435, 501
enthalpy, 238, 239
equation of state, 165, 213
equatorial anomaly, 371
equatorial fountain, 30, 371
equatorial F region, 371
equilibrium potential, 517
equivalent depth, 302
error function, 558
escape flux, 311, 322–324, 453, 456
escape velocity, 322, 453
Euler equations, 63, 144, 291–292
Europa, 40, 503
EUV solar flux, 254 258, 368, 489, 584
EUVAC solar flux model, 259, 584
evanescent wave, 303
exobase, 28, 34, 321
exosphere, 28, 321
exothermic reaction, 238
expansion phase (of a substorm), 449
exponential interaction potentials, 80
extraordinary waves (X-mode), 179, 202, 533

F_1 region, 30, 347, 349
F_2 region, 30, 347, 349, 351, 353
$F_{10.7}$ radio flux, 259, 485
Faraday's law, 68, 561
fast MHD wave, 224
fast plasma jet, 432
Fick's law, 115

field aligned current (see Birkeland current)
floating potential, 518
fluxgate magnetometer, 529
Fokker–Planck collision term, 571
forbidden transition, 248
forward shock, 19
fossil bubbles, 373
frictional heating, 274, 414–416, 428, 454, 463
frozen in magnetic field, 216
fully ionized plasma, 113, 125, 144–146, 358, 362, 582

Galilean satellites, 503
Galileo spacecraft, 21, 37, 496, 502, 503
Ganymede, 40, 503
Gamma function, 558
Gauss' law, 68, 160, 561
Gaussian pillbox, 227
general transport equations, 55–58, 62–63, 89–92, 97–102
generalized Ohm's law, 211 (see also Ohm's law)
geographic coordinates, 341
geographic pole, 342, 402, 404
geomagnetic field, 341
geomagnetic indices, 345
geomagnetic pole, 341, 424
geomagnetic storms, 345, 359, 445
geomagnetic variations, 344
geopotential height (see reduced height)
Giotto spacecraft, 509
global positioning system (GPS), 538
granules, 11
gravitational drift, 138, 376
gravity waves, **295**, 375–379, 445, 502
ground state, 241, 243, 247
group velocity, 163, 178, 299, 304, 533
growth phase (of a geomagnetic storm), 345, 445
GSM co-ordinate system, 38
guiding center, 264
gyrofrequency (see electron and ion cyclotron frequencies)
gyroradius, 45, 264

H, 28, 90, 105, 107, 109, 180, 207, 232, 239, 242, 247, 261, 270, 272, 311–313, 318, 326, 463, 470, 496, 497, 508, 595–598
H theorem, 575
H^+, 240, 247, 323, 357, 450, 453, 469, 496, 499
H^+–O charge exchange, 232, 240, 326, 357
H_2, 38, 239, 247, 277, 280, 496, 508, 592
H_2^+, 245, 496
H_2O, 239, 245, 261, 278, 281, 354, 499, 508, 589
H_2O^+, 245, 499, 508

H_3^+, 245, 497
H_3O^+, 245, 499, 508
half-thicknesses, 347
Hall conductivity, 141
Hall current, 213, 345
hard precipitation, 419–422
hard sphere collisions, 73, 96, 98, 109
$HCNH^+$, 502, 505
He, 278 33, 34, 239, 261, 263, 456, 483, 589, 595–598
He^+, 242, 360, 381, 450, 455
heat flow, 54, 61–65, 118, 122, 143–149, 207, 293, 577
heat flow equation, 58, 62, 63, 91, 100, 101, 122, 144–153
heat of formation, 239
heat sources:
    ionospheric, 122, 272–276, 254–263, 414, 416, 429, 430, 445, 452, 493, 507
    thermospheric, 238, 244, 270, 304, 315–318, 323, 326
heating efficiency, 273, 314
heliospheric current sheet, 17
helium magnetometer, 530
heterogeneous reaction, 231
heterosphere, 306
homogeneous reaction, 231
homopause, 306, 497
homosphere, 306, 312
hot atoms, **325**
hot plasma, 12, 449
Hough function, 302
hydration, 354
hydrocarbon molecules, 497
hydrodynamic equations (see Euler and Navier–Stokes)
hydrodynamic shocks, 191, 202
hydrostatic equilibrium, 296, 307

IMAGE, 533
impact parameter, 74, 77–79, 567
inclination, 344
incoherent scatter, **538**
incompressible flow, 213, 339, 405
inelastic collisions:
    electronic excitation, 283
    fine structure excitation, 282
    rotational excitation, 277
    vibrational excitation, 278
inertial reference frame, 290
in-situ measurement techniques, 517
integrated column density, 258
interaction potential, 76, 80

intermediate layers, 379
intermediate species, 232
internal field, 337
internal gravity waves, 300
International Geomagnetic Reference Field (IGRF), 342
interplanetary magnetic field (IMF), 17, 21, 34, 389, 399, 405–411, 414, 423, 430
intrinsic magnetic field, 22, 485, 495
inverse collisions, 569
inverse-power interaction potentials, 80, 102
Io, 39, 502
ion-acoustic Mach number, 134
ion-acoustic speed, 20, 133, 202
ion-acoustic waves, 170, 201, 537
ion current, 139–141
ion cyclotron frequency, 45, 122, 170
ion-cyclotron waves, 175, 201
ion–ion recombination, 246, 354
ion line, 540
ion mass spectrometers:
    Bennett, 525
    magnetic deflection, 525
    quadrupole, 527
    retarding potential analyzer, 523
    time of flight, 528
ion–molecule reaction rate, 242
ion–neutral cooling rate, 97, 106, 107, 283
ion–neutral thermal coupling, 363, 414, 428
ion outflow (see polar wind and energetic ion outflow)
ion production rate, 262, 269
ion thermal conductivity, 150, 153
ion velocity distribution (see distribution function)
ionization cross section, 262, 269, 584
ionization energy, 270
ionization frequency, 262
ionization threshold potential, 261
ionization-stripping, 270
ionogram, 532
ionopause, 33, 42, 215, 486, 496
ionosheath, 34
ionosonde, 522
ionospheric critical frequencies (see critical frequency)
ionospheric decay, 352
ionospheric features:
    light ion trough, 360
    mid-latitude trough, 427
    polar hole, 427
    propagating plasma patches, 430
    temperature hot spots, 429
    tongue of ionization, 428

ionospheric half-thicknesses (see half-thickness)
ionospheric holes, 489
ionospheric layers:
    E, F_1, F_2, 346
    F_3 layer, 381
    He^+ layer, 381
ionospheric peak densities, 44, 351, 492, 497, 498, 502, 503
ionospheric peak heights, 44, 260, 351, 483, 492, 497, 498, 503
ionospheric regions:
    D region, 31, 245, 353
    E region, 31, 347, 417, 432, 445
    F_1 region, 347, 349
    F_2 region, 347, 349, 356, 357
ionospheric sounder, 532
ionospheric storms, **386**
ionospheric variations:
    diurnal, 365
    seasonal, 367
    solar cycle, 368
irrotational flow, 214

Jacobian, 556
Jeans escape flux, 322
Jicamarca incoherent scatter radar, 373, 543
Joule heating, 445, 491
Jupiter, 37, **496**

K index, 346
kinetic pressure, 33, 215
kinetic transport equation, 50–52, 264, 563
kinetic viscosity, 294
K_p index, 346
Krook collision model, **572**

L waves, 184, 202
Langevin model, 241
Langmuir condition, 200
Langmuir probe, 517, **519–522**
Laplace's tidal equation, 302
large-scale ionospheric features, 425
Lavalle nozzle, 135
Lennard-Jones interaction potential, 80
limiting flux, 311
linear collision term (13-moment), 99
linear polarization, 180, 202
linearization technique, 167, 222
Liouville's theorem, 324, 326
local approximation, 267
local drifting Maxwellian, 59, 576
longitudinal mode, 160
loss frequency, 353, 370, 452
loss function, 268

lower hybrid oscillations, 174, 201
lunar influence, 300, 344
Lyman α radiation, 353

Mach number:
  ion-acoustic, 134, 452
  sonic, 120, 194
magnetic barrier, 34
magnetic cloud, 19
magnetic deflection ion mass spectrometer, 525
magnetic diffusion, 216
magnetic dip, 343
magnetic equator, 340, 342, 371
magnetic field (see geomagnetic field)
magnetic field divergence, 338
magnetic flux tube, 338
magnetic moments of solar system bodies, 23
magnetic pile-up, 36, 507
magnetic pressure, 20, 22, 34, 215, 486
magnetic Reynolds number, 216
magnetic scalar potential, 337, 341
magnetic storms (see geomagnetic storms)
magnetized plasma, 160, 168, 185, 217, 219,
      400, 543
magnetohydrodynamic (MHD) equations, 206, 213
magnetohydrodynamic (MHD) waves, 221
magnetometers, **529**
magnetopause, 23, 31, 37, 399
magnetosheath, 22, 22, 31, 34, 36, 37, 43, 399, 505
magnetosonic waves, 185, 202, 221
magnetosphere, 22, 31, 38, 345, 470, 505
magnetospheric tail, 24, 400
main phase (of a geomagnetic storm), 345, 445
major ion diffusion equation, 126
Mariner 5, 488
Mariner 6, 537
Mariner 10, 32
Mars, 36, **492**
Mars 4 and *5*, 495
Mars Global Surveyor (*MGS*), 36, 494, 496
Mars Express, 36, 494, 496, 533
Maunder Minimum Period, 15
Maxwell–Boltzmann velocity distribution function,
      58, **575**
Maxwell equations of electricity and magnetism,
      68, 561
Maxwell molecule collisions, 82, 90
Maxwell speed distribution, 579
Maxwell transfer equations, 60, 562
mean-free-path, 73, 115, 321
Mercury, 31, 482
meridional wind, 350–352, 366, 371, 375,
      380, 445

mesopause, 27, 304
mesosphere, 27, 304, 309, 311
Messenger, 32, 482
metallic ions, 379, 502
metastable state, 247, 248, 484
MHD discontinuities, 225, 227
mid-latitude trough, 427
migrating tide, 295, 300
Millstone Hill incoherent scatter radar, 543
minor ion diffusion equation, 130
minor ion scale height, 131
Mitra–Rowe six-ion model, 354
mixed distribution, 306, 308
mixing ratio, 312
mobility coefficient, 121, 139
molecular diffusion, 305, 307
moments of distribution:
  density, 53, 207
  heat flow, 54, 207
  pressure, 54, 207
  stress, 55, 207
  velocity, 53, 207
  temperature, 54, 207
momentum equation, 57, 62, 91, 97, 99, 119, 125,
      126, 133, 136, 139, 141, 151, 165, 213, 225,
      294, 301, 350, 370, 452, 458
momentum transfer collision frequency (see
      collision frequency)
Monte Carlo methods, 67, 321, 326, **614**
most probable speed, 579
MSIS (mass spectrometer incoherent scatter model),
      318, 320, 597–599, 299

N^+, 506, 508
N_2, 28, 30, 34, 36, 39, 40, 41, 83, 97, 99, 232, 239,
      242, 247, 261, 263, 277, 278, 311, 318, 347,
      367, 415, 447, 483, 506, 585, 595–598
N_2^+, 247, 347, 380, 506, 508
Navier–Stokes equations, 63, 144, 292–294
NBZ currents, 424
negative ionospheric storm, 387
negative ions, 245, 354–356
Neptune, 38, 501
neutral current sheet, 25
neutral density structures, 437
neutral gas heating efficiency, 314, 315
neutral gas polarizability, 90, 107, 241
neutral gas scale height, 257
neutral polar wind, 470
neutral wind, 28, 138, 140, 350, 353, 366, 370, 380,
      413, 443
NH_3, 42
NO, 36, 90, 239, 242, 247, 261, 354, 415, 483

NO^+, 247, 347, 354, 355, 380, 415
nonresonant ion–neutral collisions, 105
normal shock, 192, 227
Nozomi, 522
NRLMSIS, 319
numerical solution for F-region density, 608

O, 28, 30, 34, 36, 90, 105, 107, 232, 239, 228, 243,
    249, 261, 263, 282, 283, 310, 311, 318, 323,
    326, 347, 355, 365, 368, 447, 483, 586,
    595–598
O^+, 31, 35, 37, 232, 239, 240, 243, 244, 326, 347,
    351, 357, 415, 456, 461–464, 467–469, 483,
    493
O_2, 28, 30, 36, 40, 90, 105, 107, 232, 239, 245, 261,
    263, 277, 279, 310, 311, 318, 347, 355, 415,
    585, 597, 598
O_2^+, 232, 243, 244, 347, 354, 380, 483, 484, 493
O_2^-, 245, 246, 354, 355
oblique Alfvén wave, 224
oblique shock, 227
Ohm's law, 142, 211, 213, 216
OI 130.4 nm airglow, 328
open field lines, 29, 132, 400, 450
optical depth, 256, 258, 262, 348, 349
optical thickness, 256
order of a reaction, 233
orbital motion limited condition, 521
ordinary waves (O mode), 179, 202, 533
orthogonal expansions, 60
oxygen fine structure cooling, 276, 282, 284, 361
oxygen red line emission, 233, 249, 364

parallel electrical conductivity, 142, 212
parallel propagation, 160
parallel shock, 228
parallel temperature, 64–66, 458
partial pressure, 55, 207
partially ionized plasma, 31, 113, 125, 152, 154,
    360, 572, 583
particle precipitation:
    characteristic energy, 419
    diffuse auroral patches, 419
    diffuse auroral precipitation, 419
    electron precipitation, 419, 429, 433, 434,
        463, 489
    energy flux, 419
    Io, 504, 505
    ion precipitation, 421, 497
    Jupiter, 496
    polar rain, 419, 464
    Sun-aligned arc, 434
    theta (Θ) aurora (see Sun-aligned arcs)

    Titan, 505
    Venus, 489
partition function, 240
Pedersen conductivity, 141
Pedersen current, 141, 344
perpendicular propagation, 160
perpendicular shock, 228
perpendicular temperature, 64, 66, 458
perturbation technique, 143, 148, 467, 291–294, 296
phase space, 51, 567
phase velocity, 162, 300, 304
physical parameters of planets, 23
physical parameters of satellites, 24
Phobos, 36
photoabsorption, **254**, 584
photochemical equilibrium (see chemical
    equilibrium)
photodetachment, 245
photodissociation, 246, 309, 314, 347, 354, 483, 496
photoelectron calculations:
    continuous loss approximation, 268
    local approximation, 267
    two-stream approximation, 265
photoelectron heating rate, 274, 362, 367, 430
photoelectron production rate, 262
photoemission, 518
photoionization, **260**, 266, 351, 367, 496, 505,
    584–593
photon flux, 255
photosphere, 12
Pioneer Venus Orbiter (PVO), 483, 489, 490, 522,
    524, 525
Pioneers 10 and *11*, 21, 496, 498, 502, 503
pitch angle, 265, 266
plane waves, 161, 168, 297
planetary parameters, 23
planetary waves, 294
plasma β, 20, 215
plasma bubbles, 373
plasma convection (see convection)
plasma expansion, 135
plasma frequency, 45, 170, 174, 179, 202, 532, 560
plasma oscillations, 170, 201
plasma parameters (ionospheric), 44
plasma scale height, 128, 350
plasma sheet, 24, 449, 465, 466
plasma temperature, 358, 360, 370, 489–492,
    494, 507
plasma thermal structure, **360**, 489–492, 494, 507
plasmapause, 26
plasmasphere, 26, 356, 359, 388
plasmoid, 19
Pluto, 39, 502

Poisson equation, 129, 160
polar cap, 26, 197, 400
polar cusp, 24, 420, 428
polar hole, 427
polar rain, 419, 465
polar wind, 29, 132, 195, **450**
polarization electric field, 128, 137, 150, 452
positive ion sheath, 518
positive ionospheric storm, 387
Poynting vector, 162, 226
precipitation (see particle precipitation)
predawn effect, 364
pressure balance, 214
pressure tensor, 54, 208, 219, 577
prominence, 15
propagating plasma patches, 430
propagation constant (vector), 160, 161
protonosphere, 31, 357

quadrupole mass spectrometer, 527
quenching, 248

R waves, 184, 202
radar (incoherent) backscatter stations, 543
radiative recombination, 243, 244, 496
radiative zone, 11
radio frequency (Bennett) ion mass spectrometer, 525
radio occultation technique, **534**
radio wave spectrum, 605
random current, 518
random flux across a plane, 579
random velocity, 54, 207, 562, 577
Rankine–Hugoniot relations, 194, 228
Rayleigh–Taylor (R–T) instability, 375
reaction rates, 233, 236, 242
recombination rate (see dissociative and radiative)
reconnection, 449
recovery phase (of a geomagnetic storm), 345, 447
reduced height, 257
reduced mass, 79, 86, 93
reduced temperature, 93
refractive bending angle, 535
refractive index, 532
refractivity, 536
Region 1 current, 423
Region 2 current, 423
relative velocity, 74
resonance (wave), 160
resonant ion–neutral collisions, 107
retarding potential analyzer (RPA), 517, **522–525**
reverse shock, 19
reversible reaction, 232

ring current, 26, 345, 365
Rosetta, 43
rotating reference frame, 290–292
rotational axis, 337
rotational excitation, 277
Rutherford scattering cross section, 83

Saturn, 37, 498–500
Saturn electrostatic discharge (SED), 499
scale height:
    minor ion, 131
    mixed gas, 307
    neutral gas, 257, 296, 306
    plasma, 128, 350
scattering angle, 74, 78
Schumann–Runge continuum, 314
seasonal anomaly, 367
secular variation, 342
self-similar solution, 136
semi-diurnal tide, 300, 304, 379
shock waves, 191, 202, 225
simplified MHD equations, **213**
single fluid MHD equations, 206
skin depth, 179
slow MHD wave, 225
SO_2, 502, 591
Sodankylä ion chemistry (SIC) model, 354
soft precipitation (see particle precipitation)
solar activity, 14, 259
solar constant, 15
solar flares, 15
solar fluxes (see EUV solar flux)
solar magnetic field, 12
solar wind, 12, 217
solar wind parameters, 20
solar zenith angle, 256
Sondrestrom incoherent scatter radar, 543
sound speed, 20, 172, 221, 298
South Atlantic anomaly, 344
spacecraft (see specific spacecraft)
spacecraft potential, **517–519**
specific heat, 165, 191, 292, 294, 296
speed of light, 161, 179
spiral angle, 17, 218
Spitzer conductivity, 362
spontaneous de-excitation, 233, 248
sporadic E layer, 379, 502
spread F, 373
stationary tide, 295
statistical weight, 240
stoichiometric equation, 232
Stokes' theorem, 227, 554
stopping cross section, 268

storm-enhanced density (SED), 387
stratopause, 27
stratosphere, 27, 295, 313, 381
streaming instabilities, 188, 196, 418
stress tensor, 55, 61, 63, 122, 144, 148, 292, 452
strong shocks, 195
sub-auroral ion drift (SAID), 449
sub-auroral polarization stream (SAPS), 387
sub-auroral red arcs (SARARCS), 358, 364
subsonic flow, 120, 135, 191
substorms, 448
sudden storm commencement (SSC), 345, 445
Sun, **11**
Sun-aligned arcs, 434
sunspots, 13
supersonic flow, 17, 22, 63, 132, 154, 191, 450, 453
supersonic neutral winds, 443

tail rays, 35, 489
tangential discontinuity, 33, 42, 228, 486
temperature anisotropy, 64, 65, 350, 459
temperature hot spots, 425–430
termolecular reaction, 232, 235
terrestrial thermosphere empirical model (see MSIS)
TGCM (see thermosphere–ionosphere general
    circulation model)
thermal conduction:
    electron, 145–147, 362, 490, 494, 581
    ion, 150, 153, 581
    neutral gas, 118, 293, 317
thermal diffusion, 101, 152, 358
thermal electron heating rate, 272, 362, 367
thermal escape flux, 322
thermal potential, 196
thermal velocity, 54, 207, 562, 577
thermoelectric coefficient, 150, 583
thermoelectric effect, 101, 145, 150
thermosphere, 27, 36, 38, 314, 315, 318
thermosphere–ionosphere general circulation model
    (TIGCM), 318, 382
thermospheric composition:
    Earth, 304, 310, 314, 597–598
    Jupiter, 38
    Mars, 36
    Titan, 40
    Venus, 34
thermospheric temperatures:
    Earth, 27, 320, 597–598
    Jupiter, 23, 38, 318
    Mars, 37
    Saturn, 38
    Venus, 34
thermospheric wind, 28, 318, 413, 445

theta aurora (Sun-aligned arc), 419, 434
Thomson scatter, 538
three-body recombination, 232
tides, 294, **300,** 344, 381
TIGCM (see thermosphere–ionosphere general
    circulation model)
tilted dipole, 38, 337, 341
time constants, 234, 235, 308, 352–353, 365,
    411, 573
TIMED, 259
time-of-flight spectrometer, 528
Titan, 40, 503–507
tongue of ionization, 427, 428, 431
topside ionosphere, 31, 132, 349, 356, 450, 457,
    486, 497
total scattering cross section, 82
trace of a tensor, 61, 603
transfer collision integrals, 85, 92, 99
transport equations:
    5-moment set, 119, 165, 291
    13-moment set, 62, 98–102
    ambipolar diffusion, 127, 150, 151, 350
    continuity, 56, 62, 119, 124, 133, 136, 165, 191,
        208, 292, 301, 452, 458
    diffusion, 115, 126, 127, 130, 131, 151, 216,
        350, 370
    energy, 57, 122, 125, 292, 294, 315, 361, 416, 452
    Euler, 63, 144, 191
    heat flow, 58, 63–65, 118, 122, 143, 148, 207,
        293, 577
    momentum, 57, 62, 91, 97, 99, 119, 125, 126,
        133, 136, 139, 141, 151, 165, 213, 225, 294,
        301, 350, 370, 452, 458
    Navier–Stokes, 63, 144, **292**
    pressure tensor, 57, 62
    self-similar, 136
    stress tensor, 63, 122, 125, 144, 148, 292
    thermal conduction, 118, 146, 150, 151, 152, 293,
        362, 581
transport properties:
    ambipolar diffusion, 127, 131, 151, 350
    diffusion, 115, 126, 130, 131, 151, 216, 350, 370
    diffusion-thermal heat flow, 150, 581
    electrical conduction, 141, 142, 345
    thermal conduction, 118, 144–147, 152, 293, 317,
        362, 489, 491, 581
    thermal diffusion, 150, 151, 582
    thermoelectric effect, 101, 145, 146, 150
    viscosity, 116, 145, 292, 294
transverse mode, 160
traveling ionospheric disturbance (TID), 385,
    390, 445
Triton, 42, 508

tropopause, 27
troposphere, 27, 381
turbopause, 306
two-stream approximation (see photoelectron
    calculations)
two-stream instability, 188, 196, 418

Ulysses, 21
unit dyadic, 55, 88, 119, 219, 225, 578, 603
unmagnetized plasma, 168, 178, 196
upper hybrid oscillations, 172, 201
Uranus, 38, 501

Van Allen radiation belt, 26
velocity-dependent correction factors, 98
velocity moments:
    density, 53
    drift velocity: 53
        heat flow for ∥ energy, 64
        heat flow for ⊥ energy, 64
        heat flow tensor, 54
        heat flow vector, 54
        higher-order pressure tensor, 54
        parallel temperature, 64
        perpendicular temperature, 64
        pressure tensor, 54

    stress tensor, 55
    temperature, 54
velocity space, 51, 56, 556
Venus, 33, **482**
vertical column density, 257
vibrational excitation, 247, 279–282, 496
Viking, 356 492–495
VIRA (Venus international reference atmosphere),
    319, 595, 596
virtual height, 533
viscosity, 117, 145, 292–294
Vlasov equation, 52
*Voyager*s, 21, 38, 496–498, 501, 503, 508

water cluster ions, 30, 354
wave modes, 201
wave number-4 pattern, 383
wave–particle interactions, 365, 418, 464, 467
weak shocks, 195, 507
weakly ionized plasma, 30, 113, 120, 154
westward traveling surge, 448
whistler waves, 202
wind filtering, 384
winter helium bulge, 457

zenith angle, 256

Printed in the United States
By Bookmasters